T0281566

CONVERGENCE
FOUNDATIONS
OF TOPOLOGY

CONVERGENCE FOUNDATIONS OF TOPOLOGY

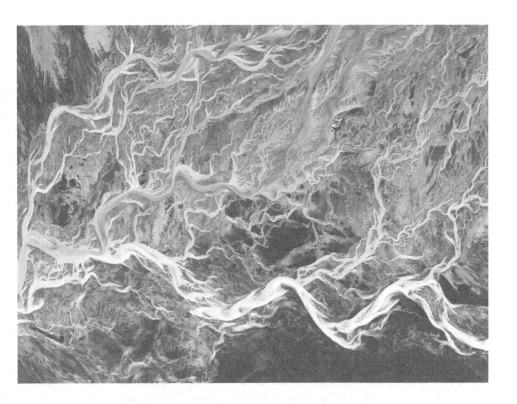

SZYMON DOLECKI
Mathematical Institute of Burgundy, France

FRÉDÉRIC MYNARD
New Jersey City University, USA

World Scientific

NEW JERSEY · LONDON · SINGAPORE · BEIJING · SHANGHAI · HONG KONG · TAIPEI · CHENNAI · TOKYO

Published by

World Scientific Publishing Co. Pte. Ltd.
5 Toh Tuck Link, Singapore 596224
USA office: 27 Warren Street, Suite 401-402, Hackensack, NJ 07601
UK office: 57 Shelton Street, Covent Garden, London WC2H 9HE

Library of Congress Cataloging-in-Publication Data
Names: Dolecki, Szymon. | Mynard, Frédéric, 1973–
Title: Convergence foundations of topology / by Szymon Dolecki (Mathematical Institute of
    Burgundy, France), Frédéric Mynard (New Jersey City University, USA).
Description: New Jersey : World Scientific, 2016. | Includes bibliographical references and index.
Identifiers: LCCN 2016003761| ISBN 9789814571517 (hardcover : alk. paper) |
    ISBN 9789814571524 (pbk. : alk. paper)
Subjects: LCSH: Topology--Textbooks. | Convergence.
Classification: LCC QA611 .D65 2016 | DDC 514--dc23
LC record available at http://lccn.loc.gov/2016003761

British Library Cataloguing-in-Publication Data
A catalogue record for this book is available from the British Library.

Cover image with courtesy of André Ermolaev.

Printed in Singapore

In memory of Gustave Choquet (1915-2006),
founder of convergence theory,
on the centenary of his birth.

Preface

The aim of this book is twofold: an elementary original introduction to *topology* and an advanced reference on *convergence theory*.

Many introductory topology books claim to emphasize the convergence-theoretic viewpoint. This is fully justified, as one of the key advantages of considering abstract topologies is to offer a unified framework for many (but unfortunately, not all) notions of convergence encountered in Analysis.

In this book, we take this viewpoint to its logical conclusion by considering the collection of convergent objects on a set as the primary notion. The resulting *convergence structures* are more general than the traditional topological structures, and can encompass most notions of convergence that may be found in Analysis ([1]). Yet, the conceptual effort required from a student to grasp this type of structures is not really greater - and for many students smaller - than what is required to get acquainted with topology in the usual sense.

Hence, we introduce topological spaces only as a particular case of convergence spaces, as part of a first course in *Foundations of Analysis* based on convergence structures, substituting for an introductory course in *Topology*. We hope to reach senior undergraduates getting acquainted with Topology for the first time, graduate students seeking a deeper understanding of foundations of Analysis, as well as experienced mathematicians. Used as an undergraduate introductory course in Topology, some of the more advanced material, for example, Chapters XIII to XVIII, should be omitted. Nevertheless, students acquainted with a proof-based approach to Analysis in the metric space setting, but not necessarily in the general topological setting, can learn point-set topology *from the convergence viewpoint*.

---

[1] Even more general are those defined by *limitoids* [50] in particular, Γ-*convergences* [49] where convergent objects are isotone families.

The book also intends to serve as a useful reference for mathematicians, including topologists, as it gathers for the first time in a textbook format a wealth of basic results and examples on convergence spaces ([2]).

Exercises are scattered throughout the text. They are an integral part of the text, and each exercise should be attempted by the reader.

Solutions to these exercises are provided in a companion document available on the book's website at

http : //www.worldscientific.com/worldscibooks/10.1142/9012.

It also contains additional exercises and problems for each chapter, without solutions. It will also be regularly updated with complements.

We have tried to avoid lengthy preliminaries, and to rely only on what can reasonably be expected from an undergraduate student. All the necessary set-theoretic background is however gathered in an Appendix at the end of the book.

Ordinal numbers and cardinal arithmetic deserve a special mention. Due to the dual nature of our intended audience, we have included results formulated in terms of ordinals, for the more mathematically mature reader. Consequently, the Appendix treats ordinal numbers and cardinal arithmetic. While we encourage students to study this more advanced material, we understand that this part goes at times beyond the scope of an introductory course. For the most part, the student can formulate these results in terms of the simple distinction between countable and uncountable.

Historical remarks

A *topological space* is usually defined as a set and a family of its subsets fulfilling the axioms of *open* sets ([3]). Notions of *closed* set, *closure*, *interior*, *neighborhood* are derived from that of open set, and in turn, each of these notions determines open sets. Therefore any of them can be used to define a topological space. Georg Cantor introduced a concept of *closed* set in

---

[2]with the exception of W. Gähler's two volumes monograph, *Grunstrukturen der Analysis*, [44] which is in German, out of print, and seems directed to a more mathematically mature audience. Other monographs, like [9] and [8], focus on the convergence-theoretic approach to Functional Analysis. G. Preuss books [89] and [90] are more similar in spirit but focus primarily, even though not only, on the uniform setting. It includes basic material on convergence spaces, that is not developed in a way comparable to the present book.

[3]A family that contains the empty set and the whole space, every union and each finite intersection of its elements.

Euclidean space in [14, 1884]; in [86, 1887] Giuseppe Peano defined the *interior* and the *closure* of a set in Euclidean space and related them to closed sets of Cantor.

In [41, 1906] Maurice Fréchet developed a theory of convergence of *sequences* in abstract *metric* spaces extending the notion of Euclidean distance ([4]). He generalized all the mentioned topological notions to metric spaces. The theory of metric spaces was universally adopted in spite of a more general approach (in terms of *accumulation* points) proposed by Frigyes Riesz in [95, 1907]. Felix Hausdorff in [53, 1914], Leopold Vietoris, Kazimierz Kuratowski and others initiated an axiomatic theory of topological spaces. *Metrizable spaces* constitute a (narrow but important) subclass of topological spaces; the search for necessary and sufficient conditions of metrizability (Chapter XVIII) was one of the main quests at the beginning of the development of topology.

Eduard Čech, Wacław Sierpiński and Felix Hausdorff considered also a more general concept of *non-idempotent closure (adherence),* which defines a larger class of *pretopological spaces* ([5]). This concept remained rather marginal in their writings.

A conceptual turnover however was operated by Gustave Choquet in [18, 1947-1948] who studied pretopologies and introduced *pseudotopologies* as a subclass of *convergence spaces*, where convergence is defined in terms of *filters.* Convergences and pseudotopologies turn out to be *exponential* ([6]) *categories* that include topologies. They are for topologies what complex numbers are for real numbers, that is, extensions of a class, in which certain operations carry out of the class, to classes stable for these operations.

Filters were formally introduced by Henri Cartan in [15, 16, 1937], but the idea can be traced back to earlier papers, for example, by Vietoris. Choquet studied a dual notion of *filter grill* in [17, 1947]. This concept was however already recognized by Cantor in [14, 1884] and formalized, under the name of *distributive family,* by Peano in [86, 1887], who considered also a third element of the triad, called *antidistributive families,* that is, *ideals* of sets, and used such families as *covers.*

---

[4]that he called *écart.*

[5]The term *pretopology* is due to Choquet. Hausdorff used the term *mehrstuffige Topologie.*

[6]Commonly called *Cartesian closed.*

Terminology and references

For the sake of comprehension we seek to use standard terminology. However often either standard terminology does not exist or there exist several terms with the same meaning, some more frequently used than others ([7]).

As a rule, we avoid employing personal names as mathematical terms, although this rule cannot be implemented systematically. On one hand, it is recurrent that notions and theorems were baptized not after those who contributed most to their introduction and study. On the other, a well chosen common name may indicate the sense or the role of the notion. If we feel that comprehension could be enhanced by a non-standard name, we adopt it ([8]).

Generally speaking, we did not track down the first appearance of a particular result, and thus we usually do not attribute specific theorems. However, we have included sources in the references. Our primary source for standard results and arguments of General Topology is [37], and occasionally [48] and [19].

Acknowledgements

We are most grateful to Gabriele Greco (Trento, Italy) and to Robert Leek (University of Oxford, UK) for a perceptive perusal of large parts of the book and their precious suggestions. Cordial thanks are due to Iwo Labuda (University of Mississippi, Oxford, MS) and to Alois Lechicki (Fürth, Germany) for their helpful remarks.

We express our gratitude to Jérôme Laurens (Mathematical Institute of Burgundy, Dijon, France) who deployed his impressive expertise of LaTeX to help the formatting of this book.

We are obliged to André Ermolaev for the kind permission to use one of his splendid photos of Iceland volcanic rivers for the cover of this book.

---

[7]For example, filters generated by sequences were often called *elementary filters* or *Fréchet filters*, but we have chosen the term *sequential filters*, which is more evocative than *elementary filter*, while we reserve the term *Fréchet filter* for a filter that is an intersection of sequential filters, because *Fréchet pretopologies* can be characterized in terms of such filters.

[8]An important class of topologies, introduced by Urysohn in [102, 1925] have been most frequently called *completely regular topologies*, but also *Tikhonov topologies* or else $T_{3\frac{1}{2}}$-*topologies*. We are convinced, however, that the term (Hausdorff) *functionally regular topologies* is more appropriate, because it evokes the essence of the property, and thus should be adopted.

We would appreciate any report of eventual misprints and errors to fmynard@njcu.edu. Errata would be posted on the book's page referenced above, or at http : //is.gd/fmynard, as needed.

December 2015

Szymon Dolecki
Mathematical Institute of Burgundy
University of Bourgogne Franche Comté, Dijon, France

Frédéric Mynard
Department of Mathematics
New Jersey City University, USA

Contents

Chapter I

Introduction

The most basic ingredient needed to develop Calculus is a notion of limit.

Continuous maps arise naturally as those that preserve limits, derivatives are defined as limits of difference-quotients ([1]), Riemann integrals are defined with the aid of a limiting process.

The purpose of this book is to provide an introduction to structures defining limits, that is, to *convergence*. Of course, they form the foundation for Calculus, hence Analysis. On the other hand, they also provide a framework for "modern geometry". To explain what is meant, consider the picture below:

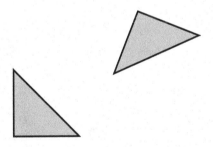

Figure I.1

You are accustomed to seeing these two triangles as identical, because they have the same sides. In other words, the two objects are identified because they are image of each other under an isometry of the plane, that is, a bijective transformation of the plane that preserves distances. Classical Euclidean geometry is therefore geometry *modulo isometries*, which in the

---

[1]which requires additional algebraic structure compatible with limits.

1

case of planar geometry means modulo translations, rotations, reflections and glide reflections.

What if geometric objects were seen modulo a larger class of transformations?

You may have already encountered such an example if you have studied *projective geometry,* where geometric objects are studied up to projective transformations. *Topological geometry,* often referred to as rubber sheet geometry, studies geometric objects modulo *continuous deformations,* or *homeomorphisms,* that is, bijective maps that are continuous with continuous inverse.

In the case of the plane, the meaning of continuity of $f : \mathbb{R}^2 \to \mathbb{R}^2$ is the usual one:

$$\lim_{n\to\infty} \|(x_n, y_n) - (x_0, y_0)\| = 0 \implies \lim_{n\to\infty} \|f(x_n, y_n) - f(x_0, y_0)\| = 0,$$

where $\| \cdot \|$ is the Euclidean norm ([2]).

It turns out that each one of the figures below can be mapped onto one another under a continuous deformation.

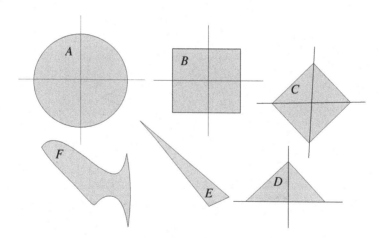

Figure I.2 All these figures are *homeomorphic,* that is, for any two of them, there exists a continuous one-to-one map such that the inverse map is also continuous.

---

[2]This book is about convergence, and we will introduce various notion of limits under the symbol lim. However, we use freely the symbol

$$\lim_{n\to\infty}$$

to denote the limit of a sequence in the usual sense.

We can easily give explicit formulas for homeomorphic transformations of some of them. For example, the ball centered at $(0,0)$ with radius 1 (figure A) is transformed onto the square (figure B) with the vertices at

$$(1,1),(-1,1),(-1,-1),(1,-1)$$

by the mapping

$$g\left(x,y\right) := \begin{cases} \dfrac{\sqrt{x^2+y^2}}{\max\left(|x|,|y|\right)}(x,y) & \text{if } (x,y) \neq (0,0) \\ (0,0) & \text{if } (x,y) = (0,0), \end{cases}$$

which is a homeomorphism. If we now rotate and scale the square so that its vertices become

$$(1,0),(0,1),(-1,0),(0,-1),$$

(figure C), then the homeomorphism

$$f\left(x,y\right) := \left(x, \frac{y+1-|x|}{2}\right)$$

transforms it into the triangle (figure D) with the vertices at

$$(1,0),(0,1),(-1,0).$$

An affine homeomorphic transformation maps D onto the triangle E. It would be more tedious to write an explicit formula for a homeomorphism that transforms E onto F, but it could be done.

In other words, in the eye of the topologist, they are the same. You may wonder what the purpose of identifying so many different objects is. The point is to focus on what is *really* different between two geometric objects. If a certain quality of an object is left unchanged by homeomorphism, then two objects that do not share this property *cannot be* image of one another under homeomorphism. In other words, they are different, even in the eyes of the topologist.

Such properties that distinguish objects are called *invariants*. What we call "modern geometry" is in part concerned with finding and studying such invariants, in order to classify geometric objects up to continuous deformations.

The foundational concept needed to embark on such a program is ultimately a notion of continuity (for geometric transformations). Continuity, as observed before, is simply preservation of limits. Hence, once again, the key concept is that of convergence. This introductory chapter shows that, as far as convergence is concerned, an adequate way of viewing a sequence is in terms of the *isotone family of sets* (Section II.1) that contains the tails of the sequence, that is, as the *filter* it generates (Section II.2).

Even in the classical metric setting, the collection of convergent sequences naturally induces a notion of convergence for families of sets that are not directly generated by sequences.

Furthermore, standard notions of convergence from Analysis, such as *pointwise convergence* of functions, cannot be modeled by a metric (Section I.7). Moreover, we will see that an adequate description of pointwise convergence cannot be restricted to convergence of sequences and necessarily involves a notion of convergence for filters.

Similarly, the classical definition of Riemann integral involves limits that are not limits of sequences, but of filters (Section I.7). As a result, we will set out to define and study convergence for general filters in the subsequent chapters.

I.1 Preliminaries and conventions

We denote by $f : X \to Y$ a *map* (or *function*) from X to Y. A *partial map*

$$f : X \rightarrowtail Y,$$

from X to Y is a map to Y from a subset of X, called the *domain* of f and denoted by $\operatorname{dom} f$ (see Appendix A for formal definitions). If $\operatorname{dom} f = X$, then a partial map becomes a map.

We denote by Y^X the set of all maps from X to Y. In the special case, where $Y = \{0, 1\}$, then $\{0, 1\}^X$ is isomorphic to the set 2^X of all subsets of X. Two standard bijections to associate to each subset A of X, a map from X to $\{0, 1\}$, are the *characteristic function* χ_A of A and the *indicator function* ψ_A of A, that are defined respectively by

$$\chi_A(x) := \begin{cases} 1 \text{ if } x \in A, \\ 0 \text{ otherwise,} \end{cases} \qquad \psi_A(x) := \begin{cases} 0 \text{ if } x \in A, \\ 1 \text{ otherwise.} \end{cases}$$

Of course, $\chi_A(x) + \psi_A(x) = 1$ for each $x \in X$ and $A \subset X$. When the ambient set is clear from context we use the notation $A^c := X \setminus A$ for the complement of A in X. Of course, $\chi_{A^c} = \psi_A$.

Given a map $f : X \to Y$, $A \subset X$, and $B \subset Y$, the *image* of A by f is denoted by

$$f(A) := \{f(x) : x \in A\},$$

and the *preimage* of B by f is denoted by

$$f^-(B) := \{x \in X : f(x) \in B\}.$$

In particular, $f(\{x\}) = \{f(x)\}$ and $f^-(\{y\})$ are sets ([3]).

One denotes by $\mathbb{R}$, the set of real numbers and by $\mathbb{R}_+$ the set of positive real numbers, that is, $\mathbb{R}_+ := \{x \in \mathbb{R} : x \geq 0\}$. Note that we use the term *positive*, rather than *non-negative*, for "greater or equal to 0". Similarly, a *positive real-valued function* is a function $f : X \to \mathbb{R}$ with $f(x) \geq 0$ for all $x \in X$.

In the context of real-valued functions on a given set, when the domain considered is clear and can be made implicit, we use the following notational conventions: If $f : X \to \mathbb{R}$ and $g : X \to \mathbb{R}$, and $r \in \mathbb{R}$, we denote by $\bar{r} : X \to \mathbb{R}$ the function $\bar{r}(x) = r$ for all $x \in X$, and

$$\begin{aligned}
\{f \leq g\} = \{g \geq f\} &:= \{x \in X : f(x) \leq g(x)\} \\
\{f < g\} = \{g > f\} &:= \{x \in X : f(x) < g(x)\} \\
\{f = g\} &:= \{x \in X : f(x) = g(x)\} \qquad \text{(I.1.1)} \\
\{f = r\} &:= \{f = \bar{r}\} = f^-(r) \\
\{f \leq r\} &:= \{f \leq \bar{r}\},
\end{aligned}$$

and $\{f < r\}$, $\{f \geq r\}$ and $\{f > r\}$ are defined similarly.

If X, Y are sets, then $X \times Y := \{(x, y) : x \in X, y \in Y\}$ is the *product* of X and Y. The *projection maps* $p_X : X \times Y \to X$, $p_Y : X \times Y \to Y$ are defined by

$$p_X(x, y) := x, \ p_Y(x, y) := y.$$

If $\mathcal{A}$ is a family of subsets of X and $\mathcal{B}$ is a family of subsets of Y, then we denote

$$\mathcal{A} \times \mathcal{B} := \{A \times B : A \in \mathcal{A}, B \in \mathcal{B}\} \qquad \text{(I.1.2)}$$

the *product family* on $X \times Y$.

A map $f : X \to Y$ is called *one-to-one (injective)* (or *injective*) if $f(x_0) = f(x_1)$ implies that $x_0 = x_1$; *onto* (or *surjective*) if $f(X) = Y$; *bijective* if it is both one-to-one and onto.

Two sets are *equipotent* if there exists a bijection between them. A set X is *finite* if each injection $f : X \to X$ is a surjection (for details, see Appendix A). A set is *infinite* if it is not finite.

---

[3] Often $f^-(\{y\})$ is denoted by $f^-(y)$, in spite of possible confusion with the value at y of the inverse map of f in case where f is inversible.

A *cardinal number* is an equivalence class of all equipotent sets. A detailed exposition of cardinal and ordinal numbers can be found in Appendices A.2 and A.9. We denote by $\operatorname{card} X$ the cardinal (number) of X. If κ, λ are cardinal numbers, then $\kappa \leq \lambda$ if there exists an injection $f : X \to Y$, where $\operatorname{card} X = \kappa$ and $\operatorname{card} Y = \lambda$ ([4]). The symbol λ^κ represents $\operatorname{card}\left(Y^X\right)$, where $\operatorname{card} X = \kappa$ and $\operatorname{card} Y = \lambda$.

In particular, $0 := \operatorname{card} \varnothing$ is smaller than any other cardinal ([5]). We denote by $\mathbb{N}$, the set of natural numbers, the first element of $\mathbb{N}$ being 0 (see Appendix A).

$$\aleph_0 := \operatorname{card} \mathbb{N} \text{ and } \mathfrak{c} := \operatorname{card} \mathbb{R}.$$

Recall that $\aleph_0 < \mathfrak{c} = 2^{\aleph_0}$, that is, $\aleph_0 \leq \mathfrak{c}$ but $\aleph_0 \neq \mathfrak{c}$.

A set X is said to be *countable* if it is the image of $\mathbb{N}$ under a map ([6]). In particular, it follows that a set is countable whenever it is equipotent with $\mathbb{N}$, or finite and non-empty.

For each $k \in \mathbb{N}$, we write

$$\mathbb{N}_k := \{n \in \mathbb{N} : n \geq k\}.$$

Finally, the family of all finite subsets of a given set X is denoted by

$$[X]^{<\omega} := \{A \subset X : \operatorname{card} A < \infty\}.$$

I.2   Premetrics and balls

A function $d : X \times X \to \mathbb{R}_+$ is called a *premetric* if

$$d(x,y) = d(y,x) \qquad\qquad \text{(symmetry)}$$
$$d(x,y) = 0 \iff x = y \qquad\qquad \text{(separation)}$$

for every $x, y \in X$. A premetric d is called a *metric* if

$$d(x,z) \leq d(x,y) + d(y,z) \qquad\qquad \text{(triangular inequality)}$$

for every $x, y, z \in X$. The couple (X,d) is called a *premetric* space *if d is a premetric* on X or a *metric space* if *d is a metric* on X.

---

[4]We omit quantifiers, because here "there exist X and Y" and "for every X and Y" are equivalent.

[5]Because, for each set X, the *empty map* $f : \varnothing \to X$ is injective (for precisions see Appendix A.2).

[6]Sometimes, a *countable* set is defined as a set of cardinality not greater than that of $\mathbb{N}$. This includes the empty set $\varnothing$, which is not countable in the sense of our definition.

The *diameter* diam A of a subset A of a premetric space is defined by
$$\text{diam } A := \sup \{d(x,y) : x, y \in A\}.$$
A subset A of a premetric space is said to be *bounded* if diam $A < \infty$.

If d is a premetric on X, then
$$B(x,r) := \{y \in X : d(x,y) < r\}$$
is called a *ball* centered at x with *radius* $r > 0$. We denote by
$$\mathcal{B}(x) := \{B(x,r) : r > 0\}$$
the family of all balls centered at x. If several premetrics are considered, we may use an index, as in $B_d(x,r)$ or $\mathcal{B}_d(x)$, to indicate with respect to which premetric the balls are considered.

Here is an example of a premetric that is not a metric.

Example I.2.1 (Féron cross). Consider $d : \mathbb{R}^2 \times \mathbb{R}^2 \to \mathbb{R}_+$ defined by
$$d\left((x_0,y_0),(x_1,y_1)\right) := \begin{cases} |x_0 - x_1| & \text{if } y_0 = y_1 \\ |y_0 - y_1| & \text{if } x_0 = x_1 \\ 3 & \text{if } x_0 \neq x_1 \text{ and } y_0 \neq y_1. \end{cases}$$
It is clear that d is a premetric on $\mathbb{R}^2$ but it is not a metric: $d\left((0,0),(1,1)\right) = 3 > d\left((0,0),(0,1)\right) + d\left((0,1),(1,1)\right) = 2$.

The ball for this premetric, centered at (x_0,y_0) with radius r is the following cross
$$B\left((x_0,y_0),r\right) = \{x_0\} \times (y_0 - r, y_0 + r) \cup (x_0 - r, x_0 + r) \times \{y_0\}$$
provided that $0 < r < 3$.

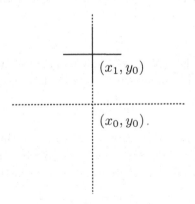

Figure I.3 The ball $B\left((x_0,y_0),r\right)$ is represented with dotted contours. Notice that if $(x,y) \in B\left((x_0,y_0),r\right)$ and $(x,y) \neq (x_0,y_0)$ then no ball $B\left((x,y),s\right)$ with $s > 0$, is a subset of $B\left((x_0,y_0),r\right)$.

Exercise I.2.2. Show that the following functions from $\mathbb{R} \times \mathbb{R}$ to $\mathbb{R}_+$ are all metrics:

$$d(x,y) := |y - x|, \tag{I.2.1}$$

$$a(x,y) := \arctan(|y - x|), \tag{I.2.2}$$

$$i(x,y) := \begin{cases} 0, & \text{if } x = y, \\ 1, & \text{if } x \neq y. \end{cases} \tag{I.2.3}$$

Example I.2.3. It is easy to check that the function $d : \mathbb{R}^2 \times \mathbb{R}^2 \to \mathbb{R}_+$ defined by $d((x_0, y_0), (x_1, y_1)) := \max(|x_0 - x_1|, |y_0 - y_1|)$, is a metric.

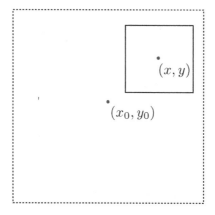

Figure I.4   The ball $B((x_0, y_0), r)$ is represented with dotted contours. In this case, for each element (x, y) of $B((x_0, y_0), r)$, where $r > 0$, there exists $t > 0$ such that $B((x, y), t) \subset B((x_0, y_0), r)$.

Exercise I.2.4. Let (X, d) be a metric space. Show that for each $x \in X$ and each $y \in B(x, r)$, there exists $t > 0$ such that $B(y, t) \subset B(x, r)$.

Exercise I.2.5. Consider the metrics d, a and i on $\mathbb{R}$ of Exercise I.2.2.

(1) Show that $a(x, y) \leq d(x, y)$ for every x and y in $\mathbb{R}$ and that there is no $c > 0$ such that $d(x, y) \leq c \cdot a(x, y)$ for all x and y in $\mathbb{R}$.
(2) Show that for each $r > 0$ there exist $s > 0$ such that for each $x \in \mathbb{R}$,

$$B_d(x, s) \subset B_a(x, r) \subset B_d(x, r). \tag{I.2.4}$$

(3) Show that $B_i(x, r) = \{x\}$ whenever $0 < r \leq 1$ and $B_i(x, r) = \mathbb{R}$ whenever $r > 1$.

Definition I.2.6. A subset V of a premetric space is called a *vicinity* of x if there is $r > 0$ such that $B(x, r) \subset V$. We denote by $\mathcal{V}(x)$ the set of vicinities of x.

Exercise I.2.7. Let X be a premetric space and $x \in X$. Check that

$$V \in \mathcal{V}(x) \Longrightarrow x \in V, \tag{I.2.5}$$

$$W \supset V \in \mathcal{V}(x) \Longrightarrow W \in \mathcal{V}(x), \tag{I.2.6}$$

$$V_0, V_1 \in \mathcal{V}(x) \Longrightarrow V_0 \cap V_1 \in \mathcal{V}(x). \tag{I.2.7}$$

Definition I.2.8. A subfamily $\mathcal{B}$ of $\mathcal{V}(x)$ is called a *vicinity base* of x if for every $V \in \mathcal{V}(x)$ there is $B \in \mathcal{B}$ such that $B \subset V$.

Notice that $\mathcal{V}(x)$ and $\left\{ B\left(x, \frac{1}{n}\right) : n \in \mathbb{N}_1 \right\}$ are vicinity bases of x. It follows that each element of a premetric space has a countable vicinity base.

If d and g are two premetrics on X, then (X, d) is called *coarser* than (X, g) ((X, g) is *finer* than (X, d))

$$(X, d) \leq (X, g)$$

if $\mathcal{V}_d(x) \subset \mathcal{V}_g(x)$ for every $x \in X$.

Observe that if $d(x_0, x_1) \leq g(x_0, x_1)$ for each $x_0, x_1 \in X$, then (X, d) is coarser than (X, g). However,

Example I.2.9. Let d, a, i be the metrics from Exercises I.2.2 and I.2.5. Observe $(\mathbb{R}, d)$ is coarser than $(\mathbb{R}, i)$, but $d \nleq i$. On the other hand, $a(x, y) \leq d(x, y)$, hence $(\mathbb{R}, a)$ is coarser than $(\mathbb{R}, d)$. By (I.2.4), $(\mathbb{R}, a)$ is also finer than $(\mathbb{R}, d)$, that is, $(\mathbb{R}, d)$ and $(\mathbb{R}, a)$ are equivalent.

I.3 Sequences

Following Gabriele H. Greco in [52], we say that:

Definition I.3.1. A *sequence* on X is a map to X from any infinite subset N of natural numbers.

Let $f \in X^N$ be a sequence, that is, $f(n) \in X$ for each $n \in N$ ([7]). As one usually denotes a generic element of X by x, it has become habitual

---

[7]Traditionally, $N = \mathbb{N}$. In this context, the image $f(n)$ of n by f is traditionally called the *n-th term* of the sequence f. This terminology might be confusing when $N \neq \mathbb{N}$.

to denote by x_n, the value of n by a sequence under the identification $x_n := f(n)$. Therefore a sequence on X is often indicated by

$$\{x_n\}_{n \in N} \text{ or even by } \{x_n\}_n,$$

because, in most situations, it is irrelevant to know which particular infinite subset N of $\mathbb{N}$ is the domain of the sequence.

Remark I.3.2. Every infinite subset N of $\mathbb{N}$ ([8]) is order-isomorphic with $\mathbb{N}$. In particular, if $s_m : \mathbb{N} \to \mathbb{N}$ is defined by $s_m(n) := n + m$, then s_m is an order isomorphism of $\mathbb{N}$ onto

$$s_m(\mathbb{N}) := \mathbb{N}_m.$$

Although Definition I.3.1 and the traditional one describe the same object (up to an order isomorphism of the domains), the present one turns out to be more flexible. For instance, traditionally, one defines a sequence on X as a map from $\mathbb{N}$ to X, but soon after a sequence $\{\frac{1}{n}\}_{n \in \mathbb{N}_1}$ is defined from $\mathbb{N}_1$ to X, which is more practical than to keep $\mathbb{N}$ as a domain and write $\{\frac{1}{n+1}\}_{n \in \mathbb{N}}$.

Definition I.3.3. A *tail* of a sequence $\{x_n\}_{n \in N}$ is the image by the sequence of the elements of N greater than a (given) natural number. The set of all its tails is denoted by $(x_n)_n$, that is,

$$(x_n)_n := \{\{x_k : N \ni k \geq n\} : n \in \mathbb{N}\}.$$

The *filter associated with a sequence* $\{x_n\}_{n \in N}$ on X is defined as the family of the subsets F of X, for which there is $n \in \mathbb{N}$ such that $\{x_k : N \ni k \geq n\} \subset F$. This family is also called *sequential filter* of $\{x_n\}_{n \in N}$. We denote it by

$$(x_n)_n^\uparrow.$$

Exercise I.3.4. Let $\{x_n\}_n$ be a sequence on X. Check that

$$F \in (x_n)_n^\uparrow \implies F \neq \varnothing, \tag{I.3.1}$$

$$G \supset F \in (x_n)_n^\uparrow \implies G \in (x_n)_n^\uparrow, \tag{I.3.2}$$

$$F_0, F_1 \in (x_n)_n^\uparrow \implies F_0 \cap F_1 \in (x_n)_n^\uparrow. \tag{I.3.3}$$

It follows from the definition that if $X \neq \varnothing$, the *range* $f(N)$ of a sequence $f : N \to X$ is countable. Notice that if the range of a sequence f on X is finite, then there exists $x \in X$ such that $\{n \in \mathbb{N} : f(n) = x\}$ is infinite.

---

[8]Considered with the order inherited from $\mathbb{N}$.

A sequence $f : N \to X$ is said to be *one-to-one* (or *injective*) if f is injective, that is, if $\{n \in \mathbb{N} : f(n) = x\}$ is either a singleton or empty for each $x \in X$. A sequence f on X is said to be *finite-to-one* if $\{n \in \mathbb{N} : f(n) = x\}$ is finite for each $x \in X$.

Example I.3.5. Consider the following sequences

$$1, \frac{1}{2}, \frac{1}{3}, \ldots \frac{1}{n}, \ldots \tag{I.3.4}$$

$$1, \frac{1}{2}, 1, \frac{1}{2}, 1, \frac{1}{2}, \ldots \tag{I.3.5}$$

$$1, \frac{1}{2}, \frac{1}{2}, \frac{1}{3}, \frac{1}{3}, \frac{1}{3}, \ldots, \underbrace{\frac{1}{n}, \ldots, \frac{1}{n}}_{n \text{ times}}, \ldots \tag{I.3.6}$$

$$1, 1, \frac{1}{2}, 1, \frac{1}{2}, \frac{1}{3}, 1, \frac{1}{2}, \frac{1}{3}, \frac{1}{4} \ldots, \underbrace{1, \ldots, \frac{1}{n}}_{n \text{ terms}}, \ldots \tag{I.3.7}$$

on the set $\mathbb{R}$ (of real numbers). The sequence (I.3.4) is one-to-one, while (I.3.6) is finite-to-one. The remaining ones are not finite-to-one. The sequence (I.3.5) has finite range and the preimage of every element of the range is infinite. The range of (I.3.7) is $\{\frac{1}{n} : n \in \mathbb{N}_1\}$ and the preimage of an element $\frac{1}{n}$ of the range is infinite.

Definition I.3.6. A sequence $g \in X^M$ is a *subsequence of* $f \in X^N$ if there exists an increasing map $h : M \to N$ such that $g = f \circ h$.

Consequently, a sequence $\{y_k\}_{k \in M}$ is a subsequence of a sequence $\{x_n\}_{n \in N}$ if there exists an increasing map $h : M \to N$ such that

$$y_k = x_{h(k)}$$

for each $k \in M$.

Notice that h is an order isomorphism between M and $h(M)$, so that we can identify $\{y_k\}_{k \in M}$ with $\{x_n\}_{n \in h(M)}$. Our Definition I.3.1 enables a more direct approach: On identifying order isomorphic domains, each subsequence of a sequence $\{x_n\}_{n \in N}$ can be determined by $M \subset N$, where M is infinite, so that,

$$\{\{x_n\}_{n \in M} : M \subset N, \operatorname{card} M = \infty\}$$

represent all the subsequences of $\{x_n\}_{n \in N}$.

Definition I.3.7. We say that a sequence $\{x_n\}_{n \in N}$ *converges* to x,

$$x \in \lim_{n \in N} x_n,$$

or that x is a *limit* of $\{x_n\}_{n \in N}$ if for each $B \in \mathcal{B}(x)$ there is n_B such that $x_n \in B$ for each $n \geq n_B$ and $n \in N$.

Remark I.3.8. It follows from the definition that $\lim_{n \in N} x_n$, is a (possibly empty) set. As we shall see, in premetric spaces, it need not be a singleton (Example III.4.9). Here we depart from a common usage in the context of metric spaces, where the limit of sequence either is an element of a space or does not exists.

Example I.3.9. Consider the premetric space of Example I.2.1. Observe that if a sequence converges to (x_0, y_0) then a tail of the sequence lies on

$$\{x_0\} \times \mathbb{R} \cup \mathbb{R} \times \{y_0\}.$$

Exercise I.3.10. Let (X, d) be a premetric space and $\{x_n\}_{n \in N}$ a sequence on X. Show that $x \in \lim_{n \in N} x_n$ if and only if one of the following conditions holds:

$$\underset{\varepsilon > 0}{\forall} \, \underset{n_\varepsilon}{\exists} \, \underset{N \ni n \geq n_\varepsilon}{\forall} d(x_n, x) < \varepsilon \tag{I.3.8}$$

$$\underset{B \in \mathcal{B}(x)}{\forall} \operatorname{card} \{n \in N : x_n \notin B\} < \infty, \tag{I.3.9}$$

$$\underset{V \in \mathcal{V}(x)}{\forall} \, \underset{n_V}{\exists} \, \{x_n : n_V \leq n \in N\} \subset V, \tag{I.3.10}$$

$$\mathcal{V}(x) \subset (x_n)_n^\uparrow. \tag{I.3.11}$$

The characterization of convergence (I.3.10) can be reformulated more concisely if we introduce a preorder (that is, a reflective and transitive relation) on the set of families of subsets of a given set.

If $\mathcal{A}, \mathcal{D}$ are two families of subsets of X, then we say that $\mathcal{D}$ is *finer* than $\mathcal{A}$, in symbols

$$\mathcal{A} \leq \mathcal{D}, \tag{I.3.12}$$

if for each $A \in \mathcal{A}$ there exists $D \in \mathcal{D}$ such that $D \subset A$. They are said *equivalent* if $\mathcal{A} \leq \mathcal{D}$ and $\mathcal{D} \leq \mathcal{A}$, in which case we write

$$\mathcal{A} \approx \mathcal{D}. \tag{I.3.13}$$

In these terms, (I.3.10) becomes

$$\mathcal{V}(x) \leq (x_n)_n. \tag{I.3.14}$$

As $(x_n)_n^\uparrow$ consists of those sets that include an element of $(x_n)_n$, (I.3.14) implies (I.3.11).

Since convergence of the sequence $\{x_n\}_{n \in N}$ only depends on the filter $(x_n)_n^{\uparrow}$ that it generates, or even only on its collection of tails $(x_n)_n$, we alternatively use the notation

$$\lim(x_n)_n := \lim_{n \in N} x_n.$$

Of course,

$$(x_n)_n \approx (y_k)_k \implies \lim(x_n)_n = \lim(y_k)_k.$$

The characterization of convergence (I.3.9) makes no use of the order of the domain of the sequence. It says that $\{x_n\}_{n \in N}$ converges to x whenever the preimage (by a sequence) of every strict ball centered at x is the complement of a finite set, that is, a *cofinite* set.

Exercise I.3.11. Let $\{x_n\}_{n \in \mathbb{N}}$ be a sequence on a premetric space and let $\varphi : \mathbb{N} \to \mathbb{N}$ be a bijection. Then $x \in \lim_{n \in \mathbb{N}} x_n$ if and only if $x \in \lim_{k \in \mathbb{N}} x_{\varphi(k)}$.

From the point of view of convergence the order on the set of indices of sequences is irrelevant. The only thing that matters are cofinite subsets of the set of indices.

Therefore, sequences can be freed from order, hence defined as maps from arbitrary infinitely countable sets. We call the resulting objects *quences* (see I.5).

What counts for convergence, are cofinite subsets of the set of indices. We shall investigate this aspect in the next section.

I.4   Cofiniteness

Definition I.4.1. A subset M of X is *cofinite* if $X \setminus M$ is finite. Let $(X)_0$ denote the set of all cofinite subsets of X.

Of course, if X is finite, each subset of X is cofinite; in particular, $\varnothing \in (X)_0$. Observe that if X is infinite, then

$$\varnothing \notin (X)_0 \,,$$
$$M \supset N \in (X)_0 \implies M \in (X)_0 \,,$$
$$N_0, N_1 \in (X)_0 \implies N_0 \cap N_1 \in (X)_0 \,.$$

Definition I.4.2. A partial map $h : X \rightarrowtail Y$ is *cofinitely continuous* if $h^-(M)$ is cofinite in X for each cofinite subset M of Y.

It follows that the domain of a cofinitely continuous partial map is cofinite, because $\operatorname{dom} h = h^-(Y)$.

Lemma I.4.3. *A partial map between infinite sets is cofinitely continuous if and only if the preimages of finite sets are finite and the domain is cofinite.*

Proof. Let X and Y be infinite and let $h : X \rightarrowtail Y$. We have already noticed that $\operatorname{dom} h$ is cofinite. Suppose that h is cofinitely continuous and B is finite. Then $h^-(Y \setminus B)$ is cofinite in X, and *a fortiori* in $h^-(Y)$, so that $h^-(B) = h^-(Y) \setminus h^-(Y \setminus B)$ is finite.

Conversely, if f is finite-to-one with cofinite domain and M is cofinite in Y then $Y \setminus M$ is finite, hence $h^-(M) = h^-(Y) \setminus h^-(Y \setminus M)$ is cofinite in $h^-(Y)$ which is cofinite in X, thus $h^-(M)$ is cofinite in X. $\qquad\square$

Exercise I.4.4. Check that the following statements are equivalent:

(1) $f : X \rightarrowtail Y$ is cofinitely continuous,
(2) for each $B \in (Y)_0$ there exists $A \in (X)_0$ such that $f(A) \subset B$,
(3) for each $B \in (Y)_0$ there exists $A \in (X)_0$ such that $A \subset f^-(B)$.

If $\mathcal{A}$ is a family of subsets of X and $\mathcal{B}$ is a family of subsets of Y, then we denote

$$f[\mathcal{A}] := \{f(A) : A \in \mathcal{A}\}, \tag{I.4.1}$$

$$f^-[\mathcal{B}] := \{f^-(B) : B \in \mathcal{B}\}. \tag{I.4.2}$$

With this notation, Exercise I.4.4 reformulates as:

Corollary I.4.5. *A partial map $f : X \rightarrowtail Y$ is cofinitely continuous if and only if $f[(X)_0] \geq (Y)_0$, equivalently, $(X)_0 \geq f^-[(Y)_0]$.*

I.5   Quences

Definition I.5.1. A *quence* on a set X is a map from a countably infinite set to X.

Every sequence is a quence. On the other hand, if $f : N \to X$ is a quence on X and $h : \mathbb{N} \to N$ is a bijection, then $f \circ h$ is a sequence.

Definition I.5.2. A quence $g : B \to X$ is called a *subquence* of a quence $f : A \to X$, in symbols,

$$g \succ f,$$

if there exists a cofinitely continuous partial map h from B to A such that $g = f \circ h$ on dom h. Two quences f and g are called *equivalent* if $g \succ f$ and $f \succ g$.

Exercise I.5.3. Show that the sequences (I.3.4) and (I.3.6) are equivalent (as quences).

It follows from the definitions that a sequence $g : M \to X$ is a subquence of a sequence $f : N \to X$ if and only if there exists a partial map $h : M \rightarrowtail N$ such that $g(k) = (f \circ h)(k)$ for $k \in$ dom h and $\lim_{k \to \infty} h(k) = \infty$.

Proposition I.5.4. *A quence* $g : M \to X$ *is a subquence of a quence* $f : N \to X$ *if and only if for each* $A \in (N)_0$ *there exists* $B \in (M)_0$ *such that* $g(B) \subset f(A)$, *that is, whenever*

$$f[(N)_0] \leq g[(M)_0]. \tag{I.5.1}$$

Proof. Let h be as in Definition I.5.2. Each element of $f[(N)_0]$ is of the form $f(A)$ where A is a cofinite subset of N. Consequently $h^-(A)$ is cofinite in M and thus $g(h^-(A)) \in g[(M)_0]$. As $g = f \circ h$ on dom h,

$$g(h^-(A)) = f(h(h^-(A))) \subset f(A),$$

which proves (I.5.1).

Conversely, suppose that the (I.5.1) holds. Up to a bijection, $N = \mathbb{N}$. By (I.5.1), there exists a sequence of cofinite subsets $\{M_n\}_{n \in \mathbb{N}}$ of M such that $g(M_n) \subset f(\mathbb{N}_n)$. By an immediate induction we can assume that $M_n \supsetneq M_{n+1}$ for all n and $\bigcap_{n \in \mathbb{N}} M_n = \varnothing$. For each $m \in M_n$, let

$$h_n(m) := \min\{k \in \mathbb{N}_n : g(m) = f(k)\},$$

and let

$$h(m) := h_n(m) \text{ if } m \in M_n \setminus M_{n+1}.$$

Then, for each $n \in \mathbb{N}$ the set $h^-(\mathbb{N}_n) \supset M_n$, hence is cofinite, and thus h is cofinitely continuous. On the other hand, $g(m) = f(h(m))$ for each $m \in M_0$. $\square$

Let us observe that (I.5.1) amounts to

$$(g^- \circ f)[(N)_0] \leq (M)_0. \tag{I.5.2}$$

Corollary I.5.5. *Two quences* $f : N \to X$ *and* $g : M \to X$ *are equivalent if and only if* $f[(N)_0] \approx g[(M)_0]$ *(in the sense of (I.3.12)).*

Of course, each subsequence of a sequence f is a subquence of f. On the other hand,

Proposition I.5.6. *For each subquence g of a sequence f there is subquence s of g that is a subsequence of f.*

Proof. Let $f \in X^N$ (where N is an infinite subset of $\mathbb{N}$) and let $g \in X^B$ be a subquence of f. Let $h : B \rightarrowtail N$ be a cofinitely continuous partial map such that $g = f \circ h$ on $\operatorname{dom} h$. Accordingly $h(B)$ is infinite and thus a map $s : h(B) \to X$ such that $s(n) = f(n)$ for each $n \in h(B)$, is a subsequence of f. For each $n \in h(B)$, let $j(n)$ be any element of $h^-(n)$. Then $j : h(\operatorname{dom} h) \to B$ is cofinitely continuous and $s = g \circ j$, hence s is a subquence of g. $\qquad\square$

Remark I.5.7. The sequence (I.3.4) is a subsequence of (I.3.6). Conversely, (I.3.6) is a subquence of (I.3.4) because of Lemma I.4.3, but it is not a subsequence of (I.3.4).

Of course, two sequences $\{y_k\}_{k \in \mathbb{N}}$ and $\{x_n\}_{n \in \mathbb{N}}$ are equivalent as quences if and only if their families of tails $(y_k)_k$ and $(x_n)_n$ are equivalent families (in the sense of (I.3.13)). Thus:

Corollary I.5.8. *If $\{x_n\}_{n \in \mathbb{N}}$ and $\{y_k\}_{k \in \mathbb{N}}$ are equivalent (as quences), then $x \in \lim_n x_n$ if and only if $x \in \lim_k y_k$.*

Definition I.5.9. In a premetric space X, a quence $f : N \to X$ converges to x (or x is a *limit* of f), in symbols,

$$x \in \lim f \qquad\qquad (I.5.3)$$

if $f^-(B(x,r))$ is cofinite for every $r > 0$.

In particular, a sequence $f \in X^N$ converges to $x \in X$ if and only if the preimage by f of every ball centered at x is cofinite in N, which is equivalent to (I.3.9). Thus if $f \in X^N$ is a sequence

$$x \in \lim f \text{ as in (I.5.3)} \iff x \in \lim(f(n))_n \text{ as in Definition I.3.7.}$$

By definition, N is an infinite subset of $\mathbb{N}$, but convergence does not require any order on N. Therefore, the definition uses merely the fact that f is a quence. In terms of (I.4.1) and (I.4.2), (I.5.3) is equivalent to

$$\mathcal{V}(x) \leq f[(N)_0].$$

Notice that if

$$(X,g) \geq (X,d) \implies \lim_g f \subset \lim_d f.$$

Proposition I.5.10. *If a quence converges to x then its every subquence converges to x.*

Proof. Let $f : A \to X$ be a quence and $x \in \lim f$, that is, $f^- (B(x,r))$ is cofinite in A for each $r > 0$. If $g : B \to X$ is a subquence of f, that is, there is a cofinitely continuous map $h : B \rightarrowtail A$ such that $g = f \circ h$ on $\operatorname{dom} h$, then

$$g^- (B(x,r)) = (f \circ h)^- (B(x,r)) = h^- (f^- (B(x,r)))$$

is cofinite, so that $x \in \lim g$. □

By Proposition I.5.10,

Corollary I.5.11. *If a sequence converges to x then its every subquence, hence its every subsequence, converges to x.*

We have introduced the concept of quence, because, from the point of view of convergence, the notion of sequence is too restrictive. Although the new concept enables us to better understand convergence of sequences, we did not yet prove that its introduction was necessary. We shall see now that it is however the case. In the next section we shall demonstrate (Corollary I.6.5 and Example I.6.6) that a usual diagonal procedure employing infinite sequences necessitates in fact the notion of quence.

I.6   Almost inclusion

In this section, we assume that all sets considered are subsets of a fixed set X. We say that a set A is *almost included* in a set B, in symbols,

$$A \subset_0 B,$$

if $\operatorname{card}(A \setminus B) < \infty$. We notice that, for each A, B and C,

$$A \subset_0 A, \tag{reflexive}$$

$$A \subset_0 B \text{ and } B \subset_0 C \Longrightarrow A \subset_0 C. \tag{transitive}$$

We say that two sets A and B are *almost equal* (in symbols, $A \approx_0 B$) if $A \subset_0 B$ and $B \subset_0 A$. Note that $\approx_0$ is an equivalence relation. We denote by $[A]_0$ the set of subsets of X that are almost equal to the set A.

Exercise I.6.1. Check that

(1) the relation $\subset_0$ on the family of almost equal subsets of X (defined by $[A]_0 \subset_0 [B]_0$ if $A \subset_0 B$) is well-defined and is an order;

(2) $A \approx_0 B$ if and only if the symmetric difference $A \triangle B := (A \setminus B) \cup (B \setminus A)$ is finite;

(3) Every two finite sets are almost equal (considered, for instance, as subsets of their union).

The concept of almost inclusion enables us to rephrase Definition I.5.9 as follows:

Proposition I.6.2. *A quence $\{x_n\}_{n \in N}$ converges to x if and only if the image $\{x_n : n \in N\}$ of the quence is almost included in every ball containing x.*

Definition I.6.3. A set C is an *almost intersection* of a family $\mathcal{A}$ of sets whenever $C \subset_0 A$ for every $A \in \mathcal{A}$.

Of course, if C is an almost intersection of $\mathcal{A}$ and $D \approx_0 C$ then C is also an almost intersection of $\mathcal{A}$.

Theorem I.6.4. *A family $\{A_n : n \in \mathbb{N}\}$ of infinite sets such that $A_{n+1} \subset_0 A_n$ for each $n \in \mathbb{N}$ admits an infinite almost intersection.*

Proof. Let $x_0 \in A_0$. By assumption, there exists $x_1 \in A_0 \cap A_1 \setminus \{x_0\}$. If we have already chosen distinct $x_0, \ldots, x_n$ such that $x_k \in A_k$ for each $0 \le k \le n$, then there exists $x_{n+1} \in \bigcap_{0 \le k \le n+1} A_k \setminus \{x_0, \ldots, x_n\}$, because $\bigcap_{0 \le k \le n+1} A_k$ is infinite. Then $A_\infty := \{x_n : n \in \mathbb{N}\}$ has the property that $A_\infty \setminus A_n \subset \{x_0, \ldots, x_n\}$ for each n. $\square$

Corollary I.6.5. *If f_{n+1} is a subsequence of a sequence f_n for each $n \in \mathbb{N}$, then there is a sequence f_∞ that is a subquence of f_n for each $n \in \mathbb{N}$.*

Proof. Let $f_n : N_n \to X$ be such that $N_{n+1} \subset N_n$ and $f_{n+1} = f_n|_{N_{n+1}}$. By Theorem I.6.4, there is N_∞ (included in N_0) such that $N_\infty \subset_0 N_n$ for each $n \in \mathbb{N}$. Then $f_\infty := f_0|_{N_\infty}$ is a subsequence of each sequence f_n. Indeed, if we define a partial map h_n from N_∞ to N_n as the identity on $N_\infty \cap N_n$, then f_n coincides with f_∞ on $\mathrm{dom}\, h_n = N_\infty \cap N_n$. $\square$

In general, in Corollary I.6.5, it is not possible to find f_∞ which is a common subsequence of each f_n.

Example I.6.6. Let X be an (infinite) set and $f : \mathbb{N} \to X$ an injective map. Define $f_n := f|_{\mathbb{N}_n}$ for each $n \in \mathbb{N}$. Suppose that $f_\infty : M \to X$ is a common subsequence of all f_n for $n \in \mathbb{N}$. Let $h_0 : M \to \mathbb{N}_0$ be an increasing map such that $f_\infty = f_0 \circ h_0$. If $n > h_0 (\min M)$, then $f_\infty (\min M) \notin f_n (\mathbb{N}_n)$, hence there is no map $h_n : M \to \mathbb{N}_n$ such that $f_\infty (\min M) = (f_n \circ h_n) (\min M)$.

I.7 When premetrics and sequences do not suffice

I.7.1 *Pointwise convergence*

In Analysis, you have encountered various notions of convergence for sequences of functions. For instance, a sequence $\{f_n\}_{n \in N}$ of real-valued functions on a set X *converges pointwise* to a function f if for every $x \in X$ the numerical sequence $\{f_n(x)\}_{n \in N}$ converges to $f(x)$, that is, by definition

$$f \in \lim_p (f_n)_n \iff \underset{x \in X}{\forall} \; f(x) \in \lim(f_n(x))_n.$$

Of course, $f \in \lim_p(f_n)_n$ whenever the constant zero function $\overline{0} \in \lim_p(f_n - f)_n$, so that we can restrict ourselves to pointwise convergence to the zero function.

For our purpose of illustrating how (pre)metrics and sequences are inadequate to describe pointwise convergence, we will focus on sequences of characteristic functions of subsets.

Lemma I.7.1. *If $\{A_n\}_n$ is a sequence of subsets of X then*

$$\overline{0} \in \lim_p(\chi_{A_n})_n \iff \overline{0} \in \lim_p \left(\chi_{\bigcup_{n \geq m} A_n} \right)_m. \qquad (I.7.1)$$

Proof. By definition $\overline{0} \in \lim_p(\chi_{A_n})_n$ if and only if for every x, there is m such that for every $n \geq m$, $\chi_{A_n}(x) = 0$, that is, $x \notin \bigcup_{n \geq m} A_n$, or equivalently, $\chi_{\bigcup_{n \geq m} A_n}(x) = 0$. In other words, (I.7.1). $\qquad \square$

Note that if $\{A_n\}_n$ is a sequence of subsets of X, the sequence $\left\{ \bigcup_{n \geq m} A_n \right\}_m$ is a decreasing sequence of subsets of X, and thus the corresponding sequence $\left\{ \chi_{\bigcup_{n \geq m} A_n} \right\}_m$ of characteristic functions is decreasing as well (for the pointwise order on $\mathbb{R}^X$: $f \leq g$ if $f(x) \leq g(x)$ for all $x \in X$). Moreover,

Proposition I.7.2. *A decreasing sequence $\{f_n\}_{n \in N}$ of non-negative real-valued functions converges to $\overline{0}$ if and only if for each $k \in \mathbb{N}_1$*

$$\left\{ x \in X : \inf_{n \in N} f_n(x) \geq \tfrac{1}{k} \right\} = \varnothing. \qquad (I.7.2)$$

Proof. The condition (I.7.2) means that for each $x \in X$ and for every $k \in \mathbb{N}_1$ there is n such that $|f_n(x)| < \frac{1}{k}$, and since the sequence is decreasing, $|f_m(x)| < \frac{1}{k}$ for each $m \geq n$. □

Thus, given a sequence $\{A_n\}_n$ of subsets of X, Proposition I.7.2 applies to the sequence $\{\chi_{\bigcup_{n \geq m} A_n}\}_n$ of characteristic functions, to the effect that $\overline{0} \in \lim_p \left(\chi_{\bigcup_{n \geq m} A_n}\right)_m$ if and only if for each $x \in X$ there is m such that $x \notin \bigcup_{n \geq m} A_n$. In other words, in view of Lemma I.7.1:

Corollary I.7.3. *If $\{A_n\}_n$ is a sequence of subsets of X, then*

$$\overline{0} \in \lim_p (\chi_{A_n})_n \iff \bigcap_{m \in \mathbb{N}} \bigcup_{n \geq m} A_n = \varnothing.$$

Corollary I.7.4. *Let X be an infinite set. Then there exists a sequence $\{A_n\}_n$ of cofinite subsets of X with*

$$\overline{0} \in \lim_p (\chi_{A_n})_n$$

if and only if X is countable.

Proof. If X is countable, say $X = \{x_n : n \in \mathbb{N}\}$ is an injective representation, then $A_n := \{x_k : k \geq n\}$ satisfies the required conditions (and is decreasing). Conversely, if there is a sequence $\{A_n\}_n$ of cofinite subsets of X with $\overline{0} \in \lim_p (\chi_{A_n})_n$, then by Corollary I.7.3, $\bigcap_{m \in \mathbb{N}} \bigcup_{n \geq m} A_n = \varnothing$ so that $X = \bigcup_{m \in \mathbb{N}} \bigcap_{n \geq m} (X \setminus A_n)$ is a countable union of finite sets, hence is countable. □

Although we have not defined any premetric on $\mathbb{R}^X$, we can extend the notion of vicinity to this case. A subset V of $\mathbb{R}^X$ is a *vicinity* of $f \in \mathbb{R}^X$ if there exists a finite subset F of $\mathbb{R}$ and $r > 0$ such that

$$\max_{x \in F} |g(x) - f(x)| < r \implies g \in V.$$

We denote by $\mathcal{V}_p(f)$ the set of all vicinities of f. Notice that $\mathcal{V}_p(f)$ has the same properties as $\mathcal{V}(x)$ from Exercise I.2.7.

Proposition I.7.5. *A sequence $\{f_n\}_n$ in $\mathbb{R}^X$ converges pointwise to a function f_∞ if and only if*

$$\mathcal{V}_p(f_\infty) \subset (f_n)_n^{\uparrow}. \tag{I.7.3}$$

Proof. If (I.7.3) holds, then, in particular,

$$V_{x,r} := \{g : |g(x) - f_\infty(x)| < r\} \in (f_n)_n^\uparrow$$

for each $x \in \mathbb{R}$, that is, there is n_r such that $|f_n(x) - f_\infty(x)| < r$ for each $n \geq n_r$.

Conversely, if $\{f_n\}_n$ converges to f_∞ pointwise, then for each $r > 0$ and each $x \in \mathbb{R}$,

$$V_{x,r} \in (f_n)_n^\uparrow.$$

If F is a finite set then, by (I.3.3),

$$V_{F,r} := \bigcap\nolimits_{x \in F} V_{x,r} \in (f_n)_n^\uparrow,$$

and, by (I.2.7), $V_{F,r} \in \mathcal{V}_p(f_\infty)$. Hence, by (I.2.6) and (I.3.2), $\mathcal{V}_p(f_\infty) \subset (f_n)_n^\uparrow$. $\square$

Let us notice that:

Proposition I.7.6. *A function $f \in \mathbb{R}^X$ has a countable (infinite) vicinity base if and only if X is countable.*

Proof. We have already observed that it is enough to study the pointwise convergence to the constant zero function. Indeed, if X is countable, then the sets

$$\left\{f : \max\nolimits_{x \in F} |f(x)| < \tfrac{1}{n}\right\},$$

where F is a finite subset of X and $n \in \mathbb{N}_1$, is a vicinity base of $\overline{0}$.

Suppose that X is uncountable and $\{V_n : n \in \mathbb{N}\}$ is a vicinity base of $\overline{0}$. Then for each n there is a finite subset F_n of X and $r_n \in (0,1)$ such that $\{f : \max_{x \in F_n} |f(x)| < r_n\} \subset V_n$. The set $\bigcup_{n \in \mathbb{N}} F_n$ is countable, hence there exists $x_0 \in X \setminus \bigcup_{n \in \mathbb{N}} F_n$. Of course, $\{f : |f(x_0)| < 1\} \in \mathcal{V}_p(\overline{0})$, but

$$\psi_{\{x_0\}} \in \{f : |f(x_0)| < 1\} \setminus V_n$$

for each $n \in \mathbb{N}$. $\square$

As each point of a premetrizable space admits a countable vicinity base, we conclude:

Proposition I.7.7. *If X is uncountable, then there is no premetric d on $\mathbb{R}^X$, in particular no metric, such that $\{f_n\}_n$ converges pointwise to f if and only if $\{f_n\}_n$ converges to f for the premetric d.*

We shall see in Corollary XVIII.3.5 that the converse also holds.

Theorem I.7.8. *The pointwise convergence on $\mathbb{R}^X$ is (pre)metrizable if and only if X is countable.*

Therefore, to provide a general framework to consider convergence of sequences that includes notions arising from Analysis, structures more general than (pre)metrics need to be introduced. Moreover, we shall see that not only premetrics are not adequate to describe pointwise convergence (on uncountable sets), but also convergent sequences are not sufficient to specify vicinities for this convergence.

Consider the family

$$\mathbb{E}_0 := \left\{ (f_n)_n^{\uparrow} : \overline{0} \in \lim_p (f_n)_n \right\}, \tag{I.7.4}$$

of sequential filters of sequences that converge pointwise to the zero function. If we set $\mathcal{F} := \bigcap_{\mathcal{E} \in \mathbb{E}_0} \mathcal{E}$, then it is straightforward that

$$\varnothing \notin \mathcal{F}, \tag{I.7.5}$$

$$D \supset F \in \mathcal{F} \Longrightarrow D \in \mathcal{F}, \tag{I.7.6}$$

$$F_0, F_1 \in \mathcal{F} \Longrightarrow F_0 \cap F_1 \in \mathcal{F}. \tag{I.7.7}$$

We have already observed that various objects, like vicinities, sequential filters, the set of cofinite subsets of infinite set, share properties (I.7.5)-(I.7.7).

Definition I.7.9. *A family $\mathcal{F}$ of subsets of a given set fulfilling (I.7.5)-(I.7.7) is called a* filter.

Proposition I.7.5 implies that if $\mathcal{F} = \bigcap_{\mathcal{E} \in \mathbb{E}_0} \mathcal{E}$ then

$$\mathcal{V}_p \left(\overline{0} \right) \subset \mathcal{F}. \tag{I.7.8}$$

This condition is analogous to the characterization of pointwise convergence of sequences in Proposition I.7.5. We shall use it to extend the concept of convergence from (se)quences to arbitrary filters: If $\mathcal{F}$ is a filter on $\mathbb{R}^X$ then

$$f \in \lim_p \mathcal{F} \iff \mathcal{V}_p(f).$$

It follows that the infimum of all sequences that pointwise converge to $f \in \mathbb{R}^X$, also converges to f. Moreover, if X is infinite, then this infimum is not a sequential filter, and if X is uncountable, this infimum is not the least filter that converges to f:

Proposition I.7.10. *If X is uncountable then*

$$\left(\bigcap_{\mathcal{E} \in \mathbb{E}_0} \mathcal{E} \right) \setminus \mathcal{V}_p \left(\overline{0} \right) \neq \varnothing. \tag{I.7.9}$$

Proof. Consider the set

$$A := \{\chi_D : D \in (X)_0\}$$

and its complement $B := \mathbb{R}^X \setminus A$. Then $B \in \bigcap_{\mathcal{E} \in \mathbb{E}_0} \mathcal{E}$. Indeed, if there were $(f_n)_n^\uparrow \in \mathbb{E}_0$ with $B \notin (f_n)_n^\uparrow$, then for every $n \in \mathbb{N}$, $\{f_k : k \geq n\} \cap A \neq \varnothing$. Therefore $(f_n)_n^\uparrow$ would have a subsequence of elements of A, which would converge to $\overline{0}$, because $(f_n)_n^\uparrow$ converges to $\overline{0}$. On the other hand, by Corollary I.7.4, no sequence of elements of A converges to $\overline{0}$, which yields a contradiction.

On the other hand, for each finite subset F of X and each $r > 0$,

$$\chi_{X \setminus F} \in \left\{ g \in \mathbb{R}^X : \max_{x \in F} |g(x)| < r \right\} \setminus B,$$

so that $B \notin \mathcal{V}_p(\overline{0})$. We have proved that $B \in \bigcap_{\mathcal{E} \in \mathbb{E}_0} \mathcal{E} \setminus \mathcal{V}_p(\overline{0})$. $\square$

In other words, sequences do not suffice to reconstruct $\mathcal{V}_p(\overline{0})$.

I.7.2 *Riemann integrals*

More concretely, many constructions involve a limiting process that does not straightforwardly correspond to the limit of a sequence, but such limits can usually easily be interpreted as limits of filters.

For instance, when defining Riemann integral of a bounded function $f : \mathbb{R} \to \mathbb{R}$ on a closed interval $[a, b]$, one usually proceeds with one variant or another of the following:

A finite set of points $x_0, x_1, \ldots, x_n$ such that

$$x_0 = a \leq x_1 \leq x_2 \leq \ldots \leq x_n = b$$

is a *subdivision of* $[a, b]$ and partitions $[a, b]$ into subintervals $[x_k, x_{k+1}]$ of width Δx_k, for $k \in \{0, 1 \ldots, n - 1\}$. Picking a sample point x_k^* in each interval $[x_k, x_{k+1}]$, we obtain *a marked partition*

$$\Pi = \left\{ x_0 = a \leq x_0^* \leq x_1 \leq x_1^* \leq \ldots \leq x_{n-1} \leq x_{n-1}^* \leq x_n = b \right\}.$$

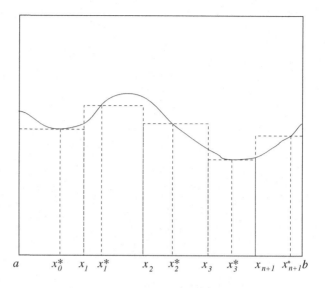

Figure I.5   A marked partition.

To such a partition, we associate the *Riemann sum*

$$S_\Pi(f) := \sum_{k=0}^{n-1} f(x_k^*)\Delta x_k.$$

Let $p(\Pi) = \max_{i \in \{0,\dots,n-1\}} \Delta x_i$ be the *parameter of the partition* Π. The function f is *Riemann integrable on* $[a,b]$ if there is a finite number I such that for every $\varepsilon > 0$ there is $\delta > 0$ such that $|S_\Pi(f) - I| < \varepsilon$ whenever Π is a marked partition with $p(\Pi) < \delta$.

The number I is usually denoted $\int_a^b f(x)\,dx$ and called *Riemann integral of f on $[a,b]$*.

Hence the integral is defined, roughly speaking, as the limit of Riemann sums of f over marked partitioned, as the parameter of the partition goes to 0. However, this does not straightforwardly correspond to the limit of a sequence, because there are a lot more marked partitions to consider than simply a sequence of them. On the other hand, the collection

$$\mathcal{F} := \{\{S_\Pi(f) : \Pi \text{ marked partition of } [a,b],\, p(\Pi) < r\} : r > 0\}^\uparrow$$

of subsets of $\mathbb{R}$ is a filter. The function f is integrable if and only if for some number l (also denoted $\int_a^b f(x)\,dx$)

$$\mathcal{F} \supset \mathcal{V}(l) \qquad\qquad (\text{I.7.10})$$

where $\mathcal{V}(l)$ is the collection of vicinities (in the sense of Definition I.2.6) of the real number l for the usual Euclidean metric of $\mathbb{R}$. In view of the characterization (I.3.11) of convergence of sequences in a premetric space, it is natural to say that a filter $\mathcal{F}$ converges to an element l of a premetric space if (I.7.10) holds. In these terms, Riemann integrability is formulated in terms of the convergence of a filter rather than a sequence.

Having demonstrated that the class of convergent objects need be extended to filters, we shall study filters and other related families of subsets in Chapter II.

Chapter II

Families of sets

A *family* of subsets of X is defined as a set of subsets of X. We use the term *family* in order to distinguish subsets of X from subsets of 2^X. For a similar reason, a *collection* on X is defined as a set of families of subsets of X, that is, as a subset of $2^{2^X} := 2^{(2^X)}$.

II.1 Isotone families of sets

Recall from (I.3.12) that a family $\mathcal{A}$ is said to be *coarser* than $\mathcal{B}$ ($\mathcal{B}$ is *finer* than $\mathcal{A}$), in symbols, $\mathcal{A} \leq \mathcal{B}$, if for each $A \in \mathcal{A}$ there exists $B \in \mathcal{B}$ such that $B \subset A$.

The relation $\leq$ on the set of all families of subsets of a given set X ([1]) is *transitive* and *reflexive*

$$(\mathcal{A} \leq \mathcal{B} \text{ and } \mathcal{B} \leq \mathcal{C}) \Longrightarrow \mathcal{A} \leq \mathcal{C}, \qquad \text{(transitive)}$$

$$\mathcal{A} \leq \mathcal{A}. \qquad \text{(reflexive)}$$

It is not antisymmetric, hence it is not an order relation.

Exercise II.1.1. Give an example of two families $\mathcal{A}, \mathcal{B}$ of subsets of $\{0, 1\}$ such that $\mathcal{A} \leq \mathcal{B} \leq \mathcal{A}$ and $\mathcal{A} \neq \mathcal{B}$.

Recall from (I.3.13) that two families $\mathcal{A}$ and $\mathcal{D}$ are said to be *equivalent* $\mathcal{A} \approx \mathcal{D}$ if $\mathcal{A} \leq \mathcal{D}$ and $\mathcal{D} \leq \mathcal{A}$. The relation $\leq$ can be quotiented to the class of equivalent families, where it becomes antisymmetric, hence an order.

Definition II.1.2. A family $\mathcal{A}$ of subsets of X is called *isotone* if $A \in \mathcal{A}$ and $A \subset B$ imply that $B \in \mathcal{A}$. The least isotone family that includes a given family $\mathcal{A}$ is called the *isotonization* of $\mathcal{A}$ and is denoted by $\mathcal{A}^\uparrow$.

---

[1]that is, on 2^{2^X}.

It is straightforward that

$$\mathcal{A}^\uparrow := \{B \subset X : \underset{A \in \mathcal{A}}{\exists} \; A \subset B\}. \tag{II.1.1}$$

When it is unclear from context in what ambient set X the isotonization of a family is performed, we may specify $\mathcal{A}^{\uparrow X}$. We denote by $\mathbb{I}X$ the set of all isotone families of subsets of X. Dually, if $\mathcal{A} \subset 2^X$ we define

$$\mathcal{A}^\downarrow := \{B \subset X : \underset{A \in \mathcal{A}}{\exists} \; B \subset A\}, \tag{II.1.2}$$

and we use $\mathcal{A}^{\downarrow X}$ instead when the underlying set needs to be specified.

Exercise II.1.3. Let $X := \{1, 2, 3\}$ and $\mathcal{A} := \{\{1, 2\}, \{3\}\}$. Describe $\mathcal{A}^\uparrow$ and $\mathcal{A}^\downarrow$.

Of course, $\mathcal{A} \subset \mathcal{A}^\uparrow$ hence $\mathcal{A} \leq \mathcal{A}^\uparrow$ for each family $\mathcal{A}$ of subsets of X. On the other hand, by the very definition, $\mathcal{A}^\uparrow \leq \mathcal{A}$. Thus $\mathcal{A}^\uparrow \approx \mathcal{A}$. In fact, by definition

$$\mathcal{A} \leq \mathcal{B} \iff \mathcal{A} \subset \mathcal{B}^\uparrow. \tag{II.1.3}$$

Thus

Corollary II.1.4. *If $\mathcal{A}$ and $\mathcal{B}$ are families of subsets of X and $\mathcal{B}$ is isotone, then $\mathcal{A} \leq \mathcal{B}$ if and only if $\mathcal{A} \subset \mathcal{B}$.*

Proposition II.1.5. *Two families $\mathcal{A}$ and $\mathcal{B}$ are equivalent if and only if $\mathcal{A}^\uparrow = \mathcal{B}^\uparrow$.*

Corollary II.1.6. *The relation $\leq$ coincides with the inclusion $\subset$ on the set of isotone families, hence is a relation of order therein.*

Definition II.1.7. $\mathcal{B}$ *refines* $\mathcal{A}$ (or $\mathcal{B}$ is a *refinement* of $\mathcal{A}$), in symbols

$$\mathcal{B} \lhd \mathcal{A},$$

if for every $B \in \mathcal{B}$ there is $A \in \mathcal{A}$ with $B \subset A$.

If $\mathcal{D}$ is a family of subsets of X, then we denote by $\mathcal{D}_c$ the set of the complements in X of the elements of $\mathcal{D}$, that is,

$$\mathcal{D}_c := \{X \setminus D : D \in \mathcal{D}\}. \tag{II.1.4}$$

Proposition II.1.8. *Let $\mathcal{A}, \mathcal{B}$ be two families of subsets of a given set. Then*

$$\mathcal{B} \lhd \mathcal{A} \iff \mathcal{A}_c \geq \mathcal{B}_c. \tag{II.1.5}$$

Proof. Indeed $\mathcal{A}_c \geq \mathcal{B}_c$ means that for every $B \in \mathcal{B}$ there is $A \in \mathcal{A}$ such that $B^c \supset A^c$, equivalently, $A \supset B$. $\qquad\qquad\square$

Proposition II.1.9. *If $h : X \to Y$, $\mathcal{F}$ is a family of subsets of X and $\mathcal{G}$ is a family of subsets of Y, then*

$$h\,[\mathcal{F}] \geq \mathcal{G} \Longleftrightarrow \mathcal{F} \geq h^-\,[\mathcal{G}]\,.$$

Proof. If $h\,[\mathcal{F}] \geq \mathcal{G}$, that is, for each $G \in \mathcal{G}$ there exists $F \in \mathcal{F}$ such that $h\,(F) \subset G$, then $F \subset h^-\,(h\,(F)) \subset h^-\,(G)$, hence $\mathcal{F} \geq h^-\,[\mathcal{G}]$. Conversely, if the latter holds, that is, for each $G \in \mathcal{G}$ there exists $F \in \mathcal{F}$ such that $F \subset h^-\,(G)$, then $h\,(F) \subset h\,(h^-\,(G)) \subset G$, hence $h\,[\mathcal{F}] \geq \mathcal{G}$. $\qquad\square$

II.2 Filters

One of the main objects of this book is *filter* that was introduced in Definition I.7.9 as a family $\mathcal{F}$ of subsets of X such that

$$\varnothing \notin \mathcal{F}, \qquad\qquad\qquad\qquad\qquad\text{(proper)}$$
$$D \supset F \in \mathcal{F} \Longrightarrow D \in \mathcal{F}, \qquad\qquad\text{(isotone)}$$
$$F_0, F_1 \in \mathcal{F} \Longrightarrow F_0 \cap F_1 \in \mathcal{F}. \qquad\text{(finite intersection)}$$

We denote by $\mathbb{F}X$ the set of all filters on X.

Remark II.2.1. It is isotony (isotone) that makes filters a little bit counter intuitive at first: the finer the filter, the bigger it is as a family of subsets of X. In other words, a filter is large (or fine) if it contains small (but non-empty) sets. Exercises II.2.6 and II.2.7, as well as Proposition II.2.11 below will illustrate this fact.

A family $\mathcal{B}$ of non-empty subsets of X is called a *filter-base* if $\mathcal{B}^\uparrow$ is a filter on X. We say that the filter-base $\mathcal{B}$ *generates* the filter $\mathcal{B}^\uparrow$.

Exercise II.2.2. Show that $\mathcal{B}$ is a filter-base if and only if for every B_0 and B_1 in $\mathcal{B}$, there is $B \in \mathcal{B}$ such that $B \subset B_0 \cap B_1$ and $\varnothing \notin \mathcal{B}$.

Exercise II.2.3. Show that for a given set X,

(1) the relation $\approx$ defined by (I.3.13) ([2]) is an equivalence relation on the set of filter-bases on X;
(2) each equivalence class contains exactly one filter, generated by any of the filter-bases in that class.

---
[2]that is, $\mathcal{A} \approx \mathcal{B}$ if $\mathcal{A} \geq \mathcal{B}$ and $\mathcal{B} \geq \mathcal{A}$

In other words, filters on X are in one-to-one correspondence with equivalent classes.

Example II.2.4. For each element x of a premetric space, the family $\mathcal{B}(x)$ of balls centered at x is a filter-base, and the family $\mathcal{V}(x) = \mathcal{B}(x)^{\uparrow}$ of all vicinities of x, is a filter. We call $\mathcal{V}(x)$ the *vicinity filter* of x. Similarly, if $\{r_n\}_n$ is a sequence of $\mathbb{R}_+$ with $\lim_{n\to\infty} r_n = 0$ then $\mathcal{B}' := \{B(x, r_n) : n \in \mathbb{N}\}$ is another (countable!) filter-base of $\mathcal{V}(x)$.

Recall (Section I.5) that the family $(X)_0$ of cofinite subsets of a set X is a filter if and only if X is infinite.

Exercise II.2.5. Let X be an uncountable set and
$$(X)_1 := \{A \subset X : \operatorname{card}(X \setminus A) \leq \aleph_0\}.$$
Show that:

(1) $(X)_1$ is a filter on X. We call $(X)_1$ the *cocountable filter on X*.
(2) $\{F_n : n \in \mathbb{N}\} \subset (X)_1$ implies $\bigcap_{n\in\mathbb{N}} F_n \in (X)_1$.

A filter $\mathcal{F}$ is called *countably deep* if whenever $\mathcal{A}$ is a countable family with $\mathcal{A} \subset \mathcal{F}$ then $\bigcap_{A\in\mathcal{A}} A \in \mathcal{F}$. According to Exercise II.2.5 (2), the cocountable filter of an uncountable set is countably deep.

II.2.1 *Order*

Exercise II.2.6. Verify that if $\mathbb{D} \subset \mathbb{F}X$ is a set of filters on X, then $\bigcap_{\mathcal{F}\in\mathbb{D}} \mathcal{F}$ is a filter on X and that $\left\{\bigcup_{\mathcal{F}\in\mathbb{D}} s(\mathcal{F}) : s : \mathbb{D} \to 2^X, s(\mathcal{F}) \in \mathcal{F}\right\}$ is a filter-base for $\bigcap_{\mathcal{F}\in\mathbb{D}} \mathcal{F}$.

As each filter is an isotone family, that is,
$$\mathbb{F}X \subset \mathbb{I}X,$$
the partial order $\leq$ on $\mathbb{I}X$ in Corollary II.1.6 can be restricted to $\mathbb{F}X$.

Exercise II.2.7 (infimum of filters). Verify that in the situation of Exercise II.2.6, the filter $\bigcap_{\mathcal{F}\in\mathbb{D}} \mathcal{F}$ is the *greatest lower bound* (or the *infimum*) of the set $\mathbb{D}$ in the partially ordered set $(\mathbb{F}X, \leq)$.

As we use $\bigwedge$ to denote infimum, we will write $\bigwedge_{\mathcal{F}\in\mathbb{D}} \mathcal{F}$ and $\bigcap_{\mathcal{F}\in\mathbb{D}} \mathcal{F}$ interchangeably.

Remark II.2.8. In contrast, the *least upper bound* of a family of filters on a set X does not always exist. In other, words, the supremum (within isotone families) of a set of filters is not necessarily a filter.

Definition II.2.9. Two families $\mathcal{A}$ and $\mathcal{B}$ of subsets of a set X *mesh*, in symbols

$$\mathcal{A}\#\mathcal{B},$$

if $A \cap B \neq \varnothing$ for every $A \in \mathcal{A}$ and $B \in \mathcal{B}$. Two families that do not mesh are called *dissociated*.

We write $A\#\mathcal{B}$ for $\{A\}\#\mathcal{B}$ and similarly $A\#B$ for $\{A\}\#\{B\}$, that is, when $A \cap B \neq \varnothing$.

Exercise II.2.10. Let $\mathcal{A}$ denote an isotone family of subsets of X, and let $H \subset X$. Then

$$H \notin \mathcal{A} \iff (X \setminus H)\#\mathcal{A}. \tag{II.2.1}$$

We shall denote by $\mathcal{F} \vee \mathcal{G}$ a least upper bound (supremum) of two filters $\mathcal{F}$ and $\mathcal{G}$ on the same set, that is, the least filter that is finer than $\mathcal{F}$ and $\mathcal{G}$.

Proposition II.2.11. *Two filters $\mathcal{F}$ and $\mathcal{G}$ admit a least upper bound (supremum) if and only if they mesh, in which case*

$$\mathcal{F} \vee \mathcal{G} = \{F \cap G : F \in \mathcal{F}, G \in \mathcal{G}\}.$$

Proof. If $\mathcal{F}$ and $\mathcal{G}$ do not mesh, then there is $F \in \mathcal{F}$ and $G \in \mathcal{G}$ such that $F \cap G = \varnothing$, and the pair $\{\mathcal{F}, \mathcal{G}\}$ cannot have an upper bound in $\mathbb{F}X$, for a filter finer than both $\mathcal{F}$ and $\mathcal{G}$ would have to contain $F \cap G = \varnothing$, in contradiction with (proper).

Conversely, if $\mathcal{F}\#\mathcal{G}$ then

$$\{F \cap G : F \in \mathcal{F}, G \in \mathcal{G}\} \tag{II.2.2}$$

is a filter.

In fact, if H_0, H_1 belong to (II.2.2), then there exist $F_0, F_1 \in \mathcal{F}$ and $G_0, G_1 \in \mathcal{G}$ such that $H_0 = F_0 \cap G_0$ and $H_1 = F_1 \cap G_1$. Therefore

$$H_0 \cap H_1 = (F_0 \cap G_0) \cap (F_1 \cap G_1) = (F_0 \cap F_1) \cap (G_0 \cap G_1)$$

belongs to (II.2.2). If $F \in \mathcal{F}$, $G \in \mathcal{G}$ and $F \cap G \subset H$, then $F \cup H \in \mathcal{F}$ and $G \cup H \in \mathcal{G}$, hence

$$(F \cup H) \cap (G \cup H) = (F \cap G) \cup H = H$$

belongs to (II.2.2). Finally, $\varnothing$ does not belong to (II.2.2), because $\mathcal{F}\#\mathcal{G}$.

Since $F \cap G \subset F$ and $F \cap G \subset G$ for every $F \in \mathcal{F}$ and every $G \in \mathcal{G}$, it is clear that this filter is finer than both $\mathcal{F}$ and $\mathcal{G}$.

On the other hand, if a filter is finer than both $\mathcal{F}$ and $\mathcal{G}$ it must contain $\{F \cap G : F \in \mathcal{F}, G \in \mathcal{G}\}$. Therefore

$$\mathcal{F} \vee \mathcal{G} = \{F \cap G : F \in \mathcal{F}, G \in \mathcal{G}\}.$$

$\square$

It is often handy to abbreviate

$$\mathcal{F} \vee A := \mathcal{F} \vee \{A\}^{\uparrow}.$$

We will however avoid to write $F \vee A$ for

$$\{F\}^{\uparrow} \vee \{A\}^{\uparrow} = \{F \cap A\}^{\uparrow},$$

because this might lead to a confusion.

Note that the family 2^X of all subsets of X is the only family satisfying (isotone) and (finite intersection) but not (proper), and that (proper) is the only property of filters preventing the existence of suprema for general families of filters.

For this reason, 2^X is also called *the degenerate filter on X.* The set

$$\overline{\mathbb{F}}X := \mathbb{F}X \cup \{2^X\}$$

of (possibly) degenerate filters on X is then a complete lattice under the partial order of Corollary II.1.6, in which $\mathcal{F} \vee \mathcal{G} = 2^X$ whenever $\mathcal{F}$ and $\mathcal{G}$ are dissociated.

Remark II.2.12. We have already noticed that the supremum of two families $\mathcal{F}$ and $\mathcal{G}$ in $\mathbb{I}X$ is equal to $\mathcal{F} \cup \mathcal{G}$. Therefore if $\mathcal{F}, \mathcal{G} \in \mathbb{F}X \subset \overline{\mathbb{F}}X \subset \mathbb{I}X$, then it follows from general properties of suprema that

$$\mathcal{F} \cup \mathcal{G} \leq \mathcal{F} \vee \mathcal{G},$$

where $\vee$ denotes the supremum in $\overline{\mathbb{F}}X$. Of course, this fact can be easily checked directly. In particular, if $\mathcal{F}$ and $\mathcal{G}$ are dissociated, then $\mathcal{F} \cup \mathcal{G}$ is strictly included in $\mathcal{F} \vee \mathcal{G} = 2^X$.

Definition II.2.13. A family $\mathcal{A}$ of subsets of X has the *finite intersection property* if $\bigcap_{A \in \mathcal{B}} A \neq \varnothing$ for each finite subset $\mathcal{B}$ of $\mathcal{A}$.

Given a family $\mathcal{A}$ of subsets of X, we denote by

$$\mathcal{A}^{\cap} := \{\bigcap_{A \in \mathcal{B}} A : \mathcal{B} \in [\mathcal{A}]^{<\omega}\} \tag{II.2.3}$$

the family of intersections of finitely many elements of $\mathcal{A}$, and by

$$\mathcal{A}^{\cup} := \{\bigcup_{A \in \mathcal{B}} A : \mathcal{B} \in [\mathcal{A}]^{<\omega}\} \tag{II.2.4}$$

the family of unions of finitely many elements of $\mathcal{A}$.

Proposition II.2.14. *If a family $\mathcal{A}$ of subsets of X has the finite intersection property, then $\mathcal{A}^\cap$ is a filter-base on X.*

Proposition II.2.15. *A set $\mathbb{D} \subset \mathbb{F}X$ of filters admits a least upper bound $\bigvee_{\mathcal{F} \in \mathbb{D}} \mathcal{F}$ if and only if $\bigcup_{\mathcal{F} \in \mathbb{D}} \mathcal{F}$ has the finite intersection property. In this case,*

$$\bigvee_{\mathcal{F} \in \mathbb{D}} \mathcal{F} = \left(\bigcup_{\mathcal{F} \in \mathbb{D}} \mathcal{F} \right)^{\cap\uparrow}. \tag{II.2.5}$$

Proof. If $\bigcup_{\mathcal{F} \in \mathbb{D}} \mathcal{F}$ has the finite intersection property, then $(\bigcup_{\mathcal{F} \in \mathbb{D}} \mathcal{F})^\cap$ is a filter-base and $(\bigcup_{\mathcal{F} \in \mathbb{D}} \mathcal{F})^{\cap\uparrow}$ is a filter that is finer than each $\mathcal{F} \in \mathbb{D}$. If $\mathcal{G}$ is another such filter, $\mathcal{G}$ also contains $(\bigcup_{\mathcal{F} \in \mathbb{D}} \mathcal{F})^{\cap\uparrow}$, so that (II.2.5) holds.

Conversely, if $\bigcup_{\mathcal{F} \in \mathbb{D}} \mathcal{F}$ does not have the finite intersection property, then there is a finite subset $\mathbb{B}$ of $\mathbb{D}$ and elements $F_\mathcal{F} \in \mathcal{F}$ for each $\mathcal{F} \in \mathbb{B}$ with $\bigcap_{\mathcal{F} \in \mathbb{B}} F_\mathcal{F} = \varnothing$. Thus no filter can contain all filters of $\mathbb{B}$ for otherwise it would also contain $\bigcap_{\mathcal{F} \in \mathbb{B}} F_\mathcal{F} = \varnothing$. $\qquad\square$

II.2.2 *Free and principal filters*

The intersection

$$\ker \mathcal{F} := \bigcap_{F \in \mathcal{F}} F$$

of all elements of a filter $\mathcal{F}$ is called the *kernel* of $\mathcal{F}$.

Definition II.2.16. A filter $\mathcal{F}$ is called *principal* if $\ker \mathcal{F} \in \mathcal{F}$. A filter $\mathcal{F}$ is called *free* if $\ker \mathcal{F} = \varnothing$.

We use the convention that $\mathbb{F}$ denotes the class of all filters while $\mathbb{F}X$ denotes the set of all filters on X. The class of *principal filters* is denoted by $\mathbb{F}_0$ and of *free filters* by $\mathbb{F}_*$, with the convention that $\mathcal{F} \in \mathbb{F}_0 X$ whenever $\mathcal{F}$ is a principal filter on X, and $\mathcal{F} \in \mathbb{F}_* X$ whenever $\mathcal{F}$ is a free filter on X.

Exercise II.2.17. Check that a filter $\mathcal{F}$ on X is principal if and only if there exists $\varnothing \neq A \subset X$ such that $\mathcal{F} = \{A\}^\uparrow$.

As a shorthand which should not cause confusion, we will often write $A^\uparrow$ instead of $\{A\}^\uparrow$ for the principal filter of $A \subset X$. Of course, filter cannot be simultaneously free and principal.

Exercise II.2.18. Show that each filter on a finite set is principal.

Exercise II.2.19. Show that each cofinite filter (on an infinite set) is free.

Proposition II.2.20. *If $\mathcal{G}$ is a free filter that contains A, then $\mathcal{G} \geq (A)_0$.*

Proof. Let $\mathcal{G}$ be a free filter on X. Then for every $x \in X$, there is $G_x \in \mathcal{G}$ with $x \notin G_x$. Thus, if $F = \{x_1 \dots x_n\}$ is a finite subset of X, then for each $i \in \{1 \dots n\}$ there is $G_i \in \mathcal{G}$ such that $x_i \notin G_i$, so that $F \cap (\bigcap_{i=1}^{n} G_i) = \varnothing$, equivalently, $\bigcap_{i=1}^{n} G_i \subset F^c \in$, hence $F^c \in \mathcal{G}$. Thus $A \cap F^c \in \mathcal{G}$ because $A \in \mathcal{G}$. This means $\mathcal{G}$ is finer than the cofinite filter of A. $\qquad\square$

In other words, on a given infinite set, the cofinite filter is the coarsest free filter. As a result:

Corollary II.2.21. *A filter is free if and only if it is finer than the cofinite filter of its (necessarily infinite) underlying set. Thus, an infimum of free filters is free.*

Exercise II.2.22. Let X be an uncountable non-empty set. Show that the cocountable filter $(X)_1$ is the coarsest free countably deep filter on X.

The degenerate filter $\{\varnothing\}^{\uparrow}$ is the only filter that is both free and principal.

Principal and free filters are the opposite ends of a spectrum.

Theorem II.2.23. *For every filter $\mathcal{F}$ on X, there exists a unique pair of (possibly degenerate) filters $\mathcal{F}^*$ and $\mathcal{F}^{\bullet}$ such that $\mathcal{F}^*$ is free, $\mathcal{F}^{\bullet}$ is principal, and*

$$\mathcal{F} = \mathcal{F}^* \wedge \mathcal{F}^{\bullet} \text{ and } \mathcal{F}^* \vee \mathcal{F}^{\bullet} = 2^X. \qquad (II.2.6)$$

The filter $\mathcal{F}^*$ is called the *free part* of $\mathcal{F}$, and $\mathcal{F}^{\bullet}$ is called the *principal part* of $\mathcal{F}$. Of course, $\mathcal{F}^* \vee \mathcal{F}^{\bullet} = 2^X$ means that $\mathcal{F}^*$ and $\mathcal{F}^{\bullet}$ are dissociated.

Proof. If $\mathcal{F}$ is principal, $\mathcal{F} = \mathcal{F}^{\bullet}$ and $\mathcal{F}^*$ is the degenerate filter 2^X. If $\mathcal{F}$ is not principal, then $\ker \mathcal{F} \notin \mathcal{F}$, equivalently, $(\ker \mathcal{F})^c \# \mathcal{F}$ (by (II.2.1)). Then $\mathcal{F}^* := \mathcal{F} \vee \{(\ker \mathcal{F})^c\}^{\uparrow}$ is a non-degenerate filter, which is free because

$$\ker \mathcal{F}^* = \bigcap_{F \in \mathcal{F}} (F \cap (\bigcap \mathcal{F})^c) = \bigcap \mathcal{F} \cap (\bigcap \mathcal{F})^c = \varnothing.$$

Let $\mathcal{F}^{\bullet}$ be the principal filter of $\ker \mathcal{F}$ (which is the degenerate filter $\{\varnothing\}^{\uparrow} = 2^X$ if $\mathcal{F}$ is free). Then, $\mathcal{F}^* \vee \mathcal{F}^{\bullet} = 2^X$ because $\mathcal{F}^*$ and $\mathcal{F}^{\bullet}$ do not mesh, and $\mathcal{F} = \mathcal{F}^* \wedge \mathcal{F}^{\bullet}$ because

$$F = (F \cap (\ker \mathcal{F})^c) \cup \ker \mathcal{F}$$

for each $F \in \mathcal{F}$, and sets of the form $(F \cap (\ker \mathcal{F})^c) \cup \ker \mathcal{F}$ for $F \in \mathcal{F}$ form a filter-base for $\mathcal{F}^* \wedge \mathcal{F}^\bullet$ by Exercise II.2.6.

If now $A^\uparrow$ is a principal filter finer than $\mathcal{F}$, then $A \subset \ker \mathcal{F}$, hence $\mathcal{F} \vee A^c$ is free if and only if $A = \ker \mathcal{F}$, which shows the uniqueness of the decomposition. $\qquad\qquad\qquad\qquad\qquad\qquad\qquad\qquad\qquad\qquad\qquad \square$

II.2.3 *Sequential filters*

Definition II.2.24. A filter $\mathcal{F}$ is called *sequential* if it is the sequential filter of a sequence, that is, if there exists a sequence $\{x_n\}_{n \in N}$ such that $\mathcal{F} = (x_n)_n^\uparrow$. We say that a sequence $\{x_n\}_{n \in N}$ is *free* if $(x_n)_n^\uparrow$ is free, and *principal* if $(x_n)_n^\uparrow$ is principal.

Accordingly, we may say the *kernel of a sequence* $\{x_n\}_{n \in N}$ for the kernel of $(x_n)_n^\uparrow$:

$$\ker\{x_n\}_n := \ker(x_n)_n^\uparrow = \bigcap_{k \in \mathbb{N}} \{x_n : k \le n \in N\}. \qquad (\text{II.2.7})$$

Example II.2.25. The kernels of (I.3.4) and (I.3.6) are empty so that the corresponding sequences are free. That of (I.3.5) is finite and that of (I.3.7) is infinite. The sequences (I.3.5) and (I.3.7) are principal: In the case of (I.3.5), the kernel is $\{-1, 1\}$ and is equal to any tail; in the case case of (I.3.7), the kernel is $\{\frac{1}{n} : n \in \mathbb{N}_1\}$ and is equal to any tail.

On the other hand, The sequence $x_n := \max\left\{\frac{(-1)^n}{n}, 0\right\}$ for $n \in \mathbb{N}$ is neither free nor principal. In fact, $\ker\{x_n\}_n = \{0\}$, but $\{n : x_n \ne 0\}$ is infinite.

Theorem II.2.23 particularizes for sequential filters to:

Theorem II.2.26. *For every sequence $\{x_n\}_{n \in \mathbb{N}}$ there exist a subset A of $\mathbb{N}$ such that $\{x_n\}_{n \in A}$ is a (possibly degenerate) free sequence and $\{x_n\}_{n \in \mathbb{N} \setminus A}$ is a (possibly degenerate) principal sequence. If $\{x_n\}_{n \in \mathbb{N}}$ is neither free nor principal, then A and $\mathbb{N} \setminus A$ are infinite.*

Let $\mathbb{E}$ denote the class of sequential filters, that is, $\mathcal{E} \in \mathbb{E}(X)$ if there is a sequence $\{x_n\}_{n \in N}$ on X such that $\mathcal{E} = (x_n)_n^\uparrow$.

Proposition II.2.27. *Each principal filter of a countable set is sequential.*

Proof. If card $A = k < \infty$, say $A = \{x_0, \ldots, x_{k-1}\}$, then let $f : \mathbb{N} \to X$ be defined by $f(n) := x_{n([k])}$, where $n([k])$ is the remainder in the division of n by k. Then $A = \{f(m) : m \geq n\}$ for each n, and thus $A^\uparrow = (f(n))_n^\uparrow$. If A is infinite, say $A = \{x_k : k \in \mathbb{N}\}$ with distinct x_k, then let

$$f(n) := x_{n-k!} \text{ if } k! < n \leq (k+1)!$$

It follows that, $A = \{f(m) : m \geq n\}$ for each n, and thus $A^\uparrow = (f(n))_n^\uparrow$. □

Proposition II.2.28. *Each cofinite filter of a countably infinite set is sequential.*

Proof. If $A = \{x_k : k \in \mathbb{N}\}$ with distinct x_k, then $(x_k)_k^\uparrow = (A)_0$. □

Exercise II.2.29. Show that

(1) a sequence is free if and only if it is finite-to-one;
(2) if $\{x_n\}_{n \in N}$ is free, then there exists a countably infinite set A such that $(x_n)_n^\uparrow = (A)_0$;
(3) if A is a countably infinite set and $\{x_n\}_{n \in N}$ is such that $\{x_n : n \in N\} = A$ and for each $x \in A$, the set $\{n \in N : x = x_n\}$ is finite, then $(A)_0 = (x_n)_n^\uparrow$.

Remark II.2.30. While quences and sequences are different objects, it makes no difference to use one or the other from the standpoint of convergence, which only depends on the cofinite filter on the (countable) image: If $f \in X^A$ is a quence, then $f[(A)_0]$ is a filter base and, of course, if X is a premetric space, then $x \in \lim f$ whenever $f[(A)_0] \geq \{B(x, r) : r > 0\}$. If $\{x_n\}_{n \in N}$ is a sequence and $f(n) := x_n$, then $f[(N)_0] \approx (x_n)_n$.

Lemma II.2.31. *If $\mathcal{F}_0, \mathcal{F}_1$ are sequential filters, then $\mathcal{F}_0 \cap \mathcal{F}_1$ is a sequential filter.*

Proof. Let $f_0, f_1 : \mathbb{N} \to X$ be such that $\mathcal{F}_0 = (f_0(n))_n^\uparrow$ and $\mathcal{F}_1 = (f_1(n))_n^\uparrow$. Then

$$f(n) := \begin{cases} f_1\left(\frac{n}{2}\right), & \text{if } n \text{ is even,} \\ f_0\left(\frac{n+1}{2}\right), & \text{if } n \text{ is odd,} \end{cases}$$

fulfills $\mathcal{F}_0 \cap \mathcal{F}_1 = (f(n))_n^\uparrow$. □

Proposition II.2.32. *A filter is sequential if and only if it contains a countable set and admits a countable base consisting of almost equal sets.*

Proof. Each sequential filter obviously fulfills the condition. Conversely, consider a filter $\mathcal{F}$ on X, for which the condition holds. Then there is a decreasing base $\{B_n : n \in \mathbb{N}\}$ of $\mathcal{F}$ such that B_0 is countable and $B_{n+1} \setminus B_n$ is finite. Of course, $B_\infty := \bigcap_{n \in \mathbb{N}} B_n$ is either countable or empty and $B_0 \setminus B_\infty$ is either countable or empty. Of course, B_∞ and $B_0 \setminus B_\infty$ cannot be empty simultaneously.

If $B_0 \setminus B_\infty$ is finite, then B_∞ is non-empty and $\{B_\infty\}$ is a base of $\mathcal{F}$. Then $\mathcal{F}$ is sequential by Proposition II.2.27.

If $B_0 \setminus B_\infty$ is infinite, then $\{B_n \setminus B_\infty : n \in \mathbb{N}\}$ is a base of the cofinite filter of $B_0 \setminus B_\infty$. Then $(B_0 \setminus B_\infty)_0$ is sequential by Proposition II.2.28. If $B_\infty = \varnothing$, then $\mathcal{F} = (B_0)_0$.

Finally, if $B_0 \setminus B_\infty$ is infinite and $B_\infty \neq \varnothing$, then $\mathcal{F} = (B_0 \setminus B_\infty)_0 \cap B_\infty^\uparrow$ is a sequential filter by Lemma II.2.31. $\qquad\square$

Corollary II.2.33. *A filter $\mathcal{F}$ is sequential if and only if it contains a countable set F_0 such that $F_0 \subset_0 F$ for each $F \in \mathcal{F}$.*

Proof. This condition is obviously necessary. Suppose that it holds. Then $\{F \cap F_0 : F \in \mathcal{F}\}$ is a base of $\mathcal{F}$. If $x \notin F_\infty = \bigcap_{F \in \mathcal{F}} F$, then there is $F \in \mathcal{F}$ such that $x \notin F$. Hence either $\mathcal{F} = F_\infty^\uparrow$ or $\mathcal{B} := \{F \cap F_0 \setminus F_\infty : F \in \mathcal{F}\}$ consists of cofinite subsets of $F_0 \setminus F_\infty$ and thus is a base of a sequential filter. If $F_\infty = \varnothing$, then $\mathcal{B}^\uparrow = \mathcal{F}$; otherwise, $\mathcal{F} = \mathcal{B}^\uparrow \cap F_\infty^\uparrow$. $\qquad\square$

II.2.4 *Images, preimages, products*

Exercise II.2.34 (image filter). Let $f : X \to Y$ be a function and $\mathcal{F}$ be a filter on X. Recall that (I.4.1)

$$f[\mathcal{F}] := \{f(F) : F \in \mathcal{F}\}.$$

(1) Show that the family $f[\mathcal{F}]$ is a filter-base.
(2) Show that the filter $f[\mathcal{F}]^\uparrow$ it generates is

$$f[\mathcal{F}]^\uparrow = \{B \subset Y : f^-(B) \in \mathcal{F}\}.$$

(3) Show that $f[\mathcal{F}]$ is a filter if and only if f is surjective.

If $R \subset X \times Y$ is a binary relation, $F \subset X$, $G \subset Y$, then

$$R(F) = \left\{y \in Y : \underset{x \in F}{\exists}\, (x, y) \in R\right\}$$

and alike for the inverse relation $R^- = \{(y, x) : (x, y) \in R\}$ (see Appendix A).

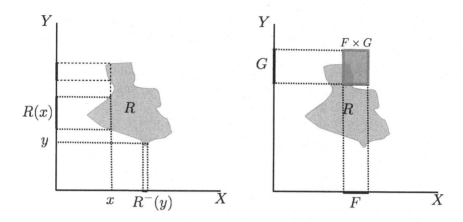

Figure II.1   A relation.

Then the following expressions are equivalent:

$$(F \times G) \# R \iff R(F) \# G \iff F \# R^-(G). \qquad (\text{II.2.8})$$

Indeed, $(x, y) \in R \iff y \in R(x) \iff x \in R^-(y)$.
If $R \subset X \times Y$, then let

$$R[\mathcal{F}] := \{R(F) : F \in \mathcal{F}\}.$$

Exercise II.2.35. Check that $R[\mathcal{F}]$ is a filter-base of a (possibly degenerate) filter on Y.

Accordingly, the image filter defined in Exercise II.2.34 is a particular case of the image of a filter under a relation, namely the graph relation.
In view of (II.2.8),

Proposition II.2.36. *If* $R \subset X \times Y, \mathcal{F} \in \mathbb{F}X$ *and* $\mathcal{G} \in \mathbb{F}Y$, *then*

$$(\mathcal{F} \times \mathcal{G}) \# R \iff R[\mathcal{F}] \# \mathcal{G} \iff \mathcal{F} \# R^-[\mathcal{G}]. \qquad (\text{II.2.9})$$

If now $\mathcal{R}$ is a filter on $X \times Y$, each of its elements can be seen as a relation, and we can consider the (possibly degenerate) filter

$$\mathcal{R}[\mathcal{F}] := \{R(F) : R \in \mathcal{R}, F \in \mathcal{F}\}^{\uparrow}.$$

Of course, we can define similarly the possibly degenerate filter $\mathcal{R}^-[\mathcal{G}]$ on X. It follows from (II.2.9) that

$$(\mathcal{F} \times \mathcal{G}) \# \mathcal{R} \iff \mathcal{R}[\mathcal{F}] \# \mathcal{G} \iff \mathcal{F} \# \mathcal{R}^-[\mathcal{G}]. \qquad (\text{II.2.10})$$

Exercise II.2.37. Let $\mathcal{F} \in \mathbb{F}X$ and $\mathcal{G} \in \mathbb{F}Y$. Verify that

(1) $\mathcal{F} \times \mathcal{G} = \{F \times G : F \in \mathcal{F}, G \in \mathcal{G}\}$ is a filter-base on $X \times Y$;
(2) $p_X^-[\mathcal{F}]$ and $p_Y^-[\mathcal{G}]$ are filter-bases, and

$$\mathcal{F} \times \mathcal{G} \approx p_X^-[\mathcal{F}]^\uparrow \vee p_Y^-[\mathcal{G}]^\uparrow. \qquad (\text{II.2.11})$$

Products of infinitely many filters are defined similarly. To this end, let $\mathcal{X}$ be a set of sets, and, for each $X \in \mathcal{X}$, let $\mathcal{F}_X \in \mathbb{F}X$. We denote by $\prod_{X \in \mathcal{X}} X$ the Cartesian product of all the sets X in $\mathcal{X}$ and by $p_{X_0} : \prod_{X \in \mathcal{X}} X \to X_0$ the projection on $X_0 \in \mathcal{X}$.

Recall from Proposition II.2.15 that a family of filters admits a least upper bound whenever their union has the finite intersection property. Since the family $\bigcup_{X \in \mathcal{X}} p_X^-[\mathcal{F}_X]$ has the finite intersection property, we can define the *polyhedral filter* $\prod_{X \in \mathcal{X}} \mathcal{F}_X$ on $\prod_{X \in \mathcal{X}} X$ by

$$\prod_{X \in \mathcal{X}} \mathcal{F}_X := \bigvee_{X \in \mathcal{X}} p_X^-[\mathcal{F}_X]. \qquad (\text{II.2.12})$$

Exercise II.2.38. Verify that

$$\left\{ \prod_{X \in \mathcal{X}} F_X : \underset{X \in \mathcal{X}}{\forall} F_X \in \mathcal{F}_X, \{X \in \mathcal{X} : F_X \neq X\} \in [\mathcal{X}]^{<\omega} \right\}$$

is a filter-base for $\prod_{X \in \mathcal{X}} \mathcal{F}_X$.

II.3 Grills

If $\mathcal{A}$ is a family of subsets of X we denote by $\mathcal{A}^\#$ *the grill of* $\mathcal{A}$, that is, the family of subsets of X that intersect every element of $\mathcal{A}$. In other words,

$$\mathcal{A}^\# := \left\{ H \subset X : \underset{A \in \mathcal{A}}{\forall} A \cap H \neq \varnothing \right\}. \qquad (\text{II.3.1})$$

Exercise II.3.1. Let $X := \{1, 2, 3, 4\}$ and let $\mathcal{A} := \{\{1, 2\}, \{3, 4\}\}$. Describe explicitly the family $\mathcal{A}^\#$.

Exercise II.3.2. Show that $\mathcal{A} \subset \mathcal{B} \implies \mathcal{B}^\# \subset \mathcal{A}^\#$.

If $\mathcal{A}$ is an isotone family of subsets of X then (II.2.1) reformulates as

$$H \in \mathcal{A}^\# \iff H^c \notin \mathcal{A}. \qquad (\text{II.3.2})$$

We use the convention that $\mathcal{A}^{\#\#} := \left(\mathcal{A}^\#\right)^\#$.

Exercise II.3.3 (grill and isotonization). Let $\mathcal{A}$ be a family of subsets of X. Show that:

(1) $\mathcal{A}^{\#}$ is isotone;
(2) $\mathcal{A}^{\#\#} = \mathcal{A}^{\uparrow}$;
(3) $\mathcal{A} = \mathcal{A}^{\#\#}$ if and only if $\mathcal{A}$ is isotone;
(4) If $\mathcal{A}, \mathcal{B} \in \mathbb{I}X$ then

$$\mathcal{A} \subset \mathcal{B} \iff \mathcal{B}^{\#} \subset \mathcal{A}^{\#}.$$

Note that (finite intersection) implies in particular that

Proposition II.3.4. *For every filter $\mathcal{F}$,*

$$\mathcal{F} \subset \mathcal{F}^{\#}.$$

Proposition II.3.5. *If $\mathbb{D} \subset \mathbb{I}X$ is a collection of isotone families of subsets of X then*

$$\left(\bigcap_{\mathcal{F} \in \mathbb{D}} \mathcal{F} \right)^{\#} = \bigcup_{\mathcal{F} \in \mathbb{D}} \mathcal{F}^{\#}$$

$$\left(\bigcup_{\mathcal{F} \in \mathbb{D}} \mathcal{F} \right)^{\#} = \bigcap_{\mathcal{F} \in \mathbb{D}} \mathcal{F}^{\#}.$$

Proof. Because $\mathcal{F} \geq \bigcap_{\mathcal{F} \in \mathbb{D}} \mathcal{F}$ for each $\mathcal{F} \in \mathbb{D}$, it follows that $\bigcup_{\mathcal{F} \in \mathbb{D}} \mathcal{F}^{\#} \subset \left(\bigcap_{\mathcal{F} \in \mathbb{D}} \mathcal{F} \right)^{\#}$. Conversely, if $A \notin \bigcup_{\mathcal{F} \in \mathbb{D}} \mathcal{F}^{\#}$ then for every $\mathcal{F} \in \mathbb{D}$ there exists $F_{\mathcal{F}} \in \mathcal{F}$ such that $F_{\mathcal{F}} \cap A = \varnothing$. Therefore, $A \cap \bigcup_{\mathcal{F} \in \mathbb{D}} F_{\mathcal{F}} = \varnothing$. But $\bigcup_{\mathcal{F} \in \mathbb{D}} F_{\mathcal{F}} \in \bigcap_{\mathcal{F} \in \mathbb{D}} \mathcal{F}$ by isotony. Hence $A \notin \left(\bigcap_{\mathcal{F} \in \mathbb{D}} \mathcal{F} \right)^{\#}$.

The second formula is immediate, as $A \in \left(\bigcup_{\mathcal{F} \in \mathbb{D}} \mathcal{F} \right)^{\#}$ if $A \cap F \neq \varnothing$ for all $F \in \mathcal{F}$ and all $\mathcal{F} \in \mathbb{D}$. $\qquad\square$

Proposition II.3.6. *If $\mathbb{G}$ is a finite collection of filters on X and $\mathcal{H} \in \mathbb{F}X$, then*

$$\mathcal{H} \# \bigwedge_{\mathcal{G} \in \mathbb{G}} \mathcal{G} \iff \underset{\mathcal{G} \in \mathbb{G}}{\exists} \mathcal{H} \# \mathcal{G}. \tag{II.3.3}$$

Proof. If $\mathcal{H} \# \mathcal{G}$ for some $\mathcal{G} \in \mathbb{G}$, then $\mathcal{H}$ meshes *a fortiori* the coarser filter $\bigwedge_{\mathcal{G} \in \mathbb{G}} \mathcal{G}$. Conversely, if $\mathcal{H}$ is dissociated from each each $\mathcal{G} \in \mathbb{G}$, then for each $\mathcal{G} \in \mathbb{G}$, there is $G_{\mathcal{G}} \in \mathcal{G}$ with $G_{\mathcal{G}} \notin \mathcal{H}^{\#}$, that is, $G_{\mathcal{G}}^{c} \in \mathcal{H}$ by (II.3.2). Because $\mathbb{G}$ is finite, $\bigcap_{\mathcal{G} \in \mathbb{G}} G_{\mathcal{G}}^{c} \in \mathcal{H}$, that is, $\bigcup_{\mathcal{G} \in \mathbb{G}} G_{\mathcal{G}} \notin \mathcal{H}^{\#}$. In view of Exercise II.2.6, $\mathcal{H}$ and $\bigwedge_{\mathcal{G} \in \mathbb{G}} \mathcal{G}$ are dissociated. $\qquad\square$

Exercise II.3.7. Find an example showing that (II.3.3) may fail if $\mathbb{G} \subset \mathbb{F}X$ is infinite.

II.4   Duality between filters and grills

Proposition II.4.1. *A non-empty family $\mathcal{G}$ of subset of X is the grill of a filter on X if and only if it is isotone and*

$$A \cup B \in \mathcal{G} \Longrightarrow A \in \mathcal{G} \text{ or } B \in \mathcal{G}. \tag{II.4.1}$$

Proof. Indeed, if $\mathcal{F}$ is a filter, $A \notin \mathcal{F}^{\#}$ and $B \notin \mathcal{F}^{\#}$, there is F_A and F_B in $\mathcal{F}$ such that $F_A \cap A = \varnothing$ and $F_B \cap B = \varnothing$. Since $\mathcal{F}$ is a filter, $F_A \cap F_B \in \mathcal{F}$ and $(F_A \cap F_B) \cap (A \cup B) = \varnothing$. Thus $A \cup B \notin \mathcal{F}^{\#}$.

Conversely, if $\mathcal{G}$ is an isotone family such that $A \in \mathcal{G}$ or $B \in \mathcal{G}$ whenever $A \cup B \in \mathcal{G}$, we need to show that there is a filter $\mathcal{F}$ such that $\mathcal{F}^{\#} = \mathcal{G}$.

In view of Exercise II.3.3, $\mathcal{G} = \mathcal{G}^{\#\#}$ because $\mathcal{G}$ is isotone. Hence, we only need to show that $\mathcal{G}^{\#}$ is a filter. Assume $A \in \mathcal{G}^{\#}$ and $B \in \mathcal{G}^{\#}$. In view of (II.3.2), $A^c \notin \mathcal{G}$ and $B^c \notin \mathcal{G}$, so that $A^c \cup B^c \notin \mathcal{G}$. By (II.3.2), it means that

$$(A^c \cup B^c)^c = A \cap B \in \mathcal{G}^{\#},$$

thus $\mathcal{G}^{\#}$ is a filter. $\qquad\square$

Remark II.4.2. Notice that $(2^X)^{\#} = \varnothing$, that is, the empty family of subsets of X is the grill of the degenerate filter on X.

Definition II.4.3. A non-empty family $\mathcal{G}$ of subsets of X is called a *filter-grill* provided that it is isotone and fulfills (II.4.1).

Proposition II.4.1 establishes that if $\mathcal{F}$ is a filter, then $\mathcal{F}^{\#}$ is a filter-grill and that if $\mathcal{G}$ is a filter-grill then $\mathcal{G}^{\#}$ is a filter. In other words, the set $\mathbb{F}X$ of filters on X is in bijection (via the operator $\cdot^{\#} : 2^X \to 2^X$) with the set $\mathbb{F}_{\#}X$ of filter-grills on X. In view of Remark II.4.2, this bijection extends to $\overline{\mathbb{F}}X = \mathbb{F}X \cup \{2^X\}$ and

$$\overline{\mathbb{F}}_{\#}X := \mathbb{F}_{\#}X \cup \{\varnothing\}.$$

The classes of filters and filter-grills are distinct:

Example II.4.4. If $\{x_n\}_{n\in\mathbb{N}}$ is an injective sequence, the associated sequential filter is not a filter-grill. Indeed,

$$\{x_n : n \in \mathbb{N}\} = \{x_{2n} : n \in \mathbb{N}\} \cup \{x_{2n-1} : n \in \mathbb{N}\} \in (x_n)_n^{\uparrow},$$

but neither $\{x_{2n} : n \in \mathbb{N}\}$ nor $\{x_{2n-1} : n \in \mathbb{N}\}$ belongs to $(x_n)_n^{\uparrow}$ because neither contains any tail $\{x_n : n \geq k\}$ for any k. As a consequence, $((x_n)_n^{\uparrow})^{\#}$ is a filter-grill, but not a filter, because we have noticed that $(x_n)_n^{\uparrow} = ((x_n)_n^{\uparrow})^{\#\#}$ is not a filter-grill.

Are there any filters that are also filter-grills?

A simple, in some sense trivial, example is the principal filter $\{x\}^\uparrow$ of a singleton: it is a filter, and it is a filter-grill because $x \in A \cup B$ means that $x \in A$ or $x \in B$. Before we establish the existence of many more such filters, let us examine their properties:

Theorem II.4.5. *Let $\mathcal{F}$ be a filter on a set X. The following are equivalent:*

(1) $\mathcal{F}$ is a filter-grill;
(2) $A \in \mathcal{F}$ or $X \setminus A \in \mathcal{F}$ for every subset A of X;
(3) $\mathcal{F} = \mathcal{F}^{\#}$;
(4) for each $\mathcal{G} \in \mathbb{F}X$,

$$\mathcal{G} \geq \mathcal{F} \Longrightarrow \mathcal{G} = \mathcal{F}.$$

Proof. Since $A \cup A^c = X \in \mathcal{F}$, (2) follows readily from (1).

Assume (2). Since $\mathcal{F}$ is a filter, $\mathcal{F} \subset \mathcal{F}^{\#}$. On the other hand if $H \in \mathcal{F}^{\#}$ but $H \notin \mathcal{F}$ then by (2), $H^c \in \mathcal{F}$, which is incompatible with $H \in \mathcal{F}^{\#}$, and we obtain (3).

(3) $\Longrightarrow$ (4). Let $\mathcal{G}$ be a filter finer than $\mathcal{F}$. If $\mathcal{F} \not\geq \mathcal{G}$ then there is $G \in \mathcal{G}$ with $G \notin \mathcal{F}$. However, this is not possible, because $G \in \mathcal{G} \geq \mathcal{F}$ ensures that $G \in \mathcal{F}^{\#} = \mathcal{F}$.

(4) $\Longrightarrow$ (1). Assume that $\mathcal{F}$ is maximal and that there is A and B with $A \cup B \in \mathcal{F}$ but $A \notin \mathcal{F}$ and $B \notin \mathcal{F}$. Then $A^c \in \mathcal{F}^{\#}$ and $B^c \in \mathcal{F}^{\#}$ by (II.3.2). Therefore $\mathcal{F} \vee \{A^c\}^\uparrow$ and $\mathcal{F} \vee \{B^c\}^\uparrow$ are proper filters finer than $\mathcal{F}$. By maximality of $\mathcal{F}$, we conclude that $\mathcal{F} = \mathcal{F} \vee \{A^c\}^\uparrow = \mathcal{F} \vee \{B^c\}^\uparrow$ so that A^c and B^c belong to $\mathcal{F}$. Therefore, $A^c \cap B^c = (A \cup B)^c \in \mathcal{F}$, which contradicts $A \cup B \in \mathcal{F}$. $\square$

Note that the fourth property in Theorem II.4.5 states that $\mathcal{F}$ is a maximal element in $(\mathbb{F}X, \leq)$.

II.5   Triad: filters, filter-grills and ideals

Notice that a (possibly empty) family $\mathcal{F}$ of subsets of X is a (possibly degenerate) filter if and only if

$$F_0 \in \mathcal{F} \text{ and } F_1 \in \mathcal{F} \Longleftrightarrow F_0 \cap F_1 \in \mathcal{F}. \tag{F}$$

Analogously, a family $\mathcal{H}$ of subsets of X is a filter-grill if and only if

$$H_0 \cup H_1 \in \mathcal{H} \Longleftrightarrow H_0 \in \mathcal{H} \text{ or } H_1 \in \mathcal{H}. \tag{D}$$

The definition above admits the grill of degenerate filter. Filter-grills are also called *distributive families*.

A family $\mathcal{A}$ of subsets of X is called *antidistributive* or an *ideal* if

$$A_0 \cup A_1 \in \mathcal{A} \Longleftrightarrow A_0 \in \mathcal{A} \text{ and } A_1 \in \mathcal{A}. \tag{A}$$

Recall that the class of distributive families (or possibly degenerate filter-grills) is $\overline{\mathbb{F}}_\#$, while the class of possibly degenerate filter is denoted $\overline{\mathbb{F}}$. Let us denote by $\mathbb{A}$ the class of antidistributive families. In order to relate the properties (D), (A) and (F), consider the unitary operations $2^X \to 2^X$ defined in (II.3.1) and (II.1.4)).

We have already seen that the operation $^\#$ is an involution between $\overline{\mathbb{F}}X$ and $\overline{\mathbb{F}}_\#X$. On the other hand:

Proposition II.5.1. *The operation $_c$ is an involution between $\overline{\mathbb{F}}X$ and $\mathbb{A}X$.*

Proof. Let $\mathcal{F} \in \overline{\mathbb{F}}X$ and let $\mathcal{A} := \mathcal{F}_c$. We need to show that (F) implies (A). To this end, let $A_1 \cup A_2 \in \mathcal{A}$, equivalently, $(A_1 \cup A_2)^c \in \mathcal{F}$, that is, $A_1^c \cap A_2^c \in \mathcal{F}$. By (F), A_1^c and A_2^c belong to $\mathcal{F}$, that is, $A_0 \in \mathcal{A}$ and $A_1 \in \mathcal{A}$. On the other hand, if $A_0 \in \mathcal{A}$ and $A_1 \in \mathcal{A}$ then A_1^c and A_2^c belong to $\mathcal{F}$, hence $A_1^c \cap A_2^c \in \mathcal{F}$, that is, $A_1 \cup A_2 \in \mathcal{A}$. The converse is similar. $\square$

In summary,

$$\mathbb{A}X \xleftrightarrow{c} \overline{\mathbb{F}}X \xleftrightarrow{\#} \overline{\mathbb{F}}_\#X.$$

In particular, as we have already said, if $\varnothing \in 2^X$, then $\varnothing^\# = 2^X$ and $\varnothing_c = \varnothing$.

Of course, composing the arrows, we see that the image of $\mathbb{A}X$ under $_c{}^\#$ is $\overline{\mathbb{F}}_\#X$ and that of $\overline{\mathbb{F}}_\#X$ under $^\#{}_c$ is $\mathbb{A}X$.

II.6 Ultrafilters

A filter $\mathcal{U}$ on a set X is said to be an *ultrafilter* if satisfy one, and therefore all, of the equivalent properties in Theorem II.4.5. In particular, ultrafilters on X are maximal elements of $\mathbb{F}X$, and a filter $\mathcal{F}$ is an ultrafilter if and only if $\mathcal{F} = \mathcal{F}^\#$.

Proposition II.6.1. *An ultrafilter is either principal or free.*

Proof. If an ultrafilter $\mathcal{U}$ is not principal, then $\mathcal{U}^*$ is a non-degenerate filter finer than $\mathcal{U}$. By maximality, $\mathcal{U} = \mathcal{U}^*$ is free. $\square$

Principal ultrafilters are exactly the principal filters of singletons, for if A contains two different points, the corresponding principal filters of singletons are strictly finer than $\{A\}^{\uparrow}$. By Example II.4.4,

Proposition II.6.2. *An ultrafilter is sequential if and only if it is principal.*

The *Zorn-Kuratowski Lemma* (Theorem A.9.10) asserts that every partially ordered set whose totally ordered subsets have upper bounds have maximal elements. In fact, the Zorn-Kuratowski Lemma is equivalent to the Axiom of Choice.

Corollary II.6.3. *Each filter is coarser than an ultrafilter.*

Proof. Let $\mathcal{F}$ be a fixed filter on X. Consider the subset $S := \{\mathcal{G} \in \mathbb{F}X : \mathcal{G} \geq \mathcal{F}\}$ of $(\mathbb{F}X, \leq)$ with the induced order. It is a partially ordered set, and if a subset $\{\mathcal{G}_i : i \in I\}$ of S is totally ordered, then $\mathcal{G} := \bigcup_{i \in I} \mathcal{G}_i$ is a filter, and is therefore an upper bound in S for $\{\mathcal{G}_i : i \in I\}$.

Indeed, $\mathcal{G}$ is isotone as a union of isotone family, and if A and B are elements of $\mathcal{G}$ there is i_A and i_B in I such that $A \in \mathcal{G}_{i_A}$ and $B \in \mathcal{G}_{i_B}$. Since, $\{\mathcal{G}_i : i \in I\}$ is totally ordered, one of these two filters contains the other, say, $\mathcal{G}_{i_A} \geq \mathcal{G}_{i_B}$. Then $A \cap B \in \mathcal{G}_{i_A} \subset \mathcal{G}$, and $\mathcal{G}$ is a filter.

By the Zorn-Kuratowski Lemma, S admits a maximal element $\mathcal{U}$, which is also maximal in $\mathbb{F}X$, for if $\mathcal{G} \geq \mathcal{U}$ then $\mathcal{G} \in S$. $\square$

Note that in view of Proposition II.2.14:

Corollary II.6.4. *Each family of subsets of X with the finite intersection property is contained in an ultrafilter on X.*

Let $\mathbb{F}_{\#}X$ denote the set of filter-grills on X and $\mathbb{U}X$ denote the set of ultrafilters on X. Further, given a filter-base $\mathcal{F}$ on X, we denote

$$\beta(\mathcal{F}) := \{\mathcal{U} \in \mathbb{U}X : \mathcal{U} \geq \mathcal{F}\}. \qquad (\text{II.6.1})$$

Corollary II.6.3 shows that $\beta(\mathcal{F})$ is always non-empty, by the Axiom of Choice.

Proposition II.6.5. *If for each $\mathcal{U} \in \beta(\mathcal{F})$, one selects $U_{\mathcal{U}} \in \mathcal{U}$, then there exists a finite sub-collection $\mathbb{D}$ of $\beta(\mathcal{F})$ such that $\bigcup_{\mathcal{U} \in \mathbb{D}} U_{\mathcal{U}} \in \mathcal{F}$.*

Proof. If this is not true, then $\bigcup_{\mathcal{U} \in \mathbb{G}} U_{\mathcal{U}} \notin \mathcal{F}$ for every finite sub-collection $\mathbb{D}$ of $\beta(\mathcal{F})$, hence $\mathcal{B} = \{\bigcap_{\mathcal{U} \in \mathbb{D}} U_{\mathcal{U}}^c : \mathbb{D} \in [\beta(\mathcal{F}]^{<\omega}\}$ is a filter-base that meshes with $\mathcal{F}$. If $\mathcal{W}$ is an ultrafilter finer than $\mathcal{B} \vee \mathcal{F}$, then there is $U_{\mathcal{W}} \in \mathcal{W}$ such that $U_{\mathcal{W}}^c \in \mathcal{B} \subset \mathcal{W}$ yielding a contradiction. $\square$

Lemma II.6.6. *If* $\mathcal{F} \in \mathbb{F}X$ *and* $f : X \to Y$, *then for every* $\mathcal{G} \# f[\mathcal{F}]$, *there exists* $\mathcal{U} \in \beta(\mathcal{F})$ *such that*

$$f[\mathcal{U}] \geq \mathcal{G} \vee f[\mathcal{F}].$$

Proof. If $\mathcal{G} \# f[\mathcal{F}]$, equivalently, $f^-[\mathcal{G}] \# \mathcal{F}$, then there exists an ultrafilter $\mathcal{U}$ finer than $f^-[\mathcal{G}] \vee \mathcal{F}$. This ultrafilter $\mathcal{U}$ has the required properties, because $f[\mathcal{U}] \geq f[f^-[\mathcal{G}]] \geq \mathcal{G}$ and $f[\mathcal{U}] \geq f[\mathcal{F}]$. $\qquad\square$

Therefore:

Corollary II.6.7. *If* $\mathcal{F} \in \mathbb{F}X$ *and* $f : X \to Y$, *then for every* $\mathcal{W} \in \beta(f[\mathcal{F}])$, *there exists* $\mathcal{U} \in \beta(\mathcal{F})$ *such that* $\mathcal{W} \approx f[\mathcal{U}]$.

It follows from Corollary II.6.7 that the images of ultrafilters by maps are ultrafilters (or rather, ultrafilter bases), because if $\mathcal{F}$ is an ultrafilter, then $\beta(\mathcal{F}) = \{\mathcal{F}\}$.

Ultrafilters generate all filters and all filter-grills. Indeed,

Proposition II.6.8. *Every filter is the intersection of all finer ultrafilters:*

$$\mathcal{F} \in \mathbb{F}X \Longrightarrow \mathcal{F} = \bigcap_{\mathcal{U} \in \beta(\mathcal{F})} \mathcal{U};$$

and every filter-grill is the union of coarser ultrafilters:

$$\mathcal{G} \in \mathbb{F}_{\#}X \Longrightarrow \mathcal{G} = \bigcup_{\mathcal{U} \in \mathbb{U}X, \mathcal{U} \subset \mathcal{G}} \mathcal{U}.$$

Proof. Of course, $\mathcal{F} \leq \bigwedge_{\mathcal{U} \in \beta(\mathcal{F})} \mathcal{U}$. On the other hand, if $\mathcal{F} \not\geq \bigwedge_{\mathcal{U} \in \beta(\mathcal{F})} \mathcal{U}$ there is $A \in \bigwedge_{\mathcal{U} \in \beta(\mathcal{F})} \mathcal{U}$ with $A \notin \mathcal{F}$, equivalently, $A^c \in \mathcal{F}^{\#}$. Then $\mathcal{F} \vee \{A^c\}^{\uparrow}$ is a filter and, in view of Corollary II.6.3, there is an ultrafilter $\mathcal{W}$ finer than $\mathcal{F} \vee \{A^c\}^{\uparrow}$. Then $\mathcal{W} \in \beta(\mathcal{F})$ and $A^c \in \mathcal{W}$, so that $A \notin \mathcal{W}$; a contradiction with $A \in \bigwedge_{\mathcal{U} \in \beta(\mathcal{F})} \mathcal{U}$.

Because $\mathcal{U}^{\#} = \mathcal{U}$ for each ultrafilter $\mathcal{U}$, the second part of the theorem follows from the first part and Proposition II.3.5. $\qquad\square$

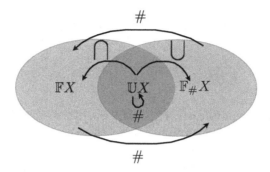

Figure II.2 Duality between filter-grills and filters via #, and relation between filters, filter-grills and ultrafilters.

II.7 Cardinality of the set of ultrafilters

Theorem II.7.1. *If X is an infinite set then*

$$\operatorname{card} \mathbb{U}X = 2^{2^{\operatorname{card} X}}.$$

In particular, card $\mathbb{U}\mathbb{N} = 2^{\mathfrak{c}}$. In order to prove this theorem, we consider:

Definition II.7.2. Let X be an infinite set. A family $\mathcal{A}$ of subsets of X is *independent* if

$$\bigcap_{B \in \mathcal{B}} B \cap \bigcap_{C \in \mathcal{C}} (X \setminus C) \neq \varnothing$$

whenever $\mathcal{B}$ and $\mathcal{C}$ are disjoint finite subsets of $\mathcal{A}$.

The key is to establish:

Theorem II.7.3 (Hausdorff). *Let X be an infinite set. There is an independent family of subsets of X of cardinality $2^{\operatorname{card} X}$.*

Proof. We construct an independent family of the desired cardinality, not on X, but on the set

$$I := \left\{ (F, \mathcal{F}) : F \in [X]^{<\omega}, \mathcal{F} \subset 2^F \right\},$$

which has the same cardinality, by Exercise II.7.4. This is sufficient, as it is easily verified that if $\mathcal{A}$ is an independent family on I and $f : I \to X$ is a bijection, then $f(\mathcal{A}) := \{f(A) : A \in \mathcal{A}\}$ is an independent family on X.

For each $A \subset X$, consider

$$A' := \{(F, \mathcal{F}) \in I : A \cap F \in \mathcal{F}\}.$$

The family $\mathcal{A} := \{A' : A \in 2^X\}$ is the desired independent family. It is of cardinality $2^{\operatorname{card} X}$ because if $A \neq B$ then $A' \neq B'$, for if there is $x \in A \setminus B$ then $(\{x\}, \{\{x\}\}) \in A' \setminus B'$.

To see that it is independent, let $\mathcal{B}$ and $\mathcal{C}$ be two disjoint finite subsets of $\mathcal{A}$. Because $\mathcal{B}$ and $\mathcal{C}$ are disjoint, for every $B' \in \mathcal{B}$ and $C' \in \mathcal{C}$ there is $x_{B,C} \in (B \setminus C) \cup (C \setminus B)$. Consider now

$$F := \{x_{B,C} : B' \in \mathcal{B}, C' \in \mathcal{C}\}$$

$$\mathcal{F} := \{B \cap F : B' \in \mathcal{B}\}.$$

By definition, $(F, \mathcal{F}) \in I$, $F \cap B \in \mathcal{F}$ for every $B' \in \mathcal{B}$, and $F \cap C \notin \mathcal{F}$ for every $C' \in \mathcal{C}$, so that

$$(F, \mathcal{F}) \in \bigcap_{B \in \mathcal{B}} B' \cap \bigcap_{C' \in \mathcal{C}} (I \setminus C') \neq \varnothing.$$

$\square$

Exercise II.7.4. Show that the set I defined in the proof of Theorem II.7.3 satisfies $\operatorname{card} I = \operatorname{card} X$.

Proof of Theorem II.7.1. Of course $\operatorname{card} \mathbb{U}X \leq 2^{2^{\operatorname{card} X}}$ because $\mathbb{U}X \subset 2^{2^X}$. To see the reverse inequality, let $\mathcal{A}$ be an independent family of cardinality $2^{\operatorname{card} X}$.

Consider for each $f : \mathcal{A} \to \{0, 1\}$ and $A \in \mathcal{A}$ the set

$$A_f := \begin{cases} A & \text{if } f(A) = 1 \\ X \setminus A & \text{if } f(A) = 0, \end{cases}$$

and define

$$\mathcal{F}_f := \Big\{ \bigcap_{A \in \mathcal{B}} A_f : \mathcal{B} \in [\mathcal{A}]^{<\omega} \Big\}.$$

Because $\mathcal{A}$ is independent, $\mathcal{F}_f$ has the finite intersection property, and is therefore contained in an ultrafilter $\mathcal{U}_f$ on X, by Corollary II.6.4.

Note that if f and g are two distinct maps of $\{0, 1\}^{\mathcal{A}}$, then there is $A \in \mathcal{A}$ with $f(A) \neq g(A)$, so that $A_f \cap A_g = \varnothing$. But $A_f \in \mathcal{U}_f$ and $A_g \in \mathcal{U}_g$ so that $\mathcal{U}_f \neq \mathcal{U}_g$ and $F : \{0, 1\}^{\mathcal{A}} \to \mathbb{U}X$ defined by

$$F(f) := \mathcal{U}_f$$

is injective. Thus $\operatorname{card} \mathbb{U}X \geq 2^{2^{\operatorname{card} X}}$. $\square$

Definition II.7.5. The *robustness* of a filter $\mathcal{F}$ is defined as

$$\|\mathcal{F}\| := \min\{\operatorname{card} F : F \in \mathcal{F}\}.$$

A filter $\mathcal{F}$ on X is called *uniform* (or *robust*) if $\|\mathcal{F}\| = \operatorname{card} X$.

Remark II.7.6. If $\mathcal{U}$ is a uniform ultrafilter on X and $A \subset X$ is such that $\operatorname{card} A < \operatorname{card} X$, then $X \setminus A \in \mathcal{U}$. Indeed, $(X \setminus A) \cup A \in \mathcal{U}$ hence either $A \in \mathcal{U}$ or $X \setminus A \in \mathcal{U}$, because $\mathcal{U}$ is an ultrafilter. But $A \notin \mathcal{U}$ because $\operatorname{card} A < \|\mathcal{U}\|$.

Proposition II.7.7. *Let X be an infinite set. For any $\mathcal{F} \in \mathbb{F}X$, there is $\mathcal{U} \in \beta(\mathcal{F})$ with*

$$\|\mathcal{F}\| = \|\mathcal{U}\|.$$

In particular, there are uniform ultrafilters on each infinite set.

Proof. Of course, $\|\mathcal{U}\| \leq \|\mathcal{F}\|$ for every $\mathcal{U} \in \beta(\mathcal{F})$, because $\mathcal{F} \subset \mathcal{U}$. To see that there is $\mathcal{U} \in \beta(\mathcal{F})$ with $\|\mathcal{U}\| \geq \|\mathcal{F}\|$, assume to the contrary that each ultrafilter $\mathcal{U} \in \beta\mathcal{F}$ contains a set $U_\mathcal{U}$ with $\operatorname{card} U_\mathcal{U} < \|\mathcal{F}\|$. By Proposition II.6.5, there is a finite subset $\mathbb{D}$ of $\mathbb{U}X$ with $\bigcup_{\mathcal{U} \in \mathbb{D}} U_\mathcal{U} \in \mathcal{F}$, which is a contraction because $\operatorname{card} \bigcup_{\mathcal{U} \in \mathbb{D}} U_\mathcal{U} < \|\mathcal{F}\|$.

Applying this result to $\mathcal{F} = \{X\}$ gives that there is a uniform ultrafilter on X whenever X is infinite. $\square$

II.8 Remarks on sequential filters

II.8.1 *Countably based and Fréchet filters*

Let $\mathbb{F}_1$ denote the class of *countably based filters*. Accordingly, $\mathbb{F}_1(X)$ denotes the set of filters on X that admit a filter-base that is countable. Recall that $\mathbb{E}$ denotes the class of sequential filters. Since the collection of tails of a sequence is countable, every sequential filter is countably based, symbolically, $\mathbb{E} \subset \mathbb{F}_1$. On the other hand, there are countably based filters that are not sequential.

Example II.8.1. On the real line with its canonical metric $d(x,y) := |x - y|$, a vicinity filter $\mathcal{V}(x)$ is not a sequential filter, because all of its elements are uncountable, but it is countably based. For instance $\{B(x, \frac{1}{n}) : n \in \mathbb{N}_1\}$ is a countable filter-base.

Exercise II.8.2. Show that if $\mathcal{H} \in \mathbb{F}_1$ then $\mathcal{H}$ also admits a *decreasing* filter-base (i.e., such that $H_{n+1} \subset H_n$ for each n). Deduce that every countably based filter admits finer sequential filters.

In view of Proposition II.2.15:

Proposition II.8.3. *If a countable collection of countably based filters on X admits a supremum (in $\mathbb{F}X$), it is countably based too.*

However,

Theorem II.8.4. *Let $\mathcal{E}_n$ be a free sequential filter on a set X such that $\mathcal{E}_n \leq \mathcal{E}_{n+1}$ for each $n \in \mathbb{N}$. If $\bigvee_{n \in \mathbb{N}} \mathcal{E}_n$ is a sequential filter, then there is n_0 such that $\mathcal{E}_{n_0} = \mathcal{E}_n$ for every $n \geq n_0$.*

Proof. Suppose that, contrary to the claim $\mathcal{E}_n < \mathcal{E}_{n+1}$ for each $n \in \mathbb{N}$. Then there is a sequence $\{E_n\}_n$ of subsets of X such that $\mathcal{E}_n$ is the cofinite filter of E_n and $E_n \setminus E_{n+1}$ is infinite for every n. We shall see that $\{E_n : n \in \mathbb{N}\}$ is a base of $\mathcal{F} := \bigvee_{n < \omega} \mathcal{E}_n$, and thus by Proposition II.2.32, $\mathcal{F}$ is not sequential. If $\{E_n : n \in \mathbb{N}\}$ is not a base of $\mathcal{F}$, then there is $F \in \mathcal{F}$ and a sequence $\{x_n\}_n$ such that $x_n \in E_n \setminus F$ for each $n \in \mathbb{N}$. Then the set $\{x_n : n < \omega\}$ is almost included in E_n, hence $\{x_n\}_n \geq \mathcal{E}_n$ for every $n \in \mathbb{N}$, and thus $\{x_n\}_n \geq \mathcal{F}$. But on the other hand, $\{x_n : n \in \mathbb{N}\} \cap F = \varnothing$, which yields a contradiction. $\qquad\square$

Definition II.8.5. A filter $\mathcal{F}$ on a set X is called *Fréchet* if ([3])

$$\mathcal{F} = \bigwedge_{\mathcal{E} \in \mathbb{E}, \, \mathcal{E} \geq \mathcal{F}} \mathcal{E}.$$

Exercise II.8.6. Show that a filter is Fréchet if and only if it is an intersection of sequential filters.

Proposition II.8.7. *For each element x of a premetric space, the vicinity filter $\mathcal{V}(x)$ is Fréchet. Namely,*

$$\mathcal{V}(x) = \bigwedge_{x \in \lim(x_n)_n} (x_n)_n^{\uparrow}.$$

Remark II.8.8. In particular, vicinity filters on the real line (for the usual metric) are Fréchet but not sequential, as seen in Example II.8.1. Thus an intersection of sequential filters need not be a sequential filter.

---

[3]Note that, in contrast to the present terminology, some authors use the term Fréchet filter for sequential filters.

Note also that convergence of sequences depends on a filter $\mathcal{V}(x)$ which is generally not a sequential filter.

Exercise II.8.9. Prove Proposition II.8.7.

In view of Proposition I.7.10, Proposition II.8.7 gives an alternative proof that pointwise convergence on a non-countable set is not premetrizable.

Exercise II.8.10. Show that the following are equivalent:

(1) $\mathcal{F}$ is a Fréchet filter on X;
(2) For all $A \subset X$,

$$A \in \mathcal{F}^{\#} \implies \underset{\mathcal{E} \in \mathbb{E}}{\exists} \ \mathcal{E} \geq \mathcal{F} \vee A;$$

(3) For all $A \subset X$,

$$A \in \mathcal{F}^{\#} \implies \underset{\mathcal{H} \in \mathbb{F}_1}{\exists} \ \mathcal{H} \geq \mathcal{F} \vee A.$$

Exercise II.8.11. Show that countably based filters are Fréchet.

Example II.8.12 (A Fréchet filter that is not countably based). On $\mathbb{N} \times \mathbb{N}$ consider for each $n \in \mathbb{N}$, $\mathcal{E}_n := \{\{(n, m) : m \geq k\} : k \in \mathbb{N}\}^{\uparrow}$ and let

$$\mathcal{F} := \bigcap_{n \in \mathbb{N}} \mathcal{E}_n.$$

As an intersection of sequential filters, $\mathcal{F}$ is Fréchet by Exercise II.8.6. On the other hand, $\mathcal{F}$ does not have a countable filter-base. Suppose to the contrary that there is a filter-base $\{F_n : n \in \mathbb{N}\} \subset \mathcal{F}$. Since $F_n \in \mathcal{F}$, there is for each $j \in \mathbb{N}$, $E_j(n) \in \mathcal{E}_j$ with $E_j(n) \subset F_n$, and we can assume $E_j(n+1) \subsetneq E_j(n)$ for all n and j in $\mathbb{N}$, by Exercise II.8.2. Consider now

$$A := \bigcup_{n \in \mathbb{N}} E_n(n) \in \mathcal{F}.$$

We claim that $F_k \not\subset A$ for all $k \in \mathbb{N}$, that is, $\{F_n : n \in \mathbb{N}\}$ is not a filter-base. Indeed, we can assume that $F_k = \bigcup_{j \in \mathbb{N}} E_j(k)$, but $E_{k+1}(k+1) \subsetneq E_{k+1}(k)$ so that there are elements of F_k (in $E_{k+1}(k) \setminus E_{k+1}(k+1)$) that are not in A.

Exercise II.8.13. Show that free ultrafilters are not Fréchet.

II.8.2 *Infima and products of filters*

Note that if $\mathcal{F}$ is a filter on X and $(\mathcal{G}_i)_{i \in I}$ are filters on Y then

$$\bigwedge_{i \in I}(\mathcal{F} \times \mathcal{G}_i) \geq \mathcal{F} \times (\bigwedge_{i \in I} \mathcal{G}_i), \qquad (\text{II.8.1})$$

because $\mathcal{F} \times \mathcal{G}_i \geq \mathcal{F} \times (\bigwedge_{i \in I} \mathcal{G}_i)$ for each $i \in I$.

If I is finite, then (II.8.1) becomes equality:

Lemma II.8.14. *If $\mathcal{F}$ is a filter on X and $\mathcal{G}_1 \ldots \mathcal{G}_n$ are filters on Y then*

$$\bigwedge_{i=1}^{n}(\mathcal{F} \times \mathcal{G}_i) \approx \mathcal{F} \times (\bigwedge_{i=1}^{n} \mathcal{G}_i).$$

Proof. In view of (II.8.1), it suffices to show that $\bigwedge_{i=1}^{n}(\mathcal{F} \times \mathcal{G}_i) \leq \mathcal{F} \times (\bigwedge_{i=1}^{n} \mathcal{G}_i)$. To this end, pick an element of $\bigwedge_{i=1}^{n}(\mathcal{F} \times \mathcal{G}_i)$ of the form $\bigcup_{i=1}^{n}(F_i \times G_i)$ where each F_i is chosen in $\mathcal{F}$ and each G_i in chosen in $\mathcal{G}_i$. Then

$$(\bigcap_{i=1}^{n} F_i) \times (\bigcup_{i=1}^{n} G_i) \subset \bigcup_{i=1}^{n}(F_i \times G_i)$$

and $\bigcap_{i=1}^{n} F_i \in \mathcal{F}$, $\bigcup_{i=1}^{n} G_i \in \bigwedge_{i=1}^{n} \mathcal{G}_i$. $\qquad\square$

Remark II.8.15. It should be apparent from the proof that the above Lemma can be improved for particular filters. For instance if $\mathcal{F}$ is countably deep, then the conclusion of the Lemma remains true for countably many filters $\mathcal{G}_i$.

However, the reverse inequality in (II.8.1) is in general not true:

Example II.8.16. Let $X = Y = \mathbb{N}$, $\mathcal{F} := \{\mathbb{N}_n : n \in \mathbb{N}\}^{\uparrow}$ ([4]) (the cofinite filter) and, for each $n \in \mathbb{N}$, let $\mathcal{G}_n := \{n\}^{\uparrow}$ (the principal ultrafilter of n). Then $\bigwedge_{n \in \mathbb{N}} \mathcal{G}_n = \{\mathbb{N}\}$, so that a base of $\mathcal{F} \times \bigwedge_{n \in \mathbb{N}} \mathcal{G}_n$ is of the form $\{\mathbb{N}_n \times \mathbb{N} : n \in \mathbb{N}\}$. Therefore the set

$$\bigcup_{i \in \mathbb{N}} \mathbb{N}_i \times \{i\}$$

belongs to $\bigwedge_{n \in \mathbb{N}}(\mathcal{F} \times \mathcal{G}_n)$ but not to $\mathcal{F} \times \bigwedge_{n \in \mathbb{N}} \mathcal{G}_n$.

Here is an alternative example that, while more complicated, will turn out to be useful in understanding other examples later on.

---

[4]Recall that $\mathbb{N}_n = \{k \in \mathbb{N} : k \geq n\}$.

Example II.8.17. Let $X := \{\frac{1}{n} : n \in \mathbb{N}_1\}$ and $Y := \{x_{k,i} : k, i \in \mathbb{N}\}$ where all $x_{k,i}$ are distinct.

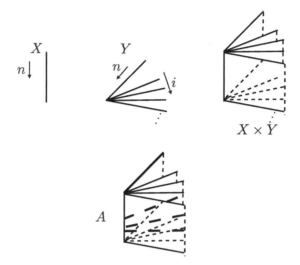

Figure II.3   For the sake of the picture $X = \{\frac{1}{n} : n \in \mathbb{N}\}$ is drawn vertically, as the "spine of a book" and the parts $Y_i := \{x_{k,i} : k \in \mathbb{N}\}$ of Y are drawn as attached to each other at one point, as to form edges of a page. The "book" has countably many pages. The set A is the union of the $\{\frac{1}{i}\} \times Y_i$, that is, the top edge of the first page, and the full length of the page in the middle of the second page, the full length of the page at one third of the height of the third page, and so on.

Consider on X the cofinite filter $\mathcal{F} = (\frac{1}{n})_n^\uparrow$ of X and on Y the cofinite filter $\mathcal{G}_i := (x_{k,i})_k^\uparrow$ of $\{x_{k,i} : k \in \mathbb{N}\}$ for each $i \in \mathbb{N}$. Then the subset

$$A := \bigcup_{n \in \mathbb{N}} \left(\{\tfrac{1}{n}\} \times \{x_{k,n} : k \in \mathbb{N}\} \right)$$

of $X \times Y$ meshes $\mathcal{F} \times \bigwedge_{i \in \mathbb{N}} \mathcal{G}_i$ but not $\bigwedge_{i \in \mathbb{N}} (\mathcal{F} \times \mathcal{G}_i)$, so that

$$\bigwedge_{i \in \mathbb{N}} (\mathcal{F} \times \mathcal{G}_i) \not\leq \mathcal{F} \times \bigwedge_{i \in \mathbb{N}} \mathcal{G}_i.$$

Indeed, $A \in (\mathcal{F} \times \bigwedge_{i \in \mathbb{N}} \mathcal{G}_i)^\#$ because each $\{x_{k,i} : k \in \mathbb{N}\} \in (\bigwedge_{i \in \mathbb{N}} \mathcal{G}_i)^\#$. On the other hand for each $i \in \mathbb{N}$

$$B_i := \left(\{\tfrac{1}{n} : n \geq i+1\} \times \{x_{k,i} : k \in \mathbb{N}\} \right) \in \mathcal{F} \times \mathcal{G}_i$$

but $B_i \cap A = \varnothing$. Therefore $\bigcup_{i \in \mathbb{N}} B_i$ is an element of $\bigwedge_{i \in \mathbb{N}} (\mathcal{F} \times \mathcal{G}_i)$ disjoint from A.

II.9   Contours and extensions

Let X be a non-empty subset. Denote by

$$\S X := 2^{2^X}$$

the set of families of subsets of X. Then we abridge $\S\S X := \S(\S X)$. Accordingly $\S\S\S X := \S(\S\S X)$.

Let $\mathcal{H}$ be a family of subsets of $\S X$, that is, $\mathcal{H} \in \S\S X$. The *contour* $\mathcal{H}^\star$ of $\mathcal{H}$ is defined as a family of subsets of X, that is, $\mathcal{H}^\star \subset 2^X$, given by

$$\mathcal{H}^\star := \bigcup_{H \in \mathcal{H}} \bigcap_{\mathcal{A} \in H} \mathcal{A}. \qquad (\text{II.9.1})$$

In other words, $A \in \mathcal{H}^\star$ if and only there exists $H \in \mathcal{H}$ such that $A \in \mathcal{A}$ for each $\mathcal{A} \in H$.

In particular, if $H \subset \S X$, then $H^\star := \{H\}^\star = \left(\{H\}^\uparrow\right)^\star$, that is,

$$H^\star = \bigcap_{\mathcal{A} \in H} \mathcal{A}.$$

In practice, one often considers a parametrized collection of families

$$\{\mathcal{A}(y) : y \in Y\}$$

of subsets of X, and a family $\mathcal{G}$ on Y. Then the *contour* $\mathcal{A}(\mathcal{G})$ of $\mathcal{A}(\cdot)$ along $\mathcal{G}$ is a family of subsets of X defined by

$$\mathcal{A}(\mathcal{G}) := \bigcup_{G \in \mathcal{G}} \bigcap_{y \in G} \mathcal{A}(y).$$

In particular, for $G \subset Y$, we abridge $\mathcal{A}(G) := \bigcap_{y \in G} \mathcal{A}(y)$.

In order to relate the two notions of contour introduced so far, consider the family $\widehat{\mathcal{G}}$ of subsets of $\S X$ such that $H \in \widehat{\mathcal{G}}$ if there exists $G \in \mathcal{G}$ such that $\{\mathcal{A}(y) : y \in G\} \subset H$. Then it turns out that

$$\mathcal{A}(\mathcal{G}) = \widehat{\mathcal{G}}^\star. \qquad (\text{II.9.2})$$

Exercise II.9.1. Prove (II.9.2).

Of course, for a family $\mathcal{G}$ of subsets of $\S\S X$, the contour $\mathcal{G}^\star$ is a family of subsets of $\S X$. The set X has a natural injection $j : X \to \S\S X$, namely,

$$j(x) := \left\{\{x\}^{\uparrow x}\right\}^{\uparrow \S X} = \left\{\mathcal{A} \subset \S X : \{x\}^{\uparrow x} \in \mathcal{A}\right\}.$$

Therefore, one can represent a family $\mathcal{G}$ on X as a family $j[\mathcal{G}]$ on $\S\S X$ and consider the contour of $j[\mathcal{G}]$, which is a family on $\S X$. Namely, we define a family $\mathcal{G}^\Uparrow$ of subsets of $\S X$, by

$$\mathcal{G}^\Uparrow := (j[\mathcal{G}])^\star$$

and call it the *extension* of $\mathcal{G}$ to $\S X$.

In particular, if $G \subset X$, then $G^{\Uparrow} := \{G\}^{\Uparrow} = (\{G\}^{\uparrow X})^{\Uparrow}$, that is,

$$G^{\Uparrow} = \{\mathcal{A} \in \S X : G \in \mathcal{A}\}. \tag{II.9.3}$$

Therefore, $H \in \mathcal{G}^{\Uparrow}$ if and only if there exists $G \in \mathcal{G}$ such that $G^{\Uparrow} \subset H$, that is,

$$\mathcal{G}^{\Uparrow} = \left\{ G^{\Uparrow} : G \in \mathcal{G} \right\}^{\uparrow \S X}.$$

Theorem II.9.2. *For each* $\mathcal{H} \in \S\S X$ *and* $\mathcal{G} \in \S X$,

$$(\mathcal{H}^{\star})^{\Uparrow} = \mathcal{H}^{\uparrow \S X}, \ (\mathcal{G}^{\Uparrow})^{\star} = \mathcal{G}^{\uparrow X}.$$

Proof. By definition, $B \in (\mathcal{H}^{\star})^{\Uparrow}$ whenever there exists $G \in \mathcal{H}^{\star}$ such that $G^{\Uparrow} \subset B$. Now, $G \in \mathcal{H}^{\star}$ whenever there exists $H \in \mathcal{H}$ such that $G \in H^{\star}$. Gathering the two conditions, $B \in (\mathcal{H}^{\star})^{\Uparrow}$ if and only if there exists $H \in \mathcal{H}$ such that $G \in H^{\star}$ and $G^{\Uparrow} \subset B$. Now, $G \in H^{\star} = \bigcap_{\mathcal{A} \in H} \mathcal{A}$ means that $G \in \mathcal{A}$, by (II.9.3) equivalently, $H \subset G^{\Uparrow} \subset B$, that is, $B \in \mathcal{H}^{\uparrow}$.

Now, $D \in (\mathcal{G}^{\Uparrow})^{\star}$ if and only if there exists $H \in \mathcal{G}^{\Uparrow}$ such that $D \in \mathcal{A}$ for each $\mathcal{A} \in H$, equivalently, there exists $G \in \mathcal{G}$ such that if $G \in \mathcal{A}$ then $\mathcal{A} \in H \subset D$, that is, $D \in \mathcal{G}^{\uparrow}$. $\square$

Chapter III

Convergences

III.1 Definitions and first examples

Recall that $\mathbb{F}X$ denotes the set of filters on a (non-empty) set X. If ξ is an arbitrary relation between $\mathbb{F}X$ and X, then we write

$$x \in \lim_\xi \mathcal{F} \tag{III.1.1}$$

if $(\mathcal{F}, x) \in \xi$, and we say that the filter $\mathcal{F}$ *converges* to x with respect to ξ (equivalently, x is a *limit* of $\mathcal{F}$ with respect to ξ).

Of course, the notation (III.1.1) is redundant with respect to the standard way $x \in \xi(\mathcal{F})$ of expressing the image of $\mathcal{F}$ under the relation ξ, but is more suggestive of its interpretation as a convergence. Similarly, we denote the image of x under the inverse relation of ξ by

$$\lim_\xi^{-1}(x) := \xi^-(x) = \{\mathcal{F} \in \mathbb{F}X : (\mathcal{F}, x) \in \xi\}.$$

A relation ξ is called a *preconvergence* if it is *isotone*, that is, if

$$\mathcal{F} \leq \mathcal{G} \implies \lim_\xi \mathcal{F} \subset \lim_\xi \mathcal{G}. \tag{III.1.2}$$

In other words, ξ is a preconvergence if $(x, \mathcal{F}) \in \xi$ and $\mathcal{F} \leq \mathcal{G}$ implies $(x, \mathcal{G}) \in \xi$.

As we have seen in Exercise II.2.3, a filter is an equivalence class of filter-bases with respect to the equality $\approx$ modulo the preorder $\leq$ used in (III.1.2). Therefore a natural extension of the relation ξ to filter-bases via

$$\lim_\xi \mathcal{B} := \lim_\xi \mathcal{B}^\uparrow$$

is compatible with the property (III.1.2).

A preconvergence ξ is said to be a *convergence* if it is *centered*, that is, if for every $x \in X$,

$$x \in \lim_\xi \{x\}^\uparrow, \tag{centered}$$

55

where $\{x\}^{\uparrow}$ is the principal filter determined by x.

If ξ is a convergence on a set X, then the couple (X, ξ) is called a *convergence space* ([1]). The set on which a convergence ξ is defined, is called the *underlying set* of ξ and is denoted by $|\xi|$.

A convergence determines the corresponding convergence space thanks to (centered). Therefore we will use the terms *convergence* and *convergence space* interchangeably.

A (pre)convergence ξ is called *finitely deep* ([2]) if

$$\lim_\xi(\mathcal{F} \wedge \mathcal{G}) \supset \lim_\xi \mathcal{F} \cap \lim_\xi \mathcal{G} \qquad \text{(finitely deep)}$$

for every filters $\mathcal{F}$ and $\mathcal{G}$ on $|\xi|$. The inclusion above is an equality, because the reverse inclusion follows from (III.1.2).

Example III.1.1 (Standard convergence of the real line). The *standard convergence ν on* $\mathbb{R}$ is defined by

$$x \in \lim_\nu \mathcal{F} \iff \mathcal{F} \geq \left\{ \left(x - \tfrac{1}{n}, x + \tfrac{1}{n}\right) : n \in \mathbb{N}_1 \right\}.$$

It is easy to verify that this is a finitely deep convergence.

Example III.1.2 (Premetric convergence). More generally, if d is a premetric on X, then we define a convergence $\tilde{d}$ on X by $x \in \lim_{\tilde{d}} \mathcal{F}$ whenever $B_d(x, r) \in \mathcal{F}$ for each $r > 0$. Such convergences are always finitely deep. Example III.1.1 is the particular case where $X = \mathbb{R}$ and $d(x, y) = |x - y|$.

Example III.1.3 (Empty preconvergence). On an arbitrary non-empty set X one defines the *empty preconvergence* $\varnothing_X$ by

$$\lim_{\varnothing_X} \mathcal{F} = \varnothing$$

for all $\mathcal{F} \in \mathbb{F}X$. This is not a convergence because (centered) is not satisfied.

Here is an example of a convergence that is not finitely deep.

Example III.1.4. Let X be an infinite set and let x_∞ be a point outside of X. Let $\mathcal{F}$ and $\mathcal{G}$ be two free filters on X that do not mesh. By Corollary II.2.21, $\mathcal{F} \wedge \mathcal{G}$ is a free filter, and it is strictly coarser than both $\mathcal{F}$ and $\mathcal{G}$ because $\mathcal{F}$ and $\mathcal{G}$ are dissociated.

---

[1]Various authors (e.g., [91]) call it *generalized convergence*.

[2]Some authors call it *limit space* (after the German term *Limesraüme* introduced by Kowalsky [71] and Fischer [40]). Others use the name "convergence" for it. The intermediate class of convergences satisfying

$$x \in \lim_\xi \mathcal{F} \implies x \in \lim_\xi \mathcal{F} \wedge \{x\}^{\uparrow}$$

is often (e.g., [91]) called a *Kent space* after D.C. Kent's work in e.g., [60].

Define on $X \cup \{x_\infty\}$ a convergence ξ so that $x \in \lim_\xi \mathcal{H}$ if and only if $\mathcal{H} = \{x\}^\uparrow$ for each $x \in X$, and $x_\infty \in \lim_\xi \mathcal{H}$ if ([3])

$$\mathcal{H} \geq \mathcal{F} \wedge \{x_\infty\}^\uparrow \text{ or } \mathcal{H} \geq \mathcal{G} \wedge \{x_\infty\}^\uparrow.$$

This convergence is not finitely deep because $x_\infty \in \lim_\xi \mathcal{F} \cap \lim_\xi \mathcal{G}$ but neither $\mathcal{F} \wedge \mathcal{G} \geq \mathcal{F}$ nor $\mathcal{F} \wedge \mathcal{G} \geq \mathcal{G}$ and thus $x_\infty \notin \lim_\xi (\mathcal{F} \wedge \mathcal{G})$.

Example III.1.5 (Discrete convergence). On an arbitrary non-empty set X one defines the *discrete preconvergence* ι (also denoted ι_X if there might be ambiguity on the set on which we consider the discrete convergence) to the effect that

$$x \in \lim_\iota \mathcal{F} \iff \mathcal{F} = \{x\}^\uparrow.$$

The discrete preconvergence is a finitely deep convergence.

Example III.1.6 (Chaotic convergence). On an arbitrary non-empty set X one defines the *chaotic preconvergence* o (also denoted o_X if there might be ambiguity on the set on which we consider the chaotic convergence) so that $\lim_o \mathcal{F} = X$ for every filter $\mathcal{F}$ on X. The chaotic preconvergence is a finitely deep convergence.

Definition III.1.7. If ζ and ξ are preconvergences on the same underlying set, then ζ is *finer* than ξ, or that ξ is *coarser* than ζ (in symbols, $\zeta \geq \xi$), if

$$\lim_\zeta \mathcal{F} \subset \lim_\xi \mathcal{F}$$

for each filter $\mathcal{F}$.

It is immediate that for every preconvergence ξ on a set X

$$o_X \leq \xi \leq \varnothing_X.$$

Because a preconvergence is a convergence if and only if it satisfies (centered):

Proposition III.1.8. *A preconvergence ξ on X is a convergence if and only if*

$$\xi \leq \iota_X.$$

Definition III.1.9. A preconvergence is called *Hausdorff*, or T_2, if each filter has at most one limit point.

---

[3]Where we identify $\mathcal{F}$ and $\mathcal{G}$ with the filters they generate on Y, as they are filter-bases on Y.

In other words, a preconvergence is Hausdorff if and only if

$$x_0 \neq x_1 \Longrightarrow \lim^{-1}(x_0) \cap \lim^{-1}(x_1) = \varnothing,$$

if and only if

$$x_0 \neq x_1, \mathcal{F}_0 \in \lim^{-1}(x_0) \text{ and } \mathcal{F}_1 \in \lim^{-1}(x_1),$$

then $\mathcal{F}_0$ and $\mathcal{F}_1$ are dissociated (see Definition II.2.9).

Definition III.1.10. A preconvergence ξ is called T_1 if $\lim\{x\}^{\uparrow} \subset \{x\}$ for each $x \in |\xi|$, and T_0 if points are distinguishable, that is, if

$$x_0 \neq x_1 \Longrightarrow \lim^{-1}(x_0) \neq \lim^{-1}(x_1).$$

Of course, a *convergence* is T_1 if and only if $\lim\{x\}^{\uparrow} = \{x\}$ for each x, so that each T_2-convergence is also T_1, but the converse is not true (see Example III.4.9 or Example IV.9.3). Similarly, each T_1-convergence is T_0 for if $x \neq y$ then $\{x\}^{\uparrow} \in \lim^{-1}(x) \setminus \lim^{-1}(y)$, but the converse is not true. For instance, Example III.1.14 below (or the Sierpiński space of (III.2.1)) is T_0 but not T_1. On the other hand, the chaotic convergence on a set with at least two points is not T_0.

Proposition III.1.11. *Each T_1-preconvergence on a finite set is T_2. Each T_1-convergence on a finite set is discrete.*

Proof. Let ξ be a preconvergence on a finite set X. If ξ is not T_2 there exist a filter $\mathcal{F}$ on X and two distinct points a and b such that $\{a, b\} \subset \lim_\xi \mathcal{F}$. As the filter $\mathcal{F}$ is principal, by Exercise II.2.18, there exists $x \in \bigcap \mathcal{F}$, so that $\{x\}^{\uparrow} \geq \mathcal{F}$ and $\{a, b\} \subset \lim_\xi \{x\}^{\uparrow}$. Thus, ξ is not T_1.

If ξ is a convergence and $x \in \lim_\xi A^{\uparrow}$ then $x \in \lim_\xi \{a\}^{\uparrow}$ for each $a \in A$. Hence $x = a$ for each $a \in A$, so that $A = \{x\}$. $\qquad\qquad\square$

Proposition III.1.12. *Let ξ be a T_1-convergence and let $x \in \lim_\xi \mathcal{F}$. Then $\ker \mathcal{F} \subset \{x\}$.*

Proof. Suppose that there exists $x_0 \neq x$ such that $x_0 \in \ker \mathcal{F}$. Then $\mathcal{F} \leq \{x_0\}^{\uparrow}$ and thus $x \in \lim_\xi \{x_0\}^{\uparrow}$. But as $\{x_0\}$ is closed, $x = x_0$, which is a contradiction. $\qquad\qquad\square$

Exercise III.1.13. Show that if $\xi \leq \tau$ and ξ is T_1 (respectively T_2) then so is τ.

In contrast, this property does not hold for T_0:

Example III.1.14 (A T_0-convergence coarser than a non-T_0 convergence). On $X = \{0, 1, 2\}$ define the finitely deep convergence ξ by

$$\lim_\xi \{0\}^\uparrow = \{0, 2\} = \lim_\xi \{2\}^\uparrow, \lim_\xi \{1\}^\uparrow = \{1, 2\}.$$

This completely determines the convergence by finite depth. Explicitly, $\lim_\xi \{0, 1\}^\uparrow = \lim_\xi \{0, 1, 2\}^\uparrow = \varnothing$, $\lim_\xi \{0, 2\}^\uparrow = \{0, 2\}$ and $\lim_\xi \{1, 2\}^\uparrow = \{2\}$. Of course, ξ is not T_1. On the other hand,

$$\lim_\xi^{-1}(0) = \{\{0\}^\uparrow, \{2\}^\uparrow, \{0, 2\}^\uparrow\}, \lim_\xi^{-1}(1) = \{\{1\}^\uparrow\},$$

$$\lim_\xi^{-1}(2) = \{\{0\}^\uparrow, \{2\}^\uparrow, \{0, 2\}^\uparrow, \{1\}^\uparrow, \{1, 2\}^\uparrow\},$$

so that ξ is T_0. On the other hand, the finer finitely deep convergence τ defined by

$$\lim_\tau \{0\}^\uparrow = \lim_\tau \{2\}^\uparrow = \{0, 2\}, \text{ and } \lim_\tau \{1\}^\uparrow = \{1\}$$

is not T_0 for $\lim_\tau^{-1}(0) = \lim_\tau^{-1}(2)$.

Example III.1.15 (prime cofinite convergence). Let X be a non-empty set and let $x \in X$ be a distinguished point. We denote by

$$\pi[x, (X)_0]$$

the following convergence on X:

$t \in \lim_{\pi[x,(X)_0]} \{t\}^\uparrow$ for each $t \in X$ and $x \in \lim_{\pi[x,(X)_0]} \mathcal{F}$ for $\mathcal{F} \neq \{x\}^\uparrow$ if $X \setminus A \in \mathcal{F}$ for each finite subset A of X. This convergence is called *prime cofinite (at x)*. This is a T_2-convergence.

(1) If X is finite, then $\pi[x, (X)_0]$ is the discrete convergence of X.
(2) If X is infinite then there exists a coarsest filter that converges to x in $\pi[x, (X)_0]$. This filter is given by

$$\mathcal{V}_{\pi[x,(X)_0]}(x) = \{x\}^\uparrow \wedge (X)_0, \tag{III.1.3}$$

where $(X)_0$ is the cofinite filter on X. Notice that for every infinite set X,

$$\text{card} \{X \setminus A : \text{card } A < \infty\} = \text{card } X. \tag{III.1.4}$$

(3) If X is countably infinite, then the filter $\mathcal{V}_{\pi[x,(X)_0]}(x)$ is *countably based*.
(4) If X is uncountable, then $\mathcal{V}_{\pi[x,(X)_0]}(x)$ is not countably based.

Definition III.1.16. If ξ is a preconvergence on X, then for each $x \in X$,

$$\mathcal{V}_\xi(x) := \bigwedge_{\mathcal{F} \in \lim_\xi^{-1}(x)} \mathcal{F} \tag{III.1.5}$$

is called the *vicinity filter* of ξ at x. A subset V of X is called a *vicinity* of x for ξ if $V \in \mathcal{V}_\xi(x)$.

The vicinity filter of ξ at x need not converge to x in ξ. In fact, in Example III.1.4, $\mathcal{V}_\xi(x_\infty) = \mathcal{F} \wedge \mathcal{G} \wedge \{x_\infty\}^\uparrow$ and $x_\infty \notin \lim_\xi \mathcal{V}_\xi(\infty)$, because we have seen that $x_\infty \notin \lim_\xi (\mathcal{F} \wedge \mathcal{G})$.

Here is an example of a convergence in which no vicinity filter converges.

Example III.1.17. Let ν denote the standard convergence on $\mathbb{R}$. Let

$$x \in \lim_{\mathrm{Seq}\,\nu} \mathcal{F}$$

if there exists a sequential filter $\mathcal{E}$ such that $x \in \lim_\nu \mathcal{E}$ and $\mathcal{E} \leq \mathcal{F}$. This is a convergence, because if $x \in \lim_{\mathrm{Seq}\,\nu} \mathcal{F}$ and $\mathcal{F} \leq \mathcal{G}$ then there is a sequential filter $\mathcal{E}$ such that $x \in \lim_\nu \mathcal{E}$ and $\mathcal{E} \leq \mathcal{F} \leq \mathcal{G}$. On the other hand, the principal filter $\{x\}^\uparrow$ of x is sequential. It follows that each filter that converges for $\mathrm{Seq}\,\nu$ contains a countable set!

Observe that $\mathrm{Seq}\,\nu \geq \nu$. Recall that $\mathbb{E}X$ denotes the sets of all sequential filters on X. It is obvious that the vicinity filter of $\mathrm{Seq}\,\nu$ at x is given by

$$\mathcal{V}_{\mathrm{Seq}\,\nu}(x) = \bigwedge_{\mathcal{V}_\nu(x) \leq \mathcal{E} \in \mathbb{E}X} \mathcal{E}.$$

By Proposition II.8.7,

$$\mathcal{V}_{\mathrm{Seq}\,\nu}(x) = \mathcal{V}_\nu(x).$$

We notice that $x \notin \lim_{\mathrm{Seq}\,\nu} \mathcal{V}_\nu(x)$, because every vicinity of x is uncountable (it includes an open interval containing x), while all filters convergent in $\mathrm{Seq}\,\nu$ contain a countable set.

III.2   Preconvergences on finite sets

III.2.1   *Preconvergences on two-point sets*

It may seem of little importance to dwell on trivial examples such as spaces with two or three points only. We will see that some of these very simple spaces "generate" (in a sense to be specified later) important classes of convergence spaces, and as such, they will play an important role.

If $X := \{0,1\}$ then there are only three filters on X: $\{0\}^\uparrow$, $\{1\}^\uparrow$, $\{0,1\}^\uparrow = \{\{0,1\}\}$. The *discrete convergence* $\iota = \iota_{\{0,1\}}$ is finitely deep, because

$$\lim_\iota \{0\}^\uparrow = \{0\}, \lim_\iota \{1\}^\uparrow = \{1\}, \lim_\iota \{0,1\}^\uparrow = \varnothing.$$

Hence $\lim_\iota \{0\}^\uparrow \cap \lim_\iota \{1\}^\uparrow = \varnothing = \lim_\iota \{0,1\}^\uparrow$.

If a *finitely deep convergence* ξ is not discrete, then either $1 \in \lim_\xi\{0\}^\uparrow$ hence $1 \in \lim_\xi\{0,1\}^\uparrow$ or $0 \in \lim_\xi\{1\}^\uparrow$ hence $0 \in \lim_\xi\{0,1\}^\uparrow$. Therefore there are three possibilities ([4]):

The *Sierpiński convergence* $\$_0$

$$\lim_{\$_0}\{0\}^\uparrow = \{0,1\}, \lim_{\$_0}\{1\}^\uparrow = \{1\}, \lim_{\$_0}\{0,1\}^\uparrow = \{1\}. \qquad \text{(III.2.1)}$$

The *Sierpiński convergence* $\$_1$

$$\lim_{\$_1}\{0\}^\uparrow = \{0\}, \lim_{\$_1}\{1\}^\uparrow = \{0,1\}, \lim_{\$_1}\{0,1\}^\uparrow = \{0\}. \qquad \text{(III.2.2)}$$

The *chaotic convergence* o

$$\lim_o\{0\}^\uparrow = \{0,1\}, \lim_o\{1\}^\uparrow = \{0,1\}, \lim_o\{0,1\}^\uparrow = \{0,1\}.$$

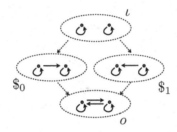

Figure III.1 In each oval a finitely deep convergence on $\{0,1\}$ is defined by solid arrows. The principal filter of that point at the tail of a solid arrow converges to the point at the head of the arrow. Left points represent 0, and right ones represent 1. A dotted arrow indicates that the convergence at the tail of the arrow is finer than that at the arrowhead.

The Sierpiński convergences $\$_0$ and $\$_1$ are equivalent, in the sense that the bijective map $f(0) = 1$, $f(1) = 0$ transforms one into another. Thus, if we only consider convergences on $\{0,1\}$ up to this type of equivalence, there are only three possibilities: the discrete, the chaotic, and the Sierpiński convergence. Note that the discrete convergence is T_2, the Sierpiński convergence is T_0 but not T_1, and the chaotic convergence is not even T_0.

For each convergence considered above, removing the condition (centered) at one point, at the other, or at both, yields three new preconvergences. Of course, because of this process, they cease to be convergences.

---

[4]For $\$_0$ and $\$_1$, if we removed the condition of finite depth and keep the first two conditions, then $\{0,1\}^\uparrow$ might not converge. For the chaotic convergence, if we removed the condition of finite depth, then $\{0,1\}^\uparrow$ might not converge, converge only to 0, or only to 1.

In each group of four preconvergences below, the eastern and western preconvergences are equivalent via the transformation defined by $f(0) = 1$, $f(1) = 0$. Additionally, each of the four preconvergences in the eastern group is equivalent to its counterpart in the western group via the same map.

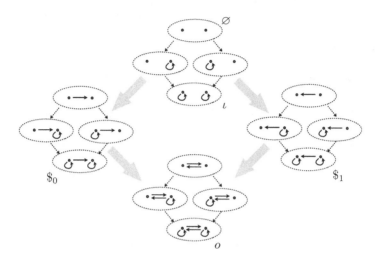

Figure III.2   In each oval a finitely deep preconvergence on $\{0, 1\}$ is defined by solid arrows. The principal filter of the point at the tail of a solid arrow converges to the point at the head of the arrow. Left points represent 0, and right ones represent 1. A dotted arrow indicates that the preconvergence at the tail is finer than that at the arrowhead. A large grey arrow represents the same relation of being finer from spaces in the tail group to spaces in the corresponding positions in the head group.

III.2.2   *Preconvergences on three-point sets*

A *finitely deep convergence* on a three-point set $\{0, 1, 2\}$ is determined by the convergence of the three principal filters of points, as other filters are finite infima of such filters. In a convergence each principal filter of a point converges to that point. It can also converge to none of the other two points, both, or one and not the other. Thus there are 4 possibilities at each of the three points. Hence there are $4^3 = 64$ finitely deep convergences on $\{0, 1, 2\}$.

Identifying convergences that are transformed into one another under a permutation of $\{0, 1, 2\}$, we are left with 16 possibilities represented below

(^{5}). In this figure, we apply the same conventions as in Figures III.1 and III.2, except that the loops representing convergence of principal filters of points to the defining point are omitted, as we only consider convergences.

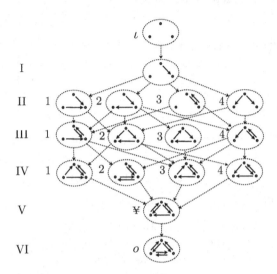

Figure III.3 There are 16 equivalence classes of finitely deep convergences on a three-point set. In each oval a finitely deep convergence representing such a class is defined by solid arrows. The principal filter of the point at the tail of a solid arrow converges to the point at the head of the arrow. A dotted arrow indicates the that the convergence at the tail is finer than that at the head. The 16 classes are organized in rows that correspond to the number of solid arrows.

Exercise III.2.1. Show that the convergence IV(3) in Figure III.3 is T_0 but that the space ¥ of row V in the same figure is not.

If we remove the constraint that the principal filter of a point converges to that point, we have now 8 possibilities at each of the three points. Indeed, $\lim \{x\}^{\uparrow}$ can be any subset of $\{0, 1, 2\}$, which accounts for $2^3 = 8$ possibilities. As there are 3 (principal) ultrafilters, there are $8^3 = 512$ finitely deep preconvergences on $\{0, 1, 2\}$. There are many more preconvergences.

---

[5]Explicitly, the discrete and chaotic convergences are invariant under bijection and thus represent only one convergence each. On the other hand, each of the ovals I, II(2), III(1), III(2), III(4), IV(3) and V represent 6 convergences, for each permutation of $\{0, 1, 2\}$ yields an equivalent convergence. The ovals II(1), II(3), II(4), IV(1), IV(2) and IV(4) each represent 3 different convergences for only the orientation preserving transformations yield equivalent convergences. Finally, the cycle III(3) represents 2 convergences, corresponding to two possible orientations.

Exercise III.2.2. Let $X := \{0, 1, 2\}$.

(1) How many relations are there in $X \times \mathbb{F}X$?

(2) How many preconvergences are there on X?

III.3 Induced (pre)convergence

Definition III.3.1. If (X, ξ) is a preconvergence space and $A \subset X$, let $i_X^A : A \to X$ denote the *inclusion map* of A in X, that is,

$$i_X^A(x) := x$$

for each $x \in A$. Then the *preconvergence* $\xi_{|A}$ *induced on* A *by* ξ is defined by

$$x \in \lim_{\xi_{|A}} \mathcal{F} \iff x \in \lim_{\xi} i_X^A[\mathcal{F}], \tag{III.3.1}$$

whenever $x \in A$ and $\mathcal{F} \in \mathbb{F}A$.

Note that if $A \subset X$ and $\mathcal{F} \in \mathbb{F}A$ then $A \in (i_X^A[\mathcal{F}])^\uparrow \in \mathbb{F}X$ and that conversely, if $A \in \mathcal{G} \in \mathbb{F}X$ then $\mathcal{F} := \{G \cap A : G \in \mathcal{G}\}$ is a filter of $\mathbb{F}A$ with $(i_X^A[\mathcal{F}])^\uparrow = \mathcal{G}$. Therefore, we can make the identification

$$\mathbb{F}A \approx \{\mathcal{F} \in \mathbb{F}X : A \in \mathcal{F}\}. \tag{III.3.2}$$

Clearly, if $A \subset X$ and (X, ξ) satisfies (centered), so does $\xi_{|A}$. In other words, if (X, ξ) is a convergence space, so is $(A, \xi_{|A})$.

Definition III.3.2. A (pre)convergence space (A, τ) is a *subspace* of a (pre)convergence space (X, ξ) if $A \subset X$ and $\tau = \xi_{|A}$.

Exercise III.3.3. Show that a subspace of a discrete convergence is a discrete convergence and that a subspace of a chaotic convergence is a chaotic convergence.

Exercise III.3.4. Show that the convergence induced on $\mathbb{Z}$ by the standard convergence on $\mathbb{R}$ is the discrete convergence.

Exercise III.3.5. Show that (each of the two variants of) the Sierpiński space \$ is (homeomorphic to) a subspace of each of the convergences of Figure III.3 except for the discrete convergence, the chaotic convergence, and the convergence is ovals II(3) and IV(2).

Since for every $A \subset X$ and every $\mathcal{F}, \mathcal{G} \in \mathbb{F}A$,

$$(i_X^A[\mathcal{F} \wedge \mathcal{G}])^{\uparrow x} = (i_X^A[\mathcal{F}])^{\uparrow x} \wedge (i_X^A[\mathcal{G}])^{\uparrow x},$$

we have:

Proposition III.3.6. *A subspace of a finitely deep (pre)convergence is finitely deep.*

III.4 Premetrizable convergences

Recall from Example III.1.2 that if d is a premetric on a set X, then we define on X a convergence $\tilde{d}$ by

$$x \in \lim_{\tilde{d}} \mathcal{F} \iff \mathcal{V}_d(x) \leq \mathcal{F}, \qquad (\text{III.4.1})$$

where $\mathcal{V}_d(x)$ is the vicinity filter for the premetric d. Recall that $V \in \mathcal{V}_d(x)$ whenever there is $r > 0$ such that $B_d(x, r) \subset V$. This is a convergence, because if $x \in \lim_{\tilde{d}} \mathcal{F}$ and $\mathcal{F} \leq \mathcal{G}$ then $\mathcal{V}_d(x) \leq \mathcal{F} \leq \mathcal{G}$, and $x \in N$ for each $N \in \mathcal{V}_d(x)$, that is, $\mathcal{V}_d(x) \leq \{x\}^\uparrow$, thus $x \in \lim_{\tilde{d}} \{x\}^\uparrow$. Therefore $\mathcal{V}_d(x)$ is the coarsest filter that converges to x in $\tilde{d}$.

In particular, the *standard convergence* ν *on* $\mathbb{R}$ is the convergence fulfills $\nu = \tilde{d}$, where $d(x, y) = |x - y|$.

We call a convergence ξ on X *premetrizable* if there exists a premetric d on X such that $\xi = \tilde{d}$, and *metrizable* if d can be chosen to be a metric.

Exercise III.4.1. Show that each subspace of a premetrizable (respectively metrizable) convergence space is premetrizable (respectively metrizable).

The metric (I.2.2) of the real line is an instance of a family of metrics that can be defined on any set:

Example III.4.2. If X is an arbitrary non-empty set, then the function

$$i_X(x_0, x_1) := \begin{cases} 0 \text{ if } x_0 = x_1, \\ 1 \text{ if } x_0 \neq x_1, \end{cases} \qquad (\text{III.4.2})$$

is a metric. It is clear that $\{x\} \in \mathcal{V}_{i_X}(x)$ for every $x \in X$. Therefore, $x \in \lim_{\widetilde{i_X}} \mathcal{F}$ whenever $\mathcal{F} = \{x\}^\uparrow$.

Therefore:

Proposition III.4.3. *The discrete convergence on each non-empty set is metrizable.*

Example III.4.4. The convergence induced on the set $X := \{\frac{1}{n} : n \in \mathbb{N}\}$ by the standard convergence ν of $\mathbb{R}$ is discrete. Therefore it is metrizable by the usual metric, but also by (III.4.2).

Thus a convergence may be premetrizable by different premetrics. Two premetrics d_1 and d_2 on X are called *equivalent* if $\widetilde{d_1} = \widetilde{d_2}$.

Exercise III.4.5. Show that the Euclidean metric d_2 on $\mathbb{R}^2$, defined by $d_2(x, y) = \sqrt{(x_1 - y_1)^2 + (x_2 - y_2)^2}$, and the *taxicab metric* d_1 defined by $d_1(x, y) = |x_1 - y_1| + |x_2 - y_2|$ are equivalent, and that they are not equivalent to the discrete metric of $\mathbb{R}^2$.

Example III.4.6. If $\mathcal{F}$ is a filter on $\mathbb{R}^{\mathbb{R}}$ and $f \in \mathbb{R}^{\mathbb{R}}$, let

$$f \in \lim_p \mathcal{F} \iff \mathcal{F} \geq \mathcal{V}_p(f),$$

where $\mathcal{V}_p(f)$ is defined as in Proposition I.7.5. It is easy to verify that this defines a convergence on $\mathbb{R}^{\mathbb{R}}$, which is not premetrizable by Theorem I.7.8.

Proposition III.4.7. *Each metrizable convergence is Hausdorff and each premetrizable convergence is T_1.*

Proof. If ξ is metrizable by a metric d and if x and y are two points in $\lim_\xi \mathcal{F}$ then $\mathcal{F} \geq \mathcal{V}_d(x)$ and $\mathcal{F} \geq \mathcal{V}_d(y)$ so that $\mathcal{V}_d(x) \# \mathcal{V}_d(y)$. In particular, for every $\varepsilon > 0$ there is $z_\varepsilon \in B(x, \varepsilon) \cap B(y, \varepsilon)$ so that, by (triangular inequality),

$$d(x, y) \leq d(x, z_\varepsilon) + d(z_\varepsilon, y) \leq 2\varepsilon.$$

Therefore, $d(x, y) = 0$ and $x = y$ by (separation).

If ξ is premetrizable by d and $y \in \lim_\xi \{x\}^\uparrow$ then $x \in \bigcap_{n \in \mathbb{N}} B(y, \frac{1}{n})$ so that $d(x, y) = 0$. In view of (separation), $x = y$. □

By Proposition III.1.11:

Proposition III.4.8. *Each premetrizable convergence on a finite set is discrete.*

The chaotic convergence on a set with at least two points is not T_1 and therefore not premetrizable. Since premetrizable convergences are T_1, there are T_1-convergences that are not T_2:

Example III.4.9 (A non-Hausdorff premetrizable (hence T_1) convergence). Let $X := \left\{0, \infty, \frac{1}{n} : n \in \mathbb{N}\right\}, d\left(0, \frac{1}{n}\right) = d\left(\infty, \frac{1}{n}\right) := \frac{1}{n}$ and $d\left(\frac{1}{n}, \frac{1}{k}\right) := \left|\frac{1}{n} - \frac{1}{k}\right|$ for each $n, k \in \mathbb{N}_1$ and, say, $d(\infty, 0) := 1$. This is a non-Hausdorff premetric space. In view of Proposition III.4.7, $\widetilde{d}$ is premetrizable but not metrizable.

Proposition III.4.10. *For each premetrizable convergence, for every point, there is a countably based filter that is the coarsest filter convergent to that point.*

Proof. If d is a premetric on X, then the coarsest filter convergent to $x \in X$ is $\mathcal{V}_d(x) = \left\{ B_d\left(x, \frac{1}{n}\right) : n \in \mathbb{N}_1 \right\}^{\uparrow}$. $\qquad \square$

Accordingly the convergences of Examples III.1.4, III.1.15(4) and III.1.17 are not premetrizable.

III.5 Adherence and cover

Definition III.5.1. For a given convergence ξ the *adherence* of a family $\mathcal{A}$ of subsets of $|\xi|$ is defined by

$$\text{adh}_\xi\, \mathcal{A} = \bigcup_{\mathbb{F}X \ni \mathcal{H} \# \mathcal{A}} \lim_\xi \mathcal{H}. \qquad (\text{III.5.1})$$

We call a family *adherent* if its adherence is non-empty, and *non-adherent* if its adherence is empty.

As $\mathcal{A}^\# = \left(\mathcal{A}^\uparrow\right)^\#$, it is clear that $\text{adh}_\xi\, \mathcal{A} = \text{adh}_\xi\, \mathcal{A}^\uparrow$ for every family $\mathcal{A}$ of sets. Notice that

$$\xi \geq \theta \Longrightarrow \text{adh}_\xi\, \mathcal{A} \subset \text{adh}_\theta\, \mathcal{A}.$$

For a convergence ξ and a filter $\mathcal{F}$ on $|\xi|$,

$$\ker \mathcal{F} \subset \text{adh}_\xi\, \mathcal{F}, \qquad (\text{III.5.2})$$

because $\ker \mathcal{F} = \text{adh}_\iota\, \mathcal{F}$ and $\iota \geq \xi$.

Exercise III.5.2. Show that if $\mathcal{F}$ is a filter on $|\xi|$ then

$$\text{adh}_\xi\, \mathcal{F} = \bigcup_{\mathcal{U} \in \beta(\mathcal{F})} \lim_\xi \mathcal{U}. \qquad (\text{III.5.3})$$

Proposition III.5.3. *Let X be a premetric space and let $\mathcal{F} \in \mathbb{F}X$ and $x \in X$. Then*

$$x \in \text{adh}\, \mathcal{F} \iff \mathcal{V}(x)\#\mathcal{F} \iff \underset{n \in \mathbb{N}_1}{\forall}\, B(x, \tfrac{1}{n})\#\mathcal{F}.$$

Proof. In a premetric space, convergence is defined modulo (III.4.1), so that $x \in \text{adh}\, \mathcal{F}$ means that there is a filter $\mathcal{H}\#\mathcal{F}$ with $\mathcal{H} \geq \mathcal{V}(x)$, and thus $\mathcal{F}\#\mathcal{V}(x)$. Conversely, if $\mathcal{F}$ is a filter and $\mathcal{F}\#\mathcal{V}(x)$, then $\mathcal{F} \vee \mathcal{V}(x)$ is a filter that meshes with $\mathcal{F}$ and converges to x. Since $\{B\left(x, \frac{1}{n}\right) : n \in \mathbb{N}_1\}$ is a filter-base for $\mathcal{V}(x)$, the second equivalent follows. $\qquad \square$

In particular:

Corollary III.5.4. *In a premetric space,*

$$x \in \mathrm{adh}\,(x_n)_n^\uparrow$$

if and only if there exists a subsequence $\{x_{n_k}\}_k$ of $\{x_n\}_n$ such that $x = \lim(x_{n_k})_k$.

Proof. If $\{x_{n_k}\}_{k \in \mathbb{N}}$ is a subsequence of $\{x_n\}_{n \in \mathbb{N}}$ that converges to x, then $(x_{n_k})_k \# (x_n)_n$ and thus $x \in \mathrm{adh}(x_n)_n^\uparrow$.

Conversely, by Proposition III.5.3, $x \in \mathrm{adh}(x_n)_n^\uparrow$ if and only if $B_d(x, \frac{1}{k}) \cap \{x_n : n > k\} \neq \varnothing$ for each k, hence there is $n_k > k$ such that $x_{n_k} \in B_d(x, \frac{1}{k})$. Therefore $x \in \lim (x_{n_k})_k$ and $\{n_k\}_{n \in \mathbb{N}}$ tends to ∞, hence $\{x_{n_k}\}_k$ is a subsequence of $\{x_n\}_n$. $\square$

In view of Proposition II.3.6:

Proposition III.5.5. *If $\mathbb{G}$ is a finite collection of filters on $|\xi|$ then*

$$\mathrm{adh}_\xi \Big(\bigwedge_{\mathcal{G} \in \mathbb{G}} \mathcal{G} \Big) = \bigcup_{\mathcal{G} \in \mathbb{G}} \mathrm{adh}_\xi \, \mathcal{G}. \qquad (\mathrm{III.5.4})$$

Exercise III.5.6. Prove Proposition III.5.5 and find an example showing that (III.5.4) may fail if $\mathbb{G}$ is an infinite family.

Definition III.5.7. A non-empty family $\mathcal{P}$ of subsets of X is a *cover* of a subset A of a convergence space (X, ξ), or *ξ-cover of A*, in symbols, $\mathcal{P} \succ_\xi A$ if every filter converging to a point of A for ξ contains an element of $\mathcal{P}$:

$$\mathcal{P} \succ_\xi A \iff (\lim_\xi \mathcal{F} \cap A \neq \varnothing \implies \mathcal{F} \cap \mathcal{P} \neq \varnothing).$$

In particular, a *cover* of the space (X, ξ) is a ξ-cover of X.

It follows that a family $\mathcal{P}$ of X is a cover of X with respect to the discrete convergence ι whenever $X \subset \bigcup_{P \in \mathcal{P}} P$.

Exercise III.5.8. Let ξ be a premetrizable convergence on X. Check that $\mathcal{P}$ is a cover of a subset A of X if and only if for each $x \in A$ there is $r > 0$ and $P \in \mathcal{P}$ such that $B(x, r) \subset P$.

Exercise III.5.9. Show:

(1) If $\mathcal{R}$ refines $\mathcal{P}$ (see Definition II.1.7) and $\mathcal{R}$ is a cover, then $\mathcal{P}$ is also a cover, in symbols

$$\mathcal{P} \triangleright \mathcal{R} \text{ and } \mathcal{R} \succ A \implies \mathcal{P} \succ A.$$

(2) If $\mathcal{P}$ is a cover of A for ξ and $\xi \leq \zeta$ then $\mathcal{P}$ is a cover of A for ζ.

Proposition III.5.10. *If $\mathcal{P}$ is a cover of Y and X is a subspace of Y, then $\mathcal{P}_X := \{P \cap X : P \in \mathcal{P}\}$ is a cover of X.*

Proof. If $x \in X$ and $\mathcal{F}$ is a filter on X such that $x \in \lim_X \mathcal{F} = X \cap \lim_Y \mathcal{F}$, then there is $P \in \mathcal{F} \cap \mathcal{P}$, hence $P \cap X \in \mathcal{F} \cap \mathcal{P}_X$. $\qquad\square$

Lemma III.5.11. *If $\mathcal{P}$ and $\mathcal{R}$ are two covers of ξ, then*

$$\mathcal{P} \vee \mathcal{R} := \{P \cap R : P \in \mathcal{P}, R \in \mathcal{R}, P \cap R \neq \varnothing\} \qquad (\text{III.5.5})$$

is also a cover of ξ.

Proof. If $\mathcal{P}$ and $\mathcal{R}$ are two covers of ξ and $\varnothing \neq \lim_\xi \mathcal{F}$, then there exist $P \in \mathcal{P}$ and $R \in \mathcal{R}$ such that $P \in \mathcal{F}$ and $R \in \mathcal{F}$. As $\mathcal{F}$ is a filter, also $P \cap R \in \mathcal{F}$ showing that $\mathcal{P} \vee \mathcal{R}$ is a cover of ξ. $\qquad\square$

Of course, $\mathcal{P} \vee \mathcal{R}$ is a common refinement of $\mathcal{P}$ and $\mathcal{R}$.

The notion of cover is dual to that of adherence. Recall (from page 28) that if $\mathcal{P}$ is a family of subsets of X,

$$\mathcal{P}_c := \{X \setminus P : P \in \mathcal{P}\}.$$

Theorem III.5.12. *The following are equivalent:*

$$\mathcal{P} \succ_\xi A; \qquad (\text{III.5.6})$$

$$\text{adh}_\xi \, \mathcal{P}_c \cap A = \varnothing. \qquad (\text{III.5.7})$$

Proof. By definition, (III.5.7) means that if a filter $\mathcal{F}$ converges in ξ to an element of A, then $\mathcal{F}$ does not mesh $\mathcal{P}_c$: there exist $P \in \mathcal{P}$ and $F \in \mathcal{F}$ such that $P^c \cap F = \varnothing$, that is, $F \subset P$ and equivalently $\mathcal{P} \cap \mathcal{F} \neq \varnothing$, since $\mathcal{F}$ is isotone. This is equivalent to (III.5.6). $\qquad\square$

Notice that in general $\mathcal{P}_c$ in (III.5.7) is not a filter. More precisely, $\mathcal{P}_c$ is a filter-base if and only if for each $P_1, P_2 \in \mathcal{P}$ there is $P \in \mathcal{P}$ with $P_1 \cup P_2 \subset P$, that is, if $\mathcal{P}$ is an ideal-base.

For a filter $\mathcal{G} \in \mathbb{F}X$ and $A \subset X$, we have, in view of Theorem III.5.12 and (III.5.3),

$$\text{adh}\,\mathcal{G} \cap A = \varnothing \iff \mathcal{G}_c \succ A \iff \beta(\mathcal{G}) \cap \xi^- A = \varnothing. \qquad (\text{III.5.8})$$

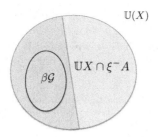

Figure III.4 The disc represents the set of all ultrafilters on X (the ultrafilters ξ-convergent to A are on the right). The filter $\mathcal{G}$ is non-adherent to A.

III.6 Lattice of convergences

A preordered set in which every pair of elements x, y admits a supremum $x \vee y$ and an infimum $x \wedge y$ is called a *lattice*. A lattice X is called *complete* if every $A \subset X$ admits a supremum $\bigvee A = \bigvee_{a \in A} a$ and an infimum $\bigwedge A = \bigwedge_{a \in A} a$ (6).

The sets $\mathbf{J}(X)$ and $\mathbf{I}(X)$ of all preconvergences, and of all convergences, are complete lattices: Every family Ξ of (pre)convergences on X admits a supremum $\bigvee \Xi = \bigvee_{\xi \in \Xi} \xi$ and an infimum $\bigwedge \Xi = \bigwedge_{\xi \in \Xi} \xi$ given by

$$\lim_{\bigvee \Xi} \mathcal{F} = \bigcap_{\xi \in \Xi} \lim_{\xi} \mathcal{F}; \qquad (\text{III.6.1})$$

$$\lim_{\bigwedge \Xi} \mathcal{F} = \bigcup_{\xi \in \Xi} \lim_{\xi} \mathcal{F}. \qquad (\text{III.6.2})$$

The chaotic (pre)convergence o_X is the smallest element of $\mathbf{J}(X)$ and of $\mathbf{I}(X)$; the discrete convergence ι_X is the largest element of $\mathbf{I}(X)$, while the empty preconvergence $\varnothing_X$ is the largest element of $\mathbf{J}(X)$.

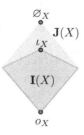

---

[6] See Appendix A.7 for details.

Exercise III.6.1. Verify that (III.6.1) and (III.6.2) define indeed the supremum and infimum respectively of a non-empty family Ξ of (pre)convergences on X.

III.7 Finitely deep modification

The structure of the lattice $\mathbf{I}(X)$ of convergences on X will play an important role in understanding various types of convergences. Let us consider the set $\mathbf{L}(X)$ of all finitely deep convergences on X.

Observe that $\mathbf{L}(X)$ is non-empty as it contains at least the discrete and chaotic convergences. Moreover, if $\Xi \subset \mathbf{L}(X)$ is a collection of finitely deep convergences on X then the supremum $\bigvee \Xi$ (in $\mathbf{I}(X)$) is also a finitely deep convergence: if $\mathcal{F}$ and $\mathcal{G}$ are two filters converging to x in $\bigvee \Xi$, they converge in each of the finitely deep convergences $\xi \in \Xi$ and therefore, so does $\mathcal{F} \wedge \mathcal{G}$.

Now given any convergence ξ on X the set of finitely deep convergences that are coarser than ξ is non-empty because it contains the chaotic convergence, and has a finitely deep supremum, which is the finest finitely deep convergence on X that is coarser than ξ:

Definition III.7.1. Given a convergence ξ on X the finest finitely deep convergence on X that is coarser than ξ is called *finitely deep modification of* ξ and is denoted by $\mathrm{L}\,\xi$.

Exercise III.7.2. Let ξ be a convergence. Show that $x \in \lim_{\mathrm{L}\,\xi} \mathcal{F}$ if and only if there are finitely many filters $\mathcal{G}_1, \mathcal{G}_2, \ldots, \mathcal{G}_n$ on $|\xi|$ such that $x \in \bigcap_{i=1}^{n} \lim_{\xi} \mathcal{G}_i$ and $\mathcal{F} \geq \bigwedge_{i=1}^{n} \mathcal{G}_i$.

Thus,

$$\xi \geq \mathrm{L}\,\xi$$

for every convergence ξ.

Exercise III.7.3. Show that

(1) a convergence ξ is finitely deep if and only if $\mathrm{L}\,\xi \geq \xi$, and deduce that L is *idempotent*, that is, $\mathrm{L}(\mathrm{L}\,\xi) = \mathrm{L}\,\xi$ for all convergence ξ;
(2) if $\xi \leq \tau$ then $\mathrm{L}\,\xi \leq \mathrm{L}\,\tau$.

Example III.7.4. We have seen that the convergence ξ of Example III.1.4 is not finitely deep. Moreover, $x \in \lim_{\mathrm{L}\,\xi} \mathcal{H}$ if and only if $\mathcal{H} = \{x\}^{\uparrow}$ for each $x \in X$, and $x_{\infty} \in \lim_{\mathrm{L}\,\xi} \mathcal{H}$ if $\mathcal{H} \geq \mathcal{F} \wedge \mathcal{G} \wedge \{x_{\infty}\}^{\uparrow}$.

III.8   Pointwise properties of convergence spaces

Given a preconvergence ξ on a set X, it is often insightful to study the collection

$$\lim_\xi^{-1}(x) = \xi^-(x) \qquad \qquad \text{(III.8.1)}$$

of all filters that converge to x in ξ, separately for each $x \in X$. In keeping with our convention for inverse relation, we often use the shorthand

$$\lim_\xi^-(x) := \lim_\xi^{-1}(x)$$

In fact, certain properties of preconvergence spaces can be reformulated in terms of some properties of the preimages of singletons, taken one by one.

For example, a preconvergence ξ is a convergence if and only if $\{x\}^\uparrow \in \lim_\xi^-(x)$ for each $x \in |\xi|$. Another example is that of discrete convergence on a (non-empty) set X; in fact, $\xi = \iota_X$ if and only if $\mathcal{F} \in \lim_\xi^-(x)$ implies that $\mathcal{F} = \{x\}^\uparrow$. Still another example is that of chaotic convergence on X that can be characterized by the fact that the preimage of every singleton consists of $\mathbb{F}X$.

We shall call the *bundle* of ξ at x the preimage of x by ξ, that is, (III.8.1).

Definition III.8.1. A property P of preconvergences is called *pointwise* if there exists a property $\mathfrak{P}$ of classes of filters such that $\xi \in P$ if and only if $\xi^-(x) \in \mathfrak{P}$ for each $x \in |\xi|$.

Example III.8.2. Finite depth is a pointwise property. Indeed, ξ is finitely deep if and only if for each $x \in |\xi|$, $n \in \mathbb{N}_1$ and $\mathcal{F}_1, \ldots, \mathcal{F}_n \in \lim_\xi^-(x)$ implies that $\mathcal{F}_1 \wedge \ldots \wedge \mathcal{F}_n \in \lim_\xi^-(x)$. The property of classes of filters is the stability for finite infima.

Example III.8.3. Hausdorffness is not a pointwise property. Indeed, ξ is Hausdorff whenever $\lim_\xi^-(x_0) \cap \lim_\xi^-(x_1) = \varnothing$ for each couple $x_0 \neq x_1$ of elements of $|\xi|$. Therefore one cannot decide if a preconvergence is Hausdorff by considering separately its bundles.

Definition III.8.4. A *pavement* of a convergence ξ at a point x of $|\xi|$ is a family $\mathbb{P} \subset \lim_\xi^-(x)$ of filters that converge to x such that whenever $\mathcal{F} \in \lim_\xi^-(x)$, there is $\mathcal{H} \in \mathbb{P}$ such that $\mathcal{H} \leq \mathcal{F}$.

We call the *paving number at* x of ξ the least cardinal $\mathfrak{p}(\xi, x)$ such that there is a pavement at x of cardinality $\mathfrak{p}(\xi, x)$. The *paving number of* ξ is defined as $\mathfrak{p}(\xi) := \sup_{x \in |\xi|} \mathfrak{p}(\xi, x)$. A convergence is κ-*paved* if $\mathfrak{p}(\xi) \leq \kappa$.

We observe that each convergence is determined by all its pavements. Notice that being κ-paved is a pointwise property.

In the simplest case, each pavement is composed of a single filter, that is, the space is 1-paved. For instance, the discrete convergence, the chaotic convergence, the Sierpiński convergences are all 1-paved. By the very definition, premetrizable convergences, in particular the standard convergence ν of the real line, are 1-paved too.

In contrast, the paving number of $\operatorname{Seq}\nu$ is infinite, even uncountable:

Example III.8.5. $\operatorname{Seq}\nu$ is not countably paved. Indeed, consider a countable collection

$$\mathbb{P} = \{(x_{n,k})_k : n \in \mathbb{N}\}$$

of sequences that converge to a point $x \in \mathbb{R}$ for ν. Then $\mathbb{P}$ cannot be a pavement at x for $\operatorname{Seq}\nu$, for $S := \{x_{n,k} : n, k \in \mathbb{N}\}$ is a countable set, so that for each $p \in \mathbb{N}$ there is $y_p \in B(x, \frac{1}{p}) \setminus S$. Then $(y_p)_p$ converges to x for ν, but is not finer than any element of $\mathbb{P}$. By Corollary A.5.4, there are $\mathfrak{c}$ many sequential filters on $\mathbb{R}$, so that $\mathfrak{p}(\operatorname{Seq}\nu) \leq \mathfrak{c}$ ([7]).

This pointwise perspective leads us to consider pointwise properties of (pre)convergences at individual points, independently of other points. For example, we noticed that the discrete convergence is characterized as a preconvergence, for which the bundle at x consists of $\{x\}^{\uparrow}$ for each x. This property contemplated at each point separately gives rise to a notion of isolated point.

Definition III.8.6. A point x of a convergence space (X, ξ) is *isolated* if $\lim_{\xi}^{-1}(x) = \{\{x\}^{\uparrow}\}$.

Obviously, a convergence space is discrete if and only if each point is isolated.

Prime spaces

Definition III.8.7. A convergence with at most one non-isolated point is called *prime*. A non-isolated point of a prime convergence is said to be *distinguished*.

Hence the discrete convergence of Example III.1.5 (in which all points are isolated) and the convergence spaces of Examples III.1.4 and III.1.15 are examples of prime spaces.

---

[7]In fact, $\mathfrak{p}(\operatorname{Seq}\nu) = \mathfrak{c}$.

To a non-empty family $\mathbb{G}$ of filters on X we can associate a prime convergence

$$\pi[x, \mathbb{G}]$$

on X by picking a point $x \in X$ and declaring all points of $X \setminus \{x\}$ isolated and $\mathbb{G} \wedge \{x\}^\uparrow := \{\mathcal{G} \wedge \{x\}^\uparrow : \mathcal{G} \in \mathbb{G}\}$ a pavement at x.

Clearly, this convergence is finitely deep if and only if for every finite subset $\mathbb{G}_0$ of $\mathbb{G}$, there is $\mathcal{G}_0 \in \mathbb{G}$ with

$$\bigwedge_{\mathcal{G} \in \mathbb{G}_0} \mathcal{G} \geq \mathcal{G}_0.$$

Conversely, if X is a prime convergence space with distinguished point x, then $\lim^{-1}(x)$ characterizes the convergence. In the case where $\mathbb{G} = \{\mathcal{G}\}$ consists of a single filter, we abridge

$$\pi[x, \mathcal{G}] := \pi[x, \{\mathcal{G}\}].$$

Exercise III.8.8. Is primeness a pointwise property?

Definition III.8.9. A 1-paved convergence is also called a *pretopology*.

Notice that pretopologicity is a pointwise property. We can say that a convergence ξ is *pretopological at* x if the bundle $\xi^-(x)$ contains a least element; this is, by definition, the vicinity filter $\mathcal{V}_\xi(x)$ of ξ at x.

Notice that each convergence is pretopological at each of its isolated points.

Pretopologies will be studied systematically in Chapter V. We observe right now that:

Proposition III.8.10. *Each finitely deep convergence is an infimum of prime pretopologies.*

Proof. If ξ is a finitely deep convergence on a set X, then $x \in \lim_\xi(\mathcal{F} \wedge \{x\}^\uparrow)$ for every x and $\mathcal{F}$ such that $x \in \lim_\xi \mathcal{F}$. Let $\pi[x, \mathcal{F}]$ be the prime pretopology on X such that $\{\mathcal{F} \wedge \{x\}^\uparrow\}$ is a pavement at x and $y \neq x$ is isolated. Then

$$\xi = \bigwedge_{x \in X} \bigwedge_{\mathcal{F} \in \lim_\xi^-(x)} \pi[x, \mathcal{F}],$$

because $x \in \lim_\xi \mathcal{F}$ if and only if $x \in \lim_{\pi[x,\mathcal{F}]} \mathcal{F}$. $\qquad\square$

Exercise III.8.11. Show that a finitely deep convergence that is finitely paved at each point is a pretopology.

As we have seen, a filter $\mathcal{F}$ on a set X together with a specified point $\infty_X \in X$ defines uniquely a prime pretopology $\pi[\infty_X, \mathcal{F}]$ on X by

$$\infty_X \in \lim \mathcal{G} \iff \mathcal{G} \geq \mathcal{F} \wedge \{\infty_X\}^\uparrow.$$

This pretopology is Hausdorff if and only if $\ker \mathcal{F} \subset \{\infty_X\}$. This is the case in particular when $\mathcal{F}$ is free.

Conversely, a prime pretopology determines a unique filter

$$\mathcal{V}(\infty_X) := \bigwedge_{\infty_X \in \lim \mathcal{G}} \mathcal{G}$$

on X which is the only pavement at ∞_X of cardinality 1.

Hence, filters and prime pretopologies correspond to each other, and prime pretopologies determine finitely deep convergences, via Proposition III.8.10.

III.9 Convergences on a complete lattice

See Appendix A.7 for definitions concerning lattices.

Definition III.9.1. If $(X, \leq)$ is a complete lattice, the associated *Scott convergence*, or *lower convergence*, on X is defined by

$$x \in \lim_s \mathcal{F} \iff x \leq \bigvee_{F \in \mathcal{F}} \bigwedge_{a \in F} a. \tag{III.9.1}$$

The Scott convergence s^* for the reverse order can be expressed in terms of the original order by

$$x \in \lim_{s^*} \mathcal{F} \iff x \geq \bigwedge_{F \in \mathcal{F}} \bigvee_{a \in F} a, \tag{III.9.2}$$

and is also called *upper convergence*.

Proposition III.9.2. *If $(X, \leq)$ is a complete lattice with at least two points, then the lower and upper convergences are T_0 but not T_1.*

Proof. It is enough to check this for the lower convergence s. If $x \neq y$ are two points of X then either x and y are incomparable or either $x \leq y$ or $y \leq x$ but not both. Say, x and y are incomparable or $x \leq y$ but $y \not\leq x$. Then $\{x\}^\uparrow \in \lim_s^{-1}(x) \setminus \lim_s^{-1}(y)$. On the other hand, s is not T_1 for if 1 denotes the greatest element of X then $X = \lim_s \{1\}^\uparrow$. $\qquad\square$

Traditionally, one denotes

$$\liminf \mathcal{F} := \bigvee_{F \in \mathcal{F}} \bigwedge_{a \in F} a,$$

$$\limsup \mathcal{F} := \bigwedge_{F \in \mathcal{F}} \bigvee_{a \in F} a,$$

the lower and upper limits.

In general, for every filter $\mathcal{F}$,

$$\liminf \mathcal{F} \le \limsup \mathcal{F}$$

because $F_1 \cap F_2 \ne \varnothing$ for each F_1 and F_2 in $\mathcal{F}$, so that $\bigvee F_1 \ge \bigwedge F_2$. Hence if a filter $\mathcal{F}$ converges both in s and s^*, then $\limsup \mathcal{F} = \liminf \mathcal{F}$. In general, for a filter $\mathcal{F}$,

$$\bigvee_{H \in \mathcal{F}^\#} \bigwedge_{a \in H} a \le \bigwedge_{F \in \mathcal{F}} \bigvee_{a \in F} a. \tag{III.9.3}$$

A lattice is called *completely distributive* if the equality holds in (III.9.3) for each family $\mathcal{F}$.

Definition III.9.3. The convergence $s \vee s^*$ on a complete lattice is called *convergence in order*.

$$x \in \lim_{s \vee s^*} \mathcal{F} \iff x = \limsup \mathcal{F} = \liminf \mathcal{F}.$$

Hence, unlike Scott convergence, convergence in order on a complete lattice is Hausdorff.

Example III.9.4. In the extended real line $\overline{\mathbb{R}}$, for every $x \in \mathbb{R}$,

$$x \in \lim_s \mathcal{F} \iff \mathcal{F} \ge \{(a, \infty] : a < x\}$$
$$x \in \lim_{s^*} \mathcal{F} \iff \mathcal{F} \ge \{[-\infty, a) : x < a\}$$
$$x \in \lim_{s \vee s^*} \mathcal{F} \iff \mathcal{F} \ge \{(a, b) : a < x < b\}.$$

Proposition III.9.5. *Let X be a complete lattice. If X is a frame ([8]) then Scott convergence on X is finitely deep. Conversely, if X is distributive and its Scott convergence is finitely deep, then X is a frame.*

---

[8]In Appendix A.7, a complete lattice is called a *frame* if $\wedge$ distributes with $\bigvee$.

Proof. Let $(X, \leq)$ be a frame and let $x \in \lim_s \mathcal{F} \cap \lim_s \mathcal{G}$. As $x \leq \bigvee_{F \in \mathcal{F}} \bigwedge F$ and $x \leq \bigvee_{G \in \mathcal{G}} \bigwedge G$,

$$
\begin{aligned}
x = x \wedge \bigvee_{F \in \mathcal{F}} \bigwedge F &= \bigvee_{F \in \mathcal{F}} (x \wedge \bigwedge F) \\
= x \wedge \bigvee_{G \in \mathcal{G}} \bigwedge G &= \bigvee_{G \in \mathcal{G}} (x \wedge \bigwedge G) \\
&= \left(\bigvee_{F \in \mathcal{F}} (x \wedge \bigwedge F) \right) \wedge \left(\bigvee_{G \in \mathcal{G}} (x \wedge \bigwedge G) \right) \\
&= \bigvee_{F \in \mathcal{F}, G \in \mathcal{G}} x \wedge \bigwedge F \wedge \bigwedge G \\
&= \bigvee_{F \in \mathcal{F}, G \in \mathcal{G}} x \wedge \bigwedge (F \cup G) \\
&= x \wedge \bigvee_{F \in \mathcal{F}, G \in \mathcal{G}} \bigwedge (F \cup G).
\end{aligned}
$$

Therefore, $x \leq \limsup(\mathcal{F} \wedge \mathcal{G})$ and $x \in \lim_s \mathcal{F} \wedge \mathcal{G}$.

Conversely, assume that Scott convergence is finitely deep and let $x \in X$ and $A \subset X$. Denote by $A^{\vee}$ the closure of A under all finitary joins and $\uparrow a := \{x \in X : x \geq a\}$. With these notations $\{\uparrow a : a \in A^{\vee}\}$ is a filter-base for a filter $\mathcal{F}$ on X and

$$
\bigvee_{F \in \mathcal{F}} \bigwedge F = \bigvee_{a \in A^{\vee}} \bigwedge \uparrow a = \bigvee A^{\vee} = \bigvee A.
$$

Hence $x \wedge \bigvee A \in \lim_s \mathcal{F}$ and $x \wedge \bigvee A \in \lim_s \{x\}^{\uparrow}$. Since s is finitely deep, we conclude that $x \wedge \bigvee A \in \lim_s \mathcal{F} \wedge \{x\}^{\uparrow}$, that is,

$$
\begin{aligned}
x \wedge \bigvee A \leq \bigvee_{F \in \mathcal{F}} \bigwedge (F \wedge x) &= \bigvee_{a \in A^{\vee}} (a \wedge x) \\
&= \bigvee_{a, b \in A} ((a \vee b) \wedge x) \\
&= \bigvee_{a, b \in A} (a \wedge x) \vee (b \wedge x) \text{ by distributivity} \\
&= \bigvee_{a \in A} a \wedge x.
\end{aligned}
$$

The reverse inequality $\bigvee_{a \in A} (a \wedge x) \leq x \wedge \bigvee A$ is always true, so we conclude that X is a frame. $\qquad\square$

Chapter IV

Continuity

In this chapter we shall study continuous maps and several fundamental operations on convergence spaces that are performed with the aid of continuous maps.

IV.1 Continuous maps

Definition IV.1.1. Let ξ be a convergence on X and τ be a convergence on Y. A map $f : X \to Y$ is said to be *continuous* (from ξ to τ) if $x \in \lim_\xi \mathcal{F}$ implies $f(x) \in \lim_\tau f[\mathcal{F}]$ for every filter $\mathcal{F}$ on X. In other words, f is continuous if and only if

$$f(\lim_\xi \mathcal{F}) \subset \lim_\tau f[\mathcal{F}], \tag{IV.1.1}$$

for every filter $\mathcal{F}$ on X ([1]).

We denote by $C(\xi, \tau)$, or $C(X, Y)$ if no ambiguity on the convergences can arise, the set of continuous maps from (X, ξ) to (Y, τ).

We will adopt the convention that

$$f : |\xi| \to |\tau|$$

denotes the map $f : X \to Y$ where X is endowed with the convergence ξ and Y with the convergence τ ([2]).

---

[1] Recall that $f[\mathcal{F}]$ is a filter-base.

[2] Even though one may argue that this is improper because $|\xi| = X$ for any convergence on X, the notation $f : |\xi| \to |\tau|$ enables us to speak about the continuity of the map without having to repeat what convergence structures are considered on the domain and codomain of the map.

Exercise IV.1.2. Show that for every convergences ξ and τ and a set Σ of convergences on the same set

$$C(\xi, \bigvee \Sigma) = \bigcap_{\sigma \in \Sigma} C(\xi, \sigma),$$

$$C(\bigwedge \Sigma, \tau) = \bigcap_{\sigma \in \Sigma} C(\sigma, \tau).$$

In contrast, the inclusions

$$\bigcup_{\sigma \in \Sigma} C(\xi, \sigma) \subset C(\xi, \bigwedge \Sigma),$$

$$\bigcup_{\sigma \in \Sigma} C(\sigma, \tau) \subset C(\bigvee \Sigma, \tau),$$

may be strict.

Example IV.1.3. Consider the convergences $o, \$_0, \$_1$ and ι on a two points set $\{0, 1\}$ as in Section III.1. Let $\Sigma := \{\$_0, \$_1\}$. Then the identity map

$$i : |o| \to |\$_0 \wedge \$_1|$$

is continuous because $\$_0 \wedge \$_1 = o$, but neither $i : |o| \to |\$_0|$ nor $i : |o| \to |\$_1|$ is continuous. Thus $i \in C(o, \$_0 \wedge \$_1) \setminus (C(o, \$_0) \cup C(o, \$_1))$.

Similarly, the identity map

$$i : |\$_0 \vee \$_1| \to |\iota|$$

is continuous because $\$_0 \vee \$_1 = \iota$, but neither $i : |\$_0| \to |\iota|$ nor $i : |\$_1| \to |\iota|$ is continuous. Thus $i \in C(\$_0 \vee \$_1, \iota) \setminus (C(\$_0, \iota) \cup C(\$_0, \iota))$.

Order between convergences is characterized in terms of continuity via

$$\xi \leq \tau \iff i \in C(\tau, \xi), \tag{IV.1.2}$$

where i the identity map of $|\xi| = |\tau|$.

Observe also that every map from a discrete convergence is continuous, and so is every map to a chaotic convergence. In other words, if $|\xi| = X$ and $|\tau| = Y$, then

$$C(\iota_X, \tau) = Y^X = C(\xi, o_Y).$$

Exercise IV.1.4. Show that if $f \in C(\xi, \tau)$ and $g \in C(\tau, \sigma)$, then $g \circ f \in C(\xi, \sigma)$.

Exercise IV.1.5. Show that if $f \in C(\xi, \tau)$, $\theta \geq \xi$ and $\sigma \leq \tau$, then $f \in C(\theta, \sigma)$.

Recall from Example III.1.2 that a premetric d on a set X induces a convergence $\tilde{d}$ on X. A convergence Seq $\tilde{d}$ can then be defined on X just like the convergence Seq ν defined in Example III.1.17 on $\mathbb{R}$: $x \in \lim_{\text{Seq} \, \tilde{d}} \mathcal{F}$ if $\mathcal{F} \geq \mathcal{H}$ for some sequential filter $\mathcal{H}$ satisfying $x \in \lim_{\tilde{d}} \mathcal{H}$.

Exercise IV.1.6. Let (X, d_X) and (Y, d_Y) be premetric spaces and $f : X \to Y$. Show that the following are equivalent:

$$\underset{x \in X}{\forall} \; \underset{\varepsilon > 0}{\forall} \; \underset{\delta > 0}{\exists} \; d_X(x, t) < \delta \implies d_Y(f(x), f(t)) < \varepsilon,$$

$$f \in C\left(\widetilde{d_X}, \widetilde{d_Y}\right),$$

$$f \in C\left(\text{Seq} \, \widetilde{d_X}, \text{Seq} \, \widetilde{d_Y}\right).$$

Exercise IV.1.7. Show that if $f : |\xi| \to |\tau|$ is continuous then $f : |\mathsf{L}\,\xi| \to |\mathsf{L}\,\tau|$ is also continuous.

Definition IV.1.8. A bijective map f is called a *homeomorphism* if both f and f^- are continuous. If $f : |\xi| \to |\sigma|$ is a homeomorphism, then the convergences ξ and σ are called *homeomorphic*. We denote by $H(\xi, \sigma)$ the set of *homeomorphisms* from ξ to σ.

As explained in the Introduction, objects that are homeomorphic are often identified. Roughly speaking, two homeomorphic convergence spaces correspond to two presentations of the same space. The homeomorphism provides a way to "change the point of view" on this object.

Exercise IV.1.9. Show that if $\mathbb{R}$ is equipped with its standard convergence ν (induced by the usual metric $d(x, y) = |y - x|$) then the identity map $i : |\text{Seq}\,\nu| \to |\nu|$ is a continuous bijective map that is not an homeomorphism.

Exercise IV.1.10. Show that the Sierpiński spaces $\$_0$ and $\$_1$ described by (III.2.1) and (III.2.2) in Example III.2.1, are homeomorphic ([3]).

Exercise IV.1.11. Let $X := \left(-\frac{\pi}{2}, \frac{\pi}{2}\right)$ with the induced convergence from the standard convergence of $\mathbb{R}$. Show that $\tan : \left(-\frac{\pi}{2}, \frac{\pi}{2}\right) \to \mathbb{R}$ is a homeomorphism. Deduce that any non-empty open interval (a, b) of $\mathbb{R}$ (with the induced convergence) is homeomorphic to $\mathbb{R}$.

A convergence ξ on X is said to be *homogeneous* if for every couple $x_0, x_1 \in X$ there exists a homeomorphism $h : X \to X$ such that $h(x_0) = x_1$.

---

[3] Therefore there are exactly three non mutually homeomorphic finitely deep convergences on $\{0, 1\}$.

Exercise IV.1.12. Show that:

(1) discrete convergences are homogeneous;
(2) chaotic convergences are homogeneous;
(3) the standard convergence ν of the real line is homogeneous;
(4) $\operatorname{Seq} \nu$ is homogeneous;
(5) the Sierpiński space is not homogeneous.

If a convergence admits an isolated point, then it is homogeneous if and only if it is discrete. In particular, a prime convergence is homogeneous if and only if it is discrete.

IV.2 Initial and final convergences

Claim IV.2.1. For every map $f : X \to Y$ and each (pre)convergence τ on Y, there exists the coarsest among the (pre)convergences ξ on X for which f is continuous (from ξ to τ).

Proof. Let

$$S_I := \{\xi \in \mathbf{I}(X) : f \in C(\xi, \tau)\},$$
$$S_J := \{\xi \in \mathbf{J}(X) : f \in C(\xi, \tau)\}.$$

If τ is a convergence then the set S_I is non-empty because it contains the discrete convergence. So is the set S_J because it contains the empty preconvergence.

Moreover, if S is either S_I or S_J then $\bigwedge_{\xi \in S} \xi \in S$. Indeed, $x \in \lim_{\bigwedge_{\xi \in S} \xi} \mathcal{F}$ if and only if there is $\xi \in S$ with $x \in \lim_{\xi} \mathcal{F}$, by (III.6.2). Since $\xi \in S$, $f(x) \in \lim_{\tau} f[\mathcal{F}]$ so that $\bigwedge_{\xi \in S} \xi \in S$. Hence $\bigwedge_{\xi \in S} \xi$ is the desired (pre)convergence. $\qquad\square$

Note that we do not need to distinguish between the coarsest convergence and the coarsest preconvergence making f continuous, because if τ is a convergence, then both S_I and S_J contain the discrete convergence and $\bigwedge_{\xi \in S_I} \xi = \bigwedge_{\xi \in S_J} \xi$.

Definition IV.2.2. Given $f : X \to Y$ and a (pre)convergence τ on Y, the *initial (pre)convergence* for (f, τ), denoted $f^{-}\tau$, is the coarsest (pre)convergence on X making f continuous (to τ).

It follows from (IV.1.1) that if f is continuous from ξ to τ, then $\lim_\xi \mathcal{F} \subset f^-(\lim_\tau f[\mathcal{F}])$. Therefore

$$\lim{}_{f-\tau} \mathcal{F} = f^-(\lim{}_\tau f[\mathcal{F}]), \qquad\qquad (IV.2.1)$$

is the initial (pre)convergence.

Exercise IV.2.3. Verify (IV.2.1).

Dually to Claim IV.2.1, we have:

Claim IV.2.4. For every map $f : X \to Y$ and each (pre)convergence ξ on X, there exists the finest among the (pre)convergences τ on Y for which f is continuous (from ξ to τ).

Exercise IV.2.5. Show Claim IV.2.4.

Definition IV.2.6. Given $f : X \to Y$ and a preconvergence ξ on X, the *final preconvergence for* (f, ξ), denoted $\widehat{f\xi}$, is the finest preconvergence on Y making f continuous (from ξ).

Continuity of $f : |\xi| \to |\widehat{f\xi}|$ imposes that $\bigcup_{f[\mathcal{F}] \le \mathcal{G}} f(\lim_\xi \mathcal{F}) \subset \lim_{\widehat{f\xi}} \mathcal{G}$. Therefore

$$\lim{}_{\widehat{f\xi}} \mathcal{G} = \bigcup_{f[\mathcal{F}] \le \mathcal{G}} f(\lim{}_\xi \mathcal{F}).$$

Definition IV.2.7. Given $f : X \to Y$ and a convergence ξ on X, the *final convergence for* (f, ξ), denoted $f\xi$, is the finest convergence on Y making f continuous (from ξ).

Since a preconvergence ξ is a convergence if and only if it is coarser than the discrete convergence, we conclude that

$$f\xi = \iota \wedge \widehat{f\xi}.$$

In other words,

$$\lim{}_{f\xi} \mathcal{G} = \bigcup_{f[\mathcal{F}] \le \mathcal{G}} f(\lim{}_\xi \mathcal{F}) \cup \lim{}_\iota \mathcal{G}. \qquad\qquad (IV.2.2)$$

Note also that $\lim_\iota \mathcal{G} \subset \bigcup_{f[\mathcal{F}] \le \mathcal{G}} f(\lim_\xi \mathcal{F})$ whenever $f(X) \in \mathcal{G}^\#$, so that (IV.2.2) above becomes

$$\lim{}_{f\xi} \mathcal{G} = \bigcup_{f[\mathcal{F}] \le \mathcal{G}} f(\lim{}_\xi \mathcal{F}) \qquad\qquad (IV.2.3)$$

if f is onto.

Exercise IV.2.8. Verify (IV.2.2).

Note that initial and final constructions preserve order: if $\tau \leq \sigma$ then $f^-\tau \leq f^-\sigma$ and $f\tau \leq f\sigma$.

Moreover,

$$f\left(f^-\tau\right) \geq \tau, \tag{IV.2.4}$$

because f is continuous from $f^-\tau$ to τ and $f\left(f^-\tau\right)$ is the finest convergence making f continuous from $f^-\tau$. Similarly,

$$f^-\left(f\xi\right) \leq \xi, \tag{IV.2.5}$$

because f is continuous from ξ to $f\xi$ and $f^-\left(f\xi\right)$ is the coarsest convergence making f continuous to $f\xi$.

Proposition IV.2.9. *Let $f : |\xi| \to |\tau|$ be a function. The following are equivalent:*

$$f \in C\left(\xi, \tau\right);$$
$$\xi \geq f^-\tau;$$
$$f\xi \geq \tau.$$

Proof. If f is continuous then $\xi \geq f^-\tau$ by definition of the initial convergence $f^-\tau$. If $\xi \geq f^-\tau$ then $f\xi \geq ff^-\tau \geq \tau$ by (IV.2.4). Finally, if $f\xi \geq \tau$ then $f : |\xi| \to |f\xi|$ is continuous and the identity map $i : |f\xi| \to |\tau|$ is continuous, so that the composite $f : |\xi| \to |\tau|$ is continuous as well. $\quad\square$

Proposition IV.2.10. *If f is surjective, then*

$$f(f^-\tau) = \tau. \tag{IV.2.6}$$

Proof. Because of (IV.2.4), we only need to show that $\tau \geq f(f^-\tau)$. To this end, let $y \in \lim_\tau \mathcal{G}$. Then $f^-(y) \subset \lim_{f^-\tau} f^-[\mathcal{G}]$, because by (IV.2.1) $\lim_{f^-\tau} f^-[\mathcal{G}] = f^-\left(\lim_{f^-\tau} ff^-[\mathcal{G}]\right) = f^-(\lim_\tau \mathcal{G})$. Hence $y \in \lim_{ff^-\tau} ff^-[\mathcal{G}] = \lim_{ff^-\tau} \mathcal{G}$. $\quad\square$

Proposition IV.2.11. *If f is injective, then*

$$f^-\left(f\xi\right) = \xi.$$

Proof. In view of (IV.2.5), we only need to show that $f^-(f\xi) \geq \xi$. If $x \in \lim_{f^-(f\xi)} \mathcal{F}$ then $f(x) \in \lim_{f\xi} f[\mathcal{F}]$, that is, by (IV.2.2), there is a filter $\mathcal{F}_0$ such that $x \in \lim_\xi \mathcal{F}_0$ and $f[\mathcal{F}] = f[\mathcal{F}_0]$ and, since f is injective, $\mathcal{F} = \mathcal{F}_0$. $\quad\square$

Initial and final constructions are also well-behaved with respect to composition: consider two maps f and g such that

$$X \xrightarrow{f} Y \xrightarrow{g} Z,$$

and let ξ be a convergence on X. A final convergence on Z can be defined via $g \circ f$ or by taking the final convergence under g of the final convergence $f\xi$ on Z. In fact, that makes no difference:

$$(g \circ f)\xi = g(f\xi). \tag{IV.2.7}$$

Similarly, if τ is a convergence on Z, an initial convergence on X can be defined via $g \circ f$ or by taking the initial convergence under f of the initial convergence $g^-\tau$ on Y. We have:

$$(g \circ f)^-\tau = f^-(g^-\tau). \tag{IV.2.8}$$

Exercise IV.2.12. Verify (IV.2.7) and (IV.2.8).

Exercise IV.2.13. Use Exercise IV.1.7 and Proposition IV.2.9 to show:

(1) $\mathrm{L}(f^-\tau) \geq f^-(\mathrm{L}\,\tau)$, equivalently $f(\mathrm{L}\,\xi) \geq \mathrm{L}\,(f\xi)$, for every map $f :$ $X \to Y$ and every convergence ξ on X and τ on Y;
(2) Deduce that if τ is a finitely deep convergence, so is $f^-\tau$;
(3) Show that moreover

$$\mathrm{L}(f^-\tau) = f^-(\mathrm{L}\,\tau). \tag{IV.2.9}$$

We have already encountered a standard construction involving initial convergence: *induced convergence*. It is immediate from (III.3.1) that:

Proposition IV.2.14. *If* (X,ξ) *is a (pre)convergence and A is a subset of X with inclusion map $i_X^A : A \to X$, then*

$$\xi_{|A} = (i_X^A)^-\xi.$$

Proposition IV.2.15. *Let* $f : (X,\xi) \to (Y,\tau)$ *be a continuous map, and let $A \subset X$ and $B \subset Y$. Then the restrictions*

$$f_{|A} : |\xi_{|A}| \to |\tau_{|f(A)}|$$
$$f^{|B} : |\xi_{|f^-B}| \to |\tau_{|B}|$$

are continuous.

Proof. If $\mathcal{F} \in \mathbb{F}A$ and $x \in \lim_{\xi_{|A}} \mathcal{F}$ then $x \in \lim_\xi i_X^A[\mathcal{F}]$. Thus, $f(x) \in \lim_\tau f[i_X^A[\mathcal{F}]]$ by continuity of f. Moreover, $f(A) \in f[i_X^A[\mathcal{F}]]$ because $A \in i_X^A[\mathcal{F}]$, so that, in view of (III.3.2),

$$i_Y^{f(A)}[f_{|A}[\mathcal{F}]] = f[i_X^A[\mathcal{F}]].$$

Thus, $f(x) \in \lim_{\tau_{|f(A)}} f_{|A}[\mathcal{F}]$ and $f_{|A}$ is continuous. $\square$

Exercise IV.2.16. Prove the second part of Proposition IV.2.15, namely, that

$$f^{|B} : |\xi_{|f^{-}B}| \to |\tau_{|B}|$$

is continuous if $f : |\xi| \to |\tau|$ is continuous.

In view of Exercise IV.2.13:

Proposition IV.2.17. *A subspace of a finitely deep convergence is finitely deep. Moreover if ξ is a convergence and A is a subset of $|\xi|$ then*

$$(L\,\xi)_{|A} = L(\xi_{|A}).$$

Definition IV.2.18. A map $f : |\xi| \to |\tau|$ is an *embedding* if it is one-to-one and $f : |\xi| \to |\tau_{|f(X)}|$ is a homeomorphism.

IV.3   Initial and final convergences for multiple maps

More generally, consider the situation where F is a collection of maps $f : X \to Y$, where X is the same for each $f \in F$ and Y depends on f. In other words, for each $f \in F$ there exists Y_f such that

$$f : X \to Y_f.$$

If υ_f is a convergence on Y_f for each $f \in F$ and if $Y := \{\upsilon_f : f \in F\}$, then we denote by

$$F^{-}Y := \bigvee_{f \in F} f^{-}\upsilon_f$$

the coarsest convergence on X, for which every $f \in F$ is continuous. We call it the *initial convergence* with respect to $\{(f, \upsilon_f) : f \in F\}$. Accordingly,

$$x \in \lim_{F^{-}Y} \mathcal{F} \tag{IV.3.1}$$

if and only if $f(x) \in \lim_{\upsilon_f} f[\mathcal{F}]$ for every $f \in F$.

Note that even if F is not a set but a class, the supremum is taken over a subset of the set of convergences on X.

Definition IV.3.1. A class of convergences Θ is *initially dense* in another class Π of convergences, if for each $\pi \in \Pi$ there exists a subclass S of Θ and, a collection F of maps $f : |\pi| \to |\tau_f|$ where $\tau_f \in S$ for every $f \in F$, such that

$$\pi = F^{-}S = \bigvee_{f \in F} f^{-}\tau_f.$$

For instance, we will see (Theorem V.1.14) that there is a convergence that is initially dense in the class of pretopologies.

Dually, consider a collection F of maps $f : |\xi_f| \to Y$ with codomain Y. Let $X := \{\xi_f : f \in F\}$. Then

$$FX := \bigwedge_{f \in F} f\xi_f,$$

is the finest convergence on Y making each $f \in F$ continuous.

Definition IV.3.2. A class Ξ of convergences is said to be *finally dense* (respectively, *surjectively finally dense*) in a class Σ if for each $\sigma \in \Sigma$ there is a subclass S of Ξ and a collection F of maps (respectively, onto maps) $f : |\xi_f| \to |\sigma|$ for where $|\xi_f| \in S$ for every $f \in F$ such that

$$\sigma = FS = \bigwedge_{f \in F} f\xi_f.$$

Corollary IV.3.3. *Prime pretopologies are surjectively finally dense in the class of finitely deep convergences.*

Proof. Following the notations of the proof of Proposition III.8.10, let $I := \{(x, \mathcal{F}) : x \in X, \mathcal{F} \in \mathbb{F}X, x \in \lim_\xi \mathcal{F}\}$ and for each $i = (x, \mathcal{F}) \in I$, let f_i be the identity map of X from $\pi[x, \mathcal{F}]$ to ξ. Then $f_i(\pi[x, \mathcal{F}]) = \pi[x, \mathcal{F}]$ and the result follows from Proposition III.8.10. $\square$

The importance of finally dense and initially dense classes (in some other classes of convergences) will become apparent, for example, in the context of convergences on spaces of mappings.

Exercise IV.3.4. Show that:

(1) If $f : X \to Y$ and Ξ is a set of convergences on X, then

$$f\left(\bigwedge_{\xi \in \Xi} \xi\right) = \bigwedge_{\xi \in \Xi} f\xi. \tag{IV.3.2}$$

(2) If $f : X \to Y$ and Θ is a set of convergences on Y, then

$$f^-\left(\bigvee_{\tau \in \Theta} \tau\right) = \bigvee_{\tau \in \Theta} f^-\tau. \tag{IV.3.3}$$

(3) Deduce that if $g : Y \to Z$ and $f_i : |\xi_i| \to Y$ for every $i \in I$, then

$$\bigwedge_{i \in I} (g \circ f_i)\xi_i = g\left(\bigwedge_{i \in I} f_i\xi_i\right), \tag{IV.3.4}$$

and that if $f : X \to Y$ and $g_i : Y \to |\tau_i|$, then

$$\bigvee_{i \in I} (g_i \circ f)^- \tau_i = f^-\left(\bigvee_{i \in I} g_i^-\tau_i\right). \tag{IV.3.5}$$

Initial and final convergences can alternatively be characterized in terms of the following universal properties:

Proposition IV.3.5. *Let $F = \{f : X \to Y_f\}$, let v_f be a convergence on Y_f for each $f \in F$, let $Y = \{v_f : f \in F\}$, and let ξ be a convergence on X. Then*

$$\xi = F^- Y$$

if and only if the continuity of a map $g : |\theta| \to |\xi|$ is equivalent to that of every composite $f \circ g : |\theta| \to |v_f|$, for $f \in F$.

$$
\begin{array}{ccc}
|\xi| & \xrightarrow{\ f\ } & |v_f| \\
{\scriptstyle g}\big\uparrow & \nearrow & \\
& {\scriptstyle f \circ g} & \\
|\theta| & &
\end{array}
$$

Proof. Assume that $\xi = F^- Y$. If g is continuous, so is $f \circ g$ for every $f \in F$, because each $f \in F$ is continuous. Conversely, if $f \circ g : |\theta| \to |v_f|$ is continuous for each $f \in F$, then

$$\theta \geq \bigvee_{f \in F} (f \circ g)^- v_f \overset{\text{(IV.3.5)}}{=} g^- \left(\bigvee_{f \in F} f^- v_f \right) = g^- \xi,$$

that is, $g : |\theta| \to |\xi|$ is continuous. Assume now that ξ has the universal property and let $\theta := F^- Y$ on X. Then $f \circ i_X : |\theta| \to |v_f|$ is continuous for each $f \in F$, thus the identity map $i_X : |\theta| \to |\xi|$ is continuous, that is, $\theta \geq \xi$. Moreover, $\xi \geq \theta$ for ξ makes each $f \in F$ continuous. Thus $\xi = \theta$. $\qquad\square$

Dually,

Proposition IV.3.6. *Let $F = \{f : |\xi_f| \to Y\}$, let $X = \{\xi_f : f \in F\}$, and let τ be a convergence on Y. Then*

$$\tau = F X$$

if and only the continuity of a map $g : |\tau| \to |\sigma|$ is equivalent to that of every composite $g \circ f : |\xi_f| \to |\sigma|$, for $f \in F$.

$$
\begin{array}{ccc}
|\xi_f| & \xrightarrow{\ f\ } & |\tau| \\
& {\scriptstyle g \circ f}\searrow & \big\downarrow {\scriptstyle g} \\
& & |\sigma|
\end{array}
$$

Exercise IV.3.7. Prove Proposition IV.3.6.

Inclusion maps were used to define induced convergence as an initial convergence. Dually, they can be used for a standard final construction:

Definition IV.3.8. If Ξ is a collection of convergences, their *sum*

$$\bigoplus_{\xi \in \Xi} \xi$$

is the convergence whose underlying set is the disjoint union $X := \coprod_{\xi \in \Xi} |\xi|$ equipped with the final convergence $\bigwedge_{\xi \in \Xi} i_X^{|\xi|} \xi$ for the inclusion maps $i_X^{|\xi|}$: $|\xi| \to X$. In other words, a filter $\mathcal{F}$ on X converges to $x \in |\xi| \subset X$ if $|\xi| \in \mathcal{F}$ and $x \in \lim_\xi \mathcal{F}$ when $\mathcal{F}$ is considered as a filter on $|\xi|$.

Exercise IV.3.9. Show that a sum of pretopologies is a pretopology.

IV.4 Product convergence

Definition IV.4.1. Let Ξ be a set of convergence spaces. The *product convergence* $\prod_{\xi \in \Xi} \xi$ is the coarsest convergence on the Cartesian product $X := \prod_{\xi \in \Xi} |\xi|$ making each projection map $p_\xi : X \to |\xi|$ continuous (for ξ). In other words,

$$\prod_{\xi \in \Xi} \xi := \bigvee_{\xi \in \Xi} p_\xi^- \xi. \tag{IV.4.1}$$

Accordingly,

$$x \in \lim_{\prod_{\xi \in \Xi} \xi} \mathcal{F} \iff \bigvee_{\xi \in \Xi} \left(p_\xi(x) \in \lim_\xi p_\xi [\mathcal{F}] \right). \tag{IV.4.2}$$

We shall first focus on finite convergence products and, in particular, on products of two convergence spaces.

IV.4.1 *Finite product*

Let (X, ξ) and (Y, τ) be two convergence spaces. By Definition IV.4.1, the *product convergence* $\xi \times \tau$ is the coarsest convergence on the Cartesian product $X \times Y$ making both projections $p_X : X \times Y \to X$ and $p_Y : X \times Y \to Y$ continuous.

In particular, (IV.4.1) yields

$$\xi \times \tau := p_X^- \xi \vee p_Y^- \tau, \tag{IV.4.3}$$

and (IV.4.2) becomes

$$(x, y) \in \lim_{\xi \times \tau} \mathcal{F} \iff x \in \lim_{\xi} p_X[\mathcal{F}] \text{ and } y \in \lim_{\tau} p_Y[\mathcal{F}].$$

Thus if $(x, y) \in \lim_{\xi \times \tau} \mathcal{H}$, there is a filter $\mathcal{F} = p_X[\mathcal{H}]$ on X with $x \in \lim_{\xi} \mathcal{F}$ and a filter $\mathcal{G} = p_Y[\mathcal{H}]$ on Y with $y \in \lim_{\tau} \mathcal{G}$ such that $\mathcal{H} \geq \mathcal{F} \times \mathcal{G}$.

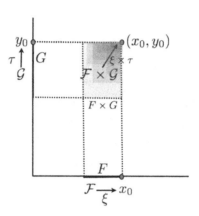

Figure IV.1   The product of two convergent filters converges in the product convergence.

Product filter-bases $\mathcal{F} \times \mathcal{G}$ where $\mathcal{F}$ is convergent in ξ and $\mathcal{G}$ is convergent in τ are sufficient to determine the product convergence $\xi \times \tau$. Indeed every convergent filter in $\xi \times \tau$ is convergent precisely because it is finer than such a filter-base. In other words, the product convergence has pavements composed of product filters, also called polyhedral filters.

Proposition IV.4.2. *Each finite product of finitely deep convergences is finitely deep.*

Proof. Let ξ be a finitely deep convergence on X and τ be a finitely deep convergence on Y. By Exercise IV.2.13, $p_X^{-}\xi$ and $p_X^{-}\tau$ are finitely deep, and we have seen that a supremum of finitely deep convergences is finitely deep. Hence $\xi \times \tau = p_X^{-}\xi \vee p_Y^{-}\tau$ is finitely deep. We conclude by elementary induction. $\qquad \square$

Corollary IV.4.3.

$$\mathrm{L}\,\xi \times \mathrm{L}\,\tau = \mathrm{L}(\xi \times \tau)$$

for every convergences ξ and τ.

Proof. By Proposition IV.4.2, a product of finitely deep convergences is finitely deep, so that $L\xi \times L\tau$ is finitely deep. Moreover, $L\xi \times L\tau \leq \xi \times \tau$ because $L\xi \leq \xi$ and $L\tau \leq \tau$. But $L(\xi \times \tau)$ is the finest finitely deep convergence coarser than $\xi \times \tau$. Therefore

$$L\xi \times L\tau \leq L(\xi \times \tau).$$

To see the reverse inequality, first note that

$$\xi \times L\tau \geq L(\xi \times \tau) \qquad\qquad (IV.4.4)$$

for each ξ and τ. Indeed, if $(x,y) \in \lim_{\xi \times L\tau} (\mathcal{F} \times \mathcal{H})$ then there are finitely many filters $\mathcal{G}_1 \ldots \mathcal{G}_n$ with $\mathcal{H} \geq \bigwedge_{i=1}^n \mathcal{G}_i$ and $y \in \bigcap_{i=1}^n \lim_\tau \mathcal{G}_i$. Moreover,

$$\mathcal{F} \times \mathcal{H} \geq \mathcal{F} \times (\bigwedge_{i=1}^n \mathcal{G}_i) \geq \bigwedge_{i=1}^n (\mathcal{F} \times \mathcal{G}_i)$$

by Lemma II.8.14. Since $(x,y) \in \bigcap_{i=1}^n \lim_{\xi \times \tau} (\mathcal{F} \times \mathcal{G}_i)$ we conclude that $(x,y) \in \lim_{L(\xi \times \tau)} (\mathcal{F} \times \mathcal{H})$.

Therefore

$$L\xi \times L\tau \geq L(\xi \times L\tau) \geq L(\xi \times \tau)$$

by applying the finitely deep modification L on both sides of (IV.4.4) and apply (IV.4.4) to the result, with ξ playing the role of τ. $\qquad\square$

Exercise IV.4.4. Show that a product of two premetrizable (respectively metrizable) convergence spaces is premetrizable (respectively metrizable).

IV.4.2 *Infinite product*

Recall from (II.2.12) that, given a set $\mathcal{X}$ of sets and, for each $X \in \mathcal{X}$, filters $\mathcal{F}_X \in \mathbb{F}X$, the product filter, or polyhedral filter, is given by

$$\prod_{X \in \mathcal{X}} \mathcal{F}_X := \bigvee_{X \in \mathcal{X}} p_X^-[\mathcal{F}_X],$$

and that

$$\left\{ \prod_{X \in \mathcal{X}} F_X : \bigvee_{X \in \mathcal{X}} F_X \in \mathcal{F}_X, \{X \in \mathcal{X} : F_X \neq X\} \in [\mathcal{X}]^{<\omega} \right\} \qquad (IV.4.5)$$

is a filter-base for $\prod_{X \in \mathcal{X}} \mathcal{F}_X$, by Exercise II.2.38.

In view of (IV.4.2), it is immediate that if $\mathcal{F} \in \mathbb{F}\left(\prod_{\xi \in \Xi} |\xi| \right)$ then

$$x \in \lim_{\prod \Xi} \mathcal{F} \iff x \in \lim_{\prod \Xi} \prod_{\xi \in \Xi} p_\xi[\mathcal{F}], \qquad (IV.4.6)$$

because

$$\mathcal{F} \geq \prod_{\xi \in \Xi} p_\xi[\mathcal{F}].$$

In particular,

$$\bigvee_{\xi \in \Xi} x_\xi \in \lim_\xi \mathcal{F}_\xi \Longrightarrow (x_\xi)_{\xi \in \Xi} \in \lim_{\prod \Xi} \prod_{\xi \in \Xi} \mathcal{F}_\xi,$$

that is, a filter is convergent in the product convergence exactly if it is finer than a product of convergent filters on each component. In other words, there is a pavement of product convergence consisting of very coarse filters: polyhedral filters. In fact, by (IV.4.5), each element of such a filter includes a cube whose all the edges but finitely many are the whole of the component.

It turns out that product convergence on infinite products is intimately related to some convergence on function spaces.

IV.5   Functional convergences

Exercise IV.5.1 (pointwise convergence and product). Show that

(1) the convergence space of Example III.4.6 is homeomorphic to the product convergence space $\prod_{x \in \mathbb{R}} \mathbb{R}_x$ where each copy $\mathbb{R}_x$ of $\mathbb{R}$ carries the standard convergence, in such a way that the projections correspond via this homeomorphism to point evaluation maps $e_x : \mathbb{R}^{\mathbb{R}} \to \mathbb{R}$ defined by $e_x(f) = f(x)$.

(2) the pointwise convergence is the coarsest convergence on $\mathbb{R}^{\mathbb{R}}$ making each map e_x continuous.

Exercise IV.5.2. Use the preceding exercise to show that a product of metrizable convergence spaces may fail to be premetrizable.

Example IV.5.3 (convergence of joint continuity on $\mathbb{R}^{\mathbb{R}}$). Consider on $\mathbb{R}^{\mathbb{R}}$ the relation

$$f \in \lim \mathcal{F} \iff \bigvee_{x \in \mathbb{R}} f(x) \in \lim_{\mathbb{R}} \langle \mathcal{V}_{\mathbb{R}}(x), \mathcal{F} \rangle, \qquad \text{(IV.5.1)}$$

where $e = \langle \cdot, \cdot \rangle : \mathbb{R} \times \mathbb{R}^{\mathbb{R}} \to \mathbb{R}$ is the evaluation map defined by

$$e(x, f) := \langle x, f \rangle := f(x).$$

Note that

$$\langle \mathcal{V}_{\mathbb{R}}(x), \mathcal{F} \rangle = e[\mathcal{V}_{\mathbb{R}}(x) \times \mathcal{F}],$$

according to two alternate notations.

This is a preconvergence because $\mathcal{G} \geq \mathcal{F}$ implies $\langle \mathcal{V}_{\mathbb{R}}(x), \mathcal{G} \rangle \geq \langle \mathcal{V}_{\mathbb{R}}(x), \mathcal{F} \rangle$ for each x. Moreover, $f \in \lim\{f\}^{\uparrow}$ if and only if $f(x) \in \lim_{\mathbb{R}} f[\mathcal{V}_{\mathbb{R}}(x)]$ for each x, that is, if and only if f is continuous. In other words, this preconvergence is centered only at continuous functions. Hence it defines a convergence on the set $C(\mathbb{R}, \mathbb{R})$ of real-valued continuous functions on $\mathbb{R}$, called *convergence of joint continuity*. This convergence turns out to be of fundamental interest and will be studied more systematically as we advance in the book. It is also called *continuous convergence* (e.g., [9], [8]), *power convergence* (e.g., [28]), or *natural convergence* (e.g., [38], [34]). This last term will be our preferred terminology when studying this convergence systematically in Chapter XVI.

Exercise IV.5.4. Considering Example IV.5.3,

(1) Show that the convergence of joint continuity is the coarsest convergence on $C(\mathbb{R}, \mathbb{R})$ that makes the evaluation map $e : \mathbb{R} \times C(\mathbb{R}, \mathbb{R}) \to \mathbb{R}$ continuous (for the product convergence);
(2) Deduce that the convergence of joint continuity is finer than the pointwise convergence on $C(\mathbb{R}, \mathbb{R})$.

Remark IV.5.5. Let us denote by ν the standard convergence on $\mathbb{R}$ and by $[\mathbb{R}, \mathbb{R}]$ the convergence of joint continuity. We have seen that $[\mathbb{R}, \mathbb{R}]$ is the coarsest of the convergences θ on $C(\mathbb{R}, \mathbb{R})$ making the evaluation $e : |\nu \times \theta| \to |\nu|$ continuous. In other words, in view of Proposition IV.2.9, it is the coarsest convergence θ such that

$$\nu \times \theta \geq e^- \nu. \tag{IV.5.2}$$

It is interesting to note that with the properties of the finitely deep modification L already established (in Exercises IV.1.7 and IV.2.13, and Corollary IV.4.3), we can prove algebraically that $[\mathbb{R}, \mathbb{R}]$ is finitely deep, without using the explicit description of $[\mathbb{R}, \mathbb{R}]$ or of the finitely deep modification L: Applying L to (IV.5.2) for $\theta = [\mathbb{R}, \mathbb{R}]$, we have

$$\mathrm{L}(\nu \times [\mathbb{R}, \mathbb{R}]) \geq \mathrm{L}(e^- \nu) \geq e^-(\mathrm{L}\nu).$$

Since ν is finitely deep and L commutes with product, we obtain:

$$\nu \times \mathrm{L}[\mathbb{R}, \mathbb{R}] \geq e^- \nu.$$

Hence $\mathrm{L}[\mathbb{R}, \mathbb{R}]$ satisfies (IV.5.2). But $[\mathbb{R}, \mathbb{R}]$ is the coarsest such convergence, so that $\mathrm{L}[\mathbb{R}, \mathbb{R}] \geq [\mathbb{R}, \mathbb{R}]$, which amount to equality. Thus $[\mathbb{R}, \mathbb{R}]$ is finitely deep.

IV.6   Diagonal and product maps

IV.6.1   *Diagonal map*

Given a set F of maps $f : X \to Y_f$ with domain X, the *diagonal map*

$$\Delta F : X \to \prod_{f \in F} Y_f$$

is defined by

$$\left((\Delta F)(x) \right)(f) = p_{Y_f}\left((\Delta F)(x) \right) := f(x) \qquad \text{(IV.6.1)}$$

for each $f \in F$.

Of course, by the very definition of injectivity,

Proposition IV.6.1. *The diagonal map $\Delta F : X \to \prod_{f \in F} Y_f$ is injective if and only if for each couple x_0, x_1 of distinct elements of X, there exists $f \in F$, for which $f(x_0) \neq f(x_1)$.*

In other words, $\Delta F : X \to \prod_{f \in F} Y_f$ is injective whenever F *separates the points of X.*

If $\prod_{f \in F} Y_f$ is endowed with the product convergence

$$\upsilon := \prod_{f \in F} \upsilon_f,$$

then $h \in \lim_\upsilon \mathcal{H}$ if and only if

$$p_{Y_f}(h) = h(f) \in \lim_{\upsilon_f} p_{Y_f} [\mathcal{H}]$$

for every projection $p_{Y_f} : \prod_{j \in F} Y_j \to Y_f$. Therefore

Proposition IV.6.2. *The initial convergence $F^- Y$ is the coarsest convergence, for which the diagonal map ΔF is continuous.*

Proof. Indeed, $x \in \lim_{F^- Y} \mathcal{F}$ whenever $f(x) \in \lim_{\upsilon_f} f [\mathcal{F}]$ for each $f \in F$. On the other hand, $(\Delta F)(x)(f) = f(x) = p_{Y_f}(\Delta F(x))$ and

$$(\Delta F)[\mathcal{F}](f) = f [\mathcal{F}] = \left(p_{Y_f} \circ \Delta F \right) [\mathcal{F}]$$

for each $f \in F$, that is,

$$(\Delta F)(x) \in \lim_{\prod_{f \in F} \upsilon_f} (\Delta F)[\mathcal{F}],$$

whenever $f(x) \in \lim_{\upsilon_f} f [\mathcal{F}]$ for each $f \in F$.   □

Corollary IV.6.3. *If F separates the elements of X, then the diagonal map from X to $\prod_{j \in F} Y_j$ is an embedding of $F^- Y$ to the product convergence.*

IV.6.2 *Product map*

If $f : X \to Y$ and $g : W \to Z$ then the *product map*

$$f \times g : X \times W \to Y \times Z$$

is defined by $(f \times g)(x, w) = (f(x), g(w))$.

Proposition IV.6.4. *The product map of two continuous maps is continuous.*

Proof. In view of Section IV.4, it is enough to note that if $x \in \lim_\xi \mathcal{F}$ and $w \in \lim_\pi \mathcal{G}$ then

$$(f \times g)[\mathcal{F} \times \mathcal{G}] = f[\mathcal{F}] \times g[\mathcal{G}]$$

converges to $(f(x), g(w))$ in $\sigma \times \tau$ by continuity of f and g. $\square$

More generally, given a set $\boldsymbol{F}$ of maps $f : X_f \to Y_f$, we define the *product map*

$$\bigotimes \boldsymbol{F} : \prod_{f \in \boldsymbol{F}} X_f \to \prod_{f \in \boldsymbol{F}} Y_f$$

by

$$\left(p_{Y_f} \circ \bigotimes \boldsymbol{F} \right)(x) := f \circ p_{X_f}(x),$$

where as usual $p_{Y_f} : \prod_{f \in \boldsymbol{F}} Y_f \to Y_f$ and $p_{X_f} : \prod_{f \in \boldsymbol{F}} X_f \to X_f$ are the respective projections on the f-th factors.

$$\begin{array}{ccc} \prod_{f \in \boldsymbol{F}} X_f & \xrightarrow{\;\otimes \boldsymbol{F}\;} & \prod_{f \in \boldsymbol{F}} Y_f \\ {\scriptstyle p_{X_f}} \downarrow & & \downarrow {\scriptstyle p_{Y_f}} \\ X_f & \xrightarrow[\;f\;]{} & Y_f \end{array}$$

Proposition IV.6.5. *If $\boldsymbol{F}$ is a set of continuous maps $f : |\xi_f| \to |\tau_f|$, then*

$$\bigotimes \boldsymbol{F} : \left| \prod_{f \in \boldsymbol{F}} \xi_f \right| \to : \left| \prod_{f \in \boldsymbol{F}} \tau_f \right|$$

is continuous.

Exercise IV.6.6. Prove Proposition IV.6.5.

IV.7   Initial and final convergences for product maps

Let $f : X \to Y$ and $g : W \to Z$. If σ is a convergence on Y and τ is a convergence on Z then we can either consider first the product $\sigma \times \tau$ on $Y \times Z$ and then the initial convergence $(f \times g)^-(\sigma \times \tau)$ on $X \times W$ or, the other way round, consider the initial convergences $f^-\sigma$ on X and $g^-\tau$ on Y, and then their product $f^-\sigma \times g^-\tau$ on $X \times W$. As it turns out, it makes no difference.

Proposition IV.7.1.

$$f^-\sigma \times g^-\tau = (f \times g)^-(\sigma \times \tau). \qquad (IV.7.1)$$

Proof. Since $f : |f^-\sigma| \to |\sigma|$ and $g : |g^-\tau| \to |\tau|$ are continuous, so is

$$f \times g : |f^-\sigma \times g^-\tau| \to |\sigma \times \tau|$$

by Proposition IV.6.4. Therefore, $f^-\sigma \times g^-\tau \geq (f \times g)^-(\sigma \times \tau)$ because the latter is the coarsest convergence with this property.

Conversely, the projections $p_X : |(f \times g)^-(\sigma \times \tau)| \to |f^-\sigma|$ and $p_W : |(f \times g)^-(\sigma \times \tau)| \to |g^-\tau|$ are continuous, which proves the reverse inequality $f^-\sigma \times g^-\tau \leq (f \times g)^-(\sigma \times \tau)$. Indeed, the continuity of projections follows from the observation that $(x, w) \in \lim_{(f \times g)^-(\sigma \times \tau)} \mathcal{H}$ implies $(f(x), g(w)) \in \lim_{\sigma \times \tau}(f \times g)[\mathcal{H}]$, that is, $f(x) \in \lim_\sigma f[p_X[\mathcal{H}]]$ and $g(w) \in \lim_\tau g[p_W[\mathcal{H}]]$. Thus, $x \in \lim_{f^-\sigma} p_X[\mathcal{H}]$ and $w \in \lim_{g^-\tau} p_W[\mathcal{H}]$. $\square$

Similarly,

Proposition IV.7.2. *If ξ is a convergence on X and π is a convergence on W, then*

$$(f \times g)(\xi \times \pi) = f\xi \times g\pi. \qquad (IV.7.2)$$

Proof. By Proposition IV.6.4, $f \times g : |\xi \times \pi| \to |f\xi \times g\pi|$ is continuous, so that

$$(f \times g)(\xi \times \pi) \geq f\xi \times g\pi,$$

by definition of $(f \times g)(\xi \times \pi)$.

Conversely, assume that $(y, z) \in \lim_{f\xi \times g\pi} \mathcal{H}$, that is, $y \in \lim_{f\xi} p_Y[\mathcal{H}]$ and $z \in \lim_{g\pi} p_Z[\mathcal{H}]$. Let us assume first that $y \in f(X)$ and $z \in g(W)$. Then there are filters $\mathcal{F} \in \mathbb{F}X$ and $\mathcal{G} \in \mathbb{F}W$ and points $x \in f^-(y)$ and $w \in g^-(z)$ such that $x \in \lim_\xi \mathcal{F}$, $w \in \lim_\pi \mathcal{G}$, $p_Y[\mathcal{H}] \geq f[\mathcal{F}]$ and $p_Z[\mathcal{H}] \geq g[\mathcal{G}]$. Hence

$$\mathcal{H} \geq p_Y[\mathcal{H}] \times p_Z[\mathcal{H}] \geq f[\mathcal{F}] \times g[\mathcal{G}] = (f \times g)[\mathcal{F} \times \mathcal{G}]. \qquad (IV.7.3)$$

Since $(y, z) \in \lim_{(f \times g)(\xi \times \pi)} (f \times g)[\mathcal{F} \times \mathcal{G}]$, we conclude that $\mathcal{H}$ converges to (y, z) in $(f \times g)(\xi \times \pi)$.

If on the other hand $y \notin f(X)$ or $z \notin g(W)$ or both, then $p_Y[\mathcal{H}] = \{y\}^\uparrow$ or $p_Z[\mathcal{H}] = \{z\}^\uparrow$ or both. Adapting (IV.7.3) accordingly, we also conclude that $\mathcal{H}$ converges to (y, z) in $(f \times g)(\xi \times \pi)$ in this case, which proves the reverse inequality. $\qquad\square$

The results of this section generalize to infinite products:

Theorem IV.7.3. *If F is a set of maps $f : |\xi_f| \to |\tau_f|$ then*

$$\left(\bigotimes F\right)^- \left(\prod_{f \in F} \tau_f\right) = \prod_{f \in F} f^- \tau_f \qquad (\text{IV.7.4})$$

and

$$\left(\bigotimes F\right) \left(\prod_{f \in F} \xi_f\right) = \prod_{f \in F} f \xi_f. \qquad (\text{IV.7.5})$$

Exercise IV.7.4. Prove Theorem IV.7.3.

IV.8 Quotient

We have been using the concept of equivalence relation as a matter of fact. We have done the same with the notion of partition. Let us emphasize now their interconnection.

An *equivalence relation* $\sim$ on X is a binary relation on X, that is,

$$x \sim x, \qquad \text{(reflexive)}$$
$$x \sim y \implies y \sim x, \qquad \text{(symmetric)}$$
$$(x \sim y \text{ and } y \sim z) \implies x \sim z. \qquad \text{(transitive)}$$

For each $x \in X$, the *equivalence class* $\sim (x)$ of x is, as usual, the image of x by the relation $\sim$. Traditionally, it is often denoted $[x]$, hence

$$[x] := \sim (x) = \{y \in X : x \sim y\}.$$

Of course, for any $x, y \in X$, either $[x] \cap [y] = \varnothing$ or $[x] = [y]$.

A family $\mathcal{P}$ of subsets of X is called a *partition* of X if $X = \bigcup_{P \in \mathcal{P}} P$ and if $P_0, P_1 \in \mathcal{P}$ then

$$P_0 \neq P_1 \implies P_0 \cap P_1 = \varnothing.$$

In other words,

Proposition IV.8.1. *Each partition of a set X defines an equivalence relation $\sim_{\mathcal{P}}$ on X by $x \sim_{\mathcal{P}} y$ if there is $P \in \mathcal{P}$ such that $x, y \in P$.*

Conversely, each equivalence relation $\sim$ on X defines a partition, which is the set of equivalence classes of $\sim$.

Let us add that if $\mathcal{P}_{\approx}$ is the partition of X defined by an equivalence relation $\approx$, then $\approx$ is equal to $\sim_{\mathcal{P}_{\approx}}$.

The set of equivalence classes for $\sim$ is denoted $X/\sim$ and called the *quotient of X by $\sim$*. The map $f : X \to X/\sim$ defined by

$$f(x) := [x]$$

is onto, and is called the *canonical surjection*.

If $\mathcal{P}$ is a partition of X, the *quotient of X by $\mathcal{P}$* is $X/\sim_{\mathcal{P}}$ and can be identified with $\mathcal{P}$.

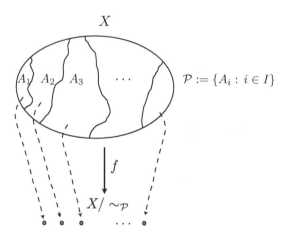

Figure IV.2   A partition, and the associated quotient.

If $f : X \to Y$ is onto, then $\{f^-(y) : y \in Y\}$ is a partition of X and defines therefore an equivalence relation on X:

$$x \sim_f t \iff f(x) = f(t).$$

Equivalent classes for $\sim_f$ are the fibers $f^-(y)$ so that Y can be identified with the quotient set $X/\sim_f$ and f is the canonical surjection. Conversely, any equivalence relation $\sim$ on a set X determines a surjection $f : X \to X/\sim$ that associates to each $x \in X$ its equivalence class, and we have $\sim = \sim_f$.

Definition IV.8.2. If (X, ξ) is a convergence space and $f : X \to Y$ is onto, we call the final convergence $f\xi$ the *quotient convergence* on Y.

A surjective map $f : |\xi| \to |\tau|$ is called a *convergence quotient map* if $\tau = f\xi$.

Example IV.8.3. Consider $[0, 1]$ with the standard convergence (induced $(\mathbb{R}, \nu)$), denoted here ξ. In this convergence, $x \in \lim_\xi \mathcal{F}$ if $\mathcal{F} \geq \mathcal{V}_{[0,1]}(x)$ where $\mathcal{V}_{[0,1]}(x)$ has a filter-base of the form

$$\left\{ B\left(x, \tfrac{1}{n}\right) := \left\{ y \in [0, 1] : |x - y| < \tfrac{1}{n} \right\} : n \in \mathbb{N} \right\}.$$

Let $S_1 := \{ z \in \mathbb{C} : |z| = 1 \}$ be the unit circle in the complex plane. Define $f : [0, 1] \to S_1$ by $f(t) = e^{2\pi i t}$. Note that f is onto, that the restriction of f to $(0, 1)$ is an homeomorphism onto $S_1 \setminus \{1\}$ and that $f(0) = f(1) = 1$. Geometrically, this quotient map identifies (that is, glues together) the two ends of the interval $[0, 1]$ to form the circle.

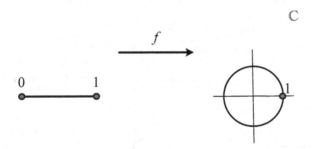

The quotient convergence $f\xi$ on S_1 is finer than the convergence induced on S_1 by the standard (metrizable) convergence on $\mathbb{C}$. Indeed, the vicinity filter $\mathcal{V}_{\mathbb{C}}(1)$ of 1 in this convergence has a filter-base formed by sets $\{z = e^{i\theta} : -\frac{\pi}{2n} < \theta < \frac{\pi}{2n}\}$ for $n \in \mathbb{N}$. Therefore this filter is not finer than the image under f of a filter converging either to 0 or to 1 in $[0, 1]$ because such a filter has a filter-base composed of sets either all included in the upper half-circle (case of 0) or all included in the lower half-circle (case of 1).

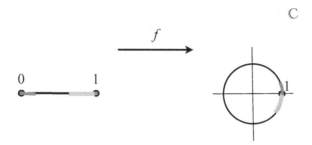

Figure IV.3   An open arc around 1 is obtained by gluing together vicinities of 0 and of 1 in $[0,1]$.

On the other hand, the finitely deep modification $\mathrm{L}\,(f\xi)$ coincides with the convergence induced by the standard convergence of $\mathbb{C}$ on S_1 because

$$f[\mathcal{V}_{[0,1]}(0)] \wedge f[\mathcal{V}_{[0,1]}(1)] = \mathcal{V}_{\mathbb{C}}(1).$$

Example IV.8.4. Consider the unit square $[0,1] \times [0,1]$ in the plane $\mathbb{R}^2$ with the convergence (or metric) inherited by the standard convergence (or metric) of $\mathbb{R}^2$. Consider the equivalence relation $\sim$ of $[0,1] \times [0,1]$ that coincides with the identity in the interior of the square and

$$(0,y) \sim (1,y) \text{ for each } y \in [0,1], \tag{IV.8.1}$$
$$(x,0) \sim (x,1) \text{ for each } x \in [0,1]. \tag{IV.8.2}$$

This equivalence relation identifies opposite sides of the square (see Figure IV.4).

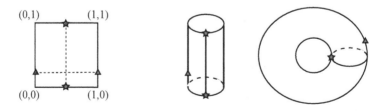

Figure IV.4   The condition (IV.8.1) glues together the vertical edges (the two triangles in the square are identified in the cylinder), the condition (IV.8.2) glues the horizontal edges of the square (the two stars in the cylinder are identified in the torus).

If ξ denotes the standard convergence on the square (inherited from the plane) and $f : [0,1]^2 \to [0,1]^2/\sim$ is the canonical quotient surjection from the square to the torus, the final convergence $f\xi$ on the torus is finer than the convergence induced by the standard convergence of $\mathbb{R}^3$. Indeed, a filter converges to, say, the triangle point of the torus on Figure IV.4, if it is the image of a filter converging to one of the two triangle points on the square. A filter-base for such a filter is formed by "half-balls" around the triangle, as shown on Figure IV.5.

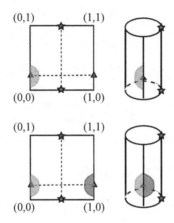

Figure IV.5   If we take a ball centered at the left triangle, then it becomes a half ball. Two balls in the square are glued to yield a ball centered at the identified triangles.

On the other hand, the vicinity filter of the triangle point in the convergence induced on the torus by $\mathbb{R}^3$ has a filter-base formed by balls drawn on the torus around the triangle point, obtained for instance by gluing together the images of the half balls around each of the triangle points in the square, as shown on Figure IV.5. In other words, this vicinity filter is obtained as the infimum of the images of the vicinity filters of the two triangle points.

Note that the four corners of the square have the same image x_0 under f and that a ball around x_0 is obtained by gluing together quarter balls around each one of the corner points. In any case, the vicinity in the convergence induced on the torus by $\mathbb{R}^3$ is obtained as the infimum of the vicinity filters of the finitely many points in the preimage, so that the finitely deep modification $\mathrm{L}\,(f\xi)$ coincides with the convergence induced on the torus by $\mathbb{R}^3$.

We can modify the argument of Corollary IV.3.3 to obtain that every finitely deep convergence is the convergence quotient of a pretopology:

Proposition IV.8.5. *If σ is a finitely deep convergence and for each $x \in |\sigma|$, a pavement $\mathbb{P}_x$ at x for σ is given, let*

$$\xi := \bigoplus_{x \in |\sigma|, \mathcal{F} \in \mathbb{P}_x} \pi[x, \mathcal{F}].$$

Then the map $f : |\xi| \to |\sigma|$ that sends a point of $|\xi|$ to the corresponding point of $|\sigma|$ is a convergence quotient map, that is, $\sigma = f\xi$.

Proof. The map f is obviously onto. If $x \in \lim_\sigma \mathcal{H}$, then there is $\mathcal{F} \in \mathbb{P}_x$ with $x \in \lim_\sigma \mathcal{F}$ and $\mathcal{F} \leq \mathcal{H}$. Of course, $x \in \lim_{\pi[x, \mathcal{F}]} \mathcal{F}$ and thus, denoting by $i_{x, \mathcal{F}}$ the inclusion map of $|\pi[x, \mathcal{F}]|$ into $|\xi|$, $i_{x, \mathcal{F}}(x) \in \lim_\xi i_{x, \mathcal{F}}[\mathcal{F}]$. By definition, $f^-(x) = \{i_{x, \mathcal{F}}(x) : \mathcal{F} \in \mathbb{P}_x\}$ and $f \circ i_{x, \mathcal{F}}[\mathcal{F}] = \mathcal{F} \leq \mathcal{H}$ and thus $x \in \lim_{f\xi} \mathcal{H}$. As f is easily seen to be continuous, $\sigma = f\xi$. □

In view of Exercise IV.3.9, we obtain:

Corollary IV.8.6. *Every finitely deep convergence is the convergence quotient of a pretopology.*

IV.9 Convergence invariants

As explained in the Introduction, objects that seem very different at first sight may turn out to be "essentially the same", that is, homeomorphic. For example, the set $\mathbb{R} \setminus \mathbb{Q}$ of irrational numbers (with the standard convergence inherited from $\mathbb{R}$) is homeomorphic to the set $\mathbb{N}^{\mathbb{N}}$ with the pointwise convergence, equivalently, $\prod_{n \in \mathbb{N}} \mathbb{N}$ with the product of the discrete convergences.

In order to decide that two spaces are not homeomorphic, one examines those properties that are invariant under homeomorphisms. Of course, if one space has such a property and another does not, these two spaces cannot be homeomorphic. In particular, a property invariant under continuous maps is invariant under homeomorphisms.

In this section we will introduce a first sample of such invariants, and illustrate how they can be used to distinguish spaces we have encountered so far, as well as new examples introduced to illustrate the notions at hand.

Maybe the most obvious invariant is the *cardinality* of the underlying space ([4]). As a homeomorphism is in particular bijective, it preserves the cardinality of the space. For instance, no convergence on $\mathbb{R}$ can be homeomorphic to a convergence on $\mathbb{N}$ simply because one of the underlying sets is countable, and the other is not.

Since each homeomorphism is, in particular, a bijection, the cardinality is *invariant*, a shortcut for *preserved by homeomorphisms* ([5]). Actually, every set-theoretic property is invariant.

Moreover, two homeomorphic convergence spaces are identified. In other words, each property defined solely in terms of convergences is, by definition, preserved by homeomorphisms. For example, the T_0, T_1 and Hausdorff properties are invariants.

IV.9.1 *Premetrizability, metrizability*

Proposition IV.9.1. *If $f : |\xi| \to |\sigma|$ is a homeomorphism and ξ is premetrizable, so is σ. If ξ is metrizable, so is σ.*

Exercise IV.9.2. Prove Proposition IV.9.1.

We have seen after Example III.1.17 that in the case of the usual metric d on the real line, $\operatorname{Seq} \widetilde{d}$ is not premetrizable. In view of Proposition IV.9.1, a consequence is that $\operatorname{Seq} \widetilde{d}$ is not homeomorphic to $\widetilde{d}$, nor to any other premetrizable convergence. Note that Exercise IV.1.9 did not show that $\operatorname{Seq} \widetilde{d}$ and $\widetilde{d}$ are not homeomorphic, but only that the identity map is not a homeomorphism, which does not prevent the existence of another homeomorphism.

Similarly, we have observed (Proposition I.7.7) that the *pointwise convergence of real-valued functions on* $\mathbb{R}$ of Example III.4.6 is not premetrizable.

Note that Proposition IV.9.1 presents *two different* invariants: premetrizability and metrizability. Metrizability implies premetrizability, but not conversely, as illustrated for instance by Example III.4.9.

---

[4]We discuss rather thoroughly the notion of cardinality in the Appendix A.2.

[5]Of course, the cardinality is not preserved by continuous maps, not even by quotient maps.

IV.9.2 *Isolated points, paving number, finite depth*

Example IV.9.3 (cofinite space). Let X be an infinite set. The *cofinite convergence* γ *on* X is defined by the pavement $\{(X)_0 \wedge \{x\}^\uparrow\}$ at x, for each $x \in X$ (where $(X)_0$ is the cofinite filter on X). In other words, $x \in \lim_\gamma \mathcal{F}$ if and only if for every finite subset F of $X \setminus \{x\}$ the complement $X \setminus F$ belongs to $\mathcal{F}$.

It is clear that this defines a pretopology, in particular a convergence. Moreover, by Proposition II.2.20, $\lim_\gamma \mathcal{F} = X$ if and only if $\mathcal{F}$ is free (in particular for the cofinite filter). Therefore γ is not Hausdorff and without isolated point.

Exercise IV.9.4. Consider the cofinite convergence γ of Example IV.9.3.

(1) Explain why γ is not homeomorphic to the prime cofinite convergence of Example III.1.15.
(2) Check that the cofinite convergence on X of Example IV.9.3 is the infimum of the set of prime cofinite convergences $\pi[x, (X)_0]$ on X of Example III.1.15:

$$\gamma = \bigwedge_{x \in X} (\pi[x, (X)_0]).$$

Exercise IV.9.5. Show that the prime convergences of Examples III.1.5, III.1.4 and III.1.15 are not homeomorphic with one another.

Recall from Definition III.8.4 that a pavement of a convergence ξ at a point x of $|\xi|$ is a family $\mathbb{H}$ of filters that converge to x in ξ such that whenever $x \in \lim_\xi \mathcal{F}$, there is $\mathcal{H} \in \mathbb{H}$ such that $\mathcal{H} \leq \mathcal{F}$, and that a convergence is κ-paved if it admits a pavement of cardinality at most κ at each point. Recall that the *paving number* $\mathfrak{p}(\xi)$ of a convergence ξ is the smallest cardinality κ for which ξ is κ-paved. As a consequence, the paving number is an invariant. In particular, being a pretopology is an invariant.

Example IV.9.6. Let ν denote the standard convergence on $\mathbb{R}$. It is a pretopology with pavement $\{\mathcal{V}_d(x)\}$ at x. Let $\mathbb{U}^\cap$ stand for the class of finite infima of ultrafilters. A convergence ξ on $\mathbb{R}$ defined by

$$x \in \lim_\xi \mathcal{F} \iff x \in \lim_\nu \mathcal{F} \text{ and } \mathcal{F} \in \mathbb{U}^\cap(\mathbb{R}).$$

Then ξ is a finitely deep convergence that is not a pretopology. Indeed, each ultrafilter $\mathcal{U}$ of $\mathcal{V}_d(x)$ converges to x in ξ but $\bigwedge_{\mathcal{U} \in \beta(\mathcal{V}_d(x))} \mathcal{U}$ does not, because $\beta(\mathcal{V}_d(x))$ is infinite. However, in a pretopology, an infimum of filters converging to a point x also converges to x.

As a consequence ξ and ν are not homeomorphic.

IV.9.3   *Characters and weight*

Definition IV.9.7. Let $\mathcal{B}$ be a family of sets. A filter $\mathcal{F}$ is called *based in* $\mathcal{B}$ if there exists a filter-base $\mathcal{F}_0$ of $\mathcal{F}$ with $\mathcal{F}_0 \subset \mathcal{B}$.

Definition IV.9.8. A family $\mathcal{B}$ is *a base of ξ at x* if there exists a pavement $\mathbb{H}$ of ξ at x such that each $\mathcal{H}$ from $\mathbb{H}$ is based in $\mathcal{B}$.

If $\mathcal{B}$ is a base of ξ at every $x \in |\xi|$, then it is said to be *a base of ξ*.

Definition IV.9.9. The *weight* $\mathrm{w}(\xi)$ of ξ is the least cardinal of a base of ξ.

Example IV.9.10. Let X be a non-empty set.

(1) The family $\{X^\uparrow\}$ is a pavement of the chaotic convergence o_X at every $x \in X$, hence $\{X\}$ is a base of o_X at every $x \in X$ and thus a base of o_X. Therefore the weight $\mathrm{w}(o_X)$ is 1.

(2) The family $\left\{\{x\}^\uparrow\right\}$ is a pavement of ι_X at x, hence $\{\{x\}\}$ is a base of ι_X at x and thus $\{\{x\} : x \in X\}$ is a base of ι_X. In fact, it is the only base. Therefore $\mathrm{w}(\iota_X) = \operatorname{card} X$.

Example IV.9.11. The standard convergence ν of the real line is 1-paved and $\mathcal{V}_\nu(r) \approx \left\{\left(r - \frac{1}{n}, r + \frac{1}{n}\right) : n \in \mathbb{N}_1\right\}$ is the single element of a pavement at r, so that $\left\{\left(r - \frac{1}{n}, r + \frac{1}{n}\right) : n \in \mathbb{N}_1\right\}$ is a base of ν at r. It follows (exercise below) that $\{(s,t) : s < t, s, t \in \mathbb{Q}\}$ is a base of ν, so that $\mathrm{w}(\nu) \leq \aleph_0$. In fact, $\mathrm{w}(\nu) = \aleph_0$, because there is no finite base of ν.

Exercise IV.9.12. Show that $\{(s,t) : s < t, s, t \in \mathbb{Q}\}$ is a base of ν.

Example IV.9.13. The sequential modification $\operatorname{Seq}\nu$ of ν is $\mathfrak{c}$-paved as seen in Example III.8.5. As each sequential filter is based in countable sets, the set of all countable sets is a base of $\operatorname{Seq}\nu$. As each countable subset of $\mathbb{R}$ is the image of $\mathbb{N}$, there are $\operatorname{card}(\mathbb{R}^\mathbb{N}) = \mathfrak{c}^{\aleph_0} = 2^{\aleph_0} = \mathfrak{c}$ countable subsets. Therefore $\mathrm{w}(\operatorname{Seq}\nu) \leq \mathfrak{c}$.

Definition IV.9.14. The *character* $\chi(\mathcal{F})$ of a filter $\mathcal{F}$ is the least cardinality of a filter-base of $\mathcal{F}$.

The character of each principal filter is 1, because $\{A\}$ is a base of $A^\uparrow$, and conversely if $\chi(\mathcal{F}) = 1$ then there exists $A \in \mathcal{F}$ such that $A \subset F$ for $F \in \mathcal{F}$, that is, $\mathcal{F} = A^\uparrow$.

Moreover, if the character $\chi(\mathcal{F})$ is finite then it is 1. Indeed, if $\mathcal{B}$ is a finite filter-base of $\mathcal{F}$, then $B_0 := \bigcap_{B \in \mathcal{B}} B \in \mathcal{F}$ thus $B_0 \subset B$ for each $B \in \mathcal{B}$, hence $\{B_0\}$ is a filter-base of $\mathcal{F}$.

Therefore, each non-principal filter has infinite character.

Definition IV.9.15. If $\mathbb{G}$ is a collection of filters, then the *character* $\chi(\mathbb{G})$ is defined by $\chi(\mathbb{G}) := \sup\{\chi(\mathcal{G}) : \mathcal{G} \in \mathbb{G}\}$.

Definition IV.9.16. The *character* $\chi(x, \xi)$ of a convergence ξ at x is the least cardinal κ such that there is a pavement $\mathbb{H}$ of ξ at x with $\chi(\mathbb{H}) = \kappa$. *The character $\chi(\xi)$ of a convergence ξ is defined as*

$$\chi(\xi) := \sup\{\chi(x, \xi) : x \in |\xi|\}.$$

Proposition IV.9.17. *Each sequentially based convergence has countable character.*

Of course, if ξ is a pretopology, then the character of ξ at x is equal to the character of the vicinity filter of x for ξ, that is,

$$\chi(x, \xi) = \chi(\mathcal{V}_\xi(x)).$$

Therefore, by Proposition III.4.10,

Proposition IV.9.18. *Each premetrizable space is of countable character.*

Definition IV.9.19. The *strong character* $\chi_*(x, \xi)$ of a convergence ξ at x is the least cardinal κ such that there exists a base of ξ at x of cardinality κ. The *strong character* $\chi_*(\xi)$ of a convergence ξ is defined by

$$\chi_*(\xi) := \sup\{\chi_*(x, \xi) : x \in |\xi|\}.$$

Observe that

Proposition IV.9.20. *The character and the strong character coincide for pretopologies.*

More generally, since each base $\mathcal{B}$ of ξ at x determines a pavement $\mathbb{H}$ with $\chi(\mathbb{H}) \leq \operatorname{card} \mathcal{B}$, we conclude that $\chi(x, \xi) \leq \chi_*(x, \xi)$. On the other hand, every base of ξ is also a base of ξ at x, so that $\chi_*(x, \xi) \leq \mathrm{w}(\xi)$. Consequently,

$$\chi(\xi) \leq \chi_*(\xi) \leq \mathrm{w}(\xi), \qquad\qquad (\text{IV.9.1})$$

for every convergence ξ.

Character does not increase under convergence quotient maps.

Proposition IV.9.21. *If* $f : X \to Y$ *is map,* (X, ξ) *is a convergence space, and* Y *carries the final convergence* $f\xi$ *then* $\chi(f\xi) \le \chi(\xi)$.

Proof. By definition of $f\xi$, a non-isolated point $y \in Y$ is a limit point of a filter $\mathcal{G}$ on Y if $\mathcal{G} \ge f[\mathcal{F}]$ for some filter $\mathcal{F}$ on X with $f^-(y) \cap \lim_\xi \mathcal{F} \ne \varnothing$. By definition of the character, there is a filter $\mathcal{L} \le \mathcal{F}$ with a filter-base of cardinality at most $\chi(\xi)$ with $f^-(y) \cap \lim_\xi \mathcal{L} \ne \varnothing$. Then $\mathcal{G} \ge f[\mathcal{L}]$, $y \in \lim_{f\xi} f[\mathcal{L}]$ and $f[\mathcal{L}]$ has a filter-base of cardinality at most $\chi(\xi)$. Therefore $\chi(f\xi) \le \chi(\xi)$. $\qquad\qquad\square$

Exercise IV.9.22. What are the weight, the strong character and the character of the prime cofinite space of Example III.1.15 on an infinite set X?

Exercise IV.9.23. What are the character, the strong character of

(1) the standard convergence ν of the real line?
(2) its sequential modification $\mathrm{Seq}\,\nu$?

The following two examples illustrate that the inequalities in (IV.9.1) may be strict.

Example IV.9.24 (A convergence of countable strong character and uncountable weight). The *Sorgenfrey line convergence* is defined on the real line by the pavement $\{\mathcal{V}_s(x)\}$ at each x, where

$$\mathcal{V}_s(x) \approx \{[x, r) : r > x\}.$$

Since $\{[x, x + \frac{1}{n}) : n \in \mathbb{N}\}$ is a base at x, the Sorgenfrey line has countable strong character (hence countable character).

It is finitely deep, and it is Hausdorff because if $x < y$ then there is n such that $y \notin [x, x + \frac{1}{n})$ so that $\mathcal{V}_s(x)$ and $\mathcal{V}_s(y)$ do not mesh. As a consequence, no filter can be finer than both $\mathcal{V}_s(x)$ and $\mathcal{V}_s(y)$.

Its weight is uncountable. To see that, assume to the contrary that there is a countable base $\mathcal{B} = \{B_n : n \in \mathbb{N}\}$. We can assume without loss of generality that each B_n is bounded below ([6]). For each $n \in \mathbb{N}$, let $p_n := \inf B_n$.

---

[6]Indeed, if there is $B_0 \in \mathcal{B}$ with $-\infty$ as greatest lower bound, then $\mathcal{B} \setminus \{B_0\}$ is still a basis, because for each $x \in B_0$ and each $n \in \mathbb{N}$, there is $B_{x,n} \in \mathcal{B} \setminus \{B_0\}$ with $x \in B_{x,n} \subset [x, x + \frac{1}{n})$.

As $P := \{p_n : n \in \mathbb{N}\}$ is countable, there is $x \in \mathbb{R} \setminus P$. Then $\mathcal{B}$ is not a base at x. Indeed, if $p_n < x$, then no set of the form $[x, r)$ includes B_n, and if $x < p_n$, then $[x, p_n)$ is disjoint from B_n.

Example IV.9.25 (*Uncountable sequential fan*: a convergence of countable character with uncountable strong character). Consider a countable set N with distinguished point ∞_N, endowed with the prime cofinite convergence at ∞_N. Let N_x denote a copy of that convergence for each $x \in \mathbb{R}$ and let

$$S := \bigoplus_{x \in \mathbb{R}} N_x$$

be the disjoint sum of these copies of N, endowed with the sum convergence ξ. Let

$$X = \{(x, n) : x \in \mathbb{R}, n \in N_x \setminus \{\infty_{N_x}\}\} \cup \{\infty_X\}$$

and define $f : S \to X$ by

$$f(x, n) = \begin{cases} (x, n), & n \neq \infty_{N_x} \\ \infty_X, & n = \infty_{N_x}. \end{cases}$$

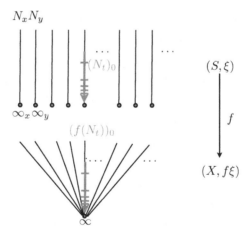

Figure IV.6 In Example IV.9.25 all points ∞_{N_x} are identified to a single point ∞_X under $\sim$. Yet, only filters carried by one of the sets $f(N_x)$ converge in the quotient convergence.

The map f is onto, and we endow X with the quotient convergence $f\xi$. All points but ∞_X are isolated, and a filter converges to ∞_X if and only if it is finer than $\{\infty_X\}^\uparrow \wedge (f(N_t))_0$ for some $t \in \mathbb{R}$, where $(f(N_t))_0$ denotes the cofinite filter on $f(N_t)$. In particular, free filters converging to ∞_X have a filter-base composed of subsets of a given branch $f(N_t)$.

The convergence $f\xi$ is prime and of countable character. Indeed, in view of Exercise IV.9.22, the character of each N_x is countable, and therefore, so is the character of S. In view of Proposition IV.9.21, the character of X is countable.

On the other hand, the strong character is uncountable. Indeed, a base $\mathcal{B}$ at ∞_X in X contains a union of images of bases $\mathcal{B}_x$ at ∞_{N_x} in each copy N_x of N under f.

Example IV.9.25 shows that the analog of Proposition IV.9.21 for strong character does not hold, as S has countable strong character.

Note also that while the convergence of Example IV.9.25 is not finitely deep, its finitely deep modification has the same character and strong character. More generally:

Proposition IV.9.26. *The finitely deep modification preserves infinite weight, character, and strong character, that is,*

$$\mathrm{w}(\xi) \geq \mathrm{w}(\mathrm{L}\,\xi)$$
$$\chi(\xi) \geq \chi(\mathrm{L}\,\xi)$$
$$\chi_*(\xi) \geq \chi_*(\mathrm{L}\,\xi)$$

whenever $\mathrm{w}(\xi)$, $\chi(\xi)$ *and* $\chi_*(\xi)$ *are infinite.*

Exercise IV.9.27. Prove Proposition IV.9.26.

Example III.1.4 is a (non-finitely deep) 2-paved convergence. But, in view of Exercise III.8.11, a finitely deep convergence is either κ-paved for an infinite cardinal κ, or 1-paved, that is, a pretopology. The next chapter will investigate in details pretopologies.

IV.9.4 *Density and separability*

Definition IV.9.28. A subset D of a convergence space (X, ξ) is *dense* if every point of X is a limit of a filter on D, that is,

$$\underset{x \in X}{\forall} \; \underset{\mathcal{F} \in \mathbb{F}X}{\exists} \; D \in \mathcal{F}, \; x \in \lim_\xi \mathcal{F}.$$

The least cardinal κ such that there exists a ξ-dense subset of cardinality κ is called the *density* of ξ and is denoted by $\mathrm{d}\,(\xi)$.

Proposition IV.9.29. *If ξ is Hausdorff, then*

$$\text{card}\,|\xi| \leq 2^{2^{\mathrm{d}(\xi)}}. \tag{IV.9.2}$$

Proof. Let D be a subset of $|\xi|$ such that $\text{card}\,D = \mathrm{d}\,(\xi)$ and for every $x \in |\xi|$ there is a filter $\mathcal{F}_x \in \mathbb{F}D$ with $x \in \lim \mathcal{F}_x$. Since ξ is Hausdorff, the map $x \mapsto \mathcal{F}_x$ is injective, which shows that $\text{card}\,|\xi| \leq \text{card}\,(\mathbb{F}D)$. As there are $2^{2^{\text{card}\,D}}$ filters on D, the inequality (IV.9.2) follows. $\qquad\square$

Definition IV.9.30. A subset D is called *strongly dense* if there is a base $\mathcal{B}$ for ξ for which $D \in \mathcal{B}^\#$. This means that D intersects every element of a base. The least cardinal κ such that there exists a strongly ξ-dense subset of cardinality κ is called the *strong density* of ξ and is denoted by $\mathrm{d}^*\,(\xi)$.

Definition IV.9.31. A convergence is called *separable* if it has a countable dense subset, and *strongly separable* if it has a countable strongly dense subset.

In other words, a convergence is separable if its density is countable; strongly separable if its strong density is countable.

Exercise IV.9.32. Show that a strongly dense subset of a convergence space is dense and that a strongly separable space is separable.

The converse of the statement in Exercise IV.9.32 is not true in general, but true among pretopologies:

Example IV.9.33 (A separable convergence that is not strongly separable). Let ξ be a convergence on $\mathbb{R}$ such that

$$x \in \lim_\xi \mathcal{F} \iff x \in \lim_\nu \mathcal{F} \text{ and } \mathcal{F} \in \mathbb{U}\,(\mathbb{R}),$$

where ν is the standard convergence of the real line. Then ξ is separable, because $\mathbb{Q} \subset \mathbb{R}$ is dense for ν so that for every $x \in \mathbb{R}$, there is an ultrafilter $\mathcal{U}$ with $\mathbb{Q} \in \mathcal{U}$ and $x \in \lim_\nu \mathcal{U} = \lim_\xi \mathcal{U}$.

On the other hand, ξ is not strongly separable. Indeed, by Remark II.7.6, for any countable subset N of $\mathbb{R}$, every uniform ultrafilter contains $\mathbb{R} \setminus N$. Moreover, in view of Proposition II.7.7, for every $x \in \mathbb{R}$, there exists a uniform ultrafilter $\mathcal{U}_x$ finer than $\mathcal{V}_\nu(x)$ and thus $x \in \lim_\xi \mathcal{U}_x$. Thus every base $\mathcal{B}$ for ξ contains subsets of $\mathbb{R} \setminus N$ and thus $N \notin \mathcal{B}^\#$.

Proposition IV.9.34. *A dense subset of a pretopology is strongly dense.*

Proof. Let D be a dense subset of a pretopological space X, with pavement $\{\mathcal{V}(x)\}$ at x. Then $\mathcal{B} := \bigcup_{x \in X} \mathcal{V}(x)$ is a base for the convergence. By density of D, for each $x \in X$, there is a filter $\mathcal{F}$ with $D \in \mathcal{F}$ and $x \in \lim \mathcal{F}$. Then $\mathcal{F} \geq \mathcal{V}(x)$ because $\{\mathcal{V}(x)\}$ is a pavement at x, so that $D \in \mathcal{V}(x)^{\#}$. Hence $D \in \mathcal{B}^{\#}$ and D is strongly dense. $\qquad \square$

Corollary IV.9.35. *A pretopology is separable if and only if it is strongly separable.*

Of course, a convergence on a countable set is always strongly separable. The usual convergence on the real line, which is a pretopology, is (strongly) separable, because the set of rationals meshes with the family of non-empty intervals, which is a base for the convergence. On the other hand, the (pretopological) discrete convergence on the real line is not separable, as the only base is $\{\{x\} : x \in \mathbb{R}\}$ and cannot be meshed by a countable set.

Proposition IV.9.36. *The image of a dense subset under a continuous surjective map is dense.*

Proof. If D is a dense subset of ξ and $f : |\xi| \to |\sigma|$ is continuous, then $f(D)$ is a dense subset of $|\sigma|$. Indeed, for each $y \in Y$ there is $\mathcal{F} \in \mathbb{F}X$ with $D \in \mathcal{F}$ and $f^{-}(y) \cap \lim_{\xi} \mathcal{F} \neq \varnothing$. Then $y \in \lim_{\sigma} f[\mathcal{F}]$ and $f(D) \in f[\mathcal{F}]$. $\qquad \square$

In particular, the continuous image of a separable convergence is separable.

Exercise IV.9.37. Show that a convergence of countable weight is strongly separable.

Proposition IV.9.38. *If $f, g : |\xi| \to |\sigma|$ are two continuous functions, σ is Hausdorff, and the restrictions $f_{|D}$ and $g_{|D}$ of f and g to a dense subset D of (X, ξ) coincide, then $f = g$.*

Exercise IV.9.39. Prove Proposition IV.9.38.

We have seen that every metrizable convergence is premetrizable, and that every premetrizable convergence has countable character. On the other hand, every convergence of countable weight has countable character. But not every metrizable convergence has countable weight. For instance, the discrete pretopology on an uncountable set is metrizable and has uncountable weight. However, we have:

Theorem IV.9.40. *If a convergence space is both metrizable and separable, then it has countable weight, and cardinality at most $\mathfrak{c}$.*

Note that if a convergence is metrizable, it is a pretopology, so that separable and strongly separable are equivalent.

Proof. Let D be a countable dense subset of (X, ξ) where ξ is metrizable by a metric d, and let $\mathcal{A}$ be a base for ξ satisfying $D \in \mathcal{A}^{\#}$. Then the family $\mathcal{B} := \left\{ B(x, \frac{1}{n}) : x \in D, \, n \in \mathbb{N}_1 \right\}$ is countable. Moreover, it is a base for ξ. Indeed, for every $y \in X$ and every $\varepsilon \in (0, 1)$ there is $x \in D$ and $n \in \mathbb{N}$ such that $y \in B(x, \frac{1}{n}) \subset B(y, \varepsilon)$, which shows that $\mathcal{V}_d(y)$ has a filter-base composed of elements of $\mathcal{B}$.

To see how to obtain $x \in D$ and $n \in \mathbb{N}$ as above, take n sufficiently large to have $\frac{1}{n} < \frac{\varepsilon}{2}$. As $\varepsilon < 1$ we have

$$\frac{\varepsilon}{n} < \frac{1}{n} < \frac{\varepsilon}{2}. \tag{IV.9.3}$$

Since $\mathcal{A}$ is a base, there is $A \in \mathcal{A}$ with $A \subset B(y, \frac{\varepsilon}{n})$ so that, by density of D, there is $x \in D \cap B(y, \frac{\varepsilon}{n})$. Then $y \in B(x, \frac{1}{n})$ because $d(x, y) < \frac{\varepsilon}{n} < \frac{1}{n}$. Moreover if $z \in B(x, \frac{1}{n})$ then

$$d(z, y) \le d(z, x) + d(x, y) < \frac{1}{n} + \frac{\varepsilon}{n} < \frac{\varepsilon}{2} + \frac{\varepsilon}{2} = \varepsilon,$$

by (IV.9.3). Hence $B(x, \frac{1}{n}) \subset B(y, \varepsilon)$.

To see that $\operatorname{card} X \le \mathfrak{c}$, note that D is dense and ξ is metrizable, so that for every $x \in X$, there is a sequence $\{d_n^x\}_n$ on D, hence an element of $D^{\mathbb{N}}$, with $x = \lim_{\xi} (d_n^x)_n$. The map $x \mapsto \{d_n^x\}_n$ is injective from X into $D^{\mathbb{N}}$, for sequences in a metrizable space have unique limits. Thus $\operatorname{card} X \le \operatorname{card} D^{\mathbb{N}} = \aleph_0^{\aleph_0} = \mathfrak{c}$ (see Section A.5). $\qquad \square$

Corollary IV.9.41. *The Sorgenfrey line of Example IV.9.24 is a strongly separable convergence of countable character that is not metrizable.*

Proof. Since $\{[x, a) : x \in \mathbb{R}, \, a > x\}$ is a base for the convergence, and the set of rational numbers, which is countable, meshes with this base, the Sorgenfrey line is separable. We have seen that it is also of countable (strong) character. If it were metrizable, it would have countable weight, according to Theorem IV.9.40. But we have seen in Example IV.9.24 that the Sorgenfrey line has uncountable weight. Therefore, it is not metrizable. $\qquad \square$

In particular, metrizability cannot be relaxed to countable character in Theorem IV.9.40.

Exercise IV.9.42. Show that the product of finitely many separable convergence spaces is separable.

Many examples of convergence spaces introduced so far are pretopologies, and we have seen that several important notions find simpler expressions in the context of pretopologies. We explore this important class of convergence spaces in more details in the next chapter. However, it cannot be stressed enough that even starting from a pretopological space, standard and useful constructions can lead to convergence spaces outside of this class. For instance, a quotient (in convergence spaces) of a pretopology need not be a pretopology (see Example IV.9.25), the sequentially based modification $\mathrm{Seq}\,\xi$ of a pretopology ξ need not be a pretopology, etc. We will also see, for example when discussing function space structures, that natural examples of convergences that are not pretopologies abound.

Below is another natural example, for the reader familiar with measure theory.

Example IV.9.43. Let (X, μ) be a measure space in which singletons have measure 0. A sequence $f_n : X \to \mathbb{R}$ *converges almost everywhere to* $f : X \to \mathbb{R}$ if $f(x) \in \lim_n f_n(x)$ for all $x \in X \setminus F$ where $\mu(F) = 0$. More generally, a filter $\mathcal{F}$ on $\mathbb{R}^X$ *converges almost everywhere to* f, in symbols $f \in \lim_{ae} \mathcal{F}$, if $f(x) \in \lim_{\mathbb{R}}\langle x, \mathcal{F}\rangle$ for each $x \in X \setminus F$ where $\mu(F) = 0$. It is easily verified that convergence almost everywhere is indeed a convergence. However, in general, it is not a pretopology.

To see that, consider for each $a \in X$ a filter $\mathcal{F}_a$ on $\mathbb{R}^X$ such that $\langle x, \mathcal{F}_a\rangle$ converges to $f_0(x)$ for all $x \in X \setminus \{a\}$ but $f_0(a) \notin \lim_{\mathbb{R}}\langle a, \mathcal{F}_a\rangle$. Then $f_0 \in \bigcap_{a \in X} \lim_{ae} \mathcal{F}_a$ but $f_0 \notin \lim_{ae} \bigwedge_{a \in X} \mathcal{F}_a$. Indeed, for each $x \in X$,

$$\langle x, \bigwedge_{a \in X} \mathcal{F}_a\rangle \leq \langle x, \mathcal{F}_x\rangle$$

and $f_0(x) \notin \lim_{\mathbb{R}}\langle x, \mathcal{F}_x\rangle$ so that $f_0(x) \notin \lim_{\mathbb{R}}\langle x, \bigwedge_{a \in X} \mathcal{F}_a\rangle$.

Chapter V

Pretopologies

We have already introduced pretopologies in Section III.8 and seen several examples of pretopological convergences. In this chapter we shall study in detail this important class of convergences.

V.1   Definition and basic properties

Recall that a convergence is called a *pretopology* if it is 1-paved (Definition III.8.9), in other words, if for each x, there is a coarsest filter that converges to x. Because of (III.1.2), this filter is the infimum of all filters converging to x.

Recall that the *vicinity filter* of a convergence ξ at x is defined as the intersection of all filters that converge to x with respect to ξ, as defined in (III.1.5),

$$\mathcal{V}_\xi(x) := \bigwedge_{x \in \lim_\xi \mathcal{F}} \mathcal{F}.$$

We have given several examples of convergences, for instance Example III.1.17, for which the vicinity filter of x does not converge to x.

It follows from the considerations above that if π is a pretopology, then $\{\mathcal{V}_\xi(x)\}$ is a pavement at x for each $x \in |\pi|$. Therefore

Proposition V.1.1. *A convergence π on X is a pretopology if and only if*

$$x \in \lim_\pi \mathcal{F} \Longrightarrow \mathcal{F} \geq \mathcal{V}_\pi(x), \qquad (\text{V.1.1})$$

equivalently,

$$x \in \lim_\pi \mathcal{V}_\pi(x),$$

for every $x \in X$.

Remark V.1.2. The vicinity filter only depends on convergent ultra-filters:

$$\mathcal{V}_\xi(x) = \bigwedge_{x \in \lim_\xi \mathcal{F}} \mathcal{F} = \bigwedge_{\mathcal{U} \in \lim_\xi^{-1}(x) \cap \mathbb{U}X} \mathcal{U}. \qquad (V.1.2)$$

Indeed, each $\mathcal{F}$ with $x \in \lim_\xi \mathcal{F}$ can be written as $\mathcal{F} = \bigwedge_{\mathcal{U} \in \beta(\mathcal{F})} \mathcal{U}$ by Proposition II.6.8, so that

$$\mathcal{V}_\xi(x) = \bigwedge_{x \in \lim_\xi \mathcal{F}} \bigwedge_{\mathcal{U} \in \beta(\mathcal{F})} \mathcal{U},$$

and clearly $x \in \lim_\xi \mathcal{U}$ if and only if there is $\mathcal{F} \leq \mathcal{U}$ with $x \in \lim_\xi \mathcal{F}$.

Most of the convergences that we have considered so far are pretopologies. Each premetrizable convergence is a pretopology. Hence, the discrete convergence on any set is a pretopology. The chaotic convergence on any set is a pretopology. It is immediate that if X is a non-empty set, $\iota = \iota_X$ is the discrete pretopology on X and $o = o_X$ is the chaotic pretopology on X, then

$$\mathcal{V}_\iota(x) = \left\{ \{x\}^\uparrow \right\} \text{ and } \mathcal{V}_o(x) = \{\{X\}\} \qquad (V.1.3)$$

for every $x \in X$.

Exercise V.1.3. Show that

(1) each pretopology is finitely deep,
(2) each finitely deep convergence on a finite set is a pretopology.

It follows that the Sierpiński space of Example III.2.1 is a pretopology. On the other hand, the sequential modification $\text{Seq}\,\nu$ of the standard convergence of real line of Example III.1.17 is a finitely deep convergence that is not a pretopology, because its vicinity filters do not converge.

Proposition V.1.4. *A convergence ξ is a pretopology if and only if*

$$\lim_\xi \bigwedge_{\mathcal{F} \in \mathbb{D}} \mathcal{F} = \bigcap_{\mathcal{F} \in \mathbb{D}} \lim_\xi \mathcal{F}, \qquad (V.1.4)$$

for any family $\mathbb{D}$ of filter on $|\xi|$.

Proof. By (III.1.2), $\lim_\xi \bigwedge_{\mathcal{F} \in \mathbb{D}} \mathcal{F} \subset \bigcap_{\mathcal{F} \in \mathbb{D}} \lim_\xi \mathcal{F}$ is true for any convergence ξ. If $x \in \bigcap_{\mathcal{F} \in \mathbb{D}} \lim_\xi \mathcal{F}$ then $\mathcal{F} \geq \mathcal{V}_\xi(x)$ for all $\mathcal{F} \in \mathbb{D}$ so that $\bigwedge_{\mathcal{F} \in \mathbb{D}} \mathcal{F} \geq \mathcal{V}_\xi(x)$ and $x \in \lim_\xi \bigwedge_{\mathcal{F} \in \mathbb{D}} \mathcal{F}$ whenever ξ is a pretopology.

Conversely, if (V.1.4) holds, then in particular

$$x \in \bigcap_{\mathcal{F} \in \lim_\xi^{-1}(x)} \lim_\xi \mathcal{F} \subset \lim_\xi \mathcal{V}_\xi(x),$$

because $\mathcal{V}_\xi(x) = \bigwedge_{\mathcal{F} \in \lim_\xi^{-1}(x)} \mathcal{F}$, hence ξ is a pretopology by Proposition V.1.1. □

Corollary V.1.5. *A supremum of pretopologies is a pretopology.*

Proof. Let Ξ be a set of pretopologies on X and let $\mathbb{D}$ be a set of filters on X. By (III.6.1),

$$\lim_{\vee \Xi} \bigwedge_{\mathcal{F} \in \mathbb{D}} \mathcal{F} = \bigcap_{\xi \in \Xi} \lim_\xi \bigwedge_{\mathcal{F} \in \mathbb{D}} \mathcal{F} = \bigcap_{\xi \in \Xi} \bigcap_{\mathcal{F} \in \mathbb{D}} \lim_\xi \mathcal{F}$$

$$= \bigcap_{\mathcal{F} \in \mathbb{D}} \bigcap_{\xi \in \Xi} \lim_\xi \mathcal{F} = \bigcap_{\mathcal{F} \in \mathbb{D}} \lim_{\vee \Xi} \mathcal{F}.$$

□

It follows from Corollary V.1.5 that:

Proposition V.1.6. *For every convergence ξ there exists the finest pretopology $S_0 \xi$ that is coarser than ξ.*

Proof. Let ξ be a convergence on a set X. Let Π_ξ be the set all the pretopologies π on X such that $\xi \geq \pi$. The set Π_ξ is not empty, because the chaotic pretopology o is in Π_ξ. By Corollary V.1.5,

$$S_0 \xi := \bigvee \Pi_\xi$$

is a pretopology, and is coarser than ξ. Thus $S_0 \xi$ is the finest pretopology that is coarser than ξ. □

The map S_0 associating to each convergence the finest pretopology coarser than that convergence is called the *pretopologizer*, and $S_0 \xi$ is the *pretopological modification* of ξ.

By definition of $S_0 \xi$,

$$S_0 \xi \leq \xi. \tag{V.1.5}$$

If ξ is a pretopology, then ξ itself is the finest pretopology coarser than ξ, that is, $\xi = S_0 \xi$. In view of (V.1.5)

$$\xi \text{ is a pretopology} \iff \xi \leq S_0 \xi.$$

Moreover S_0 is idempotent, that is,

$$S_0(S_0\,\xi) = S_0\,\xi;$$

and

$$\zeta \leq \xi \Rightarrow S_0\,\zeta \leq S_0\,\xi,$$

because $S_0\,\zeta$ is a pretopology coarser than ξ whenever $\zeta \leq \xi$.

An explicit description of $S_0\,\xi$ is given by

$$x \in \lim_{S_0\,\xi} \mathcal{F} \iff \mathcal{F} \geq \mathcal{V}_\xi(x). \qquad (\text{V.1.6})$$

Exercise V.1.7. Prove (V.1.6).

Moreover, S_0 preserves continuity:

Proposition V.1.8. *If* $f : |\xi| \to |\tau|$ *is continuous, so is* $f : |S_0\,\xi| \to |S_0\,\tau|$.

Proof. In view of (V.1.6), it is enough to show

$$f[\mathcal{V}_\xi(x)] \geq \mathcal{V}_\tau(f(x)),$$

equivalently (by Exercise II.3.3(4)),

$$(f[\mathcal{V}_\xi(x)])^\# \subset \mathcal{V}_\tau(f(x))^\#.$$

Let $A \# f[\mathcal{V}_\xi(x)]$, equivalently, $f^- A \# \mathcal{V}_\xi(x)$. Since $\mathcal{V}_\xi(x) = \bigwedge_{x \in \lim_\xi \mathcal{F}} \mathcal{F}$, by Proposition II.3.5, there is a filter $\mathcal{F}$ with $x \in \lim_\xi \mathcal{F}$ and $f^- A \in \mathcal{F}^\#$. The latter amounts to $A \in f[\mathcal{F}]^\#$. On the other hand, $f(x) \in \lim_\tau f[\mathcal{F}]$ by continuity of f, hence $f[\mathcal{F}] \geq \mathcal{V}_\tau(f(x))$ and thus $A \in \mathcal{V}_\tau(f(x))^\#$. $\qquad \square$

Summarizing:

Theorem V.1.9. *The pretopologizer is isotone, contractive and idempotent, i.e.,*

$$\zeta \leq \xi \Rightarrow S_0\,\zeta \leq S_0\,\xi; \qquad\qquad (\text{isotone})$$
$$S_0\,\xi \leq \xi; \qquad\qquad (\text{contractive})$$
$$S_0(S_0\,\xi) = S_0\,\xi. \qquad\qquad (\text{idempotent})$$

Moreover, for each map f *and convergence* τ *on the range set of* f,

$$S_0\left(f^- \tau\right) \geq f^-\left(S_0\,\tau\right). \qquad (\text{V.1.7})$$

Proof. To see (V.1.7), note that f is continuous from $f^- \tau$ to τ and therefore from $S_0(f^- \tau)$ to $S_0\,\tau$ by Proposition V.1.8. In view of Proposition IV.2.9, this fact rephrases as (V.1.7). $\qquad \square$

Corollary V.1.10. *An initial convergence with respect to a pretopology is a pretopology.*

Proof. If $f : X \to |\tau|$ and τ is a pretopology, that is, $S_0\,\tau \geq \tau$ then $f^-(S_0\,\tau) \geq f^-\tau$. By (V.1.7), we conclude that

$$S_0(f^-\tau) \geq f^-\tau,$$

that is, that $f^-\tau$ is a pretopology. □

Corollaries V.1.10 and V.1.5 immediately give:

Corollary V.1.11. *Subspaces and products of pretopologies are pretopologies.*

On the other hand, a quotient of a pretopology need not be a pretopology:

Example V.1.12. Consider the construction of *uncountable sequential fan* of Example IV.9.25. For each $x \in \mathbb{R}$ consider a countably infinite set N_x with a distinguished point ∞_{N_x}. All points of each N_x but ∞_{N_x} are isolated and

$$\mathcal{V}(\infty_{N_x}) = (N_x)_0 \wedge \{\infty_{N_x}\}^\uparrow.$$

This is a prime pretopology. The space $S = \bigoplus_{x\in\mathbb{R}} N_x$ with the sum convergence ξ also is pretopological. On the other hand, the quotient space $(X, f\xi)$ equipped with the quotient convergence is *not* a pretopology. Indeed,

$$\mathcal{V}_{f\xi}(\infty_X) = (\bigwedge_{x\in\mathbb{R}} (f(N_x))_0) \wedge \{\infty_X\}^\uparrow \qquad (V.1.8)$$

does not converge to ∞_X, because filters converging to ∞_X in the quotient convergence have a filter-base composed of subsets of $f(N_x)$ for some x and elements of $\mathcal{V}_{f\xi}(\infty_X)$ intersect every branch $f(N_x)$.

Clearly, the notion of vicinity is pretopological, that is,

$$\mathcal{V}_{S_0\,\xi}(x) = \mathcal{V}_\xi(x) \qquad (V.1.9)$$

for each convergence ξ and every $x \in |\xi|$.

Proof. Of course, $\mathcal{V}_{S_0\,\xi}(x) \leq \mathcal{V}_\xi(x)$, because $S_0\,\xi \leq \xi$. On the other hand, if we define $x \in \lim_\pi \mathcal{F}$ if $\mathcal{V}_\xi(x) \leq \mathcal{F}$, then π is a pretopology and $\pi \leq \xi$, hence $\pi \leq S_0\,\xi$ and thus $\mathcal{V}_\pi(x) = \mathcal{V}_\xi(x) \leq \mathcal{V}_{S_0\,\xi}(x)$. □

Recall from Proposition III.8.10 that prime pretopologies are finally dense in the class of finitely deep convergences. Example IV.9.3 illustrates an instance of this construction. In that case, the resulting cofinite convergence is a pretopology, and can be obtained as the infimum of the prime cofinite pretopologies on that set. Another illustration is given by Example V.1.12.

We will see that the following pretopology on a three-point set, denoted ¥ in Figure III.3, plays an exceptional role among all pretopologies.

Example V.1.13 (Bourdaud pretopology). We denote by ¥ the following pretopology on a three-point set, say $Y := \{0, 1, 2\}$:

$$\mathcal{V}_{¥}(0) := \{Y\}, \ \mathcal{V}_{¥}(1) := \{Y\}, \ \mathcal{V}_{¥}(2) := \{\{1, 2\}, Y\}. \qquad (\text{V.1.10})$$

We notice that $\{0, 1\} \subset \lim_{¥} \mathcal{F}$ for each filter $\mathcal{F}$ on Y; on the other hand, $2 \in \lim_{¥} \mathcal{F}$ if and only if $\{1, 2\} \in \mathcal{F}$.

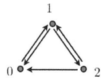

Figure V.1   The arrows indicate to which points each principal filter of a singleton converges for ¥. Loops at each point are omitted.

Theorem V.1.14. *The Bourdaud pretopology ¥ is initially dense in the class of pretopologies.*

Proof. Consider an arbitrary pretopology π on a set X. A map $f : X \to \{0, 1, 2\}$ is continuous at $x \in X$ from π to ¥ whenever $f[\mathcal{V}_\pi(x)] \geq \mathcal{V}_{¥}(f(x))$. Therefore, by (V.1.10), it is continuous at every element of $f^-\{0, 1\}$, while it is continuous at $x \in f^-(2)$ provided that $f^-\{1, 2\} \in \mathcal{V}_\pi(x)$.

Accordingly, for each $x \in X$ and every $V \in \mathcal{V}_\pi(x)$, the map

$$f_{x,V}(v) := \begin{cases} 2 \text{ if } v = x, \\ 1 \text{ if } v \in V \setminus \{x\}, \\ 0 \text{ if } v \notin V, \end{cases}$$

is continuous. Moreover

$$V = f_{x,V}^- \{1, 2\} \qquad (\text{V.1.11})$$

for each $V \in \mathcal{V}_\pi(x)$ and every $x \in X$. Therefore

$$\pi = \bigvee_{x \in X, V \in \mathcal{V}_\pi(x)} f_{x,V}^- \Psi.$$

Indeed,

$$x_0 \in \lim\nolimits_{\bigvee_{x \in X, V \in \mathcal{V}_\pi(x)} f_{x,V}^- \Psi} \mathcal{F}$$

if and only if for each $x \in X$ and $V \in \mathcal{V}_\pi(x)$,

$$f_{x,V}(x_0) \in \lim\nolimits_\Psi f_{x,V}[\mathcal{F}]. \qquad \text{(V.1.12)}$$

If $x \neq x_0$ then $f_{x,V}(x_0) \in \{0,1\}$, so that (V.1.12) holds for each filter $\mathcal{F}$. If $x = x_0$ then $f_{x_0,V}(x_0) = 2$, and thus by (V.1.12), $\{1,2\} \in f_{x,V}[\mathcal{F}]$, that is, by (V.1.11), $\mathcal{F} \geq \mathcal{V}_\pi(x_0)$.

Consequently, each pretopology is initial with respect to a set of maps valued in the Bourdaud space. $\qquad \square$

V.2 Principal adherences and inherences

Definition V.2.1. If A is a subset of a convergence space ξ, then the *(principal) adherence* $\mathrm{adh}\, A$ *(of a set A)* is defined by

$$\mathrm{adh}_\xi A := \bigcup_{\mathcal{F} \# A} \lim\nolimits_\xi \mathcal{F}.$$

Remark V.2.2. Of course, Definition V.2.1 specializes Definition III.5.1 to principal families.

Exercise V.2.3. Show that for any subset A of a convergence space

$$\mathrm{adh}\, A = \bigcup_{A \in \mathcal{F}^\#} \lim \mathcal{F} = \bigcup_{A \in \mathcal{G}} \lim \mathcal{G} = \bigcup_{\mathcal{U} \in \mathrm{U}X, A \in \mathcal{U}} \lim \mathcal{U}.$$

By definition, for every convergence ξ and subsets A and B of $|\xi|$,

$$\mathrm{adh}_\xi \varnothing = \varnothing;$$
$$A \subset B \Longrightarrow \mathrm{adh}_\xi A \subset \mathrm{adh}_\xi B;$$
$$A \subset \mathrm{adh}_\xi A.$$

Moreover,

$$\mathrm{adh}_\xi(A \cup B) = \mathrm{adh}_\xi A \cup \mathrm{adh}_\xi B. \qquad \text{(V.2.1)}$$

Exercise V.2.4. Prove (V.2.1) using Proposition II.4.1.

Definition V.2.5. A point x is called an *accumulation point* of a subset A of $|\xi|$ if $x \in \mathrm{adh}_\xi (A \setminus \{x\})$.

Exercise V.2.6. Show that in a T_1-convergence space, x is an accumulation point of A if and only if there is a free filter $\mathcal{F}$ such that $A \in \mathcal{F}$ and $x \in \lim \mathcal{F}$.

Example V.2.7. The adherence for the discrete pretopology ι on X fulfills

$$\mathrm{adh}_\iota A = A,$$

since the only filter that converges to x is the principal filter $\{x\}^\uparrow$ of x.

The adherence for the chaotic pretopology o on X fulfills

$$\mathrm{adh}_o A = X$$

if $A \neq \varnothing$, because every filter converges to each point.

Definition V.2.8. A family $\mathcal{A}$ of subsets of a convergence space is called *discrete* if for each convergent filter $\mathcal{F}$ there exists $F \in \mathcal{F}$ such that

$$\mathrm{card}\, \{A \in \mathcal{A} : A \# F\} \leq 1.$$

It follows that the elements of a discrete family $\mathcal{A}$ are pairwise disjoint, because if $x \in A \in \mathcal{A}$, then $x \in \lim \{x\}^\uparrow$ and $\{x\} \cap A \neq \varnothing$.

Definition V.2.9. A family $\mathcal{A}$ of subsets of a convergence space is called *locally finite* if for each convergent filter $\mathcal{F}$ there exists $F \in \mathcal{F}$ such that

$$\mathrm{card}\, \{A \in \mathcal{A} : A \# F\} < \infty.$$

Of course, each discrete family is locally finite. Easy examples show that the converse does not hold.

Proposition V.2.10. *If $\mathcal{A}$ is a locally finite family of subsets of a convergence space, then*

$$\mathrm{adh} \left(\bigcup_{A \in \mathcal{A}} A \right) = \bigcup_{A \in \mathcal{A}} \mathrm{adh}\, A.$$

Proof. The inclusion $\supset$ holds for an arbitrary family $\mathcal{A}$. If $x \in \mathrm{adh}(\bigcup_{A \in \mathcal{A}} A)$, then there exists a filter $\mathcal{F}$ on $\bigcup_{A \in \mathcal{A}} A$ such that $x \in \lim \mathcal{F}$. As $\mathcal{A}$ is locally finite, there exist $F \in \mathcal{F}$ and a finite subfamily $\mathcal{A}_0$ of $\mathcal{A}$ such that $F \cap \bigcup_{A \in \mathcal{A} \setminus \mathcal{A}_0} A = \varnothing$. Therefore, $x \in \mathrm{adh}(\bigcup_{A \in \mathcal{A}_0} A) = \bigcup_{A \in \mathcal{A}_0} \mathrm{adh}\, A$ by (V.2.1). $\square$

Example V.2.11. Consider the Bourdaud pretopology ¥ and $A \subset \{0, 1, 2\}$. Then $\{0, 1\} \subset \mathrm{adh}_{¥} A$ for each $A \neq \varnothing$, and $2 \in \mathrm{adh}_{¥} A$ if and only if $\{1, 2\} \cap A \neq \varnothing$.

The following proposition relates principal adherences to vicinity filters.

Proposition V.2.12.

$$x \in \mathrm{adh}_\xi A \Longleftrightarrow A \in \mathcal{V}_\xi(x)^{\#}. \qquad (V.2.2)$$

Proof. If $x \in \mathrm{adh}_\xi A$ then there is a filter $\mathcal{F}$ with $A \in \mathcal{F}^{\#}$ and $x \in \lim_\xi \mathcal{F}$. Then $A \in \mathcal{V}_\xi(x)^{\#}$ because $\mathcal{F} \geq \mathcal{V}_\xi(x)$. Conversely, if

$$A \in \mathcal{V}_\xi(x)^{\#} = (\bigwedge_{x \in \lim_\xi \mathcal{F}} \mathcal{F})^{\#},$$

then there is a filter $\mathcal{F}$ with $A \in \mathcal{F}^{\#}$ and $x \in \lim_\xi \mathcal{F}$ by Proposition II.3.5, so that $x \in \mathrm{adh}_\xi A$. □

Example V.2.13 (Radial pretopology). Let $X := \mathbb{R}^2$. A subset V of X is a vicinity of x if for each $h \in X$ there is $t_h > 0$ such that $x + th \in V$ for each $0 \leq t < t_h$. Accordingly, $x \in \mathrm{adh}\, A$ if there exists $h \in X$ and a sequence $(t_n)_n$ of positive reals converging to 0 such that $x + t_n h \in A$ for each n.

Proposition V.2.12 shows that adherence is a pretopological notion and thus

$$\mathrm{adh}_{S_0\,\xi} A = \mathrm{adh}_\xi A \qquad (V.2.3)$$

for every convergence ξ and each $A \subset |\xi|$ (because of (V.1.9)).

Definition V.2.14. The *(principal) inherence* $\mathrm{inh}\, A$ (of a set A) is defined by

$$x \in \mathrm{inh}_\xi A \Longleftrightarrow A \in \mathcal{V}_\xi(x). \qquad (V.2.4)$$

Exercise V.2.15. Show that

$$\mathrm{inh}_\xi A = (\mathrm{adh}_\xi A^c)^c. \qquad (V.2.5)$$

Exercise V.2.16. Show that

$$x \in \mathrm{inh}_\xi A \iff (x \in \lim_\xi \mathcal{F} \Longrightarrow A \in \mathcal{F}).$$

The following properties are dual to those of adherences:

$$\mathrm{inh}_\xi \, X = X;$$
$$A \subset B \Longrightarrow \mathrm{inh}_\xi \, A \subset \mathrm{inh}_\xi \, B;$$
$$\mathrm{inh}_\xi (A \cap B) = \mathrm{inh}_\xi A \cap \mathrm{inh}_\xi \, B;$$
$$\mathrm{inh}_\xi \, A \subset A,$$

for every A and B.

Proposition V.2.17. *The principal adherences of the initial convergence of τ by a map f, and of the final convergence of ξ by f are, respectively,*

$$\mathrm{adh}_{f^-\tau} H = f^-(\mathrm{adh}_\tau \, f(H)), \qquad (V.2.6)$$
$$\mathrm{adh}_{f\xi} H = H \cup f(\mathrm{adh}_\xi \, f^-(H)). \qquad (V.2.7)$$

Proof. By definition, $x \in \mathrm{adh}_{f^-\tau} H$ whenever there is a filter $\mathcal{F}$ such that $H \in \mathcal{F}^\#$ and $x \in \lim_{f^-\tau} \mathcal{F}$, equivalently $f(x) \in \lim_\tau f[\mathcal{F}]$ and $f(H) \in f[\mathcal{F}]^\#$, thus $f(x) \in \mathrm{adh}_\tau \, f(H)$, that is, $x \in f^-(\mathrm{adh}_\tau \, f(H))$.

On the other hand, if $f(x) \in \mathrm{adh}_\tau \, f(H)$ then there is a filter $\mathcal{G}$ such that $f(x) \in \lim_\tau \mathcal{G}$ and $f(H) \in \mathcal{G}^\#$, which amounts to $H \# f^-[\mathcal{G}]$, by (II.2.10). As $\mathcal{G} \leq f[f^-[\mathcal{G}]]$,

$$x \in f^- (\lim_\tau \mathcal{G}) \subset f^- \left(\lim_\tau f[f^-[\mathcal{G}]]\right) = \lim_{f^-\tau} f^-[\mathcal{G}],$$

thus $x \in \mathrm{adh}_{f^-\tau} H$.

By definition, $y \in f(\mathrm{adh}_\xi \, f^-(H))$ whenever there exists a filter $\mathcal{F}$ such that $f^-(H) \in \mathcal{F}^\#$ and $y \in f(\lim_\xi \mathcal{F}) \subset \lim_{f\xi} f[\mathcal{F}]$. But $H \in f[\mathcal{F}]^\#$ so that $y \in \mathrm{adh}_{f\xi} H$.

Conversely, $y \in \mathrm{adh}_{f\xi} H$ means that there is a filter $\mathcal{G}$ such that $y \in \lim_{f\xi} \mathcal{G}$ and $H \in \mathcal{G}^\#$. If $y \notin f(X)$ then $\mathcal{G} = \{y\}^\uparrow$ and $y \in H$. Otherwise, there is a filter $\mathcal{F}$ on X such that $\mathcal{G} = f[\mathcal{F}]$ and $y \in f(\lim_\xi \mathcal{F})$. Of course, $f^-(H) \in \mathcal{F}^\#$ by (II.2.10), hence $y \in f(\mathrm{adh}_\xi \, f^-(H))$. $\square$

Proposition V.2.18. *Let (X, ξ) and (Y, τ) be two pretopological spaces. The following are equivalent for a map $f : |\xi| \to |\tau|$:*

(1) $f \in C(\xi, \tau)$;
(2) for all $x \in X$,

$$f([\mathcal{V}_\xi(x)] \geq \mathcal{V}_\tau(f(x)));$$

(3) for all $A \subset X$,

$$f(\mathrm{adh}_\xi \, A) \subset \mathrm{adh}_\tau \, f(A);$$

(4) for all $B \subset Y$,

$$f^-(\mathrm{inh}_\tau \, B) \subset \mathrm{inh}_\xi(f^-(B)).$$

Proof. (1) $\Longrightarrow$ (2) by definition, because $x \in \lim_\xi \mathcal{V}_\xi(x)$ and τ is a pretopology.

(2) $\Longrightarrow$ (3) because if $x \in \mathrm{adh}_\xi \, A$ then $A \# \mathcal{V}_\xi(x)$, so that $f(A) \# f[\mathcal{V}_\xi(x)]$. In view of (2), $f(A) \# \mathcal{V}_\tau(f(x))$ so that $f(x) \in \mathrm{adh}_\tau \, f(A)$.

(3) $\Longrightarrow$ (4). Let $x \in f^-(\mathrm{inh}_\tau \, B)$, that is,

$$f(x) \in \mathrm{inh}_\tau \, B = (\mathrm{adh}_\tau \, B^c)^c.$$

Let $A := f^-(B^c)$. Then $f(A) \subset B^c$ so that $f(x) \notin \mathrm{adh}_\tau \, f(A)$. In view of (3), $x \notin \mathrm{adh}_\xi \, A$, that is, $x \in \mathrm{inh}_\xi \, A^c$. Since $A^c = f^-(B)$, we obtain (4).

(4) $\Longrightarrow$ (1). In view of (V.2.4), $B \in \mathcal{V}_\tau(f(x))$ means that $f(x) \in \mathrm{inh}_\tau \, B$. By (4), $x \in \mathrm{inh}_\xi \, f^-(B)$, that is, $f^-(B) \in \mathcal{V}_\xi(x)$ and $f(f^-(B)) \subset B$ so that $f([\mathcal{V}_\xi(x)] \geq \mathcal{V}_\tau(f(x))$. $\square$

Proposition V.2.19. *A convergence π is a pretopology if and only if*

$$\lim_\pi \mathcal{F} \supset \bigcap_{H \in \mathcal{F}^\#} \mathrm{adh}_\pi H. \qquad (V.2.8)$$

Of course, the reverse inclusion $\lim_\pi \mathcal{F} \subset \bigcap_{H \in \mathcal{F}^\#} \mathrm{adh}_\pi H$ is true for every convergence, so that (V.2.8) is in fact an equality.

Proof. Suppose that π is a pretopology and let $x \in \mathrm{adh}_\pi \, H$ for every $H \in \mathcal{F}^\#$. Since, in view of (V.2.2), $x \in \mathrm{adh}_\pi \, H$ amounts to $H \in \mathcal{V}_\pi(x)^\#$, we infer that $\mathcal{F}^\# \subset \mathcal{V}_\pi(x)^\#$, which, in view of Exercises II.3.2 and II.3.3, means $\mathcal{F} \geq \mathcal{V}_\pi(x)$, that is, $x \in \lim_\pi \mathcal{F}$.

Conversely, suppose that $\mathcal{F} \geq \mathcal{V}_\pi(x)$ and $x \notin \lim_\pi \mathcal{F}$, hence by (V.2.8), there exists $H \in \mathcal{F}^\#$ such that $x \notin \mathrm{adh}_\pi \, H$. The latter means that $H \notin \mathcal{V}_\pi^\#(x)$, which yields a contradiction. $\square$

Corollary V.2.20.

$$\lim_{S_0 \, \xi} \mathcal{F} = \bigcap_{H \in \mathcal{F}^\#} \mathrm{adh}_\xi H. \qquad (V.2.9)$$

Proof. Since $\mathrm{adh}_\xi H = \mathrm{adh}_{S_0 \, \xi} H$ for each H, by (V.2.8) $\lim_{S_0 \, \xi} \mathcal{F} \supset \bigcap_{H \in \mathcal{F}^\#} \mathrm{adh}_\xi H$. On the other hand, if $x \in \lim_{S_0 \, \xi} \mathcal{F}$ and $H \in \mathcal{F}^\#$ then $x \in \mathrm{adh}_{S_0 \, \xi} H = \mathrm{adh}_\xi H$. $\square$

We have seen that the quotient of a pretopology need not be pretopological. It is instructive to examine the pretopological modification of such a quotient.

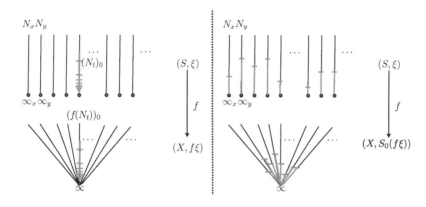

Figure V.2   In Example V.2.21 a filter can converge to ∞ in $f\xi$ only along one of the branches. In the pretopological modification $S_0(f\xi)$ some convergent filters, for instance $\mathcal{V}_{f\xi}(\infty)$, mesh every single branch.

Example V.2.21. Recall the *uncountable sequential fan* of Example IV.9.25, in which (S,ξ) is a pretopological space and $f : S \to X$ is a surjection such that the final convergence $f\xi$ on X is not a pretopology. As noticed in Example V.1.12,

$$\mathcal{V}_{f\xi}(\infty) = \bigwedge_{x\in\mathbb{R}} (f(N_x))_0 \wedge \{\infty\}^{\uparrow},$$

so that a base of $\mathcal{V}_{f\xi}(\infty)$ consists of $\{\infty\} \cup \bigcup_{x\in\mathbb{R}} f(A_x)$, where A_x is a cofinite subset of N_x for each $x \in \mathbb{R}$.

Exercise V.2.22. Show that in the situation of Example IV.9.25 as depicted above, $\mathcal{V}_{f\xi}(\infty)$ is a Fréchet filter that does not have a countable filter-base.

Proposition V.2.23.

$$S_0\left(f^-\tau\right) = f^-\left(S_0\,\tau\right).$$

Proof. We have seen in Theorem V.1.9 that $S_0\left(f^-\tau\right) \geq f^-\left(S_0\,\tau\right)$. Conversely, $x \in \lim_{f^-(S_0\,\tau)} \mathcal{F}$ if and only if $f(x) \in \lim_{S_0\,\tau} f[\mathcal{F}]$. By (V.2.9), $f(x) \in \mathrm{adh}_\tau H$ for every $H \in f[\mathcal{F}]^{\#}$. If $A \in \mathcal{F}^{\#}$ then $f(A) \in f[\mathcal{F}]^{\#}$, hence $f(x) \in \mathrm{adh}_\tau f(A)$, which, in view of (V.2.6), means that, $x \in f^-(\mathrm{adh}_\tau f(A)) = \mathrm{adh}_{f^-\tau} A$, hence $x \in \lim_{S_0(f^-\tau)} \mathcal{F}$. $\square$

Corollary V.2.24. *If (X,ξ) is a convergence space and $A \subset X$ then the pretopological modification of the induced convergence coincides with the convergence induced on A by the pretopological modification, that is,*

$$S_0(\xi_{|A}) = (S_0\,\xi)_{|A}.$$

Proposition V.2.25. *Let (X,ξ) and (Y,τ) be two convergence spaces, and let $A \subset X$ and $B \subset Y$. Then*

$$\mathrm{adh}_{\xi\times\tau}(A \times B) = \mathrm{adh}_\xi A \times \mathrm{adh}_\tau B.$$

Proof. $(x,y) \in \mathrm{adh}_\xi A \times \mathrm{adh}_\tau B$ if and only if there is $\mathcal{F}$ with $A \in \mathcal{F}$ and $x \in \lim_\xi \mathcal{F}$, and there is $\mathcal{G}$ with $B \in \mathcal{G}$ and $y \in \lim_\tau \mathcal{G}$. This is equivalent to $(x,y) \in \mathrm{adh}_{\xi\times\tau}(A \times B)$ because basic convergent filters in $\xi \times \tau$ are of the form $\mathcal{F} \times \mathcal{G}$ for $\mathcal{F}$ ξ-convergent and $\mathcal{G}$ τ-convergent. $\square$

Since the pretopological modification is determined by principal adherences (e.g., via (V.2.9)), one might be misled to believe that Proposition V.2.25 states that (finite) product and pretopological modification commute. No such conclusion can be drawn, as the formula for $\mathrm{adh}_{\xi\times\tau}$ applies only to the very special sets of the form $A \times B$.

In fact, the pretopological modification of a product does *not coincide* with the product of the pretopological modification. This might look surprising as product is an initial construction: $\xi \times \tau = p_{|\xi|}^-\xi \vee p_{|\tau|}^-\tau$. The problem is that suprema do not commute with S_0.

More specifically, we have

$$S_0(\xi \vee \tau) \geq S_0\,\xi \vee S_0\,\tau$$

for every convergence ξ and τ on a set X, because $S_0\,\xi \vee S_0\,\tau$ is a pretopology as a supremum of pretopologies, and is coarser than $\xi \vee \tau$. However, the reverse inequality may badly fail as shown in the following example.

Example V.2.26 ($S_0\,\sigma \vee \tau \not\geq S_0(\sigma \vee \tau)$ with $\tau = S_0\,\tau$). Let X be the uncountable sequential fan of Examples IV.9.25, V.1.12 and V.2.21. The convergence $\sigma := f\xi$ defined on X in Example IV.9.25 is prime. We define on X another prime pretopology τ with the same distinguished point ∞_X. To this end, consider a family

$$\left\{ \bigcup_{x \geq r} f(N_x) : r \in \mathbb{R} \right\}$$

of subsets of X and notice that it is a filter-base that meshes with $\mathcal{V}_\sigma(\infty_X)$. Therefore, there is a free ultrafilter

$$\mathcal{U} \in \beta\left(\mathcal{V}_\sigma(\infty_X) \vee \left\{\bigcup_{x \geq r} f(N_x) : r \in \mathbb{R}\right\}^\uparrow\right).$$

Let $\mathcal{V}_\tau(\infty_X) := \{\infty_X\}^\uparrow \wedge \mathcal{U}$.

Notice that $\sigma \vee \tau$ is the discrete pretopology on X: the only free filter converging to ∞_X in τ is $\mathcal{U}$, but $\infty_X \notin \lim_\sigma \mathcal{U}$ because

$$\mathcal{U} \geq \left\{\bigcup_{x \geq r} f(N_x) : r \in \mathbb{R}\right\}^\uparrow$$

and therefore $\mathcal{U}$ does not have a filter-base composed of subsets of a single branch $f(N_x)$. Accordingly, $\infty_X \in \lim_{S_0 \sigma \vee \tau} \mathcal{U}$ because $\mathcal{U} \geq \mathcal{V}_\sigma(\infty_X)$, but $\infty_X \notin \lim_{S_0(\sigma \vee \tau)} \mathcal{U}$. Hence

$$S_0\,\sigma \vee \tau \not\geq S_0(\sigma \vee \tau).$$

Example V.2.27 ($S_0\,\sigma \times \tau \not\geq S_0(\sigma \times \tau)$). If X, σ and τ are as in Example V.2.26, then $S_0\,\sigma \times \tau \not\geq S_0(\sigma \times \tau)$, because

$$(\infty_X, \infty_X) \in \lim_{S_0 \sigma \times \tau}(\mathcal{U} \times \mathcal{U}) \setminus \lim_{S_0(\sigma \times \tau)}(\mathcal{U} \times \mathcal{U}).$$

In fact, by (V.2.9) the subset $\Delta := \{(x, x) : x \in X, x \neq \infty_X\}$ of $X \times X$ meshes with $\mathcal{U} \times \mathcal{U}$ but $(\infty_X, \infty_X) \notin \operatorname{adh}_{\sigma \times \tau} \Delta$. Indeed, if a filter $\mathcal{H}$ on $X \times X$ contains Δ then it has the same projection $p_X[\mathcal{H}]$ on each factor, and if $(\infty_X, \infty_X) \in \lim_{\sigma \times \tau} \mathcal{H}$ then $\infty_X \in \lim_{\sigma \vee \tau} p_X[\mathcal{H}]$. But we have seen that $\sigma \vee \tau$ is discrete; a contradiction, because $\mathcal{H}$ has to be free.

V.3   Open and closed sets, closures, interiors, neighborhoods

Definition V.3.1. A subset O of a convergence space (X, ξ) is called *ξ-open* if

$$\lim_\xi \mathcal{F} \cap O \neq \varnothing \implies O \in \mathcal{F}, \qquad (V.3.1)$$

for every filter $\mathcal{F}$.

Let $\mathcal{O}_\xi$ stand for the family of all ξ-open sets ([1]). We will also occasionally use the notation

$$\mathcal{O}_\xi(x) := \{O \in \mathcal{O}_\xi : x \in O\}.$$

---

[1]If no ambiguity is probable on the convergence, we may talk of *open* sets, and denote by $\mathcal{O}$ the family of open subsets of a convergence space.

Exercise V.3.2. Let $\mathcal{O}$ be the family of open sets of a convergence. Verify that:

$$\mathcal{G} \in [\mathcal{O}]^{<\omega} \implies \bigcap_{G \in \mathcal{G}} G \in \mathcal{O}; \tag{V.3.2}$$

$$\mathcal{G} \subset \mathcal{O} \implies \bigcup_{G \in \mathcal{G}} G \in \mathcal{O}; \tag{V.3.3}$$

$$\varnothing, X \in \mathcal{O}. \tag{V.3.4}$$

In particular, $\varnothing$ is open, because the intersection of $\varnothing$ with any set is empty so that the implication holds for each filter $\mathcal{F}$; the whole underlying space is open, because it belongs to every filter.

The finer a convergence, the less convergent filters; therefore, $\xi \leq \zeta$ implies that $\mathcal{O}_\xi \subset \mathcal{O}_\zeta$.

Exercise V.3.3. Show that if (X, d) is a premetric space, then $O \subset X$ is open for $\tilde{d}$ if and only if

$$\underset{x \in O}{\forall} \ \underset{r > 0}{\exists} \ B(x, r) \subset O.$$

Hence a subset O of a premetric space (X, d) is open if and only O is a union of balls, i.e., for each $x \in O$ there is $r_x > 0$ such that

$$O = \bigcup_{x \in O} B(x, r_x).$$

Note in particular that, in view of Exercise I.2.4, balls $B(x, r)$ are open in the convergence $\tilde{d}$ whenever d is a metric.

As a shorthand, we will often refer to a cover composed of open sets as an *open cover*.

Definition V.3.4. A subset C of a convergence space ξ is called ξ-*closed* if it is the complement of a ξ-open set.

The family of ξ-closed subsets of $|\xi|$ is denoted by $\mathcal{C}_\xi$, or $\mathcal{C}$ if no confusion can arise.

Exercise V.3.5. Show that the following are equivalent for a subset C of a convergence space:

(1) C is closed;
(2) $\lim \mathcal{H} \subset C$ for each filter $\mathcal{H}$ for which $C \in \mathcal{H}^\#$;
(3) $\lim \mathcal{H} \subset C$ for each filter $\mathcal{H}$ for which $C \in \mathcal{H}$.

Of course, properties of closed sets are dual to those of open sets. Indeed, if $\mathcal{C}$ is the family of closed sets of a convergence, then

$$\mathcal{H} \in [\mathcal{C}]^{<\omega} \implies \bigcup_{H \in \mathcal{H}} H \in \mathcal{C}; \qquad (V.3.5)$$

$$\mathcal{H} \subset \mathcal{C} \implies \bigcap_{H \in \mathcal{H}} H \in \mathcal{C}; \qquad (V.3.6)$$

$$\varnothing, X \in \mathcal{C}. \qquad (V.3.7)$$

The set of all ξ-open subsets of a given set A is non-empty because it contains $\varnothing$; its union is an open set that is called the *interior* of A and is denoted by $\mathrm{int}_\xi A$. Hence, $\mathrm{int}_\xi A$ is the largest ξ-open set contained in A.

As a consequence, if ξ is a convergence on X, then

$$\mathrm{int}_\xi X = X,$$
$$A \subset B \implies \mathrm{int}_\xi A \subset \mathrm{int}_\xi B,$$
$$\mathrm{int}_\xi A \subset A,$$
$$\mathrm{int}_\xi(\mathrm{int}_\xi A) = \mathrm{int}_\xi A.$$

Exercise V.3.6. Show that additionally

$$\mathrm{int}_\xi A_0 \cap \mathrm{int}_\xi A_1 = \mathrm{int}_\xi(A_0 \cap A_1),$$

for every subsets A_0 and A_1 of X.

Note that if $x \in \mathrm{int}_\xi A$ then $A \in \mathcal{F}$ whenever $x \in \lim_\xi \mathcal{F}$, because $\mathrm{int}\, A$ is open, so that $A \in \bigwedge_{x \in \lim_\xi \mathcal{F}} \mathcal{F} = \mathcal{V}_\xi(x)$. In other words, $x \in \mathrm{inh}_\xi A$. Hence

$$\mathrm{int}_\xi A \subset \mathrm{inh}_\xi A, \qquad (V.3.8)$$

for all subset A of a convergence space.

The dual notion is that of *closure*: $\mathrm{cl}_\xi A$ is the intersection of all ξ-closed sets that include A, that is, the smallest ξ-closed subset of X that contains A.

Exercise V.3.7. Let ξ be a convergence ξ. Show that

$$\mathrm{cl}_\xi A = (\mathrm{int}_\xi A^c)^c. \qquad (V.3.9)$$

In view of (V.3.8), we conclude that

$$\mathrm{adh}_\xi A \subset \mathrm{cl}_\xi A, \qquad (V.3.10)$$

for each subset A of $|\xi|$.

The properties of the closure are dual to those of the interior:

$$\mathrm{cl}_\xi \, \varnothing = \varnothing,$$

$$A \subset B \Longrightarrow \mathrm{cl}_\xi \, A \subset \mathrm{cl}_\xi \, B,$$

$$A \subset \mathrm{cl}_\xi \, A,$$

$$\mathrm{cl}_\xi \, (\mathrm{cl}_\xi \, A) = \mathrm{cl}_\xi \, A,$$

$$\mathrm{cl}_\xi (A_0 \cup A_1) = \mathrm{cl}_\xi \, A_0 \cup \mathrm{cl}_\xi \, A_1.$$

Exercise V.3.8. If (X, d) is a metric space and $A \subset X$, let

$$\mathrm{dist}_A : X \to [0, \infty)$$

be the map defined by $\mathrm{dist}_A(x) = \inf_{a \in A} d(x, a)$. Let ξ be the convergence induced by d.

(1) Show that $\mathrm{dist}_A : |\xi| \to |\nu_{[0,\infty)}|$ is continuous;
(2) Show that $x \in \mathrm{cl}_\xi \, A$ if and only if $\mathrm{dist}_A(x) = 0$.

Definition V.3.9. A subset W of $|\xi|$ is a *neighborhood* of an element x of $|\xi|$ if there exists a ξ-open set O such that

$$x \in O \subset W.$$

The set of all the ξ-neighborhoods of x is denoted by $\mathcal{N}_\xi(x)$.

In other words, $W \in \mathcal{N}_\xi(x)$ if and only if $x \in \mathrm{int}_\xi \, W$. On the other hand,

Exercise V.3.10. Show that

$$x \in \mathrm{cl}_\xi \, A \Longleftrightarrow A \in \mathcal{N}_\xi(x)^\#. \tag{V.3.11}$$

The properties of neighborhoods follow from those of interior (or of closure). Let $\mathcal{N}(x)$ stand for the set of all neighborhoods of x with respect to a convergence on X. By definition, the *contour* of $\mathcal{N}(\cdot)$ along $\mathcal{N}(x)$ is

$$\mathcal{N}(\mathcal{N}(x)) = \bigcup_{V \in \mathcal{N}(x)} \bigcap_{v \in V} \mathcal{N}(v).$$

We shall study more in detail the notion of contour in Section VI.1. Then for every $x \in X$ and for each $V, W, V_0, V_1 \subset X$,

$$X \in \mathcal{N}(x),$$

$$W \supset V \in \mathcal{N}(x) \Longrightarrow W \in \mathcal{N}(x),$$

$$V_0, V_1 \in \mathcal{N}(x) \Longrightarrow V_0 \cap V_1 \in \mathcal{N}(x), \tag{V.3.12}$$

$$V \in \mathcal{N}(x) \Longrightarrow x \in V,$$

$$\mathcal{N}(x) \subset \mathcal{N}(\mathcal{N}(x)).$$

The first property means that $\mathcal{N}(x) \neq \varnothing$ for each $x \in X$, that is, each point has a neighborhood. The next three properties say that $\mathcal{N}(x)$ is a *filter* coarser than the principal filter of x. We shall refer to $\mathcal{N}(x)$ as to the *neighborhood filter of* x. The penultimate property of neighborhoods rephrases as $x \in \ker \mathcal{N}(x)$.

The last property corresponds to the idempotency of the interior (and of the closure). Therefore the last property can be reformulated as

$$W \in \mathcal{N}(x) \Longrightarrow \underset{V \in \mathcal{N}(x)}{\exists} \underset{v \in V}{\forall} W \in \mathcal{N}(v).$$

Exercise V.3.11. Prove the properties (V.3.12) of neighborhoods.

Exercise V.3.12. Show that in a convergence space, the vicinity filter of a point is finer than the neighborhood filter:

$$\mathcal{N}(x) \leq \mathcal{V}(x).$$

Exercise V.3.13. Show that

(1) the following are equivalent for a subset O of a convergence space:

(a) O is open;
(b) O is a neighborhood of each of its points:

$$O \in \bigcap_{x \in O} \mathcal{N}(x);$$

(c) O is a vicinity of each of its points:

$$O \in \bigcap_{x \in O} \mathcal{V}(x);$$

(d) $\operatorname{inh} O \supset O$;
(e) $\operatorname{int} O \supset O$.

(2) the following are equivalent for a subset C of a convergence space:

(a) C is closed;
(b) $\operatorname{cl} C \subset C$;
(c) $\operatorname{adh} C \subset C$.

Example V.3.14. We have seen that the discrete convergence ι on a set X of Example III.1.5 is a pretopology. By (V.1.3), $A \in \mathcal{V}_\iota(x)$ for A and $x \in A$, so that each set is open. Therefore

$$\mathcal{O}_{\iota_X} = 2^X = \mathcal{C}_{\iota_X}.$$

Example V.3.15. We have also noticed that the chaotic convergence $o = o_X$ on a set X of Example III.1.6 is a pretopology, for which X is the only vicinity of every $x \in X$. Accordingly, X is the only non-empty open set with respect to o; the empty set $\varnothing$ is another open set for o. Therefore

$$\mathcal{O}_{o_X} = \{\varnothing, X\} = \mathcal{C}_{o_X}.$$

Example V.3.16. If $\$_0$ is the Sierpiński pretopology (see Example III.2.1) on $\{0, 1\}$, then $\{1\}$ is open, because $\{1\} \in \{1\}^\uparrow$ and the only filter that converges to 1 is $\{1\}^\uparrow$, and $\{0\}^\uparrow$ is not, as $0 \in \lim_\$ \{0, 1\}^\uparrow$ but $\{0\} \notin \{0, 1\}^\uparrow$. Hence $\varnothing, \{1\}$ and $\{0, 1\}$ are all the $\$_0$-open sets. Of course, $\{1\}^c = \{0\}$ is closed.

In the foregoing examples we have encountered sets that are simultaneously open and closed.

Definition V.3.17. A set that is open and closed is called *clopen*.

Of course, the empty set and the whole underlying set are clopen for every convergence. These are the only clopen sets, for example, for the Sierpiński convergence and for the standard convergence of the real line. On the other hand, all sets are clopen for the discrete convergence.

Exercise V.3.18. Let $r, s \in \mathbb{R}$ equipped with the standard convergence, with $r < s$. Show that (r, s) is open, that $[r, s]$ is closed and that $[r, s)$ is neither open nor closed.

Exercise V.3.19. Describe the open subsets and the closed subsets of the cofinite convergence of Example IV.9.3.

Exercise V.3.20. Let ν be the standard convergence of $\mathbb{R}$ and let $\nu_{|[-1,1)}$ be its restriction to $[-1, 1)$. Are the following subsets of $[-1, 1)$ closed, open, neither closed nor open, or clopen with respect to $\nu_{|[-1,1)}$?

$$(0, 1), \ [0, 1), \ (-1, 0), \ [-1, 0], \ [-1, 0).$$

Exercise V.3.21. Consider the Sorgenfrey line of Example IV.9.24.

(1) Show that if $x < a$ then the set $[x, a)$ is open;
(2) Show that $O \subset \mathbb{R}$ is open in the Sorgenfrey line convergence if and only if for each $x \in O$ there is $a_x > x$ such that $O = \bigcup_{x \in O} [x, a_x)$;
(3) Show that if $x < a$ then $[x, a)$ is also closed.

Note from the examples and exercises above that a subset of a convergence space may be open but not closed, closed but not open, neither open nor closed, or both open and closed.

Exercises V.3.3 and V.3.21 suggest the following definition:

Definition V.3.22. A family $\mathcal{B}$ of open subsets of a convergence space (X, ξ) is a *base of open sets* if every open subset of X can be obtained as a union of elements of $\mathcal{B}$.

Exercise V.3.23. Show that a family $\mathcal{B}$ of open subsets of (X, ξ) is a base of open sets if and only if

$$\underset{x \in X}{\forall} \; \underset{O \in \mathcal{O}_\xi(x)}{\forall} \; \underset{B \in \mathcal{B}}{\exists} \; x \in B \subset O.$$

Definition V.3.24. A convergence space is called *zero-dimensional* if it has a base of open sets composed of sets that are also closed.

Example V.3.25. In view of Exercise V.3.21, $\mathcal{B} := \{[x, a) : x < a\}$ is a base of open sets for the Sorgenfrey line that is composed of sets that are also closed. Hence the Sorgenfrey line is zero-dimensional.

On the other hand, $\mathbb{R}$ with its usual convergence is at the other end of the spectrum: the only open subsets that are both open and closed are $\varnothing$ and $\mathbb{R}$, as we will see in Chapter XII when investigating in more details the role of closed and open subsets. In particular $\mathbb{R}$ is not zero-dimensional for its usual convergence.

Proposition V.3.26. *A set H is $f\xi$-closed if and only if $f^-(H)$ is ξ-closed.*

Proof. By (V.2.7) and Exercise V.3.13(2), a set H is $f\xi$-closed if and only if $f(\mathrm{adh}_\xi f^-(H)) \subset H$. This implies

$$\mathrm{adh}_\xi f^-(H) \subset f^- f(\mathrm{adh}_\xi f^-(H)) \subset f^-(H),$$

in other words, $f^-(H)$ is ξ-closed. Conversely, $\mathrm{adh}_\xi f^-(H) \subset f^-(H)$ implies that $f(\mathrm{adh}_\xi f^-(H)) \subset f f^-(H) \subset H$, that is, H is $f\xi$-closed. $\square$

It follows that:

Corollary V.3.27. *The preimage of a closed set by a continuous map is closed. The preimage of an open set by a continuous map is open.*

Proof. If f is a continuous map from ξ to τ, that is, $f\xi \geq \tau$, then each τ-closed set H is $f\xi$-closed, hence by Proposition V.3.26, $f^-(H)$ is ξ-closed. If P is τ-open, then $f^-(P) = (f^-(P^c))^c$, hence is ξ-open. $\qquad\square$

Recall that χ_A denotes the characteristic function of A, as defined on page 4.

Exercise V.3.28. Let (X, ξ) be a convergence space and let $A \subset X$. Show that:

(1) A is ξ-open if and only if

$$\chi_A \in C\left(\xi, \$_1\right);$$

(2) A is ξ-closed if and only if

$$\chi_A \in C\left(\xi, \$_0\right).$$

Exercise V.3.29. Let A be a subset of a T_1-convergence space X and consider the family $\mathcal{A} := \{\{x\} : x \in A\}$. Show that the following are equivalent:

(1) $\mathcal{A}$ is a locally finite family;
(2) $\mathcal{A}$ is a discrete family;
(3) A is a closed and discrete subspace of X.

V.4 Topologies

A pretopology is called a *topology* if the adherence is idempotent, that is, if

$$\text{adh}_\xi(\text{adh}_\xi A) = \text{adh}_\xi A$$

for each $A \subset |\xi|$ (equivalently, if

$$\text{inh}_\xi(\text{inh}_\xi A) = \text{inh}_\xi A$$

for each $A \subset |\xi|$).
   We will often use $\text{adh}^2(\cdot)$ and $\text{inh}^2(\cdot)$ as shorthands for $\text{adh}(\text{adh}(\cdot))$ and $\text{inh}(\text{inh}(\cdot))$.

Example V.4.1. The discrete pretopology ι is a topology on a set X, because $\text{adh}_\iota A = A$.

Example V.4.2. The chaotic pretopology o on X is a topology, because $\text{adh}_o A = X$ for each $A \neq \varnothing$ and $\text{adh}_o \varnothing = \varnothing$.

Example V.4.3. The Sierpiński pretopologies $\$_0$ and $\$_1$ on $\{0,1\}$ are topologies. For example,

$$\mathrm{adh}_{\$_1} \varnothing = \varnothing, \; \mathrm{adh}_{\$_1}\{0\} = \{0\},$$
$$\mathrm{adh}_{\$_1}\{1\} = \{0,1\} = \mathrm{adh}_{\$_1}\{0,1\}.$$

Proposition V.4.4. *A pretopology ξ is a topology if and only if one of the following equalities holds*

$$\mathrm{adh}_\xi A = \mathrm{cl}_\xi A$$
$$\mathrm{inh}_\xi A = \mathrm{int}_\xi A$$
$$\mathcal{V}_\xi(x) = \mathcal{N}_\xi(x)$$

for each $x \in X$ and each $A \subset |\xi|$.

Proof. If ξ is a topology, then adh_ξ is idempotent, so that, for each subset A of $|\xi|$, $\mathrm{adh}_\xi A$ is a closed subset containing A. As $\mathrm{adh}_\xi A \subset \mathrm{cl}_\xi A$ and $\mathrm{cl}_\xi A$ is the smallest set with this property, we conclude that $\mathrm{adh}_\xi A = \mathrm{cl}_\xi A$. In view of (V.2.5) and (V.3.9), we conclude that $\mathrm{inh}_\xi A = \mathrm{int}_\xi A$. In view of the definitions of $\mathrm{inh}_\xi A$ and of $\mathcal{N}_\xi(x)$, this means that $\mathcal{V}_\xi(x) = \mathcal{N}_\xi(x)$, which in turn, in view of (V.2.2) and (IV.1.1), means that $\mathrm{adh}_\xi A = \mathrm{cl}_\xi A$, so that adh_ξ is idempotent. $\qquad\square$

Corollary V.4.5. *A convergence ξ is a topology if and only if*

$$x \in \lim_\xi \mathcal{N}_\xi(x),$$

for each $x \in |\xi|$.

Example V.4.6 (A non-topological pretopology: the bisequence pretopology). Let $X = \{x_\infty\} \cup \{x_n : n < \infty\} \cup \{x_{n,k} : n,k < \infty\}$, where all the elements are distinct. We define a convergence π by $x_{n,k} \in \lim_\pi \mathcal{F}$ if $\mathcal{F} = \{x_{n,k}\}^\uparrow$, $x_n \in \lim_\pi \mathcal{F}$ if $\{x_n\}^\uparrow \wedge (x_{n,k})_k^\uparrow \leq \mathcal{F}$, and $x_\infty \in \lim_\pi \mathcal{F}$ provided that $\{x_\infty\}^\uparrow \wedge (x_n)_n^\uparrow \leq \mathcal{F}$.

The convergence π is a pretopology, because $x \in \lim \mathcal{V}_\pi(x)$ for each $x \in X$. Indeed, the elements of the form $x_{n,k}$ are isolated, hence $\{x_{n,k}\} \in \mathcal{V}_\pi(x_{n,k})$ for each $n,k < \infty$; $\mathcal{V}_\pi(x_n) = \{x_n\}^\uparrow \wedge (x_{n,k})_k^\uparrow$ for every $n < \infty$; $\mathcal{V}_\pi(x_\infty) = \{x_\infty\}^\uparrow \wedge (x_n)_n^\uparrow$.

The pretopology π is not a topology, because if O such that $x_\infty \in O$ is an open set, then there is n_0 such that $x_n \in O$ for every $n \geq n_0$ because $x_\infty \in \lim(x_n)_n^\uparrow$. For every $n \geq n_0$, $x_n \in O \cap \lim(x_{n,k})_k^\uparrow$, so that there is

$\kappa(n) < \infty$ such that $x_{n,k} \in O$ for each $k > \kappa(n)$. The neighborhood filter $\mathcal{N}_\pi(x_\infty)$, which is generated by sets of the form

$$V = \{x_\infty\} \cup \{x_n : n > n_0\} \cup \{x_{n,k} : k > \kappa(n), n > n_0\},$$

where $n_0 < \infty$ and $\kappa : \mathbb{N} \to \mathbb{N}$, does not converge to x_∞ in π.

Exercise V.4.7. Show that:

(1) The Féron cross pretopology induced by the premetric of Example I.2.1 is not a topology;
(2) The radial pretopology of Example V.2.13 is not a topology.

Exercise V.4.8. Show that the Sorgenfrey line convergence of Example IV.9.24 is a topology.

Proposition V.4.9. *Each prime pretopology is topological.*

Proof. We know that if an isolated point $x \in \operatorname{adh} A$ then $x \in A$. Let $x \in \operatorname{adh}^2 A$. If x is isolated then $x \in \operatorname{adh} A$. Suppose that $x_\infty \in \operatorname{adh}^2 A \setminus \operatorname{adh} A$, then $\mathcal{V}(x_\infty) \# \operatorname{adh} A$ but there is $V \in \mathcal{V}(x_\infty)$ with $V \cap A = \varnothing$, that is, $A \subset X \setminus V$. Thus A is composed of isolated points and thus $\operatorname{adh} A = A \cup \{x_\infty\}$ is closed. $\square$

Exercise V.4.10. Show that a prime topology on X with $\mathcal{N}(x_\infty) \geq (X)_0 \wedge \{x_\infty\}^\uparrow$ is zero-dimensional.

Example V.4.11 (ordinals as prime topological spaces). Let γ be an ordinal (see Appendix A.9). Define on $\gamma + 1 = \{\alpha : \alpha \leq \gamma\}$ a canonical pretopology, where each $\alpha < \gamma$ is isolated and V is a vicinity of γ if there exists $\alpha_V < \gamma$ such that $\{\alpha : \alpha_V < \xi \leq \gamma\} \subset V$. In view of Proposition V.4.9, this is a topology.

Example V.4.12 (*order topology* induced by a total order). Define on a totally order set $(X, \leq)$ the pretopology in which

$$\mathcal{V}(x) := \{(a, b) : a < x < b\}^\uparrow,$$

where $(a, b) := \{t \in X : a < t < b\}$. By definition, $(a, b) \in \bigcap_{t \in (a,b)} \mathcal{V}(t)$ so that $\mathcal{V}(x) = \mathcal{N}(x)$ for all $x \in X$. Thus the pretopology is in fact a topology.

Of course, the standard convergence of $\mathbb{R}$ is an instance of order topology.

Example V.4.13 (ω_1 with the order topology). Denote by ω_1 the first uncountable ordinal (see Appendix A.9). By Proposition A.9.3, we can view ω_1 as the set

$$\omega_1 := \{\alpha \in \mathrm{Ord} : \alpha < \omega_1\}$$

of countable ordinals. As a totally ordered set, it can be endowed with the order topology as in Example V.4.12. Note that for every $\alpha \in \omega_1$, the initial segment $\{\beta \in \mathrm{Ord} : \beta < \alpha\}$ is countable, and that countable subsets of ω_1 admit a supremum in ω_1. Thus any non-decreasing sequence of ω_1 converges (to its supremum) for the order topology. Note that in a well-ordered set, every sequence has a non-decreasing subsequence (Exercise A.8.2). Thus every sequence of ω_1 has a convergent subsequence.

Because every pretopology $\pi\,[x, \mathcal{F}]$ form the proof of Proposition III.8.10 is prime, hence topological by Proposition V.4.9, we get:

Corollary V.4.14. *Each finitely deep convergence is an infimum of topologies.*

Corollary V.4.15. *Prime topologies are surjectively finally dense in the class of finitely deep convergences.*

Exercise V.4.16. Show that a sum of topologies is a topology.

Thus, in view of Exercise V.4.16, Corollary IV.8.6 becomes:

Corollary V.4.17. *Every finitely deep convergence is the (convergence) quotient of a topology.*

Proposition V.4.18. *The Sierpiński topology is initially dense in the class of topologies.*

Proof. If τ is a topology, then A is τ-open if and only if the characteristic function χ_A is continuous from τ to $\$_1$ (see Exercise V.3.28). Hence

$$\tau = \bigvee_{f \in C(\tau, \$_1)} f^- \$_1,$$

because $x \in \lim_{\bigvee_{A \in \mathcal{O}_\tau} \chi_A^- \$_1} \mathcal{F}$ if $\chi_A(x) \in \lim_{\$_1} \chi_A[\mathcal{F}]$ for every open subset A of $|\tau|$, that is, if for every open set A with $x \in A$, $A \in \mathcal{F}$. In other words, $x \in \lim_{\bigvee_{A \in \mathcal{O}_\tau} \chi_A^- \$_1} \mathcal{F} \iff \mathcal{F} \geq \mathcal{N}_\tau(x)$. $\qquad\square$

Proposition V.4.19. *If (X, ξ) is a topological space then whenever a family $\mathcal{B}$ of subsets of X is a base for the convergence, the family ([2])*

$$\text{int}^\natural \mathcal{B} := \{\text{int } B : B \in \mathcal{B}\}$$

is another base for the convergence, composed of open sets.

Moreover, a family of open subsets is a base for the convergence if and only if it is a base of open sets.

Proof. Since ξ is a topology, $\{\mathcal{N}(x)\}$ is the only pavement at x. Hence, if $\mathcal{B}$ is a base for the convergence, $\mathcal{N}(x) \cap \mathcal{B}$ is a filter-base for $\mathcal{N}(x)$. Since $\text{int } B \in \mathcal{N}(x)$ whenever $B \in \mathcal{N}(x)$, $\mathcal{N}(x) \cap \text{int}^\natural \mathcal{B}$ is also a filter-base of $\mathcal{N}(x)$ and $\text{int}^\natural \mathcal{B}$ is a base for the convergence.

Assume $\mathcal{B}$ is a base of open sets. Since ξ is a topology, $\{\mathcal{N}(x)\}$ is a pavement at x, for each $x \in X$. Moreover $\mathcal{B} \cap \mathcal{N}(x)$ is a filter-base of $\mathcal{N}(x)$. Indeed for each $V \in \mathcal{N}(x)$ there is an open set O such that $x \in O \subset V$, and therefore, there is $B \in \mathcal{B}$ with $x \in B \subset O \subset V$, and $B \in \mathcal{N}(x)$. Hence $\mathcal{B}$ is a base for the convergence.

Conversely, if $\mathcal{B}$ is a base for the convergence then each $\mathcal{N}(x)$ has a filter-base composed of elements of $\mathcal{B}$. In particular, for each open set O and $x \in O$, there is $B \in \mathcal{B} \cap \mathcal{N}(x)$ with $x \in B \subset O$. Hence $\mathcal{B}$ is a base of open sets. $\square$

Exercise V.4.20. Deduce that a topology has weight at most κ if and only if it has a base of open sets of cardinality at most κ.

V.4.1 *Topological modification*

Proposition V.4.21. *A supremum of a non-empty set of topological convergences on the same set is a topology.*

Proof. Let $\mathcal{T}$ be a non-empty family of topologies on X. The convergence $\bigvee \mathcal{T}$ is a pretopology, because each topology is a pretopology and thus Corollary V.1.5 applies. If $x \in \text{adh}_{\bigvee \mathcal{T}} A$, then there is a filter $\mathcal{F}$ such that $A \in \mathcal{F}$ and $x \in \lim_\tau \mathcal{F}$ for each $\tau \in \mathcal{T}$, hence $x \in \text{adh}_\tau A = \text{cl}_\tau A$ for each $\tau \in \mathcal{T}$, showing that $\text{adh}_{\bigvee \mathcal{T}} A$ is closed for $\bigvee \mathcal{T}$. Hence $\bigvee \mathcal{T}$ is a topology. $\square$

---

[2]More generally if $\mathcal{B} \subset 2^X$ and $m : 2^X \to 2^X$ then

$$m^\natural \mathcal{B} := \{m(B) : B \in \mathcal{B}\}.$$

Corollary V.4.22. *For every convergence ξ there exists the finest topology on $|\xi|$, that is coarser than ξ.*

Proof. Let $\mathcal{T}$ be the set of topologies on X such that $\tau \leq \xi$ for each $\tau \in \mathcal{T}$. Then $\mathcal{T}$ is not empty, because the chaotic convergence o is a topology and therefore $o \in \mathcal{T}$. By Proposition V.4.21, $\bigvee \mathcal{T}$ and is the finest topology coarser than ξ.   □

The map T is called the *topologizer*. It is clear that open and closed sets, closure, interior and neighborhood filters are topological notions, that is, they are the same for a convergence ξ and $\mathrm{T}\,\xi$.

It follows from Proposition V.3.26 that T preserves continuity, that is,

$$C(\xi, \tau) \subset C(\mathrm{T}\,\xi, \mathrm{T}\,\tau). \qquad (\mathrm{V}.4.1)$$

Exercise V.4.23. Show that the following are equivalent:

(1) (V.4.1) holds for every convergences ξ and τ;
(2) For every map f and convergence τ on the codomain

$$\mathrm{T}\left(f^- \tau\right) \geq f^-(\mathrm{T}\,\tau);$$

(3) For every map f and convergence ξ on the domain

$$f(\mathrm{T}\,\xi) \geq T(f\xi). \qquad (\mathrm{V}.4.2)$$

On summarizing,

Proposition V.4.24. *The topologizer T is isotone, contractive and idempotent, i.e.,*

$$\zeta \leq \xi \Rightarrow \mathrm{T}\,\zeta \leq \mathrm{T}\,\xi; \qquad \text{(isotone)}$$
$$\mathrm{T}\,\xi \leq \xi; \qquad \text{(contractive)}$$
$$\mathrm{T}(\mathrm{T}\,\xi) = \mathrm{T}\,\xi. \qquad \text{(idempotent)}$$

Moreover, for each map f and a convergence τ on the range space of f,

$$\mathrm{T}\left(f^- \tau\right) \geq f^-(\mathrm{T}\,\tau). \qquad (\mathrm{V}.4.3)$$

Since each topology is in particular a pretopology, and $\mathrm{S}_0\,\xi$ is the finest pretopology coarser than ξ, it follows that

$$\xi \geq \mathrm{S}_0\,\xi \geq \mathrm{T}\,\xi \qquad (\mathrm{V}.4.4)$$

for every convergence ξ.

In view of Propositions V.4.24 and V.4.4, we conclude that closure, interior and neighborhood are topological notions:

$$\mathrm{cl}_\xi = \mathrm{cl}_{\mathrm{T}\xi}$$
$$\mathrm{int}_\xi = \mathrm{int}_{\mathrm{T}\xi}$$
$$\mathcal{N}_\xi(\cdot) = \mathcal{N}_{\mathrm{T}\xi}(\cdot).$$

Proposition V.4.25. *Let ξ be a convergence. Then*

$$x \in \lim_{\mathrm{T}\xi} \mathcal{F} \iff \mathcal{F} \geq \mathcal{N}_\xi(x)$$
$$\iff x \in \bigcap_{A \in \mathcal{F}^{\#}} \mathrm{cl}_\xi A.$$

Proof. The convergence σ defined by

$$x \in \lim_\sigma \mathcal{F} \iff \mathcal{F} \geq \mathcal{N}_\xi(x),$$

is a pretopology that satisfies $\mathcal{V}_\sigma(x) = \mathcal{N}_\xi(x)$ and is therefore a topology with $\mathcal{V}_\sigma(x) = \mathcal{N}_\sigma(x) = \mathcal{N}_\xi(x)$. Hence $\sigma = \mathrm{T}\,\xi$.

As a result of Exercise II.3.3(4), $x \in \lim_{\mathrm{T}\xi} \mathcal{F}$ if and only if $\mathcal{F}^{\#} \subset \mathcal{N}_\xi(x)^{\#}$. In view of (V.3.11), this is equivalent to $x \in \bigcap_{A \in \mathcal{F}^{\#}} \mathrm{cl}_\xi A$. □

In particular, the collection of open subsets of a convergence space (X, ξ) completely determines its topological modification $\mathrm{T}\,\xi$. Hence, notions defined solely in terms of open sets, like zero-dimensionality, are topological notions, in the sense that a convergence has the property if and only if its topological modification does.

Similarly, the collection of closed subsets, the closure operator cl_ξ, the interior operator int_ξ and the collection of neighborhood filters $\mathcal{N}_\xi(x)$ all completely characterize the topology $\mathrm{T}\,\xi$ and can be seen as equivalent descriptions of the topology.

Example V.4.26. The Bourdaud pretopology of Example V.1.13 is not a topology. Indeed, $\mathrm{adh}_{\Upsilon} \{0\} = \{0, 1\}$ and $\mathrm{adh}_{\Upsilon} \{0, 1\} = \{0, 1, 2\}$, so that $\mathrm{adh}_{\Upsilon}^2 \{0\} \neq \mathrm{adh}_{\Upsilon} \{0\}$.

Exercise V.4.27. Show that the topological modification $\mathrm{T}\,\Upsilon$ of the Bourdaud pretopology Υ as well as that of the convergence IV(3) of Figure III.3 are the chaotic topology o of $\{0, 1, 2\}$.

In view of Exercises V.4.27 and III.2.1, a convergence may be T_0 while its topological modification is not. In fact,

Proposition V.4.28. *If ξ is a convergence, then $\mathrm{T}\,\xi$ is T_0 if and only if for every pair of different points, there is an open subset of $|\xi|$ that contains one but not the other.*

Proof. Since $x \in \lim_{\mathrm{T}\,\xi} \mathcal{F}$ if and only if $\mathcal{F} \geq \mathcal{N}_\xi(x)$, we conclude that $\mathrm{T}\,\xi$ is T_0 if and only if

$$x \neq y \implies \mathcal{N}_\xi(x) \neq \mathcal{N}_\xi(y),$$

which is equivalent to the desired condition, for neighborhood filters of a filter-base composed of open sets. $\square$

On the other hand,

Exercise V.4.29. Show that $\mathrm{T}\,\xi$ is Hausdorff if and only if for every $x \neq y$ in $|\xi|$, there are two *disjoint* open sets U and V with $x \in U$ and $y \in V$.

Despite Example III.1.14, we have:

Exercise V.4.30. Show that if $S_0\,\xi$ is T_0, so is ξ.

We will see that the same holds for the topological modification T.

In view of Proposition V.2.18 and the results of the last two sections, the following should come as no surprise:

Exercise V.4.31. Show that the following are equivalent, for any pair of convergences ξ and τ and $f : |\xi| \to |\tau|$:

(1) $f \in C\,(\mathrm{T}\,\xi, \mathrm{T}\,\tau)$;
(2) for all $x \in |\xi|$,

$$f[\mathcal{N}_\xi(x)] \geq \mathcal{N}_\tau(f(x));$$

(3) for all $A \subset |\xi|$,

$$f(\mathrm{cl}_\xi\, A) \subset \mathrm{cl}_\tau(f(A));$$

(4) for all $B \subset |\tau|$,

$$f^-(\mathrm{int}_\tau\, B) \subset \mathrm{int}_\xi(f^-(B));$$

(5) $f^-(O)$ is ξ-open whenever O is τ-open;

(6) $f^-(C)$ is ξ-closed whenever C is τ-closed.

Proposition V.4.32. *The initial convergence of a topology is a topology.*

Proof. If $\tau = \mathrm{T}\tau$ then (V.4.3) gives

$$\mathrm{T}(f^-\tau) \geq f^-(\mathrm{T}\tau) = f^-\tau,$$

which gives topologicity of $f^-\tau$, as the reverse inequality is always true. $\square$

Since a supremum of topologies is a topology, we obtain:

Corollary V.4.33. *Subspaces and products of topologies are topologies.*

V.4.2 Induced topology

Corollary V.4.33 shows that the convergence induced by a topology is a topology, classically called *induced topology*.

Exercise V.4.34 (Induced topology). Let (X, ξ) be a topological space, and let $A \subset X$.

(1) Show that $U \subset A$ is $\xi_{|A}$-open if and only if there is a ξ-open subset O of X with $U = O \cap A$;
(2) Show that $F \subset A$ is $\xi_{|A}$-closed if and only if there is a ξ-closed subset C of X with $F = C \cap A$;
(3) Show that if $B \subset A$ then $\mathrm{cl}_{\xi_{|A}} B = \mathrm{cl}_\xi B \cap A$.

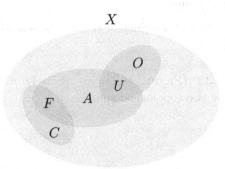

In contrast to Corollary V.2.24 however, the convergence induced by the topological modification may fail to coincide with the topological modification of the induced convergence. Of course, since a subspace of a topological space is topological, $(\mathrm{T}\,\xi)_{|A}$ is a topology on A coarser than $\xi_{|A}$, so that

$$(\mathrm{T}\,\xi)_{|A} \leq \mathrm{T}(\xi_{|A}),$$

but the reverse inequality may fail:

Example V.4.35 (A subset A of a pretopological space π such that $(\mathrm{T}\,\pi)_{|A} \not\geq \mathrm{T}(\pi_{|A})$). Let π denote the bisequence pretopology of Example V.4.6, and let

$$A := \{x_{n,k} : n, k \in \mathbb{N}\} \cup \{x_\infty\}.$$

Since $A \in \mathcal{N}_\pi(x_\infty)^\#$ the topology $(\mathrm{T}\,\pi)_{|A}$ is not discrete, and $\mathcal{N}_{(\mathrm{T}\,\pi)_{|A}}(x_\infty) = \mathcal{N}_\pi(x_\infty) \vee A$. On the other hand, $\pi_{|A} = \mathrm{T}(\pi_{|A})$ is the discrete topology.

On the other hand,

Proposition V.4.36. *If A is a closed subset of a convergence space (X, ξ) then $(\mathrm{T}\,\xi)_{|A} = \mathrm{T}(\xi_{|A})$.*

Exercise V.4.37. Prove Proposition V.4.36.

V.4.3 *Product topology*

By Corollary V.4.33, each product of topologies is a topology, classically called *product topology*.

Exercise V.4.38 (Product topology (finitely many factors)). Let (X, ξ) and (Y, τ) be two topological spaces.

(1) Show that

$$\{U \times V : U \in \mathcal{O}_\xi, V \in \mathcal{O}_\tau\}$$

form a base for open sets in $\xi \times \tau$;

(2) Show that if C is ξ-closed and F is τ-closed then $C \times F$ is $(\xi \times \tau)$-closed.

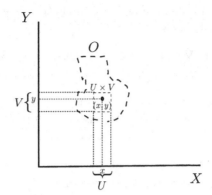

Exercise V.4.39 (product topology (infinitely many factors)). Let Ξ be an infinite set of topological spaces. Show that

$$\left\{ \prod_{X\in\Xi} U_X : U_X \in \mathcal{O}_X, \{X \in \Xi : U_X \neq X\} \in [\Xi]^{<\omega} \right\}$$

form a base for open sets in the product topology $\prod_{X\in\Xi} X$.

Observe that

$$\prod_{\xi\in\Xi} \mathrm{T}\,\xi \leq \mathrm{T}(\prod_{\xi\in\Xi} \xi),$$

because $\prod_{\xi\in\Xi} \mathrm{T}\,\xi$ is a topology, as a product of topologies, and is coarser than $\prod_{\xi\in\Xi} \xi$, as $\mathrm{T}\,\xi \leq \xi$ for each ξ.

The converse however is not valid even for two convergences, one of which is a topology.

Example V.4.40 ($\mathrm{T}\,\sigma \times \tau \not\geq \mathrm{T}(\sigma \times \tau)$.). Consider the situation of Example V.2.26. Note that in view of Proposition V.4.9, τ and $\mathrm{S}_0\,\sigma$ are both topologies. In particular $\mathrm{S}_0\,\sigma = \mathrm{T}\,\sigma$. The desired inequality therefore follows from the observation that Δ was shown to be $(\sigma \times \tau)$-closed, and from Proposition V.4.25.

Let us revisit density in the context of topological spaces, and product thereof.

Lemma V.4.41. *Let (X,ξ) be a convergence space. Then D is a dense subset for $\mathrm{T}\,\xi$ if and only if*

$$\mathrm{cl}\,O = \mathrm{cl}\,(O \cap D),$$

for every non-empty ξ-open subset O of X.

Proof. As D is $\top\xi$-dense if and only if $\mathrm{cl}_\xi D = X$, that is, $\mathcal{N}(x)\#D$ for all $x \in X$, the result follows. $\qquad\square$

Proposition V.4.42. *If D is a dense subspace of a topological space X and C is a dense subspace of D, then C is a dense subspace of X.*

Proof. We apply Lemma V.4.41. If O is a non-empty open subset of X, then $O \cap D$ is a non-empty (by density) open subset of D and thus $\mathrm{cl}_D(O \cap D) = \mathrm{cl}_D(O \cap C)$. Thus, $\mathrm{cl}_X(O \cap C) = \mathrm{cl}_X(O \cap D) = \mathrm{cl}_X O$. $\qquad\square$

Lemma V.4.43. *If X is a discrete space of cardinality κ, then the density of $\prod_{t \in 2^\kappa} X$ (with the product topology) is κ.*

Proof. The product $\prod_{t \in \kappa} \{0,1\}$, where $\{0,1\}$ is discrete, admits a base open sets $\mathcal{B}$ of cardinality κ, for example, the one that consists of $\{f \in \prod_{t \in \kappa} \{0,1\} : f(t_i) = \alpha_i, i = 1, \ldots, n\}$, where $n < \omega, \alpha_i \in \{0,1\}$ and $t_i \in \kappa$. Let $\mathcal{T}$ the set of finite sequences $\{B_1, \ldots, B_n\}$ of disjoint elements of $\mathcal{B}$ ($n < \omega, B_k \in \mathcal{B}$ if $1 \le k \le n$). Then card $\mathcal{T} = \kappa$. Let $\mathcal{A}$ be the set of elements F of

$$\prod_{t \in 2^\kappa} X \approx X^{2^\kappa}$$

such that there exists $n < \omega$ and $\{B_1, \ldots, B_n\} \in \mathcal{T}$, such that F is constant on B_k for $1 \le k \le n$ as well as on $2^\kappa \backslash \bigcup_{k=1}^n B_k$. Then $\mathcal{A}$ is dense in $\prod_{t \in 2^\kappa} X$.

Indeed, a base of open subsets of $\prod_{t \in 2^\kappa} X$ consists of

$$\mathcal{F} := \left\{ F \in \prod_{t \in 2^\kappa} X : \underset{k=1,\ldots,n}{\forall} F(t_k) = x_k \right\},$$

where $n < \omega, t_1, \ldots, t_n \in 2^\kappa$ and $x_1, \ldots, x_n \in X$. As 2^κ is Hausdorff, there exists $\{B_1, \ldots, B_n\} \in \mathcal{T}$ such that $t_k \in B_k$ for $1 \le k \le n$. Hence there exists $F \in \prod_{t \in 2^\kappa} X$ such that $F^{-1}(x_k) = B_k$ pour $1 \le k \le n$, and thus $F \in \mathcal{A} \cap \mathcal{F}$. $\qquad\square$

Theorem V.4.44 (Hewitt-Marczewski-Pondiczery). *Let X_t be a topological space for each $t < 2^\kappa$, where κ is infinite. If $\mathrm{d}(X_t) \le \kappa$ for each $t \in 2^\kappa$, then $\mathrm{d}(\prod_{t \in 2^\kappa} X_t) \le \kappa$.*

Proof. For each $t \in 2^\kappa$ there exists a dense subset A_t of X_t such that card $A_t = \kappa$. Of course, $\prod_{t \in 2^\kappa} A_t$ is dense in $\prod_{t \in 2^\kappa} X_t$. It suffices to show that $\prod_{t \in 2^\kappa} A_t$ has a dense subset of cardinality κ. If Y_t is a topological space of cardinality κ for each $t \in 2^\kappa$, then there exists a continuous bijection of $\prod_{t \in 2^\kappa} Y_t$ onto $\prod_{t \in 2^\kappa} A_t$. But the density of $\prod_{t \in 2^\kappa} Y_t$ is κ, by Lemma V.4.43. $\qquad\square$

In particular,

Corollary V.4.45. *The Cantor cube* $\prod_{t \in \mathfrak{c}} \{0, 1\}$ *(of dimension* $\mathfrak{c}$*) is separable. Its weight is* $\mathfrak{c}$ *and its cardinality is* $2^{\mathfrak{c}}$.

V.5   Open maps and closed maps

Definition V.5.1. A map $f : |\xi| \to |\tau|$ is *open* if $f(U)$ is τ-open whenever U is ξ-open and *closed* if $f(C)$ is τ-closed whenever C is ξ-closed.

By definition, f is open (respectively, closed) if and only if $f : |T\,\xi| \to |T\,\tau|$ is. Therefore, we can restrict ourselves to topologies when studying such maps.

Exercise V.5.2. Let $f : |\xi| \to |\tau|$. Show that

(1) f is open if and only if

$$f(\mathrm{int}_\xi A) \subset \mathrm{int}_\tau f(A)$$

for every $A \subset |\xi|$.
(2) f is closed if and only if

$$\mathrm{cl}_\tau (f(A)) \subset f(\mathrm{cl}_\xi A)$$

for every $A \subset |\xi|$.

Exercise V.5.3. Show that a continuous bijection f between topological spaces is an homeomorphism if and only if f is open if and only if f is closed.

However, the notions of open, of closed, and of continuous map are independent in general. It is not surprising, because continuity (between topological spaces) can be expressed in terms of the preimages of open sets, equivalently closed sets, while open and closed maps are defined in terms of their images.

As the preimage of a complement of a set is equal to the complement of the preimage, continuity conditions using open sets are equivalent to the corresponding ones for closed sets. As the image of the complement is not equal to the complement of the image even for surjections, the concepts of open and of closed map are different.

Example V.5.4 (A continuous map that is neither open nor closed). Let $f : \mathbb{R} \to \mathbb{R}$ where $\mathbb{R}$ carries its usual topology be defined by

$$f(x) := \frac{1}{1+x^2}.$$

This is clearly a continuous map. Then its range $f(\mathbb{R}) = (0,1]$ is neither open nor closed, even though the whole space $\mathbb{R}$ is both open and closed. Thus f is neither open nor closed.

Example V.5.5 (A map that is both open and closed, but not continuous). Recall that the unit circle can be represented as $S_1 := \{e^{2i\pi t} : t \in [0,1)\}$. Consider the map $f : S_1 \to [0,1)$ defined by $f(e^{2i\pi t}) = t$, where S_1 is endowed with the topology induced by the natural topology of $\mathbb{R}^2$ and $[0,1)$ carries the topology induced by the natural topology of $\mathbb{R}$. This map is not continuous because $f^{-1}\left([0,\frac{1}{2})\right)$ is not open in S_1 even though $[0,\frac{1}{2})$ is an open subset of $[0,1)$. But f is closed and open because f^{-1} is continuous, so that $f(C) = \left(f^{-1}\right)^{-1}(C)$ is closed (respectively open) for every closed (respectively open) subset C of S_1.

Exercise V.5.6. Show that a function $f : |\xi| \to |\iota_Y|$ is always open and closed, but not always continuous.

Proposition V.5.7. *A projection* $p : |\xi \times \tau| \to |\xi|$, *defined by* $p(x,y) = x$, *is always open.*

Exercise V.5.8. Prove Proposition V.5.7.

Example V.5.9 (A map that is open but not closed). In view of Proposition V.5.7, the projection $p : \mathbb{R} \times \mathbb{R} \to \mathbb{R}$ defined by $p(x,y) = x$ is open. However, it is not closed, for

$$A := \left\{ \left(x, \frac{1}{x}\right) : x \neq 0 \right\}$$

is closed in $\mathbb{R} \times \mathbb{R}$ but $p(A) = \mathbb{R} \setminus \{0\}$ is not.

Example V.5.10 (A map that is closed but not open). Let $f : \mathbb{R} \to \mathbb{R}$ where $\mathbb{R}$ carries its usual topology be defined by $f(x) = x^2$. It is not open for $f((-1,1)) = [0,1)$ is not open. On the other hand, it is closed, because if F is a closed subset of $\mathbb{R}$, and $y \in \operatorname{cl} f(F)$, then there exists a sequence $\{y_n\}_n$ of elements of $f(F)$ such that $y \in \lim (y_n)_n$ then there exists a sequence $\{x_n\}_n$ of elements of F such that $y_n = f(x_n)$. As each convergent sequence in a metric space is bounded, there is $r > 0$ such that $0 \leq y_n \leq r$ and thus $|x_n| \leq \sqrt{r}$ for each $n \in \mathbb{N}$. Therefore there exists a subsequence $\{x_{n_k}\}_n$ and $x \in \mathbb{R}$ with $x \in \lim (x_{n_k})_k$. As F is closed, $x \in F$ and, since f is continuous, $y = f(x) \in f(F)$ proving that $f(F)$ is closed.

Let us revisit Corollary IV.6.3 in the case of topologies, in light of Exercise V.5.3.

We say that a family $\boldsymbol{F}$ of maps with common topological domain X *separates points from closed sets* if for each closed $A \subset X$ and $x \notin A$ there exists $f \in \boldsymbol{F}$ such that $f(x) \notin \mathrm{cl}_{Y_f} f(A)$. We say that f *separates points from closed sets* if $\boldsymbol{F} = \{f\}$ does.

Lemma V.5.11. *If $\boldsymbol{F}$ is a family of maps with common topological domain X that separates points from closed sets, then the diagonal map $\{\Delta \boldsymbol{F}\}$ does too.*

Proof. Indeed, if $(\Delta \boldsymbol{F})(x) \in \mathrm{cl}_{\prod_{f \in \boldsymbol{F}} v_f}(\Delta \boldsymbol{F})(A)$ then for each $f \in \boldsymbol{F}$,

$$f(x) \in p_{Y_f}(\mathrm{cl}_{\prod_{f \in \boldsymbol{F}} v_f}(\Delta \boldsymbol{F})(A))$$
$$\subset \mathrm{cl}_{v_f} p_{Y_f}(\Delta \boldsymbol{F})(A)) = \mathrm{cl}_{v_f} f(A).$$

$\square$

Proposition V.5.12. *If v is a topology, ζ is a T_0-topology and $f \in C(\zeta, v)$ separates points from closed sets, then f is an embedding.*

Proof. If ζ is T_0-topology and f separates points from closed sets, then f separates points and thus is injective. Indeed, if $x_0 \neq x_1$ then there is a ζ-closed set A such that either $x_0 \in A$ and $x_1 \notin A$ or vice versa. In the first case for example, $f(x_1) \notin \mathrm{cl}_v f(A)$ and $f(x_0) \in f(A) \subset \mathrm{cl}_v f(A)$, so that $f(x_1) \neq f(x_0)$.

By Exercise V.5.3, we only need to show that f is closed. To this end, let $A = \mathrm{cl}_\zeta A$ and let $y \notin f(A)$, that is, $f^-(y) \cap A = \varnothing$. Because f is injective, there is $x \in X$ such that $\{x\} = f^-(y)$. By assumption, $y = f(x) \notin \mathrm{cl}_v f(A)$, which proves that $f(A)$ is v-closed. $\square$

V.6 Topological defect and sequential order

This section uses few basic facts on ordinal numbers that can be found in Appendix A.9.

V.6.1 *Iterated adherence and topological defect*

If ξ is a convergence on X, then for every $A \subset X$,

$$A \subset \mathrm{adh}_\xi A \subset \mathrm{cl}_\xi A,$$

which means that the adherence and the closure are expansive maps (on 2^X), thus can be iterated.

Given an ordinal γ we define the γ-*iterate* of the adherence $\mathrm{adh}_\xi\, A$ by $\mathrm{adh}_\xi^0\, A = A, \mathrm{adh}_\xi^1\, A = \mathrm{adh}_\xi\, A$ and if $\gamma > 1$,

$$\mathrm{adh}_\xi^\gamma\, A = \mathrm{adh}_\xi(\bigcup_{\alpha<\gamma} \mathrm{adh}_\xi^\alpha\, A). \tag{V.6.1}$$

In particular, $\mathrm{adh}_\xi^n\, A = \mathrm{adh}_\xi(\mathrm{adh}_\xi^{n-1}\, A)$ for every natural number n. The γ-iterate of the closure is defined in the same way, but $\mathrm{cl}^\gamma\, A = \mathrm{cl}\, A$ for each $\gamma > 0$, because the closure is idempotent. It follows that, for two ordinal numbers $\alpha \le \gamma$,

$$0 \le \alpha \le \gamma \Longrightarrow \mathrm{adh}_\xi^\alpha\, A \subset \mathrm{adh}_\xi^\gamma\, A \subset \mathrm{cl}_\xi\, A. \tag{V.6.2}$$

Example V.6.1 (Iterated adherence in the Féron cross pretopology). Consider the *Féron cross pretopology* induced on the plane by the premetric of Example I.2.1, and let

$$A := \{(x,y) \in \mathbb{R}^2 : y = x,\, x \in (0,1)\}.$$

Note that A is closed because "crosses" centered at (0,0) or at (1,1) do not intersect A. On the other hand

$$B := \{(x,y) \in \mathbb{R}^2 : 0 < x < y < 1\}$$

is not closed, and

$$\mathrm{adh}\, B = \{(x,y) \in \mathbb{R}^2 : 0 \le x \le y \le 1\} \setminus \{(0,0),(0,1),(1,1)\}$$
$$\mathrm{adh}^2\, B = \{(x,y) \in \mathbb{R}^2 : 0 \le x \le y \le 1\} \text{ is closed.}$$

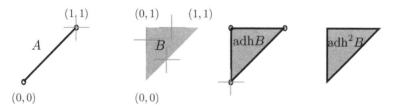

It follows from (V.6.2) that

Proposition V.6.2. *Let ξ be a convergence on X. For every $A \subset X$, there exists a least ordinal δ such that*

$$\mathrm{adh}_\xi^\delta\, A = \mathrm{adh}_\xi^{\delta+1}\, A. \tag{V.6.3}$$

Proof. If (V.6.3) did not hold for $\delta >$ card(X), then for each $1 < \gamma < \delta$, there would exist $x_\gamma \in \mathrm{adh}_\xi^\gamma A \setminus \bigcup_{\alpha < \gamma} \mathrm{adh}_\xi^\alpha A$, which is impossible because the cardinality of $\{x_\gamma \in X : \gamma < \delta\}$ is greater than card(X). $\qquad\square$

Of course, if (V.6.3) holds then $\mathrm{adh}_\xi^\delta A = \mathrm{adh}_\xi^{\delta+\eta} A$ for every ordinal η; therefore $\mathrm{adh}_\xi^\delta A$ is ξ-closed, because

$$\mathrm{adh}_\xi(\mathrm{adh}_\xi^\delta A) = \mathrm{adh}_\xi^{\delta+1} A \subset \mathrm{adh}_\xi^\delta A.$$

Definition V.6.3. The least ordinal δ for which (V.6.3) holds for every $A \subset X$, is called the *topological defect* of ξ and is denoted by $\mathrm{t}(\xi)$.

Of course, a pretopology is a topology if and only if its topological defect is at most 1. Incidentally, a pretopology has the topological defect 0 if and only if it is a discrete topology. More generally:

Proposition V.6.4. *A convergence ξ has topological defect at most 1 if and only if $\mathrm{S}_0\,\xi = \mathrm{T}\,\xi$. In particular, a convergence is a topology if and only if it is a pretopology and has topological defect at most 1.*

Note that while the set B of Example V.6.1 is closed after two iterations of the adherence, this does not mean that the topological defect of the Féron cross pretopology is 2 ([3]).

The notion of topological defect is global. Sometimes it is insightful to consider a related pointwise concept.

The *topological defect* $\mathrm{t}(x, \xi)$ *of* ξ *at* x is the supremum of ordinals γ such that there exists a set A for which $x \in \mathrm{cl}_\xi A \setminus \mathrm{adh}_\xi^\alpha A$ for every $\alpha < \gamma$. We notice that

$$\mathrm{t}(\xi) = \sup_{x \in |\xi|} \mathrm{t}(x, \xi).$$

The topological defect of ξ is said to be *attained* if the supremum above is attained, that is, if there is $x_0 \in |\xi|$ for which $\mathrm{t}(\xi) = \mathrm{t}(x_0, \xi)$. Of course, the topological defect can be non-attained only if it is a limit ordinal.

Example V.6.5. The discrete topology has the (attained) topological defect 0, and every other topology is of (attained) topological defect 1.

Proposition V.6.6. *If ξ is a convergence, then the topological defect $\mathrm{t}(\xi)$ is the least γ such that*

$$\mathrm{cl}_\xi A = \mathrm{adh}_\xi^\gamma A$$

for every A.

---

[3]In fact, it can be shown [51] to be ω_1.

Proof. As $A \subset \text{cl}_\xi A$, also $\text{adh}_\xi^\alpha A \subset \text{cl}_\xi A$ for every α, and thus in particular $\text{adh}_\xi^{t(\xi)} A \subset \text{cl}_\xi A$. But $\text{adh}_\xi^{t(\xi)} A$ is closed, because $\text{adh}_\xi^{t(\xi)+1} A = \text{adh}_\xi(\text{adh}_\xi^{t(\xi)} A) \subset \text{adh}_\xi^{t(\xi)} A$, and thus $\text{cl}_\xi A \subset \text{adh}_\xi^{t(\xi)} A$. $\square$

Example V.6.7. Consider the bisequence pretopology of Example V.4.6. The topological defect at each isolated point is 0. The topological defect at each point x_n is 1 because $x_n \in \text{cl} A \setminus A$ if and only if $x_n \in \text{adh} A \setminus A$. On the other hand, the topological defect at x_∞ is 2.

Indeed, $x_\infty \in \text{cl} A \setminus A$ if and only if

$$\underset{n \in \mathbb{N}}{\forall} \underset{\kappa:\mathbb{N}\to\mathbb{N}}{\forall} A \cap \left(\bigcup_{p \geq n} \{x_{p,k} : k \geq \kappa(p)\} \right) \neq \varnothing. \tag{V.6.4}$$

In particular, A intersects infinitely many branches $I_n := \{x_{n,k} : k \in \mathbb{N}\}$ along an infinite set. Otherwise, the cardinality of

$$J := \{n \in \mathbb{N} : \text{card}(A \cap I_n) = \infty\}$$

is finite, so that, picking $n > \max J$, there is, for each $p \geq n$, an integer $\kappa(p)$ such that $A \cap \{x_{p,k} : k \geq \kappa(p)\} = \varnothing$, which would contradict (V.6.4).

Hence the cardinality of J is infinite, and for each $n \in J$, $x_n \in \text{adh} A$ because $A \cap I_n$ is infinite and therefore $(x_{n,k})_k$ has a subsequence on A. Since $\text{adh} A$ contains infinitely points of $\{x_n : n \in \mathbb{N}\}$, $x_\infty \in \text{adh}(\text{adh} A)$. If A is for instance $\{x_{n,k} : n, k \in \mathbb{N}\}$ then $x_\infty \in \text{adh}^2 A \setminus \text{adh} A$ because $A \notin (x_n)_n^\#$.

Proposition V.6.8. *For every ordinal γ, there exists a pretopology with the attained topological defect γ.*

Proof. We proceed by transfinite induction (see Theorem A.9.9). By Example V.6.5, there exist pretopologies of topological defect 0 and 1. Let γ be an ordinal greater than 1, and assume that for every $\alpha < \gamma$, there exists a pretopology π_α on X_α with the topological defect γ attained at $x_\alpha \in X_\alpha$. Consider a disjoint union

$$X_\gamma := \bigcup_{\alpha < \gamma} X_\alpha \cup \{x_\gamma\}.$$

We define on X_γ a pretopology π_γ so that $\mathcal{V}_{\pi_\alpha}(x)$ is a filter-base of $\mathcal{V}_{\pi_\gamma} x$ if $x \in X_\alpha$, and $W \in \mathcal{V}_{\pi_\gamma}(x_\gamma)$ if there exists $\alpha < \gamma$ such that

$$\{x_\beta : \alpha < \beta < \gamma\} \cup \{x_\gamma\} \subset W.$$

Then the topological defect of π_γ is γ and is attained at x_γ. Indeed, let A_α be a subset of X_α such that $x_\alpha \in \mathrm{cl}_{\pi_\alpha} A_\alpha \setminus \mathrm{adh}_{\pi_\alpha}^\delta A_\alpha$ (^{4}) for every $\delta < \alpha$ and set $A = \bigcup_{\alpha < \gamma} A_\alpha$. If $\beta < \gamma$, then

$$\mathrm{adh}_{\pi_\gamma}^\beta A \cap (\{x_\delta : \beta < \delta < \gamma\} \cup \{x_\gamma\}) = \varnothing$$

and thus $x_\gamma \notin \mathrm{adh}_{\pi_\gamma}^\beta A$. On the other hand, $\bigcup_{\beta < \gamma} \mathrm{adh}_{\pi_\gamma}^\beta A \supset \{x_\beta : \beta < \gamma\}$, hence

$$\mathrm{adh}_{\pi_\gamma}^\gamma A = \mathrm{adh}_{\pi_\gamma} (\bigcup_{\beta < \gamma} \mathrm{adh}_{\pi_\gamma}^\beta A) = X_\gamma.$$

$\square$

It is interesting to notice that despite Proposition V.6.8, the following procedure produces an open set in countably many steps.

Proposition V.6.9. *If π is a pretopology, $x_0 \in |\pi|$, $W_1 \in \mathcal{V}_\pi(x_0)$ and*

$$W_{k+1} \in \mathcal{V}_\pi(W_k) := \bigcap_{x \in W_k} \mathcal{V}_\pi(x)$$

for each $k \geq 1$, then $\bigcup_{k \in \mathbb{N}} W_k$ is open.

Proof. If $x \in \bigcup_{k \in \mathbb{N}} W_k$, then there is $k < \omega$ such that $x \in W_k$, hence $x \in \mathrm{inh}\, W_{k+1} \subset \mathrm{inh}(\bigcup_{k \in \mathbb{N}} W_k)$. $\square$

Of course, each open set could be obtained via this procedure (in fact if O is open and we pick $W_1 = O$ the procedure stops immediately). However, it should be noted that to produce all open subsets containing a given point x_0, we have to repeat this procedure for every possible choice of $W_{k+1} \in \mathcal{V}(W_k)$ at each step k of the iteration.

V.6.2 *Sequentially based convergence and sequential order*

Following Example III.1.17, we associate to each convergence τ its *sequentially based modification* $\mathrm{Seq}\,\tau$ defined by $x \in \lim_{\mathrm{Seq}\,\tau} \mathcal{F}$ if $\mathcal{F} \geq (x_n)_n$ for some sequence $\{x_n\}_n$ such that $x \in \lim_\tau (x_n)_n^\uparrow$.

Definition V.6.10. A convergence τ is called *sequentially based* if $\tau = \mathrm{Seq}\,\tau$, in other words, if at each point x the collection $\mathbb{E} \cap \lim_\tau^{-1}(x)$ of sequential filters converging to x is a pavement of τ at x.

Exercise V.6.11. Show that:

---

^{4}Our construction could be made with a precision that each X_α is a disjoint union $\{x_\alpha\} \cup A_\alpha$.

(1) the map Seq is isotone, idempotent and expansive ([5]);
(2) for every function f and convergence ξ on the domain of f,

$$f(\operatorname{Seq}\xi) \geq \operatorname{Seq}(f\xi); \qquad (V.6.5)$$

(3) Deduce that $f : |\operatorname{Seq}\xi| \to |\operatorname{Seq}\tau|$ is continuous whenever $f : |\xi| \to |\tau|$ is;
(4) Deduce from (2) that $f\xi$ is sequentially based whenever ξ is.

Definition V.6.12. The adherence associated to $\operatorname{Seq}\tau$ is called the *sequential adherence* of τ. The topological defect of $\operatorname{Seq}\tau$ is called the *sequential order* $\operatorname{so}(\tau)$ of τ.

Note that the sequential adherence $\operatorname{adh}_{\operatorname{Seq}\tau} A$ of a subset A is obtained as the set of all limits (for τ) of sequences (or rather sequential filters) on A.

Definition V.6.13. A subset of $|\tau|$ is *sequentially closed* if it is $\operatorname{Seq}\tau$-closed; *sequentially open* if it is $\operatorname{Seq}\tau$-open.

Of course, a set A is τ-closed whenever $\operatorname{adh}_\tau A \subset A$, and $\operatorname{Seq}\tau$-closed whenever $\operatorname{adh}_{\operatorname{Seq}\tau} A \subset A$. Since $\operatorname{Seq}\tau \geq \tau$ for every convergence τ, for every $A \subset |\tau|$,

$$\operatorname{adh}_{\operatorname{Seq}\tau} A \subset \operatorname{adh}_\tau A.$$

Therefore each τ-closed set is sequentially τ-closed.

As $\operatorname{adh}_{\operatorname{Seq}\tau}$ is not in general idempotent, $\operatorname{adh}_{\operatorname{Seq}\tau} A$ is not necessarily sequentially closed for a given $A \subset |\tau|$. The sequential order $\operatorname{so}(\tau)$ of τ is thus the least ordinal number γ such that $\operatorname{adh}^\gamma_{\operatorname{Seq}\tau} A$ is closed for each $A \subset |\tau|$. In other words, $\operatorname{so}(\tau)$ is the least ordinal for every subset A of $|\tau|$,

$$\operatorname{cl}_{\operatorname{Seq}\tau} A = \operatorname{adh}^{\operatorname{so}(\tau)}_{\operatorname{Seq}\tau} A.$$

Definition V.6.14. A convergence is *sequential* if sequentially closed sets are closed.

It follows immediately from the definition that

---

[5] That is,

$$\xi \leq \tau \Longrightarrow \operatorname{Seq}\xi \leq \operatorname{Seq}\tau, \qquad \text{(isotone)}$$
$$\operatorname{Seq}\operatorname{Seq}\xi = \operatorname{Seq}\xi, \qquad \text{(idempotent)}$$
$$\xi \leq \operatorname{Seq}\xi. \qquad \text{(expansive)}$$

Proposition V.6.15. *A convergence ξ is sequential if and only if* $\mathrm{cl}_\xi = \mathrm{cl}_{\mathrm{Seq}\,\xi}$, *that is, if* $\mathrm{T}\,\xi = \mathrm{T}\,\mathrm{Seq}\,\xi$, *which can be rephrased as*

$$\xi \geq \mathrm{T}\,\mathrm{Seq}\,\xi. \tag{V.6.6}$$

Exercise V.6.16. Show that the topological modification $\mathrm{T}\,\pi$ of the bisequence pretopology of Example V.4.6 is sequential of sequential order 2.

It turns out that a subspace of a sequential topology need not be a sequential topology. A topology is called *subsequential* if it is a subspace topology of a sequential topology.

Example V.6.17 (Arens topology). Let π be the bisequence pretopology of Example V.4.6. Consider on $A := \{x_{n,k} : n, k \in \mathbb{N}\} \cup \{\infty\}$ the induced topology $(\mathrm{T}\,\pi)_{|A}$. The topological space $(A, (\mathrm{T}\,\pi)_{|A})$ is often called the *Arens space*.

The subset $B := \{x_{n,k} : n, k \in \mathbb{N}\}$ of the topological space $(A, (\mathrm{T}\,\pi)_{|A})$ is sequentially closed, because it was shown in Example V.4.6 that the only free $\mathrm{T}\,\pi$-converging sequences on B converge to an element of the form x_n.

On the other hand, $\mathcal{N}_{(\mathrm{T}\,\pi)_{|A}}(\infty) = \{V \cap A : V \in \mathcal{N}_\pi(\infty)\}$, so that $B \in \mathcal{N}_{(\mathrm{T}\,\pi)_{|A}}(\infty)^{\#}$. In other words, $\infty \in \mathrm{cl}_{(\mathrm{T}\,\pi)_{|A}} B \setminus B$, hence B is not closed. In particular, $(\mathrm{T}\,\pi)_{|A}$ is a subsequential non-sequential topology.

Sequential convergences form a natural class to study for several reasons: by definition, it is the class of convergences for which the closure operator can be reconstructed from convergent sequences. They are also the convergences for which sequentially continuous functions to topological spaces are continuous.

Recall that a map f from a convergence ξ to another convergence τ is *sequentially continuous* if $f(x) \in \lim_\tau (f(x_n))_n^\uparrow$ whenever $x \in \lim_\xi (x_n)_n^\uparrow$.

Exercise V.6.18. Show that $f : |\xi| \to |\tau|$ is sequentially continuous if and only if $f : |\mathrm{Seq}\,\xi| \to |\mathrm{Seq}\,\tau|$ is continuous if and only if $f : |\mathrm{Seq}\,\xi| \to |\tau|$ is continuous.

Proposition V.6.19. *The following statements are equivalent for a convergence ξ:*

(1) ξ is sequential;
(2) for every convergence $\tau \leq \mathrm{T}\,\mathrm{Seq}\,\tau$, each sequentially continuous map from ξ to τ is continuous;

(3) for every topology τ each sequentially continuous map from ξ to τ is continuous.

Proof. (1) $\Longrightarrow$ (2): If f is sequentially continuous, that is, $f : |\operatorname{Seq}\xi| \to |\operatorname{Seq}\tau|$ is continuous, then $f : |\operatorname{T}\operatorname{Seq}\xi| \to |\operatorname{T}\operatorname{Seq}\tau|$ is also continuous by Exercise V.4.23. If ξ is sequential, that is, $\xi \geq \operatorname{T}\operatorname{Seq}\xi$, and $\tau \leq \operatorname{T}\operatorname{Seq}\tau$, then $f : |\xi| \to |\tau|$ is continuous by Exercise IV.1.5.

(2) $\Longrightarrow$ (3) follows from the fact that $\tau \leq \operatorname{Seq}\tau$ implies $\tau = \operatorname{T}\tau \leq \operatorname{T}\operatorname{Seq}\tau$ whenever τ is a topology.

(3) $\Longrightarrow$ (1): Let $\tau := \operatorname{T}\operatorname{Seq}\xi$. Then the identity map $i : |\xi| \to |\tau|$ is sequentially continuous because $\operatorname{Seq}\xi \geq \operatorname{T}\operatorname{Seq}\xi$ and therefore $\operatorname{Seq}\xi \geq \operatorname{Seq}\operatorname{T}\operatorname{Seq}\xi$. Hence $i : |\xi| \to |\operatorname{T}\operatorname{Seq}\xi|$ is continuous, that is, $\xi \geq \operatorname{T}\operatorname{Seq}\xi$. $\square$

Note that in view of Proposition V.6.19 and Exercise IV.1.6, premetrizable spaces are sequential. More generally:

Exercise V.6.20. Show that premetrizable topologies are (sequential and) of sequential order at most 1.

With Proposition II.8.7 in mind, show:

Exercise V.6.21. Let ξ be a topology. Show that ξ is sequential of sequential order at most 1 if and only if

$$\xi \geq \operatorname{S}_0 \operatorname{Seq}\xi.$$

We call a convergence as in Exercise V.6.21 a *Fréchet convergence*.

Exercise V.6.22. Show that a pretopology is Fréchet if and only if each of its vicinity filters is a Fréchet filter in the sense of Definition II.8.5.

Proposition V.6.23. *For every convergence, the sequential order is not greater than ω_1.*

Proof. If $x \in \operatorname{adh}_{\operatorname{Seq}\tau}^{\omega_1} A$, then, by definition, there exist a sequence $(\alpha_n)_{n \in \mathbb{N}}$ of ordinals $\alpha_n < \omega_1$ and a sequence $\{x_n\}_{n \in \mathbb{N}}$ such that $x_n \in \operatorname{adh}_{\operatorname{Seq}\tau}^{\alpha_n} A$ and $x \in \lim_\tau (x_n)_n^\uparrow$. It follows that $\alpha = \sup_{n \in \mathbb{N}_0} \alpha_n < \omega_1$, hence $\{x_n : n \in \mathbb{N}\} \subset \operatorname{adh}_{\operatorname{Seq}\tau}^{\alpha} A$, thus $x \in \lim_{\operatorname{Seq}\tau} (x_n)_n^\uparrow \subset \operatorname{adh}_{\operatorname{Seq}\tau}^{\alpha+1} A$ and $\alpha + 1 < \omega_1$. Thus $\bigcup_{\alpha < \omega_1} \operatorname{adh}_{\operatorname{Seq}\tau}^{\alpha} A = \operatorname{adh}_{\operatorname{Seq}\tau}^{\omega_1} A$, and consequently,

$$\operatorname{adh}_{\operatorname{Seq}\tau}^{\omega_1} A := \operatorname{adh}_{\operatorname{Seq}\tau}\left(\bigcup_{\alpha < \omega_1} \operatorname{adh}_{\operatorname{Seq}\tau}^{\alpha} A\right) = \operatorname{adh}_{\operatorname{Seq}\tau}^{\omega_1+1} A.$$

Hence $\operatorname{adh}_{\operatorname{Seq}\tau}^{\omega_1} A = \operatorname{adh}_{\operatorname{Seq}\tau}^{\gamma} A$ for each $\gamma \geq \omega_1$. $\square$

Example V.6.24 (The radial topology is of sequential order ω_1). A subset C of $\mathbb{R}^2$ is closed for the *radial topology* ρ whenever $C \cap L$ is closed in L for every straight line L in $\mathbb{R}^2$. We shall see that the sequential order of ρ is ω_1.

Observe that a sequence $\{x_n\}_{n \in \mathbb{N}}$ converges to x in ρ, if and only if it converges to x in the Euclidean topology, and there is $m \in \mathbb{N}$ and straight lines $L_1, L_2, \ldots, L_m$ such that

$$\{x_n : n \in \mathbb{N}\} \subset L_1 \cup L_2 \cup \ldots \cup L_m.$$

The sufficiency follows from the definition of ρ. To prove the necessity, it is enough to prove this for the sequences of distinct terms. If this were not the case, then we could construct by induction a subsequence $\{x_{n_k}\}_k$ such that for each $k_0 \in \mathbb{N}$, the intersection of each straight line with $\{x_{n_k} : k < k_0\}$ has at most two points.

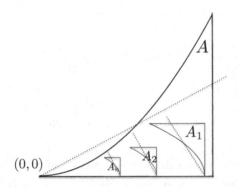

Figure V.3 The distance along any straight line between $(0,0)$ and the set A is strictly positive, as is that between $(\frac{1}{n}, 0)$ and A_n along straight lines.

If this is the case for k_0 then, by the inductive assumption, there is n_{k_0+1} such that $x_{n_{k_0+1}}$ does not belong to the (finite) union of all the straight lines that join the elements of $\{x_{n_k} : k < k_0\}$. Therefore, for every straight line and each subset of $\{x_{n_k} : k \in \mathbb{N}\} \cap L$ has at most two points, thus is closed in L, consequently closed for ρ. Therefore,

$$\bigcap_{l \in \mathbb{N}} \mathrm{cl}\{x_{n_k} : l < k\} = \varnothing$$

and thus $\{x_{n_k}\}_k$ is not convergent, hence $\{x_n\}_n$ is not convergent, which is a contradiction.

Let us show now that there exists a subset A of $\mathbb{R}^2$ such that $(0,0) \in$ adh$^2_{\text{Seq}\,\rho}\,A \setminus$ adh$_{\text{Seq}\,\rho}\,A$. Set

$$A := \left\{ \left(\tfrac{1}{n}, \tfrac{1}{k}\right) : \tfrac{1}{k} < \tfrac{1}{n^2},\, n,k \in \mathbb{N}_1 \right\}.$$

Then adh$_{\text{Seq}\,\rho}\,A \setminus A = \left\{ \left(\tfrac{1}{n},0\right) : n \in \mathbb{N} \right\}$. The point $(0,0) \notin$ adh$_{\text{Seq}\,\rho}\,A$, because for each straight line passing by $(0,0)$ given parametrically by $\{(rt,st) : t \in \mathbb{R}\}$, where $r,s \in \mathbb{R}$, there is $\delta > 0$ such that $\{(\alpha t, \beta t) : |t| < \delta\} \cap A = \varnothing$. On the other hand, $(0,0) \in$ adh$^2_{\text{Seq}\,\rho}\,A$, because the sequence $\left\{\left(\tfrac{1}{n},0\right)\right\}_{n \in \mathbb{N}_1}$ converges to $(0,0)$.

Let $\gamma < \omega_1$ and assume that for each $\delta < \gamma$ there is a bounded subset A_δ of $\left\{(x,y) : 0 < y < x^2, x > 0\right\}$ such that

$$(0,0) \in \text{adh}^\delta_{\text{Seq}\,\rho}\,A \setminus \bigcup_{\alpha < \delta} \text{adh}^\alpha_{\text{Seq}\,\rho}\,A.$$

As γ is countable, there is a non-decreasing sequence $\{\delta_n\}_n$ such that $\sup_{n \in \mathbb{N}}(\delta_n + 1) = \gamma$. Because homotheties, translations and rotations are homeomorphisms for ρ, there is a sequence of subsets $\{A_n\}_{n \in \mathbb{N}_1}$ of $\left\{(x,y) : 0 < y < x^2, x > 0\right\}$ such that

$$\left(\tfrac{1}{n},0\right) \in \text{adh}^{\delta_n}_{\text{Seq}\,\rho}\,A_n \setminus \bigcup_{\alpha < \delta_n} \text{adh}^\alpha_{\text{Seq}\,\rho}\,A$$

and the diameter of A_n is less than $\tfrac{1}{n^2}$; it follows that these sets are disjoint. For each $\alpha < \gamma$, $\left(\tfrac{1}{n},0\right) \in$ adh$^\alpha_{\text{Seq}\,\rho}\,A$ if and only if $\delta_n \leq \alpha$, hence $\left\{\left(\tfrac{1}{n},0\right) : n \in \mathbb{N}\right\} \subset \bigcup_{\alpha < \gamma} \text{adh}^\alpha_{\text{Seq}\,\rho}\,A$, while $(0,0) \in$ adh$^\gamma_{\text{Seq}\,\rho}\,A_n \setminus \bigcup_{\alpha < \gamma} \text{adh}^\alpha_{\text{Seq}\,\rho}\,A$.

Anticipating Section XV.1.2, we call a continuous onto map $f : |\xi| \to |\tau|$ *topologically quotient* or simply *quotient*, if $\tau \geq \text{T}(f\xi)$. This is to be contrasted with a *convergence quotient* map, for which $\tau = f\xi$. Of course, a convergence quotient map is quotient for T is contractive, but the converse may fail. In fact if ξ is a non-topological convergence then the identity map $i : |\xi| \to |\text{T}\,\xi|$ is quotient but not convergence quotient. Sequentiality is preserved under quotient maps:

Proposition V.6.25. *If ξ is a sequential convergence and $f : |\xi| \to |\tau|$ is a quotient map, then τ is sequential.*

Proof. If $\xi \geq \text{T}\,\text{Seq}\,\xi$ then

$$f\xi \geq f(\text{T}\,\text{Seq}\,\xi) \geq \text{T}\,\text{Seq}(f\xi)$$

by (V.4.2) and (V.6.5). Moreover, $f\xi \geq \tau$ by continuity of f, so that $f\xi \geq \text{T}\,\text{Seq}\,\tau$. Applying the topological modifier yields $\text{T}(f\xi) \geq \text{T}\,\text{Seq}\,\tau$ and thus $\tau \geq \text{T}\,\text{Seq}\,\tau$ because f is quotient. $\square$

Another reason why the class of sequential convergences is fundamental is:

Theorem V.6.26. *A convergence is sequential if and only it is the image of a metrizable topology under a quotient map.*

Proof. We have already observed that premetrizable, in particular metrizable, topologies are sequential. Thus, by Proposition V.6.25, the image of a metrizable topology under a quotient map is sequential. Conversely, if $\tau \geq \mathrm{T}\,\mathrm{Seq}\,\tau$ is sequential, let $\sigma := \mathrm{Seq}\,\tau$, and consider at each $x \in |\sigma|$ the pavement $\mathbb{E}_x$ for σ consisting of sequential filters that τ-converges to x. Let ξ be the topology defined in Proposition IV.8.5 for which $f : |\xi| \to |\sigma|$ is convergence quotient. Then $f : |\xi| \to |\tau|$ is quotient, for

$$\tau \geq \mathrm{T}\,\sigma = \mathrm{T}(f\xi).$$

It remains to see that ξ is metrizable. By definition, $\xi = \bigoplus_{x \in |\tau|, \mathcal{E} \in \mathbb{E}_x} \pi[x, \mathcal{E}]$. Each $\pi[x, \mathcal{E}]$ is metrizable with the distance $d_{x,\mathcal{E}}$ defined as follows. If $\mathcal{E} = \{x\}^\uparrow$ then $d_{x,\mathcal{E}}$ is the discrete metric. Otherwise there is a sequence $\{x_n\}_{n \in \mathbb{N}}$ on $|\tau| \setminus \{x\}$ such that $\mathcal{E} = (x_n)_n^\uparrow$. Let $d_{x,\mathcal{E}}(x_n, x) = d_{x,\mathcal{E}}(x, x_n) = \frac{1}{2^n}$, $d_{x,\mathcal{E}}(x, t) = 1$ if $t \in |\tau| \setminus (\{x_n : n \in \mathbb{N}\} \cup \{x\})$, $d_{x,\mathcal{E}}(t, y) = 0$ if $t = y$, and $d_{x,\mathcal{E}}(t, y) = 1$ if $t \neq y$ are in $|\tau| \setminus \{x\}$. It is easily verified (see Exercise V.6.27 below) that $d_{x,\mathcal{E}}$ is a metric and that $\widetilde{d_{x,\mathcal{E}}} = \pi[x, \mathcal{E}]$. Now define on $|\xi|$ the distance d by $d(s, t) = d_{x,\mathcal{E}}(s, t)$ if $\{s, t\} \subset |\pi[x, \mathcal{E}]|$ and $d(s, t) = 1$ if s and t do not lie in the same component $\pi[x, \mathcal{E}]$. It is easily verified that d is a metric and that $\tilde{d} = \xi$. $\square$

Exercise V.6.27. Verify that in the proof of Theorem V.6.26, $d_{x,\mathcal{E}}$ is a metric such that $\widetilde{d_{x,\mathcal{E}}} = \pi[x, \mathcal{E}]$ and that d is a metric such that $\tilde{d} = \xi$.

Chapter VI

Diagonality and regularity

Diagonality is akin to topologicity, while regularity is a separation property. Both diagonality and regularity can be defined with the aid of contours. The role of the contour operation in these definitions can however be qualified as opposite or complementary. It will turn out that in some minimal classes of convergences the two properties become equivalent (Theorem IX.2.1).

VI.1 More on contours

We revisit the notion of contour introduced in Section II.9. Recall:

Definition VI.1.1. If $\mathcal{F}$ is a family of subsets of a set A, and $\mathcal{G}(a)$ is a family of subsets of X for every $a \in A$, then the *contour* of $\mathcal{G}(\cdot)$ along $\mathcal{F}$ is the following family of subsets of X:

$$\mathcal{G}(\mathcal{F}) = \bigcup_{F \in \mathcal{F}} \bigcap_{a \in F} \mathcal{G}(a). \tag{VI.1.1}$$

In the particular case where $F^{\uparrow}$ is the principal filter of a set F, then we abridge

$$\mathcal{G}(F) = \bigcap_{a \in F} \mathcal{G}(a). \tag{VI.1.2}$$

Consequently, $\mathcal{G}(\mathcal{F}) = \bigcup_{F \in \mathcal{F}} \mathcal{G}(F)$. We notice that:

Lemma VI.1.2. *If $\mathcal{F}_0 \leq \mathcal{F}_1$ and $\mathcal{G}_0(a) \subset \mathcal{G}_1(a)$ for every $a \in A$ then*

$$\mathcal{G}_0(\mathcal{F}_0) \subset \mathcal{G}_1(\mathcal{F}_1).$$

Proof. Indeed, if $A \in \mathcal{G}_0(\mathcal{F}_0)$, then there is $F_0 \in \mathcal{F}_0$ such that $A \in \bigcap_{a \in F_0} \mathcal{G}_0(a)$. Because $\mathcal{F}_0 \leq \mathcal{F}_1$ there is $F_1 \in \mathcal{F}_1$ such that $F_1 \subset F_0$ and therefore $A \in \bigcap_{a \in F_1} \mathcal{G}_0(a) \subset \bigcap_{a \in F_1} \mathcal{G}_1(a)$. $\square$

Proposition VI.1.3. *If $\mathcal{F}$ and $\mathcal{G}(x)$ are filters for each x, then $\mathcal{G}(\mathcal{F})$ is a filter.*

Proof. Indeed, if $\mathcal{G}(x)$ is isotone for each x, then $\mathcal{G}(\mathcal{F})$ is isotone. Assume that $\mathcal{F}$ and $\mathcal{G}(x)$ for each x are closed under finite intersections, and let $H_0, H_1 \in \mathcal{G}(\mathcal{F})$, that is, there exist $F_0, F_1 \in \mathcal{F}$ such that $H_0 \in \bigcap_{x \in F_0} \mathcal{G}(x)$ and $H_1 \in \bigcap_{x \in F_1} \mathcal{G}(x)$. It follows that $H_0 \cap H_1 \in \mathcal{G}(x)$ every $x \in F_0 \cap F_1$, hence $H_0 \cap H_1 \in \mathcal{G}(\mathcal{F})$. $\qquad\square$

Proposition VI.1.4. *If $\mathcal{G} : A \to \mathbb{F}X$ and $\mathcal{F} \in \mathbb{F}A$ let $\mathcal{G}^{\#} : A \to \mathbb{F}_{\#}X$ be defined by $\mathcal{G}^{\#}(x) = (\mathcal{G}(x))^{\#}$. Then*

$$(\mathcal{G}(\mathcal{F}))^{\#} = \mathcal{G}^{\#}(\mathcal{F}^{\#}). \qquad (VI.1.3)$$

Proof. If $H \in (\mathcal{G}(\mathcal{F}))^{\#}$ then $H \in (\bigcap_{x \in F} \mathcal{G}(x))^{\#}$ for all $F \in \mathcal{F}$. In view of Proposition II.3.5, there is $x_F \in F$ with $H \in \mathcal{G}(x_F)^{\#}$.

In other words, there is $R := \{x_F : F \in \mathcal{F}\} \in \mathcal{F}^{\#}$ such that $H \in \bigcap_{x \in R} \mathcal{G}^{\#}(x)$. Hence $H \in \mathcal{G}^{\#}(\mathcal{F}^{\#})$.

Conversely, if $H \notin (\mathcal{G}(\mathcal{F}))^{\#}$ there is $F_0 \in \mathcal{F}$ with $H \notin (\bigcap_{x \in F_0} \mathcal{G}(x))^{\#}$, that is, $H \notin \mathcal{G}(x)^{\#}$ for all $x \in F_0$. Hence for each $R \in \mathcal{F}^{\#}$ there is $x \in R \cap F_0$ so that $H \notin \bigcap_{x \in R} \mathcal{G}^{\#}(x)$. In other words, $H \notin \mathcal{G}^{\#}(\mathcal{F}^{\#})$. $\qquad\square$

By virtue of Propositions VI.1.3 and VI.1.4,

Proposition VI.1.5. *If $\mathcal{F}$ and $\mathcal{G}(x)$ are filter-grills for each x, then $\mathcal{G}(\mathcal{F})$ is a filter-grill.*

Since ultrafilters are filters that are also filter-grills,

Corollary VI.1.6. *If $\mathcal{F}$ and $\mathcal{G}(x)$ are ultrafilters for each x, then $\mathcal{G}(\mathcal{F})$ is an ultrafilter.*

Moreover, there is a converse of sorts.

Definition VI.1.7. A family of filters $\{\mathcal{G}_i : \in I\}$ on X is *discrete* if there is a family $\{X_i : i \in I\}$ of pairwise disjoint subsets of X such that $X_i \in \mathcal{G}_i$ for each $i \in I$.

With this definition in mind, we have:

Proposition VI.1.8. *If a family $\{\mathcal{G}(x) : x \in X\}$ is discrete and $\mathcal{F}$ is not an ultrafilter, then $\mathcal{G}(\mathcal{F})$ is not an ultrafilter.*

Proof. Let $\{X_x : x \in X\}$ be the family of pairwise disjoint subsets of X such that $X_x \in \mathcal{G}(x)$ for all x. If $\mathcal{F}$ is not an ultrafilter, there is $A \in \mathcal{F}^\#$ with $A^c \in \mathcal{F}^\#$. If $\mathcal{G}(\mathcal{F})$ was an ultrafilter, then by maximality, we would have

$$\mathcal{G}(\mathcal{F}) = \mathcal{G}(\mathcal{F} \vee A) = \mathcal{G}(\mathcal{F} \vee A^c).$$

But $\bigcup_{x \in A} X_x \in \mathcal{G}(\mathcal{F} \vee A)$, $\bigcup_{x \in A^c} X_x \in \mathcal{G}(\mathcal{F} \vee A^c)$ and $(\bigcup_{x \in A} X_x) \cap (\bigcup_{x \in A^c} X_x) = \varnothing$. $\qquad\square$

Lemma VI.1.9. *If $\mathcal{S} : A \to \mathbb{U}X$ then*

$$\beta(\mathcal{S}(\mathcal{F})) = \{\mathcal{S}(\mathcal{U}) : \mathcal{U} \in \beta(\mathcal{F})\}$$

for every $\mathcal{F} \in \mathbb{F}A$.

Proof. $\beta(\mathcal{S}(\mathcal{F})) \supset \{\mathcal{S}(\mathcal{U}) : \mathcal{U} \in \beta(\mathcal{F})\}$ follows from Corollary VI.1.6 and Lemma VI.1.2.

Consider now $\mathcal{W} \in \beta(\mathcal{S}(\mathcal{F}))$. If $\mathcal{W}$ was not equal to any of the ultrafilters $\mathcal{S}(\mathcal{U})$, $\mathcal{U} \in \beta(\mathcal{F})$, then for each $\mathcal{U} \in \beta(\mathcal{F})$ there would be $W_\mathcal{U} \in \mathcal{W}$ with $W_\mathcal{U}^c \in \mathcal{S}(\mathcal{U})$. Hence, there is $U_\mathcal{U} \in \mathcal{U}$ with $W_\mathcal{U}^c \in \bigcap_{a \in U_\mathcal{U}} \mathcal{S}(a)$.

By Proposition II.6.5, there is a finite subset $\mathbb{D}$ of $\beta(\mathcal{F})$ with $\bigcup_{\mathcal{U} \in \mathbb{D}} U_\mathcal{U} \in \mathcal{F}$. Hence, $\bigcap_{a \in \bigcup_{\mathcal{U} \in \mathbb{D}} U_\mathcal{U}} \mathcal{S}(a) \subset \mathcal{S}(\mathcal{F})$ and

$$\bigcup_{\mathcal{U} \in \mathbb{D}} W_\mathcal{U}^c = (\bigcap_{\mathcal{U} \in \mathbb{D}} W_\mathcal{U})^c \in \bigcap_{a \in \bigcup_{\mathcal{U} \in \mathbb{D}} U_\mathcal{U}} \mathcal{S}(a).$$

Thus, there would be $W_0 := \bigcap_{\mathcal{U} \in \mathbb{D}} W_\mathcal{U} \in \mathcal{W}$ with $W_0^c \in \mathcal{S}(\mathcal{F})$, in contradiction to $\mathcal{W} \in \beta(\mathcal{S}(\mathcal{F}))$. $\qquad\square$

VI.2   Diagonality

If ξ is a pretopology, we denote by ξ^2 the pretopology whose principal adherence is adh_ξ^2 and more generally, for each ordinal γ, we denote by ξ^γ the pretopology whose principal adherence is

$$\mathrm{adh}_{\xi^\gamma} := \mathrm{adh}_\xi^\gamma$$

as defined in (V.6.1). By convention, $\xi^0 = \iota$ and $\xi^1 = \xi$. Since a pretopology is completely determined by its principal adherence (e.g., modulo (V.2.8)), ξ^γ is well defined.

Of course, $\alpha < \beta$ implies $\mathrm{adh}_\xi^\alpha \subset \mathrm{adh}_\xi^\beta$ and therefore $\xi^\alpha \geq \xi^\beta$ so that the transfinite sequence $(\xi^\alpha)_{\alpha \in \mathrm{Ord}}$ is a decreasing sequence of pretopologies on $|\xi|$, which, in view of Proposition V.6.6, terminates for $\alpha = \mathrm{t}(\xi)$ at $\mathrm{T}\xi$.

The formula (V.2.4) linking inherence and vicinity enables us to write down the vicinity filter of these iterates in terms of the vicinity filter for ξ. Indeed,

Proposition VI.2.1. $V \in \mathcal{V}_{\xi^2}(x)$ *if and only if there exists* $W \in \mathcal{V}_\xi(x)$ *such that* $V \in \mathcal{V}_\xi(w)$ *for each* $w \in W$.

Proof. By definition, $V \in \mathcal{V}_{\xi^2}(x)$ whenever $x \in \operatorname{inh}_{\xi^2} V = \operatorname{inh}_\xi(\operatorname{inh}_\xi V)$, that is, $\operatorname{inh}_\xi V \in \mathcal{V}_\xi(x)$. By setting $W := \operatorname{inh}_\xi V$, we get that $W \in \mathcal{V}_\xi(x)$ and $V \in \mathcal{V}_\xi(w)$ for each $w \in W$.   □

Recall (from Proposition V.6.9) that V is a *vicinity of a set* A if V is a vicinity of x for each $x \in A$. We denote by $\mathcal{V}(A)$ the set of all vicinities of A. Of course,

$$\mathcal{V}(A) = \bigcap_{x \in A} \mathcal{V}(x).$$

If $\mathcal{A}$ is a family of subsets of X, then the contour $\mathcal{V}(\mathcal{A})$ of $\mathcal{V}(\cdot)$ along $\mathcal{A}$ introduced in Definition VI.1.1 is equal to

$$\mathcal{V}(\mathcal{A}) = \bigcup_{A \in \mathcal{A}} \mathcal{V}(A).$$

In these terms,

$$\mathcal{V}_{\xi^2}(x) = \mathcal{V}_\xi(\mathcal{V}_\xi(x)).$$

Proposition VI.2.2. *A pretopology* ξ *is a topology if and only if for each* $x \in |\xi|$,

$$\mathcal{V}_\xi(x) \subset \mathcal{V}_{\xi^2}(x) \tag{VI.2.1}$$

if and only if $\mathcal{V}_\xi(x) = \mathcal{N}_\xi(x)$ *for each* $x \in |\xi|$.

Let us explicate (VI.2.1): if $V \in \mathcal{V}_\xi(x)$ then there is $W \in \mathcal{V}_\xi(x)$ such that $V \in \mathcal{V}_\xi(w)$ for each $w \in W$. This is precisely the last property of the neighborhood filter in (V.3.12).

Similarly,

Proposition VI.2.3.

$$\mathcal{V}_{\xi^\gamma}(x) = \bigcup_{V \in \mathcal{V}_\xi(x)} \bigcap_{v \in V, \, \alpha < \gamma} \mathcal{V}_{\xi^\alpha}(v). \tag{VI.2.2}$$

Proof. Notice that $A \in \mathcal{V}_{\xi^\gamma}(x)^\#$ if and only if

$$x \in \mathrm{adh}_\xi^\gamma A = \mathrm{adh}_\xi\Big(\bigcup_{\alpha < \gamma} \mathrm{adh}_\xi^\alpha A\Big),$$

equivalently, $\mathcal{V}_\xi(x) \# (\bigcup_{\alpha < \gamma} \mathrm{adh}_\xi^\alpha A)$, that is, for every $V \in \mathcal{V}_\xi(x)$ there is $v \in V$ and $\alpha < \gamma$ with $\mathcal{V}_{\xi^\alpha}(v) \# A$. In other words, $A \in \mathcal{V}_{\xi^\gamma}(x)^\#$ whenever

$$A \in \Big(\bigcup_{V \in \mathcal{V}_\xi(x)} \bigcap_{v \in V,\, \alpha < \gamma} \mathcal{V}_{\xi^\alpha}(v)\Big)^\#,$$

which complete the proof by Exercise II.3.3(4). $\qquad\square$

VI.2.1 *Various types of diagonality*

Definition VI.2.4. A *selection* of a convergence space (X, ξ) is a map $\mathcal{S} : X \to \mathbb{F}X$ such that $x \in \lim_\xi \mathcal{S}(x)$ for each $x \in X$.

Definition VI.2.5. We call a convergence ξ *diagonal* if for every selection $\mathcal{S}$ of ξ and every filter $\mathcal{F}$ on $|\xi|$ with $x \in \lim_\xi \mathcal{F}$,

$$x \in \lim_\xi \mathcal{S}(\mathcal{F}). \qquad (\text{VI.2.3})$$

Remark VI.2.6. Diagonality (hence topologicity) is not a pointwise property.

Exercise VI.2.7. Show that a diagonal convergence such that ([1])

$$x \in \lim \mathcal{F} \Longrightarrow x \in \lim \mathcal{F} \wedge \{x\}^\uparrow$$

is finitely deep.

Definition VI.2.8. We say that ξ is *diagonal for ultrafilters* if (VI.2.3) holds for every selection $\mathcal{S} : X \to \mathbb{U}X$ of ξ and every ultrafilter $\mathcal{F}$ on $|\xi|$ with $x \in \lim_\xi \mathcal{F}$.

Definition VI.2.9. We say that a convergence ξ is *weakly diagonal* if $\mathrm{adh}_\xi \mathcal{F}$ is closed for every filter $\mathcal{F}$.

Proposition VI.2.10. *If ξ is diagonal for ultrafilters then it is weakly diagonal.*

Proof. Let $x_0 \in \mathrm{adh}(\mathrm{adh}_\xi \mathcal{F})$. In other words, there is $\mathcal{U}$ with $\mathrm{adh}_\xi \mathcal{F} \in \mathcal{U}$ and $x_0 \in \lim_\xi \mathcal{U}$. Therefore, for each $x \in \mathrm{adh}_\xi \mathcal{F}$ there is $\mathcal{S}(x) \in \beta(\mathcal{F})$ with $x \in \lim_\xi \mathcal{S}(x)$. If $x \notin \mathrm{adh}_\xi \mathcal{F}$, let $\mathcal{S}(x) = \{x\}^\uparrow$.

---
[1]that is, a diagonal Kent space.

By diagonality for ultrafilters, $x_0 \in \lim_\xi \mathcal{S}(\mathcal{U})$. Moreover, $\mathcal{S}(\mathcal{U}) \geq \mathcal{F}$ because $\bigcap_{x \in \mathrm{adh}_\xi \mathcal{F}} \mathcal{S}(x) \geq \mathcal{F}$. Hence $x_0 \in \mathrm{adh}_\xi \mathcal{F}$ and thus $\mathrm{adh}_\xi \mathcal{F}$ is closed. $\qquad\square$

In particular, if ξ is diagonal then $\mathrm{adh}_\xi A = \mathrm{adh}_\xi A^\uparrow$ is closed for every $A \subset |\xi|$ so that $\mathrm{S}_0\, \xi = \mathrm{T}\, \xi$. Hence we have seen that each of the following property of a convergence ξ implies the next:

diagonal $\implies$ diagonal for ultrafilters $\implies$ weakly diagonal $\implies \mathrm{S}_0\, \xi = \mathrm{T}\, \xi$.

$$(\mathrm{VI.2.4})$$

Of course, the last property of (VI.2.4) amounts to the closedness of principal adherences, that is, to the topological defect being at most 1.

Recall that $\mathbb{U}^* X$ denote the set of free ultrafilters on X.

Example VI.2.11. Let X be an infinite set, $\infty_X \in X$ and $\mathcal{U}_0$ a robust ultrafilter on X. We define the following convergence ξ on X: each $x \neq \infty_X$ is isolated, while a free filter $\mathcal{F}$ converges to ∞_X in ξ whenever $\mathcal{U}_0 \notin \beta\mathcal{F}$. This is a Hausdorff prime convergence satisfying $\mathrm{S}_0\, \xi = \mathrm{T}\, \xi$.

As $\bigwedge \{\mathcal{U} \in \mathbb{U}^* X : \mathcal{U} \neq \mathcal{U}_0\}$ is the cofinite filter $(X)_0$ of X, the vicinity filter of ∞_X is $(X)_0 \wedge \{\infty_X\}^\uparrow$. Accordingly, $\mathrm{S}_0\, \xi = \mathrm{T}\, \xi$ is the prime cofinite topology $\pi[\infty_X, (X)_0]$.

In general, none of the implications can be reversed:

Example VI.2.12 (A non-weakly diagonal convergence ξ with $\mathrm{S}_0\, \xi = \mathrm{T}\, \xi$). Consider a free ultrafilter $\mathcal{U}_0$ on $\mathbb{N}$. Define now on $\mathbb{N}$ the convergence ξ in which $n \in \lim \mathcal{F}$ if and only $\mathcal{F} \geq (\mathbb{N})_0 \wedge \{n\}^\uparrow$ if $n \neq 1$ and $1 \in \lim \mathcal{F}$ if $\mathcal{F} \geq (\mathbb{N})_0 \wedge \{1\}^\uparrow$ and $\mathcal{U}_0 \notin \beta(\mathcal{F})$.

Then $\mathrm{S}_0\, \xi$ is the cofinite topology, because, as each $U \in \mathcal{U}_0$ belongs to some free ultrafilter different from $\mathcal{U}_0$, we have

$$\bigcap_{\mathcal{W} \in \mathbb{U}^* X, \mathcal{W} \neq \mathcal{U}_0} \mathcal{W} = \bigcap_{\mathcal{W} \in \mathbb{U}^* X} \mathcal{W} = (\mathbb{N})_0,$$

where $\mathbb{U}^* X$ is the set of free ultrafilters on X, so that $\mathcal{V}_\xi(1) = (\mathbb{N})_0 \wedge \{1\}^\uparrow$. Therefore $\mathrm{S}_0\, \xi = \mathrm{T}\, \xi$ (that is, principal adherences are closed).

On the other hand, $\mathrm{adh}_\xi \mathcal{U}_0 = \lim_\xi \mathcal{U}_0 = \mathbb{N} \setminus \{1\}$ is not closed.

Example VI.2.13 (A weakly diagonal convergence that is not diagonal for ultrafilters). Consider two disjoint copies $\mathbb{N}$ and $\mathbb{N}'$ of the natural numbers, and denote by n' the copy in $\mathbb{N}'$ of $n \in \mathbb{N}$, by A' the copy in $\mathbb{N}'$ of a subset of A of $\mathbb{N}$ and by $\mathcal{F}'$ the copy on $\mathbb{N}'$ of a filter $\mathcal{F}$ on $\mathbb{N}$. Pick a free ultrafilter $\mathcal{U}_0$ on $\mathbb{N}$.

Define on $X := \mathbb{N} \cup \mathbb{N}'$ the convergence in which each point $n' \in \mathbb{N}'$ is isolated, the only filters converging to $n \neq 1$ are $\{n\}^{\uparrow}$ and $\{n'\}^{\uparrow}$ and

$$1 \in \lim \mathcal{F} \iff \beta(\mathcal{F}) \subset \{\{1\}^{\uparrow}\} \cup \{\mathcal{U}_0\}.$$

The convergence space X is not diagonal for ultrafilters: Let $\mathcal{S} : X \to \mathbb{U}X$ be defined by $\mathcal{S}(n) = \{n'\}^{\uparrow}$ for all $n \neq 1$, $\mathcal{S}(1) = \{1\}^{\uparrow}$, $\mathcal{S}(n') = \{n'\}^{\uparrow}$. Then $\mathcal{S}$ is a selection, but $\mathcal{S}(\mathcal{U}_0) = \mathcal{U}_0'$ does not converge to 1 even though $\mathcal{U}_0$ does.

Adherences are closed in X. Indeed, let $\mathcal{F}$ be a filter on X with $x \in$ adh(adh $\mathcal{F}$). If x is isolated ($x = n'$ for some n), $x \in$ adh $\mathcal{F}$.

If $x = n \neq 1$ then either n or n' belongs to adh $\mathcal{F}$. If $n \in$ adh $\mathcal{F}$ we are done. If $n' \in$ adh $\mathcal{F}$, then $\{n'\}^{\uparrow} \geq \mathcal{F}$ and $n \in \lim \{n'\}^{\uparrow}$, so $n \in$ adh $\mathcal{F}$.

If $x = 1$ and $1 \notin$ adh $\mathcal{F}$ then adh $\mathcal{F} \in \mathcal{U}_0$. Moreover, for each $n \in$ adh $\mathcal{F} \cap \mathbb{N}$, either $n \in \ker \mathcal{F}$ or $n' \in \ker \mathcal{F} \subset$ adh $\mathcal{F}$. In other words,

$$\text{adh } \mathcal{F} \cap \mathbb{N} \subset (\ker \mathcal{F} \cap \mathbb{N}) \cup (\text{adh } \mathcal{F} \setminus \ker \mathcal{F})' \in \mathcal{U}_0.$$

Since $\mathcal{U}_0$ is an ultrafilter with a filter-base on $\mathbb{N}$, we conclude that $\ker \mathcal{F} \cap \mathbb{N} \in \mathcal{U}_0$ so that $\mathcal{U}_0 \geq \mathcal{F}$, and $1 \in$ adh $\mathcal{F}$.

Example VI.2.14 (A non-diagonal convergence that is diagonal for ultrafilters). Let X be an infinite set and let $\mathcal{F}$ and $\mathcal{G}$ be two distinct free filters, that are not finite infima of ultrafilters. Let $x_0 \in X$. Define on X the prime convergence in which $x_0 \in \lim \mathcal{H}$ if there is a finite subset $\mathbb{D}$ of $\beta(\mathcal{F})$ with $\mathcal{H} \geq \mathcal{G} \wedge \{x_0\}^{\uparrow} \bigwedge_{W \in \mathbb{D}} W$ **or** there is a finite subset $\mathbb{H}$ of $\beta(\mathcal{G})$ with $\mathcal{H} \geq \mathcal{F} \wedge \{x_0\}^{\uparrow} \bigwedge_{\mathcal{U} \in \mathbb{H}} \mathcal{U}$. For this convergence,

$$x \in \lim \mathcal{H} \implies x \in \lim \mathcal{H} \wedge \{x\},$$

but the convergence in not finitely deep, because $\mathcal{F}$ and $\mathcal{G}$ converge to x_0 but $\mathcal{F} \wedge \mathcal{G}$ does not. By Exercise VI.2.7, the convergence is not diagonal.

On the other hand, if $\mathcal{S} : X \to \mathbb{U}X$ is a selection then $\mathcal{S}(x) = \{x\}^{\uparrow}$ for all $x \neq x_0$ and $\mathcal{S}(x_0)$ is either finer than $\mathcal{G} \wedge \{x_0\}^{\uparrow} \wedge \bigwedge_{W \in \mathbb{D}} W$ for some finite subset $\mathbb{D}$ of $\beta(\mathcal{F})$ or than $\mathcal{F} \wedge \{x_0\}^{\uparrow} \wedge \bigwedge_{\mathcal{U} \in \mathbb{H}} \mathcal{U}$ for some finite subset $\mathbb{H}$ of $\beta(\mathcal{G})$. Let us assume that we are in the first case. In view of Exercise VIII.1.1(1), $\mathcal{S}(x_0)$ is either $\{x_0\}^{\uparrow}$ or in $\beta(\mathcal{G})$ or in $\beta(\mathcal{F})$ because $\bigwedge_{W \in \mathbb{D}} W \geq \mathcal{F}$. Of course, a similar argument applies to the second case, so that

$$\mathcal{S}(x_0) \in \beta(\mathcal{F}) \cup \beta(\mathcal{G}) \cup \{\{x_0\}^{\uparrow}\}.$$

If now $\mathcal{U}_0$ is an ultrafilter with $\mathcal{U}_0 \geq \mathcal{G} \wedge \{x_0\}^{\uparrow} \wedge \bigwedge_{W \in \mathbb{D}} W$ for some finite subset $\mathbb{D}$ of $\beta(\mathcal{F})$, then

$$\mathcal{S}(\mathcal{U}_0) \geq \mathcal{S}(x_0) \wedge \mathcal{G} \wedge \{x_0\}^{\uparrow} \wedge \bigwedge_{W \in \mathbb{D}} W.$$

The right-hand side is equal to $\mathcal{G} \wedge \{x_0\}^\uparrow \wedge \bigwedge_{\mathcal{W} \in \mathbb{D}} \mathcal{W}$ if $\mathcal{S}(x_0) \in \beta(\mathcal{G}) \cup \{\{x_0\}^\uparrow\}$ and to $\mathcal{G} \wedge \{x_0\}^\uparrow \wedge \bigwedge_{\mathcal{W} \in \mathbb{D}'} \mathcal{W}$ for $\mathbb{D}' = \mathbb{D} \cup \{\mathcal{S}(x_0)\}$ finite subset of $\beta(\mathcal{F})$ otherwise.

A similar argument applies to the case $\mathcal{U}_0 \geq \mathcal{F} \wedge \{x_0\}^\uparrow \wedge \bigwedge_{\mathcal{U} \in \mathbb{H}} \mathcal{U}$ for some finite subset $\mathbb{H}$ of $\beta(\mathcal{G})$, to the effect that $x_0 \in \lim \mathcal{S}(\mathcal{U}_0)$.

However, all four notions in (VI.2.4) coincide among pretopologies. Indeed,

Proposition VI.2.15. *Every topology is diagonal.*

Proof. The last property of $\mathcal{N}(x)$ in (V.3.12) translates to $\mathcal{N}(\mathcal{N}(x)) = \mathcal{N}(x)$, as observed after Proposition VI.2.2. Moreover, if $\mathcal{S}$ is a selection, then $\mathcal{S}(x) \geq \mathcal{N}(x)$ for each x and if $x_0 \in \lim \mathcal{F}$ then $\mathcal{F} \geq \mathcal{N}(x_0)$. Hence, by Lemma VI.1.2, $\mathcal{S}(\mathcal{F}) \geq \mathcal{N}(\mathcal{N}(x_0)) = \mathcal{N}(x_0)$ and $x_0 \in \lim \mathcal{S}(\mathcal{F})$. $\square$

In particular:

Corollary VI.2.16. *A convergence is a topology if and only if it is a diagonal pretopology.*

In particular each pretopology that is not a topology, such as the bisequence pretopology, the Féron cross pretopology and the radial pretopology provide examples of a non-diagonal convergences. On the other hand, there are non-topological diagonal convergences:

Example VI.2.17 (A diagonal non-topological convergence). On $[0,1]$, pick a free ultrafilter $\mathcal{U}$ and consider the prime topology defined by $\mathcal{N}(0) = \mathcal{U} \wedge \{0\}^\uparrow$. Let $\{I_n : n \in \mathbb{N}\}$ denote countably many copies of this space and let ξ denote the sum convergence. The quotient convergence $f\xi$, obtained by identifying all the zeros, is not pretopological, hence not topological, because $f[\mathcal{N}_{I_n}(0)]$ are the only filters that converge to 0 in $f\xi$. On the other hand, $f\xi$ is diagonal, because a selection $\mathcal{G}$ for $f\xi$ is necessarily $\mathcal{G}(x) = \{x\}^\uparrow$ for all $x \neq 0$ and $\mathcal{G}(0)$ is either $\{0\}^\uparrow$ or $f[\mathcal{U}_n]$ for some n, where $\mathcal{U}_n$ denotes the copy of $\mathcal{U}$ in I_n. Since a filter $\mathcal{F}$ convergent in $f\xi$ is either a principal ultrafilter or $\mathcal{U}_n$ for some n, $\mathcal{G}(\mathcal{F})$ converges to the limit of $\mathcal{F}$ either way.

Remark VI.2.18. Note that a convergence is pretopological if for any collection $(\mathcal{F}(i))_{i \in I}$ with $x \in \bigcap_{i \in I} \lim \mathcal{F}(i)$ we have $x \in \lim \bigcap_{i \in I} \mathcal{F}(i)$, in other words, if x is a limit of the contour of $\mathcal{F}(\cdot)$ along the principal filter of I.

This is a property akin to diagonality, except that the map playing the role of the selection allows for many filters converging to the same point, and the base filter of the contour is principal.

Hence diagonality and pretopologicity can be combined into a diagonal property equivalent to topologicity, using an appropriate indexing device:

Theorem VI.2.19. *The following statements are equivalent:*

(1) a convergence on X is a topology;

(2) for every set A, every pair of maps $s : A \to X$ and $S : A \to \mathbb{F}X$ such that $s(a) \in \lim S(a)$ for each $a \in A$,

$$\lim s[\mathcal{F}] \subset \lim S(\mathcal{F}), \qquad (\text{VI.2.5})$$

for every $\mathcal{F} \in \mathbb{F}A$;

(3) for every set A, every pair of maps $s : A \to X$ and $S : A \to \mathbb{U}X$ such that $s(a) \in \lim S(a)$ for each $a \in A$,

$$\lim s[\mathcal{F}] \subset \lim S(\mathcal{F}), \qquad (\text{VI.2.6})$$

for every $\mathcal{F} \in \mathbb{F}A$.

Proof. $(1) \implies (2)$. If the convergence is topological then it is pretopological and diagonal. In particular, letting $\mathcal{G}(x) := \{x\}^{\uparrow}$ if $s^{-}(x) = \varnothing$ and $\mathcal{G}(x) := \bigcap_{a \in s^{-}(x)} S(a)$ otherwise defines a selection.

By diagonality, if $x \in \lim s[\mathcal{F}]$ then $x \in \lim \mathcal{G}(s[\mathcal{F}])$. Moreover, $\mathcal{G}(s[\mathcal{F}]) = S(\mathcal{F})$ because $H \in \mathcal{G}(s[\mathcal{F}])$ if and only if there is $F \in \mathcal{F}$ with $H \in \bigcap_{x \in s(F)} \mathcal{G}(x)$, that is, $H \in \bigcap_{a \in F} S(a)$.

$(2) \implies (3)$ is obvious.

$(3) \implies (1)$. Assume that (VI.2.6) holds for every A, s, S and $\mathcal{F}$ as in the statement. In particular, consider $A := \{(x, \mathcal{W}) : x \in \lim \mathcal{W}\} \subset X \times \mathbb{U}X$ and let $s(x, \mathcal{W}) = x$ and $S(x, \mathcal{W}) = \mathcal{W}$.

Then by definition $s(x, \mathcal{W}) \in \lim S(x, \mathcal{W})$ for each $(x, \mathcal{W}) \in A$. Let $\mathcal{F}$ be the (principal) filter generated on A by $\{(x_0, \mathcal{W}) : x_0 \in \lim \mathcal{W}\}$. Then $s[\mathcal{F}] = \{x_0\}^{\uparrow}$ so that $x_0 \in \lim S(\mathcal{F})$. But

$$S(\mathcal{F}) = \bigcap_{x_0 \in \lim \mathcal{W}} \mathcal{W} = \mathcal{V}(x_0),$$

so that the convergence is pretopological. Moreover,

$$\mathcal{V}(\mathcal{V}(x_0)) = S(\mathcal{G})$$

where $\mathcal{G}$ is the filter generated on A by

$$\{\{(x, \mathcal{W}) : x \in \lim \mathcal{W}, : x \in V\} : V \in \mathcal{V}(x_0)\}.$$

Since $s[\mathcal{G}] = \mathcal{V}(x_0)$ converges to x_0 so does $\mathcal{V}(\mathcal{V}(x_0))$ and the convergence is topological. $\qquad\square$

VI.2.2 *Diagonal modification*

Proposition VI.2.20. *A supremum of diagonal convergences is diagonal.*

Proof. Let Ξ be a family of diagonal convergences on a set X and let $\mathcal{G} : X \to \mathbb{F}X$ be a selection for $\bigvee \Xi$ and $x_0 \in \lim_{\bigvee \Xi} \mathcal{F}$. Since $\mathcal{G}$ is also a selection for each $\xi \in \Xi$ and $x_0 \in \lim_\xi \mathcal{F}$ for each $\xi \in \Xi$, we have

$$\underset{\xi \in \Xi}{\forall}\, x_0 \in \lim_\xi \mathcal{G}(\mathcal{F}),$$

by diagonality of each ξ. Thus $x_0 \in \lim_{\bigvee \Xi} \mathcal{G}(\mathcal{F})$ and $\bigvee \Xi$ is diagonal. $\quad\square$

Corollary VI.2.21. *For each convergence ξ there is a finest diagonal convergence $\mathrm{D}\,\xi$ coarser than ξ.*

Proof. The set of diagonal convergences coarser than ξ is non-empty, because it contains the chaotic topology, and closed under supremum. $\quad\square$

$\mathrm{D}\,\xi$ is called the *diagonal modification* of ξ, and the map D satisfies:

$$\mathrm{D}\,\xi \leq \xi$$
$$\xi \leq \tau \Longrightarrow \mathrm{D}\,\xi \leq \mathrm{D}\,\tau$$
$$\mathrm{D}(\mathrm{D}\,\xi) = \mathrm{D}\,\xi.$$

Since each topology is diagonal and pretopological,

$$\mathrm{D} \geq \mathrm{T} \quad \text{and} \quad \mathrm{S}_0 \geq \mathrm{T}. \tag{VI.2.7}$$

Example VI.2.17 shows that $\mathrm{D}\,\xi > \mathrm{T}\,\xi$ is possible, and Example V.6.1 shows that $\mathrm{S}_0\,\xi > \mathrm{T}\,\xi$ may occur. However, we have seen (e.g., (VI.2.4)) that $\mathrm{S}_0\,\xi = \mathrm{T}\,\xi$ whenever ξ is diagonal, so that

$$\mathrm{T} = \mathrm{S}_0\,\mathrm{D}.$$

Proposition VI.2.22. *If the topological defect of ξ is finite, then*

$$\mathrm{D}\,\mathrm{S}_0\,\xi = \mathrm{T}\,\xi.$$

Proof. It is clear that $\mathrm{D}\,\mathrm{S}_0\,\xi \geq \mathrm{T}\,\xi$, because $\mathrm{S}_0\,\xi \geq \mathrm{T}\,\xi$ and $\mathrm{T}\,\xi$ is diagonal, so $\mathrm{D}\,\mathrm{T}\,\xi = \mathrm{T}\,\xi$.

To see the reverse inequality, let $d\xi$ be the convergence on $|\xi|$ defined by

$$x \in \lim_{d\xi} \mathcal{F}$$

if there is a selection $\mathcal{G}(\cdot)$ for ξ and a filter $\mathcal{H}$ with $x \in \lim_\xi \mathcal{H}$ such that $\mathcal{F} \geq \mathcal{G}(\mathcal{H})$. Using the discrete selection $\mathcal{G}(x) = \{x\}^\uparrow$ for all $x \in |\xi|$, we have $\mathcal{G}(\mathcal{H}) = \mathcal{H}$ for each filter $\mathcal{H}$, so that $\xi \geq d\xi$. On the other hand, ξ is diagonal if and only if $d\xi \geq \xi$. Therefore

$$d\xi \geq d(\mathrm{D}\,\xi) \geq \mathrm{D}\,\xi. \qquad (\mathrm{VI.2.8})$$

If ξ is a pretopology, so is $d\xi$. Indeed, $d\xi$ is the pretopology ξ^2: A map $\mathcal{G} : |\xi| \to \mathbb{F}|\xi|$ is a selection for ξ if and only if $\mathcal{G}(x) \geq \mathcal{V}_\xi(x)$ for all x so that

$$x \in \lim_{d\xi} \mathcal{F} \iff \mathcal{F} \geq \mathcal{V}_\xi(\mathcal{V}_\xi(x)).$$

Defining $d^1\xi = d\xi$ and $d^n\xi = d(d^{n-1}\xi)$ for $n \geq 2$, we obtain by induction that $d^n\xi$ is the pretopology ξ^n for all $n \in \mathbb{N}$.

Since the topological defect of ξ is finite, there is $n_0 \in \mathbb{N}$ with $\mathrm{T}\xi = (\mathrm{S}_0\,\xi)^{n_0}$. Hence,

$$\mathrm{T}\,\xi = d^{n_0}(\mathrm{S}_0\,\xi) \geq \mathrm{D}\,\mathrm{S}_0\,\xi,$$

because, in view of (VI.2.8), $d^n \geq \mathrm{D}$ for all $n \in \mathbb{N}$. $\qquad\square$

Unlike S_0 and T, the diagonalizer D does not preserve continuity:

Example VI.2.23 (A map $f \in C(\xi, \tau)\backslash C(\mathrm{D}\,\xi, \mathrm{D}\,\tau)$). Let (X, ξ) denote the bisequence pretopology of Example V.4.6 and let f be the convergence quotient map identifying $\{x_n : n \in \mathbb{N}\} \cup \{x_\infty\}$ to a single point $\{\infty\}$. Hence $Y := \{x_{n,k} : n, k \in \mathbb{N}\} \cup \{\infty\}$ carries the quotient convergence $\tau := f\xi$. Note that a selection $\mathcal{G}(\cdot)$ for τ satisfies $\mathcal{G}(x_{n,k}) = \{x_{n,k}\}^\uparrow$ for each n and k, and $\mathcal{G}(\infty) \geq \{\infty\}^\uparrow \wedge (x_{n,k})_k^\uparrow$ for some $n \in \mathbb{N}$. Therefore τ is diagonal $(\tau = \mathrm{D}\,\tau)$.

On the other hand, since the bisequence pretopology has topological defect 2, $x_\infty \in \lim_{\mathrm{T}\xi} \mathcal{N}_\xi(x_\infty) = \lim_{\mathrm{D}\xi} \mathcal{N}_\xi(x_\infty)$ by Proposition VI.2.22. But $f(x_\infty) = \infty \notin \lim_\tau f[\mathcal{N}_\xi(x_\infty)]$ because $f[\mathcal{N}_\xi(x_\infty)]$ meshes with every $(x_{n,k})_k$ and is not finer than any of them. Thus $f : |\mathrm{D}\xi| \to |\mathrm{D}\tau|$ is not continuous.

VI.3 Self-regularity

Definition VI.3.1. Given a self-map $a : 2^X \to 2^X$ of the powerset of a set X and a family $\mathcal{A}$ of subsets of X, we denote by

$$a^\natural\mathcal{A} := \{a(A) : A \in \mathcal{A}\}, \qquad (\mathrm{VI.3.1})$$

the family of images of elements of $\mathcal{A}$ under a.

Note that if $\mathcal{F}$ is a filter on X and $a : 2^X \to 2^X$ is isotone, that is,

$$A \subset B \Longrightarrow a(A) \subset a(B),$$

then $a^\natural \mathcal{F}$ is a (possibly degenerate) filter-base on X.

Definition VI.3.2. A convergence ξ is called *regular* (or *self-regular*) at x if

$$x \in \lim_\xi \mathcal{F} \Longrightarrow x \in \lim_\xi \mathrm{adh}_\xi^\natural \mathcal{F}. \qquad (VI.3.2)$$

A convergence ξ is said to be *regular* (or *self-regular*) if it is regular at each $x \in |\xi|$.

Remark VI.3.3. Regularity is not a pointwise property.

Proposition VI.3.4. *For each convergence ξ there exists the finest regular convergence $\mathrm{Reg}\,\xi$ that is coarser than ξ.*

Proof. Let Ξ be the set of all regular convergences that are coarser than ξ. The chaotic convergence o belongs to Ξ, because each filter o-converges to every point. If $x \in \lim_{\bigvee \Xi} \mathcal{F}$, that is, $x \in \lim_\theta \mathcal{F}$ for each $\theta \in \Xi$, and since θ is regular, $x \in \lim_\theta \mathrm{adh}_\theta^\natural \mathcal{F}$. As $\mathrm{adh}_{\bigvee \Xi} F \subset \mathrm{adh}_\theta F$ for each $\theta \in \Xi$ (and for every set F), $\mathrm{adh}_\theta^\natural \mathcal{F} \le \mathrm{adh}_{\bigvee \Xi}^\natural \mathcal{F}$ and thus

$$x \in \lim_\theta \mathrm{adh}_{\bigvee \Xi}^\natural \mathcal{F}$$

for each $\theta \in \Xi$, that is,

$$x \in \lim_{\bigvee \Xi} \mathrm{adh}_{\bigvee \Xi}^\natural \mathcal{F}.$$

It follows, that is, $\bigvee \Xi \in \Xi$, so that $\mathrm{Reg}\,\xi = \bigvee \Xi$. $\qquad \square$

The map Reg associating to each convergence the finest regular convergence coarser than that convergence is called the *regularizer*, and $\mathrm{Reg}\,\xi$ is the *regular modification* of ξ.

Proposition VI.3.5. *The regularizer is isotone, contractive and idempotent, i.e.,*

$$\zeta \le \xi \Rightarrow \mathrm{Reg}\,\zeta \le \mathrm{Reg}\,\xi; \qquad \text{(isotone)}$$
$$\mathrm{Reg}\,\xi \le \xi; \qquad \text{(contractive)}$$
$$\mathrm{Reg}(\mathrm{Reg}\,\xi) = \mathrm{Reg}\,\xi. \qquad \text{(idempotent)}$$

Moreover, let f be a map valued in $|\tau|$. Then

Proposition VI.3.6. *If τ is regular, then $f^-\tau$ is regular.*

Proof. Let $x \in \lim_{f^- \tau} \mathcal{F}$, that is,

$$f(x) \in \lim_\tau f[\mathcal{F}] \subset \lim_\tau \mathrm{adh}_\tau^\natural f[\mathcal{F}],$$

because τ is regular. Thus $x \in \lim_{f^- \tau} f^- \left[\mathrm{adh}_\tau^\natural f[\mathcal{F}] \right]$. Moreover, in view of (V.2.6),

$$f^- \left[\mathrm{adh}_\tau^\natural f[\mathcal{F}] \right] = \mathrm{adh}_{f^- \tau}^\natural \mathcal{F},$$

which concludes the proof. $\qquad\square$

Corollary VI.3.7. $C(\xi, \tau) \subset C(\mathrm{Reg}\,\xi, \mathrm{Reg}\,\tau).$

Proof. By continuity of f and contractivity of Reg,

$$\xi \geq f^- \tau \geq f^- \mathrm{Reg}(\tau),$$

and, by Proposition VI.3.6, $f^-(\mathrm{Reg}\,\tau)$ is a regular convergence coarser than ξ, hence $\mathrm{Reg}\,\xi \geq f^-(\mathrm{Reg}\,\tau)$, which proves the corollary. $\qquad\square$

In particular, by setting $\xi = f^- \tau$, we get

$$\mathrm{Reg}(f^- \tau) \geq f^-(\mathrm{Reg}\,\tau). \qquad (VI.3.3)$$

As a result,

Corollary VI.3.8. *Each subspace of a regular convergence is regular.*

Corollary VI.3.9. *Each product of regular convergences is regular.*

It follows from the definition that a topology τ on X is regular if and only if

$$\mathcal{N}_\tau(x) \leq \mathrm{cl}_\tau^\natural \mathcal{N}_\tau(x) \qquad (VI.3.4)$$

for each $x \in X$, that is, the neighborhood filter $\mathcal{N}_\tau(x)$ has a filter-base formed of τ-closed sets.

Proposition VI.3.10. *A topology is regular if and only if for each closed set F and $x \notin F$ there exist two disjoint open sets O_0, O_1 such that $x \in O_0$ and $F \subset O_1$.*

Proof. Let F is a τ-closed set and $x \notin F$, thus $X \backslash F \in \mathcal{N}_\tau(x)$. By regularity, there exists $V \in \mathcal{N}_\tau(x)$ such that $\mathrm{cl}_\tau V \cap F = \varnothing$. Equivalently,

$$\mathrm{int}_\tau V \cap (X \backslash \mathrm{cl}_\tau V) = \varnothing,$$

$x \in \mathrm{int}_\tau V$ and $F \subset X \backslash \mathrm{cl}_\tau V$. It is enough to set $O_0 := \mathrm{int}_\tau V$ and $O_1 := X \backslash \mathrm{cl}_\tau V$.

Conversely, let O be an open neighborhood of x. Then $X \setminus O$ is a closed set with $x \notin X \setminus O$ and, by the condition, there exist disjoint open sets O_0, O_1 such that $x \in O_0$ and $X \setminus O \subset O_1$, that is,

$$O_0 \subset X \setminus O_1 \subset O.$$

As $X \setminus O_1$ is closed, $x \in O_0 \subset \mathrm{cl}_\tau O_0 \subset O$, so that $O \in \mathrm{cl}_\tau^\natural \mathcal{N}_\tau(x)$. □

Corollary VI.3.11. *A metrizable topology is regular.*

Exercise VI.3.12. Show Corollary VI.3.11, for instance using Proposition VI.3.10 and Exercise V.3.8.

It follows from Proposition VI.3.10 that

Corollary VI.3.13. *A topology τ is not regular at x if and only if there exists a closed set F such that $x \notin F$ and*

$$\mathcal{N}_\tau(x) \# \mathcal{N}_\tau(F).$$

Proposition VI.3.14. *If ξ is a regular topology, then*

$$\mathrm{w}(\xi) \leq 2^{\mathrm{d}(\xi)}.$$

Proof. If ξ is a regular topology then $\mathcal{B} := \{\mathrm{cl}_\xi O : O \in \mathcal{O}_\xi\}$ is a base of ξ-open sets. If D is a dense subset of $|\xi|$ such that $\mathrm{card}\, D = \mathrm{d}(\xi)$, then by Lemma V.4.41, $\mathcal{B} = \{\mathrm{cl}_\xi (O \cap D) : O \in \mathcal{O}_\xi\}$, that is, $\mathrm{card}\, \mathcal{B} \leq 2^{\mathrm{card}\, D} = 2^{\mathrm{d}(\xi)}$. □

Corollary VI.3.15. *The weight of a separable regular topology is not greater than $\mathfrak{c}$.*

Exercise VI.3.16. (1) Show that a T_0 regular topology is Hausdorff; (2) Find an example of a (non T_0) regular topology that is not Hausdorff.

On the other hand, there exist Hausdorff non-regular topologies, as shows the example below:

Example VI.3.17 (A Hausdorff non-regular topology). Let $X = \{x_\infty\} \cup \{x_n : n \in \mathbb{N}\} \cup \{x_{n,k} : n, k \in \mathbb{N}\}$. Define on X the following pretopology: $V \in \mathcal{V}(x_\infty)$ if there is $n_V \in \mathbb{N}$ such that

$$\{x_\infty\} \cup \{x_{n,k} : n \geq n_V, k \in \mathbb{N}\} \subset V,$$

and $V \in \mathcal{V}(x_n)$ if there exists k_V such that $\{x_n\} \cup \{x_{n,k} : k \geq k_V\} \subset V$; finally $V \in \mathcal{V}(x_{n,k})$ if $x_{n,k} \in V$, that is, $x_{n,k}$ is isolated for each n and k.

This is a Hausdorff topology.

It is not regular. Indeed $F := \{x_n : n \in \mathbb{N}\}$ is closed and $x_\infty \notin F$. Each open set G containing x_∞, includes $\{x_\infty\} \cup \{x_{n,k} : n \geq n_G, k \in \mathbb{N}\}$ for some $n_G \in \mathbb{N}$. On the other hand, each open set H that includes F, includes also $\{x_n\} \cup \{x_{n,k} : n \in \mathbb{N}, k \geq k_{H,n}\}$. It follows that $G \cap H \neq \varnothing$.

The following result shows, in light of Theorem VI.2.19, that regularity and topologicity are dual properties.

Theorem VI.3.18. *The following are equivalent:*

(1) A convergence on X is regular;

(2) For every set A, every pair of maps $s : A \to X$ and $\mathcal{S} : A \to \mathbb{F}X$ with $s(a) \in \lim \mathcal{S}(a)$ for each $a \in A$,

$$\lim \mathcal{S}(\mathcal{F}) \subset \lim s[\mathcal{F}], \qquad (\text{VI.3.5})$$

for each $\mathcal{F} \in \mathbb{F}A$;

(3) For every set A, every pair of maps $s : A \to X$ and $\mathcal{S} : A \to \mathbb{U}X$ with $s(a) \in \lim \mathcal{S}(a)$ for each $a \in A$, (VI.3.5) for each $\mathcal{F} \in \mathbb{F}A$.

Proof. $(1) \Longrightarrow (2)$: Assume the convergence is regular and consider A, s, $\mathcal{S}$ and $\mathcal{F}$ as in the statement. For each $H \in \mathcal{S}(\mathcal{F})$ there is $F \in \mathcal{F}$ such that $H \in \bigcap_{a \in F} \mathcal{S}(a)$ so that $s(F) \subset \operatorname{adh} H$ because $s(a) \in \lim \mathcal{S}(a)$ for each $a \in F$. Hence

$$s[\mathcal{F}] \geq \operatorname{adh}^\natural \mathcal{S}(\mathcal{F}),$$

and

$$\lim s[\mathcal{F}] \supset \lim \operatorname{adh}^\natural \mathcal{S}(\mathcal{F}) \supset \lim \mathcal{S}(\mathcal{F}),$$

where the second inclusion follows from regularity.

$(2) \Longrightarrow (3)$ is clear. $(3) \Longrightarrow (1)$: Assume that (VI.3.5) holds for every A, s, $\mathcal{S} : X \to \mathbb{U}X$ with $x \in \lim \mathcal{S}(x)$ and $\mathcal{U}$ as in the statement, and that $x_0 \in \lim \mathcal{G}$.

For each $G \in \mathcal{G}$ and $x \in \operatorname{adh} G$ there is an ultrafilter $\mathcal{U}_{x,G}$ with $G \in \mathcal{U}_{x,G}$ and $x \in \lim \mathcal{U}_{x,G}$. Of course, $G \in \bigcap_{x \in \operatorname{adh} G} \mathcal{U}_{x,G}$.

Let $A := X \times \mathcal{G}$, let $s : A \to X$ be defined by $s(x, G) = x$ and $\mathcal{S} : A \to \mathbb{U}X$ be defined by $\mathcal{S}(x, G) = \mathcal{U}_{x,G}$. Consider the filter

$$\mathcal{F} := \{\operatorname{adh} G \times \{B \in \mathcal{G} : B \subset G\} : G \in \mathcal{G}\}^\uparrow$$

on A.

Then $s[\mathcal{F}] = \operatorname{adh}^\natural \mathcal{G}$ and $\mathcal{S}(\mathcal{F}) \geq \mathcal{G}$ because if $x \in \operatorname{adh} G$ and $B \in \mathcal{G}$ with $B \subset G$ then $G \in \mathcal{U}_{x,B}$. Hence

$$x_0 \in \lim \mathcal{S}(\mathcal{F}) \subset \lim s[\mathcal{F}] = \lim \operatorname{adh}^\natural \mathcal{G},$$

and the convergence is regular. $\qquad\square$

VI.4   Topological regularity

A convergence ξ is said to be *topologically regular* at x if

$$x \in \lim_{\xi} \mathcal{F} \implies x \in \lim_{\xi} \mathrm{cl}_{\xi}^{\natural} \mathcal{F}, \qquad\qquad (\text{VI.4.1})$$

and *topologically regular* if it is regular at each $x \in |\xi|$. Of course, a topologically regular convergence is self-regular ([2]) and the two properties coincide for topologies. Since the principal adherence of a convergence satisfies $\mathrm{adh}_{\xi} = \mathrm{adh}_{\mathrm{S}_0\,\xi}$ and cl_{ξ} is the principal adherence for $\mathrm{T}\,\xi$, we have:

Proposition VI.4.1. *If ξ is regular and $\mathrm{S}_0\,\xi = \mathrm{T}\,\xi$ then ξ is topologically regular.*

In particular, if a convergence has any of the variants of diagonality in (VI.2.4) and is regular, then it is topologically regular.

However, a regular pretopology does not need to be topologically regular:

Example VI.4.2. Let Σ denote the *sequential tree*, that is, the set of finite sequences of natural numbers: $s \in \Sigma$ if there is $k \in \mathbb{N}$ and $n_1, \ldots, n_k \in \mathbb{N}$ such that $s = (n_1, \ldots, n_k)$. The *length* of s is denoted by $l(s)$. The only element of Σ of zero length is the *empty sequence*. If $s \in \Sigma$ and $n \in \mathbb{N}$ then $s \frown n := (n_1, \ldots, n_k, n)$.

The natural pretopology of Σ can be defined by: $V \in \mathcal{V}(s)$ whenever $s \in V$ and $\{n \in \mathbb{N} : s \frown n \notin V\}$ is finite.

Let ∞ be an additional point and $X := \Sigma \cup \{\infty\}$. By definition, $V \in \mathcal{V}(\infty)$ if there exists $n \in \mathbb{N}$ such that $V_n \subset V$, where

$$V_n := \{s \in \Sigma : l(s) > n\} \cup \{\infty\}.$$

Of course, $\{V_n : n \in \mathbb{N}\}$ is a vicinity base of ∞. If $s \in \Sigma$, then a vicinity base of s consists of closed sets $\{s \frown m : m \geq n\} \cup \{\infty\}$, where $n \in \mathbb{N}$. Therefore the pretopology of X is topologically regular at each $s \in \Sigma$.

However it is regular but not topologically regular at ∞. Indeed, $V_n = \mathrm{adh}\,V_{n+1}$, hence $X = \mathrm{cl}\,V_n = \mathrm{adh}^n\,V_n$.

In an analogous way as for regularity, one proves that

Proposition VI.4.3. *For each convergence ξ there exists the finest topologically regular convergence $\mathrm{Reg}_T\,\xi$ that is coarser than ξ.*

---

[2]because $\mathrm{cl}^{\natural}\,\mathcal{F} \leq \mathrm{adh}^{\natural}\,\mathcal{F} \leq \mathcal{F}$.

Of course, the *topological regularizer* Reg_T is isotone, contractive and idempotent and, moreover,

$$\text{Reg}_T \left(f^- \tau \right) \geq f^- \left(\text{Reg}_T \tau \right). \tag{VI.4.2}$$

Therefore:

Corollary VI.4.4. *Each subspace of a topologically regular convergence is topologically regular.*

Corollary VI.4.5. *Each product of topologically regular convergences is topologically regular.*

Exercise VI.4.6. Prove Proposition VI.4.3 and (VI.4.2).

VI.5 Regularity with respect to another convergence

Topological regularity is an instance of regularity with respect to another convergence. A convergence ξ on X is *regular* (at x) with respect to a convergence θ on X (shortly, *θ-regular*) if

$$\lim_\xi \mathcal{F} \subset \lim_\xi \text{adh}_\theta^\natural \mathcal{F} \tag{VI.5.1}$$

for every filter $\mathcal{F}$ on X (such that $x \in \lim_\xi \mathcal{F}$).

Sure enough, a convergence ξ is regular whenever ξ is ξ-regular; ξ is topologically regular whenever ξ is $\text{T}\,\xi$-regular.

Proposition VI.5.1. *A pretopology ξ is θ-regular if and only if for each $x \in |\xi|$,*

$$\mathcal{V}_\xi(x) \leq \text{adh}_\theta^\natural \mathcal{V}_\xi(x). \tag{VI.5.2}$$

Proof. Indeed, if ξ is a pretopology, then $x \in \lim_\xi \mathcal{V}_\xi(x)$, so that if moreover ξ is θ-regular, then $x \in \lim_\xi \text{adh}_\theta^\natural \mathcal{V}_\xi(x)$, that is, $\mathcal{V}_\xi(x) \leq \text{adh}_\theta^\natural \mathcal{V}_\xi(x)$. Conversely, if (VI.5.2) and $x \in \lim_\xi \mathcal{F}$, then $\mathcal{V}_\xi(x) \leq \mathcal{F}$ hence $\mathcal{V}_\xi(x) \leq \text{adh}_\theta^\natural \mathcal{V}_\xi(x) \leq \text{adh}_\theta^\natural \mathcal{F}$, thus $x \in \text{adh}_\theta^\natural \mathcal{F}$. $\square$

Here is an observation that is often useful in connection with regularity.

Proposition VI.5.2. *If $\mathcal{A}$ and $\mathcal{B}$ are isotone families of subsets of a convergence space (X, ξ) then*

$$(\text{adh}_\xi^\natural \mathcal{A}) \# \mathcal{B} \iff \mathcal{A} \# \mathcal{V}_\xi(\mathcal{B}),$$

where $\mathcal{V}_\xi(\mathcal{B})$ is the contour of $\mathcal{V}_\xi(\cdot)$ along $\mathcal{B}$.

Proof. $\mathrm{adh}_\xi^\natural \mathcal{A} \# \mathcal{B}$ means that $\mathrm{adh}_\xi A \cap B \neq \varnothing$ for every $A \in \mathcal{A}$ and $B \in \mathcal{B}$, which, in view of (V.2.2), means that there is $b \in B$ with $A \in (\mathcal{V}_\xi(b))^\#$. In other words, taking Proposition II.3.5 into account,

$$A \in \bigcup_{b \in B} \mathcal{V}_\xi(b)^\# = (\bigcap_{b \in B} \mathcal{V}_\xi(b))^\# = \mathcal{V}_\xi(B)^\#$$

for every $A \in \mathcal{A}$ and $B \in \mathcal{B}$, that is, $\mathcal{A} \# \mathcal{V}_\xi(\mathcal{B})$. $\qquad\square$

An analogue of Theorem VI.3.18 for θ-regularity is easily obtained by adapting the proof:

Theorem VI.5.3. *Let ξ and θ be two convergences on the same set X. The following are equivalent:*

(1) ξ *is θ-regular;*
(2) *For every set A, every pair of maps $s : A \to X$ and $\mathcal{S} : A \to \mathbb{F}X$ with $s(a) \in \lim_\theta \mathcal{S}(a)$ for each $a \in A$,*

$$\lim_\xi \mathcal{S}(\mathcal{F}) \subset \lim_\xi s[\mathcal{F}], \qquad\qquad (VI.5.3)$$

for each $\mathcal{F} \in \mathbb{F}A$;
(3) *For every set A, every pair of maps $s : A \to X$ and $\mathcal{S} : A \to \mathbb{U}X$ with $s(a) \in \lim_\theta \mathcal{S}(a)$ for each $a \in A$, (VI.5.3) for each $\mathcal{F} \in \mathbb{F}A$.*

Exercise VI.5.4. Prove Theorem VI.5.3.

Note that Theorem VI.5.3 characterizes in particular topological regularity, when taking $\theta = \mathrm{T}\,\xi$.

Chapter VII

Types of separation

In this chapter we consider convergences which distinguish certain objects, like points or closed sets. We have already discussed *Hausdorff convergences* and *regular convergences*. It turns out that the former can be characterized in terms of separation of points by convergent filters, and the latter in terms of separation by convergent filters of points from closed sets. In this chapter we study also *normal convergences*, in which convergent filters separate disjoint closed sets.

Separation turns out to play a key role in the problem, which we consider in the later part of this chapter, of finding a continuous extension of a continuous partial map.

VII.1   Convergence separation

Definition VII.1.1. We say that a convergence ξ *separates* two subsets A_0 and A_1 of $|\xi|$ (or A_0 and A_1 are ξ-*separated*) if for every selection $\mathcal{F}(\cdot)$ of ξ (Definition VI.2.4), the contours (in the sense of Definition VI.1.1) $\mathcal{F}(A_0)$ and $\mathcal{F}(A_1)$ are dissociated.

If a convergence separates A_0 and A_1, then $A_0 \cap A_1 = \varnothing$, because $\mathcal{F}(x) := \{x\}^{\uparrow}$ is a selection of every convergence. Of course,

Proposition VII.1.2. *A pretopology ξ separates A_0 and A_1 if and only if $\mathcal{V}_\xi(A_0)$ and $\mathcal{V}_\xi(A_1)$ are dissociated.*

Proposition VII.1.3. *A topology ξ separates A_0 and A_1 if and only if $\mathcal{N}_\xi(A_0)$ and $\mathcal{N}_\xi(A_1)$ are dissociated.*

Exercise VII.1.4. Prove Propositions VII.1.2 and VII.1.3.

It follows that a topology ξ separates A_0 and A_1 if and only if there exist disjoint $O_0, O_1 \in \mathcal{O}_\xi$ such that $A_0 \subset O_0$ and $A_1 \subset O_1$. More generally,

Definition VII.1.5. Two subsets A_0 and A_1 of a convergence space X are called *topologically separated* if there are two disjoint open subsets U_0 and U_1 such that $A_0 \subset U_0$ and $A_1 \subset U_1$.

Proposition VII.1.6. *A convergence is Hausdorff if and only if it separates each pair of distinct singletons.*

Definition VII.1.7. A convergence ξ is *pretopologically Hausdorff* if $S_0\, \xi$ is Hausdorff, *topologically Hausdorff* if $T\, \xi$ is Hausdorff.

Of course, *topologically Hausdorff* implies *pretopologically Hausdorff*, which implies *Hausdorff*, because $\xi \geq S_0\, \xi \geq T\, \xi$. None of these implications can be reversed.

Example VII.1.8 (A Hausdorff convergence that is not pretopologically Hausdorff). Let $\{X_n : n \in \mathbb{N}\}$ be a set of disjoint infinitely countable sets, $0, \infty \notin \bigcup_{n\in\mathbb{N}} X_n$ and let

$$X := \{\infty\} \cup \{0\} \cup \bigcup_{n\in\mathbb{N}} X_n.$$

Consider a filter-base

$$\mathcal{X} := \left\{ \bigcup_{p>n} X_p : p \in \mathbb{N} \right\}. \tag{VII.1.1}$$

Define a convergence ζ on X by assuming that each $x \in \bigcup_{n\in\mathbb{N}} X_n$ is isolated and

$$\underset{n\in\mathbb{N}}{\forall} \; \{0\} = \lim_\zeta (X_n)_0 \text{ and } \{\infty\} = \lim_\zeta \mathcal{X}.$$

This is a Hausdorff convergence, because $X_n \cap \bigcup_{p>n} X_p = \varnothing$ for each $n \in \mathbb{N}$. However, ζ is not pretopologically Hausdorff. Indeed, $\mathcal{V}_\zeta(\infty) = \mathcal{X}^\uparrow$, $\mathcal{V}_\zeta(0) = \bigcap_{n\in\mathbb{N}} (X_n)_0$ and $\mathcal{V}_\zeta(x) = \{x\}^\uparrow$ if $x \in \bigcup_{n\in\mathbb{N}} X_n$. Let $n \in \mathbb{N}$ and $V \in \mathcal{V}_\zeta(0)$, that is, for each $k \in \mathbb{N}$, there is a cofinite subset V_k of X_k such that $\bigcup_{k\in\mathbb{N}} V_k \subset V$. It follows that $\bigcup_{p>n} X_p \cap V \neq \varnothing$ for each n, that is, $\mathcal{V}_\zeta(\infty) \,\#\, \mathcal{V}_\zeta(0)$.

Example VII.1.9 (A pretopologically Hausdorff convergence that is not topologically Hausdorff). Let $\{X_n : n \in \mathbb{N}\}$ be a set of disjoint infinitely countable sets, let $\{x_n : n \in \mathbb{N}\}$ be a set of distinct points and let

$$X := \{\infty\} \cup \{0\} \cup \{x_n : n \in \mathbb{N}\} \cup \bigcup_{n\in\mathbb{N}} X_n$$

be a disjoint union. Let σ be a sequentially based convergence for which each $x \in \bigcup_{n \in \mathbb{N}} X_n$ is isolated,

$$\{0\} = \lim_\sigma (x_n)_n \,, \{x_n\} = \lim_\sigma (X_n)_0 \text{ and } \{\infty\} = \lim_\sigma \mathcal{E}$$

whenever $\mathcal{E}$ is a sequential filter finer than $\mathcal{X}$, where $\mathcal{X}$ s defined in (VII.1.1). Of course,

$$\mathcal{V}_\sigma(0) = (x_n)_n^\uparrow, \mathcal{V}_\sigma(\infty) = \mathcal{X}^\uparrow, \mathcal{V}_\sigma(x_n) = (X_n)_0$$

and $\mathcal{V}_\sigma(x) = \{x\}^\uparrow$ if $x \in \bigcup_{n \in \mathbb{N}} X_n$.

Observe that σ is pretopologically Hausdorff, but not topologically Hausdorff, because $\mathcal{N}_\sigma(0)$ is the contour of $\{(X_n)_0 : n \in \mathbb{N}\}$ along $(x_n)_n^\uparrow$, so that if $N \in \mathcal{N}_\sigma(0)$, then there is $n \in \mathbb{N}$ and for each $k > n$ there is a cofinite subset V_k of X_k such that $\bigcup_{k>n} V_k \subset N$. This shows that $N \in \mathcal{V}_\sigma(\infty)^\# = \mathcal{N}_\sigma(\infty)^\#$.

Proposition VII.1.10. *A pretopology is regular if and only if it separates points from the complements of their vicinities.*

Proof. Let ξ be a pretopology on X. Suppose that $\{x\}$ is ξ-separated from $X \setminus V$ for each $V \in \mathcal{V}_\xi(x)$. In other words, $\mathcal{V}_\xi(x)$ is dissociated from $\mathcal{V}_\xi(X \setminus V)$, which by Proposition VI.5.2 means that $\text{adh}_\xi^\natural \mathcal{V}_\xi(x)$ is dissociated from $\{X \setminus V\}^\uparrow$. Equivalently, for each $V \in \mathcal{V}_\xi(x)$ there exists $W \in \text{adh}_\xi^\natural \mathcal{V}_\xi(x)$ such that $W \subset V$, that is, $\mathcal{V}_\xi(x) \leq \text{adh}_\xi^\natural \mathcal{V}_\xi(x)$. $\square$

As a consequence we recover Proposition VI.3.10.

Corollary VII.1.11. *A topology is regular if and only if it separates points from closed sets.*

Definition VII.1.12. A convergence is called *normal* if it separates each pair of disjoint closed sets. A convergence ξ is called *topologically normal* if $\mathrm{T}\,\xi$ is normal.

Example VII.1.13 (A normal convergence that is not topologically normal). Consider the convergence σ from Example VII.1.9. To see that this convergence has the desired properties, we shall review infinite σ-closed sets. Let A be an infinite set such that $A \subset \bigcup_{n \in \mathbb{N}} X_n$. Then $x_n \in \text{adh}_\sigma A$ if and only if $A \cap X_n$ is infinite; $\infty \in \text{adh}_\sigma A$ (equivalently, $\infty \in \text{cl}_\sigma A$) if and only if $A \cap X_n \neq \varnothing$ for infinitely many n; finally, $0 \in \text{cl}_\sigma A$ if and only if $\{x_n : n \in \mathbb{N}\} \cap \text{adh}_\sigma A$ is infinite.

Accordingly, if a σ-closed set F_0 has infinite intersection with $\bigcup_{n \in \mathbb{N}} X_n$ and F_1 is an infinite σ-closed set disjoint from F_0, then $F_1 \cap \bigcup_{n \in \mathbb{N}} X_n$ is finite. Therefore $H_1 := F_1 \cap \{x_n : n \in \mathbb{N}\}$ is infinite and thus $0 \in F_1$. It follows that $H_0 := F_0 \cap \{x_n : n \in \mathbb{N}\}$ is finite, because, otherwise, $0 \in F_0$, contradicting that $F_0 \cap F_1 = \varnothing$. As the elements of $\bigcup_{n \in \mathbb{N}} X_n$ are isolated, it is enough to see that $H_1 \cup \{0\}$ and $H_0 \cup \{\infty\}$ are separated. This is obviously the case, because if $x_n \in H_1$ then $X_n \cap F_0$ is finite.

On the other hand, σ is not topologically normal, because each σ-open set O including the closed set $\{0\} \cup \{x_n : n \in \mathbb{N}\}$ belongs to (VII.1.1), hence $O \in \mathcal{N}_\sigma (\infty)^\#$.

VII.2   Regularity with respect to a family of sets

Let $\mathcal{Z}$ be a family of subsets of X. A convergence ξ on X is called $\mathcal{Z}$-*regular* at x if $x \in \lim_\xi \mathcal{F}$ implies the existence of a filter base $\mathcal{H} \subset \mathcal{Z}$ such that $x \in \lim_\xi \mathcal{H}$ and $\mathcal{H} \leq \mathcal{F}$.

In other words, ξ is $\mathcal{Z}$-regular at x if there exists a pavement of ξ at x, each element of which is based in $\mathcal{Z}$. This can be rephrased as follows.

Proposition VII.2.1. *A convergence ξ is $\mathcal{Z}$-regular at x if and only if $\mathcal{Z}$ is a base of ξ at x.*

Of course, a convergence ξ is said to be $\mathcal{Z}$-*regular* if it is $\mathcal{Z}$-regular at x for every $x \in |\xi|$.

Example VII.2.2. If X is a vector space and $\mathcal{Z}$ is the set of convex sets, then ξ is called *locally convex* (at x) provided that ξ is $\mathcal{Z}$-regular (at x). Notice that ξ is locally convex whenever

$$\lim_\xi \mathcal{F} \subset \lim_\xi \mathrm{conv}^\natural \, \mathcal{F},$$

where $\mathrm{conv}\, F$ stands for the convex hull of F and $\mathrm{conv}^\natural \mathcal{F}$ is the filter generated by $\{\mathrm{conv}\, F : F \in \mathcal{F}\}$. Indeed, if $x \in \lim_\xi \mathcal{F}$ then by definition there is a convexly based filter $\mathcal{H}$ coarser than $\mathcal{F}$ such that $x \in \lim_\xi \mathcal{H}$. Accordingly $\mathcal{H} \approx \mathrm{conv}^\natural \mathcal{H} \leq \mathrm{conv}^\natural \mathcal{F}$, so that $x \in \lim_\xi \mathrm{conv}^\natural \mathcal{F}$.

Example VII.2.3. It is clear that ξ is topologically regular (or $\mathrm{T}\,\xi$-regular) if and only if it is $\mathcal{Z}$-regular, where

$$\mathcal{Z} = \{Z \subset X : \mathrm{cl}_\xi Z \subset Z\} .$$

Notice that in Example VII.2.2, $\mathcal{Z}$ is independent of ξ, while in Example VII.2.3, $\mathcal{Z}$ depends on ξ.

If $\mathcal{Z}$ is a family of subsets of X that is closed under finite intersection, that is, if $\mathcal{Z} = \mathcal{Z}^\cap$, we call $\mathcal{Z}$ a *meet-semilattice of subsets of X.*

Proposition VII.2.4. *Let $\mathcal{Z}$ be a meet-semilattice of subsets of X (independent of a convergence). Then each supremum of $\mathcal{Z}$-regular convergences is $\mathcal{Z}$-regular.*

Proof. Let Ξ be a set of $\mathcal{Z}$-regular convergences on X and let $x \in \lim_{\bigvee \Xi} \mathcal{F}$, that is, $x \in \lim_{\xi} \mathcal{F}$ for each $\xi \in \Xi$. Then for each $\xi \in \Xi$ there exists a filter base $\mathcal{H}_\xi \leq \mathcal{F}$ with $\mathcal{H}_\xi \subset \mathcal{Z}$ such that $x \in \lim_{\xi} \mathcal{H}_\xi$. In view of Proposition II.2.15, there exists $\bigvee_{\xi \in \Xi} \mathcal{H}_\xi \leq \mathcal{F}$, and since $\bigvee_{\xi \in \Xi} \mathcal{H}_\xi$ consists of finite intersections of elements of $\bigcup_{\xi \in \Xi} \mathcal{H}_\xi$, it is based in $\mathcal{Z}$, which proves that $\bigvee_{\xi \in \Xi} \xi$ is $\mathcal{Z}$-regular. $\qquad\square$

If $\mathcal{Z}(\xi)$ is a meet-semilattice of subsets of $|\xi|$ for each convergence ξ, then we call $\mathcal{Z}(\cdot)$ an *assignment*. An assignment $\mathcal{Z}(\cdot)$ is called *isotone* if

$$\xi \leq \zeta \Longrightarrow \mathcal{Z}(\xi) \leq \mathcal{Z}(\zeta).$$

Proposition VII.2.5. *If $\mathcal{Z}(\cdot)$ is an isotone assignment, then each supremum of $\mathcal{Z}$-regular convergences is $\mathcal{Z}$-regular, and thus for every convergence ξ there is the finest $\mathcal{Z}$-regular convergence $\mathrm{Reg}_{\mathcal{Z}}\, \xi$ that is coarser than ξ.*

Proof. As in the preceding proof, on observing that $\bigcup_{\xi \in \Xi} \mathcal{Z}_\xi \subset \mathcal{Z}_{\bigvee_{\xi \in \Xi} \xi}$. $\qquad\square$

Proposition VII.2.6. *If an assignment $\mathcal{Z}(\cdot)$ is isotone and if*

$$f^-\left[\mathcal{Z}(\tau)\right] \subset \mathcal{Z}(f^-\tau), \tag{VII.2.1}$$

for each convergence τ on the range of f, then $\mathrm{Reg}_{\mathcal{Z}}$ preserves continuity:

$$C(\xi, \tau) \subset C(\mathrm{Reg}_{\mathcal{Z}}\, \xi, \mathrm{Reg}_{\mathcal{Z}}\, \tau)$$

for every ξ and τ.

Proof. It is enough to show that if τ is $\mathcal{Z}(\tau)$-regular then $f^-\tau$ is $\mathcal{Z}(f^-\tau)$-regular. If $x \in \lim_{f^-\tau} \mathcal{F}$ then $f(x) \in \lim_{\tau} f\,[\mathcal{F}]$ and since τ is $\mathcal{Z}(\tau)$-regular, there is a filter base $\mathcal{H} \subset \mathcal{Z}(\tau)$ such that $\mathcal{H} \leq f\,[\mathcal{F}]$ and $f(x) \in \lim_{\tau} \mathcal{H}$. Therefore, $f^-\,[\mathcal{H}] \leq f^- f\,[\mathcal{F}] \leq \mathcal{F}$ and $x \in \lim_{f^-\tau} f^-\,[\mathcal{H}]$, because $f(x) \in \lim_{\tau} f f^-\,[\mathcal{H}]$ and $\mathcal{H} \leq f f^-\,[\mathcal{H}]$. In view of (VII.2.1), $f^-\,[\mathcal{H}] \subset \mathcal{Z}(f^-\tau)$, so that $f^-\tau$ is $\mathcal{Z}(f^-\tau)$-regular. $\qquad\square$

In particular, because $\xi \leq \zeta$ implies that $C(\xi, \$) \subset C(\zeta, \$)$ and $f \in C(\xi, \tau)$ implies that $f \in C(\mathrm{T}\, \xi, \mathrm{T}\, \tau)$, we recover (VI.4.2) for $\mathrm{Reg}_T = \mathrm{Reg}_{\mathcal{Z}}$ where $\mathcal{Z}(\cdot)$ assigns to a convergence its collection of closed subsets.

Remark VII.2.7. Preimages of convex sets are convex under linear maps. Therefore functorial aspects of locally convex convergences concern categories of convergence vector spaces.

In Examples VII.2.2 and VII.2.3, the classes $\mathcal{Z}$ of subsets of a convergence spaces could be represented as the fixed points of the convex hull conv and of the closure cl respectively. This is akin to regularity with respect to a convergence as introduced in the preceding chapter.

VII.3 Functionally induced convergences

In Section IV.3, we considered initial convergences with respect to a family of maps $\boldsymbol{F}$. When ζ is initial with respect to $\boldsymbol{F} = C(\zeta, \upsilon)$, we say that ζ is υ-*initial*. In other words, the class of all the convergences that are υ-initial is precisely that in which υ is initially dense. If we set ([1])

$$R_\upsilon \zeta := \bigvee_{f \in C(\zeta, \upsilon)} f^- \upsilon, \qquad (\text{VII.3.1})$$

then ζ is υ-initial whenever $\zeta = R_\upsilon \zeta$.

As in (IV.3.1),

$$x \in \lim_{R_\upsilon \zeta} \mathcal{F}$$

if and only if $f(x) \in \lim_\upsilon f[\mathcal{F}]$ for every $f \in C(\zeta, \upsilon)$. If in particular υ is a topology, then for any convergence ξ, $R_\upsilon \xi$ is a topology because the class of topologies is closed under initial convergences and suprema (Propositions V.4.32 and V.4.21). Moreover:

Proposition VII.3.1. *If υ is a topology and ζ a convergence, then $R_\upsilon \zeta$ is a topology and*

$$\left\{ f^-(O) : O \in \mathcal{O}_\upsilon, f \in C(\zeta, \upsilon) \right\}^\cap \qquad (\text{VII.3.2})$$

is a base of open sets for $R_\upsilon \zeta$.

Proof. In this situation, $x \in \lim_{R_\upsilon \zeta} \mathcal{F}$ if and only if $f[\mathcal{F}] \geq \mathcal{O}_\upsilon(f(x))$ for every $f \in C(\zeta, \upsilon)$, that is, if and only if $f^-(O) \in \mathcal{F}$ whenever $O \in \mathcal{O}_\upsilon$, $f \in C(\zeta, \upsilon)$ and $f(x) \in O$. Thus

$$\mathcal{N}_{R_\upsilon \zeta}(x) = \left(\left\{ f^-(O) : O \in \mathcal{O}_\upsilon(f(x)), f \in C(\zeta, \upsilon) \right\}^\cap \right)^\uparrow,$$

and the result follows. $\qquad \square$

---

[1]We will study a more general scheme in Section XIV.7.

If ζ is a convergence on X, υ is a convergence on Z, and $F \subset C(\zeta,\upsilon)$, then

$$\Delta F : X \to \prod_{f \in F} Z \approx Z^F \subset Z^{(Z^X)}$$

becomes the *point-evaluation map*, that is, $(\Delta F)(x) : F \to Z$ and, for each $h \in F$,

$$(\Delta F)(x)(h) = h(x).$$

By Proposition IV.6.2:

Corollary VII.3.2. *If* $F \subset C(\zeta,\upsilon)$ *then* ΔF *is continuous from* ζ *to* $\prod_{f \in F} \upsilon$.

In particular,

Corollary VII.3.3. *If* $F = C(\zeta,\upsilon)$ *then* ΔF *is continuous from* ζ *to* $\prod_{f \in C(\zeta,\upsilon)} \upsilon$.

By Corollary IV.6.3:

Proposition VII.3.4. *If* ζ *is* υ-initial and $F = C(\zeta,\upsilon)$ *separates points, then* ΔF *is an embedding of* ζ *to* $\prod_{f \in C(\zeta,\upsilon)} \upsilon$.

Example VII.3.5. The class of convergences that are initial with respect to the Sierpiński topology $\$$ is that of all topologies, because $\$$ is initially dense in the class of topologies (Proposition V.4.18). We observe that for each convergence ξ,

$$\mathrm{T}\xi = \mathrm{R}_\$ \, \xi$$

is the topologization of ξ.

Recall that there are two homeomorphic variants of $\$$ on $\{0,1\}$, namely $\$_0$ for which $\{0\}$ is open (and $\{1\}$ is not) and $\$_1$ for which $\{1\}$ is open (and $\{0\}$ is not).

Accordingly, $f \in C(\zeta,\$_0)$ whenever f is the indicator function of a ζ-open set. The general formula (VII.3.2) for a base of open sets of a topology ζ becomes

$$\left\{ f^-(0) : f \in C(\zeta,\$_0) \right\},$$

because $\{0\}$ is the only non-empty open (distinct from the whole space) set for $\$_0$. Therefore, if $\mathcal{B}$ is a base of open sets for ζ, then ζ is $\{\psi_B : B \in \mathcal{B}\}$-initial.

Similarly,

Example VII.3.6. The class of convergences that are initial for $\maltese$ (where $\maltese$ is the Bourdaud pretopology) is that of pretopologies. Therefore

$$S_0\,\xi = R_{\maltese}\,\xi.$$

Example VII.3.7. If ν is the usual convergence of the real line and ξ is a convergence, then $R_\nu\,\xi$ is the coarsest among the convergences θ on $|\xi|$ for which

$$C(\xi,\nu) \subset C(\theta,\nu).$$

As we have seen, if υ is a topology then each υ-initial convergence is a topology. Consequently, $R_\nu\,\xi$ is a topology for each ξ. The class of ν-initial topologies is strictly included in that of all topologies and will be studied in details in Section VII.6.

We notice that for a convergence υ, and two convergences ζ and ξ on the same set,

$$\zeta \geq \xi \Longrightarrow R_\upsilon\,\zeta \geq R_\upsilon\,\xi, \qquad\qquad \text{(isotone)}$$

$$\zeta \geq R_\upsilon\,\zeta, \qquad\qquad \text{(contractive)}$$

$$R_\upsilon(R_\upsilon\,\zeta) = R_\upsilon\,\zeta. \qquad\qquad \text{(idempotent)}$$

Moreover,

Lemma VII.3.8. *Let υ be a convergence. Then for each convergence τ and each map f valued in $|\tau|$,*

$$R_\upsilon\left(f^-\tau\right) \geq f^-\left(R_\upsilon\,\tau\right). \qquad\qquad \text{(initial)}$$

Proof. Indeed,

$$f^-(R_\upsilon\,\tau) = f^-(\bigvee_{\varphi\in C(\tau,\upsilon)} \varphi^-\upsilon) \overset{(IV.3.5)}{=} \bigvee_{\varphi\in C(\tau,\upsilon)} (\varphi\circ f)^-\upsilon.$$

Now $\varphi \in C\,(\tau,\upsilon)$ yields

$$\upsilon \leq \varphi\tau \leq \varphi(f(f^-\tau)) = (\varphi\circ f)\,(f^-\tau),$$

that is, $\varphi\circ f \in C\,(f^-\tau,\upsilon)$. Thus

$$f^-(R_\upsilon\,\tau) = \bigvee_{\varphi\in C(\tau,\upsilon)} (\varphi\circ f)^-\upsilon \leq \bigvee_{\psi\in C(f^-\tau,\upsilon)} \psi^-\upsilon = R_\upsilon\left(f^-\tau\right).$$

$\square$

Therefore, each subspace and every product of υ-initial convergences is υ-initial.

VII.4 Real-valued functions

In the rest of this chapter, real-valued continuous functions on a convergence space will play a crucial role. Therefore, we gather here some basic information on the set $C(\xi, \nu)$ of real-valued continuous functions $f : |\xi| \to |\nu|$. As usual, algebraic operations on $\mathbb{R}$ such as sum $+$, difference $-$, multiplication $\times$, absolute value $|\cdot|$, pairwise minima $\wedge$ and pairwise maxima $\vee$ are lifted to real-valued functions by defining these operations pointwise. Specifically, if $*$ is a binary operation on $\mathbb{R}$ (such as $+, -, \cdot, \div, \vee, \wedge$), we lift $*$ to $\mathbb{R}^X$ by

$$(f * g)(x) = f(x) * g(x),$$

and $|f|(x) := |f(x)|$ for all $x \in X$.

Let X be a set and $r \in \mathbb{R}$. We define $\bar{r} \in \mathbb{R}^X$ by

$$\bar{r}(x) := r$$

for each $x \in X$. We will often use the notation $\bar{r}$ without specifying its domain, when it is clear from context.

Exercise VII.4.1. Let ξ be a convergence, ν be the standard convergence of the real line and let $f, g \in C(\xi, \nu)$. Show that

(1) $|f|$ is continuous;
(2) $f + g$, $f - g$ and $f \cdot g$ are also continuous;
(3) $\dfrac{f}{g}$ is continuous at each x such that $g(x) \neq 0$.

Uniform convergence

Definition VII.4.2. Let (Y, d) be a metric space and let X be a set. A filter $\mathcal{F}$ on Y^X is said to *converge uniformly* to a function $f_0 \in Y^X$ if for every $\varepsilon > 0$, there is $F \in \mathcal{F}$ such that

$$\langle x, F \rangle \subset B(f_0(x), \varepsilon) \qquad\qquad \text{(VII.4.1)}$$

for every every $x \in X$.

This defines a convergence on Y^X called *uniform convergence*. In fact, uniform convergence is a topology.

Exercise VII.4.3. Show that uniform convergence is a topology.

Accordingly, a sequence $\{f_n\}_n$ of real-valued functions on X converges to $f \in \mathbb{R}^X$ if for every $\varepsilon > 0$, there is $k \in \mathbb{N}$ such that

$$\underset{x \in |\xi|}{\forall}\ n \geq k \Longrightarrow |f_n(x) - f(x)| < \varepsilon.$$

Exercise VII.4.4. Show that the uniform convergence on $C(\mathbb{R}, \mathbb{R})$ is finer than the convergence of joint continuity of Example IV.5.3.

Proposition VII.4.5. *Let (X, ξ) be a convergence space, (Y, d) be a metric space, $\mathcal{F}$ on filter on $C(\xi, \tilde{d})$ and $f_0 \in Y^X$. If $\mathcal{F}$ converges uniformly to f_0, then $f_0 \in C(\xi, \tilde{d})$.*

Proof. Let $\mathcal{G} \in \mathbb{F}X$ with $x_0 \in \lim_\xi \mathcal{G}$ and let $\varepsilon > 0$. By uniform convergence of $\mathcal{F}$ to f_0, there is $F \in \mathcal{F}$ such that for every $x \in X$ and every $f \in F$, $d(f(x), f_0(x)) < \frac{\varepsilon}{3}$. On the other hand, each $f \in F$ is continuous, so that there is $G_f \in \mathcal{G}$ with $f(G_f) \subset B(f(x_0), \frac{\varepsilon}{3})$. Thus, for a given $f \in F$ and $x \in G_f$,

$$d(f_0(x), f_0(x_0)) \leq d(f_0(x), f(x)) + d(f(x), f(x_0)) + d(f(x_0), f_0(x_0)),$$

where the first and last terms are less than $\frac{\varepsilon}{3}$ by uniform convergence, and the middle term is less than $\frac{\varepsilon}{3}$ by continuity of f. Thus $f_0(G_f) \subset B(f_0(x_0), \varepsilon)$ and f_0 is continuous. $\square$

In particular, the uniform limit of a sequence of real-valued continuous functions is continuous.

If $\{g_n\}_{n \in \mathbb{N}}$ is a sequence of real-valued continuous functions then each of the associated partial sums $S_k := \sum_{n=0}^{k} g_n$ is also continuous. Under the following assumptions, the sequence of partial sums is also uniformly convergent, hence has a continuous limit ([2]):

Corollary VII.4.6. *If $g_n : |\xi| \to |\nu|$, $n \in \mathbb{N}$, defines a sequence of continuous real-valued functions and for each n there is M_n with $|g_n| \leq \overline{M_n}$ and $\sum_{n=0}^{\infty} M_n$ is a convergent numerical series, then $\sum_{n=0}^{\infty} g_n$ is a continuous function.*

VII.5   Functionally closed and open sets

In the rest of this chapter, we will frequently make use of the conventions spelled out in (I.1.1). If ξ is a convergence and $f : |\xi| \to \mathbb{R}$ is *continuous*,

---

[2]This a classical result of Advanced Calculus, often referred to as Weierstraß' test, whose proof relies on Proposition VII.4.5 and completeness of the reals. We thus postpone a full justification to Chapter X.

then $f^-(0) = \{f = 0\}$ is called the *zero-set of f*; its complement is called *cozero-set of f*. A subset A of $|\xi|$ is a *zero-set* or a *functionally closed set* if there is a continuous function $f : |\xi| \to \mathbb{R}$ such that $A = \{f = 0\}$. A *cozero-set* or *functionally open set* is the complement of a zero-set.

We denote by $\mathcal{C}_\xi^\nu$ and $\mathcal{O}_\xi^\nu$ respectively the sets of all functionally ξ-closed and ξ-functionally open sets.

Of course, each functionally closed set is closed because singletons are closed in ν and the preimage of a closed set under a continuous function is closed (Corollary V.3.27), and therefore each functionally open set is open. Note that for every real-valued functions f and g, and every $n \in \mathbb{N}_1$,

$$\{f = 0\} = \{|f| = 0\}) = \{f^n = 0\},$$
$$\{f \cdot g = 0\} = \{f = 0\} \cup \{g = 0\},$$
$$\{|f| + |g|) = 0\} = \{f = 0\} \cap \{g = 0\},$$
$$\{f = 0\} = \bigcap_{n \in \mathbb{N}} \left\{ x \in X : |f(x)| < \frac{1}{2^n} \right\}.$$

Hence $\mathcal{C}_\xi^\nu$ (and thus also $\mathcal{O}_\xi^\nu$) is closed under both finite intersections and finite unions. Recall that a family of subsets of a given set is called a *lattice of subsets* if it is stable for finite unions and finite intersections.

Corollary VII.5.1. *The families $\mathcal{C}_\xi^\nu$ of functionally closed sets, and $\mathcal{O}_\xi^\nu$ of functionally open sets are lattices of subsets of $|\xi|$.*

Moreover every functionally closed set is a $G_{\aleph_0}$-set, also called G_δ-set, that is, a countable intersection of open sets. Dually, a countable union of closed sets is called an F_σ-set.

Since for each real-valued function f,

$$\{f \geq 0\} = \{f \wedge \overline{0} = 0\} = \{f - |f| = 0\}$$
$$\{f \leq 0\} = \{f \vee \overline{0} = 0\} = \{f + |f| = 0\},$$

we conclude that these sets are functionally closed sets, so that the open sets $\{f > 0\}$ and $\{f < 0\}$ are functionally open sets. Moreover, each functionally open set is of that form, since $|\xi| \setminus \{f = 0\} = \{|f| > 0\}$.

Proposition VII.5.2. *A countable intersection of functionally closed sets is functionally closed.*

Proof. If $\{f_n\}_{n \in \mathbb{N}}$ is a sequence of continuous real-valued functions, so is $\{g_n\}_{n \in \mathbb{N}}$ where $g_n := \min(|f_n|, \frac{1}{2^n})$. Moreover, in view of Corollary VII.4.5,

$g = \sum_{n=0}^{\infty} g_n$ is continuous. The conclusion thus follows from the observation that

$$\{g = 0\} = \bigcap_{n \in \mathbb{N}} \{g_n = 0\} = \bigcap_{n \in \mathbb{N}} \{f_n = 0\}.$$

$\square$

Definition VII.5.3. Two subsets A_0, A_1 of a convergence space X are called *functionally separated* if there is a continuous function $f : X \to [0,1] \subset \mathbb{R}$ (where $\mathbb{R}$ carries its standard convergence) with $f(A_0) = \{0\}$ and $f(A_1) = \{1\}$.

Of course, every functionally separated pair of sets is also topologically separated, for if f is as in the definition of functionally separated sets, then $\{f < \frac{1}{2}\}$ and $\{f > \frac{1}{2}\}$ are two disjoint open sets containing A_0 and A_1 respectively.

Note that to see that A_0 and A_1 are functionally separated, it is enough to find $g \in C(X, \mathbb{R})$ with

$$\sup_{x \in A_0} g(x) < \inf_{x \in A_1} g(x).$$

Indeed, if $r := \sup_{x \in A_0} g(x)$ and $s := \inf_{x \in A_1} g(x)$, then

$$f := \frac{1}{s-r}\left(\left((\overline{r} \vee g) \wedge \overline{s}\right) - \overline{r}\right)$$

has the property from the definition.

Proposition VII.5.4. *Two sets are functionally separated if and only if they are included in disjoint functionally closed sets.*

Proof. If A and B are functionally separated by f with $f(A) = \{0\}$ and $f(B) = \{1\}$ then $\{f \le \frac{1}{3}\}$ and $\{f \ge \frac{2}{3}\}$ are disjoint zero-sets containing A and B respectively.

Conversely, if $A \subset \{f = 0\}$ and $B \subset \{g = 0\}$ with $\{f = 0\} \cap \{g = 0\} = \varnothing$ then

$$h(x) := \frac{|f(x)|}{|f(x)| + |g(x)|}$$

is a continuous real-valued function satisfying $h(A) \subset h(\{f = 0\}) = \{0\}$ and $h(B) \subset h(\{g = 0\}) = \{1\}$. $\square$

As in a metric space, for every non-empty set A, the distance function dist_A defined by $\text{dist}_A(x) = \inf\{d(a,x) : a \in A\}$ is continuous and $\{\text{dist}_A(\cdot) = 0\} = \text{cl}\, A$, (Exercise V.3.8) we infer that:

Proposition VII.5.5. *In a metrizable space each closed set is functionally closed.*

Definition VII.5.6. We say that a convergence space X is *functionally Hausdorff* if for any two different points x_0, x_1, the singletons $\{x_0\}, \{x_1\}$ are functionally separated.

Of course, every functionally Hausdorff convergence is topologically Hausdorff. There exist Hausdorff (regular) topologies (of cardinality strictly greater than 1) in which each continuous function is constant (see [54] by Horst Herrlich). In particular, such a topology is Hausdorff but not functionally Hausdorff.

VII.6 Functional regularity (aka complete regularity)

In this section, we revisit the results of Section VII.3 when v is the standard convergence ν on the real line $\mathbb{R}$.

Recall that $C(\xi, \nu)$ or $C(X, \mathbb{R})$ denotes the set of continuous functions $f : |\xi| \to |\nu|$. In this case (VII.3.1) becomes

$$\mathrm{R}_\nu \, \xi = \bigvee_{f \in C(\xi,\nu)} f^- \nu.$$

Definition VII.6.1. A convergence is called *functionally regular* if it is ν-initial.

A Hausdorff functionally regular space is traditionally called *completely regular* or *Tikhonov* or $T_{3\frac{1}{2}}$.

In other words, a convergence ξ is functionally regular if it carries the initial convergence for its real-valued continuous functions, that is, if

$$\xi \leq \mathrm{R}_\nu \, \xi. \tag{VII.6.1}$$

Of course, the reverse inequality is always true, as the continuity of f rephrases, via Proposition IV.2.9, as $\xi \geq f^- \nu$.

By Proposition V.4.32, $f^- \nu$ is a topology for each $f \in C(\xi, \nu)$, and a supremum of topologies is a topology. Hence:

Proposition VII.6.2. *A functionally regular convergence is a topology.*

By virtue of Proposition VII.3.1,

Corollary VII.6.3. *The finite intersections of sets of the form*
$$\{x \in X : |f(x) - f(x_0)| < \varepsilon\},$$

where $f \in C(\xi, \nu)$ *and* $\varepsilon > 0$ *form a neighborhood base of* $\mathrm{R}_\nu \xi$ *at* x_0.

Proposition VII.6.4. *A convergence is functionally regular if and only if it is initial with respect to the usual topology of* $[0, 1]$.

Proof. As $C\left(\xi, \nu_{|[0,1]}\right) \subset C(\xi, \nu)$, the initial convergence with respect to ν is finer than that with respect to $\nu_{|[0,1]}$. But since $C\left(\xi, \nu_{|(0,1)}\right) \subset C\left(\xi, \nu_{|[0,1]}\right)$ and $\nu_{|(0,1)}$ is homeomorphic to ν, also the converse inequality holds. $\qquad\square$

Proposition VII.6.5. *A topological space is functionally regular if and only if the functionally open sets form a base for open sets.*

Proof. Assume that X is functionally regular, and denote by $\mathcal{Z}(X)$ the set of its functionally closed subsets. Let $x_0 \in O$, where O is an open subset of a functionally regular space X. The family

$$\{A \subset X : x_0 \in A, \, X \setminus A \in \mathcal{Z}(X)\}$$

is a filter-base, because

$$\{f_1 \cdot f_2 \neq 0\} \subset \{f_1 \neq 0\} \cap \{f_2 \neq 0\}.$$

Let $\mathcal{F}$ denote the filter it generates. For every $\varepsilon > 0$, the set

$$\{x \in X : |f(x) - f(x_0)| < \varepsilon\} = \{g < 0\},$$

where g is the continuous function $|f - f(x_0)| - \varepsilon$, is a functionally open set as observed in the previous section, and contains x_0, hence belongs to $\mathcal{F}$. Thus $f(x_0) \in \lim_\nu f[\mathcal{F}]$ for each $f \in C(X, \mathbb{R})$, so that

$$x_0 \in \lim_{\bigvee_{f \in C(X,\mathbb{R})} f^- \nu} \mathcal{F} \subset \lim \mathcal{F}.$$

Therefore, $O \in \mathcal{F}$, because O is open, and there is a functionally open set A with $x_0 \in A \subset O$. Hence functionally open sets form a base for open sets.

Conversely, if the functionally open sets form a base for open sets and $\mathcal{F}$ is a filter on a topological space X with $f(x_0) \in \lim_\nu f[\mathcal{F}]$ for each $f \in C(X, \mathbb{R})$, it is enough to show that each functionally open set containing x_0 is in $\mathcal{F}$ to conclude that $x \in \lim \mathcal{F}$. Let $A := \{f_A \neq 0\}$ with $f_A(x_0) \neq 0$. If $A \notin \mathcal{F}$ then $A^c \in \mathcal{F}^\#$ so that each $F \in \mathcal{F}$ contains a point x_F such that $f_A(x_F) = 0$. Hence $0 \in \ker(f_A[\mathcal{F}]) \subset \mathrm{adh}_\nu \, f_A[\mathcal{F}]$, which is incompatible with

$$0 \neq f_A(x_0) \in \lim_\nu f_A[\mathcal{F}] = \mathrm{adh}_\nu \, f_A[\mathcal{F}],$$

because ν is Hausdorff. Thus $A \in \mathcal{F}$.

We conclude that X is functionally regular. $\qquad\square$

Corollary VII.6.6. *A topology is functionally regular if and only if the functionally closed sets form a base for closed sets.*

In view of Proposition VII.5.5:

Corollary VII.6.7. *Every metrizable space is functionally regular.*

For instance, the discrete topology is metrizable, hence functionally regular. On the other hand, there are non T_1 functionally regular topologies:

Example VII.6.8. The chaotic topology o on any set X is functionally regular. In fact, $C(o, \nu)$ consists of constant functions. Of course, if f is constant, that is, if there is $r \in \mathbb{R}$ such that $f(x) = r$ for each $x \in X$, then

$$f^-(0) = \begin{cases} X, & \text{if } r = 0, \\ \varnothing, & \text{if } r \neq 0, \end{cases}$$

which give all the closed subsets of o_X.

Corollary VII.6.9. *A topology ξ is functionally regular if and only if for each closed set F and $x \notin F$ there exists $f \in C(\xi, \nu)$ such that $f(x) = 1$ and $f(F) = \{0\}$.*

Proof. By Corollary VII.6.6, for each ξ-closed set F and $x \notin F$, there exists $f_0 \in C(\xi, \nu)$ such that $F \subset \{f_0 = 0\}$ and $x \notin \{f_0 = 0\}$, hence $f_0(x) > 0$. Set $f := \min(\frac{1}{f_0(x)} f_0, 1)$, and the result follows. $\square$

Remark VII.6.10. In view of the proof, the map f in Corollary VII.6.9 can equivalently be chosen in $C(\xi, \nu_{|[0,1]})$.

As a consequence, if a functionally regular space is T_1 then *continuous functions separate points*, that is,

$$x \neq t \implies \underset{f \in C(X, \mathbb{R})}{\exists} f(x) \neq f(t),$$

because $\{x\}$ and $\{t\}$ are functionally separated. The topology is in particular Hausdorff.

In view of Proposition VI.3.10, functionally regular convergence spaces are in particular regular topological spaces, which justifies the name.

However, there are Hausdorff regular topologies that are not functionally regular. Indeed, as we have already mentioned, there exist non-trivial Hausdorff regular topologies in which each continuous real-valued function is constant [54].

Example VII.6.11 (A regular Hausdorff topology that is not functionally regular). Let X be the subset of the plane
$$X := \{(x,y) : y \geq 0\} \cup \{(0,-1)\}.$$
Points (x,y) of X with $y > 0$ are isolated. Points $z = (x_0, 0)$ have vicinity filter
$$\mathcal{V}(z) := \{A(z) \setminus F : F \in [X \setminus \{z\}]^{<\omega}\}^{\uparrow},$$
where
$$A(z) := \{(x_0, y) \in X : 0 \leq y \leq 2\} \cup \{(x_0 + y, y) : 0 \leq y \leq 2\}.$$

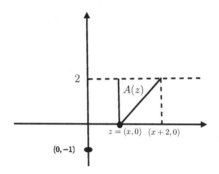

Note that at these points $\mathcal{V}(\mathcal{V}(z)) = \mathcal{V}(z)$. Moreover, since two sets of the form $A(z_1)$ and $A(z_2)$ for $z_1 \neq z_2$ intersect in at most one point, elements $A(z) \setminus F$ of the filter-base of $\mathcal{V}(z)$ are not only open, but also closed.

Finally, let $z_0 := (0, -1)$ admit the vicinity filter
$$\mathcal{V}(z_0) := \{U_i : i \in \mathbb{N}\}^{\uparrow},$$
where $U_i := \{z_0\} \cup \{(x,y) \in X : x \geq i\}$.

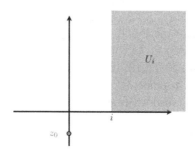

Note that the sets U_i are open, so that we have endowed X with a topology. It is easily seen to be Hausdorff.

It is also regular: since points of $X \setminus \{z_0\}$ have vicinities with a filterbase composed of open and closed sets, it is enough to check that if C is a closed subset of X and $z_0 \notin C$, we can find disjoint open sets O and V with $z_0 \in O$ and $C \subset V$.

Since $z_0 \in X \setminus C$ and $X \setminus C$ is open, there is $i_0 \in \mathbb{N}$ such that $U_{i_0} \cap C = \varnothing$. Let $O := U_{i_0+2}$ and

$$V = X \setminus (U_{i_0+2} \cup \{(x,0) : i_0 \leq x \leq i_0 + 2\}).$$

Then V is open because if $z = (x,0) \in V$ then $x < i_0$ so that $A(z) \cap U_{i_0+2} = \varnothing$. Moreover, $C \subset X \setminus U_{i_0} \subset V$.

Finally, X is not functionally regular: Note that $C := \{(x,0) : 0 \leq x \leq 1\}$ is closed, and $z_0 \notin C$. If now $f : X \to \mathbb{R}$ is continuous and $f(C) = \{0\}$, we will show that $f(z_0) = 0$.

To this end, we show that $K_i := \{x \in [i, i+1] : f((x,0)) = 0\}$ is infinite for each i, so that, $\{f = 0\} \# V(z_0)$ and $z_0 \in \{f = 0\}$ by continuity of f, because $\{f = 0\}$ is then closed.

We proceed by induction, with the case $i = 0$ taken care of, because $C = K_0$. Assume K_i is infinite.

First note that for each $z = (x_0, 0) \in K_i$, there is a countable subset $S(z)$ of $L(z) := \{(x_0 + y, y) : 0 \leq y \leq 2\}$ such that

$$f(L(z) \setminus S(z)) = \{0\}.$$

Indeed $\{s \in X : |f(s)| \geq \frac{1}{n}\}$ is closed for each n by continuity of f, so that $\{s \in L(z) : |f(s)| \geq \frac{1}{n}\}$ is closed in $L(z)$ with the induced convergence. Of course, $L(z)$ carries the prime cofinite topology at z, so that $\{s \in L(z) : |f(s)| \geq \frac{1}{n}\}$ is finite, and $S(z) := \bigcup_{n \in \mathbb{N}} \{s \in L(z) : |f(s)| \geq \frac{1}{n}\}$ is countable and has the desired property.

Consider now an infinite countable subset $\{z_n : n \in \mathbb{N}\}$ of K_i, and let A be the projection on $\{(x,0) : x \in \mathbb{R}\}$ of $\bigcup_{n \in \mathbb{N}} S(z_n)$. It is countable, and for each $t = (x,0)$ with $x \in [i+1, i+2]$ such that $t \notin A$ we have $A(t) \cap (L(z_n) \setminus S(z_n)) \neq \varnothing$ for all $n \in \mathbb{N}$. Hence $V(t) \# (\bigcup_{n \in \mathbb{N}} L(z_n) \setminus S(z_n))$ and by continuity of f we conclude that $f(t) = 0$, and $t \in K_{i+1}$, so that K_{i+1} is infinite.

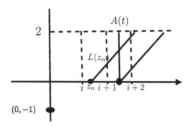

The *Hilbert cube* (indexed by Y) is defined as
$$[0,1]^Y \approx \prod_{y \in Y} [0,1].$$
The standard convergence of the Hilbert cube is the usual pointwise convergence $\nu^Y_{|[0,1]}$, homeomorphically, the product convergence ([3]). Of course, this convergence is topological. In particular, if $\boldsymbol{F} \subset C(\xi, \nu_{|[0,1]})$ then the diagonal map
$$\Delta \boldsymbol{F} : X \to [0,1]^{\boldsymbol{F}}$$
is continuous from ξ to $\nu^{\boldsymbol{F}}_{|[0,1]}$ by Proposition VII.3.2. If moreover $\xi = \boldsymbol{F}^- \nu_{|[0,1]}$, then ξ is a functionally regular topology on X. If moreover, ξ is Hausdorff, then $\Delta \boldsymbol{F}$ is an embedding by Proposition VII.3.4. We conclude:

Corollary VII.6.12. *A convergence ξ is (homeomorphically) embedded into a Hilbert cube if and only if ξ is a Hausdorff functionally regular topology.*

Note that more specifically, we have proved that ξ is embedded into $\nu^{C(\xi, \nu_{|[0,1]})}_{|[0,1]}$ by the embedding $\Delta \boldsymbol{F}$ where $\boldsymbol{F} = C(\xi, \nu_{|[0,1]})$.

It turns out that separable metric spaces can be embedded in the Hilbert cube of countable dimension:

Exercise VII.6.13. If (X, d) is a separable metric space such that d is bounded by 1 and $\{a_n : n \in \mathbb{N}\}$ is dense in X, then let $h : X \to \prod_{n \in \mathbb{N}} [0,1]$ be defined by
$$h(x)(n) := d(x, a_n)$$
for each $n \in \mathbb{N}$ and each $x \in X$. Show that

(1) h is continuous;
(2) h is injective and $h^{-1} : h(X) \to X$ is continuous;
(3) each separable metrizable space is homeomorphic to a subset of a Hilbert cube of countable dimension.

---

[3]see Exercise IV.5.1 for the particular case of real-valued functions on the reals.

VII.7 Normality

Recall that two subsets A and B of a convergence space X are called *topologically separated* if there are two disjoint open subsets U and V with $A \subset U$ and $B \subset V$.

Definition VII.7.1. Two subsets A and B of a convergence space are *topologically almost separated* if

$$A \cap \mathrm{cl}\, B = \mathrm{cl}\, A \cap B = \varnothing.$$

Of course, two topologically separated subsets are topologically almost separated; two disjoint closed subsets are topologically almost separated. Recall (Definition VII.1.12) that a convergence space is *topologically normal* if any two disjoint closed sets are topologically separated.

Definition VII.7.2. A convergence space is called *hereditarily normal* if any pair of topologically almost separated subsets is topologically separated.

Since topological normality is defined in terms of closed sets and of open sets, a convergence is topologically normal if and only if its topological modification is. In other words topological normality is a topological notion. Similarly, hereditary normality is a topological notion, because it is defined in terms of closure and of open sets. In other words, a convergence is hereditarily normal if and only if its topological modification is. Therefore results in this subsection, with a few exceptions, are formulated for topologies, in which case the adjective "topologically" can be omitted when discussing normality, separation, and almost separation.

For instance, by Proposition VI.3.10, a topology is regular if every closed subset can be separated from each disjoint singleton. Since singleton in a T_1 space are closed, we immediately obtain:

Proposition VII.7.3. *A T_1 normal topology is regular and Hausdorff.*

Exercise VII.7.4. Show that the Sierpiński topology is normal but not regular.

However, Proposition VII.7.3 does not mean that a T_1 normal *convergence* need be regular, but only that its topological modification does:

Example VII.7.5 (A T_1 hereditarily normal convergence that is not regular). Let ν denote the standard convergence on $\mathbb{R}$, and let ξ be the convergence on $\mathbb{R}$ in which $x \in \lim_\xi \mathcal{F}$ if $x \in \lim_\nu \mathcal{F}$ and $\mathcal{F}$ is an ultrafilter. Note that $\mathcal{V}_\xi(x) = \mathcal{V}_\nu(x)$ for each x, because of (V.1.2). Hence $S_0 \xi = T \xi = \nu$ is hereditarily normal, as we will see from Proposition VII.7.10, and T_1, so that ξ is as well. But ξ is not regular because $\mathrm{adh}_\xi^\natural \mathcal{U} = \mathrm{adh}_\nu^\natural \mathcal{U}$ may not be an ultrafilter when $\mathcal{U}$ is a convergent ultrafilter.

By definition, every hereditarily normal space is topologically normal. More precisely:

Proposition VII.7.6. *A topological space is hereditarily normal if and only if every subspace is normal.*

Proof. Assume that X is a hereditarily normal topological space and let $Y \subset X$ carry the induced topology. Let A and B be disjoint closed subsets of Y. Then $\mathrm{cl}_X A \cap B = A \cap \mathrm{cl}_X B = \varnothing$, and by hereditary normality of X, there are disjoint open subsets U and V of X such that $A \subset U$ and $B \subset V$. As $U \cap Y$ and $V \cap Y$ are open in Y (see Exercise V.4.34), we conclude that Y is normal.

Conversely, if X is not hereditarily normal, there are two subsets A and B of X such that $A \cap \mathrm{cl}_X B = \mathrm{cl}_X A \cap B = \varnothing$ but $\mathcal{N}_X(A) \# \mathcal{N}_X(B)$. Consider the subset $Y := X \setminus (\mathrm{cl}_X A \cap \mathrm{cl}_X B)$ of X with the induced topology. Then $A' := \mathrm{cl}_X A \cap Y$ and $B' := \mathrm{cl}_X B \cap Y$ are disjoint closed subsets of Y, but $\mathcal{N}_Y(A') \# \mathcal{N}_Y(B')$.

Indeed, if $U \in \mathcal{O}_Y$ and $A' \subset U$ then U is also open in X because Y is, and $A \subset U$ because $A \subset X \setminus \mathrm{cl}_X B$ by assumption. Similarly, if $V \in \mathcal{O}_Y$ and $B' \subset V$ then $B \subset V \in \mathcal{O}_X$. Hence $U \cap V \neq \varnothing$, and Y fails to be normal. $\qquad\square$

Exercise VII.7.7. Show that the discrete topology is hereditarily normal.

More generally:

Exercise VII.7.8. Show that every zero-dimensional topology (hence every prime topology, by Exercise V.4.10) is hereditarily normal.

On the other hand, there are Hausdorff normal topologies that are not hereditarily normal:

Example VII.7.9 (A normal topological space that is not hereditarily normal). Let X be a countable set endowed with the prime cofinite topology at $x_\infty \in X$ and let Y be an uncountable set endowed with the prime cofinite topology at $y_\infty \in Y$. The spaces X and Y are both hereditarily normal by Exercise VII.7.8. Their product $Z := X \times Y$ is normal, but its subspace $Z \setminus \{(x_\infty, y_\infty)\}$ is not, which, in view of Proposition VII.7.6, shows that Z is not hereditarily normal.

We will justify that Z is normal later[4].

To see that $Z \setminus \{(x_\infty, y_\infty)\}$ is not normal, note that the subsets $A := \{(x, y_\infty) : x \in X \setminus \{x_\infty\}\}$ and $B := \{(x_\infty, y) : y \in Y \setminus \{y_\infty\}\}$ are closed (in $Z \setminus \{(x_\infty, y_\infty)\}$) and disjoint. Yet, they cannot be separated. Indeed, if U and V are open subsets of $Z \setminus \{(x_\infty, y_\infty)\}$ with $A \subset U$ and $B \subset V$, then for each $x \in X \setminus \{x_\infty\}$ there is a finite subset F_x of $Y \setminus \{y_\infty\}$ with $\{x\} \times (Y \setminus F_x) \subset U$, and for each $y \in Y \setminus \{y_\infty\}$ there is a finite subset G_y of $X \setminus \{x_\infty\}$ with $(X \setminus G_y) \times \{y\} \subset V$. Since X is countable, so is $\bigcup_{x \in X} F_x$. Thus there is $y_0 \in Y \setminus (\{y_\infty\} \cup \bigcup_{x \in X} F_x)$ because Y is uncountable. Now if $x_0 \in X \setminus G_{y_0}$ then $(x_0, y_0) \in U \cap V$, so that $U \cap V \neq \varnothing$.

Proposition VII.7.10. *Every metrizable topology is hereditarily normal.*

Proof. Let A and B be subsets of $(X, \tilde{d})$ satisfying $A \cap \mathrm{cl}\, B = \mathrm{cl}\, A \cap B = \varnothing$. For each $a \in A$ there is $r_a > 0$ with $B(a, r_a) \subset X \setminus \mathrm{cl}\, B$ and for each $b \in B$ there is $r_b > 0$ with $B(b, r_b) \subset X \setminus \mathrm{cl}\, A$. The sets $U := \bigcup_{a \in A} B(a, \frac{r_a}{2})$ and $V := \bigcup_{b \in B} B(b, \frac{r_b}{2})$ are open because d is a metric. Moreover, $A \subset U$, $B \subset V$ and $U \cap V = \varnothing$. Indeed, if $x \in U \cap V$ then $d(a, x) < \frac{r_a}{2}$ and $d(x, b) < \frac{r_b}{2}$ for some $a \in A$ and $b \in B$, so that

$$d(a, b) < \frac{r_a + r_b}{2} \leq \max(r_a, r_b),$$

which is not possible because $B(a, r_a)$ is disjoint from B and $B(b, r_b)$ is disjoint from A. □

In particular, the real line is hereditarily normal, and the space Z of Example VII.7.9 is not metrizable. On the other hand, there are hereditarily normal spaces that are not metrizable: we have shown the Sorgenfrey line to be non-metrizable (Corollary IV.9.41), and it is hereditarily normal:

Example VII.7.11 (The Sorgenfrey line is hereditarily normal). If A and B are subsets of $\mathbb{R}$ with $A \cap \mathrm{cl}_s B = B \cap \mathrm{cl}_s A = \varnothing$, then we can pick for each $a \in A$ an $x_a > a$ such that $[a, x_a) \subset \mathbb{R} \setminus \mathrm{cl}_s B$ and for each $b \in B$

---

[4]We will see that compact Hausdorff topologies are normal (Corollary IX.2.6), that a prime cofinite topology is compact (Example IX.1.7), and that a product of compact spaces is compact (Theorem IX.1.32), which shows that Z is compact Hausdorff and therefore normal.

an $x_b > b$ such that $[b, x_b) \subset \mathbb{R} \setminus \mathrm{cl}_s A$. The sets $U := \bigcup_{a \in A}[a, x_a)$ and $V := \bigcup_{b \in B}[b, x_b)$ are open, $A \subset U$, $B \subset V$ and $U \cap V = \varnothing$. Indeed, $[a, x_a) \cap [b, x_b) = \varnothing$ for each $a \in A$ and $b \in B$, for otherwise we would have $b \in [a, x_a) \cap B$ if $a < b$, and $a \in [b, x_b) \cap A$ if $b < a$, neither of which is possible.

Note that in view of Example VII.7.9, and of Exercise VII.7.8, a product of hereditarily normal topological spaces does not need to be hereditarily normal. In fact, it does not even need to be normal, as we will see for instance in Example VII.9.5.

Exercise VII.7.12. Show that a topological space is normal if and only if for any closed set A and any open set U_1 containing A, there is an open set U_0 with

$$A \subset U_0 \subset \mathrm{cl}\, U_0 \subset U_1.$$

One of the main results about normal topologies, known for historical reasons as the *Urysohn Lemma*, states that in normal topological spaces, disjoint closed sets are not only topologically separated but functionally separated. As a consequence, every T_1 normal topology is not only regular but functionally regular. We will provide an example of a non-normal functionally regular space after we establish the Urysohn Lemma.

Lemma VII.7.13. *Let D be a dense subset of the real line $\mathbb{R}$ (with its usual convergence), and let X be a topological space. Suppose that a family $\{U_d : d \in D\}$ of open subsets of X satisfies $d < t \implies \mathrm{cl}\, U_d \subset U_t$,*

$$\bigcup_{d \in D} U_d = X \quad and \quad \bigcap_{d \in D} U_d = \varnothing.$$

Then

$$f(x) := \inf\{d \in D : x \in U_d\} \tag{VII.7.1}$$

defines a function $f : X \to \mathbb{R}$ that is continuous on X.

Proof. The conditions on $\{U_d : d \in D\}$ ensure that (VII.7.1) is a well-defined function. We show continuity of f at $a \in X$.

Since D is dense in $\mathbb{R}$, the collection of intervals

$$\{[d, t] : d < f(a) < t; d, t \in D\}$$

is a filter-base of $\mathcal{N}_{\mathbb{R}}(f(a))$. If d and t are in D and $d < f(a) < t$ then

$$V := U_t \setminus \mathrm{cl}\, U_d \in \mathcal{N}_X(a)$$

and $f(V) \subset [d, t]$, so that f is continuous at a.

To see that, note that by definition of f

$$f(a) < t \Longrightarrow a \in U_t,$$

and

$$x \in \mathrm{cl}\, U_d \Longrightarrow \underset{s > d}{\forall}\, x \in U_s \Longrightarrow f(x) \leq d.$$

Therefore a belongs to the open set $V = U_t \setminus \mathrm{cl}\, U_d$. Moreover, $f(x) \leq t$ whenever $x \in U_t$ so that $f(V) \subset [d, t]$. $\quad\square$

Theorem VII.7.14 (Urysohn Lemma). *Each couple of disjoint closed sets in a normal topological space is functionally separated.*

Proof. Given two disjoint closed subsets A and B of a normal topological space X, we define a family $\{U_r : r \in \mathbb{Q}\}$ of open subsets of X as follows:

$U_r = \varnothing$ for all $r < 0$ and $U_r = X$ for all $r > 1$. Let $U_1 := X \setminus B$, which is an open set containing A. By normality and Exercise VII.7.12, there is an open set U_0 with

$$A_0 \subset U_0 \subset \mathrm{cl}\, U_0 \subset U_1.$$

Enumerate now the rationals of $[0, 1]$ with a sequence $\{r_n\}_{n \in \mathbb{N}}$ with $r_1 = 1$ and $r_2 = 0$. We define then U_{r_n} by induction on $n > 2$ so that

$$r_k < r_n < r_l \Longrightarrow \mathrm{cl}\, U_{r_k} \subset U_{r_n} \subset \mathrm{cl}\, U_{r_n} \subset U_{r_l}$$

for any $k, l < n$.

The family $\{U_r : r \in \mathbb{Q}\}$ satisfies the conditions of Lemma VII.7.13, which provides a continuous function f satisfying $f(A) = \{0\}$ because $A \subset U_0$ and $A \cap U_r = \varnothing$ for $r < 0$, and $f(B) = \{1\}$ because $B \subset \bigcap_{r > 1} U_r$ but $B \cap U_1 = \varnothing$. $\quad\square$

Corollary VII.7.15. *A T_1 normal topology is functionally regular.*

Example VII.7.16 (the Niemytzki plane is a functionally regular topology that is not normal). Let $Y := \{(x, y) \in \mathbb{R}^2 : y > 0\}$ carry the topology induced by the usual topology of the plane. Let $L := \{(x, 0) \in \mathbb{R}^2 : x \in \mathbb{R}\}$ and $X := Y \cup L$ and define on X the pretopology defined by

$$z \in Y \Longrightarrow \mathcal{V}_X(z) := \mathcal{N}_Y(z)^\uparrow$$
$$z = (x, 0) \in L \Longrightarrow \mathcal{V}_X(z) := \{V_n(z) : n \in \mathbb{N}\}^\uparrow,$$

where $V_n(z) := \{z\} \cup B_Y((x, \frac{1}{n}), \frac{1}{n})$ is the open disk of radius $\frac{1}{n}$ in Y (for the Euclidean metric) tangent to L at z to which we adjoin z.

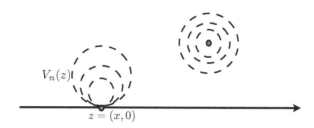

We call this convergence space *the Niemytzki plane*. It is easy to see that sets $V_n(z)$ are vicinities of each of their points and are therefore open. Hence

$$\mathcal{V}_X(z) = \mathcal{N}_X(z)$$

for each $z \in X$, and X is topological. Moreover:

Claim VII.7.17. The Niemytzki plane is functionally regular.

Proof. Let $z \in X$ and let F be a closed subset of X with $z \notin F$. If $z \in Y$, then we can separate $\{z\}$ and F by a continuous function from the closed upper-half plane $\{(x, y) : y \geq 0\}$ with the topology induced by that of the plane, because the plane is functionally regular. Since the topology of X is finer than that of the upper-half plane, the function is also continuous on X.

Hence, it is enough to check that if $z \in L$ and $n \in \mathbb{N}$ there is a continuous function $f : X \to [0, 1]$ with $f(z) = 0$ and $f(X \setminus V_n(z)) = 1$, for $X \setminus F$ is an open neighborhood of z and contains $V_n(z)$ for some n. To see that, consider for each $t \in V_n(z) \setminus \{z\}$ the intersection point t' of the half-ray joining z to t with the bounding circle of $V_n(z)$.

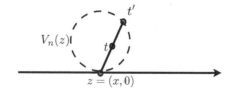

Define $f : X \to [0,1]$ by

$$f(t) = \begin{cases} 0 & \text{if } t = z \\ 1 & \text{if } t \in X \setminus V_n(z) \\ \dfrac{d(z,t)}{d(z,t')} & \text{if } t \in V_n(z) \setminus \{z\} \end{cases}.$$

The function f is continuous because $d(z,\cdot) : Y \to \mathbb{R}$ is continuous, $d(z,t') \neq 0$ for $t \in V_n(z) \setminus \{z\}$, $\lim_{d(z,t)\to 0} \frac{d(z,t)}{d(z,t')} = 0$ and $\lim_{d(t,t')\to 0} \frac{d(z,t)}{d(z,t')} = 1$. $\qquad\square$

Claim VII.7.18. The Niemytzki plane is not normal.

Proof. Since the topology induced by X on L is discrete, every subset A of L is closed in X. If X was normal, there would be, for each $A \subset L$, two disjoint open subsets U_A and V_A of X with $A \subset U_A$ and $L \setminus A \subset V_A$. Consider $C_A := U_A \cap \mathbb{Q}^2$. We will show that

$$A \neq B \implies C_A \neq C_B$$

for every subsets A and B of L, which yields a contradiction, because there are $2^{\mathfrak{c}}$ subsets of L but only $2^{\aleph_0} = \mathfrak{c} < 2^{\mathfrak{c}}$ subsets of $\mathbb{Q}^2 \cap X$.

If $A \neq B$ then $A \setminus B \neq \varnothing$ or $B \setminus A \neq \varnothing$. By symmetry, we only need to consider one case, say $A \setminus B \neq \varnothing$. By definition of U_A and V_B, we have $A \setminus B \subset U_A \cap V_B$, so that $U_A \cap V_B$ is a non-empty open subset of X, which therefore intersects the dense subset $\mathbb{Q}^2 \cap X$ of X. Hence

$$\varnothing \neq \mathbb{Q}^2 \cap U_A \cap V_B = C_A \cap V_B \subset C_A \setminus U_B \subset C_A \setminus C_B,$$

and $C_A \setminus C_B \neq \varnothing$, which concludes the proof. $\qquad\square$

Note that we have used that the Niemytzki plane is separable. On the other hand, its subspace L is a discrete subspace of cardinality $\mathfrak{c}$ and is therefore not separable. Hence *a subspace of a separable space does not need to be separable.*

Before moving on to the extension properties of normal topologies, here are two important sufficient conditions for normality:

Theorem VII.7.19. *Let X be a regular T_1 topological space.*

(1) If X has countable weight, then it is normal;
(2) If $\operatorname{card} X \leq \aleph_0$, then X is normal.

Exercise VII.7.20. Deduce Theorem VII.7.19 from Lemma VII.7.21 below.

Lemma VII.7.21. *If X is a T_1 topological space such that for every closed set F and every open set W such that $F \subset W \subset X$, there is a sequence $\{W_i\}_{i \in \mathbb{N}}$ of open subsets of X such that $\text{cl}_X W_i \subset W$ for each $i \in \mathbb{N}$ and*

$$F \subset \bigcup_{i=1}^{\infty} W_i,$$

then X is normal.

Proof. Let A and B be two disjoint closed subsets of X. Then $F = A$ and $W = X \setminus B$ are as in the assumption, so that there is a sequence $\{W_i\}_{i \in \mathbb{N}}$ of open subsets of X with $A \subset \bigcup_{i=1}^{\infty} W_i$ and $B \cap \text{cl}_X W_i = \varnothing$ for all $i \in \mathbb{N}$. If instead we pick $F = B$ and $W = X \setminus A$, we obtain a sequence $\{V_j\}_{j \in \mathbb{N}}$ of open subsets of X with $B \subset \bigcup_{j=1}^{\infty} V_j$ and $A \cap \text{cl}_X V_j = \varnothing$ for all $j \in \mathbb{N}$.

The sets

$$U_i := W_i \setminus \bigcup_{j \leq i} \text{cl}_X V_j \text{ and } O_i := V_i \setminus \bigcup_{j \leq i} \text{cl}_X W_j$$

are open, so that $U := \bigcup_{i=1}^{\infty} U_i$ and $O := \bigcup_{i=1}^{\infty} O_i$ are also open. Moreover, $A \subset U$ because $A \subset \bigcup_{i=1}^{\infty} W_i$ and $A \cap \text{cl}_X V_j = \varnothing$ for all $j \in \mathbb{N}$. Similarly, $B \subset O$. Finally, $U \cap O = \varnothing$ because $U_i \cap O_j = \varnothing$ for each i and j. Indeed, if $j \leq i$ then $U_i \cap O_j \subset U_i \cap \bigcup_{j \leq i} V_j \subset \varnothing$, and similarly if $j \geq i$. $\square$

Theorem VII.7.22 (Urysohn). *A regular T_1 topology of countable weight is metrizable.*

Proof. By Theorem VII.7.19 (1), if X is a regular T_1 topology of countable weight, then it is normal. Let $\mathcal{B}$ be a countable base and let

$$\mathcal{V} := \{(B, V) \in \mathcal{B} \times \mathcal{B} : \text{cl } B \subset V\}.$$

By the Urysohn Lemma VII.7.14, for every $(B, V) \in \mathcal{V}$ there exists $f_{B,V} \in C(X, [0, 1])$ such that $f_{B,V}(B) = \{0\}$ and $f_{B,V}(X \setminus V) = \{1\}$. The set

$$F := \{f_{B,V} : (B, V) \in \mathcal{V}\} \tag{VII.7.2}$$

is clearly countable and distinguishes points and closed sets.

Indeed, if A is closed and $x \notin A$, then there is $V \in \mathcal{B}$ such that $x \in V$ and $V \cap A = \varnothing$. By regularity, there exists $B \in \mathcal{B}$ such that $x \in B \subset \text{cl } B \subset V$, so that $(B, V) \in \mathcal{V}$ and thus $f_{B,V}(x) = 0$ and $f_{B,V}(A) = \{1\}$. It follows that (VII.7.2) distinguishes points, because X is T_1.

By Corollary VII.6.12, the embedding $\Delta F : X \to [0,1]^F$ is an (homeomorphic) embedding, and since $[0,1]$ is a metric space and F is countable, the product $\prod_{f \in F} [0,1] = [0,1]^F$ is metrizable ([5]), hence X is metrizable. $\qquad\square$

Theorem VII.7.22 combines with Theorem IV.9.40 to the effect that:

Corollary VII.7.23. *A regular T_1 topology has countable weight if and only if it is separable and metrizable.*

VII.8 Continuous extension of maps

Many topological problems boil down to extending a continuous functions $f : S \to Y$ from a subspace S of a topological space X to a continuous function $\bar{f} : X \to Y$ (where the term "extension" means that $f(s) = \bar{f}(s)$ for all $s \in S$). The Urysohn Lemma can be interpreted in those terms, even though it pertains to a rather trivial looking case: if A and B are two disjoint closed subsets, then $f : A \cup B \to \mathbb{R}$ defined by $f(a) = 0$ for all $a \in A$ and $f(b) = 1$ for all $b \in B$ is well-defined and continuous. The Urysohn Lemma states that, if X is a normal topological space, there is a continuous extension of f to X. Section VII.9 below will provide a much more far-reaching theorem, ensuring that every real-valued continuous on a closed subset of a normal topological space admits a continuous extension to the entire space. In this section, we consider the complementary problem of extending a continuous function on a non-closed subset of a convergence space to (part of) its closure.

Let (X,ξ) and (Y,τ) be two convergence spaces and let $S \subset X$ be endowed with the subspace convergence, and assume that $f : |\xi_{|S}| \to |\tau|$ is continuous. If $S \subset A \subset X$ and there is a continuous extension $\bar{f} : |\xi_{|A}| \to |\tau|$ of f to A, then for any $x \in \mathrm{adh}_\xi S \cap A$ there is a filter $\mathcal{F} \in \mathbb{F}X$ with $S \in \mathcal{F}$ and $x \in \lim_\xi \mathcal{F}$, so that, by continuity of $\bar{f}$, $\bar{f}(x) \in \lim_\tau \bar{f}[\mathcal{F}]$. But $\bar{f}[\mathcal{F}] = f[\mathcal{F}]$ because $S \subset A$ is an element of $\mathcal{F}$, so that

$$\bar{f}(x) \in \lim_\tau f[\mathcal{F}]$$

for every $\mathcal{F} \in \mathbb{F}S$ with $x \in \lim_\xi \mathcal{F}$. In particular, if there is $x \in \mathrm{adh}_\xi S \cap A$

---

[5]We have seen in Exercise IV.4.4 that a product of finitely many metrizable spaces is metrizable, and in Exercise IV.5.2 that, in general, a product of metrizable spaces may fail to be metrizable. The reader may convince himself that a countable product of metrizable spaces is metrizable; a fact that is fully proved in Corollary XVIII.3.5.

such that

$$\bigcap_{\mathcal{F}\in\mathbb{F}S,\, x\in\lim_\xi \mathcal{F}} \lim_\tau f[\mathcal{F}] = \varnothing$$

then there is **no** continuous extension of f to A, because the value $\bar{f}(x)$ cannot be picked in a way that ensures continuity of $\bar{f}$ at x. These considerations lead to the following definition:

Definition VII.8.1. Let (X, ξ) and (Y, τ) be two convergence spaces, let $S \subset X$ and $f : |\xi_{|S}| \to |\tau|$ be continuous. The *hull of extensionability of S for f* is

$$h(S, f) := \Big\{ x \in \mathrm{adh}_\xi\, S : \bigcap_{\mathcal{F}\in\mathbb{F}S,\, x\in\lim_\xi \mathcal{F}} \lim_\tau f[\mathcal{F}] \neq \varnothing \Big\}.$$

Remark VII.8.2. Note that continuity of f ensures that $S \subset h(S, f)$, because if $x \in S$ then

$$f(x) \in \bigcap_{\mathcal{F}\in\mathbb{F}S,\, x\in\lim_\xi \mathcal{F}} \lim_\tau f[\mathcal{F}].$$

Definition VII.8.3. A subspace S of a convergence space (X, ξ) is *strict* if for every filter $\mathcal{F}$ with $\mathrm{adh}\, S \in \mathcal{F}$,

$$x \in \lim_\xi \mathcal{F} \Longrightarrow \underset{\mathcal{G}\in\mathbb{F}S}{\exists}\ \mathrm{adh}^\natural\, \mathcal{G} \leq \mathcal{F} \text{ and } x \in \lim_\xi \mathcal{G}.$$

This condition should be seen as a localized weak version of diagonality. Indeed:

Proposition VII.8.4. *If (X, ξ) is a diagonal convergence space, then every subspace is strict.*

Proof. Let $\mathcal{F} \in \mathbb{F}(\mathrm{adh}_\xi\, S)$ and $x_0 \in \lim_\xi \mathcal{F}$. For each $x \in \mathrm{adh}_\xi\, S$, there is $\mathcal{S}(x) \in \mathbb{F}S$ with $x \in \lim_\xi \mathcal{S}(x)$. Defining $\mathcal{S}(x) := \{x\}^\uparrow$ if $x \notin \mathrm{adh}_\xi\, S$, the map $\mathcal{S}$ is a selection for ξ. Thus $x \in \lim_\xi \mathcal{S}(\mathcal{F})$ because ξ is diagonal. But $S \in \mathcal{S}(\mathcal{F})$ by construction, and $\mathcal{F} \geq \mathrm{adh}^\natural\, \mathcal{S}(\mathcal{F})$. Indeed for each $x \in F \subset \mathrm{adh}_\xi\, S$, $x \in \lim_\xi \mathcal{S}(x) \subset \mathrm{adh}\, \bigcup_{t\in F} S_t$ and thus

$$F \subset \mathrm{adh} \bigcup_{t\in F} S_t$$

for any choice of $S_t \in \mathcal{S}(t)$. $\qquad\qquad\square$

Theorem VII.8.5. *Let (X, ξ) be a convergence space, let (Y, τ) be a regular convergence space, and let S be a strict subspace of (X, ξ). If $f : |\xi_{|S}| \to |\tau|$ is continuous, there is a continuous extension*

$$\bar{f} : |\xi_{|h(S,f)}| \to |\tau|.$$

Proof. If $x \in S$ we define $\bar{f}(x) = f(x)$. If $x \in h(S, f) \setminus S$, we can pick

$$y = \bar{f}(x) \in \bigcap_{\mathcal{F} \in \mathbb{F}S,\, x \in \lim_\xi \mathcal{F}} \lim_\tau f[\mathcal{F}],$$

because that set is non-empty by definition of $h(S, f)$. We claim that any choice of values $\bar{f}(x)$ in the sets $\bigcap_{\mathcal{F} \in \mathbb{F}S,\, x \in \lim_\xi \mathcal{F}} \lim_\tau f[\mathcal{F}]$ results in a continuous function $\bar{f}$:

Let $\mathcal{F} \in \mathbb{F}(h(S, f))$ with $x_0 \in \lim_\xi \mathcal{F} \cap h(S, f)$. By strictness of S, there is $\mathcal{G} \in \mathbb{F}S$ with $x_0 \in \lim_\xi \mathcal{G}$ and $\mathrm{adh}_\xi^\natural \mathcal{G} \leq \mathcal{F}$. Thus $\bar{f}(x_0) \in \lim_\tau f[\mathcal{G}]$ because $\bar{f}(x_0) \in h(S, f)$. Moreover, $\bar{f}(x_0) \in \lim_\tau \mathrm{adh}_\tau^\natural f[\mathcal{G}]$, by regularity of τ. But f is continuous on $S \in \mathcal{G}$ so that

$$\mathrm{adh}_\tau^\natural f[\mathcal{G}] \leq \bar{f}[\mathrm{adh}_\xi^\natural \mathcal{G}]$$

because $\bar{f}(\mathrm{adh}_\xi G) \subset \mathrm{adh}_\tau f(G)$ for any $G \subset S$, for if $t \in \mathrm{adh}_\xi G \cap h(S, f)$, there is $\mathcal{H} \in \mathbb{F}G$ with $t \in \lim_\xi \mathcal{H}$, but then $\bar{f}(t) \in \lim_\tau f[\mathcal{H}] \subset \mathrm{adh}_\tau f(G)$. Therefore

$$\bar{f}(x_0) \in \lim_\tau \mathrm{adh}_\tau^\natural f[\mathcal{G}] \subset \lim_\tau f[\mathrm{adh}_\xi^\natural \mathcal{G}] \subset \lim_\tau \bar{f}[\mathcal{F}],$$

which concludes the proof. $\square$

Corollary VII.8.6. *Let (X, ξ) be a convergence space, let (Y, τ) be a Hausdorff regular convergence space, and let S be a strict subspace of (X, ξ). If $f : |\xi_{|S}| \to |\tau|$ is continuous, there is a unique continuous extension*

$$\bar{f} : |\xi_{|h(S,f)}| \to |\tau|.$$

Proof. If τ is Hausdorff, each set $\lim_\tau f[\mathcal{F}]$ is at most a singleton, so that for each $x \in h(S, f)$, $\bigcap_{\mathcal{F} \in \mathbb{F}S,\, x \in \lim_\xi \mathcal{F}} \lim_\tau f[\mathcal{F}]$ is a singleton and there is only one choice of $\bar{f}(x)$. $\square$

Remark VII.8.7. In the important case where (X, ξ) is topological (hence diagonal), we only need to consider $\mathcal{F} = \mathcal{N}(x) \vee S$ in the definition of $h(S, f)$, that is,

$$h(S, f) = \{x \in \mathrm{cl}_\xi S : \lim_\tau f[\mathcal{N}(x) \vee S] \neq \varnothing\}.$$

Yet, we would like to be able to extend a continuous map from S to its closure $\mathrm{cl}_\xi S$. To this end, we need to ensure that $\lim_\tau f[\mathcal{N}(x)] \neq \varnothing$ for all $x \in \mathrm{cl}_\xi S$. We will see in Chapter X that completeness of τ plays a key role in guaranteeing this condition.

It turns out that regularity is a necessary condition in Corollary VII.8.6:

Theorem VII.8.8. *A Hausdorff convergence space* (Y, τ) *is regular if and only if for every topological convergence space* (X, ξ), *every subset* S *of* X *and every continuous map* $f : |\xi_{|S}| \to |\tau|$ *there is a (unique) continuous extension* $\bar{f} : |\xi_{|h(S,f)}| \to |\tau|$.

Proof. The "only if" part is Corollary VII.8.6.

If τ is not regular, then, in view of Theorem VI.3.18, there is a set A and maps $l : A \to Y$ and $\mathcal{S} : A \to \mathbb{F}Y$ with $l(a) \in \lim_\tau \mathcal{S}(a)$ for all $a \in A$, and there is $\mathcal{F} \in \mathbb{F}A$ and $y_0 \in Y$ with $y_0 \in \lim_\tau \mathcal{S}(\mathcal{F})$ but $y_0 \notin \lim_\tau l[\mathcal{F}]$. Let

$$X := (Y \times A) \cup A \cup \{x_0\}$$

where $x_0 \notin A$, and consider on X the pretopology π in which $Y \times A$ is a discrete subspace, $\mathcal{V}_\pi(a) = \{a\}^\uparrow \wedge (\mathcal{S}(a) \times \{a\})^\uparrow$ and $\mathcal{V}_\pi(x_0) = \{x_0\}^\uparrow \wedge \mathcal{F}^\uparrow$ where $\mathcal{F}^\uparrow$ and $(\mathcal{S}(a) \times \{a\})^\uparrow$ denote the filters generated on X by the filters on A and $Y \times A$ respectively. Let $\xi := \mathrm{T}\,\pi$. Let $S := X \setminus \{x_0\}$ and $f : S \to Y$ be defined by $f((y,a)) = y$ if $(y,a) \in Y \times A$ and $f(a) = l(a)$ if $a \in A$. Then $f : |\xi_{|S}| \to |\tau|$ is a continuous map because $\xi_{|S} = \pi_{|S}$ and $f[\mathcal{V}_\pi(a)] = \{l(a)\}^\uparrow \wedge \mathcal{S}(a)$ converges to $l(a) = f(a)$ in τ. Moreover, $x_0 \in h(S, f)$ because

$$S \in \mathcal{N}_\xi(x_0) = \mathcal{V}_\pi\big(\mathcal{V}_\pi(x_0)\big) = \{x_0\}^\uparrow \wedge \Big(\bigcup_{F \in \mathcal{F}} \bigcap_{a \in F} \mathcal{V}_\pi(a) \Big),$$

and $f[\mathcal{N}_\xi(x_0)] = \mathcal{S}(\mathcal{F})$ satisfies $\lim_\tau f[\mathcal{N}_\xi(x_0)] = \{y_0\}$ because τ is Hausdorff. Hence the only potentially continuous extension of f to X is defined by $\bar{f}(x_0) = y_0$, but this map is not continuous because $f[\mathcal{F}] = l[\mathcal{F}]$ does not converge to $y_0 = \bar{f}(x_0)$. $\square$

Additionally, strictness is an essential condition in Corollary VII.8.6, as shows the following example:

Example VII.8.9 (A continuous map $f : S \to \{0,1\}$, $S \subset X$, that has no continuous extension to $h(S, f) = X$). Let

$$X := \{x_n : n \in \mathbb{N}\} \cup \{x_{n,p} : n, p \in \mathbb{N}\} \cup \{z_n : n \in \mathbb{N}\} \cup \{\infty\}$$

be equipped with the pretopology in which every point of

$$S := \{x_{n,p} : n, p \in \mathbb{N}\} \cup \{z_n : n \in \mathbb{N}\}$$

is isolated, $\mathcal{V}(x_n) := \{\{x_n\} \cup \{x_{n,p} : p \geq k\} : k \in \mathbb{N}\}^\uparrow$ and $\mathcal{V}(\infty) := \{\{\infty\} \cup \{x_n : n \geq k\} \cup \{z_n : n \geq k\} : k \in \mathbb{N}\}^\uparrow$.

Then the function $f : S \to \{0,1\}$, where $\{0,1\}$ is discrete, defined by $f(x_{n,p}) = 0$ for all n and p in $\mathbb{N}$ and $f(z_n) = 1$ for all $n \in \mathbb{N}$ is continuous. Moreover,

$$\operatorname{adh} S = h(S,f) = X,$$

because each filter on S convergent to x_n is finer than $(x_{n,p})_{p \in \mathbb{N}}$ has the image $\{0\}^\uparrow$ under f and each filter convergent on S convergent to ∞ is finer than $(z_n)_{n \in \mathbb{N}}$ and has image $\{1\}^\uparrow$ under f. Yet, f has no continuous extension to X. Indeed, if $\bar{f} : X \to \{0,1\}$ were such an extension, then $\bar{f}(x_n) = 0$ for all n so that, by continuity, $(\bar{f}(x_n))_n$ converges to $\bar{f}(\infty)$, so that $\bar{f}(\infty) = 0$. On the other hand, by continuity, $(\bar{f}(z_n))_n$ converges to $\bar{f}(\infty)$ as well. But $\bar{f}(z_n) = f(z_n) = 1$ for all n so that $\bar{f}(\infty) = 1$, which is a contradiction.

VII.9   Tietze's extension theorem

While the previous section dealt with continuous extensions of a map from a subset to (part of) its adherence, the following one deals with continuous extensions of real-valued maps defined on a closed subset.

Theorem VII.9.1 (Tietze's extension theorem). *A topological space X is normal if and only if for every closed subset A and every continuous function $f : A \to \mathbb{R}$ there is a continuous extension $\bar{f} : X \to \mathbb{R}$.*

The proof will be based on repeated use of the following:

Lemma VII.9.2. *If A is a closed subset of a normal topological space and $f : A \to \mathbb{R}$ is a continuous function bounded by c (i.e., $|f(a)| \leq c$ for all $a \in A$) then there is a continuous function $h : X \to \mathbb{R}$ with $|h(x)| \leq \frac{c}{3}$ for all $x \in X$ and*

$$|f(a) - h(a)| \leq \frac{2}{3}c, \qquad (\text{VII.9.1})$$

for all $c \in A$.

Proof. The subsets $C_1 := \{a \in A : f(a) \geq \frac{c}{3}\}$ and $C_0 := \{a \in A : f(a) \leq -\frac{c}{3}\}$ are closed disjoint subsets of X. By the Urysohn Lemma, there is a continuous map $h_1 : X \to [0,1] \subset \mathbb{R}$ with $h_1(C_1) = \{1\}$ and $h_1(C_0) = \{0\}$. The map $f : \mathbb{R} \to \mathbb{R}$ defined by $f(x) = \frac{c}{3}(2x - 1)$ is continuous and maps $[0,1]$ onto $[-\frac{c}{3}, \frac{c}{3}]$. Then $h = f \circ h_1$ is continuous on X, $h(C_1) = \{\frac{c}{3}\}$ and $h(C_0) = \{-\frac{c}{3}\}$.

Moreover, if $a \in C_1$ then $|f(a) - h(a)| = f(a) - \frac{c}{3} \leq \frac{2}{3}c$ because $f(a) \leq c$. Similarly (VII.9.1) holds if $a \in C_0$. On the other hand, if $a \in A \setminus (C_0 \cup C_1)$ then $|f(a)| < \frac{c}{3}$ and $|h(a)| \leq \frac{c}{3}$ so that (VII.9.1) holds as well. □

Proof of Theorem VII.9.1. Assume that X is a normal topological space. If $f : A \to \mathbb{R}$ is continuous and bounded by a constant c then by Lemma VII.9.2 there is a map $g_0 : X \to \mathbb{R}$ with $|f(a) - g_0(a)| \leq \frac{2}{3}c$ for all $a \in A$ and $|g_0(x)| \leq \frac{c}{3}$ for all $x \in X$. Therefore $f - g_0 : A \to \mathbb{R}$ is continuous and bounded (by $\frac{2}{3}c$) so that Lemma VII.9.2 applies to $f - g_0$ to the effect that there is a continuous map $g_1 : X \to \mathbb{R}$ with $|f(a) - g_0(a) - g_1(a)| \leq \frac{2}{3}\frac{2}{3}c$ for all $a \in A$ and $|g_1(x)| \leq \frac{1}{3}\frac{2c}{3}$ for all x.

Iterating this process yields a sequence $\{g_n\}_n$ of real-valued continuous functions on X satisfying

$$\underset{a \in A}{\forall} \left| f(a) - \sum_{i=0}^{n} g_i(a) \right| \leq \left(\tfrac{2}{3}\right)^{n+1} c \qquad (VII.9.2)$$

$$\underset{x \in X}{\forall} |g_n(x)| \leq \tfrac{c}{3}\left(\tfrac{2}{3}\right)^{n}.$$

Since $\sum_{i=0}^{\infty} \frac{1}{3}(\frac{2}{3})^i c = c$ is convergent, we conclude by Corollary VII.4.6 that

$$\bar{f}(x) := \sum_{i=0}^{\infty} g_i(x)$$

is well-defined, continuous on X, and bounded by c. Moreover, (VII.9.2) ensures that $\bar{f}(a) = f(a)$ for all $a \in A$ so that $\bar{f}$ is a continuous extension of f.

Note also that if $|f(a)| < c$ for all $a \in A$, we can ensure that $\bar{f}$ satisfies $|\bar{f}(x)| < c$ for all $x \in X$. Indeed, if F is the extension of f obtained with the construction above, then $B = \{x \in X : |F(x)| = c\}$ is a closed subset of X disjoint from A. By the Urysohn Lemma, there is a continuous function $h : X \to \mathbb{R}$ with $h(A) = \{1\}$ and $h(B) = \{0\}$. Clearly, $\bar{f}(x) := h(x)F(x)$ is a continuous extension of f with the desired property.

If now $f : A \to \mathbb{R}$ is not bounded, pick an homeomorphism $h : \mathbb{R} \to (-c, c)$, and consider the bounded map $h \circ f : A \to \mathbb{R}$. By the previous case, $h \circ f$ has a continuous extension $\overline{h \circ f} : X \to (-c, c) \subset \mathbb{R}$ and $\bar{f} := h^{-1} \circ (\overline{h \circ f})$ is a continuous extension of f.

The converse is obvious, as we already have observed that the Urysohn Lemma follows from the conclusion of Tietze's extension Theorem, and that functional separation of a pair of sets implies separation of those sets. □

Exercise VII.9.3. Show that if X is a normal topological space, A is a closed subset of X and $f : A \to \mathbb{R}^n$ is continuous, then f has a continuous extension $\bar{f} : X \to \mathbb{R}^n$.

Among the many applications of Tietze's extension theorem, let us mention the following criterion of non-normality:

Corollary VII.9.4. *A separable normal topological space does not contain any closed discrete subspace of cardinality continuum.*

Proof. Let X be a normal topological space containing a countable dense subset C and assume that it also contains a closed discrete subspace D of cardinality $\mathfrak{c}$. By Proposition IV.9.38, every real-valued continuous function on X is determined by its restriction to C, so that there are at most $\mathfrak{c}^{\aleph_0} = \mathfrak{c}$ (see Corollary A.5.4) such functions. But each of the $2^{\mathfrak{c}}$ real-valued functions on D is continuous, because the induced convergence on D is discrete, and each one of them is continuously extendable to X by Tietze's extension theorem, because D is closed. As $\mathfrak{c} < 2^{\mathfrak{c}}$ by Cantor's Theorem A.4.15, we have a contradiction. $\square$

Note that Corollary VII.9.4 provides an alternative way to see that the Niemytzki plane of Example VII.7.16 is not normal. It also provides us with examples of hereditarily normal spaces whose product is not even normal.

Example VII.9.5 (The square of the Sorgenfrey line is not normal). Let ξ_s denote the Sorgenfrey line convergence of Example IV.9.24. The set of points with rational coordinates is countable and dense in $\xi_s \times \xi_s$. Moreover, the subspace

$$D = \{(x, y) \in \mathbb{R}^2 : y = -x\}$$

is of cardinality $\mathfrak{c}$ and is discrete, because the intersection of a basic open set $[x, a) \times [y, b)$ around $(x, y) \in D$ and D is the singleton $\{(x, y)\}$. In view of Corollary VII.9.4, $\xi_s \times \xi_s$ is not normal.

Chapter VIII

Pseudotopologies

Pretopologies are determined by set adherences (principal adherences). Pseudotopologies are determined by adherences of filters. This larger class of convergences is of great importance, because of its preservation properties. In particular, the pseudotopologizer commutes with arbitrary products, a fact that has far reaching consequences.

VIII.1 Adherence, inherence

In Definition III.5.1 we introduced the adherence of a family $\mathcal{A}$ of subsets of X with respect to a convergence ξ on X, namely

$$\operatorname{adh}_\xi \mathcal{A} = \bigcup_{\mathcal{H} \# \mathcal{A}} \lim_\xi \mathcal{H}.$$

As with the symbol of limit, we often omit the reference to a convergence. For a convergence on X and two isotone families $\mathcal{F}$ and $\mathcal{G}$ of subsets of X,

$$\operatorname{adh} 2^X = \varnothing; \qquad (\text{VIII.1.1})$$

$$\mathcal{G} \geq \mathcal{F} \Longrightarrow \operatorname{adh} \mathcal{G} \subset \operatorname{adh} \mathcal{F}; \qquad (\text{VIII.1.2})$$

$$\operatorname{adh}(\mathcal{F} \wedge \mathcal{G}) \subset \operatorname{adh} \mathcal{F} \cup \operatorname{adh} \mathcal{G}, \qquad (\text{VIII.1.3})$$

where (VIII.1.3) immediately follows from Proposition II.3.6. The last inclusion is in fact an equality, because of (VIII.1.2).

Exercise VIII.1.1. Show that if $\mathcal{F}_0, \mathcal{F}_1 \in \mathbb{F}X$ and $\mathbb{D} \subset \mathbb{F}X$, then:

(1) $\beta(\mathcal{F}_0 \wedge \mathcal{F}_1) = \beta(\mathcal{F}_0) \cup \beta(\mathcal{F}_1)$;
(2)

$$\beta\left(\bigvee_{\mathcal{F} \in \mathbb{D}} \mathcal{F} \right) = \bigcap_{\mathcal{F} \in \mathbb{D}} \beta(\mathcal{F}).$$

(3) Find an example showing that, in general,

$$\beta\left(\bigwedge_{\mathcal{F}\in\mathbb{D}} \mathcal{F}\right) \not\subseteq \bigcup_{\mathcal{F}\in\mathbb{D}} \beta(\mathcal{F}).$$

Remark VIII.1.2. Notice that Proposition II.3.6 and Exercise VIII.1.1 (1) are two sides of the same coin, because $\mathcal{H}\#(\mathcal{F}\wedge\mathcal{G})$ amounts to $\beta\left(\mathcal{H}\right)\cap\left(\beta(\mathcal{F}\wedge\mathcal{G})\right) \neq \varnothing$.

Exercise VIII.1.3. Let $\mathcal{F}$ be an isotone family of subsets of a convergence space.

(1) Show that if $\mathcal{F}$ is a (possibly degenerate) filter, then

$$\text{adh}\,\mathcal{F} = \bigcup_{\mathcal{G}\#\mathcal{F}} \lim \mathcal{G} = \bigcup_{\mathcal{G}\geq\mathcal{F}} \lim \mathcal{G} = \bigcup_{\mathcal{U}\in\beta(\mathcal{F})} \lim \mathcal{U}.$$

(2) Deduce that for every ultrafilter $\mathcal{U}$,

$$\lim \mathcal{U} = \text{adh}\,\mathcal{U}. \qquad\qquad (\text{VIII.1.4})$$

(3) Show that the formula of (1) is in general not valid for a family $\mathcal{F}$ which is not a filter.

Exercise VIII.1.4. Show that if ξ is a pretopology then for every family $\mathcal{F}$ of subsets of $|\xi|$ (in particular, for every filter):

$$\text{adh}_\xi \mathcal{F} = \bigcap_{F\in\mathcal{F}} \text{adh}_\xi F. \qquad\qquad (\text{VIII.1.5})$$

In particular, $\text{adh}_\xi \mathcal{F} = \bigcap_{F\in\mathcal{F}} \text{cl}_\xi F$ whenever ξ is a topology.
Inherence is a dual notion to adherence.

Definition VIII.1.5. Given a convergence ξ on X, the *inherence* $\text{inh}_\xi \mathcal{A}$ of $\mathcal{A}$ is the set of x such that $x \in \lim_\xi \mathcal{F}$ implies that $\mathcal{F}\cap\mathcal{A} \neq \varnothing$.

In other words, the inherence $\text{inh}_\xi \mathcal{A}$ is the greatest subset of X for which $\mathcal{A}$ is a cover.
Recall that

$$\mathcal{A}_c = \{X \setminus A : A \in \mathcal{A}\}$$

denotes the family of complements of elements of $\mathcal{A}$.

Proposition VIII.1.6. *For each family $\mathcal{A}$ of subsets of $|\xi|$,*

$$\text{inh}_\xi \mathcal{A} = (\text{adh}_\xi \mathcal{A}_c)^c.$$

Proof. Let $X := |\xi|$. We notice that $x \notin \mathrm{adh}_\xi \mathcal{A}_c$ whenever $x \notin \lim_\xi \mathcal{F}$ for each filter $\mathcal{F}$ such that $\mathcal{F} \# \mathcal{A}_c$. Equivalently, if $x \in \lim_\xi \mathcal{F}$ then there is $A \in \mathcal{A}$ such that $X \setminus A \notin \mathcal{F}^\#$, which means that $A \in \mathcal{F}$, by (II.3.2). □

Properties of inherences are obviously dual to those of adherences. Namely,

$$\xi \geq \theta \implies \mathrm{inh}_\xi \mathcal{A} \supset \mathrm{inh}_\theta \mathcal{A},$$

and formulas (VIII.1.1), (VIII.1.2), (VIII.1.3) are equivalent respectively to (^{1})

$$\mathrm{inh}(2^X) = X;$$
$$\mathcal{B} \lhd \mathcal{A} \implies \mathrm{inh}\,\mathcal{B} \supset \mathrm{inh}\,\mathcal{A};$$
$$\mathrm{inh}(\mathcal{A} \cap \mathcal{B}) \supset \mathrm{inh}\,\mathcal{A} \cap \mathrm{inh}\,\mathcal{B}.$$

Recall from Section II.5 that *ideal* is a dual notion to that of *filter*.

Definition VIII.1.7. A family $\mathcal{B}$ of subsets of X is an *ideal* on X if $\mathcal{B}_c$ is a filter on X. If $\mathcal{B}_c$ is the degenerate filter on X, then $\mathcal{B} = 2^X = \mathcal{B}_c$. This is the *improper ideal*. It follows that $\mathcal{B}$ is a *(proper) ideal* if and only if

$$X \notin \mathcal{B}; \tag{VIII.1.6}$$
$$A \subset B \in \mathcal{B} \implies A \in \mathcal{B}; \tag{VIII.1.7}$$
$$B_0, B_1 \in \mathcal{B} \implies B_0 \cup B_1 \in \mathcal{B}. \tag{VIII.1.8}$$

Dually to (III.5.2), for every ideal $\mathcal{B}$ and each convergence ξ,

$$\bigcup_{B \in \mathcal{B}} B \supset \mathrm{inh}_\xi \mathcal{B}.$$

Proposition VIII.1.8. *Let $f : X \to Y$ and let ξ be a convergence on X and τ a convergence on Y. Let $\mathcal{G}$ be a family on Y and $\mathcal{H}$ a family on X. Then*

$$\mathrm{adh}_{f\xi}\,\mathcal{G} = \ker \mathcal{G} \cup f(\mathrm{adh}_\xi f^-[\mathcal{G}]), \tag{VIII.1.9}$$
$$\mathrm{adh}_{f^-\tau}\,\mathcal{H} = f^-(\mathrm{adh}_\tau f[\mathcal{H}]). \tag{VIII.1.10}$$

---

^{1}Indeed,

$$\mathrm{inh}(\mathcal{A} \cap \mathcal{B}) = (\mathrm{adh}(\mathcal{A} \cap \mathcal{B})_c)^c = (\mathrm{adh}(\mathcal{A}_c \cap \mathcal{B}_c))^c \supseteq$$
$$(\mathrm{adh}\,\mathcal{A}_c \cup \mathrm{adh}\,\mathcal{B}_c)^c = ((\mathrm{inh}\,\mathcal{A})^c \cup (\mathrm{inh}\,\mathcal{B})^c)^c = \mathrm{inh}\,\mathcal{A} \cap \mathrm{inh}\,\mathcal{B}.$$

Proof. By definition, $y \in f(\text{adh}_\xi f^-[\mathcal{H}])$ if and only if there exists $\mathcal{F} \# f^-[\mathcal{H}]$ such that $y \in f(\lim_\xi \mathcal{F}) \subset \lim_{f\xi} f[\mathcal{F}]$. It follows from (II.2.9) that $f[\mathcal{F}] \# \mathcal{H}$ and hence $y \in \text{adh}_{f\xi} \mathcal{H}$.

Conversely, if $y \in \text{adh}_{f\xi} \mathcal{H}$ and $y \notin f(X)$ then $\{y\} \# \mathcal{H}$ and $y \in \ker \mathcal{H}$. If $y \in f(X) \cap \text{adh}_{f\xi} \mathcal{H}$ then there is $\mathcal{F} \in \mathbb{F}X$ with $\lim_\xi \mathcal{F} \cap f^- y \neq \varnothing$ and $f[\mathcal{F}] \# \mathcal{H}$. Since $\mathcal{F} \# f^-[\mathcal{H}]$ we conclude that $y \in f(\text{adh}_\xi f^-[\mathcal{H}])$, thus proving (VIII.1.9).

As for (VIII.1.10), if $x \in \text{adh}_{f^-\tau} \mathcal{H}$, then there exists a filter $\mathcal{F}$ on X such that $\mathcal{F} \# \mathcal{H}$ and $x \in \lim_{f^-\tau} \mathcal{F} = f^-(\lim_\tau f[\mathcal{F}])$, by (IV.2.1). Consequently, $f[\mathcal{F}] \# f[\mathcal{H}]$ hence $x \in f^-(\text{adh}_\tau f[\mathcal{H}])$.

Conversely, if $x \in f^-(\text{adh}_\tau f[\mathcal{H}])$, then there exists a filter $\mathcal{F}$ such that $\mathcal{F} \# f[\mathcal{H}]$ and $f(x) \in \lim_\tau \mathcal{F}$. As a result, $f^-[\mathcal{F}] \# \mathcal{H}$ and as $\mathcal{F} \leq f[f^-[\mathcal{F}]]$, we have $x \in f^-(\lim_\tau \mathcal{F}) \subset f^-(\lim_\tau f[f^-[\mathcal{F}]]) = \lim_{f^-\tau} f^-[\mathcal{F}]$. Therefore, $x \in \text{adh}_{f^-\tau} \mathcal{H}$. $\qquad\square$

It follows from (VIII.1.9) and (VIII.1.10) that if $\mathcal{A}$ is a family on Y and $\mathcal{D}$ is a family on X, then

$$\text{inh}_{f\xi} \mathcal{A} = \left(f((\text{inh}_\xi f^-[\mathcal{A}])^c) \right)^c \cap \bigcup_{A \in \mathcal{A}} A,$$

$$\text{inh}_{f^-\tau} \mathcal{D} = f^-(\text{inh}_\tau (f[\mathcal{D}_c])_c).$$

VIII.2 Pseudotopologies

By the very definition of adherence, $\lim_\xi \mathcal{F} \subset \text{adh}_\xi \mathcal{H}$ provided that $\mathcal{F} \# \mathcal{H}$. A convergence ξ is a *pseudotopology* if conversely $x \in \lim_\xi \mathcal{F}$ whenever $x \in \text{adh}_\xi \mathcal{H}$ for every filter $\mathcal{H}$ such that $\mathcal{H} \# \mathcal{F}$. In other words, a convergence ξ is a pseudotopology if and only if

$$\lim_\xi \mathcal{F} \supset \bigcap_{\mathcal{H} \# \mathcal{F}} \text{adh}_\xi \mathcal{H}. \qquad\qquad (\text{VIII.2.1})$$

As a result, pseudotopologies are the convergences determined by the limits of ultrafilters. In fact,

Proposition VIII.2.1. *A convergence ξ is a pseudotopology if and only if*

$$\lim_\xi \mathcal{F} = \bigcap_{\mathcal{U} \in \beta(\mathcal{F})} \lim_\xi \mathcal{U}. \qquad\qquad (\text{VIII.2.2})$$

Proof. By (III.1.2) $\lim_\xi \mathcal{F} \subset \lim_\xi \mathcal{U}$ for each $\mathcal{U} \in \beta(\mathcal{F})$ for every convergence ξ. Let ξ be a pseudotopology and let $x \in \lim_\xi \mathcal{U}$ for each $\mathcal{U} \in \beta(\mathcal{F})$. If $x \notin \lim_\xi \mathcal{F}$ then by (VIII.2.1) there is a filter $\mathcal{H}$ such that $x \notin \mathrm{adh}_\xi \mathcal{H}$ and $\mathcal{H} \# \mathcal{F}$. Therefore, if $\mathcal{U} \in \beta(\mathcal{H} \vee \mathcal{F})$ then $x \notin \mathrm{adh}_\xi \mathcal{H} \supset \mathrm{adh}_\xi \mathcal{U} = \lim_\xi \mathcal{U}$.

Conversely, if (VIII.2.2) holds and $x \in \mathrm{adh}_\xi \mathcal{H}$ for each $\mathcal{H} \# \mathcal{F}$, then *a fortiori* $x \in \lim_\xi \mathcal{U}$ for each $\mathcal{U} \in \beta(\mathcal{F})$ and thus by (VIII.2.2) $x \in \lim_\xi \mathcal{F}$. $\square$

Exercise VIII.2.2. Show that each pseudotopology is finitely deep.

Exercise VIII.2.3. Check that pseudotopologicity is a pointwise property. Describe the corresponding property of collections of filters.

Proposition VIII.2.4. *Each pretopology is a pseudotopology.*

Proof. If ξ is a pretopology, and $\mathcal{F}$ is such that $x \in \bigcap_{\mathcal{U} \in \beta(\mathcal{F})} \lim_\xi \mathcal{U}$ then $\mathcal{U} \geq \mathcal{V}_\xi(x)$ for each $\mathcal{U} \in \beta(\mathcal{F})$ so that, by Proposition II.6.8,

$$\mathcal{F} = \bigwedge_{\mathcal{U} \in \beta(\mathcal{F})} \mathcal{U} \geq \mathcal{V}_\xi(x),$$

and thus $x \in \lim_\xi \mathcal{F}$, because ξ is a pretopology. $\square$

The class of pseudotopologies is strictly intermediate between the classes of pretopologies and of finitely deep convergences, as the following examples show.

Example VIII.2.5 (non-pseudotopological finitely deep convergence). In an infinite set X, distinguish an element x_∞, and define a prime convergence on X as follows: every $x \in X \setminus \{x_\infty\}$ is isolated, and $x_\infty \in \lim \mathcal{H}$ if there exists a *finite* subset $\mathbb{D}$ of the set of free ultrafilters on X with $\mathcal{H} \geq \bigwedge_{\mathcal{U} \in \mathbb{D}} \mathcal{U} \wedge \{x_\infty\}^\uparrow$; in particular every free ultrafilter converges to x_∞. This convergence is finitely deep, but not a pseudotopology. Indeed the cofinite filter $(X)_0$ of X does not converge to x_∞, even though every ultrafilter finer than $(X)_0$ is free and thus converges to x_∞.

Example VIII.2.6 (non-pretopological pseudotopology). Consider the sequential modification $\mathrm{Seq}\,\nu$ of the natural topology of the real line ν. Notice that each convergent filter contains a countable set, because if $r \in \lim_{\mathrm{Seq}\,\nu} \mathcal{F}$, then there is a sequential filter $\mathcal{E} \leq \mathcal{F}$ such that $r \in \lim_{\mathrm{Seq}\,\nu} \mathcal{E}$. Hence there is a countable set E such that $E \in \mathcal{E} \leq \mathcal{F}$. In view of Proposition II.8.7 the vicinity filter is $\mathcal{V}_{\mathrm{Seq}\,\nu}(r) = \mathcal{N}_\nu(r)$. It does not contain any countable set, so that $\mathcal{V}_{\mathrm{Seq}\,\nu}(r)$ does not converge. It follows that $\mathrm{Seq}\,\nu$ is not a pretopology.

To see that $\mathrm{Seq}\,\nu$ is a pseudotopology, let $\mathcal{F}$ be a free filter such that $r \in \lim_{\mathrm{Seq}\,\nu} \mathcal{U}$ for each $\mathcal{U} \in \beta(\mathcal{F})$. In view of Proposition II.2.32, for each $\mathcal{U} \in \beta(\mathcal{F})$ there exists a countable set $E_{\mathcal{U}}$ such that $E_{\mathcal{U}} \in \mathcal{U}$ and $r \in \lim_{\mathrm{Seq}\,\nu} (E_{\mathcal{U}})_0$ (where $(E_{\mathcal{U}})_0$ is the cofinite filter of $E_{\mathcal{U}}$). By Proposition II.6.5, there is a finite subset $\mathbb{D}$ of $\beta(\mathcal{F})$ such that $E := \bigcup_{\mathcal{U} \in \mathbb{D}} E_{\mathcal{U}} \in \mathcal{F}$. As $\mathcal{F}$ is free, $(E)_0$ is a free sequential filter, $(E)_0 \leq \mathcal{F}$ and $r \in \lim_{\mathrm{Seq}\,\nu} (E)_0 \subset \lim_{\mathrm{Seq}\,\nu} \mathcal{F}$.

Example VIII.2.7 (non-pretopological pseudotopology). Consider the convergence space $(X, f\xi)$ as in Example IV.9.25. We have seen that $f\xi$ is not even finitely deep, and therefore is not a pseudotopology. However, the convergence σ on X defined by

$$\lim_\sigma \mathcal{F} := \bigcap_{\mathcal{U} \in \beta(\mathcal{F})} \lim_{f\xi} \mathcal{U}$$

is easily verified to be a pseudotopology. Indeed, since $\beta(\mathcal{U}) = \{\mathcal{U}\}$ for every ultrafilter, we get that $\lim_\sigma \mathcal{U} = \lim_{f\xi} \mathcal{U}$ whenever $\mathcal{U} \in \mathbb{U}X$ so that σ satisfies the condition of Proposition VIII.2.1.

On the other hand, σ is not a pretopology. Indeed, in view of (V.1.2),

$$\mathcal{V}_\sigma(\infty) = \bigwedge_{\infty \in \lim_\sigma \mathcal{U}} \mathcal{U} = \bigwedge_{\infty \in \lim_{f\xi} \mathcal{U}} \mathcal{U} = \mathcal{V}_{f\xi}(\infty),$$

does not converge for σ because $\mathcal{V}_{f\xi}(\infty) \# \{\bigcup_{x \geq r} f(N_x) : r \in \mathbb{R}\}$ so that there is $\mathcal{U}_0 \in \beta(\mathcal{V}_{f\xi}(\infty))$ such that $\mathcal{U}_0 \geq \{\bigcup_{x \geq r} f(N_x) : r \in \mathbb{R}\}$. Clearly, $\lim_{f\xi} \mathcal{U}_0 = \lim_\sigma \mathcal{U}_0 = \varnothing$ because $\mathcal{U}_0$ does not have a filter-base composed of subsets of a fixed branch $f(N_x)$.

Example VIII.2.8 (non-pretopological pseudotopology). The Scott convergence (or lower convergence) defined in (III.9.1) is a pseudotopology. We will see (via Theorem IX.4.1) that it may fail to be pretopological.

VIII.3 Pseudotopologizer

Proposition VIII.3.1. *Each supremum of pseudotopologies is a pseudotopology.*

Proof. If Ξ is a family of pseudotopologies on a set X and $\mathcal{F}$ is a filter on X such that $x \in \lim_{\bigvee \Xi} \mathcal{U}$ for every $\mathcal{U} \in \beta(\mathcal{F})$, then $x \in \lim_\xi \mathcal{U}$ for every $\xi \in \Xi$. Therefore $x \in \lim_\xi \mathcal{F}$, because ξ is a pseudotopology for each $\xi \in \Xi$, hence $x \in \lim_{\bigvee \Xi} \mathcal{F}$. $\square$

The set of pseudotopologies on a given set is closed under arbitrary suprema and contains the chaotic topology. As a result, for every convergence ξ there exists the finest among coarser pseudotopologies, called the *pseudotopologization* $S\xi$ of ξ.

Exercise VIII.3.2. Verify that

$$\lim_{S\xi} \mathcal{F} = \bigcap_{\mathcal{U} \in \beta(\mathcal{F})} \lim_{\xi} \mathcal{U} = \bigcap_{\mathcal{H}\#\mathcal{F}} \mathrm{adh}_\xi \, \mathcal{H}. \qquad \text{(VIII.3.1)}$$

In particular,

$$\lim_{S\xi} \mathcal{U} = \lim_{\xi} \mathcal{U}$$

for every ultrafilter $\mathcal{U}$ on $|\xi|$, because then $\beta(\mathcal{U}) = \{\mathcal{U}\}$.

Example VIII.3.3. The convergence σ of Example VIII.2.7 is the pseudotopologization $S(f\xi)$ of $f\xi$.

Note that for every filter $\mathcal{H}$,

$$\mathrm{adh}_{S\xi} \, \mathcal{H} = \mathrm{adh}_\xi \, \mathcal{H}.$$

Indeed,

$$\mathrm{adh}_{S\xi} \, \mathcal{H} = \bigcup_{\mathcal{U} \in \beta(\mathcal{F})} \lim_{S\xi} \mathcal{U} = \bigcup_{\mathcal{U} \in \beta(\mathcal{F})} \lim_{\xi} \mathcal{U} = \mathrm{adh}_\xi \, \mathcal{H}.$$

The *pseudotopologizer* S, restricted to the set of all convergences with the same underlying set, fulfills

$$\zeta \leq \xi \Rightarrow S\zeta \leq S\xi; \qquad \text{(isotone)}$$
$$S\xi \leq \xi; \qquad \text{(contractive)}$$
$$S(S\xi) = S\xi. \qquad \text{(idempotent)}$$

In view of Exercise VIII.1.1 and Proposition VIII.2.4,

$$I \geq L \geq S \geq S_0 \geq T, \qquad \text{(VIII.3.2)}$$

where I the identity and L the finitely deep modifier.

Indeed, $S_0 \xi$ is a pretopology, hence a pseudotopology, coarser than ξ, so that $S\xi \geq S_0 \xi$. Similarly, $S\xi$ is a pseudotopology, hence a finitely deep convergence, coarser than ξ so that $L\xi \geq S\xi$.

Proposition VIII.3.4. *The pseudotopologizer commutes with initial structures. Namely,*

$$S(f^-\tau) = f^-(S\tau) \qquad \text{(VIII.3.3)}$$

for every convergence τ and each map f.

Proof. Indeed, if $x \in \lim_{f^-(S\tau)} \mathcal{F}$ then equivalently $f(x) \in \lim_{S\tau} f[\mathcal{F}]$, that is, $f(x) \in \lim_\tau \mathcal{U}$ for every $\mathcal{U} \in \beta(f[\mathcal{F}])$. If now $\mathcal{W} \in \beta(\mathcal{F})$, then $f[\mathcal{W}] \in \beta(f[\mathcal{F}])$ and thus $f(x) \in \lim_\tau f[\mathcal{W}]$, equivalently $x \in \lim_{f^-\tau} \mathcal{W}$, which means that $x \in \lim_{S(f^-\tau)} \mathcal{F}$.

Conversely, assume that $x \in \lim_{S(f^-\tau)} \mathcal{F}$, that is, $x \in \lim_{f^-\tau} \mathcal{U}$ for every $\mathcal{U} \in \beta(\mathcal{F})$, and let $\mathcal{W} \in \beta(f[\mathcal{F}])$. By Corollary II.6.7, there is $\mathcal{U} \in \beta(\mathcal{F})$ with $f[\mathcal{U}] = \mathcal{W}$. Since $x \in \lim_{f^-\tau} \mathcal{U}$, $f(x) \in \lim_\tau f[\mathcal{U}]$, hence $f(x) \in \lim_\tau \mathcal{W}$. In other words, $f(x) \in \lim_{S\tau} f[\mathcal{F}]$, that is, $x \in \lim_{f^-(S\tau)} \mathcal{F}$. □

Corollary VIII.3.5. *If (X, ξ) is a convergence space and $A \subset X$ then the pseudotopological modification of the induced convergence coincides with the convergence induced on A by the pseudotopological modification, that is,*

$$\mathrm{S}(\xi_{|A}) = (\mathrm{S}\,\xi)_{|A}.$$

From (VIII.3.4), we have in particular

$$\mathrm{S}(f^-\tau) \geq f^-(\mathrm{S}\,\tau) \tag{VIII.3.4}$$

for every convergence τ and every map f, hence:

Corollary VIII.3.6. *The pseudotopologizer preserves continuity.*

Proof. Indeed, if f is continuous from ξ to τ, that is, $\xi \geq f^-\tau$, then $\mathrm{S}\,\xi \geq \mathrm{S}(f^-\tau) \geq f^-(\mathrm{S}\,\tau)$ by (isotone) and (VIII.3.4), hence f is continuous from $\mathrm{S}\,\xi$ to $\mathrm{S}\,\tau$. □

Corollary VIII.3.7. *The initial convergence of a pseudotopology is a pseudotopology.*

Proof. If τ is a pseudotopology, that is, $\tau \leq \mathrm{S}\,\tau$ then $f^-\tau \leq f^-(\mathrm{S}\,\tau) \leq \mathrm{S}(f^-\tau)$ by (VIII.3.4). □

It follows that pseudotopologies, as pretopologies and topologies, are stable under subspace, suprema, and products. Commutativity of the pseudotopologizer with suprema (and thus, products), however, is in sharp contrast with the behavior of the pretopologizer and the topologizer, even for finite sets of convergences (see Examples V.2.27 and V.4.40).

Proposition VIII.3.8. *If Θ is a set of convergences on X, then*

$$\mathrm{S}(\bigvee \Theta) = \bigvee_{\theta \in \Theta} \mathrm{S}\,\theta.$$

Proof. By definition,

$$\lim_{\bigvee_{\theta \in \Theta} S \theta} \mathcal{F} = \bigcap_{\theta \in \Theta} \bigcap_{\mathcal{U} \in \beta(\mathcal{F})} \lim_\theta \mathcal{U} = \bigcap_{\mathcal{U} \in \beta(\mathcal{F})} \bigcap_{\theta \in \Theta} \lim_\theta \mathcal{U}$$

$$= \bigcap_{\mathcal{U} \in \beta(\mathcal{F})} \lim_{\vee \Theta} \mathcal{U} = S(\lim_{\vee \Theta} \mathcal{F}).$$

$\square$

Because the product is the supremum of initial convergences with respect to the projections on factor spaces, we get the following important result:

Theorem VIII.3.9.

$$S(\prod \Xi) = \prod_{\xi \in \Xi} S \xi \qquad (VIII.3.5)$$

for every set of convergences Ξ.

Exercise VIII.3.10. Show that a convergence ξ is Hausdorff if and only if $S \xi$ is.

Proposition VIII.3.11. *If a convergence ξ is regular then $S \xi$ is regular as well.*

Exercise VIII.3.12. Prove Proposition VIII.3.11 using Proposition VIII.4.1.

Definition VIII.3.13. A convergence ξ is *almost topological* if $S \xi = T \xi$.

Exercise VIII.3.14. Show that the convergence ξ of Example IV.9.6 is almost topological.

Of course, if ξ is almost topological, then in particular $S_0 \xi = T \xi$. Thus, in view of Proposition VIII.3.11 and Proposition VI.4.1:

Proposition VIII.3.15. *A regular almost topological convergence is topologically regular.*

VIII.4 Regularity and topologicity among pseudotopologies

Proposition VI.5.2 leads to the following useful characterization of regularity among pseudotopologies:

Proposition VIII.4.1. *If a convergence ξ is θ-regular then for every family $\mathcal{H}$,*

$$\mathrm{adh}_\xi \, \mathcal{V}_\theta(\mathcal{H}) \subset \mathrm{adh}_\xi \, \mathcal{H}, \qquad\qquad (VIII.4.1)$$

and the converse is true if ξ is a pseudotopology.

Proof. $x \in \mathrm{adh}_\xi \, \mathcal{V}_\theta(\mathcal{H})$ whenever there is a filter $\mathcal{F}$ such that $x \in \lim_\xi \mathcal{F}$ and

$$\mathcal{F} \# \mathcal{V}_\theta(\mathcal{H}),$$

which is equivalent to $(\mathrm{adh}_\theta^\natural \mathcal{F}) \# \mathcal{H}$ by Proposition VI.5.2, and $x \in \lim_\xi \mathrm{adh}_\theta^\natural \mathcal{F}$ by θ-regularity. Hence, $x \in \mathrm{adh}_\xi \, \mathcal{H}$.

Conversely, if ξ is a pseudotopology that satisfies (VIII.4.1) for every filter $\mathcal{H}$, and $x \notin \lim_\xi \mathrm{adh}_\theta^\natural \mathcal{F}$ then there is $\mathcal{H} \# (\mathrm{adh}_\theta^\natural \mathcal{F})$ such that $x \notin \mathrm{adh}_\xi \, \mathcal{H}$ because ξ is a pseudotopology. By Proposition VI.5.2, $\mathcal{V}_\theta(\mathcal{H}) \# \mathcal{F}$ and $x \notin \mathrm{adh}_\xi \, \mathcal{V}_\theta(\mathcal{H})$ by (VIII.4.1). Therefore, $x \notin \lim_\xi \mathcal{F}$. In other words, $\lim_\xi \mathcal{F} \subset \lim_\xi \mathrm{adh}_\theta^\natural \mathcal{F}$ and ξ is θ-regular. $\qquad\square$

Theorems VI.2.19 and VI.3.18 provide dual characterizations of topologicity and regularity. Among pseudotopologies, where convergence only depends on ultrafilters, they can be refined as follows:

Theorem VIII.4.2. *A pseudotopology is a topology if and only if for every set A, every pair of maps $s : A \to X$ and $\mathcal{S} : A \to \mathbb{U}X$ with $s(a) \in \lim \mathcal{S}(a)$ for each $a \in A$,*

$$\lim s[\mathcal{U}] \subset \lim \mathcal{S}(\mathcal{U}),$$

for each $\mathcal{U} \in \mathbb{U}A$.

Proof. In view of Theorem VI.2.19, it suffices to show that if the condition is satisfied, so is that of (3) of Theorem VI.2.19. Assume therefore that $\mathcal{F} \in \mathbb{F}A$ with $s : A \to X$ and $\mathcal{S} : A \to \mathbb{U}X$ with $s(a) \in \lim \mathcal{S}(a)$ for each $a \in A$, and $x_0 \in \lim s[\mathcal{F}]$. To show that $x_0 \in \lim \mathcal{S}(\mathcal{F})$, we only need to show that $x_0 \in \lim \mathcal{W}$ for every $\mathcal{W} \in \beta(\mathcal{S}(\mathcal{F}))$, because the convergence is a pseudotopology. In view of Lemma VI.1.9, for each such $\mathcal{W}$, there is $\mathcal{U} \in \beta\mathcal{F}$ with $\mathcal{S}(\mathcal{U}) = \mathcal{W}$. Since $s[\mathcal{U}] \geq s[\mathcal{F}]$, $x_0 \in \lim s[\mathcal{U}]$ and $\lim s[\mathcal{U}] \subset \lim \mathcal{S}[\mathcal{U}]$ by assumption. Thus $x_0 \in \bigcap_{\mathcal{W} \in \beta(\mathcal{S}(\mathcal{F}))} \lim \mathcal{W}$, which concludes the proof. $\quad\square$

Theorem VIII.4.3. *A pseudotopology is regular if and only if for every set A, every pair of maps $s : A \to X$ and $\mathcal{S} : A \to \mathbb{U}X$ with $s(a) \in \lim \mathcal{S}(a)$ for each $a \in A$,*

$$\lim \mathcal{S}(\mathcal{U}) \subset \lim s[\mathcal{U}],$$

for each $\mathcal{U} \in \mathbb{U}A$.

Proof. In view of Theorem VI.3.18 it suffices to show that if the condition is satisfied, so is that of (3) of Theorem VI.3.18. To this end, consider $\mathcal{F} \in \mathbb{F}A$ with $s : A \to X$ and $\mathcal{S} : A \to \mathbb{U}X$ with $s(a) \in \lim \mathcal{S}(a)$ for each $a \in A$, such that $x_0 \in \lim \mathcal{S}(\mathcal{F})$. To show that $x_0 \in \lim s[\mathcal{F}]$, it is enough to show that $x_0 \in \lim \mathcal{W}$ for every $\mathcal{W} \in \beta(s[\mathcal{F}])$, because the convergence is a pseudotopology. By Corollary II.6.7, there is $\mathcal{U} \in \beta \mathcal{F}$ with $s[\mathcal{U}] = \mathcal{W}$. Note that $\mathcal{S}(\mathcal{U}) \geq \mathcal{S}(\mathcal{F})$ so that $x_0 \in \lim \mathcal{S}(\mathcal{U})$. Moreover, by assumption, $\lim \mathcal{S}(\mathcal{U}) \subset \lim s[\mathcal{U}]$. Thus $x_0 \in \lim s[\mathcal{U}] = \lim \mathcal{W}$. Hence $x_0 \in \bigcap_{\mathcal{W} \in \beta(s[\mathcal{F}])} \lim \mathcal{W}$, which concludes the proof. $\square$

VIII.5   Initial density in pseudotopologies

Definition VIII.5.1. A class of convergences Θ is *simple* if there is $\tau \in \Theta$ such that $\{\tau\}$ is initially dense in Θ.

Theorem V.1.14 shows that the class of pretopologies is simple, and Proposition V.4.18 shows that the class of topologies is simple.

In contrast, we will see in this section that the class of pseudotopologies is not simple, so that, in view of Lemma VIII.5.2 below, a collection that is initially dense in pseudotopologies must be a proper class.

Lemma VIII.5.2. *If Θ is a set of convergences that is initially dense in a class Π of convergences, then the convergence*

$$\theta_0 := \prod_{\theta \in \Theta} \theta$$

is initially dense in Π.

Proof. For each $\pi \in \Pi$, there is a subset S of Θ and, for each $\tau \in S$, there are maps $(f_i : |\pi| \to |\tau|)_{i \in I_\tau}$ such that

$$\pi = \bigvee_{\tau \in S} \bigvee_{i \in I_\tau} f_i^- \tau.$$

Let $p_\tau : |\theta_0| \to |\tau|$ denote the projection on $|\tau|$, let $y_0 \in \theta_0$, and let $\Delta_S : \pi \to \pi^S = \prod_{s \in S} \pi_s$ be defined by $\Delta_S(x)(s) = x$ for all $s \in S$. Let now $e_S : |\prod_{\tau \in S} \tau| \to |\theta_0|$ be defined by $p_\tau(e_S(x)) = p_\tau(x)$ if $\tau \in S$ and $p_\tau(e_S(x)) = p_\tau(y_0)$ otherwise. Set $I := \prod_{\tau \in S} I_\tau$ and consider, for each $j = (i_\tau)_{\tau \in S} \in I$, the map

$$F_j := e_S \circ (\bigotimes_{\tau \in S} f_{i_\tau}) \circ \Delta_S : |\pi| \to |\theta_0|.$$

Then

$$\bigvee_{j \in I} F_j^- \theta_0 = \bigvee_{j \in I} (e_S \circ (\bigotimes_{\tau \in S} f_{i_\tau}) \circ \Delta_S)^- (\prod_{\tau \in \Theta} \tau)$$

$$= \bigvee_{j \in I} \Delta_S^- (\bigotimes_{\tau \in S} f_{i_\tau})^- e_S^- (\prod_{\tau \in \Theta} \tau) \text{ by (IV.2.8)}$$

$$= \bigvee_{j \in I} \Delta_S^- (\bigotimes_{\tau \in S} f_{i_\tau})^- (\prod_{\tau \in S} \tau)$$

$$= \bigvee_{j \in I} \Delta_S^- (\prod_{\tau \in S} f_{i_\tau}^- \tau) \text{ by (IV.7.4)}$$

$$= \bigvee_{\tau \in S} \bigvee_{i \in I_\tau} f_i^- \tau = \pi.$$

$\square$

Lemma VIII.5.3. *If X is infinite and* card $X >$ card Y, *then for each uniform ultrafilter $\mathcal{U}$ on X and every map $f : X \to Y$, there is a uniform ultrafilter $\mathcal{W}$ different from $\mathcal{U}$ such that $f[\mathcal{U}] = f[\mathcal{W}]$.*

Proof. Let $\mathcal{U}$ be a uniform ultrafilter on X and let $U \in \mathcal{U}$ be such that card $U =$ card X. Consider

$$\mathcal{P} := \{ f^-(y) \cap U : y \in f(U) \},$$

as well as two subfamilies of $\mathcal{P}$,

$$\mathcal{P}_0 := [\mathcal{P}]^{<\omega}$$

and $\mathcal{P}_1 := \mathcal{P} \setminus \mathcal{P}_0$. Let

$$P_0 := \bigcup_{P \in \mathcal{P}_0} P.$$

Note that card $P_0 \leq$ card $Y <$ card X, because card $\mathcal{P}_0 \leq$ card Y. It follows that $\mathcal{P}_1$ is not empty, for otherwise $P_0 = U$.

For each $P \in \mathcal{P}_1$, let A_P and B_P be such that $P = A_P \cup B_P$, $\varnothing = A_P \cap B_P$ and card $A_P =$ card B_P. We set

$$A := P_0 \cup \bigcup_{P \in \mathcal{P}_1} A_P \text{ and } B := P_0 \cup \bigcup_{P \in \mathcal{P}_1} B_P.$$

As $A \cup B = U \in \mathcal{U}$, either $A \in \mathcal{U}$ or $B \in \mathcal{U}$, say $A \in \mathcal{U}$. Let $h : A \to B$ be such that $h(x) = x$ for each $x \in P_0$ and h is a bijection from A_P onto B_P for each $P \in \mathcal{P}_1$. This function is well-defined, because elements of $\mathcal{P}$ are pairwise disjoint. Then $\mathcal{W} := h[\mathcal{U}]$ is an ultrafilter on U and $B \in \mathcal{W}$. Moreover $\mathcal{W} \neq \mathcal{U}$, for otherwise $P_0 = A \cap B \in \mathcal{U}$, even though card $P_0 < \|\mathcal{U}\|$, which yields a contradiction.

Finally, $f[\mathcal{U}] = f[\mathcal{W}] = f \circ h[\mathcal{U}]$, because $f(A_P) = f(B_p)$ for each $P \in \mathcal{P}_1$. $\square$

Proposition VIII.5.4. *The class of all convergences is not simple.*

Proof. If τ is an arbitrary convergence on a set Y, let X be an infinite set such that $\operatorname{card} X > \operatorname{card} Y$, endowed with the convergence ξ of Example VI.2.11. Then by Lemma VIII.5.3 for each $f : X \to Y$, there exists an ultrafilter $\mathcal{W}_f \neq \mathcal{U}_0$ such that $f[\mathcal{W}_f] = f[\mathcal{U}_0]$. If $f : |\xi| \to |\tau|$ is continuous, $f(\infty) \in \lim_\tau f[\mathcal{W}_f] = \lim_\tau f[\mathcal{U}_0]$. Thus

$$\infty \in \lim_{\bigvee_{f \in \mathcal{F}} f^- \tau} \mathcal{U}_0 \neq \lim_\xi \mathcal{U}_0 = \varnothing$$

for each family $\mathcal{F}$ of maps $f : X \to Y$. In other words, ξ is not τ-initial, and thus $\{\tau\}$ is not initially dense in the class of all convergences. $\square$

Because, for each pseudotopology τ, the convergence $\bigvee_{f \in \mathcal{F}} f^- \tau$ is pseudotopological (by Propositions VIII.3.1 and VIII.3.4) and $\lim_{S\xi} \mathcal{U}_0 = \varnothing$, the proof of Proposition VIII.5.4 is unchanged if we replace ξ by $S\xi$. Therefore:

Corollary VIII.5.5. *The class of pseudotopologies is not simple.*

VIII.6   Natural convergence

Following the particular case of Example IV.5.3, we examine the *natural convergence* or *convergence of joint continuity,* as a fundamental example of pseudotopology.

If (X, ξ) and (Y, σ) are two convergence spaces, we denote by $C(\xi, \sigma)$, or $C(X, Y)$ if no confusion is possible, the set of continuous function from ξ to σ.

Definition VIII.6.1. The *natural convergence* $[\xi, \sigma]$ is the coarsest among the convergences θ on $C(\xi, \sigma)$ such that the evaluation map $e :$ $|\xi \times \theta| \to |\sigma|$

$$e(x, f) := \langle x, f \rangle := f(x)$$

is continuous.

Note that the definition makes sense, that is, the coarsest convergence on $C(\xi, \sigma)$ making the evaluation continuous exists.

Proof. Indeed, the set Θ of convergences on $C(\xi, \sigma)$ for which e is continuous is non empty, because it contains the discrete topology. Moreover,

$$\xi \times \theta \geq e^- \sigma, \tag{VIII.6.1}$$

for each $\theta \in \Theta$, by Proposition IV.2.9. Hence,

$$\bigwedge_{\theta \in \Theta} (\xi \times \theta) = \xi \times (\bigwedge_{\theta \in \Theta} \theta) \geq e^{-}\sigma,$$

and $[\xi, \sigma] = \bigwedge \Theta$ is indeed the coarsest of convergences θ satisfying (VIII.6.1).   □

An explicit description follows from the definition. Indeed, if θ makes the evaluation continuous then

$$f \in \lim_\theta \mathcal{F} \implies \underset{x \in X}{\forall} \underset{\mathcal{G} \in \mathbb{F}|\xi| \cap \lim_\xi^-(x)}{\forall} f(x) \in \lim_\sigma e[\mathcal{G} \times \mathcal{F}].$$

The coarsest convergence satisfying this condition is the one for which the implication is reversed:

$$f \in \lim_{[\xi,\sigma]} \mathcal{F} \iff \underset{x \in X}{\forall} \underset{\mathcal{G} \in \mathbb{F}|\xi| \cap \lim_\xi^-(x)}{\forall} f(x) \in \lim_\sigma e[\mathcal{G} \times \mathcal{F}]. \quad (\text{VIII.6.2})$$

Theorem VIII.6.2. *Let σ be a pseudotopology. Then $[\xi, \sigma]$ is a pseudotopology.*

Proof. By definition $\xi \times [\xi, \sigma] \geq e^{-}\sigma$. Since the pseudotopologizer is isotone, we have:

$$S(\xi \times [\xi, \sigma]) \geq S(e^{-}\sigma).$$

In view of Theorem VIII.3.9 and (VIII.3.4), we obtain:

$$\xi \times S[\xi, \sigma] \geq S\xi \times S[\xi, \sigma] \geq e^{-}(S\sigma) = e^{-}\sigma.$$

Therefore $S[\xi, \sigma]$ satisfies (VIII.6.1) and is coarser than $[\xi, \sigma]$, so that $[\xi, \sigma] = S[\xi, \sigma]$.   □

In the particular case $[\mathbb{R}, \mathbb{R}]$ of Example IV.5.3 the natural convergence turns out to be topological, but in many instances, the natural convergence is not even pretopological. We will prove these facts when we return to function space structures in Chapter XVI.

VIII.7   Convergences on hyperspaces

Recall from Exercise V.3.28 that a subset A of a convergence space (X, ξ) is open if and only if $\chi_A : |\xi| \to |\$_1|$ is continuous and closed if and only if $\chi_A : |\xi| \to |\$_0|$ is continuous, where χ_A is the characteristic function defined by

$$\langle x, \chi_A \rangle = 1 \iff x \in A.$$

As a result, we can identify the set $\mathcal{O}_\xi$ of open subsets of (X, ξ) with the set $C(\xi, \$_1)$ of continuous functions from ξ to the Sierpiński space $\$_1$ and the set $\mathcal{C}_\xi$ of closed subsets with the set $C(\xi, \$_0)$ of continuous functions from ξ to the Sierpiński space $\$_0$.

In order to simplify the discourse, we do not assume in this section that an underlying convergence ξ is arbitrary but require that ξ be a topology, so that either $C(\xi, \$_1)$ or $C(\xi, \$_0)$ determines ξ. We will now examine how the natural convergence $[\xi, \$_1]$ can be characterized in terms of open sets, via the identification

$$\mathcal{O}_\xi = C(\xi, \$_1) \text{ and } \mathcal{C}_\xi = C(\xi, \$_0).$$

Proposition VIII.7.1. *Let $\mathcal{F}$ be a filter on $\mathcal{O}_\xi$ and let U be an open subset of $|\xi|$. Then*

$$U \in \lim_{[\xi, \$_1]} \mathcal{F} \iff U \subset \bigcup_{F \in \mathcal{F}} \mathrm{int}_\xi \left(\bigcap_{O \in F} O \right). \tag{VIII.7.1}$$

Proof. By definition of the natural convergence, $U \in \lim_{[\xi, \$_1]} \mathcal{F}$ if and only if $\langle x, \chi_U \rangle \in \lim_{\$_1} \langle \mathcal{G}, \mathcal{F} \rangle$ for every $x \in X$ and every filter $\mathcal{G}$ converging to x in ξ. This condition is always true if $\langle x, \chi_U \rangle = 0$, that is, if $x \notin U$. Hence $\mathcal{F}$ converges to U if and only if

$$1 \in \lim_{[\xi, \$_1]} \langle \mathcal{N}_\xi(x), \mathcal{F} \rangle$$

for every $x \in U$. As 1 is isolated in $\$_1$, that means that $\{1\} \in \langle \mathcal{N}_\xi(x), \mathcal{F} \rangle$, that is, there is $V \in \mathcal{N}_\xi(x)$ and $F \in \mathcal{F}$ with $\chi_O(V) = 1$ for each $O \in F$, that is,

$$V \subset \bigcap_{O \in F} O.$$

In other words, $U \in \lim_{[\xi, \$_1]} \mathcal{F}$ if and only if for each $x \in U$ there is $F \in \mathcal{F}$ with $x \in \mathrm{int}_\xi(\bigcap_{O \in F} O)$, which completes the proof. $\qquad\square$

A convergence space of the type $(\mathcal{C}_\xi, \theta)$ or $(\mathcal{O}_\xi, \theta)$ is called a *hyperspace for ξ*, and θ is an *hyperconvergence*.

We now turn to the incarnation of the natural convergence on the set $\mathcal{C}_\xi$ of closed subsets of a topological space (X, ξ). If $\mathcal{F}$ is a family of subsets of $\mathcal{C}_\xi$ its *reduction on X* is, by definition, the family

$$\mathrm{rdc}(\mathcal{F}) := \left\{ \bigcup_{C \in F} C : F \in \mathcal{F} \right\} \tag{VIII.7.2}$$

of subsets of X. If $\mathcal{F}$ is a filter, then $\mathrm{rdc}(\mathcal{F})$ is called the *reduced filter* of $\mathcal{F}$.

Note that

$$\mathrm{rdc} \left(\bigcup_{\mathcal{A} \in \mathfrak{A}} \mathcal{A} \right) = \bigcup_{\mathcal{A} \in \mathfrak{A}} \mathrm{rdc}(\mathcal{A})$$

for any collection $\mathfrak{A}$ of families of subsets of $\mathcal{C}_\xi$, and

$$\mathrm{rdc} \left(\bigwedge_{\mathcal{F} \in \mathbb{D}} \mathcal{F} \right) = \bigwedge_{\mathcal{F} \in \mathbb{D}} \mathrm{rdc}(\mathcal{F}) \qquad (\mathrm{VIII.7.3})$$

for any family $\mathbb{D}$ of filters on $\mathcal{C}_\xi$.

Exercise VIII.7.2. Show that a filter $\mathcal{F}$ on the set $\mathcal{C}_\xi$ of closed subsets of a topological space (X, ξ) converges for $[\xi, \$_0]$ to a closed subset A_0 of X if and only if

$$\mathrm{adh}_\xi \, \mathrm{rdc}(\mathcal{F}) \subset A_0,$$

that is,

$$A_0 \in \lim_{[\xi, \$_0]} \mathcal{F} \iff \bigcap_{F \in \mathcal{F}} \mathrm{cl}_\xi \left(\bigcup_{C \in F} C \right) \subset A_0. \qquad (\mathrm{VIII.7.4})$$

Definition VIII.7.3. The natural convergence $[\xi, \$_0]$ on $\mathcal{C}_\xi$ given by (VIII.7.4) is usually called the *upper Kuratowski convergence*.

This is an instance of Scott convergence on the lattice $(\mathcal{C}_\xi, \supset)$. Observe that if $A_1 \in \mathcal{C}_\xi$ and $A_1 \supset A_0$, then

$$A_0 \in \lim_{[\xi, \$_0]} \mathcal{F} \implies A_1 \in \lim_{[\xi, \$_0]} \mathcal{F} .$$

Of course, $[\xi, \$_0]$ is homeomorphic to $[\xi, \$_1]$ on $\mathcal{O}_\xi$ given by (VIII.7.1). The convergence $[\xi, \$_1]$ is often called the *Scott convergence*, because the Scott convergence on a complete lattice $(X, \leq)$ is given by (III.9.1) and (VIII.7.1) is the particular case of the complete lattice $(\mathcal{O}_\xi, \subset)$.

Definition VIII.7.4. The *lower Kuratowski convergence* $[\xi, \$_0]^\#$ is defined on $\mathcal{C}_\xi$ by

$$A_0 \in \lim_{[\xi, \$_0]^\#} \mathcal{F} \iff A_0 \subset \bigcap_{H \in \mathcal{F}^\#} \mathrm{cl}_\xi \left(\bigcup_{C \in H} C \right). \qquad (\mathrm{VIII.7.5})$$

Notice that, by (VIII.1.5), $A_0 \in \lim_{[\xi, \$_0]^\#} \mathcal{F}$ is equivalent to

$$A_0 \subset \mathrm{adh}_\xi \, \mathrm{rdc} \left(\mathcal{F}^\# \right) .$$

Observe that if $A_1 \in \mathcal{C}_\xi$ and $A_0 \subset A_1$, then

$$A_0 \in \lim_{[\xi, \$_0]^\#} \mathcal{F} \implies A_1 \in \lim_{[\xi, \$_0]^\#} \mathcal{F} .$$

Of course, the convergence $[\xi, \$_0]^{\#}$ is homeomorphic to the convergence in $[\xi, \$_1]^{\#}$ defined on $\mathcal{O}_\xi$ by

$$U \in \lim{}_{[\xi,\$_1]^{\#}} \mathcal{F} \iff U \supset \bigcup_{H \in \mathcal{F}^{\#}} \mathrm{int}_\xi \left(\bigcap_{O \in H} O \right).$$

Remark that if $(X, \leq)$ is a complete lattice, the Scott convergence s^* for the reverse order can be expressed in terms of the original order by (III.9.2). Hence $[\xi, \$_1]^*$ (on $\mathcal{O}_\xi$) is finer than $[\xi, \$_1]^{\#}$, and $[\xi, \$_0]^*$ (on $\mathcal{C}_\xi$) is finer than $[\xi, \$_0]^{\#}$.

Proposition VIII.7.5. *If ξ is a topology, then the lower Kuratowski convergence $[\xi, \$_0]^{\#}$ is also a topology and the collection of finite intersections of sets*

$$O^+ := \{ C \in \mathcal{C}_\xi : C \cap O \neq \varnothing \},$$

where $O \in \mathcal{O}_\xi$ form a base of its open sets.

Proof. By definition, $A_0 \in \lim_{[\xi,\$_0]^{\#}} \mathcal{F}$ if and only if $x \in \mathrm{cl}_\xi \left(\bigcup_{C \in H} C \right)$ for each $x \in A_0$ and each $H \in \mathcal{F}^{\#}$, that is, if for each $O \in \mathcal{O}_\xi(x)$ there is $C \in H$ with $O \cap C \neq \varnothing$. In other words, for each $x \in A_0$ and $O \in \mathcal{O}_\xi(x)$,

$$\underset{H \in \mathcal{F}^{\#}}{\forall} \underset{C \in H}{\exists} O \cap C \neq \varnothing,$$

which is equivalent to

$$\underset{F \in \mathcal{F}}{\exists} \underset{C \in F}{\forall} O \cap C \neq \varnothing,$$

because a power set is completely distributive, as shown in Example A.7.1. Therefore, $A_0 \in \lim_{[\xi,\$_0]^{\#}} \mathcal{F}$ if and only if for each $x \in A_0$ and $O \in \mathcal{O}_\xi(x)$

$$\mathcal{F} \geq \left\{ O^+ : O \in \mathcal{O}_\xi(x) \right\}.$$

In other words,

$$A_0 \in \lim{}_{[\xi,\$_0]^{\#}} \mathcal{F} \iff \mathcal{F} \geq \mathcal{B} := \left\{ O^+ : O \in \mathcal{O}_\xi, A_0 \in O^+ \right\}.$$

As a consequence, each O^+ is $[\xi, \$_0]^{\#}$-open. Thus $\left\{ \bigcap_{O \in T} O^+ : T \in [\mathcal{B}]^{<\omega} \right\}$ is a filter-base for $\mathcal{N}_{[\xi,\$_0]^{\#}}(A_0)$ and $[\xi, \$_0]^{\#}$ is a topology. $\square$

Definition VIII.7.6. The *Kuratowski convergence* on $\mathcal{C}_\xi$ is by definition the supremum of the upper and lower Kuratowski convergences:

$$\xi_{\$_0} := [\xi, \$_0] \vee [\xi, \$_0]^{\#}.$$

Recall that $[\xi, \$_0]$ is a pseudotopology by Theorem VIII.6.2 and that $[\xi, \$_0]^{\#}$ is a topology, hence a pseudotopology, so that $\xi_{\$_0}$ is a pseudotopology as a supremum of pseudotopologies. It follows from the definitions that

$$A_0 \in \lim_{\xi_{\$_0}} \mathcal{F} \iff \text{adh}_\xi \, \text{rdc}(\mathcal{F}) \subset A_0 \subset \text{adh}_\xi \, \text{rdc} \left(\mathcal{F}^{\#} \right). \qquad (\text{VIII.7.6})$$

As $\text{adh}_\xi \, \text{rdc} \left(\mathcal{F}^{\#} \right) \subset \text{adh}_\xi \, \text{rdc} \left(\mathcal{F} \right)$ for each filter $\mathcal{F}$, we infer that $\xi_{\$_0}$ is Hausdorff. We have proved:

Proposition VIII.7.7. *If ξ is a topology, then the Kuratowski convergence $\xi_{\$_0}$ is a Hausdorff pseudotopology.*

Of course, $\text{adh}_\xi \, \text{rdc} \left(\mathcal{F}^{\#} \right) = \text{adh}_\xi \, \text{rdc} \left(\mathcal{F} \right)$ if $\mathcal{F}$ is an ultrafilter. We will see (Section IX.4) that neither upper Kuratowski convergence $[\xi, \$_0]$ nor Kuratowski convergence $\xi_{\$_0}$ are topological in general.

Note that when ξ is T_1 there is a one-to-one map $j : |\xi| \to C(\xi, \$_0)$ defined by

$$j(x) := \{x\},$$

and X can then be identified with a subset of $C(\xi, \$_0)$. Moreover, if $\mathcal{H}$ is a filter on X then

$$\text{rdc} \left(j[\mathcal{H}] \right) = \mathcal{H}.$$

Proposition VIII.7.8. *If (X, ξ) is a Hausdorff pseudotopological space then $j : |\xi| \to |\xi_{\$_0}|$ is an embedding.*

Proof. If ξ is Hausdorff, $\mathcal{U} \in \mathbb{U}X$ and if $x \in \lim_\xi \mathcal{U}$ then $\{x\} = \lim_\xi \mathcal{U} = \text{adh}_\xi \mathcal{U}$, so that $\{x\} \in \lim_{\xi_{\$_0}} j[\mathcal{U}]$ because $\mathcal{U} = \mathcal{U}^{\#}$. Conversely, if $\mathcal{W} \in \mathbb{U}(j[X])$, and $j(x) \in \lim_{\xi_{\$_0}} \mathcal{W}$, then noting that $j^{-}[\mathcal{W}] = \text{rdc}(\mathcal{W})$, we have $\{x\} = \text{adh}_\xi j^{-}[\mathcal{W}]$, that is $x \in \lim_\xi j^{-}[\mathcal{W}]$. Thus j^{-} is continuous. $\square$

In other words, if (X, ξ) is a Hausdorff pseudotopological space, then the convergence induced on X (more precisely on $j(X)$) by the Kuratowski convergence $\xi_{\$_0}$ coincides with ξ.

Chapter IX

Compactness

Pseudotopologies are determined by convergent ultrafilters. Compactness, one of the most consequential notions of analysis and topology, can be characterized in terms of convergent ultrafilters. This is more than an analogy. Compactness turns out to be a pseudotopological notion, as a convergence is compact whenever its pseudotopologization is.

IX.1 Compact sets

A convergence space is *compact* if every filter on its underlying set is adherent, equivalently if each ultrafilter on its underlying set is convergent.

Definition IX.1.1. If A and B are subsets of X and ξ is a convergence on X, then A is called ξ-*compact at* B if for every filter $\mathcal{H}$ on X,

$$A \in \mathcal{H}^{\#} \implies \mathrm{adh}_{\xi}\mathcal{H} \cap B \neq \varnothing. \qquad (\text{IX.1.1})$$

In particular,

(1) A is called *compact* if A is compact at A;
(2) A is called *compactoid* or *relatively compact* if A is compact at X.

We denote by $\mathcal{K}(\xi)$ the set of ξ-compact sets, and by $\mathcal{K}^{o}(\xi)$ that of ξ-compactoid (relatively ξ-compact) sets.

Exercise IX.1.2. Show that

(1) each finite union of sets that are compact at B is compact at B,
(2) each subset of a set that is compact at B is compact at B.

It follows from Exercise IX.1.2 that the set of all compactoid subsets of a convergence space forms a (possibly degenerate) ideal of subsets of the underlying set (see Definition VIII.1.7). Therefore the family

$$\mathcal{K}_c^o(\xi) := (\mathcal{K}^o(\xi))_c = \{K^c : K \in \mathcal{K}^o(\xi)\} \qquad (IX.1.2)$$

of complements of compactoid sets is a (possibly degenerate) filter, called the *cocompactoid filter* of ξ.

If there is no ambiguity regarding the convergence, we write *compact* and *compactoid* without mentioning the convergence.

Proposition IX.1.3. *If a convergence ξ is T_1-convergence and A is ξ-compact at B then $A \subset B$.*

Proof. Let $x \in A \setminus B$. Then $A \# \{x\}^\uparrow$ and $\mathrm{adh}_\xi \{x\}^\uparrow = \{x\}$ because $\{x\}$ is closed. Therefore $\{x\}$ does not mesh B and thus A is not compact at B for ξ. $\qquad\square$

So if A is compact at B in a T_1 space X, then $A \subset B \subset X$, the extreme cases being $A = B$ (compactness) and $B = X$ (relative compactness). If however X is not T_1, then it can even happen that $A \cap B = \varnothing$ and A is compact at B.

A convergence is called *hypercompact* if every filter on the underlying set converges.

Example IX.1.4. The chaotic convergence o_X (on an arbitrary non-empty set X) is hypercompact.

Example IX.1.5. Consider one of the two homeomorphic Sierpiński spaces, say $\$_0$, as defined in (III.2.1). We notice that $\$_0$ is hypercompact because each filter on $\{0, 1\}$ converges in $\$_0$.

Recall that $\$_0$ is T_0 but not T_1. Observe that $\{0\}$ is compact at $\{1\}$ because $\lim_{\$_0} \{0\}^\uparrow = \{0, 1\}$, but of course, $\{0\}$ is not a subset of $\{1\}$.

Proposition IX.1.6. *If a T_1-convergence is hypercompact then its underlying set is a singleton.*

Proof. Let ξ be a T_1-convergence. Then $\lim_\xi \{x\}^\uparrow \subset \{x\}$ for each $x \in |\xi|$. If $x_0, x_1 \in |\xi|$ and $\lim_\xi \{x_0, x_1\}^\uparrow \neq \varnothing$, then

$$\varnothing \neq \lim_\xi \{x_0, x_1\}^\uparrow \subset \lim_\xi \{x_0\}^\uparrow \cap \lim_\xi \{x_1\}^\uparrow \subset \{x_0\} \cap \{x_1\},$$

thus $x_0 = x_1$. $\qquad\square$

Example IX.1.7. The prime cofinite space $\pi[x_\infty, (X)_0]$ of Example III.1.15 is compact, because every filter on X has either a non-empty kernel, hence non-empty adherence, or is free. Since every free filter on X converges to x_∞, their adherence is the singleton $\{x_\infty\}$, and is non-empty.

On the other hand, every infinite subset A of X that does not contain x_∞ is compactoid but not compact, because free filters on A have an adherent point in X but not in A.

Exercise IX.1.8. Show that if A and B are two subsets of a convergence space (X, ξ) then A is compact at B if and only if $\lim \mathcal{U} \cap B \neq \varnothing$ for every ultrafilter $\mathcal{U} \in \mathbb{U}X$ such that $A \in \mathcal{U}$.

Exercise IX.1.9. Let X be a convergence space and let $B \subset X$. Show that:

(1) Each finite subset is compact; in particular the empty set is compact;
(2) Each finite union of subsets that are compact at B is compact at B.

Example IX.1.10. A discrete convergence space is compact if and only if it is finite. Indeed, any finite space is compact in view of the previous exercise, and if the space is infinite, it has free filters. Since only principal ultrafilters converge in a discrete space, free filters have empty adherence. Thus an infinite discrete space is not compact.

Proposition IX.1.11. *A closed subset of a compactoid set is compact.*

Proposition IX.1.12. *If a subset of a Hausdorff convergence space is compact, then it is closed.*

Exercise IX.1.13. Prove Propositions IX.1.12 and IX.1.11.

Example IX.1.14. In the Sierpiński space $\$_0$, the singleton $\{0\}$ is compact, but $\mathrm{adh}_{\$_0}\{0\} = \{0, 1\}$, hence $\{0\}$ is not closed.

Exercise IX.1.15. Let $\mathcal{H}$ be a family of non-empty compact subsets of a *Hausdorff space* X such that $\bigcap_{H \in \mathcal{F}} H \neq \varnothing$ whenever $\mathcal{F}$ is a finite subfamily of $\mathcal{H}$. Show that $\bigcap_{H \in \mathcal{H}} H$ is a non-empty compact subset of X.

Example IX.1.16 (Two compact sets whose intersection is not compact). Let Y be an infinite set and let x_0 and x_1 be two different points that are not in Y. Define on $X := Y \cup \{x_0\} \cup \{x_1\}$ the topology in which all points of Y are isolated and $\mathcal{V}(x_0) = (Y)_0 \wedge \{x_0\}^\uparrow$ and $\mathcal{V}(x_1) = (Y)_0 \wedge \{x_1\}^\uparrow$. Then $X_0 := Y \cup \{x_0\}$ and $X_1 := Y \cup \{x_1\}$ with their subspace topologies are compact, but their intersection $X_0 \cap X_1 = Y$ is infinite and discrete, and therefore is not compact.

Recall that a subset A of $\mathbb{R}^n$ is *bounded* if it is bounded for one, hence for all the norm metrics of $\mathbb{R}$. A metric d is on a normed vector space $(X, \| \cdot \|)$ is said be a *norm metric* if $d(x_0, x_1) = \|x_0 - x_1\|$ for each couple x_0, x_1 of elements of X. All norms on a finite-dimensional vector space are equivalent, that is, if $\| \cdot \|_\circ$ and $\| \cdot \|_\bullet$ are such norms, then there exist two positive numbers c_0, c_1 such that for every $x \in X$,

$$\|x\|_\circ \le c_0 \|x\|_\bullet \le c_1 \|x\|_\circ.$$

Consequently, a subset A of $\mathbb{R}^n$ is bounded if and only if there is $k \ge 0$ such that

$$\max\{|a_1|, \ldots, |a_n|\} \le k$$

for all $(a_1, \ldots, a_n) \in A$. We will see that a subset of $\mathbb{R}^n$ is compact if and only if it is closed and bounded. Consider first the case of $n = 1$.

Theorem IX.1.17 (Bolzano-Weierstraß). *A subset of $\mathbb{R}$ is compact for the standard convergence if and only if it is closed and bounded.*

Proof. In view of Proposition IX.1.11, it is enough to show that closed bounded intervals $[a, b]$ are compact, since any closed bounded subset of $\mathbb{R}$ is a closed subset of such an interval.

Let $\mathcal{U}$ be an ultrafilter with $[a, b] \in \mathcal{U}$. Let $m := \frac{b+a}{2}$ be the midpoint of $[a, b]$. As $[a, b] = [a, m] \cup [m, b] \in \mathcal{U}$, either $[a, m] \in \mathcal{U}$ or $[m, b] \in \mathcal{U}$, because $\mathcal{U}$ is an ultrafilter.

By induction, $\mathcal{U}$ contains a decreasing sequence $[a_n, b_n]$ such that $b_n - a_n = \frac{b-a}{2^n}$. Therefore for each $n \in \mathbb{N}$,

$$a_n < a_{n+1} < b_{n+1} < b_n.$$

Because $(b_n - a_n)_n$ tends to 0, there is $l \in [a, b]$ such that

$$\sup_{n \in \mathbb{N}} a_n = l = \inf_{n \in \mathbb{N}} b_n.$$

As $\{[a_n, b_n] : n \in \mathbb{N}\}$ is a neighborhood base of the standard topology of $\mathbb{R}$ at l and $[a_n, b_n] \in \mathcal{U}$ for each $n \in \mathbb{N}$, we conclude that $\{l\} = \lim_{\mathbb{R}} \mathcal{U}$.

Conversely, if A is a compact subset of $\mathbb{R}$, then it is closed by Proposition IX.1.12. If A were not bounded, then there would be a strictly increasing sequence $\{a_n\}_{n \in \mathbb{N}}$ of elements of A such that for each $r \in \mathbb{R}$ there is $n \in \mathbb{N}$ for which $r < a_n$, hence $r < a_k$ for each $k \ge n$. Consequently, $\mathcal{B} := \{\{a_k : n \le k\} : n \in \mathbb{N}\}$ is a filter base on A consisting of closed sets, and $\mathrm{adh}_{\mathbb{R}} \mathcal{B} = \bigcap_{n \in \mathbb{N}} \{a_k : n \le k\} = \varnothing$. $\square$

Example IX.1.18. (a, b) is compact at $[a, b]$, hence compactoid, but not compact, because it is not closed.

Example IX.1.19 (Cantor set). Let $A_1 := [0, 1]$, $A_2 := [0, \frac{1}{3}] \cup [\frac{2}{3}, 1]$, $A_3 := [0, \frac{1}{9}] \cup [\frac{2}{9}, \frac{1}{3}] \cup [\frac{2}{3}, \frac{7}{9}] \cup [\frac{8}{9}, 1]$ and more generally, for $n \geq 2$, A_n is obtained from A_{n-1} by removing the "open middle third" out of each of the 2^{n-1} intervals of length $\frac{1}{3^{n-2}}$ forming A_{n-1}, resulting in 2^n intervals of length $\frac{1}{3^{n-1}}$.

The *Cantor set*

$$C := \bigcap_{i=1}^{\infty} A_i$$

is compact, because, as an intersection of closed sets, it a closed subset of a compact set $[0, 1]$. It is non-empty, as it contains in particular, by construction, the endpoints of each interval composing the A_n's (e.g., $0, 1, \frac{1}{3^n}, \ldots$).

In fact it contains a lot more points. An alternative description of the Cantor set in Appendix A.5 shows in particular that

$$\operatorname{card} C = 2^{\aleph_0} = \mathfrak{c}.$$

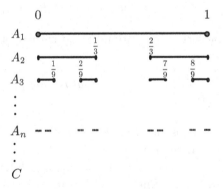

Proposition IX.1.20. *The Cantor set is homeomorphic to the Cantor cube* $\{0, 1\}^{\mathbb{N}}$ *where* $\{0, 1\}$ *is endowed with the discrete topology and* $\{0, 1\}^{\mathbb{N}}$ *carries the product topology (of pointwise convergence).*

Exercise IX.1.21. To show Proposition IX.1.20, show that the map F in the proof of Proposition A.5.2 is a homeomorphism.

Remark IX.1.22. Compactness is an absolute property, that is, if X and Y are two convergence spaces with common subspace A, then A is compact as a subset of X if and only if it is compact as a subset of Y.

Now we have two seemingly different definitions of compactness: a subset A of a convergence space (X, ξ) can be compact in the sense of Definition IX.1.1, that is, as a compact subset of X, or compact because the convergence induced by ξ on A is a compact convergence space. But *compactness is absolute*: these are equivalent properties.

Exercise IX.1.23. Show that the upper Kuratowski convergence defined by (VIII.7.4) is hypercompact.

Exercise IX.1.24. Show that the lower Kuratowski and the Kuratowski convergences, defined by (VIII.7.5) and (VIII.7.6) respectively, are compact.

Of course, if $\zeta \geq \xi$ then ζ-compactness implies ξ-compactness. On the other hand, compactness is a pseudotopological property:

Proposition IX.1.25. *Let A and B be two subsets of a convergence space (X, ξ). Then A is ξ-compact at B if and only if A is $\mathrm{S}\,\xi$-compact at B.*

Proof. Because $\mathrm{adh}_{\mathrm{S}\,\xi}\mathcal{H} = \mathrm{adh}_\xi \mathcal{H}$ for every filter $\mathcal{H}$ on $|\xi|$. $\qquad\square$

Compactness is preserved by continuity, and is therefore an invariant of homeomorphisms:

Proposition IX.1.26. *If A is compact at B, and f is continuous, then $f(A)$ is compact at $f(B)$.*

Proof. Let f be a continuous map from ξ to τ, and let A be ξ-compact at B. If $\mathcal{H}$ meshes with $f(A)$ then A meshes with $f^-[\mathcal{H}]$, hence by assumption, $\mathrm{adh}_\xi f^-[\mathcal{H}] \cap B \neq \varnothing$. It follows that there is an ultrafilter $\mathcal{U} \# f^-[\mathcal{H}]$ such that $B \cap \lim_\xi \mathcal{U} \neq \varnothing$. By continuity, $f(B) \cap \lim_\tau f[\mathcal{U}] \neq \varnothing$, thus $f(B) \cap \mathrm{adh}_\tau \mathcal{H} \neq \varnothing$ because $f[\mathcal{U}] \# \mathcal{H}$. $\qquad\square$

Compactness is a minimal property in the class of Hausdorff pseudotopologies.

Proposition IX.1.27. *If ξ is a compact pseudotopology and τ is a Hausdorff pseudotopology such that $\xi \geq \tau$, then $\xi = \tau$.*

Proof. Let ξ, τ be pseudotopologies on X. If $x \in \lim_\tau \mathcal{U}$ for an ultrafilter $\mathcal{U}$ on X, then $\lim_\xi \mathcal{U} \neq \varnothing$ because ξ is compact, hence $\varnothing \neq \lim_\xi \mathcal{U} \subset \lim_\tau \mathcal{U} = \{x\}$ because τ is Hausdorff. In other words, the pseudotopologies ξ and τ coincide on ultrafilters, hence are equal. $\qquad\square$

Corollary IX.1.28. *Let ξ be a compact pseudotopology and let τ be a Hausdorff pseudotopology. If $f : |\xi| \to |\tau|$ is a continuous bijection, then it is a homeomorphism.*

Proof. As $f \in C(\xi, \tau)$, equivalently, $f\xi \geq \tau$, hence $S(f\xi) \geq \tau$ because τ is a pseudotopology. Moreover, $S(f\xi)$ is compact by the continuity of $f : |\xi| \to |S(f\xi)|$. By Proposition IX.1.27, $\tau = S(f\xi)$. Since f is a bijection, $f^- : |\tau| \to |\xi|$ is a map. It is continuous, because

$$\lim_\tau \mathcal{W} = \lim_{f\xi} \mathcal{W} = f(\lim_\xi f^-[\mathcal{W}])$$

for each ultrafilter $\mathcal{W} \in \mathbb{U}|\tau|$. $\qquad\square$

Recall from Definition VII.1.7 that a convergence ξ is said to be *pretopologically Hausdorff* if $S_0 \xi$ is Hausdorff, *topologically Hausdorff* if $T\xi$ is Hausdorff.

By Exercise VIII.3.10, ξ is Hausdorff if and only if $S\xi$ is.

Corollary IX.1.29. *Each compact pretopologically Hausdorff pseudotopology is a pretopology. Each compact topologically Hausdorff pseudotopology is a topology.*

Exercise IX.1.30. Deduce Corollary IX.1.29 from Proposition IX.1.27.

Of course, Hausdorffness cannot be dropped in Proposition IX.1.27 and its corollaries:

Example IX.1.31. The Sierpiński topology is (hyper)compact and strictly finer than the chaotic topology.

A fundamental feature of compactness is its stability under product:

Theorem IX.1.32 (Tikhonov for sets). *Let Ξ be a family of convergence spaces. Then $\prod_{\xi \in \Xi} \xi$ is compact if and only if ξ is compact for each $\xi \in \Xi$.*

Proof. If $\prod_{\xi \in \Xi} \xi$ is compact, then each ξ is compact, because each projection

$$p_\xi : \left| \prod_{\zeta \in \Xi} \zeta \right| \to |\xi|$$

is continuous, and compactness is preserved by continuous maps (Proposition IX.1.26).

Assume conversely that each $\xi \in \Xi$ is compact. If $\mathcal{U}$ is an ultrafilter on $\prod_{\xi \in \Xi} |\xi|$ then, each projection $p_\xi[\mathcal{U}]$ is an ultrafilter on $|\xi|$ by Corollary II.6.7, and ξ is compact, so that there is $x_\xi \in \lim_\xi p_\xi[\mathcal{U}]$. Hence

$$(x_\xi)_{\xi \in \Xi} \in \lim_{\prod_{\xi \in \Xi} \xi} \prod_{\xi \in \Xi} p_\xi[\mathcal{U}] \subset \lim_{\prod_{\xi \in \Xi} \xi} \mathcal{U}$$

because $\mathcal{U} \geq \prod_{\xi \in \Xi} p_\xi[\mathcal{U}]$.   □

Corollary IX.1.33. *A subset of* $\mathbb{R}^n$ *is compact if and only if it is closed and bounded.*

Proof. Let A be a closed and bounded subset of $\mathbb{R}^n$. By the boundedness of A, there are closed intervals I_k such that $A \subset \prod_{1 \leq k \leq n} I_k$. By Proposition IX.1.17 and Theorem IX.1.32, $\prod_{1 \leq k \leq n} I_k$ is compact. Thus A is compact as a closed subset of a compact set (Proposition IX.1.11).

Conversely, by Proposition IX.1.12, a compact subset of $\mathbb{R}^n$ is closed because $\mathbb{R}^n$ is Hausdorff. On the other hand, if A is compact, then k-th projection $p_k(A)$ of A is compact in a copy of $\mathbb{R}$ for each $1 \leq k \leq n$, hence bounded. Therefore $A \subset \prod_{1 \leq k \leq n} p_k(A)$, which is bounded in $\mathbb{R}^n$.   □

IX.2 Regularity and topologicity in compact spaces

As can be seen by comparing Theorem VI.2.19 and Theorem VI.3.18, topologicity and regularity are in some sense dual properties. The minimality of compact convergences among Hausdorff pseudotopologies makes the two properties coincide in the class of compact Hausdorff pseudotopologies:

Theorem IX.2.1. *A compact Hausdorff pseudotopology is regular if and only if it is topological.*

Proof. This is a direct consequence of Theorems VIII.4.2 and VIII.4.3. Indeed, given a set A, a map $s : A \to X$ and a map $\mathcal{S} : A \to \mathbb{U}X$ with $s(a) \in \lim \mathcal{S}(a)$ for each $a \in A$, both $s[\mathcal{U}]$ and $\mathcal{S}(\mathcal{U})$ are ultrafilter bases on X whenever $\mathcal{U} \in \mathbb{U}A$. Since the convergence is compact Hausdorff, $\lim s[\mathcal{U}]$ and $\lim \mathcal{S}(\mathcal{U})$ are singletons. Thus

$$\lim s[\mathcal{U}] \subset \lim \mathcal{S}(\mathcal{U}) \iff \lim \mathcal{S}(\mathcal{U}) \subset \lim s[\mathcal{U}],$$

which, in view of Theorems VIII.4.2 and VIII.4.3, shows that the convergence is regular if and only if it is a topology. □

Corollary IX.2.2. *A compact Hausdorff topology is regular.*

Exercise IX.2.3. Show Corollary IX.2.2 directly, without Theorem IX.2.1, as follows:

(1) Show that if X is a Hausdorff topological space, K is a compact subset and $x \notin K$ then there are disjoint open subsets U and V of X with $x \in U$ and $K \subset V$.
(2) Deduce Corollary IX.2.2.

Corollary IX.2.4. *A compact Hausdorff regular convergence is almost topological, hence topologically regular.*

Proof. Let ξ be a compact Hausdorff regular convergence. In view of Exercise VIII.3.10, Proposition VIII.3.11 and Proposition IX.1.25, we conclude that $\mathsf{S}\,\xi$ is a compact regular Hausdorff pseudotopology, and is therefore a topology, by Theorem IX.2.1. Thus $\mathsf{T}\,\xi = \mathsf{S}\,\xi$ and ξ is topologically regular by Proposition VIII.3.15. □

In the situation of Theorem IX.2.1, the topology is moreover normal:

Theorem IX.2.5. *Each compact topologically regular convergence is normal.*

Proof. Suppose that a convergence on X is not normal: there exist disjoint closed sets A_0, A_1 such that $\mathcal{N}(A_0) \# \mathcal{N}(A_1)$. Let $\mathcal{U}$ be an ultrafilter finer than $\mathcal{N}(A_0) \vee \mathcal{N}(A_1)$. By compactness, there exists $x \in \lim \mathcal{U}$. If $x \notin A_0$, then A_0^c is an open set that contains x, hence by topological regularity there exists $U \in \mathcal{U}$ such that $\mathrm{cl}\, U \cap A_0 = \varnothing$, that is, $U \notin \mathcal{N}^{\#}(A_0)$, which yields a contradiction. For the same reason, $x \in A_1$, and thus $A_0 \cap A_1 \neq \varnothing$ contrary to the assumption. □

Corollary IX.2.6. *A compact Hausdorff regular convergence is normal.*

In particular:

Corollary IX.2.7. *A compact Hausdorff topology is normal.*

Unlike for topologies, a compact Hausdorff convergence need not be regular, because there exist non-topological compact Hausdorff pseudotopologies:

Example IX.2.8. The Kuratowski convergence on $\mathcal{C}_\xi$ defined by (VIII.7.6) is a Hausdorff pseudotopology, that is compact because each ultrafilter $\mathcal{U}$ on $\mathcal{C}_\xi$ converges to $\mathrm{adh}_\xi |\mathcal{U}|$. We will see (Theorem IX.4.1) that it may fail to be topological, in which case the Kuratowski convergence is compact Hausdorff but not regular.

IX.3   Local compactness

Definition IX.3.1. A convergence space is *locally compactoid* (respectively, *locally compact*) if each convergent filter contains a compactoid (respectively, compact) set.

Definition IX.3.2. A convergence is *hereditarily locally compact* if at each point there is a pavement composed of filters with a filter-base of compact sets.

Example IX.3.3. A discrete convergence space is hereditarily locally compact, because singletons are compact.

Example IX.3.4. The usual topology of $\mathbb{R}$ is hereditarily locally compact, because closed intervals centered at a point x form a filter-base for $\mathcal{N}_{\mathbb{R}}(x)$, and closed intervals are compact, by Proposition IX.1.17. Similarly, sets of the form $\prod_{i=1}^{n} I_i$ where each I_i is a closed interval centered at $x_i \in \mathbb{R}$ form a filter-base for $\mathcal{N}_{\mathbb{R}^n}((x_i)_{i \in \{1...n\}})$. Such sets are compact by Tikhonov's theorem (Theorem IX.1.32), so that $\mathbb{R}^n$ is also hereditarily locally compact.

Example IX.3.5 (A locally compact non-pretopological pseudotopology). The non-pretopological pseudotopology Seq ν from Example VIII.2.6 is locally compact, because each filter $\mathcal{F}$ such that $r \in \lim_{\mathrm{Seq}\,\nu} \mathcal{F}$ is finer than a sequential filter that converges to r. If we denote by $\{r_n\}_{n \in \mathbb{N}}$ a corresponding sequence, then $\{r\} \cup \{r_n : n \in \mathbb{N}\}$ is compact and belongs to $\mathcal{F}$.

Example IX.3.6 (A non-locally compact topology). The bisequence pretopology π of Example V.4.6 is locally compact. Indeed, $\{x_{n,k}\}$ is a compact vicinity of $x_{n,k}$ for each $n, k \in \mathbb{N}$; the set $\{x_n\} \cup \{x_{n,k} : k \in \mathbb{N}\}$ is a compact vicinity of x_n for each $n \in \mathbb{N}$; finally, $\{x_\infty\} \cup \{x_n : n \in \mathbb{N}\}$ is a compact vicinity of x_∞.

The bisequence topology $\mathrm{T}\,\pi$ obtained from π is not locally compact. Indeed, each neighborhood V of x_∞ contains one of the form

$$\{x_\infty\} \cup \{x_n : n \geq n_0\} \cup \{x_{n,k} : k \geq \kappa(n),\, n \geq n_0\},$$

where $\kappa : \mathbb{N} \to \mathbb{N}$, and the infinite subset $A := \{x_{n,\kappa(n)} : n \geq n_0\}$ is disjoint from some other neighborhood of x_∞, so that every free ultrafilter on A fails to converge. But such an ultrafilter also contains V which is therefore not compact.

As the identity i is a continuous map from π to $\mathrm{T}\,\pi$, Example IX.3.6 implies:

Proposition IX.3.7. *Local compactness is not preserved by continuous maps.*

Of course, every compact convergence space is locally compact, but not necessarily hereditarily locally compact:

Example IX.3.8. Recall that the Kuratowski convergence $\xi_{\$_0}$ defined in Definition VIII.7.6 is a compact Hausdorff pseudotopology (Proposition VIII.7.7 and Exercise IX.1.24), hence it is locally compact. We will see (Theorem IX.4.1) that if ξ is a regular Hausdorff topology, $\xi_{\$_0}$ is topological if and only if ξ locally compact. Thus, if ξ is a non-locally compact Hausdorff regular topology (such as $\mathbb{Q}$), $\xi_{\$_0}$ is a non-topological compact Hausdorff pseudotopology, hence it is not regular, by Theorem IX.2.1. Thus, it is not hereditarily locally compact. Indeed, compact subsets of a Hausdorff convergence are closed (Proposition IX.1.12), so that if the convergent had pavements composed of filters with a filter-base of compact subset, the convergence would be topologically regular, hence regular.

This distinction is not needed, however, among regular Hausdorff convergences:

Proposition IX.3.9. *A regular locally compact Hausdorff convergence is topologically regular and hereditarily locally compact.*

Proof. Let ξ be a locally compact convergence and let $x \in \lim_\xi \mathcal{F}$. Then there is a compact set $C \in \mathcal{F}$. Since a subspace of a regular convergence is regular, $\xi_{|C}$ is regular Hausdorff and compact, thus almost topological, and therefore topologically regular, by Corollary IX.2.4. Moreover, by Proposition IX.1.12, C is a closed subspace of ξ, so that $\mathrm{T}(\xi_{|C}) = (\mathrm{T}\,\xi)_{|C}$ by Proposition V.4.36. Thus $x \in \lim_\xi \mathrm{cl}^\natural_\xi \mathcal{F}$ and ξ is topologically regular.

In particular, for each $F \in \mathcal{F}$ the set $\mathrm{cl}_\xi (F \cap C)$ is a closed subset of the compact set C (which is itself closed), and is therefore compact by Proposition IX.1.11. Therefore $\mathrm{cl}^\natural_\xi \mathcal{F}$ has a filter-base composed of compact sets, $x \in \lim_\xi \mathrm{cl}^\natural_\xi \mathcal{F}$ and $\mathrm{cl}^\natural_\xi \mathcal{F} \leq \mathcal{F}$. □

In particular, Hausdorff locally compact topologies are hereditarily locally compact, because:

Proposition IX.3.10. *Each Hausdorff locally compact topology is functionally regular.*

Proof. Let ξ be a Hausdorff locally compact topology on a set X. Let $F \in \mathcal{C}_\xi$ and let $x \in X \setminus F$. By definition, there exists $K \in \mathcal{K}(\xi)$ such that $x \in \mathrm{int}_\xi K$. Then $\widehat{F} := (F \cap K) \cup (K \setminus \mathrm{int}_\xi K)$ is a closed subset of a Hausdorff compact space $\xi_{|K}$, thus there exists a continuous map f on K valued in $[0,1]$ such that $f(0) = 0$ and $\widehat{F} \subset f^-(1)$. As $K \cap (X \setminus \mathrm{int}_\xi K) = K \setminus \mathrm{int}_\xi K \subset \widehat{F}$, the function $h : X \to [0,1]$ defined by $h(x) := f(x)$ if $x \in K$ and $h(x) := 1$ if $x \in X \setminus K$ is continuous, because $h^-(1)$ is closed in X. On the other hand, $F \subset h^-(1)$. □

Proposition IX.3.11. *Each infimum of locally compact convergences is locally compact.*

Proof. Let Ξ be a family of locally compact convergences and let $x \in \lim_{\bigwedge_{\xi \in \Xi} \xi} \mathcal{F}$. It means that there is $\xi_0 \in \Xi$ with $x \in \lim_{\xi_0} \mathcal{F}$. As ξ_0 is locally compact, there is a ξ_0-compact set $C \in \mathcal{F}$. But every ξ_0-compact set is also compact for coarser convergences, in particular for $\bigwedge_{\xi \in \Xi} \xi$. Thus $\bigwedge \Xi$ is locally compact. □

A similar argument applies to hereditary local compactness.

Proposition IX.3.12. *Each infimum of hereditarily locally compact convergences is hereditarily locally compact.*

Corollary IX.3.13. *For each convergence ξ there exists a coarsest convergence $\mathrm{K}\,\xi$ among the locally compact convergences finer than ξ. More-*

over, ξ and $\mathrm{K}\xi$ have the same compact subsets, and

$$x \in \lim_{\mathrm{K}\xi} \mathcal{F} \iff \begin{cases} x \in \lim_{\xi} \mathcal{F} \\ \mathcal{K}(\xi) \cap \mathcal{F} \neq \varnothing. \end{cases} \tag{IX.3.1}$$

Proof. The discrete topology is (hereditarily) locally compact, because each singleton is compact so that the set of locally compact convergences finer than ξ is non-empty. Moreover, the infimum of this set of convergences is locally compact by Proposition IX.3.11, and is therefore the coarsest among the locally compact convergences finer than ξ. Let us temporarily denote it by σ.

The formula (IX.3.1) defines a convergence finer than ξ. Hence every $\mathrm{K}\xi$-compact set is also ξ-compact. Moreover, if C is a ξ-compact set, then every ultrafilter containing C converges to $x \in C$ for ξ, hence for $\mathrm{K}\xi$, so that C is also $\mathrm{K}\xi$-compact. In other words,

$$\mathcal{K}(\mathrm{K}\xi) = \mathcal{K}(\xi),$$

so that $\mathrm{K}\xi$ is locally compact, and, thus $\mathrm{K}\xi \geq \sigma$. On the other hand, $\sigma \geq \mathrm{K}\xi$, for a σ-convergent filter is also ξ-convergent, and contains a σ-compact, hence ξ-compact set. $\square$

The convergence $\mathrm{K}\xi$ is called *locally compact modification of ξ*. The map K on the class of convergences is called the *locally compact modifier*. By definition, $|\mathrm{K}\xi| = |\xi|$ for each convergence ξ.

Proposition IX.3.14. *The locally compact modifier fulfills*

$$\xi \leq \zeta \implies \mathrm{K}\xi \leq \mathrm{K}\zeta \qquad \text{(isotone)}$$
$$\mathrm{K}\xi \geq \xi \qquad \text{(expansive)}$$
$$\mathrm{K}(\mathrm{K}\xi) = \mathrm{K}\xi \qquad \text{(idempotent)}$$
$$\mathrm{K}(f^-\tau) \geq f^-(\mathrm{K}\tau). \qquad \text{(initial)}$$

Proof. As $\xi \leq \zeta$, by Corollary IX.3.13, $\xi \leq \mathrm{K}\zeta$, hence $\mathrm{K}\xi \leq \mathrm{K}\zeta$. Corollary IX.3.13 implies immediately $\mathrm{K}\xi \geq \xi$. Since $\mathrm{K}\xi$ is locally compact, it is the coarsest locally compact convergence finer than itself, that is, $\mathrm{K}(\mathrm{K}\xi) = \mathrm{K}\xi$.

If $x \in \lim_{\mathrm{K}(f^-\tau)} \mathcal{F}$ then $x \in \lim_{f^-\tau} \mathcal{F}$ and there is a $f^-\tau$-compact set $C \in \mathcal{F}$ by (IX.3.1). Thus $f(x) \in \lim_{\tau} f[\mathcal{F}]$ and $f(C) \in f[\mathcal{F}]$. Moreover, $f(C)$ is τ-compact by Proposition IX.1.26, so that $f(x) \in \lim_{\mathrm{K}\tau} f[\mathcal{F}]$, that is, $x \in \lim_{f^-(\mathrm{K}\tau)} \mathcal{F}$. $\square$

Similarly:

Proposition IX.3.15. *There exists a coarsest convergence* $\mathrm{Kh}\,\xi$ *among the hereditarily locally compact convergences finer than* ξ *and* Kh *is isotone, expansive, idempotent and initial.*

The convergence $\mathrm{Kh}\,\xi$ called the *hereditarily locally compact modification* and the map Kh the *hereditarily locally compact modifier*. Of course, $|\mathrm{Kh}\,\xi| = |\xi|$ for each convergence ξ and

$$\mathrm{Kh}\,\xi \geq \mathrm{K}\,\xi$$

for each convergence ξ.

Exercise IX.3.16. Show that if $f : |\xi| \to |\tau|$ is continuous, so is $f : |\mathrm{K}\,\xi| \to |\mathrm{K}\,\tau|$ and that

$$f(\mathrm{K}\,\xi) \geq \mathrm{K}(f\xi)$$

for each map $f : X \to Y$ and convergence ξ on X. Deduce that a quotient of a locally compact convergence is locally compact for the quotient convergence.

In view of Example IX.1.7, if $x \in \lim_{\xi}(x_n)_n$ then $\{x_n : n \in \mathbb{N}\} \cup \{x\}$ is a compact subspace of ξ. Thus, the convergence ξ of Theorem V.6.26 is locally compact, and therefore:

Proposition IX.3.17. *Every sequential convergence, in particular, every metrizable topology, is the image of a locally compact metrizable space under a (topologically) quotient map.*

Lemma V.4.41 that if D is a dense subset and O an open subset of a topological space, then

$$\mathrm{cl}\,(D \cap O) = \mathrm{cl}\,O. \tag{IX.3.2}$$

Proposition IX.3.18. *If X is a locally compact dense subspace of a Hausdorff topological space Y, then X is open in Y.*

Proof. Let $x \in X$. By the local compactness of X there is a $O \in \mathcal{O}_X(x)$ such that $\mathrm{cl}_X O$ is compact in X, hence compact in Y. As Y is Hausdorff, $\mathrm{cl}_X O$ is closed in Y, that is $\mathrm{cl}_X O = \mathrm{cl}_Y O \subset X$.

Let W be an open subset of Y such that $O = X \cap W$. Since X is dense in Y, by (IX.3.2),

$$x \in W \subset \mathrm{cl}_Y W = \mathrm{cl}_Y O \subset X,$$

which shows that X is open in Y. □

IX.4 Topologicity of hyperspace convergences

We revisit properties of the upper Kuratowski and Kuratowski convergences on the set

$$\mathcal{C}_\xi = C(\xi, \$_0)$$

of closed subsets of a topological space (X, ξ) introduced in Section VIII.7, examining in particular under what conditions on ξ they are topological. While no separation needs to be assumed on ξ to study this question, we present here only the simplified case where ξ is a Hausdorff regular topology. Recall that in this case the map $j : |\xi| \to |\xi_{\$_0}|$ defined by $j(x) = \{x\}$ is an embedding (Proposition VIII.7.8). Recall from (VIII.7.4) that $A \in \lim_{[\xi,\$_0]} \mathcal{F}$ if and only if $\mathrm{adh}_\xi \, \mathrm{rdc}(\mathcal{F}) \subset A$, so that if $\mathbb{D}$ is a set of filters on $C(\xi, \$_0)$ that admits a supremum $\bigvee_{\mathcal{F} \in \mathbb{D}} \mathcal{F}$ then

$$\left(\underset{\mathcal{F} \in \mathbb{D}}{\forall} \; \underset{A_\mathcal{F}}{\exists} \; A_\mathcal{F} \in \lim_{[\xi,\$_0]} \mathcal{F} \right) \Longrightarrow \bigcap_{\mathcal{F} \in \mathbb{D}} A_\mathcal{F} \in \lim_{[\xi,\$_0]} \bigvee_{\mathcal{F} \in \mathbb{D}} \mathcal{F}. \qquad (\text{IX.4.1})$$

Theorem IX.4.1. *Let ξ be a Hausdorff regular topology. The following are equivalent:*

(1) $[\xi, \$_0]$ is pretopological;
(2) $[\xi, \$_0]$ is topological;
(3) $\xi_{\$_0}$ is topological;
(4) $\xi_{\$_0}$ is pretopological;
(5) ξ is locally compact.

Proof. $(1) \Longrightarrow (2)$: It is enough to show that $A \in \lim_\sigma \mathcal{V}_\sigma(\mathcal{V}_\sigma(A))$ whenever $\sigma := [\xi, \$_0]$ is pretopological. Since σ is a pretopology, $B \in \lim_\sigma \mathcal{V}_\sigma(B)$ for each $B \in C(\xi, \$_0)$. In particular, if $\mathcal{A} \in \mathcal{V}_\sigma(A)$ then

$$\mathrm{cl}_\xi \big(\bigcup_{B \in \mathcal{A}} B \big) \in \bigcap_{B \in \mathcal{A}} \lim_\sigma \mathcal{V}_\sigma(B),$$

so that, by pretopologicity of σ, we have

$$\mathrm{cl}_\xi \big(\bigcup_{B \in \mathcal{A}} B \big) \in \lim_\sigma \big(\bigwedge_{B \in \mathcal{A}} \mathcal{V}_\sigma(B) \big).$$

In view of (IX.4.1), we conclude that

$$\bigcap_{\mathcal{A} \in \mathcal{V}_\sigma(A)} \mathrm{cl}_\xi \big(\bigcup_{B \in \mathcal{A}} B \big) \in \lim_\sigma \big(\bigvee_{\mathcal{A} \in \mathcal{V}_\sigma(A)} \bigwedge_{B \in \mathcal{A}} \mathcal{V}_\sigma(B) \big),$$

that is, $\mathrm{adh}_\xi \, \mathrm{rdc}(\mathcal{V}_\sigma(A)) \in \lim_\sigma \mathcal{V}_\sigma(\mathcal{V}_\sigma(A))$. Moreover, $\mathrm{adh}_\xi \, \mathrm{rdc}(\mathcal{V}_\sigma(A)) \subset A$ because $A \in \lim_\sigma \mathcal{V}_\sigma(A)$, so that $A \in \lim_\sigma \mathcal{V}_\sigma(\mathcal{V}_\sigma(A))$.

(2) $\Longrightarrow$ (3), because $\xi_{\$_0} = [\xi, \$_0] \vee [\xi, \$_0]^\#$ is then the supremum of two topologies, hence a topology. (3) $\Longrightarrow$ (4) is immediate.

(4) $\Longrightarrow$ (5): Since ξ is a regular Hausdorff topology, each of its neighborhood filters has a filter-base composed of closed sets. Therefore, if ξ is not locally compact, there is $x_0 \in X$ so that for each $V \in \mathcal{N}_\xi(x_0)$ there is an ultrafilter $\mathcal{U}_V$ on X with $V \in \mathcal{U}_V$ and $\lim_\xi \mathcal{U}_V = \varnothing$. Hence

$$\varnothing \in \bigcap_{V \in \mathcal{N}_\xi(x_0)} \lim_{\xi_{\$_0}} j[\mathcal{U}_V].$$

If $\xi_{\$_0}$ was a pretopology, we would then have

$$\varnothing \in \lim_{\xi_{\$_0}} \bigwedge_{V \in \mathcal{N}_\xi(x_0)} j[\mathcal{U}_V]. \qquad (\mathrm{IX}.4.2)$$

But

$$x_0 \in \mathrm{adh}_\xi \, \mathrm{rdc}\left(\bigwedge_{V \in \mathcal{N}_\xi(x_0)} j[\mathcal{U}_V] \right) \neq \varnothing$$

because $\mathrm{rdc}(\bigwedge_{V \in \mathcal{N}_\xi(x_0)} j[\mathcal{U}_V]) = \bigwedge_{V \in \mathcal{N}_\xi(x_0)} \mathcal{U}_V$ meshes with $\mathcal{N}_\xi(x_0)$, so that (IX.4.2) does not hold.

(5) $\Longrightarrow$ (1): Assume ξ is a regular locally compact, hence hereditarily locally compact topology, by Proposition IX.3.9. Let $\mathbb{D}$ be a set of filters on $C(\xi, \$_0)$ such that $A_0 \in \bigcap_{\mathcal{F} \in \mathbb{D}} \lim_{[\xi, \$_0]} \mathcal{F}$, that is, for each $\mathcal{F} \in \mathbb{D}$,

$$\mathrm{adh}_\xi \, \mathrm{rdc}\,(\mathcal{F}) \subset A_0, \qquad (\mathrm{IX}.4.3)$$

and suppose that $A_0 \notin \lim_{[\xi, \$_0]} \bigwedge_{\mathcal{F} \in \mathbb{D}} \mathcal{F}$. Then there is $x \in \mathrm{adh}_\xi \, \mathrm{rdc}(\bigwedge_{\mathcal{F} \in \mathbb{D}} \mathcal{F})$ with $x \notin A_0$. Since $X \setminus A_0$ is an open neighborhood of x, by local compactness, there exists a compact set $K \in \mathcal{N}_\xi(x)$ with $K \cap A_0 = \varnothing$. Moreover,

$$K \in \left(\mathrm{rdc}\left(\bigwedge_{\mathcal{F} \in \mathbb{D}} \mathcal{F} \right) \right)^\# = \left(\bigwedge_{\mathcal{F} \in \mathbb{D}} \mathrm{rdc}\,(\mathcal{F}) \right)^\# = \bigcup_{\mathcal{F} \in \mathbb{D}} \mathrm{rdc}(\mathcal{F})^\#$$

by (VIII.7.3) and Proposition II.3.5. Therefore, there is $\mathcal{F} \in \mathbb{D}$ with $K \in \mathrm{rdc}(\mathcal{F})^\#$, so that $\mathrm{adh}_\xi \, \mathrm{rdc}(\mathcal{F}) \cap K \neq \varnothing$ by compactness, which contradicts (IX.4.3). Thus $A_0 \in \lim_{[\xi, \$_0]} \bigwedge_{\mathcal{F} \in \mathbb{D}} \mathcal{F}$ and $[\xi, \$_0]$ is pretopological. $\qquad \square$

Note that the proof of (1) $\Longrightarrow$ (2) above does not use any separation, that is:

Proposition IX.4.2. *If ξ is a topology and $[\xi, \$_0]$ is a pretopology, then $[\xi, \$_0]$ is a topology.*

Example IX.4.3. Example IX.3.6 is a non-locally compact Hausdorff regular topology. The associated Kuratowski convergence is therefore a non-topological, compact Hausdorff pseudotopology. Thus it is not regular, by Theorem IX.2.1.

In the next section, we introduce an example of a compact Hausdorff topology on the set $\mathbb{U}X$ of ultrafilters on a set X.

IX.5 The Stone topology

Recall that, if $\mathcal{F}$ a filter on X, then $\beta(\mathcal{F})$ stands for the set of all the ultrafilters on X finer than $\mathcal{F}$, and $\beta^*(\mathcal{F})$ stands for the subset of $\beta(\mathcal{F})$ formed by free ultrafilters finer than $\mathcal{F}$. In particular, if $\mathcal{F}$ is a principal filter ([1]) on X, say, $\mathcal{F} = A^\uparrow$ where $A = \ker \mathcal{F} \subset X$, then

$$\beta(\mathcal{F}) = \beta(A) = \{\mathcal{U} \in \mathbb{U}X : A \in \mathcal{U}\}$$

denotes the set of all ultrafilters $\mathcal{U}$ on X such that $A \in \mathcal{U}$, and

$$\beta^*(A) = \beta(A) \cap \mathbb{U}^*X.$$

Note that while the set $\mathbb{U}A$ of ultrafilters on A could be identified with βA as in (III.3.2), in the present context, it is helpful to maintain a notational distinction between the set $\mathbb{U}A$ of ultrafilters *on the set* A and the *subset* $\beta(A)$ *of* $\mathbb{U}X$ of ultrafilters *on* X that contain A. The distinction between the two notations β^* and $\mathbb{U}^*$ serves a similar purpose.

Proposition IX.5.1. *If $\mathcal{F}_0, \mathcal{F}_1$ are filters on a set X, then*

$$\beta(\mathcal{F}_0 \vee \mathcal{F}_1) = \beta(\mathcal{F}_0) \cap \beta(\mathcal{F}_1); \qquad (IX.5.1)$$

$$\beta(\mathcal{F}_0 \wedge \mathcal{F}_1) = \beta(\mathcal{F}_0) \cup \beta(\mathcal{F}_1). \qquad (IX.5.2)$$

Proof. The inclusion $\subset$ in (IX.5.1) is evident. If now $\mathcal{U} \in \beta(\mathcal{F}_0) \cap \beta(\mathcal{F}_1)$, then $\mathcal{U} \geq \mathcal{F}_0$ and $\mathcal{U} \geq \mathcal{F}_1$, hence $\mathcal{U} \geq \mathcal{F}_0 \vee \mathcal{F}_1$. On the other hand (IX.5.2) was proved in Exercise VIII.1.1 (1). $\qquad \square$

Recall from Exercise VIII.1.1 that the analog of (IX.5.2) for an infinite set of filters does not hold. On the other hand, if $\mathbb{D}$ is a family of filters on X that admits a supremum (as in Proposition II.2.15) then

$$\beta\left(\bigvee_{\mathcal{F} \in \mathbb{D}} \mathcal{F}\right) = \bigcap_{\mathcal{F} \in \mathbb{D}} \beta(\mathcal{F}),$$

by the same argument as in Proposition IX.5.1.

---

[1] in the sense of Definition II.2.16

Definition IX.5.2. The *Stone convergence* β on $\mathbb{U}X$ is defined by ([2])

$$\mathcal{U} \in \lim_{\beta} \mathfrak{F} \iff \bigcup_{A \in \mathfrak{F}} \bigcap_{W \in A} W \geq \mathcal{U}. \tag{IX.5.3}$$

Note that the inequality above amounts to $\bigcup_{A \in \mathfrak{F}} \bigcap_{W \in A} W = \mathcal{U}$, because $\mathcal{U}$ is an ultrafilter.

Observe also that as $\mathfrak{F}$ is a filter on $\mathbb{U}X$ and the identity map $\mathcal{I}$ of $\mathbb{U}X$ (i.e., $\mathcal{I}(\mathcal{U}) = \mathcal{U}$) associates to each point $\mathcal{U}$ of $\mathbb{U}X$ a filter $\mathcal{U}$ on X, we can consider the contour of $\mathcal{I}$ along $\mathfrak{F}$, as in Definition VI.1.1, and

$$\mathcal{I}(\mathfrak{F}) = \bigcup_{A \in \mathfrak{F}} \bigcap_{W \in A} W.$$

As a result:

Corollary IX.5.3. *A filter $\mathfrak{F}$ on $\mathbb{U}X$ converges for β if and only if the contour filter $\mathcal{I}(\mathfrak{F})$ is an ultrafilter, in which case*

$$\lim_{\beta} \mathfrak{F} = \{\mathcal{I}(\mathfrak{F})\}.$$

Hence β is Hausdorff. Moreover $\mathcal{I}(\mathfrak{U})$ is an ultrafilter whenever $\mathfrak{U}$ is an ultrafilter on $\mathbb{U}X$, by Corollary VI.1.6. Therefore, every ultrafilter on $\mathbb{U}X$ converges for β, hence β is compact.

Let us explicit the convergence further: in view of (IX.5.3), $\mathcal{U} \in \lim_{\beta} \mathcal{F}$ if and only if

$$U \in \mathcal{U} \implies \underset{D \in \mathfrak{F}}{\exists} \ \underset{W \in D}{\forall} \ U \in W;$$

in other words, if $\mathbb{D} \subset \beta(U)$. Therefore

$$\mathcal{U} \in \lim_{\beta} \mathfrak{F} \iff \mathfrak{F} \geq \{\beta(U) : U \in \mathcal{U}\}^{\uparrow}.$$

Hence, the Stone convergence β is a pretopology with vicinity filter ([3]) at each $\mathcal{U} \in \mathbb{U}X$ given by

$$\mathcal{V}_{\beta}(\mathcal{U}) := \{\beta(U) : U \in \mathcal{U}\}^{\uparrow}. \tag{IX.5.4}$$

Note that by definition, $\beta(U) \in \bigcap_{\mathcal{U} \in \beta(U)} \mathcal{V}_{\beta}(\mathcal{U})$ so that, in view of Exercise V.3.13(1b), $\beta(U)$ is open for β. As a consequence $\mathcal{V}_{\beta}(\mathcal{U}) = \mathcal{N}_{\beta}(\mathcal{U})$ and:

---

[2]We use the symbol β to denote the Stone topology and the symbol $\beta(\mathcal{F})$ to denote the collection of ultrafilters that are finer than a filter $\mathcal{F}$. This double use should not cause any confusion.

[3]Defined this way, $\mathcal{V}_{\beta}(\mathcal{U})$ is a filter because $\beta(U) \cap \beta(V) = \beta(U \cap V)$, as in (IX.5.1).

Proposition IX.5.4. *The Stone convergence on* $\mathbb{U}X$ *is a topology and*

$$\{\beta(A) : A \subset X\}$$

form a base of its open sets.

Since for each $A \subset X$ and each $\mathcal{W} \in \mathbb{U}X$, the ultrafilter $\mathcal{W}$ contains either A or $X \setminus A$, we have

$$\mathbb{U}X \setminus \beta(A) = \beta(X \setminus A). \qquad (\text{IX.5.5})$$

As a consequence, $\beta(A)$ is both open and closed, and the topology β is zero-dimensional.

In summary, we have proved:

Theorem IX.5.5. *The Stone convergence is a Hausdorff compact and zero-dimensional topology.*

Note that as a result of Lemma IX.5.3, the assumption of discreteness is essential in Proposition VI.1.8 because of the exercise below.

Exercise IX.5.6. Show that the neighborhood filters $\mathcal{N}_\beta(\mathcal{U})$ are not ultrafilters.

Exercise IX.5.7. Show that a subset $\mathbb{D}$ of $\mathbb{U}X$ is β-closed if and only if there is $\mathcal{F} \in \mathbb{F}X$ such that $\mathbb{D} = \beta(\mathcal{F})$.

Exercise IX.5.8. Show that

$$\mathrm{cl}_\beta \mathbb{D} = \beta\Big(\bigwedge_{\mathcal{U} \in \mathbb{D}} \mathcal{U} \Big)$$

for each non-empty subset $\mathbb{A}$ of $\mathbb{U}X$.

As the function $j : X \to \mathbb{U}X$ defined by $j(x) = \{x\}^\uparrow$ is injective, X can be considered as a subset of $\mathbb{U}X$. It is natural to examine the convergence induced by β on X (that is, to be exact, on $j(X)$). Since $\beta(\{x\}) = \{\{x\}^\uparrow\}$ we see that $j(x)$ is open for each $x \in X$, that is:

Proposition IX.5.9. *The convergence induced on* X *by the Stone topology of* $\mathbb{U}X$ *is the discrete topology.*

Moreover:

Proposition IX.5.10. X *is a dense subset of* $\mathbb{U}X$.

Proof. As $\bigcup_{U \in \mathcal{U}} \bigcap_{x \in U} \{x\}^\uparrow = \mathcal{U}$, $\mathcal{U} \in \lim_\beta j[\mathcal{U}]$ for each $\mathcal{U} \in \mathbb{U}X$ so that each point of $\mathbb{U}X$ is the limit of a filter on $j(X)$. $\qquad \square$

Hence X with the discrete topology is a dense subspace of the compact space $\mathbb{U}X$, that is, $\mathbb{U}X$ is a *compactification* of the discrete space X. Moreover, it has the following universal property:

Theorem IX.5.11. *Let X be endowed with the discrete topology. Every function $f : X \to K$, where K is a compact Hausdorff topological space, has a unique continuous extension $\bar{f} : \mathbb{U}X \to K$, where $\mathbb{U}X$ is endowed with the Stone topology, defined by*

$$\bar{f}(\mathcal{U}) := \lim{}_K f[\mathcal{U}]. \tag{IX.5.6}$$

Moreover, if $Y \supset X$ is another compact Hausdorff topological space such that every function from X to a compact Hausdorff space K has a unique continuous extension to Y, then Y and $\mathbb{U}X$ are homeomorphic.

Proof. As K is compact and Hausdorff and $f[\mathcal{U}]$ is an ultrafilter, the definition (IX.5.6) of $\bar{f}$ is unambiguous. Moreover, by Proposition IV.9.38, a continuous extension of f is necessarily unique because X is a dense subset of $\mathbb{U}X$.

Finally, $\bar{f}$ is continuous because

$$\bar{f}[\mathcal{N}_\beta(\mathcal{U})] \geq \operatorname{adh}_K^\natural f[\mathcal{U}] \tag{IX.5.7}$$

and $\bar{f}(\mathcal{U}) \in \lim_K \operatorname{adh}_K^\natural f[\mathcal{U}]$, since a compact Hausdorff topology is regular, by Corollary IX.2.6.

To see (IX.5.7), note that if $y \in \operatorname{adh}_K(f(U))$ then there is an ultrafilter $\mathcal{X}$ on K with $f(U) \in \mathcal{X}$ and $y \in \lim_K \mathcal{X}$. In view of Corollary II.6.7, there is $\mathcal{W} \in \mathbb{U}X$ with $U \in \mathcal{W}$ and $f[\mathcal{W}] = \mathcal{X}$. Hence $y \in \bar{f}(\beta U)$, that is, $\bar{f}(U) \subset \operatorname{adh}_K f(U)$.

Assume now that Y is a compact Hausdorff topological space containing X (via an injective map $h : X \to Y$) such that whenever K is a compact Hausdorff topological space and $f : X \to K$ there is a unique continuous map $\tilde{f} : Y \to K$ such that $\tilde{f} \circ h = f$.

Since $\mathbb{U}X$ is a compact Hausdorff topological space, the map $j : X \to \mathbb{U}X$ has a continuous extension $\tilde{j} : Y \to \mathbb{U}X$ with $\tilde{j} \circ h = j$. Similarly, as Y is compact Hausdorff, there is a continuous map $\bar{h} : \mathbb{U}X \to Y$ with $\bar{h} \circ j = h$.

Hence $\bar{h} \circ \tilde{j} : Y \to Y$ is a continuous extension to Y of h. So is the identity map i_Y of Y. By uniqueness of the extension $\bar{h} \circ \tilde{j} = i_Y$. Similarly, $\tilde{j} \circ \bar{h} = i_{\mathbb{U}X}$ so that Y and $\mathbb{U}X$ are homeomorphic. $\qquad\square$

Corollary IX.5.12. *If $P \subset \mathbb{U}\mathbb{N}$ is countably infinite and discrete (in the Stone topology of $\mathbb{U}\mathbb{N}$), then $\operatorname{cl}_\beta P$ is homeomorphic to $(\mathbb{U}\mathbb{N}, \beta)$.*

To prove this corollary, we first observe:

Lemma IX.5.13. *If P is a discrete countably infinite subspace of $\mathbb{U}\mathbb{N}$ there is a partition of $\mathbb{N}$ into countably many pairwise disjoint sets*

$$\mathbb{N} = \coprod_{k=0}^{\infty} V_k$$

such that for each $\mathcal{W} \in P$ there is a unique k with $\{\mathcal{W}\} = P \cap \beta(V_k)$.

Proof. Let $P = \{\mathcal{W}_k : k \in \mathbb{N}\}$. As P is discrete, for every $k \in \mathbb{N}$, there is $H_k \subset \mathbb{N}$ with $\beta(H_k) \cap P = \{\mathcal{W}_k\}$. Let $W_0 := H_0$, $W_1 := H_1 \setminus H_0$ and by induction $W_{n+1} := H_{n+1} \setminus W_n$. We obtain a sequence $\{W_k\}_k$ of pairwise disjoint sets with

$$\bigcup_{k \in \mathbb{N}} H_k = \bigcup_{k \in \mathbb{N}} W_k := U,$$

and, for every k, $\beta(W_k) \cap P = \beta(H_k) = \{\mathcal{W}_k\}$. Let $\{R_k : k \in \mathbb{N}\}$ be a partition of $\mathbb{N} \setminus U$. Then $V_k := W_k \cup R_k$ defines a countable partition of $\mathbb{N}$ with the desired property. $\qquad\square$

Proof of Corollary IX.5.12. With the notations of Lemma IX.5.13, we may let $P = \{\mathcal{W}_k : k \in \mathbb{N}\}$ where for each k, $\beta(V_k) \cap P = \{\mathcal{W}_k\}$. Let $g : \mathbb{N} \to \mathbb{N}$ be defined by $g(n) = k$ if $n \in V_k$, that is, g is the canonical surjection for the quotient associated with the partition $\{V_k : k \in \mathbb{N}\}$. Let also $h : \mathbb{N} \to P$ be defined by $h(k) = \mathcal{W}_k$. We claim that if $\mathcal{U}_0 \in \mathrm{cl}_\beta P \setminus P$ then

$$\mathcal{N}_\beta(\mathcal{U}_0) \vee P = h \circ g[\mathcal{U}_0], \qquad\qquad (\mathrm{IX}.5.8)$$

so that $\mathcal{N}_\beta(\mathcal{U}_0) \vee P$ is an ultrafilter. Before we justify (IX.5.8), note that then for every continuous map $f : P \to K$ where K is a compact Hausdorff topological space, $\lim_K (\mathcal{N}_\beta(\mathcal{U}_0) \vee P) \neq \varnothing$ because $\mathcal{N}_\beta(\mathcal{U}_0) \vee P$ is an ultrafilter and K is compact. In view of Remark VII.8.7, the hull of extensionability of P for f is $h(P, f) = \mathrm{cl}_\beta P$ and, by Theorem VII.8.8, there is a unique continuous extension $\bar{f} : \mathrm{cl}_\beta P \to K$. In view of Theorem IX.5.11, $\mathrm{cl}_\beta P$ is homeomorphic to $(\mathbb{U}P, \beta)$, which in turn is homeomorphic to $\mathbb{U}\mathbb{N}$.

To see (IX.5.8), note that for each $U \in \mathcal{U}_0$,

$$g(U) = \{k \in \mathbb{N} : U \cap V_k \neq \varnothing\},$$

so that $h \circ g(U)$ is $\{\mathcal{W}_k : U \cap V_k \neq \varnothing\}$. Since $\beta(U) \cap P \neq \varnothing$ and $\beta(V_k) \cap P = \{\mathcal{W}_k\}$ we conclude that

$$\mathcal{W}_k \in \beta(U) \implies U \cap V_k \neq \varnothing,$$

that is, $\beta(U) \cap P \subset h \circ g(U)$. Thus,

$$\mathcal{N}_\beta(\mathcal{U}_0) \vee P \geq h \circ g[\mathcal{U}_0],$$

and the right-hand side is an ultrafilter, so that we have equality. $\qquad\square$

Theorem IX.5.14. *Every infinite closed subset of* $\mathbb{UN}$ *includes a subset homeomorphic to* $\mathbb{UN}$, *and thus is of cardinality* $2^{2^{\aleph_0}} = 2^{\mathfrak{c}}$.

Proof. In view of Corollary IX.5.12, it is enough to show that if H is a closed infinite subset of $\mathbb{UN}$, then it contains an infinite discrete subspace. The conclusion regarding cardinality then follows from Theorem II.7.1. Let p_0, p_1 be two distinct points in H. As the topology β of $\mathbb{UN}$ is Hausdorff and admits a base consisting of clopen sets, there exist two complementary open (hence closed) sets W_0 and W_1 such that $p_0 \in W_0$ and $p_1 \in W_1$. Either $H \cap W_0$ or $H \cap W_1$ is infinite and both are closed.

Pick an infinite one, say $H \cap W_1$, and use the same argument as before to find two open complementary subsets V_1, V_2 of W_1 such that $H \cap V_1$ and $H \cap V_2$ are non-empty. By setting $V_0 := W_0$, we get, in particular, three open disjoint sets V_0, V_1, V_2 such that $H \cap V_0, H \cap V_1$ and $H \cap V_2$ are non-void.

By induction, there exists a countable family $\{V_k : k \in \mathbb{N}\}$ of disjoint clopen subsets of $\mathbb{UN}$ such that, for each $k \in \mathbb{N}$, there exists $p_k \in H \cap V_k$. The set $P := \{p_k : k \in \mathbb{N}\}$ is discrete, hence Corollary IX.5.12 applies to yield the conclusion. $\qquad\square$

IX.6 Almost disjoint families

In Section I.6 we have considered a relation of almost inclusion and almost equality, that is, the inclusion and the equality of sets with possible exception of finite number of elements. Almost disjointedness is an akin relation.

Definition IX.6.1. Two subsets A_0 and A_1 of a set X are said to be *almost disjoint (AD)* whenever $A_0 \cap A_1$ is finite.

It follows that A_0 and A_1 are almost disjoint if and only if A_1 is almost included in $X \setminus A_0$ if and only if A_0 is almost included in $X \setminus A_1$.

Definition IX.6.2. A family $\mathcal{A}$ of infinite subsets of X is said to be *almost disjoint (AD)* if $A_0, A_1 \in \mathcal{A}$ and $A_0 \neq A_1$ then A_0 and A_1 are almost disjoint.

An almost disjoint family $\mathcal{A}$ is called *maximal almost disjoint (MAD)* if it is not strictly included in an almost disjoint family.

In other words,

Lemma IX.6.3. *An almost disjoint family $\mathcal{A}$ is maximal on a set X if and only if for every infinite subset B of X, there exists $A \in \mathcal{A}$ such that $B \cap A$ is infinite.*

Therefore, a finite almost disjoint family whose union is almost equal to X is maximal. Since the union of every chain of AD families is an AD family, for every AD family $\mathcal{A}$, there is a MAD family that includes $\mathcal{A}$ by virtue of the Zorn-Kuratowski Lemma A.9.10.

Proposition IX.6.4. *A maximal almost disjoint family is either finite or uncountable.*

Proof. If $\mathcal{A} = \{A_n : n \in \mathbb{N}\}$ is an almost disjoint family, then for each $n \in \mathbb{N}$ there is $x_n \in A_n \setminus (A_0 \cup \ldots \cup A_{n-1})$. Then $\mathcal{A} \cup \{x_n : n \in \mathbb{N}\}$ is almost disjoint, hence $\mathcal{A}$ is not maximal. $\qquad\square$

Proposition IX.6.5. *On each countably infinite set there is a MAD family of cardinality $\mathfrak{c}$.*

Proof. It is enough to to prove this statement for one countably infinite set. We will describe a MAD family of cardinality $\mathfrak{c}$ of infinite subsets of $\mathbb{Q}$. For each $r \in \mathbb{R} \setminus \mathbb{Q}$ there exists a sequence $\{q_n^r\}_{n \in N}$ in $\mathbb{Q}$ such that $r \in \lim \{q_n^r\}_{n \in N}$. Of course, $\{q_n^r : n \in N\}$ is infinite for each r and if $r \neq s$ then $\{q_n^r : n \in N\} \cap \{q_n^s : n \in N\}$ is finite. Therefore

$$\mathcal{R} := \{\{q_n^r : n \in N\} : r \in \mathbb{R}\}$$

is an almost disjoint family of cardinality of $\mathbb{R}$. By the Zorn-Kuratowski Lemma A.9.10, $\mathcal{R}$ is included in a MAD family of cardinality at least $\mathfrak{c}$. In fact, this cardinality is precisely $\mathfrak{c}$, because there are $2^{\aleph_0} = \mathfrak{c}$ subsets of $\mathbb{Q}$. $\qquad\square$

The least cardinal number such that there exists an infinite MAD family of that cardinality on a countably infinite set, is denoted by $\mathfrak{a}$. By Proposition IX.6.4,

$$\aleph_1 \leq \mathfrak{a} \leq \mathfrak{c}.$$

All the possible strict inequalities and equalities are consistent (in the sense considered on page 498) with **ZFC**. Of course, the Continuum Hypothesis (**CH**) (as considered on page 521) implies $\aleph_1 = \mathfrak{a} = \mathfrak{c}$.

If X is an infinitely countable set, recall that $\mathbb{U}^*(X)$ is the set of free ultrafilters on X. We endow it with the topology β^* induced by the Stone topology β of $\mathbb{U}X$. We have seen that X, seen as a subspace of $(\mathbb{U}X, \beta)$ by identifying X and $j(X)$, is discrete, hence open in $\mathbb{U}X$. Thus the topology of $\mathbb{U}^*(X)$ is compact. As sets of the form $\beta(H)$ for $H \subset X$ form a basis of open (actually, clopen) sets in $\mathbb{U}X$, we conclude that sets of the form

$$\beta^*(H) := \beta(H) \cap \mathbb{U}^*(X),$$

where H is an infinite subset of X, constitute a base of open (actually, clopen) sets for the topology of $\mathbb{U}^*X$. As the number of all subsets of X is $\mathfrak{c}$, the weight of the Stone topology on $\mathbb{U}^*(X)$ is not greater than $\mathfrak{c}$.

Proposition IX.6.6. *If X is an infinite countable set, then the set $\mathbb{U}^*(X)$ of free ultrafilters on X endowed with the topology β^* induced by the Stone topology of $\mathbb{U}X$ is a compact Hausdorff topological space of weight at most $\mathfrak{c}$.*

Proposition IX.6.7. *A family $\mathcal{A}$ of infinite subsets of X is almost disjoint if and only if $\{\beta^*(A) : A \in \mathcal{A}\}$ is a family of disjoint clopen subsets of $\beta^*(X)$.*

Exercise IX.6.8. Prove Proposition IX.6.7.

Let $\mathcal{A}$ be an AD family on an infinite set X. Then $\bigcup_{A \in \mathcal{A}} \beta^*(A)$ is an open subset $\mathbb{U}^*(X)$. Therefore, in view of Exercise IX.5.7, there exists a (possibly degenerate) filter $\mathcal{F}_{\mathcal{A}}$ on X such that

$$\beta^*(\mathcal{F}_{\mathcal{A}}) = \mathbb{U}^*(X) \setminus \bigcup_{A \in \mathcal{A}} \beta^*(A). \qquad (IX.6.1)$$

The filter defined by (IX.6.1) is called the *residual filter* of $\mathcal{A}$.

Proposition IX.6.9. *The residual filter $\mathcal{F}_{\mathcal{A}}$ of a MAD family $\mathcal{A}$ on an infinitely countable set is non-degenerate if and only if $\mathcal{A}$ is infinite. Then the cardinality of $\beta^*(\mathcal{F}_{\mathcal{A}})$ is infinite and thus equal to $2^{\mathfrak{c}}$.*

Proof. Let N be an infinitely countable set and let $\mathcal{A}$ be a *MAD* family on N. Then $\bigcup_{A \in \mathcal{A}} \beta^*(A) = \mathbb{U}^*(N)$ if and only if $\mathcal{A}$ is finite.

Indeed, if the equality holds, then there is a finite subfamily $\mathcal{A}_0$ of $\mathcal{A}$ such that $\bigcup_{A \in \mathcal{A}_0} \beta^*(A) = \mathbb{U}^*(N)$, because of the compactness of $\mathbb{U}^*(N)$, but since $\{\beta^*(A) : A \in \mathcal{A}\}$ is disjoint by Proposition IX.6.7, $\mathcal{A}_0 = \mathcal{A}$.

Conversely, if $\mathcal{A}$ is finite, then $\bigcup_{A \in \mathcal{A}} \beta^*(A)$ is closed, hence $\mathbb{U}^*(N) \setminus \bigcup_{A \in \mathcal{A}} \beta^*(A)$ is open, and, if it were non-empty, there would exist an infinite subset A_0 of N such that $\beta(A_0) \subset \mathbb{U}^*(N) \setminus \bigcup_{A \in \mathcal{A}} \beta^*(A)$, that is, the family $\mathcal{A} \cup \{A_0\}$ would be almost disjoint, contrary to the maximality of $\mathcal{A}$.

To prove the latter statement, consider the family $\mathcal{B}$ of those subsets B of N for which $A \cap B$ is infinite for infinitely many elements of $\mathcal{A}$. Consider a sequence $\{B_n\}_n$ of disjoint elements of $\mathcal{B}$. For every $n < \omega$ there exists an ultrafilter $\mathcal{U}_n \in \beta^*(B_n) \setminus \bigcup_{A \in \mathcal{A}} \beta^*(A)$. Indeed,

$$\mathcal{A} \vee_\infty B_n := \{A \cap B_n : A \in \mathcal{A}\},$$

is an infinite AD family, and thus can be completed to an infinite MAD family, so that the residual filter of $\mathcal{A} \vee_\infty B_n$ is non-degenerate. It follows that $\mathcal{U}_n \in \beta^*(\mathcal{F}_\mathcal{A})$ and $\mathcal{U}_n \neq \mathcal{U}_k$ for each $n, k < \omega$. $\square$

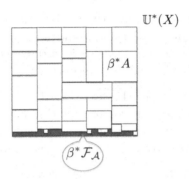

Figure IX.1 The set $\mathbb{U}^*(X)$ of all free ultrafilters on X includes the union of $\{\beta^*(A) : A \in \mathcal{A}\}$ for a MAD family $\mathcal{A}$ on X. The set $\beta^*(\mathcal{F}_\mathcal{A})$ of the ultrafilters that are not in that union is the image by β^* of the residual filter $\mathcal{F}_\mathcal{A}$ of $\mathcal{A}$.

Notice that $\mathcal{A}$ is maximal if and only if $\text{int}_{\beta^*} \beta^*(\mathcal{F}_\mathcal{A}) = \varnothing$, because otherwise there would exist an infinite subset $B \subset X$ such that $\beta^*(B) \subset \beta^*(\mathcal{F}_\mathcal{A})$ contrary to the assumption of the maximality of $\mathcal{A}$. Therefore,

Proposition IX.6.10. *An almost disjoint family $\mathcal{A}$ is maximal if and only if $\bigcup_{A \in \mathcal{A}} \beta^*(A)$ is dense in $\beta^*(X)$.*

In particular,

Proposition IX.6.11. *If $\mathcal{A}$ is an infinite MAD family on an infinite set, then there is no sequential filter finer than the residual filter $\mathcal{F}_\mathcal{A}$ of $\mathcal{A}$.*

Proof. Let N be an infinitely countable set and let $\mathcal{A}$ be a MAD family on N. Suppose that $\mathcal{E} \in \mathbb{E}$ with $\mathcal{E} \geq \mathcal{F}_\mathcal{A}$. Then $\mathcal{E}$ must be free, because $\mathcal{F}_\mathcal{A}$ is free. Hence there exists an infinite subset E of N in $\mathcal{E}$ such that $\beta^*(E) \subset \beta^*(\mathcal{F}_\mathcal{A})$. It follows that $\mathcal{A} \cup \{E\}$ is AD, contrary to the maximality of $\mathcal{A}$. $\square$

IX.7   Compact families

Compact families constitute a common generalization of compact sets and convergent filters [30] [31].

Definition IX.7.1. Let ξ be a convergence on a set X and let $\mathcal{A}$ and $\mathcal{B}$ be families of subsets of X. We say that $\mathcal{A}$ is *compact at* $\mathcal{B}$ if for every filter $\mathcal{H}$ on X,

$$\mathcal{H} \# \mathcal{A} \Longrightarrow \mathrm{adh}_\xi \, \mathcal{H} \in \mathcal{B}^\#. \tag{IX.7.1}$$

In particular, $\mathcal{A}$ is *compactoid* whenever $\mathcal{A}$ is compact at X ([4]), that is, if $\mathrm{adh}_\xi \, \mathcal{H} \neq \varnothing$ for each filter $\mathcal{H}$ that meshes with $\mathcal{A}$.

A family $\mathcal{A}$ is said to be *compact* if it is compact at itself, that is, if $\mathcal{H} \# \mathcal{A}$ implies that $\mathrm{adh}_\xi \, \mathcal{H} \in \mathcal{A}^\#$ for every filter $\mathcal{H}$.

Of course, (IX.1.1) is a special case of (IX.7.1). Indeed, if $\{A\} \subset \mathcal{A} \subset A^\uparrow$ and $\{B\} \subset \mathcal{B} \subset B^\uparrow$, then $\mathcal{A}$ is compact at $\mathcal{B}$ if and only if A is compact at B. In particular, a set A is compact if and only if the principal filter generated by A (equivalently, the family $\{A\}$) is compact.

Proposition IX.7.2. *Every convergent filter is compactoid.*

Proof. If $\mathcal{H}$ meshes with $\mathcal{F}$ and $\mathcal{F}$ is convergent, then $\varnothing \neq \lim \mathcal{F} \subset \mathrm{adh} \, \mathcal{H}$. □

Proposition IX.7.3. *Let* $\zeta \geq \xi$,

$$\mathcal{F} \leq \mathcal{G} \text{ and } \mathcal{B} \geq \mathcal{W}.$$

If $\mathcal{F}$ *is* ζ-*compact at* $\mathcal{B}$, *then* $\mathcal{G}$ *is* ξ-*compact at* $\mathcal{W}$.

Proof. If $\mathcal{H} \subset \mathcal{G}^\# \subset \mathcal{F}^\#$ then $\mathrm{adh}_\xi \, \mathcal{H} \supset \mathrm{adh}_\zeta \, \mathcal{H} \in \mathcal{B}^\# \subset \mathcal{W}^\#$. □

In other words, the coarser a convergence, the more (relatively) compact families; if a family is compact at another family, then a finer family is compact at a coarser family. In particular, if F is compactoid and $G \subset F$ then G is compactoid, because $G^\uparrow \geq F^\uparrow$.

A compact family does not need to be a filter: arbitrary unions of compact families are compact, but a union of filters may fail to be a filter.

Exercise IX.7.4. Let $(\mathcal{A}_i)_{i \in I}$ and $(\mathcal{B}_i)_{i \in I}$ be two sets of families of subsets of X such that $\mathcal{A}_i$ is compact at $\mathcal{B}_i$ for each $i \in I$. Then $\bigcup_{i \in I} \mathcal{A}_i$ is compact at $\bigcup_{i \in I} \mathcal{B}_i$.

---

[4]An abbreviation of: $\mathcal{A}$ is compact at $\{X\}$.

Example IX.7.5. The principal filter of a compact set is a compact family. Hence, if K_1 and K_2 are two disjoint compact subsets of a convergence space (X, ξ) (for instance two disjoint closed intervals in $\mathbb{R}$) the family $\mathcal{A} := \{K_1\}^\uparrow \cup \{K_2\}^\uparrow$ is compact, but there is no ultrafilter meshing with $\mathcal{A}$ ([5]) so that the analog of Proposition IX.7.8 fails.

Proposition IX.7.6. *If κ stands for the set of all isotone compact families with respect to a convergence ξ, then*

(1) $\varnothing, 2^{|\xi|} \in \kappa$,
(2) $\{\mathcal{A}_i : i \in I\} \subset \kappa \Longrightarrow \bigcup_{i \in I} \mathcal{A}_i \in \kappa$,
(3) $\{\mathcal{A}_i : i \in I\} \subset \kappa$, $\operatorname{card} I < \infty \Longrightarrow \bigcap_{i \in I} \mathcal{A}_i \in \kappa$.

Proof. Let ξ be a convergence.

(1) If $\varnothing$ is the empty family of subsets of $|\xi|$, then $\operatorname{adh}_\xi \mathcal{H} \in \varnothing^\#$ is true for every filter $\mathcal{H}$ (since there is no $A \in \varnothing$, the condition $A \cap \operatorname{adh}_\xi \mathcal{H} \neq \varnothing$ is true for each $A \in \varnothing$). On the other hand, since $\varnothing \in 2^{|\xi|}$, no filter $\mathcal{H}$ meshes with $2^{|\xi|}$, thus the condition is emptily fulfilled.

(2) If $\mathcal{H}$ is a filter such that $\mathcal{H} \# \bigcup_{i \in I} \mathcal{A}_i$, then $\mathcal{H} \# \mathcal{A}_i$ for each $i \in I$ so that $\operatorname{adh}_\xi \mathcal{H} \in \mathcal{A}_i^\#$, because $\mathcal{A}_i$ is $\mathbb{H}$-compact for each $i \in I$, that is, $\operatorname{adh}_\xi \mathcal{H} \in \left(\bigcup_{i \in I} \mathcal{A}_i\right)^\#$.

(3) Suppose that $\operatorname{adh}_\xi \mathcal{H} \notin \left(\bigcap_{i \in I} \mathcal{A}_i\right)^\# = \bigcup_{i \in I} \mathcal{A}_i^\#$, that is, $\operatorname{adh}_\xi \mathcal{H} \notin \mathcal{A}_i^\#$ for each $i \in I$. Then for each $i \in I$ there is $H_i \in \mathcal{H}$ with $H_i \notin \mathcal{A}_i^\#$. As I is finite, $\bigcap_{i \in I} H_i \in \mathcal{H}$ and $\bigcap_{i \in I} H_i \notin \bigcup_{i \in I} \mathcal{A}_i^\# = \left(\bigcap_{i \in I} \mathcal{A}_i\right)^\#$, we infer that $\mathcal{H}$ does not mesh $\bigcap_{i \in I} \mathcal{A}_i$.

$\square$

As a result, the set of all isotone compact families on a convergence space X is the set of open subsets of a topology on 2^X.

Preservation of compactness under continuous maps (Proposition IX.1.26) extends to arbitrary compact families:

Proposition IX.7.7. *Let $\mathcal{A}$ and $\mathcal{B}$ be two families of subsets of $|\xi|$ where $\mathcal{A}$ is ξ-compact at $\mathcal{B}$. If $f : |\xi| \to |\tau|$ is continuous, then $f[\mathcal{A}]$ is τ-compact at $f[\mathcal{B}]$.*

---

[5]because if $\mathcal{U} \# \mathcal{A}$ then in particular $K_1 \# \mathcal{U} \iff K_1 \in \mathcal{U}$ and similarly $K_2 \in \mathcal{U}$ which is not possible if $K_1 \cap K_2 = \varnothing$.

Proof. Let $\mathcal{H} \in \mathbb{F}|\tau|$ with $\mathcal{H} \# f[\mathcal{A}]$, equivalently, $f^-[\mathcal{H}] \# \mathcal{A}$. By compactness of $\mathcal{A}$ at $\mathcal{B}$, $\mathrm{adh}_\xi f^-[\mathcal{H}] \# \mathcal{B}$, that is, for every $B \in \mathcal{B}$, there is $\mathcal{G} \# f^-[\mathcal{H}]$ with $\lim_\xi \mathcal{G} \cap B \neq \varnothing$. Then $f[\mathcal{G}] \# \mathcal{H}$ and $\lim_\tau f[\mathcal{G}] \cap f(B) \neq \varnothing$ by continuity of f. Thus, $\mathrm{adh}_\tau \mathcal{H} \# f[\mathcal{B}]$. $\qquad\square$

On the other hand, only compactness of *filters* is preserved under product. This is because, as in the particular case of compact sets, compactness of filters depends only on ultrafilters:

Proposition IX.7.8. *A filter $\mathcal{F}$ is compact at B if and only if every ultrafilter $\mathcal{U}$ finer than $\mathcal{F}$ is convergent to an element of B.*

Proof. If a filter $\mathcal{F}$ is compact at B and $\mathcal{U} \in \beta(\mathcal{F})$, then, by definition, $\mathrm{adh}\,\mathcal{U} \cap B \neq \varnothing$, which means that $\mathcal{U}$ converges to an element of B, because $\mathrm{adh}\,\mathcal{U} = \lim \mathcal{U}$ for each ultrafilter $\mathcal{U}$. Conversely, let a filter $\mathcal{H}$ mesh with $\mathcal{F}$. Hence there exists an ultrafilter $\mathcal{U}$ finer than $\mathcal{H} \vee \mathcal{F}$. The condition implies that $\varnothing \neq B \cap \lim \mathcal{U} \subset B \cap \mathrm{adh}\,\mathcal{H}$. $\qquad\square$

Theorem IX.7.9 (Tikhonov for filters). *A filter $\mathcal{F}$ is $\prod \Xi$-compactoid if and only if $p_\xi[\mathcal{F}]$ is ξ-compactoid for every $\xi \in \Xi$.*

Proof. By Proposition IX.7.7, if $\mathcal{F}$ is $\prod \Xi$-compactoid, then $p_\xi[\mathcal{F}]$ is ξ-compactoid, because of the continuity of p_ξ. Conversely, let $p_\xi[\mathcal{F}]$ be ξ-compactoid for every $\xi \in \Xi$. Let $\mathcal{U}$ be an ultrafilter finer than $\mathcal{F}$. Then $p_\xi[\mathcal{U}]$ is an ultrafilter finer than $p_\xi[\mathcal{F}]$ so that $\lim_\xi p_\xi[\mathcal{U}] \neq \varnothing$ for every $\xi \in \Xi$. If $x_\xi \in \lim_\xi p_\xi[\mathcal{U}]$, then $(x_\xi)_{\xi \in \Xi} \in \lim_{\prod \Xi} \mathcal{U}$. In view of Proposition IX.7.8, $\mathcal{F}$ is compactoid in $\prod \Xi$. $\qquad\square$

We now relate compactness to convergence.

Proposition IX.7.10. *A filter $\mathcal{F}$ is ξ-compact at $\{x\}$ if and only if $x \in \lim_{\mathrm{S}\,\xi} \mathcal{F}$.*

Proof. If $x \in \lim_{\mathrm{S}\,\xi} \mathcal{F}$ and $\mathcal{H} \# \mathcal{F}$ then $x \in \lim_{\mathrm{S}\,\xi} \mathcal{F} \subset \mathrm{adh}_{\mathrm{S}\,\xi} \mathcal{H} = \mathrm{adh}_\xi \mathcal{H}$. Conversely, if $\mathcal{F}$ is compactoid at $\{x\}$, then $x \in \mathrm{adh}_\xi \mathcal{U} = \lim_\xi \mathcal{U}$ for every ultrafilter $\mathcal{U}$ finer than $\mathcal{F}$, hence $x \in \lim_{\mathrm{S}\,\xi} \mathcal{F}$. $\qquad\square$

Proposition IX.7.11. *A family is ξ-compact (at $\mathcal{B}$) if and only if it is $\mathrm{S}\,\xi$-compact (at $\mathcal{B}$).*

Proof. The necessity is obvious, because ξ is finer than $\mathrm{S}\,\xi$. Let $\mathcal{F}$ be $\mathrm{S}\,\xi$-compact at $\mathcal{B}$ and $\mathcal{U} \in \beta(\mathcal{F})$. Then by assumption $\mathrm{adh}_{\mathrm{S}\,\xi} \mathcal{U} \in \mathcal{B}^\#$. On the other hand, $\mathrm{adh}_{\mathrm{S}\,\xi} \mathcal{U} = \lim_{\mathrm{S}\,\xi} \mathcal{U} = \lim_\xi \mathcal{U} = \mathrm{adh}_\xi \mathcal{U}$. $\qquad\square$

The preceding observations reflect the pseudotopological nature of compactness, that is, it depends only on the pseudotopologization of a convergence. The reason for this situation is clear: compactness is defined with the aid of adherent filters and the pseudotopologizer can be expressed in terms of adherences of filters (VIII.3.1).

We will now see that in a Hausdorff convergence, compact filters find a handy characterization.

Note first that

$$\ker \mathcal{F} \subset \operatorname{adh} \mathcal{F}$$

for any filter $\mathcal{F}$ on a convergence space, because if $x \in \ker \mathcal{F}$ then $\{x\}^{\uparrow} \# \mathcal{F}$.

Definition IX.7.12. A filter $\mathcal{F}$ on a convergence space (X, ξ) is *closed* if $\operatorname{adh}_\xi \mathcal{F} = \ker \mathcal{F}$.

Of course, a subset of (X, ξ) is ξ-closed (in the sense of Definition V.3.4) if and only if its principal filter is a closed filter in the sense of Definition IX.7.12. Proposition IX.1.12 extends from principal to general filters:

Proposition IX.7.13. *A compact filter on a Hausdorff convergence space is closed.*

Proof. If $\mathcal{U} \in \beta(\mathcal{F})$ then $\operatorname{adh} \mathcal{U} = \lim \mathcal{U} \in \mathcal{F}^{\#}$ by compactness of $\mathcal{F}$. Since $\lim \mathcal{U}$ is a singleton, $\lim \mathcal{U} \subset \ker \mathcal{F}$, so that $\operatorname{adh} \mathcal{F} \subset \ker \mathcal{F}$. $\square$

Note that Proposition IX.1.11 extends as straightforwardly to filters, as long as we observe that $A \subset B$ means $B^{\uparrow} \leq A^{\uparrow}$:

Proposition IX.7.14. *If $\mathcal{F}$ is a compactoid filter and $\mathcal{G}$ is a finer closed filter, then $\mathcal{G}$ is compact.*

Proof. If $\mathcal{U} \in \beta(\mathcal{G})$ then $\mathcal{U} \in \beta(\mathcal{F})$ so that $\lim \mathcal{U} \neq \varnothing$ by relative compactness of $\mathcal{F}$. But $\lim \mathcal{U} \subset \operatorname{adh} \mathcal{G}$ and $\operatorname{adh} \mathcal{G} = \ker \mathcal{G}$ because $\mathcal{G}$ is closed. Hence $\lim \mathcal{U} \subset \ker \mathcal{G}$ and $\lim \mathcal{U} \in \mathcal{G}^{\#}$. $\square$

Corollary IX.7.15. *A filter on a Hausdorff convergence space is compact if and only if it is closed and compactoid.*

Corollary IX.7.16. *If $\mathcal{F}$ is a compact filter (at $\mathcal{B}$) on a convergence space that is either Hausdorff or regular, then $\operatorname{adh} \mathcal{F}$ is a non-empty compact set (at $\mathcal{B}$).*

Proof. Assume first that the convergence is Hausdorff. If $\mathcal{U} \in \mathbb{U}X$ and $\mathrm{adh}\,\mathcal{F} \in \mathcal{U}$ then $\ker \mathcal{F} \in \mathcal{U}$ because $\mathcal{F}$ is closed. Hence $\mathcal{U} \# \mathcal{F}$ and $\lim \mathcal{U}$ is a singleton meshing with $\mathcal{F}$, hence with $\ker \mathcal{F} = \mathrm{adh}\,\mathcal{F}$. Hence $\lim \mathcal{U} \cap \mathrm{adh}\,\mathcal{F} \neq \varnothing$. In the case where $\mathcal{F}$ is compact at $\mathcal{B}$, $\lim \mathcal{U} \in \mathcal{B}^{\#}$.

Assume now that the convergence is regular, and let $\mathcal{H}$ be a filter with $\mathrm{adh}\,\mathcal{F} \in \mathcal{H}^{\#}$. Of course, $\mathrm{adh}\,\mathcal{F} \subset \mathrm{adh}\,F$ for each $F \in \mathcal{F}$ because $\mathcal{F} \geq F^{\uparrow}$. Therefore, $\mathcal{H} \# \mathrm{adh}^{\natural}\,\mathcal{F}$ so that $\mathcal{V}(\mathcal{H}) \# \mathcal{F}$ by Proposition VI.5.2. Since $\mathcal{F}$ is compact at $\mathcal{B}$, $\mathrm{adh}\,\mathcal{V}(\mathcal{H}) \in \mathcal{B}^{\#}$. Since ξ is regular, $\mathrm{adh}\,\mathcal{V}(\mathcal{H}) \subset \mathrm{adh}\,\mathcal{H}$ by Proposition VIII.4.1, so that $\mathrm{adh}\,\mathcal{H} \in \mathcal{B}^{\#}$ and the proof is complete.   □

In the case of filters admitting a filter-base composed of open sets on a Hausdorff topological space, compactness of filters essentially reduces to compactness of sets:

Theorem IX.7.17. *Let ξ be a Hausdorff convergence on X and let $\mathcal{F}$ be a filter on X with a filter-base composed of ξ-open sets. If $\mathcal{F}$ is ξ-compact then*

$$\mathcal{F} = \mathcal{O}_{\xi}(K)^{\uparrow}$$

where $K = \mathrm{adh}_{\xi}\,\mathcal{F}$ is a ξ-compact set.

If moreover ξ is topological and there is a compact ξ-subset K of X such that $\mathcal{F} = \mathcal{O}_{\xi}(K)^{\uparrow}$ then $\mathcal{F}$ is ξ-compact and $K = \mathrm{adh}_{\xi}\,\mathcal{F}$.

Proof. If $\mathcal{F}$ is compact then $K := \mathrm{adh}\,\mathcal{F}$ is a non-empty compact set by Corollary IX.7.16. Moreover, $\mathcal{F}$ is closed by Proposition IX.7.13, so that $\mathrm{adh}\,\mathcal{F} = \ker \mathcal{F}$, and $K^{\uparrow} \geq \mathcal{F}$. Hence each open set in the filter-base of $\mathcal{F}$ contains $\ker \mathcal{F}$ and $\mathcal{F} \leq \mathcal{O}(K)$. On the other hand, if $O \in \mathcal{O}(K)$ then $\lim \mathcal{U} \cap O \neq \varnothing$ for every $\mathcal{U} \in \beta(\mathcal{F})$ by compactness of $\mathcal{F}$. Therefore $O \in \bigcap_{\mathcal{U} \in \beta(\mathcal{F})} \mathcal{U} = \mathcal{F}$ and $\mathcal{O}(K) \leq \mathcal{F}$.

Conversely, if $\mathcal{F} = \mathcal{O}(K)^{\uparrow}$ for some compact set K and ξ is a topology, then $\mathcal{F}$ is compact. Indeed $\mathcal{H} \# \mathcal{F}$ if and only if $\mathrm{cl}^{\natural}\,\mathcal{H} \# K$ so that $\mathrm{adh}(\mathrm{cl}^{\natural}\,\mathcal{H}) \cap K \neq \varnothing$ and therefore $\mathrm{adh}(\mathrm{cl}^{\natural}\,\mathcal{H}) \in \mathcal{F}^{\#}$. Since ξ is topological, $\mathrm{adh}(\mathrm{cl}^{\natural}\,\mathcal{H}) = \mathrm{adh}\,\mathcal{H}$ and the result follows. Since $\mathcal{F}$ is compact and ξ is Hausdorff,

$$\mathrm{adh}\,\mathcal{F} = \ker \mathcal{F} = \bigcap_{O \in \mathcal{O}(K)} O = K,$$

where the latter equality is true in any T_1-convergence, in particular in a Hausdorff topology. Indeed, if $x \notin K$ then $\{x\}^{c}$ is an open set containing K but not x.   □

IX.8 Conditional compactness

We have seen that compactness has a pseudotopological nature, because a convergence is compact whenever its pseudotopologization is. The link between pseudotopologies and compactness is in fact even more intimate. It hinges on the fact that $\lim_{S\xi} \mathcal{F}$ is the intersection of $\operatorname{adh}_\xi \mathcal{H}$ over all filters $\mathcal{H}$ that mesh $\mathcal{F}$, while a set F is ξ-compactoid whenever $\operatorname{adh}_\xi \mathcal{H} \neq \varnothing$ for each filter $\mathcal{H}$ that meshes F.

We shall now consider the notion of *conditional compactness*, in which not all filters are required to have non-empty adherence, but only those from a special class. This is a far reaching concept, embracing also numerous variants of completeness and of Lindelöf property.

In this section, we shall mainly consider the class of countably based filters, in which case conditional compactness becomes countable compactness. We shall see (Proposition IX.8.5) that countable compactness is paratopological in a similar way, in which compactness is pseudotopological. This is an instance of an even more general phenomenon that we will consider in full generality in Section XIV.3.

Definition IX.8.1. Let $\mathbb{H}$ be a class of filters, A and B be subsets of X and ξ be a convergence on X. Then A is $\mathbb{H}$-*compact at* B (with respect to ξ) if

$$\underset{\mathcal{H} \in \mathbb{H}}{\forall}\ A \in \mathcal{H}^{\#} \implies \operatorname{adh}_\xi \mathcal{H} \cap B \neq \varnothing. \tag{IX.8.1}$$

Proposition IX.8.2. *Let $\mathbb{H}$ be a class of filters and $f \in C(\xi, \tau)$ be surjective and such that*

$$\mathcal{H} \in \mathbb{H}(\tau) \implies f^{-}[\mathcal{H}] \in \mathbb{H}(\xi). \tag{IX.8.2}$$

Then if ξ is $\mathbb{H}$-compact, then τ is $\mathbb{H}$-compact.

Proof. Let $\mathcal{H} \in \mathbb{H}(\tau)$. Then, by assumption, $f^{-}[\mathcal{H}] \in \mathbb{H}(\xi)$ hence $\operatorname{adh}_\xi f^{-}[\mathcal{H}] \neq \varnothing$, because ξ is $\mathbb{H}$-compact. As f is surjective, $f[f^{-}[\mathcal{H}]] = \mathcal{H}$ and $\operatorname{adh}_\tau \mathcal{H} = f[f^{-}[\mathcal{H}]] \neq \varnothing$, because f is continuous. $\square$

Recall that $\mathbb{F}_1$ stands for the class of *countably based filters*, that is, filters admitting a countable filter-base, with the convention that $\mathbb{F}_1 X$ is the set of countably-based filters on the set X. It is clear that the classes $\mathbb{F}_0$ of principal filters and $\mathbb{F}_1$ of countably based filters both satisfy (IX.8.2).

Conditional compactness extends naturally from sets to families.

Definition IX.8.3. Let $\mathbb{H}$ be a class of filters, let ξ be a convergence on a set X and let $\mathcal{A}$ and $\mathcal{B}$ be families of subsets of X. We say that $\mathcal{A}$ is $\mathbb{H}$-*compact at* $\mathcal{B}$ if

$$\underset{\mathcal{H} \in \mathbb{H}}{\forall} \mathcal{H} \# \mathcal{A} \implies \mathrm{adh}_\xi \, \mathcal{H} \in \mathcal{B}^\#. \tag{IX.8.3}$$

IX.8.1 *Paratopologies*

A convergence ξ is called a *paratopology* (see Definition XIV.3.10) if for every filter $\mathcal{F}$ on $|\xi|$,

$$\lim_\xi \mathcal{F} \supset \bigcap\nolimits_{\mathbb{F}_1 \ni \mathcal{H} \# \mathcal{F}} \mathrm{adh}_\xi \, \mathcal{H}.$$

As we shall see in Section XIV.3.1 that for each convergence ξ there exists the finest paratopology $S_1 \xi$ that is coarser than ξ and that

$$\lim_{S_1 \xi} \mathcal{F} = \bigcap\nolimits_{\mathbb{F}_1 \ni \mathcal{H} \# \mathcal{F}} \mathrm{adh}_\xi \, \mathcal{H}.$$

IX.8.2 *Countable compactness*

Definition IX.8.4. Let ξ be a convergence on X. We say that a subset A of X is *countably compact at* B if it is $\mathbb{F}_1$-*compact at* B, that is, if

$$\underset{\mathcal{H} \in \mathbb{F}_1 X}{\forall} A \in \mathcal{H}^\# \implies \mathrm{adh}_\xi \, \mathcal{H} \cap B \neq \varnothing.$$

It is called *countably compactoid* (or *countably relatively compact*) if it is countably compact at the whole space, and is called *countably compact* if it is countably compact at itself.

In other words, a subset of a convergence space is countably compact if every countably based filter on it has non-empty adherence. Of course, every compact set (at B) is countably compact (at B), but not conversely, as we shall see shortly.

Proposition IX.8.5. *A convergence ξ is countably compact if and only if its paratopologization $S_1 \xi$ is.*

Proof. Of course, $\mathrm{adh}_\xi \, \mathcal{D} \subset \mathrm{adh}_{S_1 \xi} \, \mathcal{D}$ for each filter $\mathcal{D}$. If $\mathcal{D} \in \mathbb{F}_1$ and $x \in \mathrm{adh}_{S_1 \xi} \, \mathcal{D}$ then there is a filter $\mathcal{F}$ such that $\mathcal{F} \# \mathcal{D}$ and $x \in \lim_{S_1 \xi} \mathcal{F} \subset \mathrm{adh}_\xi \, \mathcal{H}$ for every $\mathcal{H} \in \mathbb{F}_1$ with $\mathcal{H} \# \mathcal{F}$, and in particular $x \in \mathrm{adh}_\xi \, \mathcal{D}$. Accordingly, $\mathrm{adh}_{S_1 \xi} \, \mathcal{D} = \mathrm{adh}_\xi \, \mathcal{D}$ for $\mathcal{D} \in \mathbb{F}_1$, which completes the proof. $\square$

Proposition IX.8.6. *A convergence ξ is countably compact if and only if $\mathrm{adh}_\xi \, \mathcal{E} \neq \varnothing$ for each sequential filter $\mathcal{E}$.*

Proof. If ξ is countably compact, then $\mathrm{adh}_\xi \mathcal{E} \neq \varnothing$ for each sequential filter $\mathcal{E}$, because it is countably based.

Let $\mathcal{H}$ be a countably based filter on $|\xi|$. As there exists a sequential filter $\mathcal{E} \geq \mathcal{H}$, we conclude that $\varnothing \neq \mathrm{adh}_\xi \mathcal{E} \subset \mathrm{adh}_\xi \mathcal{H}$. $\qquad\square$

Corollary IX.8.7. *A T_1-convergence is countably compact if and only if each infinite (countable) set has an accumulation point.*

Proof. Let A be a countably infinite subset of $|\xi|$. If ξ is countably compact, then $\mathrm{adh}_\xi (A)_0 \neq \varnothing$. If $x \in \mathrm{adh}_\xi (A)_0$ then there is a filter $\mathcal{F} \geq (A)_0$ with $x \in \lim_\xi \mathcal{F}$. As $\mathcal{F}$ is free, $A \setminus \{x\} \in \mathcal{F}$, hence $x \in \mathrm{adh}_\xi (A \setminus \{x\})$. If an infinite subset A of $|\xi|$ has no accumulation point, then no countably infinite subset F has an accumulation point. Therefore $\mathrm{adh}_\xi (F)_0 = \varnothing$, hence ξ is not countably compact. $\qquad\square$

Corollary IX.8.8. *A T_1-convergence is not countably compact if and only if it admits an infinite closed discrete subspace.*

Proposition IX.8.9. *A T_1-convergence is countably compact if and only if every locally finite (Definition V.2.9) family of non-empty sets is finite.*

Proof. Suppose that $\{A_n : n \in \mathbb{N}\}$ is an infinite ξ-locally family of non-empty sets. Then $\bigcup_{k \geq n}^\infty A_k \supset \bigcup_{k \geq n+1}^\infty A_k$ for each n, and $\bigcap_{n \in \mathbb{N}} \bigcup_{k \geq n}^\infty A_k = \varnothing$. Indeed, if $x \in \bigcap_{n \in \mathbb{N}} \bigcup_{k \geq n}^\infty A_k$, then there is $n \in \mathbb{N}$ such that $x \in A_k$ for each $k \geq n$. Hence $\{x\} \in \{x\}^\uparrow$ and $\{x\}$ intersects infinitely many elements of $\{A_n : n \in \mathbb{N}\}$, which contradicts local finiteness, because $x \in \lim_\xi \{x\}^\uparrow$.

By the countable compactness of ξ, there exists

$$ x \in \mathrm{adh}_\xi \left\{ \bigcup_{k \geq n}^\infty A_k : n \in \mathbb{N} \right\}, $$

hence there exists a free filter $\mathcal{F}$ such that $x \in \lim_\xi \mathcal{F}$ and $\bigcup_{k \geq n}^\infty A_k \in \mathcal{F}^\#$ for each $n \in \mathbb{N}$. In other words, $(\bigcup_{k \geq n}^\infty A_k) \cap F \neq \varnothing$ for every $F \in \mathcal{F}$. It follows that for each $F \in \mathcal{F}$ and each $n \in \mathbb{N}$, there exist $k_n > n$ and $x_n \in F \cap A_{k_n} \setminus \bigcup_{0 \leq k \leq n} A_k$, which contradicts local finiteness.

The converse follows from Corollary IX.8.8, because if a subset A of a T_1-convergence is discrete, then $\{\{x\} : x \in A\}$ is locally finite by Exercise V.3.29. $\qquad\square$

Definition IX.8.10. Let X be a convergence space and let $f : X \to \mathbb{R}$. We say that f is *lower semicontinuous (l.s.c.)* at $x \in X$ if for each filter $\mathcal{F}$

on X with $x \in \lim \mathcal{F}$ and for every $r < f(x)$ there exists $F \in \mathcal{F}$ such that $\inf_{w \in F} f(w) > r$.

Similarly, f is *upper semicontinuous (u.s.c.)* at $x \in X$ if for each filter $\mathcal{F}$ on X with $x \in \lim \mathcal{F}$ and for every $r > f(x)$ there exists $F \in \mathcal{F}$ such that $\inf_{w \in F} f(w) < r$.

The function f is *lower* (respectively, *upper*) *semicontinuous* if it is lower (respectively, upper) semicontinuous at each point.

Semicontinuities of real-valued functions are topological notions. Let ν_- be a topology of $\mathbb{R}$ with a base of open sets given by $\{\{r \in \mathbb{R} : r > s\} : s \in \mathbb{R}\}$, and let ν_+ be a topology of $\mathbb{R}$ with a base of open sets given by $\{\{r \in \mathbb{R} : r < s\} : s \in \mathbb{R}\}$.

Exercise IX.8.11. Let X be a convergence space and let $f : X \to \mathbb{R}$. Check that

(1) f is l.s.c. if and only if $f \in C(X, \nu_-)$,
(2) f is l.s.c. if and only if $\{f \le r\}$ is closed for each $r \in \mathbb{R}$,
(3) f is u.s.c. if and only if $f \in C(X, \nu_+)$,
(4) f is u.s.c. if and only if $\{f \ge r\}$ is closed for each $r \in \mathbb{R}$,
(5) f is continuous if and only if f is both lower and upper semicontinuous.

As $f \in C(\xi, \nu_-)$ if and only if $f \in C(\mathrm{T}\,\xi, \nu_-)$, and $f \in C(\xi, \nu_+)$ if and only if $f \in C(\mathrm{T}\,\xi, \nu_+)$, we conclude that $f : |\xi| \to \mathbb{R}$ is lower (resp. upper) semicontinuous if and only if $f : |T\xi| \to \mathbb{R}$ is.

Lemma IX.8.12. *Let f be an l.s.c. function on a convergence space X. If the filter base*

$$\{\{f \le r\} : r > \inf_X f\} \tag{IX.8.4}$$

has non-empty adherence, then there exists $x_0 \in X$ such that $f(x_0) = \inf_X f$.

Proof. Since $\{f \le r\}$ is closed for each $r \in \mathbb{R}$, and the adherence of (IX.8.4) is non-empty, there exists $x_0 \in X$ such that

$$x_0 \in \mathrm{adh}\,\{\{f \le r\} : r > \inf_X f\} \subset \bigcap\nolimits_{r > \inf_X f} \{f \le r\}.$$

Accordingly, for each $r > \inf_X f$, $f(x_0) \le r$, that is, $f(x_0) \le \inf_X f$. On the other hand, $f(x) \ge \inf_X f$ for each $x \in X$. $\qquad\square$

Proposition IX.8.13. *Each lower semicontinuous function on a countably compact convergence space attains its infimum.*

Proof. Let $\{r_n\}_{n\in\mathbb{N}}$ be a decreasing sequence such that $\inf_X f = \inf_{n\in\mathbb{N}} r_n$. Then

$$\{\{f \leq r_n\} : n \in \mathbb{N}\}$$

is another base of the filter generated by (IX.8.4). By countable compactness, $\mathrm{adh}\,\{\{f \leq r_n\} : n \in \mathbb{N}\} \neq \varnothing$, hence Lemma IX.8.12 applies. $\qquad\square$

Lemma IX.8.12 is more general than Proposition IX.8.13.

Example IX.8.14. Let X be an infinite-dimensional Hilbert space, let d denote its metric, and let B be a non-empty compact subset of X. Then X is not countably compact, but the *distance function*

$$\mathrm{dist}_B(x) := \inf\{d(x,b) : b \in B\}$$

is continuous (Exercise V.3.8), and $\mathrm{adh}\,\{\{\mathrm{dist}_B \leq r\} : r > 0\} \neq \varnothing$.

Similarly,

Remark IX.8.15. Let f be an u.s.c function on a convergence space X. If the filter base

$$\{\{f \geq r\} : r < \sup_X f\}$$

has non-empty adherence, then there exists $x_0 \in X$ such that $f(x_0) = \sup_X f$.

Therefore, we get a generalization of the Weierstraß theorem:

Corollary IX.8.16. *Each continuous function on a countably compact convergence space attains its infimum and supremum.*

IX.8.3 *Sequential compactness*

Definition IX.8.17. If A and B are two subsets of a convergence space (X,ξ) we say that A is *sequentially compact at* B if every sequence of points of A has a subsequence that converges to a point of B (for ξ).

A set is *relatively sequentially compact* (or *sequentially compactoid*) if it is sequentially compact at X and *sequentially compact* if it is sequentially compact at itself.

By Proposition IX.8.2, a continuous image of countably compact convergence is countably compact. While the class $\mathbb{E}$ of sequential filters does not satisfy (IX.8.2), sequential compactness is nevertheless preserved by continuous maps. Indeed,

Lemma IX.8.18. *A subset A of a convergence space (X,ξ) is sequentially compact (at B) if and only if it is $\mathbb{F}_1$-compact (at B) for $\mathrm{Seq}\,\xi$.*

Proof. Assume A is sequentially compact and let $\mathcal{H} \in \mathbb{F}_1 X$ with $A \in \mathcal{H}^\#$. If $\{H_n : n \in \mathbb{N}\}$ is a decreasing filter-base of $\mathcal{H}$ there is $x_n \in H_n \cap A$ for each n. This is a sequence of points x_n of A such that $(x_n)_n \geq \mathcal{H}$. By sequential compactness, there is a subsequence $\{y_k\}_{k \in N}$ of $\{x_n\}_{n \in \mathbb{N}}$ such that $\lim_\xi (y_k)_k^\uparrow \cap B \neq \varnothing$. Of course, $\lim_\xi (y_k)_k^\uparrow \subset \mathrm{adh}_{\mathrm{Seq}\,\xi} \mathcal{H}$ so that A is countably compact (at B) for $\mathrm{Seq}\,\xi$.

Conversely, if every countably based filter has non-empty adherence for $\mathrm{Seq}\,\xi$, it is true in particular of sequential filters. Hence, for every sequence $\{x_n\}_{n \in \mathbb{N}}$ on A there is a sequence $\{y_k\}_{k \in N}$ that converges (to a point of B) such that $(x_n)_n \# (y_k)_k$. Any sequential filter finer than $(x_n)_n \vee (y_k)_k$ (see Exercise II.8.2) is a subsequence of $\{x_n\}_{n \in \mathbb{N}}$ that converges (to a point of B). $\qquad\square$

Exercise IX.8.19. Show that a continuous image of a sequentially compact convergence is sequentially compact.

Corollary IX.8.20. *A sequentially compact subset of a convergence space is also countably compact.*

Hence compactness and sequential compactness overlap and both imply countable compactness:

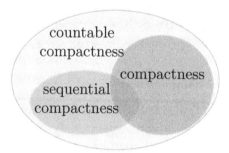

Example IX.8.21 (A sequentially compact (hence countably compact) topology that is not compact). Let $\omega_1 = \{\alpha \in \mathrm{Ord} : \alpha < \omega_1\}$ be equipped with the order topology as in Example V.4.13, where it was shown to be sequentially compact. It is not compact because the filter $\{\{\beta \in \omega_1 : \beta > \alpha\} : \alpha \in \omega_1\}^\uparrow$ has empty adherence.

Example IX.8.22 (A compact non-sequentially compact topology). Let $\{0, 1\}$ be endowed with the discrete topology. Consider

$$X = \prod_{r \in [0,1]} \{0, 1\} \approx \{0, 1\}^{[0,1]}, \qquad (\text{IX.8.5})$$

with its product topology. By the Tikhonov Theorem IX.1.32, X is compact. For each $r \in \mathbb{R}$, there exists a binary representation of r

$$r = \sum_{n=1}^{\infty} \frac{r_n}{2^n}, \qquad (\text{IX.8.6})$$

with $r_n \in \{0, 1\}$; in the case where r admits two representations, we shall consider that for which there is $n_r \in \mathbb{N}_1$ with $r_n = 0$ for each $n > n_r$ ([6]). Define a sequence $\{f_n\}_{n \in \mathbb{N}}$ of elements of X by

$$f_n(r) = r_n$$

for each $n \in \mathbb{N}_1$. There is no convergent subsequence of $\{f_n\}_{n \in \mathbb{N}}$. If on the contrary there is $f \in X$ and $\{f_{n_k}\}_{k=1}^{\infty}$ that converges to f, then for each $r \in [0, 1]$, there exists $k(r) \in \mathbb{N}$ such that $f_{n_k}(r) = f(r) \in \{0, 1\}$ for each $k \geq k(r)$, that is, $\{f_{n_k}(r)\}_{k=1}^{\infty}$ is stationary. If r is such that $r_{n_k} = 0$ for odd k and $r_{n_k} = 1$ for even k, then

$$r_{n_k} = f_{n_k}(r)$$

is not stationary, which yields a contradiction.

Example IX.8.23 (A sequentially compact non-compact topology). Let Y be a subspace of (IX.8.5) consisting of those f, for which

$$s(f) := \{r : f(r) \neq 0\}$$

is countable. Of course, $Y \neq X$ but is dense in X. Indeed, if $f_0 \in X$ and V is a neighborhood of f_0, then there exists a finite subset F of $[0, 1]$ such that

$$\{f \in X : \underset{r \in F}{\forall} f(r) = f_0(r)\} \subset V. \qquad (\text{IX.8.7})$$

Therefore if $f \in X$ is such that $s(f) = F$ and $f(r) = f_0(r)$ and for $r \in F$, then $f \in V \cap Y$. It follows that Y is not compact, because it is not a closed subset of X.

If $\{f_n\}_{n=0}^{\infty}$ is a sequence of elements of Y, then the set $S = \bigcup_{n \in \mathbb{N}} s(f_n)$ is countable. Since $\{0, 1\}$ is sequentially compact, $\prod_{r \in S} \{0, 1\}$ is sequentially compact and homeomorphic with a subset

$$\{f \in X : s(f) \subset S\}$$

of Y, which proves the sequential compactness of Y.

---

[6] For example, $\frac{1}{2}$ has two representations, one $r_1 = 1$ and $r_n = 0$ for $n > 1$, and $r_1 = 0$ and $r_n = 1$ for $n > 1$.

Example IX.8.24 (A compact topology (hence countably compact) that is not sequentially compact). The set $\mathbb{U}\mathbb{N}$ of ultrafilters on $\mathbb{N}$ with the Stone topology is compact by Theorem IX.5.5. On the other hand, if $\{X_n : n \in \mathbb{N}\}$ is a partition of $\mathbb{N}$ into infinite subsets, there is for each $n \in \mathbb{N}$ a free ultrafilter $\mathcal{U}_n$ with $X_n \in \mathcal{U}_n$. The resulting sequence $\{\mathcal{U}_n\}_{n\in\mathbb{N}}$ on $\mathbb{U}\mathbb{N}$ as well as all of its subsequences, are discrete. Thus Proposition VI.1.8 applies to the effect that, for any subsequence $\{\mathcal{U}_{n_k}\}_{k\in\mathbb{N}}$ of $\{\mathcal{U}_n\}_{n\in\mathbb{N}}$, the contour $\bigcup_{k\in\mathbb{N}} \bigcap_{p\geq k} \mathcal{U}_{n_p}$ is not an ultrafilter. Hence $\lim_\beta (\mathcal{U}_{n_k})_{n_k}^\uparrow = \varnothing$ by Lemma IX.5.3. Therefore $\mathbb{U}\mathbb{N}$ is not sequentially compact.

Theorem IX.8.25 (Tikhonov on sequential compactness of products). *A countable product of convergences is sequentially compact if and only if each component is sequentially compact.*

Proof. If the product is sequentially compact, then each component is sequentially compact as a continuous image.

Assume that X_n is sequentially compact for each $n \in \mathbb{N}$ and let $\{f_k\}_{k\in\mathbb{N}}$ be a sequence of elements of $\prod_{n\in\mathbb{N}} X_n$. Because X_0 is sequentially compact, there exists an infinite subset N_0 of $\mathbb{N}$ and $x_0 \in X_0$ such that $x_0 \in \lim_{k\in N_0} f_k(0)$. Suppose that $N_0 \supset N_1 \supset \ldots \supset N_{n-1}$ are infinite subsets of $\mathbb{N}$ such that

$$x_m \in \lim_{k\in N_m} f_k(m)$$

for each $0 \leq m \leq n - 1$. Because X_n is sequentially compact, there exists an infinite subsets N_n of N_{n-1} and $x_n \in X_m$ such that $x_n \in \lim_{k\in N_n} f_k(n)$. By Theorem I.6.4, there exists an infinite subset N_∞ of $\mathbb{N}$ such that $N_\infty \setminus N_n$ is finite for each $n \in \mathbb{N}$. Therefore, if $f(n) := x_n$, then

$$f \in \lim_{k\in N_\infty} f_k.$$

$\square$

A product of two countably compact topologies however need not be countably compact.

Example IX.8.26. If A is an infinite subset of $\mathbb{N}$, then the discrete topology on A is homeomorphic with the discrete topology on $\mathbb{N}$. Therefore $\beta(A) \subset \beta\mathbb{N}$ and $\beta(A)$ is homeomorphic with $\beta\mathbb{N}$. By Theorem IX.5.14, $\operatorname{card}(\beta A) = \operatorname{card}(\beta\mathbb{N}) = 2^{\mathfrak{c}} > \mathfrak{c}$.

For each countably infinite subset A of $\beta\mathbb{N}$, let $\alpha(A) \in \operatorname{cl}_{\beta\mathbb{N}} A \setminus A$. We shall define a transfinite sequence $(A_\gamma)_{\gamma<\omega_1}$ of subsets of $\beta\mathbb{N}$. Let $A_0 := \mathbb{N}$, and for $0 < \delta < \omega_1$, let

$$A_\delta := \bigcup_{\gamma<\delta} A_\gamma \cup \left\{\alpha(B) : B \in \left[\bigcup_{\gamma<\delta} A_\gamma\right]^{\aleph_0}\right\}.$$

Accordingly, $X := \bigcup_{\gamma < \omega_1} A_\gamma$ is countably compact, because if D is a countably infinite subset of X, then there is $\gamma < \omega_1$ such that $D \subset A_\gamma$, hence for $\gamma < \delta < \omega_1$ there is an accumulation point of D in $A_\delta \subset X$. We observe that card $A_\gamma \leq \mathfrak{c}$ for each $\gamma < \omega_1$. In fact, this true for $\gamma = 0$, for $\gamma > 0$, by induction,

$$\text{card}(A_\gamma) \leq \aleph_0 \cdot \mathfrak{c} + (\aleph_0 \cdot \mathfrak{c})^{\aleph_0} = \mathfrak{c}.$$

Therefore $\text{card}(X) \leq \aleph_1 \cdot \mathfrak{c} = \mathfrak{c}$.

The set $Y := \beta\mathbb{N} \setminus X \cup \mathbb{N}$ is countably compact, because for each countably infinite subset D of Y, the set $\text{card}(\text{cl}_{\beta\mathbb{N}} D \setminus D) > \mathfrak{c}$, hence $\text{cl}_{\beta\mathbb{N}} D \setminus D \setminus X \neq \varnothing$, so that $(\text{cl}_\beta D \setminus D) \cap Y \neq \varnothing$.

The product $X \times Y$ is not countably compact, because

$$D := \{(n, n) : n \in \mathbb{N}\} \subset X \times Y,$$

which is discrete and closed. Indeed, if $(x, y) \in \text{cl}_{X \times Y} D \setminus D$, there would be an ultrafilter $\mathcal{U}$ on $\{(n, n) : n \in \mathbb{N}\}$ such that $(x, y) \in \lim_{X \times Y} \mathcal{U} \subset \beta\mathbb{N} \times \beta\mathbb{N}$. The images $p_X(\mathcal{U})$ and $p_Y(\mathcal{U})$ by projections are copies of the same ultrafilter on $\mathbb{N}$, hence $x = y$, but $x \in X \setminus \mathbb{N}$ and $y \in Y \setminus \mathbb{N}$ and thus

$$(X \setminus \mathbb{N}) \cap (Y \setminus \mathbb{N}) = \varnothing,$$

which yields a contradiction.

Recall (from Definition V.6.14) that a convergence is sequential if every sequentially closed set is closed. In other words, ξ is sequential if

$$\lim_\xi \mathcal{F} \subset \bigcap_{H \in \mathcal{F}^\#} \text{cl}_{\text{Seq}\,\xi}\, H.$$

Proposition IX.8.27. *Countable compactness and sequential compactness coincide among sequential convergences.*

Proof. If ξ is countably compact and sequential and $\{x_n\}_{n \in \mathbb{N}}$ is a sequence on $|\xi|$ then $\text{adh}_\xi (x_n)_n^\uparrow \neq \varnothing$ because $(x_n)_n^\uparrow$ is a countably based filter. Hence there is $\mathcal{F} \geq (x_n)_n$ with $x \in \lim_\xi \mathcal{F}$ and by sequentiality, there is $F \subset \{x_n : n \in \mathbb{N}\}$ with $x \in \text{cl}_{\text{Seq}\,\xi}\, F$. In particular, $\text{adh}_{\text{Seq}\,\xi}\, F \neq \varnothing$ so that there is a subsequence of $\{x_n\}_{n \in \mathbb{N}}$ that converges. $\square$

In order to show that sequential compactness, countable compactness and compactness coincide among metric spaces, we first observe that a sequentially compact metric space is totally bounded in the following sense:

Definition IX.8.28. A metric space X is *totally bounded* if for every $\varepsilon > 0$ there is a finite subset F of X such that

$$X = \bigcup_{x \in F} B(x, \varepsilon).$$

Proposition IX.8.29. *A sequentially compact metric space is totally bounded.*

Proof. Fix $\varepsilon > 0$ and let $x_1 \in X$. If $X = B(x_1, \varepsilon)$ we are done. Otherwise, there is $x_2 \in X \setminus B(x_1, \varepsilon)$. Either $X = B(x_1, \varepsilon) \cup B(x_2, \varepsilon)$ or there is $x_3 \in X \setminus (B(x_1, \varepsilon) \cup B(x_2, \varepsilon))$.

Iterating this process, there is a finite positive integer N_ε such that $X = \bigcup_{i \in \{1 \ldots N_\varepsilon\}} B(x_i, \varepsilon)$ for otherwise the sequence $\{x_n\}_{n \in \mathbb{N}}$ would satisfy $d(x_n, x_p) > \varepsilon$ for every $n, p \in \mathbb{N}$ and would therefore have no convergent subsequence. □

Proposition IX.8.30. *Each countably compact metric space is compact.*

Proof. Under these assumptions, the space is sequentially compact by Proposition IX.8.27, hence totally bounded by Proposition IX.8.29. Therefore if $\mathcal{U}$ is an ultrafilter on X, then for every $n \in \mathbb{N}$ there is a point x_n with $B(x_n, \frac{1}{n}) \in \mathcal{U}$, because $\bigcup_{x \in F_n} B(x, \frac{1}{n}) \in \mathcal{U}$ for some finite set F_n.

Moreover, the sequence $\{x_n\}_{n \in \mathbb{N}}$ has a convergent subsequence. If x is its limit and $\varepsilon > 0$, then there is $n \in \mathbb{N}$ such that $d(x_n, x) < \frac{\varepsilon}{2}$ and $\frac{1}{n} < \frac{\varepsilon}{2}$. Then $B(x_n, \frac{1}{n}) \subset B(x, \varepsilon)$ and $B(x_n, \frac{1}{n}) \in \mathcal{U}$ so that $x \in \lim \mathcal{U}$. □

Corollary IX.8.31. *A metric space is compact if and only if it is sequentially compact if and only if it is countably compact.*

Corollary IX.8.32 (Bolzano-Weierstrass). *Each bounded sequence of $\mathbb{R}^n$ has a convergent subsequence.*

Proof. If the sequence $\{x_k\}_{k \in N}$ is bounded, the set $\{x_k : k \in N\}$ is included in $\prod_{i=1}^{n} I_i$ for some closed intervals I_i of $\mathbb{R}$. This set is compact, as a product of compact sets. In view of Corollary IX.8.31, it is also sequentially compact, and the sequence $\{x_n\}_{n \in \mathbb{N}}$ has therefore a convergent subsequence. □

IX.9 Upper Kuratowski topology

By Proposition IX.7.6, the set of all isotone compact families fulfills the properties of open sets of a topology. We shall see that its restriction to the set of open sets coincides with the Scott topology. This will enable us to characterize open sets for the upper Kuratowski topology.

A family $\mathcal{D}$ of closed subsets is called *closedly isotone* if $D \in \mathcal{D}$ and $D \subset H$ is closed implies that $H \in \mathcal{D}$.

Theorem IX.9.1. *Assume that ξ is a T_1 topology. A subset $\mathcal{D}$ of $C(\xi, \$_0)$ is $[\xi, \$_0]$-closed if and only if $\mathcal{D}$ is ξ-closedly isotone and $\mathcal{D}^\#$ is a ξ-compact family.*

Proof. Notice that if $\mathcal{D}$ is $[\xi, \$_0]$-closed, $D \in \mathcal{D}$ and $D \subset H$ then $H \in \mathcal{D}$. Indeed, $H \in \lim_{[\xi, \$_0]} \{D\}^\uparrow$, because $\operatorname{adh}_\xi \operatorname{rdc} \left(\{D\}^\uparrow \right) = D \subset H$.

Suppose that $\mathcal{D}$ is ξ-closedly isotone and such $\mathcal{D}^\#$ is ξ-compact. Let $\mathfrak{F}$ be a filter on $C(\xi, \$_0)$ such that $\mathcal{D} \in \mathfrak{F}^\#$, that is, $\mathcal{D} \cap \mathcal{F} \neq \varnothing$ for each $\mathcal{F} \in \mathfrak{F}$. Hence there is $D \in \mathcal{D}$ such that $D \cap \bigcup_{A \in \mathcal{F}} A \neq \varnothing$ for each $\mathcal{F} \in \mathfrak{F}$, which means that $\mathcal{D} \# \operatorname{rdc}(\mathfrak{F})$ and thus $\operatorname{adh}_\xi \operatorname{rdc}(\mathfrak{F}) \in \mathcal{D}^{\#\#}$, because $\mathcal{D}^\#$ is ξ-compact. As $A \in \lim_{[\xi, \$_0]} \mathfrak{F}$ whenever $\operatorname{adh}_\xi \operatorname{rdc}(\mathfrak{F}) \subset A$ and $A \in C(\xi, \$_0)$, it follows that $A \in \mathcal{D}$.

Let $\mathcal{D}$ be $[\xi, \$_0]$-closed and let $\mathcal{F}$ be a filter such that $\mathcal{F} \# \mathcal{D}^\#$, that is, $\mathcal{F} \cap C(\xi, \$_0) \subset \mathcal{D}$. Let

$$\mathfrak{F} := \{\{A \in C(\xi, \$_0) : A \subset F\} : F \in \mathcal{F}\}.$$

As $[\xi, \$_0]$ is hypercompact, $\mathfrak{F}$ is convergent. As $\mathcal{D} \in \mathfrak{F}^\#$ and since $\mathcal{D}$ is $[\xi, \$_0]$-closed, $\lim_{[\xi, \$_0]} \mathfrak{F} \subset \mathcal{D}$, hence each $A \in C(\xi, \$_0)$ such that $\operatorname{adh}_\xi \operatorname{rdc}(\mathfrak{F}) \subset A$, belongs to $\mathcal{D}$. Therefore,

$$\operatorname{cl}_\xi (\operatorname{adh}_\xi \operatorname{rdc}(\mathfrak{F})) = \operatorname{cl}_\xi (\operatorname{adh}_\xi \mathcal{F}) = \operatorname{adh}_\xi \mathcal{F} \in \mathcal{D},$$

as $\operatorname{rdc}(\mathfrak{F}) = \mathcal{F}$ because ξ is T_1, hence $\operatorname{adh}_\xi \mathcal{F} \# \mathcal{D}^\#$. $\square$

Here is an immediate corollary concerning the Scott topology. Observe first that the three conditions of Proposition IX.7.6 are still valid if we consider merely openly *isotone families* of open sets.

Corollary IX.9.2. *Assume that ξ is a T_1 topology. A subset $\mathcal{A}$ of $C(\xi, \$_1)$ is $[\xi, \$_1]$-open if and only if $\mathcal{A}$ is a ξ-openly isotone ξ-compact family.*

IX.10   More on covers

Recall (Definition III.5.7) that a family $\mathcal{P}$ of subsets of $|\xi|$ is a ξ-cover of a set A

$$\mathcal{P} \succ_\xi A$$

if $\mathcal{P} \cap \mathcal{F} \neq \varnothing$ for each filter $\mathcal{F}$ such that $\lim_\xi \mathcal{F} \cap A \neq \varnothing$.

Exercise IX.10.1. Let ξ be a pretopology. Then the following statements are equivalent:

(1) $\mathcal{P} \succ_\xi A$,
(2) $\mathcal{V}_\xi(x) \cap \mathcal{P} \neq \varnothing$ for every $x \in A$,
(3) $A \subset \bigcup_{P \in \mathcal{P}} \mathrm{inh}_\xi P$.

Proposition IX.10.2. *Each cover in a topological space has a refinement that is an open cover.*

Proof. In fact, if ξ is a topology, then by Exercise IX.10.1, $\mathcal{P}$ is a ξ-cover of A if and only if $\bigcup_{P \in \mathcal{P}} \mathrm{int}_\xi P \supset X$; the family

$$\mathrm{int}_\xi^\natural \mathcal{P} = \{\mathrm{int}_\xi P : P \in \mathcal{P}\}$$

is such a refinement, because $\bigcup_{P \in \mathcal{P}} \mathrm{int}_\xi(\mathrm{int}_\xi P) = \bigcup_{P \in \mathcal{P}} \mathrm{int}_\xi P \supset A$. $\qquad\square$

Recall from (II.2.4) that $\mathcal{P}^\cup$ denotes the family of unions of finitely many elements of $\mathcal{P}$, so that $\mathcal{P} \subset \mathcal{P}^\cup$ and $\mathcal{P}$ is closed under finite unions if and only if $\mathcal{P} = \mathcal{P}^\cup$.

Proposition IX.10.3. *Let $\mathcal{P}$ be a family of open sets in a topological space. If $\mathcal{P}^\cup$ is a cover, then $\mathcal{P}$ is a cover.*

Proof. By assumption, for each x there is a finite subfamily $\mathcal{P}_0$ of $\mathcal{P}$ such that $\bigcup_{P \in \mathcal{P}_0} P \in \mathcal{N}(x)$. Hence there exists $P \in \mathcal{P}_0 \subset \mathcal{P}$ such that $x \in P$, and since P is open, $P \in \mathcal{N}(x)$. $\qquad\square$

Recall from Definition VIII.1.7 that a family $\mathcal{R}$ is an *ideal* if $S \subset R \in \mathcal{R}$ implies $S \in \mathcal{R}$, and if $\bigcup \mathcal{T} \in \mathcal{R}$ for each finite $\mathcal{T} \subset \mathcal{R}$, that is, $\mathcal{R}$ is an ideal on X if and only if $\mathcal{R}_c = \{X \setminus R : R \in \mathcal{R}\}$ is a (possibly degenerate) filter.

For each family $\mathcal{P}$ of subsets there exists the least ideal, denoted by $\mathcal{P}^{\cup\downarrow}$, that includes $\mathcal{P}$. In fact:

Exercise IX.10.4. Show that $R \in \mathcal{P}^{\cup\downarrow}$ whenever there exists a finite subset $\mathcal{P}_R$ of $\mathcal{P}$ such that $R \subset \bigcup_{P \in \mathcal{P}_R} P$.

Note that if $\mathcal{P}$ is a ξ-cover of A then $\mathcal{P}^{\cup\downarrow}$ is also a ξ-cover of A.

Exercise IX.10.5. Find a finite family $\mathcal{P}$ of elements of $\mathbb{R}$ such that $\mathcal{P}^{\cup\downarrow}$ is a cover (with respect to the usual topology) and $\mathcal{P}$ is not.

In view of Theorem III.5.12:

Corollary IX.10.6. *Let A be a subset of a convergence space X. A family $\mathcal{P}$ of subsets of X is an ideal cover of A if and only if $\mathcal{P}_c$ is a filter on X with* $\operatorname{adh}\mathcal{P}_c \cap A = \varnothing$.

Proposition III.5.10 rephrases in terms of filters as follows:

Corollary IX.10.7. *If $\mathcal{G}$ is a non-adherent filter on Y and X is a subset of Y, then $\mathcal{G}_X$ is a non-adherent (possibly degenerate) filter on X.*

Proof. $\mathcal{G}$ is a non-adherent filter on Y if and only if $\mathcal{G}_c := \{Y \setminus G : G \in \mathcal{G}\}$ is an ideal cover of Y, hence $(\mathcal{G}_c)_X = \{X \cap (Y \setminus G) : G \in \mathcal{G}\}$ is an ideal cover of X, and $X \cap (Y \setminus G) = X \setminus G$. Therefore, $((\mathcal{G}_c)_X)_c := \mathcal{G}_X$ is a non-adherent filter on X. $\qquad\square$

Definition IX.10.8. A family $\mathcal{S}$ of non-empty subsets of $|\xi|$ is called a *pseudocover* of $A \subset |\xi|$ if each ultrafilter that converges to an element of A contains an element of $\mathcal{S}$.

Proposition IX.10.9. *If $\mathcal{P}$ is a pseudocover, then $\mathcal{P}^{\cup}$ is a cover.*

Proof. If $\lim \mathcal{F} \neq \varnothing$ then $\lim \mathcal{U} \neq \varnothing$ for each $\mathcal{U} \in \beta\mathcal{F}$, thus $\mathcal{U} \cap \mathcal{P} \neq \varnothing$. In other words, for each $\mathcal{U} \in \beta\mathcal{F}$ there exists $P_{\mathcal{U}} \in \mathcal{P}$ such that $P_{\mathcal{U}} \in \mathcal{U}$. By Proposition II.6.5, there exists a finite subset $\mathbb{A}$ of $\beta\mathcal{F}$ such that $\bigcup_{\mathcal{U} \in \mathbb{A}} P_{\mathcal{U}} \in \mathcal{F}$, which means that $\mathcal{F} \cap \mathcal{P}^{\cup} \neq \varnothing$. $\qquad\square$

We can strengthen this proposition and the proof remains almost the same.

Proposition IX.10.10. *If $\mathcal{P}$ is a pseudocover, and $\mathcal{F}$ is a compactoid filter, then* $\mathcal{F} \cap \mathcal{P}^{\cup} \neq \varnothing$.

Exercise IX.10.11. Show that

(1) every cover is a pseudocover;
(2) if $\mathcal{P}^{\cup}$ is a pseudocover, then $\mathcal{P}$ is a pseudocover;
(3) if every adherent filter contains an element of $\mathcal{P}$, then $\mathcal{P}$ is a pseudocover;

(4) each ξ-pseudocover is a S ξ-pseudocover.

It follows that $\mathcal{P}$ is a pseudocover if and only if $\mathcal{P}^\cup$ is a pseudocover if and only if $\mathcal{P}^\cup$ is a cover.

Lemma IX.10.12. *If ξ is a convergence of finite depth ([7]) and $\mathcal{A}$ is a cover of a subset B of $|\xi|$, then $\mathcal{A} \cap \{x\}^\uparrow$ is a cover of $\{x\}$, for every $x \in B$.*

Proof. Indeed, if $x \in \lim_\xi \mathcal{F} \cap B$ then $x \in \lim_\xi \mathcal{F} \wedge \{x\}^\uparrow$ so that there is $A \in \mathcal{A} \cap (\mathcal{F} \wedge \{x\}^\uparrow)$. Thus $x \in A \in \mathcal{F} \wedge \{x\}^\uparrow \subset \mathcal{F}$ and $\mathcal{A} \cap \{x\}^\uparrow$ is a cover of $\{x\}$. $\square$

Note however that this property is not true of pseudocovers, for if $\mathcal{U}$ is an ultrafilter converging to x, the filter $\mathcal{U} \wedge \{x\}^\uparrow$ is generally not an ultrafilter. For this reason, we call a pseudocover $\mathcal{A}$ of a convergence ξ *point-complete* if for every $x \in |\xi|$ and every ultrafilter $\mathcal{U}$ with $x \in \lim_\xi \mathcal{U}$, there is $A \in \mathcal{A}$ with $x \in A \in \mathcal{U}$.

Lemma IX.10.13. *A set-theoretic locally finite cover of an open subset is a pseudocover. If the set-theoretic cover is moreover composed of closed sets, it is a point-complete pseudocover.*

Proof. Let B be ξ-open, and let $\mathcal{A}$ be a locally finite family of subsets of $|\xi|$ with $B \subset \bigcup_{A \in \mathcal{A}} A$. Let $\mathcal{U}$ be an ultrafilter such that

$$x \in \lim_\xi \mathcal{U} \cap B.$$

Since B is open, $B \in \mathcal{U}$. Since $\mathcal{A}$ is locally finite, there is $U \in \mathcal{U}$ with $U \subset B$ and $\{A \in \mathcal{A} : A \cap U \neq \varnothing\}$ finite, so that $U \subset \bigcup\{A \in \mathcal{A} : A \cap U \neq \varnothing\}$. Hence $\bigcup\{A \in \mathcal{A} : A \cap U \neq \varnothing\} \in \mathcal{U}$, so that there is an element A of $\mathcal{A}$ in $\mathcal{U}$. Thus $\mathcal{A}$ is a pseudocover of B.

If $\mathcal{A}$ is composed of closed sets, then $x \in A$, for $x \in \lim_\xi \mathcal{U}$ and $A \in \mathcal{U}$. Thus, the pseudocover $\mathcal{A}$ is also point-complete. $\square$

Proposition IX.10.14 (Pasting lemma). *If ξ and τ are two convergences, ξ has finite depth, $f : |\xi| \to |\tau|$, and $\mathcal{A}$ is a cover of ξ such that for each $A \in \mathcal{A}$, the restriction $f_{|A}$ is continuous, then f is continuous. If moreover τ is a pseudotopology, then $\mathcal{A}$ only need be a point-complete pseudocover.*

---

[7]or more generally a Kent space.

Proof. If $x \in \lim_\xi \mathcal{F}$, then in view of Lemma IX.10.12, there is $A \in \mathcal{A}$ with $x \in A \in \mathcal{F}$ so that $f(x) \in \lim_\tau f_{|A}[\mathcal{F}]$ by continuity of $f_{|A}$, hence $f(x) \in \lim_\tau f[\mathcal{F}]$. Thus f is continuous.

If $\mathcal{A}$ is only a point-complete pseudocover and τ is a pseudotopology, then to check that $f(x) \in \lim_\tau f[\mathcal{F}]$ it is enough to check that $f(x) \in \lim_\tau \mathcal{W}$ for every $\mathcal{W} \in \beta(f[\mathcal{F}])$. By Corollary II.6.7, for each $\mathcal{W} \in \beta(f[\mathcal{F}])$ there is $\mathcal{U} \in \beta(\mathcal{F})$ with $f[\mathcal{U}] \approx \mathcal{W}$. Since $\mathcal{U} \in \beta(\mathcal{F})$ and $x \in \lim_\xi \mathcal{F}$, we conclude that $x \in \lim_\xi \mathcal{U}$, so that there is $A \in \mathcal{A} \cap \mathcal{U}$ with $x \in A$ because $\mathcal{A}$ is a point-complete pseudocover. The argument above applies to the effect that $f(x) \in \lim_\tau f[\mathcal{U}] = \lim_\tau \mathcal{W}$. □

In particular,

Corollary IX.10.15. *Let $\mathcal{A}$ be a locally finite family of closed subsets of a convergence ξ such that $|\xi| = \bigcup_{A \in \mathcal{A}} A$, and let τ be a pseudotopology. Assume that for each $A \in \mathcal{A}$, a continuous map $f_A : |\xi_{|A}| \to |\tau|$ is defined in such a way that $f_A(x) = f_B(x)$ whenever $A, B \in \mathcal{A}$ and $x \in A \cap B$. Then there is a unique continuous map $f : |\xi| \to |\tau|$ such that $f_{|A} = f_A$ for every $A \in \mathcal{A}$.*

Proof. Under these conditions, there is a unique map $f : |\xi| \to |\tau|$ verifying $f_{|A} = f_A$ for every $A \in \mathcal{A}$. Moreover, by Lemma IX.10.13, $\mathcal{A}$ is a point-complete pseudocover, so that Proposition IX.10.14 applies to the effect that f is continuous. □

IX.11 Cover-compactness

Compactness can be expressed in terms of covers. For example, by definition, a subset A of a convergence space (X, ξ) is compact at $B \subset X$, whenever (IX.1.1),

$$A \in \mathcal{H}^\# \implies \mathrm{adh}_\xi \, \mathcal{H} \cap B \neq \varnothing$$

for every $\mathcal{H} \in \mathbb{F}X$.

Proposition IX.11.1. *A subset A of a convergence space (X, ξ) is compact at $B \subset X$ if and only if $A \in \mathcal{P}$ for every ideal cover $\mathcal{P}$ of B.*

Proof. A is compact at B if and only if

$$\mathrm{adh}_\xi \, \mathcal{H} \cap B = \varnothing \implies A \notin \mathcal{H}^\#$$

for every filter $\mathcal{H}$ on X; equivalently, if

$$\mathcal{H}_c \succ_\xi B \Longrightarrow A \in \mathcal{H}_c$$

for every filter $\mathcal{H}$ on X, by Theorem III.5.12. The conclusion follows from Corollary IX.10.6. □

Definition IX.11.2. A subset A of a convergence space (X, ξ) is *cover-compact at* $B \subset X$ if for every ideal cover $\mathcal{P}$ of B there is $P \in \mathcal{P}$ that is a cover of A.

A subset is *cover-compactoid* or *relatively cover-compact* if it is cover-compact at the whole underlying set of the convergence; it is *cover-compact* if it is cover-compact at itself.

The only difference between compactness and cover-compactness is that, in the first case $A \in \mathcal{P}$ and in the second, $P \succ_\xi A$ for some $P \in \mathcal{P}$.

Exercise IX.11.3. Show that the following are equivalent:

(1) A is ξ-cover-compact at B;
(2) Every ξ-cover of B has a finite subfamily that is a ξ-cover of A;
(3) For every $\mathcal{H} \in \mathbb{F}X$,

$$A\#(\mathrm{adh}_\xi^\natural \mathcal{H}) \Longrightarrow \mathrm{adh}_\xi \mathcal{H} \cap B \neq \varnothing. \tag{IX.11.1}$$

Cover-compactness is formally stronger than compactness:

Exercise IX.11.4. Let A and B be two subsets of a convergence space. Show that if A is cover-compact at B then A is compact at B.

However, compactness and cover-compactness coincide in topological spaces:

Proposition IX.11.5. *If ξ is a topology on X and A and B are two subsets of X, then A is ξ-cover-compact at B if and only if A is ξ-compact at B.*

Proof. If A is compact at B and $\mathcal{H}\#\mathcal{V}_\xi(A)$ then $\mathrm{adh}_\xi^\natural \mathcal{H}\#A$ so that $\mathrm{adh}_\xi(\mathrm{adh}_\xi^\natural \mathcal{H}) \cap B \neq \varnothing$. But if ξ is a topology, then

$$\mathrm{adh}_\xi \mathcal{H} = \bigcap_{H \in \mathcal{H}} \mathrm{cl}_\xi H = \mathrm{adh}_\xi(\mathrm{adh}_\xi^\natural \mathcal{H}).$$

Therefore, the filter $\mathcal{V}_\xi(A)$ is compact at B, that is, A is cover-compact at B. □

In particular,

Corollary IX.11.6. *Let X be a topological space. The following are equivalent:*

(1) X is compact;
(2) Every ultrafilter on X is convergent;
(3) Every cover of X has a finite subcover;
(4) Every open cover of X has a finite subcover;
(5) Every ideal open cover of X has a one element subcover.

However, in a convergence space, finite subsets, and in particular singletons, are always compact, but not necessarily cover-compact.

Example IX.11.7 (If $S\xi > S_0\xi$ then there is $x \in |\xi|$ such that $\{x\}$ is not cover-compact). Indeed, if $S\xi > S_0\xi$ then there is a filter $\mathcal{F}$ and a point x with

$$x \in \lim\nolimits_{S_0\xi} \mathcal{F} \setminus \lim\nolimits_{S\xi} \mathcal{F}.$$

In view of (V.1.6) and (VIII.3.1), this means that $x \in \mathrm{adh}_\xi A$ for every $A \in \mathcal{F}^\#$, but there is a filter $\mathcal{H}\#\mathcal{F}$ such that $x \notin \mathrm{adh}_\xi \mathcal{H}$. Hence $\{x\} \cap \mathrm{adh}_\xi \mathcal{H} = \varnothing$ so that $\mathcal{H}_c$ is a (ideal) cover of $\{x\}$. If $\{x\}$ were cover-compact, there would be $H \in \mathcal{H}$ such that $\{H^c\}$ covers $\{x\}$, that is, $\mathrm{adh}_\xi H \cap \{x\} = \varnothing$, which is not possible because $H \in \mathcal{H} \subset \mathcal{F}^\#$.

Thus, compactness and cover-compactness coincide among topologies, and do not in general. We now show that they do not among pretopological spaces.

Example IX.11.8 (A compact non-cover-compact subset of a pretopological space). Let π denote the bisequence pretopology on

$$\{x_\infty\} \cup \{x_n : n \in \mathbb{N}\} \cup \{x_{n,k} : n, k \in \mathbb{N}\}$$

of Example V.4.6. The subset $A := \{x_\infty\} \cup \{x_n : n \in \mathbb{N}\}$ is compact, because $x_\infty \in \lim_\pi (x_n)_n^\uparrow$ and each free ultrafilter on A is finer than $(x_n)_n$. But A is not cover-compact. Indeed, the filter

$$\mathcal{H} := \{\{x_{n,k} : k \in \mathbb{N}, n \geq m\} : m \in \mathbb{N}\}^\uparrow$$

has the property that $\mathrm{adh}\, H \cap A \neq \varnothing$ for all $H \in \mathcal{H}$ even though $\mathrm{adh}\,\mathcal{H} = \bigcap_{H \in \mathcal{H}} \mathrm{adh}\, H = \varnothing$ is disjoint from A. In view of (IX.11.1), A is not cover-compact.

Remarkably, cover-compactness is *not* preserved by continuous maps, unlike compactness (Proposition IX.1.26):

Example IX.11.9 (A cover-compact set whose continuous image is not cover-compact). In view of Example IX.11.7, if $\xi = \mathrm{S}\,\xi > \mathrm{S}_0\,\xi$ then there is $x \in |\xi|$ such that $\{x\}$ is not cover-compact. Moreover, in view of Corollary V.4.15, there is a continuous onto map $f : |\tau| \to |\xi|$ where τ is a topology. Thus, there is $t \in |\tau|$ with $f(t) = x$. Singletons are always compact, hence, by Proposition IX.11.5, singletons are cover-compact in τ. In particular, $\{t\}$ is cover-compact but $f(\{t\}) = \{x\}$ is not, even though f is continuous.

Remark IX.11.10. In topology a basic feature of compactness is its preservation by continuous maps. Various constructions in topology and analysis depend on this feature. Accordingly, categories of compact topologies have continuous maps as morphisms. The non-preservation of cover-compactness by continuous maps suggests that the concept of compactness is principal with respect to that of cover-compactness. Importance of cover-compactness in the framework of topological spaces was due to its coincidence with compactness in that framework.

Proposition IX.11.11. *Let ξ be a convergence X and let A and B be two subsets of X. Then A is ξ-cover-compact at B if and only if $\mathcal{V}_\xi(A)$ is compact at B.*

Proof. In view of Exercise IX.11.3, A is cover-compact at B if and only if (IX.11.1) holds. But, in view of Proposition VI.5.2, $A \# \mathrm{adh}_\xi^\natural \mathcal{F}$ if and only if $\mathcal{V}_\xi(A) \# \mathcal{F}$ and the conclusion follows. $\square$

We extend the definition to families via (IX.11.1):

Definition IX.11.12. Let ξ be a convergence on X and $\mathcal{A}$ and $\mathcal{B}$ are two families of subsets of X, we say that $\mathcal{A}$ is ξ-*cover-compact at* $\mathcal{B}$ if

$$\mathrm{adh}_\xi^\natural \mathcal{H} \# \mathcal{A} \implies \mathrm{adh}_\xi \mathcal{H} \# \mathcal{B}.$$

Proposition IX.11.11 extends to families:

Proposition IX.11.13. *Let ξ be a convergence on X. A family $\mathcal{A}$ on X is ξ-cover-compact at $\mathcal{B} \subset 2^X$ if and only if $\mathcal{V}_\xi(\mathcal{A})$ is ξ-compact at $\mathcal{B}$.*

Exercise IX.11.14. Prove Proposition IX.11.13.

It turns out that coincidence of compactness and cover-compactness for all filters happens essentially only in topologies:

Theorem IX.11.15. *Let ξ be a pseudotopology on X. The following statements are equivalent:*

(1) ξ is a topology;
(2) for every filter $\mathcal{F}$ on X,

$$\lim_\xi \mathcal{F} \subset \lim_\xi \mathcal{V}_\xi(\mathcal{F});$$

(3) a filter $\mathcal{F}$ is ξ-compact at A if and only if $\mathcal{F}$ is ξ-cover-compact at A.

Proof. (1) $\Longrightarrow$ (2): (2) follows from the fact that $\mathcal{V}_\xi(\cdot)$ is a selection whenever ξ is a (pre)topology, and that a topology is diagonal.

(2) $\Longrightarrow$ (3) is proved like its particular case, Corollary IX.11.5: it is clear that cover-compactness implies compactness. If now $\mathcal{F}$ is compact at A, we need to show that $\mathcal{V}_\xi(\mathcal{F})$ is compact at A as well, which follows readily from the fact that $\mathrm{adh}_\xi^\natural \mathcal{H} \# \mathcal{F}$ if and only if $\mathcal{H} \# \mathcal{V}_\xi(\mathcal{F})$ and $\mathrm{adh}_\xi \mathcal{H} = \mathrm{adh}_\xi(\mathrm{adh}_\xi^\natural \mathcal{H})$ because of (2).

(3) $\Longrightarrow$ (1): If $x \in \lim_\xi \mathcal{F}$ then, by Proposition IX.7.10, $\mathcal{F}$ is compact at $\{x\}$, so that, by (3), $\mathcal{F}$ is also cover-compact at $\{x\}$. In view of Proposition IX.11.13, $\mathcal{V}_\xi(\mathcal{F})$ is compact at $\{x\}$. Since ξ is a pseudotopology, we conclude by Proposition IX.7.10 that $x \in \lim_\xi \mathcal{V}_\xi(\mathcal{F})$.

In the case $\mathcal{F} = \{x\}^\uparrow$, we obtain that $x \in \lim_\xi \mathcal{V}_\xi(x)$ so that ξ is pretopological. In that case, $\mathcal{V}_\xi(\cdot)$ is a selection and $\lim_\xi \mathcal{F} \subset \lim_\xi \mathcal{V}_\xi(\mathcal{F})$ amounts to diagonality of ξ. Hence ξ is topological. $\square$

IX.12 Pseudocompactness

Of course, as compactness in Section IX.8, cover compactness has conditional variants. If $\mathbb{H}$ is a class of filters, then we say that A is $\mathbb{H}$-*cover-compact at B* if (IX.11.1)

$$A \# \mathrm{adh}_\xi^\natural \mathcal{H} \Longrightarrow \mathrm{adh}_\xi \mathcal{H} \cap B \neq \varnothing$$

for each $\mathcal{H} \in \mathbb{H}X$. In particular, A is said to be *countably cover-compact at B* if A is $\mathbb{F}_1$-cover-compact at B.

Proposition IX.12.1. *Each continuous real-valued function on a countably cover compact convergence is bounded.*

Proof. If X is a countably *(cover)* compact convergence space and $f \in C(X, \mathbb{R})$, then

$$X = \bigcup_{n \in \mathbb{N}} \{|f| < n\}.$$

As $\{|f| < n\}$ is open for each $n \in \mathbb{N}$, there is $n \in \mathbb{N}$ such that $X = \{|f| < n\}$. □

Definition IX.12.2. A convergence ξ is *pseudocompact* if each $f \in C(\xi, \mathbb{R})$ is bounded.

Of course, pseudocompactness is a topological notion, because it is defined merely with the aid of topological and set-theoretic notions.

In terms of Definition IX.12.2, each countably compact convergence is pseudocompact.

Theorem IX.12.3. *Each pseudocompact normal topology is countably compact.*

Proof. If a normal topology τ is not countably compact, then by Corollary IX.8.8 there exists a τ-closed discrete countably infinite subspace $A := \{x_n : n \in \mathbb{N}\}$ of $|\tau|$. The real-valued function $f : A \to \mathbb{R}$ defined by $f(x_n) := n$, is continuous, because A is discrete, hence by Theorem VII.9.1, there exists $\widehat{f} \in C(\tau, \mathbb{R})$ such that $\widehat{f}$ coincides with f on A and thus is unbounded. Therefore τ is not pseudocompact. □

We shall provide an example of a pseudocompact (functionally regular) topology which is not countably compact. To this end, we introduce *almost disjoint topologies*:

Let X be an infinite set and $\mathcal{A}$ an infinite AD family of subsets of X. Then $\mathcal{A} \cap X = \varnothing$. We define on $Y = \mathcal{A} \cup X$ a pretopology $\zeta(\mathcal{A})$ on Y by describing vicinities of elements of Y. Let $y \in Y$ and $S \subset Y$. Then

$$ S \in \mathcal{V}_{\zeta(\mathcal{A})}(y) $$

if and only if (1) if $y \in X$ then $y \in S$, (2) if $y \in \mathcal{A}$, hence there is a unique $A \in \mathcal{A}$ such that $y = A$, then $A \setminus S$ is finite and $A \in S \cap \mathcal{A}$ ([8]). The convergence $\zeta(\mathcal{A})$ is easily seen to be a topology, called *almost disjoint topology* the AD *topology* associated to $\mathcal{A}$.

Proposition IX.12.4. *If $\mathcal{A}$ is an infinite almost disjoint family on an infinite set, $\zeta(\mathcal{A})$ is a Hausdorff locally compact topology. Moreover, if X is countable, $\zeta(\mathcal{A})$ is sequentially based and separable.*

Proof. $\zeta(\mathcal{A})$ is a Hausdorff, because if $A_0, A_1 \in \mathcal{A}$ and $A_0 \neq A_1$ then $\{A_0\} \cup A_0$ and $\{A_1\} \cup (A_1 \setminus A_0)$ are disjoint vicinities of A_0 and A_1 respectively.

---

[8]In this definition the elements of $\mathcal{A}$ are regarded in two ways: as elements of the set Y, and as subsets of Y.

The set $\{A\} \cup A$ is $\zeta(\mathcal{A})$-open and $\zeta(\mathcal{A})$-compact for each $A \in \mathcal{A}$ and hence $\zeta(\mathcal{A})$ is a locally compact topology.

On the other hand, if $\{A\} = \lim_{\zeta(\mathcal{A})} \mathcal{F}$, for a free filter $\mathcal{F}$, then $(A)_0 \leq \mathcal{F}$. Thus, if X is countable, then $\zeta(\mathcal{A})$ is a sequentially based convergence. As X is dense in Y, this topology is also separable. $\qquad\square$

If $\mathcal{A}$ is *MAD*, then ζ is called it a *MAD topology* (over the cardinal card X).

Proposition IX.12.5. *An AD topology on an infinite set is not countably compact, because $\mathcal{A}$ is a closed discrete infinite subspace, but a MAD topology is pseudocompact.*

Proof. In fact, if f were a continuous unbounded function on Y, then for every $k \in \mathbb{N}$ there would be $x_k \in X$ such that $f(x_k) > k$, because X is dense. As $\mathcal{A}$ is *MAD*, there is $A \in \mathcal{A}$ such that $\{x_k : k \in \mathbb{N}\} \cap A$ is infinite, hence there is a subsequence $\{x_{k_p}\}_p$ that converges to A. Then $\lim_p f(x_{k_p}) = f(\{A\}) < \infty$, in contradiction with the definition of $\{x_k\}_k$. $\qquad\square$

By Theorem IX.12.3, a *MAD* topology is not normal.

Chapter X

Completeness in metric spaces

Several concepts of *completeness* occur in topology and convergence theory. Each variant of completeness is specified by the definition of *fundamental filters* (called also *Cauchy filters*). A space is *complete* if each fundamental filter has non-empty adherence.

In metric spaces, fundamental filters are those having elements of arbitrarily small diameters. It is a consequence that each adherent fundamental filter in a metric space is convergent. Therefore, a metric space is complete if and only if each fundamental filter converges. Completeness in metric spaces is examined in this introductory chapter.

A topology is called *completely metrizable* if there exists a compatible metric, with respect to which the space is complete. Completely metrizable topologies form a subclass of *Čech-complete* topologies, for which fundamental filters are defined in terms of countable collections of covers. This notion of completeness corresponds to *countable completeness,* a special case of a general concept of completeness in convergence spaces, to be explored in the Chapter XI.

X.1 Complete metric spaces

Definition X.1.1. A filter $\mathcal{F}$ on a metric space (X, d) is *fundamental* or *Cauchy* if for every $\varepsilon > 0$, there is $F_\varepsilon \in \mathcal{F}$ with diam $F_\varepsilon < \varepsilon$. A metric space is *complete* if every Cauchy filter has non-empty adherence.

Note that if $\mathcal{F} \leq \mathcal{G}$ and $\mathcal{F}$ is a Cauchy filter, so is $\mathcal{G}$. In particular, every ultrafilter of a Cauchy filter is also Cauchy. Since $\mathrm{adh}\,\mathcal{U} = \lim \mathcal{U}$ whenever $\mathcal{U}$ is an ultrafilter, every Cauchy ultrafilter in a complete space converges. Moreover, if every Cauchy ultrafilter converges, then every Cauchy filter has non-empty adherence. Thus,

Proposition X.1.2. *A metric space is complete if and only if every Cauchy ultrafilter converges.*

We shall see in Corollary X.1.6 that the characterization in Proposition X.1.2 extends to Cauchy filters.

As a metrizable convergence space can admit several compatible metrics, it can have several corresponding families of fundamental filters. As a result, a metrizable space can be complete with respect to one metric and incomplete with respect to another:

Example X.1.3. Let

$$X := \left\{ \frac{1}{n} : n \in \mathbb{N}_1 \right\}$$

be endowed with the discrete convergence. In Example III.4.4 we considered on X two metrics compatible with the discrete convergence, namely the discrete metric i and the metric d inherited from $\mathbb{R}$. If $0 < \varepsilon < 1$ then diam$_i$ $A < \varepsilon$ then A is a singleton. Therefore each i-fundamental filter contains a singleton, hence is a principal ultrafilter, which is of course convergent. On the other hand, each principal ultrafilter is d-fundamental. However the cofinite filter of X is also d-fundamental, because

$$\text{diam}_d \left\{ \frac{1}{n} : n > \tfrac{1}{\varepsilon} \right\} < \varepsilon.$$

Consequently, the metric i is complete, but the metric d is not, because the cofinite filter is d-fundamental but has empty adherence, as free filters do not converge in the discrete topology.

In the sequel in this section, we shall use the term *space* to mean *metric space*, and unless specified otherwise, will denote it by (X, d).

We can particularize Definition X.1.1 to sequences: a sequence $\{x_n\}_{n \in \mathbb{N}}$ of elements of X is called *Cauchy* (or *fundamental*) if the filter $(x_n)_n^{\uparrow}$ is, that is, if

$$\lim_{n \to \infty} \text{diam}\, \{x_k : k \geq n\} = 0, \tag{X.1.1}$$

equivalently, if for each $\varepsilon > 0$ there exists n_ε such that $d(x_n, x_m) < \varepsilon$ for each $n, m \geq n_\varepsilon$. In view of (X.1.1), each Cauchy sequence is bounded, that is, diam$\{x_n : n \in \mathbb{N}\} < \infty$, and each subsequence of a Cauchy sequence is a Cauchy sequence.

Proposition X.1.4. *Each convergent filter is a Cauchy filter. In particular, each convergent sequence is a Cauchy sequence.*

Proof. If $\lim \mathcal{F} = x$, then for every $\varepsilon > 0$, there is $F \in \mathcal{F}$ with $F \subset B(x, \frac{\varepsilon}{2})$ so that

$$\operatorname{diam} F \leq \operatorname{diam} B\left(x, \frac{\varepsilon}{2}\right) < \varepsilon$$

and $\mathcal{F}$ is Cauchy. □

On the other hand,

Lemma X.1.5. *A Cauchy filter with a non-empty adherence is convergent. In particular, each adherent Cauchy sequence is convergent.*

Proof. Let $x \in \operatorname{adh} \mathcal{F}$, so that $\mathcal{F} \# \mathcal{B}(x)$, and assume that $\mathcal{F}$ is a Cauchy filter. Given $\varepsilon > 0$, there is $F_\varepsilon \in \mathcal{F}$ with $\operatorname{diam} F_\varepsilon < \frac{\varepsilon}{2}$ because $\mathcal{F}$ is Cauchy, and there is $y \in F_\varepsilon \cap B(x, \frac{\varepsilon}{2}) \neq \varnothing$ because $\mathcal{F} \# \mathcal{B}(x)$. Then

$$F_\varepsilon \subset B(x, \varepsilon),$$

for if $t \in F_\varepsilon$ then

$$d(t, x) \leq d(t, y) + d(y, x) < \frac{\varepsilon}{2} + \frac{\varepsilon}{2} = \varepsilon.$$

□

Corollary X.1.6. *The following are equivalent for a metric space* (X, d):

(1) (X, d) *is complete;*
(2) every Cauchy ultrafilter converges;
(3) every Cauchy filter converges;
(4) every Cauchy sequence converges;
(5) every Cauchy sequence has non-empty adherence.

Proof. $(1) \iff (2)$ is Proposition X.1.2. $(1) \implies (3)$ follows from Lemma X.1.5 for by definition, every Cauchy filter in a complete metric space has non-empty adherence. $(3) \implies (4) \implies (5)$ are obvious. $(5) \implies (1)$: If $\mathcal{F}$ is a Cauchy filter, then for every $n \in \mathbb{N}_1$, there is F_n with $\operatorname{diam} F_n < \frac{1}{n}$, and we can assume that $F_{n+1} \subset F_n$ by replacing F_{n+1} by $F_{n+1} \cap F_n$. Picking $x_n \in F_n$ for each n thus yields a Cauchy sequence, which has non-empty adherence by (4), and thus is convergent by Lemma X.1.5. Thus the Cauchy filter $\{F_n : n \in \mathbb{N}_1\}^\uparrow$ has non-empty adherence, and is thus convergent, so that the finer filter $\mathcal{F}$ is convergent as well. □

In view of the equivalence between (1) and (5), a metric space is complete if and only if every countably based Cauchy filter has non-empty adherence. Since in a metric space $\operatorname{adh}\mathcal{F} = \operatorname{adh}(\operatorname{cl}^{\sharp}\mathcal{F})$ and every countably based filter has a decreasing countable filter-base (Exercise II.8.2), we can restrict ourselves to Cauchy filters generated by a decreasing sequence of non-empty closed sets, that is:

Corollary X.1.7 (Cantor's characterization of completeness). *A metric space is complete if and only if for any decreasing sequence $(F_n)_{n\in\mathbb{N}}$ of non-empty closed sets with $\lim_{n\to\infty}\operatorname{diam}(F_n) = 0$,*

$$\bigcap_{n\in\mathbb{N}} F_n \neq \varnothing.$$

As observed in Example X.1.3, a space can be metrizable by two different metrics only one of which is complete. As a result, completeness is not preserved by continuous maps, or even homeomorphisms. In other words, completeness in the sense of Definition X.1.1 is a metric notion, but not a convergence or topological notion.

Theorem X.1.8 (Cauchy). *The real line with its usual metric is complete.*

Proof. Let $\{x_n\}_{n\in\mathbb{N}}$ be a Cauchy sequence. As observed before, $\{x_n\}_{n\in\mathbb{N}}$ is bounded, hence admits a convergent subsequence, that is, has non-empty adherence. In view of Corollary X.1.6, the metric is complete. $\quad\square$

Exercise X.1.9. Show that

(1) Each closed subset of a complete metric space is a complete metric space.
(2) If a subset of a metric space is complete, then it is closed.

Example X.1.10. As a result of Exercise X.1.9, $\mathbb{Q}$ and $\mathbb{R}\setminus\mathbb{Q}$ with the metric induced by that of $\mathbb{R}$ are not complete, for they are not closed.

Exercise X.1.11. Show that each metric compatible with a compact convergence is complete.

Moreover,

Proposition X.1.12. *A metric space is compact if and only if the metric is complete and totally bounded.*

Proof. Let (X, d) be a metric space. If X is compact, d is complete by Exercise X.1.11. Moreover, X is totally bounded by Proposition IX.8.29, because a compact metric space is sequentially compact by Corollary IX.8.31. If (X, d) is totally bounded, then every ultrafilter is Cauchy. Indeed, given an ultrafilter $\mathcal{U}$, for every $\varepsilon > 0$, there is a finite set F with $X = \bigcup_{x \in F} B(x, \frac{\varepsilon}{2})$ so that there is $x \in F$ with $B(x, \frac{\varepsilon}{2}) \in \mathcal{U}$ because $\mathcal{U}$ is a grill, and hence $\mathcal{U}$ is Cauchy. By completeness, every ultrafilter is convergent, that is, X is compact. $\qquad\square$

Proposition X.1.13. *The metric*

$$d(x, y) := \sum_{n \in \mathbb{N}} \frac{1}{2^n} d_n(x, y) \qquad (\text{X.1.2})$$

is complete in the product $\prod_{n \in \mathbb{N}} X_n$ *if and only if* (X_n, d_n) *is complete for each* $n \in \mathbb{N}$.

Proof. If $X := \prod_{n \in \mathbb{N}} X_n$ is complete and $\{y_k\}_k$ is a Cauchy sequence in X_n, then there exists a Cauchy sequence $\{f_k\}_k$ in X such that $f_k(n) = y_k$ for each k. Indeed, if $a \in X$ is arbitrary, then $\{f_k\}_k$ defined by

$$f_k(m) := \begin{cases} a(m) \text{ if } m \neq n \\ y_k \text{ if } m = n \end{cases}$$

has the required property. As X is complete $\{f_k\}_k$ is convergent. Therefore $\{y_k\}_k$ is convergent by the continuity of π_n, because $\pi_n(f_k) = f_k(n) = y_k$ for each k. Conversely, if $(f_k)_k$ is a Cauchy sequence in X, then by (X.1.2), $\{f_k(n)\}_k$ is a Cauchy sequence in X_n for each n, hence $x_n = \lim_{k \to \infty} (f_k(n))_k^{\uparrow}$ exists. Therefore $(\lim_{k \to \infty} f_k)(n) = x_n$ for each n. $\qquad\square$

An important classical result of Analysis that hinges on completeness in the metric setting is the Banach Fixed Point Theorem below.

Definition X.1.14. Let $k > 0$. A function $f : (X, d_X) \to (Y, d_Y)$ is *k-Lipschitz* if for every $x, t \in X$,

$$d_Y(f(x), f(t)) \leq k \cdot d_X(x, t).$$

As usual, if $f : X \to X$, we define the *iterates* of f by induction: $f^1 = f$ and $f^{n+1} = f \circ f^n$. A simple induction shows that if $f : X \to X$ is k-Lipschitz and $n \in \mathbb{N}_1$, then f^n is k^n-Lipschitz.

Theorem X.1.15 (Banach Fixed Point Theorem). *If* (X, d) *is a complete metric space,* $0 < k < 1$, *and* $f : X \to X$ *is k-Lipschitz, then* f *has a unique fixed point* x^*. *Moreover,*

$$x^* = \lim_{n \to \infty} f^n(x_0),$$

for every $x_0 \in X$ and

$$d(f^n(x_0), x^*) \leq \frac{k^n}{1-k} d(x_0, f(x_0)). \tag{X.1.3}$$

Proof. Note that for any $x, t \in X$, the triangular inequality yields

$$d(x, t) \leq d(x, f(x)) + d(f(x), f(t)) + d(f(t), t)$$
$$\leq d(x, f(x)) + k \cdot d(x, t) + d(f(t), t)$$

because f is k-Lipschitz. Solving for $d(x, t)$ gives

$$d(x, t) \leq \frac{d(x, f(x)) + d(t, f(t))}{1-k}. \tag{X.1.4}$$

Note that if x and t are two fixed points in (X.1.4), then $d(x, t) \leq 0$ so that f has at most one fixed point. To show that for any $x_0 \in X$, the sequence $\{f^n(x_0)\}_n$ converges, we only need to show that it is Cauchy, for (X, d) is complete. To this end, for $m, n \in \mathbb{N}_2$, let $x = f^n(x_0)$ and $t = f^m(x_0)$ in (X.1.4), so that, noting that $f^n \circ f = f \circ f^n$:

$$d(f^n(x_0), f^m(x_0)) \leq \frac{d(f^n(x_0), f^n(f(x_0))) + d(f^m(x_0), f^m(f(x_0)))}{1-k}.$$

Since f^n and f^m are respectively k^n-Lipschitz and k^m-Lipschitz, this yields

$$d(f^n(x_0), f^m(x_0)) \leq \frac{k^n + k^m}{1-k} d(x_0, f(x_0)). \tag{X.1.5}$$

Since $0 < k < 1$, $\lim_{n \to \infty} k^n = 0$ so that $d(f^n(x_0), f^m(x_0))$ can be made arbitrarily small by taking m, n sufficiently large, that is, $\{f^n(x_0)\}_n$ is Cauchy, hence convergent. Let $x^* := \lim_{n \to \infty} f^n(x_0)$. Since f is continuous, $f(x^*) = \lim_{n \to \infty} f \circ f^n(x_0) = x^*$ and x^* is a (unique) fixed point. Finally, letting $m \to \infty$ in (X.1.5), the continuity of d yields (X.1.3). $\qquad\square$

X.2   Completely metrizable spaces

Definition X.2.1. A space is said to be *completely metrizable* provided that there exists a compatible metric for which the space is complete.

Recall from Example X.1.3 that the same topology may admit many compatible metrics, some of which are complete some of which are not.

Theorem X.2.2 (Baire). *The intersection of a sequence of open dense subsets of a completely metrizable space is dense.*

Proof. Let d be a complete compatible metric on X, and let $(A_n)_n$ be a sequence of open sets with $\operatorname{cl} A_n = X$ for all $n \in \mathbb{N}$. To show that $\bigcap_{n \in \mathbb{N}} A_n$ is dense, we will show that for every $x \in X$ and $r > 0$, the ball $B(x, r)$ contains an element of $\bigcap_{n \in \mathbb{N}} A_n$. Since A_0 is dense and $B(x, r)$ is open, the intersection $A_0 \cap B(x, r)$ is non-empty, and open for A_0 is open. Thus, there is $x_1 \in A_0$ and $0 < r_1 < \frac{1}{2} r$ such that

$$\operatorname{cl} B(x_1, r_1) \subset A_0 \cap B(x, r).$$

Now if we have defined $r_0 := r > r_1 > \ldots > r_n > 0$ and $x = x_0, x_1, \ldots, x_n$ such that $\frac{1}{2} r_k \geq r_{k+1}$ and

$$\operatorname{cl} B(x_{k+1}, r_{k+1}) \subset A_k \cap B(x, r_k) \tag{X.2.1}$$

for all $0 \leq k < n$, there are x_{n+1} and $0 < r_{n+1} < \frac{1}{2} r_n$ satisfying (X.2.1) for $k = n$, because A_n is dense and open. Thus $F_n := \operatorname{cl} B(x_n, r_n)$ defines a decreasing sequence of non-empty closed subsets of X with

$$\lim_{n \to \infty} \operatorname{diam}(F_n) = 0.$$

By Corollary X.1.7,

$$\varnothing \neq \bigcap_{n \in \mathbb{N}} F_n \subset \bigcap_{n \in \mathbb{N}} A_n \cap B(x, r),$$

which concludes the proof. $\qquad\square$

Corollary X.2.3 (Baire). *If $(F_n)_n$ is a sequence of closed subsets of a completely metrizable space X such that*

$$\bigcup_{n \in \mathbb{N}} F_n = X,$$

then there is n such that $\operatorname{int} F_n \neq \varnothing$.

Proof. Since $X \setminus F_n$ is open for every n, and

$$\bigcap_{n \in \mathbb{N}} (X \setminus F_n) = X \setminus \bigcup_{n \in \mathbb{N}} F_n = \varnothing,$$

Theorem XI.10.1 applies to the effect that at least one of the sets $X \setminus F_n$ is not dense, so that

$$\varnothing \neq X \setminus \operatorname{cl}(X \setminus F_n) = \operatorname{int} F_n.$$

$\qquad\square$

A *Baire space* is one in which any sequence of open dense subsets has a dense intersection. In other words, Theorem XI.10.1 states that completely metrizable spaces are Baire. We will significantly generalize this result in the next chapter. The Baire property is a particularly fruitful and versatile tool that finds applications in many areas of mathematics. Some of these applications are explored in the companion online document. A more extensive account of applications can be found for instance in [12].

We already have seen that the space $\mathbb{Q}$ of rational numbers with the metric inherited from the usual metric of $\mathbb{R}$ is not complete. Moreover,

Corollary X.2.4. *The space $\mathbb{Q}$ of rational numbers is not completely metrizable.*

Proof. Since $\mathbb{Q}$ is countable, it is a countable union of singletons, which are closed non-empty subset of empty interior. In view of Corollary X.2.3, it is not completely metrizable.   □

X.3   Metric spaces of continuous functions

Recall that a map $f : X \to Y$ where (Y, d) is a metric space is *bounded* if $f(X)$ is bounded, that is, $\operatorname{diam} f(X) < \infty$. If (X, d_X) and (Y, d_Y) are metric spaces, then

$$D(f, g) := \sup_{x \in X} d_Y(f(x), g(x)) \qquad (\text{X.3.1})$$

is a metric on the set $C_b(X, Y)$ of continuous bounded maps on X.

Lemma X.3.1. *If $(f_n)_n$ is a sequence of elements of $C_b(X, Y)$ and $f_\infty : X \to Y$ is such that*

$$\lim_{n \to \infty} D(f_n, f_\infty) = 0,$$

then $f_\infty \in C_b(X, Y)$.

Proof. Let $\varepsilon > 0$ and $x_0 \in X$. Then there exists $n \in \mathbb{N}$ such that

$$\sup_{x \in X} d_Y(f_n(x), f_\infty(x)) < \varepsilon.$$

As f_n is continuous, there exists $V \in \mathcal{N}_X(x_0)$ such that $d_Y(f_n(x_0), f_n(x)) < \varepsilon$ for each $x \in V$. Therefore, if $x \in V$, then

$$\begin{aligned} d_Y(f_\infty(x_0), f_\infty(x)) &\leq d_Y(f_\infty(x_0), f_n(x_0)) \\ &+ d_Y(f_n(x_0), f_n(x)) + d_Y(f_n(x), f_\infty(x)) < 3\varepsilon, \end{aligned} \qquad (\text{X.3.2})$$

proving the continuity of f_∞. Moreover, $f_\infty \in C_b(X, Y)$ for if x and x_0 are arbitrary points in X and n is as above, (X.3.2) yields that

$$\operatorname{diam} f_\infty(X) \le \operatorname{diam} f_n(X) + 2\varepsilon.$$

$\square$

Proposition X.3.2. *The space $C_b(X, Y)$ with the metric (X.3.1) is complete if and only if (Y, d_Y) is complete.*

Proof. Let (Y, d_Y) be complete and let $\{f_n\}_n$ be a D-fundamental sequence. Then for each $\varepsilon > 0$ there exists n_ε such that

$$\sup_{x \in X} d_Y\left(f_m(x), f_n(x)\right) < \varepsilon \qquad (\text{X.3.3})$$

for $n, m \ge n_\varepsilon$. Hence, for each $x \in X$, the sequence $(f_n(x))_n$ is fundamental in (Y, d_Y), thus convergent, as d_Y is complete. Let $f_\infty(x) := \lim_{n \to \infty} f_n(x)$. It follows that

$$\sup_{x \in X} d_Y\left(f_\infty(x), f_n(x)\right) \le \varepsilon$$

for each $n \ge n_\varepsilon$, so that $\{f_n\}_n$ converges to f_∞ for D. If (Y, d_Y) is not complete and $\{y_n\}_n$ is fundamental non-convergent sequence in Y, then $\{f_n\}_n$ defined by $f_n(x) := y_n$ for each $x \in X$ and $n \in \mathbb{N}$ is fundamental and non-convergent in $C_b(X, Y)$. $\square$

Each metric space is isometric to a subspace of a complete metric space:

Theorem X.3.3. *Each metric space is isometric to a subspace of $C_b(X, \mathbb{R})$ endowed with (X.3.1).*

Proof. If (X, d) is a metric space and $a \in X$, then for each $x \in X$, the map

$$f_x(y) := d(y, x) - d(y, a)$$

is continuous. As $|d(y, x) - d(y, a)| \le d(x, a)$, the function $f_x : X \to \mathbb{R}$ is bounded. Hence $x \mapsto f_x$ maps X to $C_b(X, \mathbb{R})$. We shall see that $D(f_x, f_z) = d(x, z)$ for every $x, z \in X$. Indeed, $|d(y, x) - d(y, z)| \le d(x, z)$ for each y, x and z, so that

$$D(f_x, f_z) = \sup_{y \in X} |d(y, x) - d(y, z)| \le d(x, z).$$

On the other hand, if $y := z$ then $|d(y, x) - d(y, z)| = d(x, z)$. $\square$

A complete metric space (Y, D) is called a *completion* of a metric space (X, d) if X is densely isometrically embedded in (Y, D).

Corollary X.3.4. *Each metric space has a completion.*

Proof. Let Y denote the closure (in $C_b(X, \mathbb{R})$) of the isometric image of X in $C_b(X, \mathbb{R})$ as in Theorem X.3.3. As a closed subspace of a complete metric space, it is a complete metric space and (X, d) is densely isometrically embedded in (Y, D). □

It turns out that a completion of a metric space is unique up to isometry. The next section is in part devoted to proving this fact.

X.4   Uniform continuity, extensions, and completion

Let us return for a moment to Corollary VII.8.6. Since metrizable spaces are regular and Hausdorff (in fact, hereditarily normal, by Proposition VII.7.10), if X is a topological space, (Y, d) a metric space, and $f : X \supset S \to Y$ is a continuous partial map, then by Corollary VII.8.6, f has a unique continuous extension to

$$h(S, f) = \{x \in \operatorname{cl} S : \lim_Y f[\mathcal{N}(x) \vee S] \neq \varnothing\}.$$

In order to describe this set, we introduce the notion of *oscillation:*

Definition X.4.1. Let X be a topological space, $S \subset X$ and (Y, d) a metric space. The *oscillation* of $f : S \to Y$ at $x \in \operatorname{cl} S$ is

$$\operatorname{osc} f(x) := \inf_{U \in \mathcal{O}(x)} \operatorname{diam}(f(S \cap U)).$$

Exercise X.4.2. Show that if $f : S \to Y$ is continuous at $x \in S$, then $\operatorname{osc} f(x) = 0$.

Proposition X.4.3. *Let X be a topological space, $S \subset X$ and (Y, d) a complete metric space, and let $f : S \to Y$. Then*

$$h(S, f) = \{x \in \operatorname{cl} S : \operatorname{osc} f(x) = 0\}.$$

Proof. Clearly, $\operatorname{osc} f(x) = 0$ if and only if $f[\mathcal{N}(x) \vee S]$ is a Cauchy filter on (Y, d). Thus, if $\operatorname{osc} f(x) = 0$ then by completeness of Y, $f[\mathcal{N}(x) \vee S]$ converges and $x \in h(S, f)$. Conversely, if $\operatorname{osc} f(x) \neq 0$, then $f[\mathcal{N}(x) \vee S]$ is not Cauchy, *a fortiori* not convergent, and $x \notin h(S, f)$. □

Note that as a consequence,

$$h(S, f) = \bigcap_{n \in \mathbb{N}_1} \left\{x \in \operatorname{cl} S : \operatorname{osc} f(x) < \frac{1}{n}\right\}$$

is a G_δ-subset of X, because if osc $f(x) < \frac{1}{n}$ then there is $U \in \mathcal{O}(x)$ such that diam $f(U \cap S) < \frac{1}{n}$ so that osc $f(t) < \frac{1}{n}$ for all $t \in U$.

Proposition X.4.4. *Let X be a topological space, $S \subset X$ and (Y, d) a complete metric space. Every continuous $f : S \to Y$ has a unique continuous extension to $h(S, f)$.*

We shall investigate when $h(S, f) = \operatorname{cl} S$, that is, when osc $f(x) = 0$ for all $x \in \operatorname{cl} S$.

Definition X.4.5. A map $f : X \to Y$ between two metric spaces (X, d_X) and (Y, d_Y) is *uniformly continuous* if for every $\varepsilon > 0$, there is $\delta > 0$ such that for all $x, t \in X$,

$$d_X(x, t) < \delta \Longrightarrow d_Y(f(x), f(t)) < \varepsilon.$$

Exercise X.4.6. Verify that osc $f(x) = 0$ for all $x \in \operatorname{cl} S$ if f is uniformly continuous on S.

Every Lipschitz function (in the sense of Definition X.1.14) is uniformly continuous, and by definition, every uniformly continuous function is continuous. The three classes of functions are however distinct:

Exercise X.4.7. Verify that:

(1) The function $f(x) = x^2$ is continuous on $\mathbb{R}$ but it is not uniformly continuous;
(2) The function $f(x) = \sqrt{x}$ is uniformly continuous but not Lipschitz on $[0, \infty)$.

However, continuous and uniformly continuous maps coincide on compact subsets:

Theorem X.4.8. *Let (X, d_X) and (Y, d_Y) be two metric spaces, let K be a compact subset of X. If $f : K \to Y$ is continuous, then f is uniformly continuous.*

Proof. Let $\varepsilon > 0$. By continuity at each $x \in K$, there is δ_x with

$$f\big(B(x, \delta_x)\big) \subset B\left(f(x), \frac{\varepsilon}{2}\right). \tag{X.4.1}$$

Since $\{B(x, \frac{\delta_x}{2}) : x \in K\}$ is an open cover of the compact topological space K, there is a finite subset F of K with $K \subset \bigcup_{x \in F} B(x, \frac{\delta_x}{2})$. Let

$\delta := \min_{x \in F} \frac{\delta_x}{2}$. Let t, s be such that $d_X(t, s) < \delta$. There is $x \in F$ with $t \in B(x, \frac{\delta_x}{2})$. Then

$$d_X(x, s) \le d_X(x, t) + d_X(t, s) \le \frac{\delta_x}{2} + \delta \le \delta_x,$$

so that $t, s \in B(x, \delta_x)$ and thus

$$d_Y(f(t), f(s)) \le d_Y(f(t), f(x)) + d_Y(f(x), f(s)) \le \frac{\varepsilon}{2} + \frac{\varepsilon}{2} = \varepsilon,$$

which completes the proof. $\qquad\qquad\qquad\qquad\qquad\qquad\qquad\qquad\qquad\square$

Returning to the extension problem, we have:

Corollary X.4.9. *Let (X, d_X) and (Y, d_Y) be two metric spaces, let $S \subset X$ and let $f : S \to Y$ be uniformly continuous. If (Y, d) is complete, then f has a unique continuous extension $\bar{f} : \operatorname{cl} S \to Y$. Moreover, $\bar{f}$ is uniformly continuous.*

Proof. That there is a unique continuous extension $\bar{f} : \operatorname{cl} S \to Y$ follows from Corollary VII.8.6, Proposition X.4.3, and Exercise X.4.6. To see that $\bar{f}$ is uniformly continuous, take $\varepsilon > 0$ and $\delta > 0$ such that

$$d_X(x, t) < \delta \implies d_Y(f(x), f(t)) < \frac{\varepsilon}{2} \qquad\qquad (\text{X.4.2})$$

for every $x, t \in S$. If now $z, w \in \operatorname{cl} S$ with $d_X(z, w) < \delta$, taking $r := \frac{\delta - d_X(z,w)}{3}$ ensures that $U := B(z, r) \cup B(w, r)$ has diameter less than δ so that (X.4.2) yields $\operatorname{diam} f(U \cap S) < \frac{\varepsilon}{2}$. Thus, $\operatorname{diam} \operatorname{cl}_Y(f(U \cap S)) \le \frac{\varepsilon}{2} < \varepsilon$. Moreover, $\bar{f}(w), \bar{f}(z) \in \operatorname{cl}_Y(f(U \cap S))$ so that $d_Y(\bar{f}(w), \bar{f}(z)) < \varepsilon$. $\quad\square$

Proposition X.4.10. *If in Corollary X.4.9 both X and Y are complete and f is an isometry from S onto $f(S)$, then $\bar{f}$ is an isometry from $\operatorname{cl}_X S$ to $\operatorname{cl}_Y f(S)$.*

Proof. Let $f : S \to f(S) \subset Y$ be an isometry. Then f is uniformly continuous, so that Corollary X.4.9 applies to the effect that there is a unique (uniformly) continuous extension $\bar{f} : X \to Y$ of f. Moreover, if $y \in \operatorname{cl}_Y f(S)$, there is $\mathcal{F} \in \mathbb{F}(f(S))$ with $y \in \lim_Y \mathcal{F}$. Since $f : S \to f(S)$ is an isometry, so is the inverse map $f^- : f(S) \to S$, so that $f^-[\mathcal{F}]$ is a Cauchy filter on $S \subset X$. Thus, there is $x \in \operatorname{cl}_X S \cap \lim_X f^-[\mathcal{F}]$ by completeness of X. By definition of $\bar{f}$, $\bar{f}(x) = y$. Thus $\bar{f}(\operatorname{cl}_X S) = \operatorname{cl}_Y f(S)$. Finally, $\bar{f}$ is an isometry: if $x, t \in \operatorname{cl}_X S$,

$$d_Y(\bar{f}(x), \bar{f}(t)) = d_Y(\lim_Y f[\mathcal{N}(x) \vee S], \lim_Y f[\mathcal{N}(t) \vee S])$$
$$= \lim_{\mathbb{R}} d_Y(f[\mathcal{N}(x) \vee S] \times f[\mathcal{N}(t) \vee S])$$

by continuity of $d_Y : Y \times Y \to \mathbb{R}$. Moreover,

$$d_Y \left(f[\mathcal{N}(x) \vee S] \times f[\mathcal{N}(t) \vee S] \right) = d_X \left((\mathcal{N}(x) \vee S) \times (\mathcal{N}(t) \vee S) \right)$$

because $f : S \to f(S)$ is an isometry. Since, $d_X : X \times X \to \mathbb{R}$ is continuous, we conclude that

$$d_Y(\bar{f}(x), \bar{f}(t)) = d_X(x,t).$$

$\square$

Corollary X.4.11. *Let (X, d_X) and (Y, d_Y) be two complete metric spaces, let S be a dense subset of X and T a dense subset of Y. Every isometry from S onto T (with metrics induced from d_X and d_Y respectively) extends to an isometry from X onto Y.*

In particular,

Corollary X.4.12. *The completion of a metric space is unique up to isometry.*

Proposition X.4.13. *If a subspace of a metrizable space is completely metrizable, then it is a G_δ-subset.*

Proof. Let X be a metrizable space. If $S \subset X$ is completely metrizable, then the identity map $i : S \to (S, d)$ where d is a complete metric on S is continuous (but not necessarily uniformly). By Proposition X.4.4, there is a G_δ-subset T of $\mathrm{cl}_X S$, hence of X, and a continuous extension $\bar{\imath} : T \to S$. Moreover, $T = S$ for if $x \in T \setminus S$, then $\bar{\imath}(x) = \lim_S i[\mathcal{N}(x) \vee S]$ would not belong to S. Thus, S is a G_δ-subset of X. $\square$

Moreover,

Lemma X.4.14. *Let X be a metrizable space. Every open subset of X is homeomorphic to a closed subset of $X \times \mathbb{R}$, and every G_δ-subset of X is homeomorphic to a closed subspace $X \times \mathbb{R}^{\mathbb{N}}$.*

Proof. If $O \subset (X, d)$ and $F := X \setminus O$, then $\mathrm{dist}_F : X \to \mathbb{R}$ is a continuous function, so that the map $g : X \to X \times \mathbb{R}$ defined by $g(x) = (x, \mathrm{dist}_F(x))$, is an embedding whose image is a closed subset of $X \times \mathbb{R}$. Moreover, $g(O) = g(X) \cap X \times (0, \infty)$ is the homeomorphic image of O and is closed in $X \times (0, \infty)$, which is itself homeomorphic to $X \times \mathbb{R}$. If now $T = \bigcap_{i \in \mathbb{N}} O_i$ is a G_δ-subset of X, then $X \setminus T = \bigcup_{i \in \mathbb{N}} F_i$ with $F_i := X \setminus O_i$. Let $f : X \times \mathbb{N} \to \mathbb{R}$ be defined by $f(x, n) = \mathrm{dist}_{F_n}(x)$. The map $g : X \to X \times \mathbb{R}^{\mathbb{N}}$ defined by

$$g(x) = (x, f(x, \cdot))$$

is an embedding of X as a closed subspace of $X \times \mathbb{R}^{\mathbb{N}}$ and $g(T) = g(X) \cap X \times (0, \infty)^{\mathbb{N}}$ is the homeomorphic image of T and is closed in $X \times (0, \infty)^{\mathbb{N}}$, which is itself homeomorphic to $X \times \mathbb{R}^{\mathbb{N}}$. $\square$

Proposition X.4.13 and Lemma X.4.14 lead to:

Theorem X.4.15 (Alexandrov). *A subspace of a completely metrizable space is completely metrizable if and only if it is a G_δ-subset.*

Proof. That a completely metrizable subspace is G_δ is Proposition X.4.13. If now S is a G_δ-subset of a completely metrizable space X, by Lemma X.4.14, S is homeomorphic to closed subset of a product of countably many completely metrizable spaces, which is completely metrizable, by Proposition X.1.13. Thus S is completely metrizable. $\square$

We will also generalize this result to countable completeness in the next chapter. In view of Corollary X.2.4 and Theorem X.4.15, $\mathbb{Q}$ is not a G_δ-subset of $\mathbb{R}$.

Corollary X.4.16. *The set $\mathbb{R} \setminus \mathbb{Q}$ of irrational numbers is completely metrizable.*

Proof. $\mathbb{R} \setminus \mathbb{Q} = \bigcap_{q \in \mathbb{Q}} (\mathbb{R} \setminus \{q\})$, is a G_δ-subset of the complete metric space $\mathbb{R}$, and the result follows from Theorem X.4.15. $\square$

Chapter XI

Completeness

We shall now discuss a concept of completeness that embraces compactness, local relative compactness, metric completeness and its generalization called topological completeness or Čech completeness.

XI.1 Completeness with respect to a collection

Let $\mathbb{P}$ be a collection of families of subsets of a set X. A filter $\mathcal{F}$ on X is said to be $\mathbb{P}$-*fundamental* if

$$\bigvee_{\mathcal{P} \in \mathbb{P}} \mathcal{F} \cap \mathcal{P} \neq \varnothing. \tag{XI.1.1}$$

The set of $\mathbb{P}$-fundamental filters is denoted by $\mathbb{F}_{\mathbb{P}}$.

Definition XI.1.1. Let $\mathbb{P}$ be a collection on X. A convergence ξ on X is called $\mathbb{P}$-*complete* if $\mathrm{adh}_\xi \, \mathcal{F} \neq \varnothing$ for every $\mathbb{P}$-fundamental filter $\mathcal{F}$.

Recall that (Section IX.8) for a given collection $\mathbb{H}$ of filters on a set X, a convergence ξ on X is said to be $\mathbb{H}$-compact (that is, conditionally compact with respect to $\mathbb{H}$) if $\mathrm{adh}_\xi \, \mathcal{H} \neq \varnothing$ for every filter $\mathcal{H} \in \mathbb{H}$.

We notice that completeness with respect to a collection can be viewed as conditional compactness. Indeed, it is straightforward that

Proposition XI.1.2. *A convergence is $\mathbb{P}$-complete if and only if it is $\mathbb{F}_{\mathbb{P}}$-compact.*

Exercise XI.1.3. Let $\mathbb{P}$ be a collection on X. Show that

(1) If a filter $\mathcal{F}$ is $\mathbb{P}$-fundamental, then each $\mathcal{U} \in \beta\mathcal{F}$ is $\mathbb{P}$-fundamental.
(2) A convergence ξ on X is $\mathbb{P}$-complete if and only if each $\mathbb{P}$-fundamental ultrafilter is ξ-convergent.

(3) A convergence ξ on X is $\mathbb{P}$-complete if and only if every $\mathbb{P}$-fundamental filter $\mathcal{F}$ is ξ-compactoid (Definition IX.7.1).

In the same vein, if $\mathcal{P}$ is a family of subsets of $|\xi|$, then we say that a convergence ξ is $\mathcal{P}$-*complete* if ξ is $\{\mathcal{P}\}$-complete ([1]).

Example XI.1.4. If X is a non-empty set, then the discrete convergence ι_X is complete with respect to $\{\{x\} : x \in X\}$. Indeed, a filter is $\{\{x\} : x \in X\}$-fundamental whenever it is a principal ultrafilter, hence convergent. For the same reason, ι_X is $[X]^{<\omega}$-complete.

Example XI.1.5. If (X, d) is a metric space and

$$\mathcal{B}_n := \left\{ B_d\left(x, \frac{1}{n}\right) : x \in X \right\}$$

with $n \in \mathbb{N}_1$, then a filter $\mathcal{F}$ is $\{\mathcal{B}_n : n \in \mathbb{N}_1\}$-fundamental (or *Cauchy*) if $\inf\{\operatorname{diam}(F) : F \in \mathcal{F}\} = 0$ (see Definition X.2.1).

Example XI.1.6. A compact convergence ξ is complete with respect to every collection, because each filter is ξ-adherent.

Exercise XI.1.7. Let ξ be a convergence and let $\mathbb{P}$ be a collection on $|\xi|$. Check that the following statements are equivalent:

(1) ξ is $\mathbb{P}$-complete,
(2) ξ is $\mathbb{P}_{\cup} := \{\mathcal{P}^{\cup} : \mathcal{P} \in \mathbb{P}\}$-complete,
(3) ξ is $\mathbb{P}_{\downarrow} := \{\mathcal{P}^{\downarrow} : \mathcal{P} \in \mathbb{P}\}$-complete,
(4) ξ is $\mathbb{P}_{\cup\downarrow} := \{\mathcal{P}^{\cup\downarrow} : \mathcal{P} \in \mathbb{P}\}$-complete.

In other words, a convergence is complete with respect to a collection of families if and only if it is complete with respect to the collection of ideals generated by these families.

Recall (from Definition II.1.7) that a family $\mathcal{S}$ is a refinement of a family $\mathcal{P}$ if for every $S \in \mathcal{S}$ there is $P \in \mathcal{P}$ with $S \subset P$.

We say that a collection $\mathbb{S}$ is a *inscribed* in a collection $\mathbb{P}$ if for every $\mathcal{P} \in \mathbb{P}$ there is $\mathcal{S} \in \mathbb{S}$ such that $\mathcal{S}$ is a refinement of $\mathcal{P}$.

Proposition XI.1.8. *A convergence that is complete with respect to a collection is complete with respect to each inscribed collection.*

Proof. Let ξ be a convergence on X that is complete with respect to a collection $\mathbb{P}$ on X. Let $\mathbb{S}$ be a collection on X inscribed in $\mathbb{P}$. Let $\mathcal{H}$ be an $\mathbb{S}$-fundamental filter on X and let $\mathcal{P} \in \mathbb{P}$. Then there exists $\mathcal{S} \in \mathbb{S}$

---

[1]That is, if $\mathcal{F} \cap \mathcal{P} \neq \varnothing$ implies that $\operatorname{adh}_\xi \mathcal{F} \neq \varnothing$ for every filter $\mathcal{F}$ on X.

such that $\mathcal{S}$ is a refinement of $\mathcal{P}$. As $\mathcal{H}$ is $\mathbb{S}$-fundamental, $\mathcal{H} \cap \mathcal{S} \neq \varnothing$ hence $\mathcal{H} \cap \mathcal{P} \neq \varnothing$. Therefore $\mathcal{H}$ is $\mathbb{P}$-fundamental, and adherent by the $\mathbb{P}$-completeness of ξ. $\qquad \square$

A family $\mathcal{S}$ on a convergence space X is called a *regular refinement* of a family $\mathcal{P}$ of subsets of X if $\mathrm{adh}_X^{\natural} \mathcal{S}$ is a refinement of $\mathcal{P}$. If a convergence is regular, then each cover admits a regular refinement that is a cover. More generally,

Proposition XI.1.9. *If X is a subset of a regular convergence space Y and $\mathcal{P}$ is a cover of X, then there is a regular refinement of $\mathcal{P}$, which is a cover of X.*

Proof. Let $\mathcal{F}$ be a filter on X and let $x \in \lim_X \mathcal{F} = X \cap \lim_Y \mathcal{F}$. Then $x \in X \cap \lim_Y (\mathrm{adh}_Y^{\natural} \mathcal{F})$ by the regularity of Y.

As $\mathcal{P}$ is a cover of X, for every filter $\mathcal{F}$ on X such that $\lim_X \mathcal{F} \neq \varnothing$, there exists $F_\mathcal{F} \in \mathcal{F} \cap \mathcal{P}$ such that $\mathrm{adh}_Y F_\mathcal{F} \subset P$. Therefore $\{F_\mathcal{F} : \lim_X \mathcal{F} \neq \varnothing\}$ is a regular refinement of $\mathcal{P}$. The family $\{F_\mathcal{F} : \lim_X \mathcal{F} \neq \varnothing\}$ is a cover of X, because if $x \in \lim_X \mathcal{F}$ then $F_\mathcal{F} \in \mathcal{F}$. $\qquad \square$

A collection $\mathbb{S}$ on a convergence space is said to be *regularly inscribed* in a collection $\mathbb{P}$ if $\{\mathrm{adh}^{\natural} \mathcal{S} : \mathcal{S} \in \mathbb{S}\}$ is inscribed in $\mathbb{P}$.

Corollary XI.1.10. *A regular convergence, which is complete with respect to a collection (of covers), is complete with respect to each regularly inscribed collection (consisting of covers).*

XI.2 Cocompleteness

We shall characterize completeness in terms of collections of filters rather than in terms of collections of ideals. This way we shall get a homogeneous language of filters for completeness and convergence.

Let $\mathbb{G}$ be collection (of families of subsets) of X. A filter $\mathcal{F}$ on X is called $\mathbb{G}$-*cofundamental* if $\mathcal{F}$ is dissociated from $\mathbb{G}$, that is, if

$$\underset{\mathcal{G} \in \mathbb{G}}{\forall} \neg(\mathcal{F} \# \mathcal{G}). \qquad (XI.2.1)$$

Definition XI.2.1. We say that a convergence ξ is $\mathbb{G}$-*cocomplete* if each $\mathbb{G}$-cofundamental filter on $|\xi|$ is ξ-adherent.

In particular, if ξ is a convergence and $\mathcal{G}$ is a family of subsets of $|\xi|$, then we say that ξ is $\mathcal{G}$-*cocomplete* whenever it is $\{\mathcal{G}\}$-cocomplete.

Exercise XI.2.2. Let ξ be a convergence and let $\mathbb{G}$ be a collection on $|\xi|$. Check that the following statements are equivalent:

(1) ξ is $\mathbb{G}$-cocomplete,

(2) ξ is $\mathbb{G}_\cap := \{\mathcal{G}^\cap : \mathcal{G} \in \mathbb{G}\}$-cocomplete,

(3) ξ is $\mathbb{G}_\uparrow := \{\mathcal{G}^\uparrow : \mathcal{G} \in \mathbb{G}\}$-cocomplete.

(4) ξ is $\mathbb{G}_{\cap\uparrow} := \{\mathcal{G}^{\cap\uparrow} : \mathcal{G} \in \mathbb{G}\}$-cocomplete.

Notice that $\mathcal{G}^\cap$ is a base of a non-degenerate filter $\mathcal{G}^{\cap\uparrow}$ whenever $\mathcal{G}$ has the finite intersection property. Otherwise, $\mathcal{G}^{\cap\uparrow}$ is the family of all subsets of $|\xi|$; in particular, it contains $\varnothing$ and thus no filter meshes $\mathcal{G}^{\cap\uparrow}$. Therefore the property of being $\mathbb{G}$-cofundamental is independent of the elements of $\mathbb{G}$ that do not have the finite intersection property. Therefore, without losing generality, we can assume that the considered collections consists only of filters.

Proposition XI.2.3. *Let* $\mathbb{G} \subset \mathbb{F}X$. *A convergence* ξ *on* X *is* $\mathbb{G}$-*cocomplete if and only if*

$$\mathbb{U}(X) \setminus \xi^- X \subset \bigcup_{\mathcal{G} \in \mathbb{G}} \beta\mathcal{G}. \qquad (XI.2.2)$$

Exercise XI.2.4. Prove Proposition XI.2.3.

In other words, a convergence is $\mathbb{G}$-cocomplete if and only if for each non-convergent ultrafilter $\mathcal{U}$, there exists $\mathcal{G} \in \mathbb{G}$ such that $\mathcal{U} \geq \mathcal{G}$.

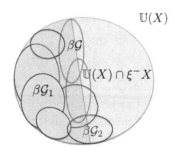

Figure XI.1   The disc represents the set of all ultrafilters on X (the ξ-convergent ultrafilters are on the right). A convergence ξ is $\mathbb{G}$-cocomplete whenever each non-convergent ultrafilter is finer than an element of $\mathbb{G}$.

Recall that for a family $\mathcal{G}$ of subsets of a given set X,

$$\mathcal{G}_c := \{X \setminus G : G \in \mathcal{G}\}$$

stands for the *complementary family*, that is, the family of complements of the elements of $\mathcal{G}$. Let us denote by

$$\mathbb{G}_\neg := \{\mathcal{G}_c : \mathcal{G} \in \mathbb{G}\}$$

the collection of the complementary families of the elements of $\mathbb{G}$.

Proposition XI.2.5. *A filter is* $\mathbb{G}$-*cofundamental if and only if it is* $\mathbb{G}_\neg$-*fundamental. A convergence is* $\mathbb{G}$-*cocomplete if and only if it is* $\mathbb{G}_\neg$-*complete.*

Exercise XI.2.6. Check Proposition XI.2.5.

Accordingly, $\mathbb{F}_{\mathbb{G}_\neg}$ is the class of cofundamental filters. Therefore, we can reduce cocompleteness to conditional compactness, as we have done in Section XI.1.

Corollary XI.2.7. *A convergence is* $\mathbb{G}$-*cocomplete if and only if it is* $\mathbb{F}_{\mathbb{G}_\neg}$-*compact.*

Example XI.2.8. It follows that if X is an infinite set, then ι_X is $(X)_0$-cocomplete (see Example XI.1.4). This means that each ultrafilter dissociated from the cofinite filter of X, is convergent. The only such ultrafilters are principal.

Let $\mathbb{G}$ and $\mathbb{H}$ be a collections on a set X. We say that a collection $\mathbb{H}$ is *rougher* than a collection $\mathbb{G}$ (or $\mathbb{G}$ is *subtler* than $\mathbb{H}$) if for every $\mathcal{G} \in \mathbb{G}$ there is $\mathcal{H} \in \mathbb{H}$ such that $\mathcal{H} \leq \mathcal{G}$.

Consider a convergence ξ on X.

Proposition XI.2.9. *If* ξ *is* $\mathbb{G}$-*cocomplete and* $\mathbb{H}$ *is rougher than* $\mathbb{G}$, *then* ξ *is* $\mathbb{H}$-*cocomplete.*

Proof. Each $\mathbb{H}$-cofundamental filter is $\mathbb{G}$-cofundamental, hence ξ-adherent by the $\mathbb{G}$-cocompleteness of ξ. $\qquad\qquad\qquad\qquad\qquad\qquad\Box$

Notice that a collection $\mathbb{H}$ is rougher than a collection $\mathbb{G}$ if and only if $\{\mathcal{H}_c : \mathcal{H} \in \mathbb{H}\}$ is inscribed in $\{\mathcal{G}_c : \mathcal{G} \in \mathbb{G}\}$. It follows that Proposition XI.2.9 is equivalent to Proposition XI.1.8.

Observe that if $\mathbb{H}$ is rougher than $\mathbb{G}$, then

$$\bigcup_{\mathcal{G} \in \mathbb{G}} \beta\mathcal{G} \subset \bigcup_{\mathcal{H} \in \mathbb{H}} \beta\mathcal{H}. \tag{XI.2.3}$$

On the other hand, (XI.2.3) implies that $\mathbb{G}$-cocompleteness entails $\mathbb{H}$-cocompletess. This suggests the following relation between collections: a collection $\mathbb{H}$ is *pseudo-rougher* than a collection $\mathbb{G}$ (or $\mathbb{G}$ is *pseudo-subtler* than $\mathbb{H}$) if for every $\mathcal{G} \in \mathbb{G}$ and each $\mathcal{U} \in \beta\mathcal{G}$ there is $\mathcal{H} \in \mathbb{H}$ such that $\mathcal{H} \leq \mathcal{U}$.

Proposition XI.2.10. *Let $\mathbb{G}$ and $\mathbb{H}$ be collections on a set X. Then $\mathbb{H}$ is pseudo-rougher than $\mathbb{G}$ if and only if every $\mathbb{G}$-cocomplete convergence on X is $\mathbb{H}$-cocomplete.*

Notice that (XI.2.3) holds if and only if:

If $\mathcal{U}$ is an ultrafilter on X such that $\mathcal{U} \cap \mathcal{H}_c \neq \varnothing$ for each $\mathcal{H} \in \mathbb{H}$ then $\mathcal{U} \cap \mathcal{G}_c \neq \varnothing$ for each $\mathcal{G} \in \mathbb{G}$. We say that $\mathbb{S}$ is *pseudo-inscribed* in $\mathbb{P}$ if every $\mathbb{S}$-fundamental ultrafilter is $\mathbb{P}$-fundamental.

Proposition XI.2.11. *$\mathbb{S}$ is pseudo-inscribed in $\mathbb{P}$ if and only if each $\mathbb{S}$-complete convergence is $\mathbb{P}$-complete.*

Proposition XI.2.12. *$\mathbb{H}$ is pseudo-rougher than $\mathbb{G}$ if and only if $\{\mathcal{H}_c : \mathcal{H} \in \mathbb{H}\}$ is pseudo-inscribed in $\{\mathcal{G}_c : \mathcal{G} \in \mathbb{G}\}$.*

Of course, in case of collections of filters, Proposition XI.2.9 follows immediately from Proposition XI.2.3 and (XI.2.3).

By Proposition XI.2.9, if ξ is $\mathcal{G}$-cocomplete and $\mathcal{H} \leq \mathcal{G}$ then ξ is $\mathcal{H}$-cocomplete.

Accordingly, recalling that $\mathcal{K}_c^o(\xi)$ denotes the family of the complements of ξ-compactoid sets, we get:

Proposition XI.2.13. *A convergence ξ is $\mathcal{G}$-cocomplete if and only if $\mathcal{G} \subset \mathcal{K}_c^o(\xi)$.*

Proof. Every convergence ξ is $\mathcal{K}_c^o(\xi)$-cocomplete, for if $\mathcal{F}$ and $\mathcal{K}_c^o(\xi)$ are dissociated, then every compactoid set K satisfies $K^c \notin \mathcal{F}^{\#}$, equivalently, $K \in \mathcal{F}$, so that $\operatorname{adh} \mathcal{F} \neq \varnothing$. Conversely, if there is $G \in \mathcal{G} \setminus \mathcal{K}_c^o(\xi)$, then G^c is not compactoid, that is, there is $\mathcal{F} \in \mathbb{F}|\xi|$ with $G^c \in \mathcal{F}$ and $\operatorname{adh} \mathcal{F} = \varnothing$. As $\mathcal{F}$ and $\mathcal{G}$ are dissociated, ξ is not $\mathcal{G}$-cocomplete. $\square$

XI.3 Completeness number

If compactness can be seen as a perfection the completeness number indicates how imperfect a convergence is. A convergence is compact if all ultrafilters converge. If only one filter is included in all non-convergent ultrafilters, then the convergence is not far from the perfection: it is locally compactoid. If there is a countable collection of non-convergent filters such that each non-convergent ultrafilter is finer than one of them, then a convergence is not too far from the perfection: it is countably complete.

Let κ be a cardinal number. A convergence ξ is called κ-*complete* if there exists a collection $\mathbb{P}$ of covers of ξ such that $\operatorname{card}\mathbb{P} \leq \kappa$ and ξ is $\mathbb{P}$-complete.

By Exercise XI.1.7, a convergence ξ is κ-complete if and only if there exists a collection $\mathbb{P}$ of ideals on $|\xi|$ such that $\operatorname{card}\mathbb{P} = \kappa$ and each $\mathcal{P} \in \mathbb{P}$ is a ξ-cover.

Recall that, by Theorem III.5.12, a family $\mathcal{P}$ on $|\xi|$ is a cover of a convergence ξ (in symbols, $\mathcal{P} \succ \xi$) if and only if $\operatorname{adh}_\xi \mathcal{P}_c = \varnothing$. Therefore, by Corollary XI.2.5,

Proposition XI.3.1. *A convergence ξ is κ-complete if and only if there exists a collection $\mathbb{G}$ of ξ-non-adherent families (equivalently, filters) such that* $\operatorname{card}\mathbb{G} \leq \kappa$ *and ξ is $\mathbb{G}$-cocomplete.*

As it was observed in (II.1.5), $\mathcal{S}$ refines $\mathcal{P}$ if and only if $\mathcal{S}_c \leq \mathcal{P}_c$.

Notice that if a refinement $\mathcal{S}$ of $\mathcal{P}$ is a cover of ξ, then $\mathcal{P}$ is also a cover of ξ. Dually, if a family is not adherent then a finer family is non-adherent.

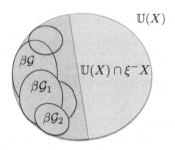

Figure XI.2   As in the preceding figure, non-convergent ultrafilters are on the left. Hence a collection $\mathbb{G}$ consists of non-adherent filters $\mathcal{G}$ whenever each set $\beta\mathcal{G}$ of ultrafilters finer than $\mathcal{G}$ is included in the set of non-convergent ultrafilters.

The filters of a collection $\mathbb{G} \subset \mathbb{F}X$ are non-adherent (with respect to a convergence ξ) if and only if

$$\bigcup_{\mathcal{G} \in \mathbb{G}} \beta\mathcal{G} \cap \xi^- X = \varnothing. \qquad (XI.3.1)$$

Thus, in view of (XI.2.2), a convergence ξ is κ-complete if there exists a collection $\mathbb{G}$ of ξ-non-adherent filters of cardinality κ such that

$$\bigcup_{\mathcal{G} \in \mathbb{G}} \beta\mathcal{G} = \mathbb{U}(X) \setminus \xi^- X.$$

The inclusion $\subset$ is equivalent to the fact that the elements of $\mathbb{G}$ are non-adherent with respect to ξ. The inclusion $\supset$ is equivalent to the $\mathbb{G}$-cocompleteness of ξ. Therefore the situations of Figures XI.2 and XI.1 are simultaneous, that is $\{\beta\mathcal{G} : \mathcal{G} \in \mathbb{G}\}$ fill the set of non-convergent ultrafilters.

The *completeness (number)* compl(ξ) of ξ is the least cardinal κ for which ξ is κ-complete.

Proposition XI.3.2. *Completeness number exists for each convergence.*

Proof. Indeed, each convergence on X is cocomplete with respect to the collection $\mathbb{S}$ of all non-convergent ultrafilters, because if $\mathcal{F}$ is $\mathbb{G}$-cofundamental, then each ultrafilter $\mathcal{U}$ finer than $\mathcal{F}$ is not in $\mathbb{S}$, hence is convergent. There exist $2^{\operatorname{card} X}$ subsets of X, hence $2^{2^{\operatorname{card} X}}$ families of subsets of X. Hence the number of non convergent ultrafilters is not greater than $2^{2^{\operatorname{card} X}}$, so that the completeness number of a convergence on X exists and is not greater than $2^{2^{\operatorname{card} X}}$. $\square$

Proposition XI.3.3. *A convergence ξ is 0-complete if and only if ξ is compact.*

Proof. It follows from the definition that ξ is 0-complete if ξ is complete with respect to the empty collection of covers of $|\xi|$. Since the condition

$$\underset{\mathcal{P} \in \varnothing}{\forall} \mathcal{F} \cap \mathcal{P} \neq \varnothing$$

is true for every filter $\mathcal{F}$ (because there is no $\mathcal{P}$ in $\varnothing$), every filter is $\varnothing$-fundamental. Therefore ξ is 0-complete whenever every filter is ξ-adherent. $\square$

If ξ is a convergence on X, then $\{X\}$ is, of course, a cover of ξ. If ξ is 0-complete, then every (non-degenerate) filter on X is ξ-adherent, hence ξ is $\{X\}$-complete. This is another characterization of a 0-complete convergence.

Corollary XI.3.4. *A convergence is 0-complete if and only if the co-compactoid filter is degenerate.*

Proof. The cocompactoid filter $\mathcal{K}_c^o(\xi)$ on a convergence ξ on X is degenerate if and only if $\varnothing \in \{X \setminus K : K \in \mathcal{K}^o(\xi)\}$, that is, $X \in \mathcal{K}^o(\xi)$, in other words, if and only if ξ is compact. □

Proposition XI.3.5. *Let Y be a Hausdorff compact pretopological space and let X be a subspace of Y. Then* $\operatorname{compl} X \leq \operatorname{card}(Y \setminus X)$.

Proof. The set $\mathbb{V} := \{\mathcal{V}_Y(y) \vee X : y \in Y \setminus X\}$ where $\mathcal{V}_Y(y)$ stands for the vicinity filter of Y at y. As Y is Hausdorff, if $\mathcal{U} \in \mathbb{U}X$ and $\mathcal{U} \# \mathcal{V}_Y(y) \vee X$ then in particular $\mathcal{U} \geq \mathcal{V}_Y(y)$, hence $\operatorname{adh}_Y \mathcal{U} = \lim_Y \mathcal{U} = \{y\}$, so that $\operatorname{adh}_X \mathcal{U} = X \cap \operatorname{adh}_Y \mathcal{U} = \varnothing$. Therefore $\mathbb{V}$ consists of X-non-adherent filters on X.

If $\mathcal{U}$ is an ultrafilter on X dissociated from $\mathbb{V}$, then $\lim_Y \mathcal{U} \neq \varnothing$, because Y is compact, and $\lim_Y \mathcal{U} \subset X$, because otherwise there would be $y \in Y \setminus X$ such that $\mathcal{V}_Y(y) \leq \mathcal{U}$. Therefore, X is $\mathbb{V}$-cocomplete and $\operatorname{card} \mathbb{V} = \operatorname{card}(Y \setminus X)$. □

XI.4 Finitely complete convergences

In particular, we say that a convergence ξ is *finitely complete* if it admits a finite collection $\mathbb{P}$ of covers of ξ such that ξ is $\mathbb{P}$-complete.

Equivalently, ξ is finitely complete if there exists a finite collection $\mathbb{G}$ of ξ-non-adherent filters such that ξ is $\mathbb{G}$-cocomplete. In view of Proposition II.3.6, ξ is $\mathbb{G}$-cocomplete if and only if ξ is $\bigwedge_{\mathcal{G} \in \mathbb{G}} \mathcal{G}$-cocomplete. On the other hand, recall from Proposition III.5.5 that

$$\operatorname{adh}_\xi \left(\bigwedge_{\mathcal{G} \in \mathbb{G}} \mathcal{G} \right) = \bigcup_{\mathcal{G} \in \mathbb{G}} \operatorname{adh}_\xi \mathcal{G},$$

because $\mathbb{G}$ is finite. Therefore, $\mathbb{G}$ consists of non-adherent filters if and only if the filter $\bigwedge_{\mathcal{G} \in \mathbb{G}} \mathcal{G}$ is non-adherent. We conclude that

Proposition XI.4.1. *A convergence is finitely complete if and only if it is 1-complete.*

By Definition IX.3.1,

Proposition XI.4.2. *A convergence is finitely complete if and only if it is locally compactoid.*

Proof. We have seen that a convergence ξ on X is finitely complete whenever there exists a ξ-non-adherent filter $\mathcal{G}$ on X such that ξ is $\mathcal{G}$-cocomplete.

Therefore for each $G \in \mathcal{G}$, every ultrafilter $\mathcal{U}$ such that $X \setminus G \in \mathcal{U}$ is dissociated from $\mathcal{G}$, hence ξ-convergent, which means that $X \setminus G$ is ξ-compactoid, that is, $\mathcal{G}$ is a ξ-non-adherent filter composed of the complements of ξ-compactoid sets, that is, $\mathcal{G}_c$ is a cover of ξ consisting of ξ-compactoid sets, which is the definition of local relative compactness.

Conversely, ξ is a locally compactoid convergence on X if and only if the ξ-cocompactoid filter $\mathcal{K}_c^o(\xi)$ is ξ-non-adherent. On the other hand, every convergence ξ is $\mathcal{K}_c^o(\xi)$-cocomplete, because each ultrafilter $\mathcal{U}$ on X that is dissociated from $\mathcal{K}_c^o(\xi)$, contains a ξ-compactoid set, hence $\lim_\xi \mathcal{U} \neq \varnothing$. $\square$

Corollary XI.4.3. *Let ξ be a convergence. The following statements are equivalent:*

(1) There exists a cover $\mathcal{P}$ of ξ such that ξ is $\mathcal{P}$-complete,
(2) $\mathcal{K}^o(\xi)$ is a cover of ξ,
(3) ξ is locally compactoid.

XI.5 Countably complete convergences

A convergence space is called *countably complete* if its completeness number is at most $\aleph_0$.

Remark XI.5.1. Traditionally, a countably complete functionally regular topological space is called *Čech-complete*.

Example XI.5.2. The real line is locally compactoid, that is, finitely complete, hence countably complete.

Example XI.5.3. The space $\mathbb{R} \setminus \mathbb{Q}$ (of irrational numbers) is countably complete, but not locally compact.

We know already that $\mathbb{R} \setminus \mathbb{Q}$ is homeomorphic with $X := (0,1) \setminus \mathbb{Q}$. The interval $Y := [0,1]$ endowed with the standard topology is compact. Therefore, by virtue of Proposition XI.3.5,

$$\mathrm{compl}\,(\mathbb{R} \setminus \mathbb{Q}) \leq \mathrm{card}\,\mathbb{Q} = \aleph_0.$$

On the other hand, it is straightforward that X is not locally compactoid ([2]).

---

[2]This follows also from the fact that $\mathbb{R} \setminus \mathbb{Q}$ is dense but not open in $\mathbb{R}$.

We have seen in Corollary X.2.4 that $\mathbb{Q}$ is not completely metrizable. We will see (Corollary XI.11.6) that $\mathbb{Q}$ is not countably complete, that is, $\aleph_0 < \text{comp}(\mathbb{Q})$.

Exercise XI.5.4. Show that $\text{compl}(\mathbb{Q}) \leq \mathfrak{c}$. ([3])

Exercise XI.5.5. Let $\{\mathcal{P}_n : n \in \mathbb{N}\}$ be a collection of families of subsets of X and let

$$\mathcal{S}_n := \mathcal{P}_0 \vee \ldots \vee \mathcal{P}_n, \tag{XI.5.1}$$

as defined in (III.5.5).

(1) Show that if $\mathcal{F}$ is a filter on X such that $\mathcal{F} \cap \mathcal{P}_n \neq \varnothing$ for each $n \in \mathbb{N}$, then there exists a decreasing sequence $(F_n)_n$ of subsets of X such that $F_n \in \mathcal{F} \cap \mathcal{S}_n$ each $n \in \mathbb{N}$.
(2) Show that if a convergence on X is $\{\mathcal{P}_n : n \in \mathbb{N}\}$-complete, then it is $\{\mathcal{S}_n : n \in \mathbb{N}\}$-complete.

XI.6 Preservation of completeness

Compactness is preserved by continuous maps (Proposition IX.1.26), but local (relative) compactness is not. For instance, in Example IX.3.6, the identity $i : |\pi| \to |\text{T}\,\pi|$ is continuous, π is locally compact, and $\text{T}\,\pi$ is not.

In other words, 0-completeness and 1-completeness behave differently under continuous maps.

Our forthcoming considerations provide an explanation of this difference and, more generally, they explain the mechanism of preservation of completeness numbers.

If $\mathbb{G}$ is a collection on X, $\mathbb{H}$ is a collection on Y and $f : X \to Y$, then we denote

$$f[\mathbb{G}] := \{f[\mathcal{G}] : \mathcal{G} \in \mathbb{G}\},$$
$$f^-[\mathbb{H}] := \{f^-[\mathcal{H}] : \mathcal{H} \in \mathbb{H}\}.$$

A map $f : |\xi| \to |\tau|$ does not increase the completeness number if

---

[3]The *dominating number* $\mathfrak{d}$ can be characterized (see [35]) as the least cardinal number so that there exists a subset G of $\mathbb{N}^{\mathbb{N}}$ of cardinality $\mathfrak{d}$ such that for each $f \in \mathbb{N}^{\mathbb{N}}$ there is $g \in G$ for which $f(n) \leq g(n)$ for each $n \in \mathbb{N}$. It is not difficult to see that $\aleph_1 \leq \mathfrak{d} \leq \mathfrak{c}$. In [29] it is shown that $\text{comp}(\mathbb{Q}) = \mathfrak{d}$.

(1) $\mathbb{G}$-cocompleteness of ξ implies $f[\![\mathbb{G}]\!]$-cocompleteness of τ for each collection $\mathbb{G}$ on $|\xi|$, and
(2) $\mathrm{adh}_\xi \mathcal{G} = \varnothing$ implies that $\mathrm{adh}_\tau f[\mathcal{G}] = \varnothing$ for each filter $\mathcal{G}$ on $|\xi|$.

The first condition is satisfied if f is continuous:

Proposition XI.6.1. *If ξ is $\mathbb{G}$-cocomplete and $f \in C(\xi, \tau)$ is surjective, then τ is $f[\![\mathbb{G}]\!]$-cocomplete.*

Proof. Let $\mathcal{H} \neg\# f[\mathcal{G}]$ for each $\mathcal{G} \in \mathbb{G}$. Equivalently, $f^-[\mathcal{H}] \neg\# \mathcal{G}$ for each $\mathcal{G} \in \mathbb{G}$, hence $\mathrm{adh}_\xi f^-[\mathcal{H}] \neq \varnothing$ by the ξ-cocompleteness of $\mathbb{G}$. If $x \in \mathrm{adh}_\xi f^-[\mathcal{H}]$ then there is $\mathcal{U} \in \beta(f^-[\mathcal{H}])$ with $x \in \lim_\xi \mathcal{U}$, thus $f(x) \in \lim_\tau f[\mathcal{U}]$ by the continuity of f. As $\mathcal{U} \# f^-[\mathcal{H}]$ amounts to $f[\mathcal{U}] \# \mathcal{H}$, we infer that $f(x) \in \mathrm{adh}_\tau \mathcal{H}$. $\qquad\square$

In particular, let $f \in C(\xi, \tau)$ and let $\mathrm{compl}(\xi) = 0$. Then ξ is $\varnothing$-complete ([4]). As $f[\![\varnothing]\!] = \varnothing$, by Proposition XI.6.1, τ is $\varnothing$-complete, hence $\mathrm{compl}(\tau) = 0$. Therefore, we recover Proposition IX.1.26.

Corollary XI.6.2. *If ξ is a compact convergence, τ is a convergence and $f : |\xi| \to |\tau|$ is surjective and continuous, then τ is compact.*

On the other hand, the second condition holds whenever f maps non-adherent filters onto non-adherent filters, equivalently, if

$$\mathrm{adh}_\tau f[\mathcal{G}] \neq \varnothing \implies \mathrm{adh}_\xi \mathcal{G} \neq \varnothing \qquad (\text{XI.6.1})$$

for each filter $\mathcal{G}$ on $|\xi|$.

Definition XI.6.3. *An onto map $f : |\xi| \to |\tau|$ is said to be perfect at y if*

$$y \in \mathrm{adh}_\tau f[\mathcal{G}] \implies f^-(y) \cap \mathrm{adh}_\xi \mathcal{G} \neq \varnothing$$

for every filter $\mathcal{G}$ on $|\xi|$. A map is called perfect if it is perfect at every point of its range.

It is obvious that each perfect map $f : |\xi| \to |\tau|$ fulfills (XI.6.1) for each filter $\mathcal{G}$ on $|\xi|$.

Theorem XI.6.4. *If $f : |\xi| \to |\tau|$ is a surjective continuous perfect map, then $\mathrm{compl}(\xi) = \mathrm{compl}(\tau)$.*

---

[4] $\varnothing$ stands for the empty collection.

Proof. Let $\mathbb{G}$ be a collection of ξ-non-adherent filters such that ξ is $\mathbb{G}$-cocomplete. Then card $(\mathbb{G}) \geq$ card $(f\,[\![\mathbb{G}]\!])$. As f is perfect, by (XI.6.1) $f\,[\![\mathbb{G}]\!]$ is a collection of τ-non-adherent filters. By Proposition XI.6.1, τ is $f\,[\![\mathbb{G}]\!]$-cocomplete. It follows that compl$(\xi) \geq$ compl(τ).

Conversely, let $\mathbb{H}$ be a collection of τ-non-adherent filters such that τ is $\mathbb{H}$-cocomplete. Then, by continuity of f the collection $f^-\,[\![\mathbb{H}]\!]$ consists of ξ-non-adherent filters. Indeed, if on the contrary, $x \in \mathrm{adh}_\xi\, f^-\,[\mathcal{H}]$ then there exists $\mathcal{U} \in \beta\,(f^-\,[\mathcal{H}])$ such that $x \in \lim_\xi \mathcal{U}$. By continuity, $f(x) \in \lim_\tau f\,[\mathcal{U}] \subset \mathrm{adh}_\tau\, \mathcal{H}$, which is a contradiction, because $f\,[\mathcal{U}]\,\#\mathcal{H}$.

To see that ξ is $f^-\,[\![\mathbb{H}]\!]$-cocomplete, let $\mathcal{F}\neg\#f^-\,[\mathcal{H}]$, equivalently $f\,[\mathcal{F}]\,\neg\#\mathcal{H}$, for every $\mathcal{H} \in \mathbb{H}$. As τ is $\mathbb{H}$-cocomplete, $\mathrm{adh}_\tau\, f\,[\mathcal{F}] \neq \varnothing$, and since f is perfect, by (XI.6.1), $\mathrm{adh}_\xi\, \mathcal{F} \neq \varnothing$. This means that compl$(\xi) \leq$ compl(τ). $\qquad\square$

Corollary XI.6.5. *If ξ is a convergence, τ is a compact convergence and $f : |\xi| \to |\tau|$ is surjective, continuous and perfect, then ξ is compact.*

Exercise XI.6.6. Show that a surjective continuous map fulfilling (XI.6.1) is perfect.

XI.7 Completeness of subspaces

Proposition XI.7.1. *If X is a closed subspace of a convergence space Y, then*

$$\mathrm{compl}\,(X) \leq \mathrm{compl}\,(Y)\,.$$

Proof. Let $\mathbb{G}$ be a collection of non-adherent filters on Y such that Y is $\mathbb{G}$-cocomplete. Let $\mathbb{G}_X := \{\mathcal{G}_X : \mathcal{G} \in \mathbb{G}\}$, where $\mathcal{G}_X$ is a restriction of $\mathcal{G}$ to X, that is, $\mathcal{G}_X := \{G \cap X : G \in \mathcal{G}\}$. It is immediate that card $(\mathbb{G}_X) \leq$ card $(\mathbb{G})$. Notice that for each filter $\mathcal{G}$ on Y,

$$\mathrm{adh}_X\, \mathcal{G}_X = X \cap \mathrm{adh}_Y\, \mathcal{G},$$

so that if $\mathcal{G}$ is non-adherent, then $\mathcal{G}_X$ is non-adherent. Finally, if Y is $\mathbb{G}$-cocomplete, then X is $\mathbb{G}_X$-cocomplete. Indeed, let $\mathcal{U} \in \mathbb{U}X$ be $\mathbb{G}_X$-cofundamental, that is, for each $\mathcal{G} \in \mathbb{G}$ there is $U \in \mathcal{U}$ and $G \in \mathcal{G}$ such that $U \cap (X \cap G) = \varnothing$. As $X \in \mathcal{U}$, it follows that $\mathcal{U}^{\uparrow Y}$ is $\mathbb{G}$-cofundamental. Therefore $\lim_Y \mathcal{U}^{\uparrow Y} \neq \varnothing$. As X is closed and $X \in \mathcal{U}$, we infer that $\lim_Y \mathcal{U}^{\uparrow Y} \subset X$ and $\lim_X \mathcal{U} = X \cap \lim_Y \mathcal{U}^{\uparrow Y} = \lim_Y \mathcal{U}^{\uparrow Y} \neq \varnothing$. $\qquad\square$

Let κ be a cardinal number. A subset X of a convergence space Y is called a G_κ-*subset* (of Y) if it is an intersection of κ many open sets. Mind that $G_{\aleph_0}$-sets are classically called G_δ-sets.

Example XI.7.2. Let $Y := \{x_\infty\} \cup \{x_n : n \in \mathbb{N}\} \cup \{x_{n,k} : n, k < \omega\}$ with the bisequence pretopology (see Example V.4.6). Then $\{x_\infty\}$ is a $G_{\aleph_0}$ non-open subset of Y.

Recall that a convergence is called *weakly diagonal* if adh $\mathcal{H}$ is closed for each filter $\mathcal{H}$ (Definition VI.2.9).

Theorem XI.7.3. *Let κ be a cardinal number. A dense κ-complete subset of a Hausdorff weakly diagonal convergence is a G_κ-subset.*

Proof. Let X be a dense κ-complete subset of Y, where Y is a Hausdorff weakly diagonal convergence space. Let $\mathbb{G}$ be a collection of non-adherent filters on X, for which X is cocomplete. Then

$$Y \backslash X = \bigcup_{\mathcal{G} \in \mathbb{G}} \mathrm{adh}_Y \, \mathcal{G}. \tag{XI.7.1}$$

Indeed, $\mathrm{adh}_X \, \mathcal{G} = X \cap \mathrm{adh}_Y \, \mathcal{G} = \varnothing$ for every $\mathcal{G} \in \mathbb{G}$, because $\mathcal{G}$ is non-adherent in X, thus $\supset$ holds in (XI.7.1). Conversely, let $y \in Y \backslash X$. As X is dense in Y and Y is Hausdorff, there exists an ultrafilter $\mathcal{U}$ on X such that $\{y\} = \lim_Y \mathcal{U} = \mathrm{adh}_Y \, \mathcal{U}$ and thus $\mathcal{U}$ is non-adherent in X. As X is $\mathbb{G}$-complete, there is $\mathcal{G} \in \mathbb{G}$ such that $\mathcal{U} \# \mathcal{G}$, thus $y \in \mathrm{adh}_Y \, \mathcal{G}$, and hence $\subset$ holds in (XI.7.1). Because Y is weakly diagonal, X is a G_κ-set. $\qquad\square$

By setting $\kappa = 1$, we get the following extension of Proposition IX.3.18:

Corollary XI.7.4. *A dense locally compactoid subspace of a Hausdorff weakly diagonal convergence is open.*

Theorem XI.7.5. *If κ, λ are equal to $0, 1$ or to an infinite cardinal, then every G_κ-subset of a regular λ-complete convergence space is $\max(\kappa, \lambda)$-complete.*

Proof. Let Y be a regular λ-complete convergence space. It is enough to prove this theorem in the case where $\kappa = \lambda$. Indeed, if $\kappa \leq \lambda$, then a G_λ-subset of Y is also a G_κ-subset of Y; if $\lambda \leq \kappa$, then Y is also κ-complete.

Let X be a G_κ-subset of Y, that is, there exists a family $\{V_\alpha : \alpha < \kappa\}$ of Y-open sets such that $X = \bigcap_{\alpha < \kappa} V_\alpha$. As Y is assumed to be κ-complete, there exists a collection $\mathbb{P}$ of covers of Y such that card $\mathbb{P} = \kappa$ and Y is $\mathbb{P}$-complete. If $\kappa = 0$, then the hypothesis is obviously fulfilled. So let $\kappa > 0$. By Proposition III.5.10,

$$\mathbb{P}_\alpha := \{\mathcal{P} \vee V_\alpha : \mathcal{P} \in \mathbb{P}\}$$

is a collection of covers of V_α. By Proposition XI.1.9, for each $\mathcal{P} \in \mathbb{P}$ there exists a regular refinement of $\mathcal{S}_{\alpha,\mathcal{P}}$ of $\mathcal{P} \vee V_\alpha$ such that $\mathcal{S}_{\alpha,\mathcal{P}}$ is a cover of V_α. Then $\mathbb{S}_\alpha := \{\mathcal{S}_{\alpha,\mathcal{P}} : \mathcal{P} \in \mathbb{P}\}$ is a collection of covers of V_α such that V_α is $\mathbb{S}_\alpha$-complete. Indeed, each $\mathbb{S}_\alpha$-fundamental ultrafilter $\mathcal{U}$ on V_α is an $\mathbb{S}_\alpha$-fundamental filter-base on Y, hence $\lim_Y \mathcal{U} \neq \varnothing$. Because $\mathbb{S}_\alpha$ is V_α-regular, $\lim_Y \mathcal{U} \subset V_\alpha$. By Proposition III.5.10, $\mathcal{S}_{\alpha,\mathcal{P}} \vee X$ is a cover of X for each $\mathcal{P} \in \mathbb{P}$ and $\alpha < \kappa$. Let

$$\mathbb{S} := \{\mathcal{S}_{\alpha,\mathcal{P}} \vee X : \mathcal{P} \in \mathbb{P}, \alpha < \kappa\}.$$

We shall show that X is $\mathbb{S}$-complete. Let $\mathcal{U}$ be an $\mathbb{S}$-fundamental ultrafilter on X, hence it is a $\mathbb{S}_\alpha$-fundamental filter-base on V_α for each $\alpha < \kappa$. Therefore $\varnothing \neq \lim_Y \mathcal{U} \subset V_\alpha$ for each $\alpha < \kappa$, hence $\varnothing \neq \lim_Y \mathcal{U} \subset X$, thus $\lim_X \mathcal{U} = X \cap \lim_Y \mathcal{U} \neq \varnothing$.

If $\kappa = 1$, then card $\mathbb{S} = 1$; if κ is infinite, then card $\mathbb{S} \leq$ card $\mathbb{P} \cdot \kappa = \kappa^2 = \kappa$. $\qquad \square$

It follows that each $G_{\aleph_0}$-subset of a regular countably complete convergence space is countably complete.

On recalling Remark XI.5.1, we infer that

Corollary XI.7.6 (Arhangel'skii-Frolík). *A functionally regular topological space is Čech-complete if and only if it is a $G_{\aleph_0}$-subspace of its any compactification.*

XI.8 Completeness of products

Lemma XI.8.1. *Let λ be an infinite cardinal. If* $\mathrm{compl}(\prod_{j \in J} \xi_j) < \lambda$, *then there is $J_0 \subset J$ such that* card $J_0 < \lambda$ *and* $\mathrm{compl}(\xi_j) = 0$ *for $j \in J \setminus J_0$.*

Proof. Let $\xi := \prod_{j \in J} \xi_j$. If $\mathrm{compl}(\xi) < \lambda$ then there exists a collection $\mathbb{P}$ of ideals that are covers of ξ such that card $\mathbb{P} = \mathrm{compl}(\xi)$ and ξ is $\mathbb{P}$-complete. Recall that each product convergence admits a pavement consisting of polyhedral filters. Let $\mathcal{H}$ be a ξ-convergent polyhedral filter. Then, for every $\mathcal{P} \in \mathbb{P}$ there exists $H_\mathcal{P} \in \mathcal{H} \cap \mathcal{P}$ and a finite subset $J_\mathcal{P}$ of J such that

$$H_\mathcal{P} = \prod_{j \in J} V_j,$$

and $V_j = X_j$ for each $j \in J \setminus J_{\mathcal{P}}$. The family

$$\mathcal{Z} := \{H_{\mathcal{P}} : \mathcal{P} \in \mathbb{P}\}^{\cap}$$

is ξ-compactoid filter base, because each $\mathcal{P} \in \mathbb{P}$ contains an element of $\mathcal{Z}$ and ξ is $\mathbb{P}$-complete. Therefore $p_j[\mathcal{Z}]$ is compactoid for each $j \in J$ by Proposition IX.7.7. Let $J_0 := \bigcup_{\mathcal{P} \in \mathbb{P}} J_{\mathcal{P}}$. Then card $J_0 < \lambda$ and $p_j[\mathcal{Z}] = \{X_j\}$ for each $j \in J \setminus J_0$, hence ξ_j is compact if $j \in J \setminus J_0$. $\qquad\square$

On setting, $\lambda = \aleph_0$, we get

Corollary XI.8.2. *If* compl$(\prod_{j \in J} \xi_j)$ *is finite, then there is a finite subset J_0 of J such that* compl $(\xi_j) = 0$ *for $j \in J \setminus J_0$.*

Theorem XI.8.3. *Let ξ_j be a convergence on X_j for each $j \in J$. Then*

$$\mathrm{compl}\left(\prod_{j \in J} \xi_j\right) \leq \sum_{j \in J} \mathrm{compl}\,(\xi_j).$$

If compl $\left(\prod_{j \in J} \xi_j\right)$ *is finite, then*

$$\sum_{j \in J} \mathrm{compl}\,(\xi_j) = \mathrm{card}\,\{j \in J : \mathrm{compl}\,(\xi_j) > 0\} < \infty. \qquad (XI.8.1)$$

If compl $\left(\prod_{j \in J} \xi_j\right)$ *is infinite, then*

$$\mathrm{compl}\left(\prod_{j \in J} \xi_j\right) = \sum_{j \in J} \mathrm{compl}\,(\xi_j). \qquad (XI.8.2)$$

Proof. Let $\xi := \prod_{j \in J} \xi_j$ be a convergence on $\prod_{j \in J} X_j$. For each $j \in J$, let $\mathbb{G}_j$ be a collection of ξ_j-non-adherent filters on X_j such that ξ_j is $\mathbb{G}_j$-cocomplete and card $\mathbb{G}_j =$ compl (ξ_j) ([5]). Let

$$\mathbb{G} := \left\{p_j^-[\mathcal{G}] : \mathcal{G} \in \mathbb{G}_j, j \in J\right\}.$$

Every element of $\mathbb{G}$ is ξ-non-adherent. Indeed, if $x \in \mathrm{adh}_\xi\, p_j^-[\mathcal{G}]$, then there is a filter $\mathcal{F}$ such that $x \in \lim_\xi \mathcal{F}$ and $\mathcal{F} \# p_j^-[\mathcal{G}]$, equivalently $p_j[\mathcal{F}] \# \mathcal{G}$. As $\mathcal{G}$ is ξ_j-non-adherent, $\lim_{\xi_j} p_j[\mathcal{F}] = \varnothing$, thus $\lim_\xi \mathcal{F} = \varnothing$.

We show that ξ is $\mathbb{G}$-cocomplete. Indeed, if $\mathcal{U}$ is a $\mathbb{G}$-cofundamental ultrafilter on $\prod_{j \in J} X_j$, then $\mathcal{U}$ does not mesh $p_j^-[\mathcal{G}]$ for each $\mathcal{G} \in \mathbb{G}_j$ and $j \in J$. Therefore, for each $j \in J$, the ultrafilter $p_j[\mathcal{U}]$ is $\mathbb{G}_j$-cofundamental, and since ξ_j is $\mathbb{G}_j$-cocomplete, there exists $x_j \in \lim p_j[\mathcal{U}]$. By the definition of product convergence, $x \in \lim_\xi \mathcal{U}$, where $p_j(x) := x_j$, which proves that ξ is $\mathbb{G}$-cocomplete. As card $(\mathbb{G}) \leq \sum_{j \in J}$ compl (ξ_j), it follows that compl $(\xi) \leq \sum_{j \in J}$ compl (ξ_j).

---

[5] If compl $(\xi_\alpha) = 0$, then let $\mathbb{G}_\alpha$ be an empty collection on X_α.

On the other hand, for each $j \in J$, the convergence ξ_j is homeomorphic to a closed subspace of ξ, hence, by Proposition XI.7.1,

$$\sup_{j \in J} \operatorname{compl}(\xi_j) \leq \operatorname{compl}(\xi). \tag{XI.8.3}$$

In general

$$\sum_{j \in J} \operatorname{compl}(\xi_j) = \sum_{\operatorname{compl}(\xi_j) > 0} \operatorname{compl}(\xi_j). \tag{XI.8.4}$$

If $\operatorname{compl}(\xi)$ is finite, then $\operatorname{compl}(\xi) \leq 1$ by Proposition XI.4.1, hence by (XI.8.3), $\operatorname{compl}(\xi_j) \leq 1$ for each $j \in J$. By Lemma XI.8.1, there exists a finite subset J_0 of J such that $\operatorname{compl}(\xi_j) = 0$ if $j \in J \setminus J_0$. Therefore (XI.8.1) holds.

If $\operatorname{compl}(\xi)$ is infinite, then (XI.8.4) is infinite by the first part of the proof. Hence by Proposition A.11.7,

$$\sum_{\operatorname{compl}(\xi_j) > 0} \operatorname{compl}(\xi_j) = \operatorname{card}\{j \in J : \operatorname{compl}(\xi_j) > 0\} \cdot \sup_{j \in J} \operatorname{compl}(\xi_j).$$

As $\{j \in J : \operatorname{compl}(\xi_j) > 0\} \subset J_0$, by Lemma XI.8.1,

$$\sum_{j \in J} \operatorname{compl}(\xi_j) \leq \operatorname{card}(J_0) \cdot \sup_{j \in J} \operatorname{compl}(\xi_j)$$

$$\leq \operatorname{compl}(\xi)^2 = \operatorname{compl}(\xi),$$

which proves that $\sum_{j \in J} \operatorname{compl}(\xi_j) \leq \operatorname{compl}(\xi)$. $\qquad\square$

As a convergence is compact whenever its completeness is 0, we recover from (XI.8.2) Tikhonov Theorem IX.1.32.

Corollary XI.8.4. *A product of convergences is compact if and only each component is compact.*

As a convergence is locally compactoid whenever its completeness is 1, from Theorem XI.8.3 and Lemma XI.8.1, we recover

Corollary XI.8.5. *A product of convergences is locally compactoid if and only if finitely many components are locally compactoid and other are compact.*

On the other hand, we get

Corollary XI.8.6. *A product of convergences is countably complete if and only if countably many components are countably complete and other are compact.*

XI.9 Conditionally complete convergences

Completeness and cocompleteness have natural weaker variants. As in the case of compactness (see Section IX.8), we shall consider conditional completeness with respect to classes of filters. Let $\mathbb{P}$ and $\mathbb{G}$ be collections (of families of subsets of a given set) and let $\mathbb{H}$ be a class of filters.

Definition XI.9.1. A convergence ξ is $\mathbb{H}$-*conditionally* $\mathbb{P}$-*complete*, if $\mathrm{adh}_\xi \mathcal{H} \neq \varnothing$ for each $\mathbb{P}$-fundamental filter $\mathcal{H} \in \mathbb{H}$.

Definition XI.9.2. A convergence ξ is $\mathbb{H}$-*conditionally* $\mathbb{G}$-*cocomplete*, if $\mathrm{adh}_\xi \mathcal{H} \neq \varnothing$ for each $\mathbb{G}$-cofundamental filter $\mathcal{H} \in \mathbb{H}$.

Recall that $\mathbb{F}_\mathbb{P}$ stands for the collection of $\mathbb{P}$-fundamental filters; hence $\mathbb{F}_{\mathbb{G}_\neg}$ represents the collection of $\mathbb{G}$-cofundamental filters. Let us denote

$$\mathbb{H}_\mathbb{P} := \mathbb{H} \cap \mathbb{F}_\mathbb{P}$$

the collection of $\mathbb{P}$-*fundamental filters belonging to* $\mathbb{H}$. This notation is coherent, because we denote by $\mathbb{F}$, the class of all filters, and by $\mathbb{F}_\mathbb{P}$, the class of all $\mathbb{P}$-fundamental filters. Consequently, $\mathbb{H}_{\mathbb{G}_\neg}$ denotes the collection of $\mathbb{G}$-*cofundamental filters belonging to* $\mathbb{H}$. It follows that:

Corollary XI.9.3. *A convergence is*

(1) $\mathbb{H}$-conditionally $\mathbb{P}$-complete if and only if it is $\mathbb{H}_\mathbb{P}$-compact.
(2) $\mathbb{H}$-conditionally $\mathbb{G}$-cocomplete if and only if it is $\mathbb{H}_{\mathbb{G}_\neg}$-compact.

Definition XI.9.4. A convergence ξ is said to be $\mathbb{H}$-*conditionally* κ-*complete* if there exists a collection $\mathbb{P}$ of covers such that $\mathrm{card}\,\mathbb{P} \leq \kappa$ and ξ is $\mathbb{H}$-conditionally $\mathbb{P}$-complete.

Remark XI.9.5. Of course, a convergence ξ is $\mathbb{H}$-conditionally κ-complete if and only if there exists a collection $\mathbb{G}$ of non-adherent filters such that $\mathrm{card}\,\mathbb{G} \leq \kappa$ and ξ is $\mathbb{H}$-conditionally $\mathbb{G}$-cocomplete.

Analogously to completeness number, we may consider an $\mathbb{H}$-*conditional completeness number* of a convergence ξ that is, the least cardinal κ such that ξ is $\mathbb{H}$-conditionally κ-complete.

Accordingly, a convergence ξ is said to be *finitely* $\mathbb{F}_1$-conditionally complete if there exists a cover $\mathcal{P}$ such that $\mathrm{adh}_\xi \mathcal{F} \neq \varnothing$ for each countably based filter $\mathcal{F}$ such that $\mathcal{F} \cap \mathcal{P} \neq \varnothing$. In other words, ξ is finitely $\mathbb{F}_1$-conditionally complete if and only if ξ is *locally countably compactoid*.

As well, a convergence ξ is said to be $\mathbb{F}_1$-*conditionally countably complete* if there exists a sequence of covers $\{\mathcal{P}_n : n \in \mathbb{N}\}$ such that $\mathrm{adh}_\xi \, \mathcal{F} \neq \varnothing$ for each countably based filter $\mathcal{F}$ such that $\mathcal{F} \cap \mathcal{P}_n \neq \varnothing$ for each $n \in \mathbb{N}$. Accordingly, the least cardinal κ such that ξ is $\mathbb{F}_1$-conditionally κ-complete is called the $\mathbb{F}_1$-*conditional completeness number* and is denoted by $\mathrm{compl}_{\mathbb{F}_1}(\xi)$.

Theorem XI.9.6. *Each product of* $\mathbb{F}_1$-*conditionally countably complete convergences is countably* $\mathbb{F}_1$-*conditionally complete.* ([6])

Proof. Let J be a set and let ξ_j be a $\mathbb{F}_1$-conditionally countably complete convergence on a set X_j. Denote by ξ their product. By assumption, for each $j \in J$, there exists a countable collection $\mathbb{P}_j := \{\mathcal{P}_j(n) : n \in \mathbb{N}\}$ of ideals on X_j such that, ξ_j is $\mathbb{F}_1$-conditionally $\mathbb{P}_j$-complete and $\mathcal{P}_j(n)$ is a cover of ξ_j for each $n \in \mathbb{N}$. By Exercise XI.5.5, we can suppose that $\mathcal{P}_j(n+1)$ is a refinement of $\mathcal{P}_j(n)$ for each $j \in J$ and $n \in \mathbb{N}$. Define

$$\mathcal{P}(n) := \left\{ \prod_{j \in J} P_j^n : P_j^n \in \mathcal{P}_j(n), \mathrm{card}\,\{P_j^n \neq X_j\} < \infty \right\}.$$

Then, by (IV.4.6) and (IV.4.5), $\mathcal{P}(n)$ is a cover of ξ. We claim that ξ is $\mathbb{F}_1$-conditionally $\{\mathcal{P}(n) : n \in \mathbb{N}\}$-complete. Indeed, if $\mathcal{F} \in \mathbb{F}_1\,|\xi|$ fulfills $\mathcal{F} \cap \mathcal{P}(n) \neq \varnothing$ for each $n \in \mathbb{N}$, then there exists a decreasing sequence $(F_n)_n$ of elements of $\mathcal{F}$ such that

$$F_n = \prod_{j \in J} P_j^n,$$

so that $P_j^n \supset P_j^{n+1}$ for each $j \in J$ and $n \in \mathbb{N}$. It follows that, for each $j \in J$, there is $x_j \in \mathrm{adh}_{\xi_j} \{P_j^n\}_n$, that is there is an ultrafilter $\mathcal{U}_j$ on X_j such that $x_j \in \lim_{\xi_j} \mathcal{U}_j$ and $\mathcal{U}_j \# \{P_j^n\}_n$. Equivalently, $p_j^-[\mathcal{U}_j] \# \mathcal{F}$ for each $j \in J$, hence $\mathcal{F} \# \bigvee_{j \in J} p_j^-[\mathcal{U}_j]$ and there is an ultrafilter $\mathcal{W}$ such that

$$\mathcal{F} \vee \bigvee_{j \in J} p_j^-[\mathcal{U}_j] \leq \mathcal{W}.$$

Thus, if $f(j) := x_j$ for each $j \in J$, then $f \in \lim_\xi \mathcal{W}$ and thus $f \in \mathrm{adh}_\xi \, \mathcal{F}$. $\square$

XI.10 Baire property

We say that a convergence ξ on X has the *Baire property* if each countably based filter $\mathcal{F}$ such that

$$\mathrm{cl}_\xi \, \mathrm{int}_\xi \, F = X$$

---

[6]This generalizes a result of [85].

for each $F \in \mathcal{F}$, fulfills

$$\mathrm{cl}_\xi\Big(\bigcap\nolimits_{F \in \mathcal{F}} F\Big) = X.$$

Of course, dually, ξ has the Baire property whenever for a countably based ideal $\mathcal{P}$ if $\mathrm{int}_\xi(\bigcup_{P \in \mathcal{P}} P) \neq \varnothing$, there exists $P \in \mathcal{P}$ such that $\mathrm{int}_\xi \mathrm{cl}_\xi P \neq \varnothing$.

Notice that Baire property is of topological nature, because it is defined in terms of set-theoretic and topological notions.

Theorem XI.10.1 (Baire). *Each $\mathbb{F}_1$-conditionally countably complete regular topology has the Baire property.*

Proof. Let ξ be a topology on a set X fulfilling the assumptions. Recall that each cover $\mathcal{P}$ of a topological space admits a cover consisting of open sets that is a refinement of $\mathcal{P}$. Therefore, there exists a countable collection $\mathbb{P} := \{\mathcal{P}_n : n \in \mathbb{N}\}$ of covers of ξ such that ξ is $\mathbb{F}_1$-conditionally $\mathbb{P}$-complete and $\mathcal{P}_n \subset \mathcal{O}_\xi$ for each $n \in \mathbb{N}$.

Let $\{F_n : n \in \mathbb{N}\}$ be a decreasing base of a filter $\mathcal{F}$ on X such that $\mathrm{cl}_\xi \mathrm{int}_\xi F_n = X$ and let $G_n := \mathrm{int}_\xi F_n$ for each $n \in \mathbb{N}$. Take an arbitrary ξ-open set O_0. Then $O_0 \cap G_n$ is non-empty and ξ-open. Let $P_0 \in \mathcal{P}_0$ be such that $P_0 \cap O_0 \cap G_0$ is non-empty. Because ξ is regular, there exists a non-empty ξ-open set O_1 such that $\mathrm{cl}_\xi O_1 \subset P_0 \cap O_0 \cap G_0$. If $O_0, O_1, \ldots, O_n$ are such non-empty ξ-open sets that

$$\mathrm{cl}_\xi O_{k+1} \subset P_k \cap O_k \cap G_k$$

for each $0 \leq k < n$, then by the same argument, there is a non-empty open set O_{n+1} such that $\mathrm{cl}_\xi O_{n+1} \subset P_n \cap O_n \cap G_n$.

Accordingly, $\{O_n : n \in \mathbb{N}\}$ is a countable base of a filter $\mathrm{int}_\xi^\natural \mathcal{F}$. Moreover $\mathrm{int}_\xi^\natural \mathcal{F} \cap \mathcal{P}_n \neq \varnothing$ for each $n \in \mathbb{N}$, so that $\mathrm{adh}_\xi \mathrm{int}_\xi^\natural \mathcal{F} \neq \varnothing$ by the $\mathbb{P}$-completeness of ξ. Therefore

$$\varnothing \neq \mathrm{adh}_\xi \mathrm{int}_\xi^\natural \mathcal{F} = \bigcap\nolimits_{n \in \mathbb{N}} \mathrm{cl}_\xi O_n \subset O_0 \cap \bigcap\nolimits_{n \in \mathbb{N}} G_n.$$

As O_0 is an arbitrary open set, this proves that $X = \mathrm{cl}_\xi(\bigcap_{n \in \mathbb{N}} G_n) \subset \mathrm{adh}_\xi \mathcal{F}$. $\qquad\qquad\square$

By Theorems XI.9.6 and XI.10.1,

Corollary XI.10.2. *Each product of regular $\mathbb{F}_1$-conditionally countably complete topologies has the Baire space.*

XI.11   Strict completeness

A collection $\mathbb{P}$ of families of subsets of a convergence space X is called *strictly complete* provided that each $\mathbb{P}$-fundamental filter is convergent. Of course, each strictly complete collection is complete.

It is important to observe that the strict completeness of a collection of families *does not entail* the strict completeness of the collection of the corresponding families or ideals!

A convergence space is said to be *strictly κ-complete* if there exists a strictly complete collection of covers of cardinality κ. Of course, a strictly κ-complete convergence is κ-complete. As we see below the converse does not hold.

Recall that a convergence is called *hypercompact* if each filter is convergent.

Proposition XI.11.1. *A convergence is strictly 0-complete if and only if it is hypercompact.*

Example XI.11.2. The chaotic convergence, the Sierpiński topology and the upper Kuratowski convergence are hypercompact, hence strictly 0-complete.

A collection $\mathbb{P}$ of families of sets is called *narrow* if whenever $\mathcal{P} \in \mathbb{P}$ we take $P_{\mathcal{P}} \in \mathcal{P}$, then the set

$$\bigcap_{\mathcal{P} \in \mathbb{P}} P_{\mathcal{P}}$$

is at most a singleton. Recall that if ξ is a topology on X, then a collection $\mathbb{P}$ on X is called *ξ-regular* if for every $\mathcal{P} \in \mathbb{P}$ and each $P \in \mathcal{P}$ there is $Q \in \mathcal{P}$ such that $\mathrm{cl}_\xi\, Q \subset P$.

It is a simple fact that a topology that admits a narrow regular collection of covers, is Hausdorff.

Example XI.11.3. Let (X, d) be a metric space and let $\mathcal{P}_n = \{B_d(x, \frac{1}{n}) : x \in X\}$. Then $\{\mathcal{P}_n : n \in \mathbb{N}\}$ is a narrow, regular sequence of covers of (X, d).

Lemma XI.11.4. *If ξ is a $\mathbb{P}$-complete topology, where $\mathbb{P}$ is a narrow regular collection of covers of ξ, then every $\mathbb{P}$-fundamental filter is ξ-convergent.*

Recall that the *pseudocharacter* of a point x of a topological T_1-space X is defined as the least cardinal κ such that there exists $\mathcal{V} \subset \mathcal{N}(x)$ such that card $\mathcal{V} \leq \kappa$ and $\{x\} = \bigcap_{V \in \mathcal{V}} V$. If moreover X is regular, then there is $\mathcal{V}$ of cardinality κ such that $\{x\} = \bigcap_{V \in \mathcal{V}} \operatorname{cl} V$. The *pseudocharacter* of X is defined as the supremum of the pseudocharacters of all its elements.

Proposition XI.11.5. *If a regular topological space of pseudocharacter κ is κ-complete, then it is strictly κ-complete.*

Proof. Let X be a topological space fulfilling the assumptions and let $\{\mathcal{P}_j : j \in J\}$ be a collection of covers of X such that X is $\{\mathcal{P}_j : j \in J\}$-complete. For each x let $\{V_j(x) : j \in J\}$ be a family of open neighborhoods of x. Then $\mathcal{V}_j := \{V_j(x) : x \in X\}$ is a cover of X, and $\{\mathcal{V}_j : j \in J\}$ is a narrow regular cover. Indeed, if $x \in \bigcap_{j \in J} V_j(x_j)$, then there is a family $\{V_{\gamma_j}(x) : j \in J\}$ such that $\operatorname{cl} V_{\gamma_j}(x) \subset V_j(x_j)$ for each $a \in J$, hence $\bigcap_{j \in J} V_{\gamma_j}(x) = \{x\}$. Consequently, $\{\mathcal{P}_j \vee \mathcal{V}_j : j \in J\}$ is a complete family of covers (as a refinement of a complete collection of covers), which is regular and narrow. $\qquad\square$

We conclude that a metrizable countably complete space is countably strictly complete. Finally,

Corollary XI.11.6. *A metrizable convergence space is completely metrizable if and only if it is countably complete.*

Chapter XII

Connectedness

A convergence space is connected if it cannot be decomposed into two disjoint non-empty closed subsets. As connectedness is defined in terms of closed subsets, it is a topological notion: a convergence is connected if and only if its topological modification is. Therefore we will restrict ourselves to topological spaces, for the most part.

XII.1   Connected spaces

Definition XII.1.1. A convergence space on a set X is *disconnected* if there exists a pair of disjoint non-empty closed subsets F_0 and F_1 of X such that $F_0 \cup F_1 = X$, that is,

$$F_0 \neq \varnothing \neq F_1, \; F_0 \cap F_1 = \varnothing \text{ and } F_0 \cup F_1 = X. \qquad (\text{XII.1.1})$$

A convergence space is *connected* if it is not disconnected, that is, if the only partition of the space by two closed subsets is $\{\varnothing, X\}$.

Since $F_1 = X \setminus F_0$, the subsets F_0 and F_1 are open and closed. Thus X is disconnected if and only if it can be partitioned by two non-empty disjoint open subsets. A subset of a convergence space that is both open and closed is also called *clopen*.

Proposition XII.1.2. *A convergence space on X is connected if and only if its only clopen subsets are $\varnothing$ and X.*

Proposition XII.1.3. *A convergence on X is connected if and only if every continuous map $f : X \to \{0,1\}$, where $\{0,1\}$ carries the discrete topology $\iota_{\{0,1\}}$, is constant.*

Proof. If there was a continuous function $f : X \to |\iota_{\{0,1\}}|$ with $f(X) = \{0,1\}$, then $f^-(0)$ and $f^-(1)$ are two disjoint non-empty closed subsets of X and $f^{-1}(0) \cup f^{-1}(1) = X$. If there exist two disjoint non-empty closed subsets F_0 and F_1 of X, then

$$f(x) := \begin{cases} 0 \text{ if } x \in F_0, \\ 1 \text{ if } x \in F_1, \end{cases}$$

is continuous and is not constant. $\qquad\square$

Example XII.1.4. A (non-empty) discrete convergence ι on X is connected if and only if $\operatorname{card} X = 1$. Indeed, if $x \in X$ and $\operatorname{card} X > 1$, then $\{x\}$ is clopen and non-empty, so that $X \setminus \{x\}$ is also clopen and non-empty. If $\operatorname{card} X = 1$ then X cannot be decomposed into two non-empty subsets.

Example XII.1.5. Every T_1-convergence space of cardinality greater than 1 that has an isolated point is disconnected. Indeed, if x is isolated then $\{x\}$ is open. It is also closed because the space is T_1. Thus $X \setminus \{x\}$ is also clopen (and non-empty card $X > 1$).

Exercise XII.1.6. Show that if τ is connected and $\tau \geq \xi$ then ξ is connected.

Definition XII.1.7. A subset A of a space (X, ξ) is *connected* if $(A, \xi_{|A})$ is a connected space.

Evidently, a convergence ξ is connected if and only if its topological modification $\operatorname{T} \xi$ is. For this reason, we will often restrict ourselves to topological spaces when discussing connectedness, even though care is needed when considering connectedness of subsets. Indeed, we know from Example V.4.35 that $\operatorname{T}(\xi_{|A})$ and $(\operatorname{T} \xi)_{|A}$ may not coincide. However,

$$(\operatorname{T} \xi)_{|A} \leq \operatorname{T}(\xi_{|A})$$

so that, in view of Exercise XII.1.6:

Proposition XII.1.8. *If A is a connected subset of ξ, it is also a connected subset of $\operatorname{T} \xi$.*

The converse, however, is not true, as we will see in Example XII.1.30.

Example XII.1.9. The space $\mathbb{Q}$ of irrational numbers is disconnected. If $r \in \mathbb{R} \setminus \mathbb{Q}$, then $F_0 := \{q \in \mathbb{Q} : q \leq r\}$ and $F_1 := \{q \in \mathbb{Q} : r \leq q\}$ are non-empty disjoint closed subsets such that $\mathbb{Q} = F_0 \cup F_1$. By a similar argument, the space $\mathbb{R} \setminus \mathbb{Q}$ of irrational numbers is disconnected.

Recall that a subset I of $\mathbb{R}$ is called an *interval* of $\mathbb{R}$ if

$$x_0, x_1 \in I,\ x_0 < t < x_1 \implies t \in I.$$

Theorem XII.1.10. *A subset of the real line (with its standard topology) is connected if and only if it is an interval.*

Proof. Suppose to the contrary that an interval I is not connected, that is, there are two disjoint closed sets A and B such that $A \cup B = I$. Let $a \in A$, let $b \in B$, and let

$$c := \sup\left\{r \in I :\ r \in [a, b] \cap A\right\}.$$

Because A is closed, $c \in A$ ([1]). Moreover $(c, b] \subset B$ because $t > c$ cannot belong to A. Since B is closed, $[c, b] \subset B$, so that $c \in A \cap B$, which is a contradiction. Conversely, if a subset A is not an interval, there are x_0 and x_1 in A with $t \in (x_0, x_1) \setminus A$. The sets $\{r \in A : r < t\}$ and $\{r \in A : r > t\}$ are disjoint non-empty open subsets of A, so that A is not connected. $\square$

In particular:

Example XII.1.11. The real line with its standard topology is connected.

Proposition XII.1.12. *If X is a connected space and $f : X \to Y$ is continuous and onto, then Y is connected.*

Proof. If $f(X) = Y$ is not connected, then there is a continuous map $g : f(X) \to \left|\iota_{\{0,1\}}\right|$ that is onto. Thus $g \circ f : X \to \left|\iota_{\{0,1\}}\right|$ is continuous and onto, hence not constant. Therefore X is not connected. $\square$

Corollary XII.1.13. *If X is connected and $f : X \to Y$ is continuous, then $f(X)$ is connected.*

Example XII.1.14. If $f : \mathbb{R} \to \mathbb{R}$ is continuous, its graph is a connected subset of $\mathbb{R}^2$, because it is the image of $\mathbb{R}$ under the continuous map $\bar{f} : \mathbb{R} \to \mathbb{R}^2$ defined by $\bar{f}(x) = (x, f(x))$.

By Corollary XII.1.13 and Theorem XII.1.10,

Theorem XII.1.15 (Bolzano). *If X is connected and $f : X \to \mathbb{R}$ is continuous, then $f(X)$ is an interval.*

---

[1] In view of the definition of c, there is an increasing sequence $(r_n)_n$ of elements of $[a, b] \cap A$ such that $\lim_n r_n = c$.

The Intermediate Value Theorem, called also the *Darboux property*, is the case $X = [a, b] \subset \mathbb{R}$:

Corollary XII.1.16 (Intermediate Value Theorem). *If $f : [a, b] \to \mathbb{R}$ is continuous and w is between $f(a)$ and $f(b)$, then there is $c \in [a, b]$ with $f(c) = w$.*

Proof. Since $f([a, b])$ is a connected subset of $\mathbb{R}$, it is an interval, so $w \in f([a, b])$, that is, there exists $c \in [a, b]$ such that $w = f(c)$. $\square$

Corollary XII.1.17. *If a connected convergence on X is such that continuous functions separate points (in particular if it is T_1 functionally regular) and* card $X > 1$, *then* card $X \geq \mathfrak{c}$.

Proof. Let x_0, x_1 be two distinct points of X. Since continuous functions separate points, there is a continuous function $f : X \to [0, 1]$ such that $f(x_0) = 0$ and $f(x_1) = 1$. By Theorem XII.1.15 f is onto, so that card $X \geq$ card $[0, 1] = \mathfrak{c}$. $\square$

Exercise XII.1.18. Let X be a topological space. We call a function $f : X \to Y$ *locally constant* if for every $x \in X$ there is $V \in \mathcal{N}(x)$ such that $f_{|V}$ is constant. Show that if X is connected and Y is a T_1-convergence space, then every locally constant function $f : X \to Y$ is constant.

Recall from Definition VII.7.1 that two subsets M and N of X are *topologically almost separated* if

$$(\operatorname{cl} M \cap N) \cup (M \cap \operatorname{cl} N) = \varnothing.$$

Proposition XII.1.19. *A subset A of a space X is connected if and only if whenever $A = M \cup N$ where M, N are topologically almost separated, $M = \varnothing$ or $N = \varnothing$.*

Proof. If $A = M \cup N$ is connected and M, N are topologically almost separated, then $\operatorname{cl}_X M \cap N = \varnothing$, so that $\operatorname{cl}_X M \cap A \subset A \setminus N = M$. Similarly $M \cap \operatorname{cl}_X N = \varnothing$, so that $\operatorname{cl}_X N \cap A \subset A \setminus M = N$. Thus M and N are closed disjoint subsets of A, so that one of them is empty, by connectedness of A. Conversely, if A is disconnected, there exists two non-empty closed disjoint subsets A_0, A_1 of A. Of course, $\operatorname{cl}_X A_0 \cap A = A_0$ (²), so that $\operatorname{cl}_X A_0 \cap A_1 = \varnothing$

---

²If X is a convergence space and A is a subspace, then a subset A_0 of A is closed if and only if there is a closed subset C of X such that $A_0 = C \cap A$. Thus $\operatorname{cl}_X A_0 \subset C$ and $\operatorname{cl}_X A_0 \cap A \subset A_0$.

and $\mathrm{cl}_X A_1 \cap A = A_1$. Therefore $A_0 \cap \mathrm{cl}_X A_1 = \varnothing$ and A_0, A_1 are non-empty and topologically almost separated in X. □

Proposition XII.1.20. *If A, B are two connected subsets of X and $A \cap B \neq \varnothing$, then $A \cup B$ is connected.*

Proof. Let $f : A \cup B \to |\iota_{\{0,1\}}|$ be continuous. As A and B are connected, f is constant on A and on B, hence constant because $A \cap B \neq \varnothing$. □

Proposition XII.1.21. *Let $\{A_t : t \in T\}$ be a family of connected subsets of X such that there is $t_0 \in T$ with $A_t \cap A_{t_0} \neq \varnothing$ for all $t \in T$. Then $\bigcup_{t \in T} A_t$ is connected.*

Proof. Consider a continuous map $f : \bigcup_{t \in T} A_t \to |\iota_{\{0,1\}}|$. Since A_t is connected, f is constant on each A_t, thus constant because $A_t \cap A_{t_0} \neq \varnothing$ for every $t \in T$. □

Corollary XII.1.22. *Let $\{A_t : t \in T\}$ be a family of connected subsets of X such that $\bigcap_{t \in T} A_t \neq \varnothing$. Then $\bigcup_{t \in T} A_t$ is connected.*

Corollary XII.1.23. *If A is a subset of X with the property that for any two points x_0 and x_1 in A there exists a connected subset C of A with $x_0, x_1 \in C$, then A is connected.*

Proof. Let $x_0 \in A$. Then for each $x \in A$ there is a connected set C_x such that $x_0, x \in C_x \subset A$. Thus $\{x_0\} \cap C_x \neq \varnothing$ for every $x \in A$ and $A = \bigcup_{x \in A} C_x$. Corollary XII.1.22 applies to the effect that A is connected. □

Exercise XII.1.24. *Let $\{A_n\}_n$ be a decreasing sequence of closed connected subsets of a topological space X. Show that $A_\infty := \bigcap_{n \in \mathbb{N}} A_n$ is connected if A_0 is compact. Show that the compactness of A_0 is necessary.*

Proposition XII.1.25. *If A is a connected subset of a convergence space and $A \subset B \subset \mathrm{adh}\, A$, then B is connected.*

Proof. If $f : B \to |\iota_{\{0,1\}}|$ is continuous, then f is constant on A because A is connected. By Proposition IV.9.38, f is also constant on $B \subset \mathrm{adh}\, A$. □

Corollary XII.1.26. *The closure of a connected subset of a topological space is connected.*

Example XII.1.27. The subset

$$A := \left\{(x, y) : |x| \neq 0, y = \sin(\tfrac{1}{x})\right\} \cup (\{0\} \times [-1, 1])$$

of the plane is connected.

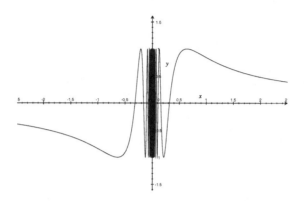

Figure XII.1   Topologist's sine curve.

Indeed, $A_+ := \left\{(x, \sin(\tfrac{1}{x})) : x > 0\right\}$ is the graph of a continuous function over an interval, hence is connected. Thus $\operatorname{cl} A_+ = A_+ \cup \{0\} \times [-1, 1]$ is connected by Corollary XII.1.26. Similarly $A_- := \left\{(x, \sin(\tfrac{1}{x})) : x < 0\right\}$ is connected, and so is $\operatorname{cl} A_- = A_- \cup \{0\} \times [-1, 1]$. Since $\operatorname{cl} A_+ \cap \operatorname{cl} A_- \neq \varnothing$, the union $A = \operatorname{cl} A_+ \cup \operatorname{cl} A_-$ is connected, by Proposition XII.1.20.

Note also that while continuous functions preserve connectedness, the converse is not true:

Example XII.1.28 (A function that preserves connected subsets but is not continuous). Let $f : \mathbb{R} \to \mathbb{R}$ be defined by $f(x) = \sin(\tfrac{1}{x})$ if $x \neq 0$ and $f(0) = 0$. The sequence defined by $x_n := \frac{2}{\pi(4n+1)}$ for $n \geq 1$ converges to 0 in $\mathbb{R}$ but $f(x_n) = 1$ does not converge to $0 = f(0)$, so that f is not continuous. On the other hand, f preserves connected sets: let C be a connected subset of $\mathbb{R}$, that is, an interval. if C does not contain 0, then $f(C)$ is connected because f is continuous at every point but 0. If $0 \in C$, then $f(C \setminus \{0\}) = [-1, 1]$ because C is an interval containing 0, so that $f(C) = [-1, 1]$ is connected.

Proposition XII.1.29. *A product of two connected spaces is connected.*

Proof. Let X and Y be connected. If $X \times Y$ is not connected, there is a continuous map $f : X \times Y \to |\iota_{\{0,1\}}|$ that is onto. Thus there is (x_0, y_0) with $f(x_0, y_0) = 0$ and (x_1, y_1) with $f(x_1, y_1) = 1$. The maps $f_{x_0} : Y \to |\iota_{\{0,1\}}|$ defined by $f_{x_0}(y) = f(x_0, y)$ and $f^{y_1} : X \to |\iota_{\{0,1\}}|$ defined by $f^{y_1}(x) = f(x, y_1)$ are constant, because they are continuous on a connected space. Thus $f_{x_0} \equiv 0$ because $f_{x_0}(y_0) = 0$ and $f^{y_1} \equiv 1$ because $f^{y_1}(x_1) = 1$. Thus

$$f(x_0, y_1) = f_{x_0}(y_1) = 0$$
$$= f^{y_1}(x_0) = 1$$

which is a contradiction. Thus $X \times Y$ is connected. $\qquad\square$

Example XII.1.30. Let

$$X := [0,1) \times [0,1] \cup \{(1,0)\}$$

be endowed with the pretopology ξ where $\mathcal{V}_\xi(x, y)$ is the neighborhood of (x, y) in the plane, whenever $y \neq 0$, $\mathcal{V}_\xi(x, 0) := \{\{x\} \times [0, \varepsilon) : 0 < \varepsilon < 1\}^\uparrow$ if $x \neq 1$ and $\mathcal{V}_\xi(1, 0) := \{(1 - \varepsilon, 1] \times \{0\} : 0 < \varepsilon < 1\}^\uparrow$. Let

$$A := [0,1) \times (0,1] \cup \{(1,0)\}.$$

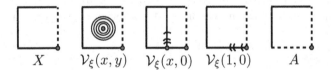

$$X \qquad \mathcal{V}_\xi(x, y) \qquad \mathcal{V}_\xi(x, 0) \qquad \mathcal{V}_\xi(1, 0) \qquad A$$

Figure XII.2

Then $\xi_{|A}$ is not connected, because $\{(1, 0)\}$ is clopen. However, $(\mathrm{T}\,\xi)_{|A}$ is the topology induced on A by the Euclidean topology of the plane. A is connected for this topology: it is the non-disjoint union of the connected product $[0, 1) \times (0, 1]$ with the connected segment $\{(x, 1 - x) : x \in [0, 1]\}$, so that Corollary XII.1.22 applies.

Theorem XII.1.31. *Let $\mathcal{X}$ be a set of topological spaces. The product $\prod_{X \in \mathcal{X}} X$ is connected if and only if each $X \in \mathcal{X}$ is connected.*

Proof. If $\prod_{X \in \mathcal{X}} X$ is connected, then each $X \in \mathcal{X}$ is connected because the projection $p_X : \prod_{X \in \mathcal{X}} X \to X$ is continuous and onto, so that Proposition XII.1.12 applies. Conversely, assume that each $X \in \mathcal{X}$ is connected. Let $\mathcal{F}$

be the family of finite subsets of $\mathcal{X}$, and let a be a fixed element of $\prod_{X \in \mathcal{X}} X$. For each $F \in \mathcal{F}$ let

$$A_F := \prod_{X \in \mathcal{X}} A_X,$$

where $A_X := X$ if $X \in F$, and $A_X := \{a(X)\}$ if $X \notin F$.

For each $F \in \mathcal{F}$, A_F is homeomorphic to $\prod_{X \in F} X$, which is connected by Proposition XII.1.29. Moreover, $a \in A_F$ for each $F \in \mathcal{F}$, so that $\bigcup_{F \in \mathcal{F}} A_F$ is connected by Corollary XII.1.22. Finally $\bigcup_{F \in \mathcal{F}} A_F$ is dense in $\prod_{X \in \mathcal{X}} X$, so that $\prod_{X \in \mathcal{X}} X = \operatorname{cl}\left(\bigcup_{F \in \mathcal{F}} A_F\right)$ is connected by Corollary XII.1.26. Indeed, if $x \in \prod_{X \in \mathcal{X}} X$ and $V \in \mathcal{N}_{\prod \mathcal{X}}(x)$ then by definition of the product topology, there is $F \in \mathcal{F}$ such that

$$V_F := \prod_{X \in F} V_X \times \prod_{X \in \mathcal{X} \setminus F} X \subset V,$$

where $V_X \in \mathcal{N}_X(x(X))$ for each $X \in F$. Of course,

$$\varnothing \neq V_F \cap A_F \subset V \cap \bigcup_{F \in \mathcal{F}} A_F.$$

$\square$

Example XII.1.32. Euclidean spaces $\mathbb{R}^n$, the space $\mathbb{R}^{\mathbb{R}}$ of real-valued functions on $\mathbb{R}$ with the topology of pointwise convergence, the space $\mathbb{R}^{\mathbb{N}}$ of sequences on $\mathbb{R}$ with the topology of pointwise convergence are thus all connected.

XII.2 Path connected and arc connected spaces

Definition XII.2.1. A *path* in X is a continuous map $p : [0,1] \to X$. The points $p(0)$ and $p(1)$ are the *endpoints* of the path. An *arc* L in a convergence space X is a subset of X that is homeomorphic to $[0,1]$, that is, such that there is an homeomorphism $l : [0,1] \to L$. The points $l(0)$ and $l(1)$ are the *endpoints of* L.

In view of Corollary IX.1.28, we conclude:

Proposition XII.2.2. *If ξ is a Hausdorff pseudotopology, the image of an injective path in ξ is an arc in ξ.*

Definition XII.2.3. A convergence space is called *arc connected* (respectively, *path connected*) if for any pair x_0, x_1 of points of X, there is an arc (respectively, a path) whose endpoints are x_0 and x_1.

Of course, every arc connected space is path connected. Moreover,

Proposition XII.2.4. *A path connected Hausdorff pseudotopology is arc connected.*

Proof. Let x_0 and x_1 be two points of $|\xi|$. There is a path $p : [0,1] \to |\xi|$ with $p(0) = x_0$ and $p(1) = x_1$. For each $x \in p([0,1])$, the set $p^-(x)$ is a closed subset of $[0,1]$ because p is continuous and $\{x\}$ is closed, as ξ is Hausdorff. Thus, $m_x := \inf p^-(x)$ belongs to $p^-(x)$. Moreover the sets $p^-(x)$ for $x \in p([0,1])$ form a partition of $[0,1]$. Thus the map $h : [0,1] \to [0,1]$ defined by $h(t) = m_{p(t)}$ is well defined, and by construction, $f := p \circ h : [0,1] \to |\xi|$ is one-to-one. Finally, f is continuous, so that $f([0,1])$ is an arc, by Proposition XII.2.2. Indeed, if $\mathcal{U} \in \mathbb{U}[0,1]$ with $t \in \lim \mathcal{U}$ then $p(t) \in \lim_\xi p[\mathcal{U}]$ by continuity of p, and $p(t) = f(t)$ and $p[\mathcal{U}] = f[\mathcal{U}]$, so that f is continuous. $\square$

The assumption that the space be Hausdorff is essential, as the following example shows.

Example XII.2.5 (A path connected topology that is not arc connected). Consider the Sierpiński space $\$_0$. It is path connected because the map $f : [0,1] \to \{0,1\}$ defined by $f(x) := 1$ if $0 < x \le 1$ and $f(0) := 0$ is continuous. Clearly, it is not arc connected.

In view of Corollary XII.1.23:

Proposition XII.2.6. *Every path connected space is connected.*

However, there are connected spaces that are not path connected:

Example XII.2.7. The topologist's sine curve of Example XII.1.27 is connected, but not path connected. Indeed, there is no path from $(\frac{1}{\pi}, 0)$ to $(0,0)$. Suppose to the contrary that there is such a path p. Then $p([0,1])$ is connected, and therefore must contain $\{(x, \sin(\frac{1}{x})) : 0 < x < \frac{1}{\pi}\}$. The sequence defined by $x_n := \frac{2}{\pi(4n+1)}$ for $n \ge 1$ converges to 0 in $[0,1]$ but $p(x_n) = (x_n, 1)$ does not converge to $(0,0)$ in the plane. Thus p cannot be continuous.

If a convergence ξ is arc connected (respectively, path connected) then its topological modification $\mathrm{T}\,\xi$ is as well. Indeed, an arc in ξ is a topological subspace, and remains a subspace in $\mathrm{T}\,\xi$. Similarly, if $p : [0,1] \to |\xi|$ is continuous, so is $p : [0,1] \to |\mathrm{T}\,\xi|$. However, the converse may be false:

Example XII.2.8 (A convergence ξ that is not path connected even though $T\xi$ is arc connected). Let ξ be the ultrafilter convergence associated with the usual topology of $[0,1]$, that is, $x \in \lim_\xi \mathcal{F}$ if and only if $\mathcal{F} \in \mathbb{U}(\mathcal{N}_{[0,1]}(x))$. Of course, $S\xi = T\xi$ is the usual topology of $[0,1]$ and is therefore arc connected. Moreover, $\mathcal{N}_\xi(\cdot) = \mathcal{N}_{[0,1]}(\cdot) := \mathcal{N}(\cdot)$. On the other hand, ξ is not path connected. Indeed, if $f : [0,1] \to |\xi|$ is continuous then

$$f[\mathcal{N}(x)] \in \mathbb{U}\left(\mathcal{N}(f(x))\right).$$

But $f[\mathcal{N}(x)] \in \mathbb{F}_1$ because $\mathcal{N}(x) \in \mathbb{F}_1$. By Exercise II.8.13, we conclude that $f[\mathcal{N}(x)]$ is a principal ultrafilter. Hence $f[\mathcal{N}(x)] = \{f(x)\}^\uparrow$ and f is locally constant. In view of Exercise XII.1.18, f is constant, and thus, is not a path between different points of $|\xi|$.

XII.3 Components and quasi-components

Definition XII.3.1. The *connected component* of an element x of a convergence space X is the union of all the connected subsets that contain x.

Proposition XII.3.2. *The connected components of a convergence space are closed, connected, and pairwise disjoint.*

Proof. Let $\mathcal{C}(x)$ denote the set of connected subsets of $|\xi|$ containing x. By definition $C(x) := \bigcup_{C \in \mathcal{C}(x)} C$ is the component of x. In view of Corollary XII.1.22, $C(x) \in \mathcal{C}(x)$ and $\mathrm{adh}\, C(x) \in \mathcal{C}(x)$ by Proposition XII.1.26. Thus $C(x)$ is closed and the largest connected subset of $|\xi|$ containing x. If $C(x_0) \cap C(x_1) \neq \varnothing$, then $C(x_0) \cup C(x_1)$ is a connected set containing x_0 and x_1, so that

$$C(x_0) \cup C(x_1) \subset C(x_0) \text{ and } C(x_0) \cup C(x_1) \subset C(x_1),$$

hence $C(x_0) = C(x_1)$. $\qquad\qquad\square$

Therefore, connected components of elements of a space X form a partition of X. Thus we call them *components of X*. Of course, X is connected if and only if X is its only component.

Definition XII.3.3. The *quasi-component* of an element x of a convergence space X is the intersection of all clopen subsets of X containing x.

Let $\mathcal{Q}(x)$ denote the family of clopen subsets of X that contain x, and let $Q(x)$ denote the quasi-component of x.

Proposition XII.3.4. *The quasi-components of the elements of a convergence space are closed and pairwise disjoint.*

Proof. Quasi-components are intersections of closed sets, hence closed. Let $x_0 \neq x_1$. If $x_1 \notin Q(x_0)$, then by definition, there is a clopen set Q such that $x_0 \in Q(x_0) \subset Q$ and $x_1 \notin Q$. Thus $X \setminus Q$ is a clopen set containing x_1, so that $Q(x_1) \subset X \setminus Q \subset X \setminus Q(x_0)$. $\qquad\square$

Therefore, quasi-components of elements of a space X form a partition of X. Thus we call them *quasi-components of X*. Of course, X is connected if and only if X is its only quasi-component.

Exercise XII.3.5. Let X be a topological space. Show that
$$C(x) \subset Q(x), \qquad\qquad (\text{XII.3.1})$$
for each $x \in X$. As a consequence:

Proposition XII.3.6. *If a quasi-component is connected, then it is a component.*

However, the reverse inclusion in (XII.3.1) may be false.

Example XII.3.7 (A quasi-component that is not a component). Let $X_n := [-1, 1] \times \{\frac{1}{n}\}$ for each $n \in \mathbb{N}_1$ and let $X_\infty := [-1, 1] \times \{0\}$. Consider $X := \bigcup_{n \in \mathbb{N}_1} X_n \cup X_\infty \setminus \{(0, 0)\}$ endowed with the induced topology from the plane. The components of X are exactly the subsets X_n for $n \in \mathbb{N}_1$, $[-1, 0) \times \{0\}$ and $(0, 1] \times \{0\}$.

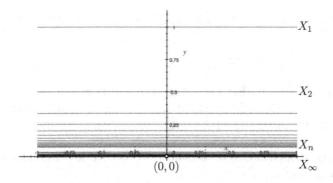

Figure XII.3

However, the quasi-component of $(1,0)$ is $X_\infty \setminus \{(0,0)\}$. Indeed, if Q is an open set that contains $(1,0)$, there exists $n_Q \in \mathbb{N}_1$ with $Q \cap X_n \neq \varnothing$ for all $n \geq n_Q$. If Q is closed as well, then $Q \cap X_n$ is a non-empty clopen subset of the connected set X_n for each $n \geq n_Q$, and we conclude that $X_n \subset Q$. Because Q is closed and

$$X_\infty \setminus \{(0,0)\} \subset \mathrm{cl}\left(\bigcup_{n \geq n_Q} X_n \right)$$

we conclude that $X_\infty \setminus \{(0,0)\} \subset Q$.

Theorem XII.3.8. *In a compact Hausdorff topological space, components and quasi-components coincide.*

Proof. Let X be a compact topological space. Assume that there is $x \in X$ such that $Q(x) \setminus C(x) \neq \varnothing$. In particular $Q(x)$ is not connected. Let F_0, F_1 be two non-empty closed disjoint subsets of $Q(x)$ such that $Q(x) = F_0 \cup F_1$. For the sake of argument, let us assume that $x \in F_0$. Since $Q(x)$ is closed, F_0 and F_1 are closed in X. In view of Corollary IX.2.6, X is normal, so that there are two disjoint open subsets O_0, O_1 of X with $F_0 \subset O_0$ and $F_1 \subset O_1$. Since $\mathcal{Q}(x)$ is a family of closed subsets of a compact space that is closed under finite intersection, we conclude that there is $Q \in \mathcal{Q}(x)$ with

$$Q(x) \subset Q \subset O := O_0 \cup O_1.$$

Otherwise, $\mathcal{Q}(x) \# O^c$, so that $\mathcal{Q}(x) \cup \{O^c\}$ is a filter-base of closed subsets of the compact space X, and has therefore non-empty intersection:

$$\varnothing \neq \bigcap_{Q \in \mathcal{Q}(x)} Q \cap O^c = Q(x) \cap O^c,$$

which is false, as $Q(x) \subset O$. The set $Q \cap O_0$ is open because Q is open. Moreover

$$\mathrm{cl}(Q \cap O_0) \subset Q \cap \mathrm{cl}\, O_0 \subset Q \cap (O_0 \cup O_1) \cap \mathrm{cl}\, O_0 = Q \cap O_0,$$

because Q is closed, so that $Q \cap O_0$ is also closed. Since $x \in F_0 \subset O_0$, we conclude that $Q \cap O_0 \in \mathcal{Q}(x)$. However, F_1 is a non-empty subset of $Q(x)$ disjoint from O_0 so that $Q(x) \not\subseteq Q \cap O_0$, which is a contradiction. $\square$

Definition XII.3.9. The *boundary* ∂A of a subset A of a topological space is by definition

$$\partial A := \mathrm{cl}\, A \setminus \mathrm{int}\, A.$$

Exercise XII.3.10. Show that if F is a closed subset of a topological space, then

$$F \setminus \partial F = \operatorname{int} F.$$

Proposition XII.3.11. *If F is a non-empty closed proper subset of a connected compact Hausdorff topological space, and C is a component of F, then $C \cap \partial F \neq \varnothing$.*

Proof. Suppose that C is a component of F, with $C \cap \partial F = \varnothing$. By Exercise XII.3.10, $F \setminus \partial F = \operatorname{int} F$ is an open set containing C. In view of Theorem XII.3.8, C is also a quasi-component, and thus is the intersection of the family $\mathcal{D}$ of clopen subsets (of F) that contain C. Thus there is $D \in \mathcal{D}$ with $C \subset D \subset F \setminus \partial F$, for otherwise, the filter-base $\mathcal{D}$ would mesh with the closed set $(F \setminus \partial F)^c$, and, by compactness, we would have $\bigcap_{D \in \mathcal{D}} D \cap (F \setminus \partial F)^c \neq \varnothing$, that is, $C \cap (F \setminus \partial F)^c \neq \varnothing$. As D is open in F, there is an open set O of X with $D = O \cap F$, so that $O \cap \partial F = \varnothing$, and $D = O \cap \operatorname{int} F$ is open in X. Moreover, D is closed in F, which is closed in X, so D is closed, hence clopen, in X. Since X is connected and $D \neq \varnothing$, $D = X$. Thus $\partial F = \varnothing$, so that F is also a clopen subset of the connected space X. But that contradicts the assumption that F is a non-empty proper closed subset. $\qquad\square$

Proposition XII.3.12. *If $X = \bigcup_{O \in \mathcal{A}} O$ where X is a topological space and $\mathcal{A}$ is a set of non-empty open, connected, pairwise disjoint subsets of X, then $\mathcal{A}$ is the set of components of X.*

Proof. If $x \in X$, there is a unique $O_x \in \mathcal{A}$ with $x \in O_x$. Then O_x is a connected set containing x so that $O_x \subset C(x)$. Moreover, for each $O \in \mathcal{A}$ with $O \cap C(x) \neq \varnothing$, $O \subset C(x)$, for O is connected. Thus

$$C(x) = \bigcup_{\mathcal{A} \ni O \cap C(x) \neq \varnothing} O = O_x \cup \left(\bigcup_{\mathcal{A} \ni O \cap C(x) \neq \varnothing, O \neq O_x} O \right).$$

By connectedness of $C(x)$, we conclude that $C(x) = O_x$ for $x \in C(x) \setminus \bigcup_{O \in \mathcal{A} \setminus \{O_x\}} O$. $\qquad\square$

The argument above applies without change to a family of closed sets if it is finite.

Proposition XII.3.13. *If $X = \bigcup_{F \in \mathcal{A}} F$ where X is a topological space and $\mathcal{A}$ is a finite set of non-empty closed, connected, pairwise disjoint subsets of X, then $\mathcal{A}$ is the set of components of X.*

The conclusion of Proposition XII.3.13 may fail if $\mathcal{A}$ is not finite.

Example XII.3.14. $\mathbb{R}^2$ (with its usual topology) can be partitioned by the pairwise disjoint non-empty connected closed subsets

$$\{\{x\} \times \mathbb{R} : x \in \mathbb{R}\},$$

but $\mathbb{R}^2$ is connected.

On the other hand, there are connected subsets of the plane $\mathbb{R}^2$ that are the disjoint union of countably many closed sets.

Example XII.3.15. ([3]) Let $S = L \cup M$, where $L := \bigcup_{k \in \mathbb{N}_1} L_k$, $M := \bigcup_{k \in \mathbb{N}_1} M_k$ and

$$L_n := \left\{ \left(x, \tfrac{1}{n} - (n-1)\, x \right) : x \geq 0 \right\}, \quad M_n := \left\{ \left(x, -\tfrac{1}{n} - (n-1)\, x \right) : x \leq 0 \right\}.$$

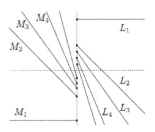

Then S is connected. Indeed,

$$L_n \cap \operatorname{cl} M \neq \varnothing \text{ et } M_n \cap \operatorname{cl} L \neq \varnothing \qquad\qquad (\text{XII.3.2})$$

for every n. Thus, if $f : S \to \{0,1\}$ is continuous, then $f(L_n)$ and $f(M_n)$ are singletons for every $n \in \mathbb{N}$. By definition of L_n and M_n, for every n there is k_n such that $f(L_n) = f(M_k)$ for every $k \geq k_n$. Therefore, $f(L_n) = f(L_p)$ for every $n, p \in \mathbb{N}$. In view of (XII.3.2), $f(L_n) = f(M_p)$ for every $n, p \in \mathbb{N}$.

However:

Theorem XII.3.16. *A compact Hausdorff connected topological space cannot be partitioned into an infinite countable family of non-empty pairwise disjoint closed sets.*

---

[3]This example was provided by Yohan Fleuriot and Louis Singrelin, students of S. Dolecki, in January 2012.

Proof. Assume to the contrary that a connected compact topological space X admits such a partition $X = \bigcup_{n \in \mathbb{N}} X_n$, where each X_n is closed and non-empty. Since X is normal by Corollary IX.2.7, and X_0, X_1 are disjoint closed subsets, there are two disjoint open sets O_0 and O_1 such that $X_0 \subset O_0$ and $X_1 \subset O_1$. Let $x \in X_1$ and let C_0 be the component of x in $\operatorname{cl} O_1$. Then $C_0 \cap X_0 = \varnothing$ and $C_0 \cap \partial O_1 \neq \varnothing$ by Proposition XII.3.11. Since $C_0 \cap X_1 \neq \varnothing$ and $X_1 \cap \partial O_1 = \varnothing$ (because $X_1 \subset O_1 = \operatorname{int} O_1$), there is $k > 1$ with $C_0 \cap X_k \neq \varnothing$. We have $C_0 = \bigcup_{n \in \mathbb{N}_1}(C_0 \cap X_n)$, where $\{C_0 \cap X_n : n \in \mathbb{N}_1\}$ are pairwise disjoint closed sets. Moreover, we have seen that $\operatorname{card}\{k \in \mathbb{N}_1 : C_0 \cap X_k \neq \varnothing\} \geq 2$. By connectedness of C_0, $\{k \in \mathbb{N}_1 : C_0 \cap X_k \neq \varnothing\}$ must be infinite. We proceed to build by induction a decreasing sequence $\{C_n\}_{n \in \mathbb{N}}$ of connected non-empty closed sets such that $X_n \cap C_n = \varnothing$ and $\{k \in \mathbb{N}_{n+1} : C_n \cap X_k \neq \varnothing\}$ is infinite, for every $n \in \mathbb{N}$. We have already built C_0, and suppose we have all C_j for $0 \leq j \leq n$. Then $C_n = \bigcup_{k \in \mathbb{N}_{n+1}}(C_n \cap X_k)$ is a compact connected space that is an infinite union of disjoint non-empty closed sets, so that the argument applied above to X to build C_0 can be applied to C_n to the effect that there is a non-empty closed connected set $C_{n+1} \subset C_n$ with $C_{n+1} \cap X_{n+1} = \varnothing$ and $\{k \in \mathbb{N}_{n+2} : C_{n+1} \cap X_k \neq \varnothing\}$ is infinite, and we conclude by induction. The resulting sequence satisfies

$$\bigcup_{n \in \mathbb{N}} X_n \cap \bigcap_{n \in \mathbb{N}} C_n = \varnothing,$$

and $\bigcap_{n \in \mathbb{N}} C_n \neq \varnothing$, for $\{C_n\}_{n \in \mathbb{N}}$ is a decreasing sequence of non-empty closed subsets of a compact space. We have a contradiction. $\qquad\square$

XII.4   Remarks on zero-dimensional spaces

We have already observed (Example XII.1.5) that a space with more than one point, at least one of which is isolated, is disconnected. The spaces of rationals and of irrationals are disconnected spaces without isolated points. In view of Theorem XII.1.31, the Cantor cube $\{0,1\}^{\mathbb{N}}$ is disconnected as a product of disconnected spaces. The Cantor set, which is homeomorphic to the Cantor cube by Proposition IX.1.20, is therefore disconnected. The *Baire space*

$$\mathbb{B} := \prod_{n \in \mathbb{N}} \mathbb{N} = \mathbb{N}^{\mathbb{N}}$$

of sequences on $\mathbb{N}$ with pointwise convergence is another important example of disconnected space, as it can also be seen as the product topology of countably many copies of the discrete, hence disconnected, space $\mathbb{N}$. Recall that a topological space is called *zero-dimensional* if each neighborhood filter has a filter base composed of clopen sets. Of course, every zero-dimensional space is disconnected. We have seen that discrete spaces are zero-dimensional, and so is the Sorgenfrey line. By Theorem IX.5.5, the Stone topology on the space $\mathbb{U}X$ of ultrafilters on a set X is also zero-dimensional.

Example XII.4.1. Since $(a, b) \cap \mathbb{Q} = [a, b] \cap \mathbb{Q}$ fro every $a, b \in \mathbb{R} \setminus \mathbb{Q}$ and $(a, b) \cap (\mathbb{R} \setminus \mathbb{Q}) = [a, b] \cap (\mathbb{R} \setminus \mathbb{Q})$ for every $a, b \in \mathbb{Q}$, the spaces $\mathbb{Q}$ of rational numbers and $\mathbb{R} \setminus \mathbb{Q}$ of irrational numbers are zero-dimensional. Similarly, the Cantor set of Example IX.1.19 is zero-dimensional.

Example XII.4.2. The following subspaces of $\mathbb{R}$:

$$\mathbb{R} \setminus \{0\}, \ \mathbb{R} \setminus \mathbb{Z},$$

are disconnected but not zero-dimensional.

Proposition XII.4.3. *A product of zero-dimensional spaces is zero-dimensional.*

Proof. In view of Exercise V.4.39, a base for open sets for $\prod_{\xi \in \Xi} \xi$ is of the form

$$\left\{ \prod_{\xi \in \Xi} U_\xi : U_\xi \in \mathcal{O}_\xi, \ \{\xi \in \Xi : U_\xi \neq |\xi|\} \in [\Xi]^{<\omega} \right\},$$

and each $U_\xi \neq |\xi|$ can be chosen clopen, so that each $\prod_{\xi \in \Xi} U_\xi$ is also closed. $\square$

As a consequence,

Example XII.4.4. The Baire space $\mathbb{B} = \mathbb{N}^\mathbb{N}$ is zero-dimensional.

Chapter XIII

Compactifications

XIII.1 Introduction

A convergence τ is called an *extension* of a convergence ξ if there exists an embedding $j : |\xi| \to |\tau|$ such that $\mathrm{adh}_\tau \, j \, (|\xi|) = |\tau|$. Up to a homeomorphism, a convergence can be seen as a dense subspace of its extension.

A convergence can be extended to another convergence in so many ways that one usually studies only some special classes of extensions. Most commonly studied classes of extensions are Hausdorff, regular extensions. Hausdorffness restricts a plethora of possible extensions.

Regularity (conjugated with density imposed by the definition) creates an interdependence between a convergence ξ and its extension τ. Indeed, let ξ be a dense subspace of a regular convergence τ. Then for each $y \in |\tau| \setminus |\xi|$ there exists $\mathcal{G}(y) \in \mathbb{F}(|\xi|)$ such that $y \in \lim_\tau \mathcal{G}(y)$. If $y_0 \in |\tau| \setminus |\xi|$ and $\mathcal{F} \in \mathbb{F}(|\tau| \setminus |\xi|)$ such that

$$y_0 \in \lim_\tau \mathcal{G}(\mathcal{F}),$$

where $\mathcal{G}(\mathcal{F})$ is the contour of $\mathcal{G}(\cdot)$ along $\mathcal{F}$ (which is a filter by Proposition VI.1.3) then $y_0 \in \lim_\tau \mathcal{F}$.

In other words, we can informally state that a filter on $|\xi|$ converges to an element $y \in |\tau| \setminus |\xi|$, then its extension to $|\tau|$ also converges to y.

Definition XIII.1.1. A compact extension τ of a convergence ξ is called a *compactification* of ξ. If $j : |\xi| \to |\tau|$ is the corresponding embedding, then $|\tau| \setminus j \, (|\xi|)$ is called the *remainder* of the compactification.

As compactness is a pseudotopological property, it is enough to study pseudotopological compactifications of pseudotopologies. How to construct a compactification of a non-compact pseudotopology? A first idea that comes to mind is to make all non-adherent filters converge to one additional point.

Definition XIII.1.2 (One-point compactification). Let ξ be a convergence on X and let $\infty_X \notin X$. Define on $Y := X \cup \{\infty_X\}$ the following convergence τ. If $x \in X$ then $x \in \lim_\tau \mathcal{F}$ if and only if $X \in \mathcal{F}$ and $x \in \lim_\xi (\mathcal{F} \vee X)$, and $\infty_X \in \lim_\tau \mathcal{F}$ if and only if $\mathrm{adh}_\xi (\mathcal{F} \vee X) = \varnothing$. Note that $\{\infty_X\}^\uparrow \vee X$ is degenerate, so that $\mathrm{adh}_\xi(\{\infty_X\}^\uparrow \vee X) = \varnothing$.

Clearly, $|\xi| = X$ is dense in $|\tau| = Y$, and τ is compact. Thus, by construction, the one-point compactification τ of a convergence ξ is indeed a compactification of ξ. We call it *one-point compactification of* ξ, and we denote it by $\overline{\omega\xi}$. Moreover ([1]),

Theorem XIII.1.3. *The one-point compactification $\overline{\omega\xi}$ of a Hausdorff pseudotopology ξ is a Hausdorff pseudotopological compactification of ξ such that*

$$\mathrm{compl}\,(\xi) \leq \mathfrak{p}\left(\overline{\omega\xi}, \infty_X\right). \tag{XIII.1.1}$$

Proof. The convergence $\overline{\omega\xi}$ is Hausdorff, for a filter on X either converges to a unique point of X, or to ∞_X in Y. To see that $\overline{\omega\xi}$ is a pseudotopology, let $x \in \bigcap_{\mathcal{U} \in \beta(\mathcal{F})} \lim_{\overline{\omega\xi}} \mathcal{U}$. If $x \in X$, then $\lim_{\overline{\omega\xi}} \mathcal{U} = \lim_\xi \mathcal{U}$ for each $\mathcal{U} \in \beta(\mathcal{F})$ and then $x \in \lim_\xi \mathcal{F} = \lim_{\overline{\omega\xi}} \mathcal{F}$ by pseudotopologicity of ξ. If $x = \infty_X$ then $\lim_\xi (\mathcal{U} \vee X) = \varnothing$ for every $\mathcal{U} \in \beta(\mathcal{F})$, so that $\mathrm{adh}_\xi(\mathcal{F} \vee X) = \varnothing$, that is, $\infty_X \in \lim_{\overline{\omega\xi}} \mathcal{F}$.

To see that $\overline{\omega\xi}$ is compact, notice that if $\mathcal{U}$ is a free ultrafilter on Y, then either $\lim_\xi (\mathcal{U} \vee X) \neq \varnothing$ or $\lim_\xi (\mathcal{U} \vee X) = \varnothing$ and thus $\lim_{\overline{\omega\xi}} \mathcal{U} = \{\infty_X\}$.

To prove (XIII.1.1), let $\mathbb{G}$ be a pavement at ∞_X for $\overline{\omega\xi}$ such that $\mathrm{card}\,\mathbb{G} = \mathfrak{p}\left(\overline{\omega\xi}, \infty_X\right)$. The collection

$$\mathbb{G}_X := \{\mathcal{G} \vee X : \{\infty_X\}^\uparrow \neq \mathcal{G} \in \mathbb{G}\}$$

consists of ξ-non-adherent filters. To see that ξ is $\mathbb{G}_X$-cocomplete, let $\mathcal{U} \in \mathbb{U}X$ that is dissociated from $\mathbb{G}_X$. This means that $\infty_X \notin \lim_{\overline{\omega\xi}} \mathcal{U}$ and thus $\lim_\xi \mathcal{U} \neq \varnothing$. This means that $\mathrm{compl}\,(\xi) \leq \mathrm{card}\,\mathbb{G} = \mathfrak{p}\left(\overline{\omega\xi}, \infty_X\right)$. $\qquad\square$

Corollary XIII.1.4. *If the one-point compactification of ξ is pretopological at the added point then ξ is locally compactoid.*

---

[1]Recall that $\mathrm{compl}\,(\xi)$ denotes the completeness number of a convergence ξ, and $\mathfrak{p}\,(\tau, x)$ denotes the paving number of a convergence τ at x.

Another idea of compactifying a Hausdorff non-compact pseudotopology, is to make converge each originally non-convergent ultrafilter $\mathcal{U}$ to an added point $y_{\mathcal{U}}$ so that $\mathcal{U} \longmapsto y_{\mathcal{U}}$ be injective. But doing so one adds a possibly infinite set to the original underlying set. In order for this construction to yield a compactification, one need define limits of all ultrafilters on the added set.

Definition XIII.1.5 (Richardson compactification). [92] Let ξ be a convergence on X. Let $q_{\xi} : \mathbb{U}X \to \mathbb{U}X$ by

$$q_{\xi}(\mathcal{U}) := \begin{cases} \{x\}^{\uparrow} & \text{if } x \in \lim_{\xi} \mathcal{U}, \\ \mathcal{U} & \text{if } \lim_{\xi} \mathcal{U} = \varnothing. \end{cases}$$

We identify X with a subset $\left\{ \{x\}^{\uparrow} : x \in X \right\}$ of $Y := q_{\xi}(\mathbb{U}X)$. Define on Y a convergence τ for ultrafilters

$$\lim_{\tau} \mathcal{W} := \begin{cases} \lim_{\xi} \mathcal{W} & \text{if } X \in \mathcal{W} \text{ and } \lim_{\xi} (\mathcal{W} \vee X) \neq \varnothing, \\ \{\mathcal{W}\} & \text{if } \lim_{\xi} \mathcal{W}^{\star} = \varnothing, \end{cases} \qquad \text{(XIII.1.2)}$$

where $\mathcal{W}^{\star}$ is the contour of $\mathcal{W}$ $(^2)$.

In the definition above, ξ-non-convergent ultrafilters on X occur both as ultrafilter-bases on Y and as elements of Y. Notice that if $X \in \mathcal{W} \in \mathbb{U}Y$, then $\mathcal{W}^{\star} = \bigcup_{\mathcal{W} \in \mathcal{W}} \bigcap_{x \in H} \{x\}^{\uparrow} = \mathcal{W}$. In view of Theorem II.9.2, this observation can be rephrased, as follows

$$\lim_{\xi} \mathcal{U} = \varnothing \implies \{\mathcal{U}\} = \lim_{\tau} \left(\mathcal{U}^{\Uparrow} \vee Y \right).$$

Proposition XIII.1.6. *The Richardson compactification of a Hausdorff pseudotopology ξ is a Hausdorff pseudotopological compactification of ξ.*

Proof. If ξ is Hausdorff, it follows from the definition that every ultrafilter is convergent to a single point, so that τ is Hausdorff and compact. As ξ is a pseudotopology and each filter that converges in τ but not in ξ, is an ultrafilter, τ is also a pseudotopology. $\square$

Proposition XIII.1.7. *If τ is the Richardson compactification of ξ and σ is a Hausdorff regular pseudotopological compactification of ξ, then there exists $f \in C(\tau, \sigma)$ which restriction to $|\xi|$ is a dense embedding.*

---

^{2}Recall that the contour $\mathcal{U}(\mathcal{W})$ is an ultrafilter on X defined by $\mathcal{W}^{\star} = \mathcal{U}(\mathcal{W}) := \bigcup_{\mathcal{W} \in \mathcal{W}} \bigcap_{\mathcal{U} \in \mathcal{W}} \mathcal{U}$.

Proof. Recall that ξ is a convergence on X and τ is defined on $Y \supset X$. Assume that ξ is a subspace of σ defined on a set $Z \supset X$. Define $f :$ $|\tau| \to |\sigma|$ so that $f(x) := x$ if $x \in X$. For each $z \in Z \setminus X$, consider $\mathbb{U}(z) := \{\mathcal{U} \in \mathbb{U}X : z \in \lim_\sigma \mathcal{U}\}$. Then let $f(\mathcal{U}) := z$ if $\mathcal{U} \in \mathbb{U}(z)$. As $\bigcup_{z \in Z \setminus X} \mathbb{U}(z) = \mathbb{U}X \setminus \xi^- X$, this defines a surjection of $|\tau|$ onto $|\sigma|$. From the construction, $f_{|X}$ is a dense embedding. By Exercise XIII.1.8 below, $f \in C(\tau, \sigma)$. □

Exercise XIII.1.8. Show that f constructed in the proof of Proposition XIII.1.7 is continuous.

Since Hausdorffness and regularity are hereditary, convergences with Hausdorff regular extensions are Hausdorff and regular. But, by Corollary IX.2.4, Hausdorff compact regular pseudotopologies are necessarily topologies. Since Hausdorffness, regularity and topologicity are hereditary properties, we shall only discuss compactifications within the class of Hausdorff regular topologies. It turns out that this class becomes automatically even narrower.

Indeed, by Corollary VII.6.12, each Hausdorff functionally regular topology can be embedded in a Hilbert cube of some dimension. As each Hilbert cube is compact, every Hausdorff functionally regular topology admits a compactification. By Corollary IX.2.7, each Hausdorff compact topology is normal, hence functionally regular. Normality is not hereditary, but functional regularity is. We have proved:

Proposition XIII.1.9. *Let ξ be a convergence. The following are equivalent:*

(1) ξ admits a regular Hausdorff pseudotopological compactification;
(2) ξ admits a Hausdorff topological compactification;
(3) ξ is a Hausdorff functionally regular topology.

Therefore, Hausdorff functionally regular topologies constitute a most natural framework to study compactifications.

XIII.2   Compactifications of functionally regular topologies

In this section we focus on compactifications that are Hausdorff topologies. In view of Proposition XIII.1.9, this means that we restrict ourselves to Hausdorff functionally regular topologies ξ and study their topological

Hausdorff compactifications. In this context, we use the term *compactification* to mean Hausdorff topological compactification, for the rest of the chapter.

If τ is a compactification of ξ, then ξ can be identified with a dense subset of τ, (up to a homeomorphism) so that in most cases we shall think of ξ as a subspace of its compactification τ ([3]). In this context, Proposition V.4.42 applies to the effect that a functionally regular Hausdorff topology and its compactifications have the same density:

$$d(\xi) = d(\tau).$$

As a result, Propositions IV.9.29 and VI.3.14 yield:

Proposition XIII.2.1. *If τ is a compactification of a functionally regular Hausdorff topology ξ, then*

$$\operatorname{card} |\tau| \leq 2^{2^{d(\xi)}} \ \text{ and } \ w(\tau) \leq 2^{d(\xi)}.$$

Nevertheless in some situations it is instrumental to consider simultaneously a compactification and the corresponding embedding, even if it is an inclusion map.

Therefore if ξ is a functionally regular Hausdorff topology and j is a dense embedding of ξ in a Hausdorff compact topology τ, then we most often write $\overline{j\xi} := \tau$.

If $\tau_0 = \overline{j_0\xi}$ and $\tau_1 = \overline{j_1\xi}$ are compactifications of ξ, then we say that $\overline{j_0\xi}$ is *bigger* than $\overline{j_1\xi}$, in symbols,

$$\overline{j_0\xi} \gtrsim \overline{j_1\xi},$$

if there exists $f \in C(\tau_0, \tau_1)$ such that $j_1 = f \circ j_0$.

It is obvious that $\gtrsim$ is a transitive and reflexive relation. Two compactifications $\overline{j_0\xi}$ and $\overline{j_1\xi}$ are said to be *equivalent*, in symbols,

$$\overline{j_0\xi} \approx \overline{j_1\xi},$$

if $\overline{j_0\xi} \lesssim \overline{j_1\xi_1}$ and $\overline{j_1\xi} \lesssim \overline{j_0\xi}$.

It is easy to see that $\gtrsim$ is antisymmetric on the equivalence classes of compactifications.

Proposition XIII.2.2. *Two compactifications $\overline{j_0\xi}$ and $\overline{j_1\xi}$ are equivalent if and only if there exists $h \in H\left(\overline{j_0\xi}, \overline{j_1\xi}\right)$ such that $j_0 = h \circ j_1$.*

---

[3]A convergence can have several dense homeomorphic embeddings in its compactification. In such cases, we consider several homeomorphic compactifications of that convergence.

Proof. If h is a homeomorphism such that $j_0 = h \circ j_1$, then, of course, $\overline{j_0\xi}$ and $\overline{j_1\xi}$ are equivalent. Suppose that there exists $f_0 \in C\left(\overline{j_0\xi}, \overline{j_1\xi}\right)$ and $f_1 \in C\left(\overline{j_1\xi}, \overline{j_0\xi}\right)$ such that $j_0 = f_0 j_1$ and $j_1 = f_1 j_0$. Then

$$j_1 = f_1 j_0 = f_1 f_0 j_1,$$

hence $j_1(x) = f_1 f_0 j_1(x)$ for each $x \in |\xi|$. Therefore, $f_1 f_0 \in C\left(\overline{j_0\xi}, \overline{j_0\xi}\right)$ coincides with the identity on a dense set. By Proposition IV.9.38, $f_1 f_0$ is the identity on $\left|\overline{j_0\xi}\right|$ because $\overline{j_0\xi}$ is Hausdorff. It follows that f_0 and f_1 are homeomorphisms and $f_1^{-1} = f_0$. $\qquad\square$

We shall see that if there exists $f \in C\left(\overline{c_0\xi}, \overline{c_1\xi}\right)$ such that $f \circ c_0 = c_1$ (in particular, $\overline{c_0\xi} \gtrsim \overline{c_1\xi}$), then f maps $c_0(|\xi|)$ on $c_1(|\xi|)$ and the remainder of $\overline{c_0\xi}$ on that of $\overline{c_1\xi}$.

Lemma XIII.2.3. *Let ξ be a Hausdorff topology on a set X, $f \in C(\xi, \tau)$ and let W be a dense subset of $|\xi|$. If $f_{|W} : |\xi_{|W}| \to |\tau_{|f(W)}|$ is a homeomorphism, then*

$$f(W) \cap f(X \setminus W) = \varnothing.$$

Proof. Suppose that there exists $x_0 \in X \setminus W$ such that $f(x_0) \in f(W)$. Consider now the restriction $\widetilde{\xi}$ of ξ to $W \cup \{x_0\}$, the restriction $\widetilde{\tau}$ of τ to $f(W)$ and $\widetilde{f} \in C(\widetilde{\xi}, \widetilde{\tau})$.

Then there exists $x_1 \in W$ such that $\widetilde{f}(x_0) = \widetilde{f}(x_1)$. As $\widetilde{\xi}$ is Hausdorff, there exist $O_0 \in \mathcal{O}_{\widetilde{\xi}}(x_0)$ and $O_1 \in \mathcal{O}_{\widetilde{\xi}}(x_1)$ such that $O_0 \cap O_1 = \varnothing$. As O_1 is open, $W \setminus O_1$ is closed in W, hence $f(W \setminus O_1)$ is closed in $f(W)$, because $f_{|W}$ is an homeomorphism. Thus $f^{-}f(W \setminus O_1) = W \setminus O_1$ is closed in $W \cup \{x_0\}$, by continuity. Thus $x_0 \notin \mathrm{cl}_{\widetilde{\xi}} O_1$ because $O_0 \cap O_1 = \varnothing$ and $x_0 \notin \mathrm{cl}_{\widetilde{\xi}} W \setminus O_1$, so that $x_0 \notin \mathrm{cl}_{\widetilde{\xi}} W$, which is a contradiction, because W is dense. $\qquad\square$

Theorem XIII.2.4. *If $\overline{c_0\xi}$ and $\overline{c_1\xi}$ are compactifications of ξ and $f \in C\left(\overline{c_0\xi}, \overline{c_1\xi}\right)$ fulfills $f \circ c_0 = c_1$, then*

$$f(c_0(|\xi|)) = c_1(|\xi|) \quad \text{and} \quad f\left(\left|\overline{c_0\xi}\right| \setminus c_0(|\xi|)\right) = \left|\overline{c_1\xi}\right| \setminus c_1(|\xi|).$$

Proof. The first equality follows from $f \circ c_0 = c_1$. As $c_0(|\xi|)$ is dense in $\left|\overline{c_0\xi}\right|$ and $f \circ c_0 = c_1$, the restriction of f to $c_0(|\xi|)$ is a homeomorphism, hence the second inequality holds by virtue of Lemma XIII.2.3. $\qquad\square$

Consider the class $\mathfrak{C}(\xi)$ of all the equivalence classes of compactifications of a Hausdorff functionally regular topology ξ. As ξ can be embedded in a

Hilbert cube of some dimension (Corollary VII.6.12), each compactification of ξ is homeomorphic to a closed subspace of that cube. Therefore $\mathfrak{C}(\xi)$ is a set.

Note that

Proposition XIII.2.5. *If ξ is compact, then $\mathfrak{C}(\xi) = \{\xi\}$.*

Proof. Indeed, if $\overline{c\xi}$ is a compactification of ξ then $c(|\xi|)$ is compact as the image of a compact set by a continuous map, hence closed, because $\overline{c\xi}$ is Hausdorff. On the other hand, by assumption, $c(|\xi|)$ is dense, hence $c(|\xi|) = |\overline{c\xi}|$, hence $c \in H(\xi, \overline{c\xi})$. Consequently, if $\overline{c_0\xi}$ and $\overline{c_1\xi}$ are compactifications of ξ, then $c_1 \circ c_0^- \in H(\overline{c_0\xi}, \overline{c_1\xi})$, that is, $\overline{c_0\xi}$ and $\overline{c_1\xi}$ are equivalent. $\square$

Theorem XIII.2.6. *Each non-empty subset of $\mathfrak{C}(\xi)$ has a least upper bound for $\gtrsim$.*

Proof. Let $\varnothing \neq T \subset \mathfrak{C}(\xi)$. For every $\tau \in T$, let $j_\tau \in C(\xi, \tau)$ be a dense embedding. Then

$$J := \Delta_{\tau \in T} j_\tau : |\xi| \to \prod_{\tau \in T} |\tau|$$

is an embedding by Corollary VII.6.12.

We denote by τ_T the restriction of $\prod_{\tau \in T} \tau$ to the closure of $J(\xi)$. We shall see that τ_T is a least upper bound of T in $(\mathfrak{C}(\xi), \lesssim)$.

For every $\tau_0 \in T$, consider the projection $p_{\tau_0} : \prod_{\tau \in T} |\tau| \to |\tau_0|$. Then $j_{\tau_0} = p_{\tau_0} \circ \Delta_{\tau \in T} j_\tau$, hence $\tau_0 \lesssim \tau_T$.

If now $\tau_1 \in \mathfrak{C}(\xi)$ is such that $\tau \lesssim \tau_1$ for each $\tau \in T$, that is, for each $\tau \in T$ there exists $f_\tau \in C(\tau_1, \tau)$ such that $j_\tau = f_\tau \circ j_{\tau_1}$. Then $\Delta_{\tau \in T} f_\tau : |\tau_1| \to \prod_{\tau \in T} |\tau|$ and

$$(\Delta_{\tau \in T} f_\tau) \circ j_{\tau_1} = \Delta_{\tau \in T} j_\tau,$$

which means that $\tau_T \leq \tau_1$. $\square$

Corollary XIII.2.7. *Every Hausdorff functionally regular topology admits a greatest compactification.*

The greatest compactification of ξ is called the *Čech-Stone compactification* of ξ and is denoted by $\overline{\beta\xi}$.

Remark XIII.2.8. Traditionally topological spaces are denoted by capital letters, say, X, Y and their Čech-Stone compactifications by $\beta X, \beta Y$ respectively. In such situations we shall follow the tradition.

Our notation $\overline{\beta\xi}$ follows the convention in which $\beta : |\xi| \to \beta\,(|\xi|) \subset |\overline{\beta\xi}|$ is a dense embedding. As $\beta\xi$ stands for the final convergence on $\beta\,(|\xi|)$ for (β, ξ), we should distinguish $\beta\xi$ and $\overline{\beta\xi}$.

In the case, when X is a discrete topological space, then the underlying set of its Čech-Stone compactification βX can be seen as the set $\mathbb{U}X$ of all ultrafilters on X endowed with the Stone topology β. This fact leads to the notation $\beta(\mathcal{F})$ for the set of all ultrafilters that are finer than a filter $\mathcal{F}$.

Theorem XIII.2.9. *A Hausdorff topology admits a least (topological Hausdorff) compactification if and only if it is locally compact.*

Proof. The one-point compactification $\overline{\omega\xi}$ of Theorem XIII.1.3 is the least compactification for $\gtrsim$, but may fail to be topological. Indeed, if $\overline{j\xi}$ is another compactification, the map $f : |\overline{j\xi}| \to |\overline{\omega\xi}|$ defined by $f(x) = x$ for $x \in |\xi|$ and $f(y) = \infty_X$ otherwise is continuous and $\omega = f \circ j$.

If ξ is a Hausdorff locally compact topology, then $\overline{\omega\xi}$ is regular, hence topological, by Theorem IX.2.1.

Indeed, under these assumptions, ξ is regular by Proposition IX.3.10. Moreover, if $x \in \lim_{\overline{\omega\xi}} \mathcal{F}$, either $x \neq \infty_X$ in which case $x \in \lim_{\xi} \mathcal{F}$ and there is a ξ-compact set $K \in \mathcal{F}$, or $x = \infty_X$ and $\lim_{\xi} \mathrm{cl}_{\xi}^{\natural}(\mathcal{F} \vee X) = \varnothing$, by regularity of ξ. In the former case $\mathrm{cl}_{\overline{\omega\xi}} K = \mathrm{cl}_{\xi} K$ because K is compact and thus $\mathrm{cl}_{\overline{\omega\xi}}^{\natural} \mathcal{F} = \mathrm{cl}_{\xi}^{\natural} \mathcal{F}$ converges for ξ hence for $\overline{\omega\xi}$ to x, by regularity of ξ.

In the latter case, $\mathrm{cl}_{\overline{\omega\xi}}^{\natural}(\mathcal{F} \vee X) = (\mathrm{cl}_{\xi}^{\natural} \mathcal{F} \vee X) \wedge \{\infty_X\}^{\uparrow}$, so that $\infty_X \in \lim_{\overline{\omega\xi}} \mathrm{cl}_{\overline{\omega\xi}}^{\natural} \mathcal{F}$.

Conversely, if $\overline{\omega\xi}$ is topological, then ξ is locally compactoid by Corollary XIII.1.4, hence locally compact, for the closure of a relatively compact set in a Hausdorff topology is compact. □

When the one-point compactification is topological Hausdorff (as in Theorem XIII.2.9), it is called the *Alexandrov compactification*.

Example XIII.2.10 (*AD* compact topology). Let $\mathcal{A}$ be an *AD* family on X, and consider the associated AD topology $\zeta(\mathcal{A})$ on $Y = X \cup \mathcal{A}$, which is locally compact and Hausdorff by Proposition IX.12.4. Of course, all the free ultrafilters that do not converge in $\zeta := \zeta(\mathcal{A})$ converge to ∞_Y in its Alexandrov compactification $\overline{\omega\zeta}$, which we call *AD compact topology associated to* $\mathcal{A}$.

Let us describe the neighborhood filter of ∞_Y. Every ultrafilter $\mathcal{U} \in \beta^*(A)$ converges to $A \in \mathcal{A} \subset Y$; each free ultrafilter on X, that does not belong to $\bigcup_{A \in \mathcal{A}} \beta^*(A)$, does not converge in ζ, hence converges to ∞_Y in $\overline{\omega\zeta}$. Therefore, the trace of $\mathcal{N}_{\overline{\omega\zeta}}(\infty_Y)$ on X is equal to the intersection of all the elements of $\beta^*(X) \setminus \bigcup_{A \in \mathcal{A}} \beta^*(A)$, that is, to the residual filter $\mathcal{F}_{\mathcal{A}}$ of $\mathcal{A}$:

$$\mathcal{N}_{\overline{\omega\zeta}}(\infty_Y) \vee X = \mathcal{F}_{\mathcal{A}}.$$

The subset $\mathcal{A}$ of (Y, ζ) is discrete, so that every ultrafilter on $\mathcal{A}$ converges to ∞_Y in $\overline{\omega\zeta}$. Therefore, the cofinite filter of $\mathcal{A}$ is the coarsest filter on $\mathcal{A}$ that converges to ∞_Y. As a result, the trace of $\mathcal{N}_{\overline{\omega\zeta}}(\infty_Y)$ on $\mathcal{A}$ is the cofinite filter of $\mathcal{A}$:

$$\mathcal{N}_{\overline{\omega\zeta}}(\infty_Y) \vee \mathcal{A} = (\mathcal{A})_0.$$

In order to describe the Čech-Stone compactification more explicitly and study its properties, we will need filters in the lattice of functionally closed sets.

XIII.3 Filters in lattices

A family $\mathcal{Z}$ of subsets of X such that $\mathcal{Z} = \mathcal{Z}^{\cap}$, that is,

$$Z_1, \ldots, Z_n \in \mathcal{Z} \implies Z_1 \cap \cdots \cap Z_n \in \mathcal{Z} \qquad \text{(XIII.3.1)}$$

is called *closed under finite intersections*. If a family $\mathcal{Z}$ of subsets of X is closed under finite intersections and finite unions, then it is called a *lattice* of subsets of X.

Example XIII.3.1. The families of closed sets, functionally closed sets, open sets and functionally open sets are lattices (of a convergence space).

For the rest of this section, $\mathcal{Z}$ is a family on X that is closed under finite intersections.

A subfamily $\mathcal{F}$ of $\mathcal{Z}$ is called a *$\mathcal{Z}$-filter* if

$$\varnothing \notin \mathcal{F},$$
$$F_0, F_1 \in \mathcal{F} \implies F_0 \cap F_1 \in \mathcal{F},$$
$$F \in \mathcal{F} \text{ and } F \subset G \in \mathcal{Z} \implies G \in \mathcal{F}.$$

We denote by $\mathbb{F}_{\mathcal{Z}}X$ the set of $\mathcal{Z}$-filters on X. If $\mathcal{F}$ is a filter on X, then $\mathcal{F} \cap \mathcal{Z}$ is a $\mathcal{Z}$-filter on X. On the other hand, each $\mathcal{Z}$-filter on X is a filter base on X. Notice that if $\mathcal{F}$ is a filter on X, then

$$(\mathcal{F} \cap \mathcal{Z})^{\uparrow} \subset \mathcal{F},$$

but in general the equality does not hold.

By the Zorn-Kuratowski lemma A.9.10, for each $\mathcal{Z}$-filter $\mathcal{F}$ there exists a maximal $\mathcal{Z}$-filter $\mathcal{U} \supset \mathcal{F}$, called a $\mathcal{Z}$-*ultrafilter*. Let $\mathbb{U}_{\mathcal{Z}}(X)$ stand for the set of all $\mathcal{Z}$-ultrafilters on X.

Proposition XIII.3.2. *A $\mathcal{Z}$-filter $\mathcal{F}$ is a $\mathcal{Z}$-ultrafilter if and only if $Z \in \mathcal{Z}$ and $Z \in \mathcal{F}^{\#}$ imply $Z \in \mathcal{F}$.*

Proof. If $Z \in \mathcal{Z} \cap \mathcal{F}^{\#}$, then $\{Z \cap F : F \in \mathcal{F}\}$ is a base of a $\mathcal{Z}$-filter $\mathcal{U}$ such that $\mathcal{F} \subset \mathcal{U}$ and thus $Z = Z \cap X \in \mathcal{U}$. Therefore if $\mathcal{F}$ is a $\mathcal{Z}$-ultrafilter, then $\mathcal{F} = \mathcal{U}$, hence $Z \in \mathcal{F}$.

Conversely if $\mathcal{F}$ is not a $\mathcal{Z}$-ultrafilter, then there exists a $\mathcal{Z}$-filter $\mathcal{U} \supset \mathcal{F}$ and $Z \in \mathcal{U} \setminus \mathcal{F}$. Then $Z \in \mathcal{U}^{\#} \subset \mathcal{F}^{\#}$, that is, $Z \in \mathcal{F}^{\#} \setminus \mathcal{F}$. $\square$

Proposition XIII.3.3. *If $\mathcal{U}$ and $\mathcal{W}$ are distinct $\mathcal{Z}$-ultrafilters, then there exist $U \in \mathcal{U}$ and $W \in \mathcal{W}$ such that $U \cap W = \varnothing$.*

Proof. If not, $\{U \cap W : U \in \mathcal{U}, W \in \mathcal{W}\}$ is a base of a $\mathcal{Z}$-filter including $\mathcal{U}$ and $\mathcal{W}$, that contradicts the maximality of $\mathcal{U}$ and of $\mathcal{W}$. $\square$

A $\mathcal{Z}$-filter $\mathcal{F}$ is called *prime* if for every $Z_0, Z_1 \in \mathcal{Z}$,

$$Z_0 \cup Z_1 \in \mathcal{F} \Longrightarrow Z_0 \in \mathcal{F} \text{ or } Z_1 \in \mathcal{F}.$$

Proposition XIII.3.4. *Every $\mathcal{Z}$-ultrafilter is prime.*

Proof. If $\mathcal{U}$ is a $\mathcal{Z}$-ultrafilter, $Z_0 \notin \mathcal{U}$ and $Z_1 \notin \mathcal{U}$, then by Proposition XIII.3.2, there exist $U_0, U_1 \in \mathcal{U}$ such that $U_0 \cap Z_0 = \varnothing$ and $U_1 \cap Z_1 = \varnothing$ and, consequently, $(U_0 \cap U_1) \cap (Z_0 \cup Z_1) = \varnothing$, that is, $Z_0 \cup Z_1 \notin \mathcal{U}^{\#}$, and since $\mathcal{U}$ is a $\mathcal{Z}$-ultrafilter, $Z_0 \cup Z_1 \notin \mathcal{U}$. $\square$

By Proposition II.4.1 and Theorem II.4.5, the converse of Proposition XIII.3.4 is true if $\mathcal{Z}$ is the lattice of all subsets.

Proposition XIII.3.5. *If $\mathcal{U}$ is an ultrafilter on X, then $\mathcal{U} \cap \mathcal{Z}$ is a prime $\mathcal{Z}$-filter.*

Proof. Let $\mathcal{U}$ be an ultrafilter and let $Z_0, Z_1 \in \mathcal{Z}$ be such that $Z_0 \cup Z_1 \in \mathcal{U} \cap \mathcal{Z}$. As $\mathcal{U}$ is prime, either $Z_0 \in \mathcal{U}$ or $Z_1 \in \mathcal{U}$ and since $Z_0, Z_1 \in \mathcal{Z}$, either $Z_0 \in \mathcal{U} \cap \mathcal{Z}$ or $Z_1 \in \mathcal{U} \cap \mathcal{Z}$. $\qquad\square$

Example XIII.3.6 (non-maximal prime $\mathcal{Z}$-filters). Let $X = \mathbb{R}$ with the natural topology and let $\mathcal{Z}$ be the lattice of closed subsets of $\mathbb{R}$. If $\mathcal{W}_+$ an ultrafilter containing $\{\frac{1}{n}, \frac{1}{n+1}, \ldots\}$ for each $n \in \mathbb{N}_1$ and $\mathcal{W}_-$ an ultrafilter containing $\{-\frac{1}{n}, -\frac{1}{n+1}, \ldots\}$ for each $n \in \mathbb{N}_1$, then $\mathcal{W}_+ \cap \mathcal{Z}$ and $\mathcal{W}_- \cap \mathcal{Z}$ are distinct prime $\mathcal{Z}$-filters, but not $\mathcal{Z}$-ultrafilters. Indeed, $0 \in Z \subset \mathbb{R}_+$ if $Z \in \mathcal{W}_+ \cap \mathcal{Z}$ and $0 \in Z \subset \mathbb{R}_-$ if $Z \in \mathcal{W}_- \cap \mathcal{Z}$. Therefore the principal $\mathcal{Z}$-ultrafilter of $\{Z \in \mathcal{Z} : 0 \in Z\}$ includes $\mathcal{W}_+ \cap \mathcal{Z}$ and $\mathcal{W}_- \cap \mathcal{Z}$.

We say that a family $\mathcal{Z}$ of subsets of X that is closed under finite intersections is *normal* if whenever $U_0, U_1 \in \mathcal{Z}$ and $U_0 \cap U_1 = \varnothing$, there exist $Z_0, Z_1 \in \mathcal{Z}$ such that

$$U_0 \cap Z_1 = \varnothing, \; U_1 \cap Z_0 = \varnothing \text{ and } Z_0 \cup Z_1 = X. \qquad (\text{XIII.3.2})$$

Proposition XIII.3.7. *If $\mathcal{Z}$ is a normal lattice, then each prime $\mathcal{Z}$-filter is included in a unique $\mathcal{Z}$-ultrafilter.*

Proof. Let $\mathcal{F}$ be a prime $\mathcal{Z}$-filter and let $\mathcal{U}_0, \mathcal{U}_1$ be $\mathcal{Z}$-ultrafilters such that $\mathcal{U}_0 \cap \mathcal{U}_1 \supset \mathcal{F}$. If $\mathcal{U}_0 \neq \mathcal{U}_1$, then, by Proposition XIII.3.3, there exist $U_0 \in \mathcal{U}_0$ and $U_1 \in \mathcal{U}_1$ such that $U_0 \cap U_1 = \varnothing$. By normality, there exist $Z_0, Z_1 \in \mathcal{Z}$ such that (XIII.3.2). As $X = Z_0 \cup Z_1 \in \mathcal{F}$ and $\mathcal{F}$ is $\mathcal{Z}$-prime, either $Z_0 \in \mathcal{F} \subset \mathcal{U}_1$ or $Z_1 \in \mathcal{F} \subset \mathcal{U}_0$. Hence either $Z_0, U_1 \in \mathcal{U}_1$ and $U_1 \cap Z_0 = \varnothing$, or $Z_1, U_0 \in \mathcal{U}_0$ and $U_0 \cap Z_1 = \varnothing$, which yields a contradiction. $\qquad\square$

XIII.4 Filters in lattices of closed and functionally closed sets

Let τ be a topology on a set X. Consider a lattice $\mathcal{Z} \subset \mathcal{C}_\tau$.

Then if $\mathcal{U}$ is a $\mathcal{Z}$-ultrafilter and $x \in \lim_\tau \mathcal{U}$, then $x \in U$ for each $U \in \mathcal{U}$. This means that $\mathcal{U} = \{Z \in \mathcal{Z} : x \in Z\}$, that is, $\mathcal{U}$ is the principal $\mathcal{Z}$-ultrafilter of x.

Recall that $\mathcal{C}_\tau^\nu$ and $\mathcal{O}_\tau^\nu$ stand respectively for the sets of all functionally closed and functionally open subsets of X for τ.

Example XIII.4.1. Let τ be a topology on X. Then $\mathcal{C}_\tau$ is normal if and only τ is normal.

In view of Proposition VII.5.4, disjoint functionally closed sets are functionally separated by a function $f \in C(\tau, [0, 1])$. As a result, the family $\mathcal{C}_\tau^\nu$

of functionally closed subsets is always normal. However, this family may be trivial unless the topology is functionally regular. Indeed, as mentioned before, there exist Hausdorff (regular) topologies (of cardinality strictly greater than 1) in which each continuous function is constant.

Proposition XIII.4.2. *Let τ be a functionally regular topology. If $\mathcal{F}$ is a $\mathcal{C}_\tau^\nu$-prime filter on X and $x \in \mathrm{adh}_\tau \mathcal{F}$ then $x \in \lim_\tau \mathcal{F}$.* ([4])

Proof. Let $x \in \mathrm{adh}_\tau \mathcal{F} = \bigcap_{F \in \mathcal{F}} F$ and let $V \in \mathcal{V}_\tau(x) \cap \mathcal{C}_\tau^\nu$. By Proposition VII.6.5, there exists $Z \in \mathcal{C}_\tau^\nu$ such that $x \in |\xi| \setminus Z \subset V$, that is, $V \cup Z = |\xi|$. As $\mathcal{F}$ is $\mathcal{C}_\tau^\nu$-prime, either $V \in \mathcal{F}$ or $Z \in \mathcal{F}$. Hence $V \in \mathcal{F}$ because $x \notin Z$. This means that $x \in \lim_\tau \mathcal{F}$. □

Lemma XIII.4.3. *If $Z \in \mathcal{C}_\tau^\nu$ and $f \in C(\xi, \tau)$, then $f^-(Z) \in \mathcal{C}_\xi^\nu$.*

Proof. If $Z \in \mathcal{C}_\tau^\nu$ then there exists $h \in C(\tau, \nu)$ such that $Z = \{h = 0\}$. Then $f^-(Z) = \{h \circ f = 0\}$ and of course $h \circ f \in C(\xi, \nu)$, so that $f^-(Z) \in \mathcal{C}_\xi^\nu$. □

Let ξ be a topology on X, τ a topology on T, $f \in C(\xi, \tau)$ and let $\mathcal{F}$ a $\mathcal{C}_\xi^\nu$-filter on X. We define

$$f^\rightarrow[\mathcal{F}] := f[\mathcal{F}]^\uparrow \cap \mathcal{C}_\tau^\nu.$$

Since $\mathcal{F}$ is a filter-base on X, $f[\mathcal{F}]^\uparrow$ is a filter on T and thus, $f^\rightarrow[\mathcal{F}]$ is a $\mathcal{C}_\tau^\nu$-filter on T.

Exercise XIII.4.4. Show that

$$f^\rightarrow[\mathcal{F}] = \left\{ Z \in \mathcal{C}_\tau^\nu : f^-(Z) \in \mathcal{F} \right\}. \qquad (\mathrm{XIII.4.1})$$

Proposition XIII.4.5. *Let $f \in C(\xi, \tau)$. If $\mathcal{F}$ is a $\mathcal{C}_\xi^\nu$-filter then $f^\rightarrow[\mathcal{F}]$ is a $\mathcal{C}_\tau^\nu$-filter; if $\mathcal{F}$ is prime, then $f^\rightarrow[\mathcal{F}]$ is prime.*

Proof. We only need to show the second statement.

If $Z_0, Z_1 \in \mathcal{C}_\tau^\nu$ and $Z_0 \cup Z_1 \in f^\rightarrow[\mathcal{F}]$, then $f^-(Z_0 \cup Z_1) = f^-(Z_0) \cup f^-(Z_1) \in \mathcal{F}$ and, as $\mathcal{F}$ is prime, $f^-(Z_0) \in \mathcal{F}$ or $f^-(Z_1) \in \mathcal{F}$, that is, $Z_0 \in f^\rightarrow[\mathcal{F}]$ or $Z_1 \in f^\rightarrow[\mathcal{F}]$, which shows that $f^\rightarrow[\mathcal{F}]$ is prime. □

---

[4]While $\mathcal{F}$ is a $\mathcal{C}_\tau^\nu$-filter and thus $\lim_\tau$ and adh_τ should not, strictly speaking, apply to $\mathcal{F}$, we have already noted that $\mathcal{F}$ is also a filter-base on X, so that $\mathrm{adh}_\tau \mathcal{F}$ and $\lim_\tau \mathcal{F}$ are in fact well-defined.

XIII.5 Maximality conditions

Theorem XIII.5.1. *Let τ be a functionally regular Hausdorff topology and let ξ be a dense subspace of τ. Then the following statements are equivalent:*

(1) Each continuous map from ξ to a compact Hausdorff topological space ζ can be extended to a continuous map from τ to ζ.

(2) Each real-valued bounded continuous map on ξ can be continuously extended to τ.

(3) If $Z_0, Z_1 \in \mathcal{C}_\xi^\nu$ and $Z_0 \cap Z_1 = \varnothing$, then $\mathrm{cl}_\tau Z_0 \cap \mathrm{cl}_\tau Z_1 = \varnothing$.

(4) If $Z_0, Z_1 \in \mathcal{C}_\xi^\nu$, then

$$\mathrm{cl}_\tau (Z_0 \cap Z_1) = \mathrm{cl}_\tau Z_0 \cap \mathrm{cl}_\tau Z_1.$$

(5) For each $y \in |\tau|$ there exists a unique $\mathcal{C}_\xi^\nu$-ultrafilter $\mathcal{U}$ on $|\xi|$ such that $y \in \lim_\tau \mathcal{U}$.

Proof. Let $X := |\xi|$ and $Y := |\tau|$.

(1) $\Longrightarrow$ (2). If $f \in C(\xi, \nu)$ is bounded then $\zeta := \mathrm{cl}_\nu f(X)$ is compact and Hausdorff.

(2) $\Longrightarrow$ (3). If $Z_0, Z_1 \in \mathcal{C}_\xi^\nu$ and $Z_0 \cap Z_1 = \varnothing$, then by Proposition VII.5.4, there exists $f \in C(\xi, \nu_{|[0,1]})$ such that $Z_0 \subset f^-(0)$ and $Z_1 \subset f^-(1)$. Let $\tilde{f}$ be a continuous extension of f:

Then $\tilde{f}^{-1}(0), \tilde{f}^{-1}(1)$ are τ-closed and disjoint, $Z_0 \subset \tilde{f}^{-1}(0)$ and $Z_1 \subset \tilde{f}^{-1}(1)$. Hence $\mathrm{cl}_\tau Z_0 \cap \mathrm{cl}_\tau Z_1 = \varnothing$.

(3) $\Longrightarrow$ (4). If $y \in \mathrm{cl}_\tau Z_0 \cap \mathrm{cl}_\tau Z_1$, then

$$V \cap Z_0 \neq \varnothing \neq V \cap Z_1$$

for all $V \in \mathcal{N}_\tau(y) \cap \mathcal{C}_\tau^\nu$. Hence $y \in \mathrm{cl}_\tau(V \cap Z_0)$ and $y \in \mathrm{cl}_\tau(V \cap Z_1)$, and by (3), $V \cap (Z_0 \cap Z_1) \neq \varnothing$ it follows that $y \in \mathrm{cl}_\tau (Z_0 \cap Z_1)$.

(4) $\Longrightarrow$ (5). As $|\xi|$ is dense in $|\tau|$, for every $y \in |\tau|$ and each $V \in \mathcal{N}_\tau(y) \cap \mathcal{C}_\tau^\nu$, the intersection $V \cap X \in \mathcal{C}_\xi^\nu$ and is non-empty. Hence

$$\{V \cap X : V \in \mathcal{N}_\tau(y) \cap \mathcal{C}_\tau^\nu\}$$

is included in an $\mathcal{C}_{\xi}^{\nu}$-ultrafilter converging to y (⁵). On the other hand, if $\mathcal{U}_0 \neq \mathcal{U}_1$ are two $\mathcal{C}_{\tau}^{\nu}$-ultrafilters on X, then there exist $U_0 \in \mathcal{U}_0$ and $U_1 \in \mathcal{U}_1$ such that $U_0 \cap U_1 = \varnothing$. By (4), $\mathrm{cl}_{\tau} U_0 \cap \mathrm{cl}_{\tau} U_1 = \varnothing$, so that two distinct $\mathcal{C}_{\xi}^{\nu}$-ultrafilters converge to distinct points.

(5) $\Longrightarrow$ (1). If ζ is a compact topology and $f \in C(\xi, \zeta)$, then by (5), for each $y \in |\zeta|$ there exists a unique $\mathcal{C}_{\xi}^{\nu}$-ultrafilter $\mathcal{U}_y$ on X converging to y. As ζ is compact, $\mathrm{adh}_{\zeta} f^{\rightarrow} [\mathcal{U}_y] \neq \varnothing$ and since the $\mathcal{C}_{\zeta}^{\nu}$-filter $f^{\rightarrow} [\mathcal{U}_y]$ is prime by Proposition XIII.4.5, it is convergent, thanks to Proposition XIII.4.2. Let

$$\{\widetilde{f}(y)\} := \lim_{\zeta} f^{\rightarrow} [\mathcal{U}_y].$$

If $x \in X$ then $\mathcal{U}_x$ is the principal $\mathcal{C}_{\xi}^{\nu}$-ultrafilter of x, thus $\widetilde{f}$ coincides with f on X. By the definition, if $Z \in \mathcal{C}_{\zeta}^{\nu}$, then $y \in \widetilde{f}^{-}(Z)$ if and only if $f^{-}(Z) \in \mathcal{U}_y$, that is, $y \in \mathrm{cl}_{\zeta} f^{-}(Z)$, thus

$$\widetilde{f}^{-}(Z) = \mathrm{cl}_{\zeta} f^{-}(Z)$$

for all $Z \in \mathcal{C}_{\xi}^{\nu}$.

It remains to show that $\widetilde{f}$ is continuous. Let $y \in |\zeta|$ and let W_0 be a functionally closed neighborhood of $\widetilde{f}(y)$. Let W_1 be a functionally closed subset of $|\zeta|$ such that $|\zeta| \setminus W_1$ is a neighborhood of $\widetilde{f}(y)$. Hence $|\zeta| = W_0 \cup W_1$ and thus $X = f^{-}(W_0) \cup f^{-}(W_1)$. As $y \notin \widetilde{f}^{-}(W_1) = \mathrm{cl}_{\zeta} f^{-}(W_1)$, thus $Y \setminus \widetilde{f}^{-}(W_1)$ is a neighborhood of y and

$$Y \setminus \widetilde{f}^{-}(W_1) = \widetilde{f}^{-}(|\zeta| \setminus W_1) \subset \widetilde{f}^{-}(W_0).$$

$\square$

Corollary XIII.2.7 shows the existence of a greatest compactification $\overline{\beta\xi}$ for each Hausdorff functionally regular topology ξ. Thus, Theorem XIII.5.1 entails

Theorem XIII.5.2. *A compactification $\overline{j\xi}$ of a Hausdorff functionally regular topology ξ is maximal, hence homeomorphic to $\overline{\beta\xi}$, if one (hence all) of the following equivalent conditions holds:*

(1) For each Hausdorff compact topology τ and each $f \in C(\xi, \tau)$ there exists an extension $\widehat{f} \in C\left(\overline{j\xi}, \tau\right)$ of f.

(2) Every bounded $f \in C(\xi, \nu)$ there exists an extension $\widehat{f} \in C\left(\overline{j\xi}, \nu\right)$.

(3) If $Z_0, Z_1 \in \mathcal{C}_{\xi}^{\nu}$ and $Z_0 \cap Z_1 = \varnothing$, then $\mathrm{cl}_{\overline{j\xi}} Z_0 \cap \mathrm{cl}_{\overline{j\xi}} Z_1 = \varnothing$.

---

⁵when considered as a filter-base on X.

(4) If $Z_0, Z_1 \in \mathcal{C}_\xi^\nu$, then

$$\mathrm{cl}_{\overline{j\xi}} (Z_0 \cap Z_1) = \mathrm{cl}_{\overline{j\xi}} Z_0 \cap \mathrm{cl}_{\overline{j\xi}} Z_1.$$

(5) For each $y \in |\overline{j\xi}|$ there exists a unique $\mathcal{C}_\xi^\nu$-ultrafilter $\mathcal{U}$ on $|\xi|$ such that $y \in \lim_{\overline{j\xi}} \mathcal{U}$.

The fifth condition in Theorem XIII.5.2 suggests a construction of one of the equivalent versions of the Čech-Stone compactification $\overline{\beta\xi}$ of a Hausdorff functionally regular topology ξ, as a topology on the set of maximal filters in the lattice of functionally closed sets. We carry it out in the next section.

XIII.6 Čech-Stone compactification

Let ξ be a Hausdorff functionally regular topology on X. Let $\mathbb{U}_\xi^\nu$ denote the set of all $\mathcal{C}_\xi^\nu$-ultrafilters on X. We shall parametrize $\mathbb{U}_\xi^\nu$ by a set Y such that $X \subset Y$,

$$\mathbb{U}_\xi^\nu = \{\mathcal{U}(y) : y \in Y\}, \tag{XIII.6.1}$$

in such a way that $y_0 \neq y_1$ entails $\mathcal{U}(y_0) \neq \mathcal{U}(y_1)$ and, for all $x \in X$,

$$\mathcal{U}(x) := \left\{ Z \in \mathcal{C}_\xi^\nu : x \in Z \right\}$$

is the principal $\mathcal{C}_\xi^\nu$-ultrafilter of x.

For each $Z \in \mathcal{C}_\xi^\nu$, let

$$\mathrm{cl}_Y Z := \{y \in Y : Z \in \mathcal{U}(y)\}. \tag{XIII.6.2}$$

We have defined an operator that associates a subset of Y with every functionally closed subset of X. Later we shall use this operator to define a topology on Y.

Of course, if $W, Z \in \mathcal{C}_\xi^\nu$ and $W \subset Z$ then $\mathrm{cl}_Y W \subset \mathrm{cl}_Y Z$.

Lemma XIII.6.1. *For every $Z_0, Z_1 \in \mathcal{C}_\xi^\nu$,*

$$\mathrm{cl}_Y (Z_0 \cup Z_1) = \mathrm{cl}_Y Z_0 \cup \mathrm{cl}_Y Z_1, \tag{XIII.6.3}$$

$$\mathrm{cl}_Y (Z_0 \cap Z_1) = \mathrm{cl}_Y Z_0 \cap \mathrm{cl}_Y Z_1. \tag{XIII.6.4}$$

Proof. If $y \in \mathrm{cl}_Y (Z_0 \cup Z_1)$ then $Z_0 \cup Z_1 \in \mathcal{U}(y)$ and since $\mathcal{U}(y)$ is a $\mathcal{C}_\xi^\nu$-ultrafilter, either $Z_0 \in \mathcal{U}(y)$ or $Z_1 \in \mathcal{U}(y)$, that is either $y \in \mathrm{cl}_Y Z_0$ or $y \in \mathrm{cl}_Y Z_1$.

Let $y \in \mathrm{cl}_Y Z_0 \cap \mathrm{cl}_Y Z_1$, that is, $Z_0 \in \mathcal{U}(y)$ and $Z_1 \in \mathcal{U}(y)$, hence $Z_0 \cap Z_1 \in \mathcal{U}(y)$, equivalently $y \in \mathrm{cl}_Y (Z_0 \cap Z_1)$. $\square$

We define on Y a topology by taking

$$\{\operatorname{cl}_Y Z : Z \in \mathcal{C}_\xi^\nu\} \tag{XIII.6.5}$$

for a base of closed sets.

Lemma XIII.6.2. *$V \in \mathcal{N}_Y(y)$ if and only there exists $U \in \mathcal{U}(y)$ such that $\operatorname{cl}_Y U \subset V$.*

Proof. If $V \in \mathcal{N}_Y(y)$, then there exists $Z \in \mathcal{C}_\xi^\nu$ such that

$$y \in Y \setminus \operatorname{cl}_Y Z \subset V.$$

But $y \notin \operatorname{cl}_Y Z$ if and only if $Z \notin \mathcal{U}(y)$ if and only if $Z \notin \mathcal{U}(y)^\#$. Hence there exists $U \in \mathcal{U}(y)$ such that $Z \cap U = \varnothing$. By virtue of Theorem XIII.5.1, $\operatorname{cl}_Y Z \cap \operatorname{cl}_Y U = \varnothing$, that is, $\operatorname{cl}_Y U \subset Y \setminus \operatorname{cl}_Y Z$. $\qquad\square$

Theorem XIII.6.3. *Let ξ be a Hausdorff functionally regular topology. Then the topology defined by (XIII.6.5) on the set of $\mathcal{C}_\xi^\nu$-ultrafilters is the Čech-Stone compactification of ξ.*

Proof. Let ξ be a Hausdorff functionally regular topology on X and let Y be endowed with the topology for which (XIII.6.5) is a base of the closed sets. By Lemma XIII.6.2, $\{y\} = \lim_Y \mathcal{U}(y)$ for all $y \in Y$, thus X is dense in Y. As $\operatorname{cl}_Y Z \cap X = Z$ for all $Z \in \mathcal{C}_\xi^\nu$, the topology of Y coincides on X with the original topology of X.

The topology of Y is Hausdorff. Indeed, if $y_0 \neq y_1$ then $\mathcal{U}(y_0)$ and $\mathcal{U}(y_1)$ are distinct, thus by Proposition XIII.3.3, there exist $Z_0, Z_1 \in \mathcal{C}_\xi^\nu$ such that $Z_0 \in \mathcal{U}(y_0), Z_1 \in \mathcal{U}(y_1)$ and $Z_0 \cap Z_1 = \varnothing$. Hence there exist functionally open subsets A_0, A_1 of X such that $Z_0 \subset A_0, Z_1 \subset A_1$ and $A_0 \cap A_1 = \varnothing$ and $Y \setminus \operatorname{cl}_Y (X \setminus A_0) \in \mathcal{N}_Y(y_0), Y \setminus \operatorname{cl}_Y (X \setminus A_0) \in \mathcal{N}_Y(y_1)$ and

$$\operatorname{cl}_Y (X \setminus A_0) \cup \operatorname{cl}_Y (X \setminus A_0) = Y,$$

that is, $\mathcal{N}_Y(y_0)$ and $\mathcal{N}_Y(y_1)$ do not mesh.

To see that the topology is compact, let us show that $\varnothing \neq \operatorname{adh}_Y \mathcal{F}$ for each filter $\mathcal{F}$ on Y. The family

$$\widehat{\mathcal{F}} := \bigcup\nolimits_{F \in \mathcal{F}} \{Z \in \mathcal{C}_\xi^\nu : F \subset \operatorname{cl}_Y Z\}$$

is a $\mathcal{C}_\xi^\nu$-filter on X, thus there exists $y \in Y$ such that $\mathcal{U}(y)$ is finer than $\widehat{\mathcal{F}}$ and thus $\lim_Y \mathcal{U}(y) = \{y\}$. Therefore

$$y \in \operatorname{adh}_Y \widehat{\mathcal{F}} = \bigcap\nolimits_{F \in \mathcal{F}} \bigcap\nolimits_{\operatorname{cl}_Y Z \supset F} \operatorname{cl}_Y Z = \bigcap\nolimits_{F \in \mathcal{F}} \operatorname{cl}_Y F = \operatorname{adh}_Y \mathcal{F}.$$

By Theorem XIII.5.1, the constructed compactification of ξ is the only one, for which every Hausdorff compact topology τ, each $f \in C(\xi, \tau)$ can be extended to a map $\tilde{f} \in C(Y, \tau)$. Therefore, if τ is a compactification of ξ and $j \in C(\xi, \tau)$ is the corresponding dense embedding, then there is an extension $\tilde{j} \in C(Y, \tau)$. This shows that the constructed topology on Y is the greatest element of $\mathfrak{C}(\xi)$. $\qquad\square$

In the following examples, $\mathbb{R}$ is endowed with its usual topology ν, $\mathbb{Q}$ is endowed as usual with the induced topology, and so is $\mathbb{N}$, which is thus discrete. In this setting, we will use the traditional notations $\beta\mathbb{R}$, $\beta\mathbb{Q}$ and $\beta\mathbb{N}$ rather than $\overline{\beta\nu}$, $\overline{\beta\nu_{|\mathbb{Q}}}$ and $\overline{\beta\iota_{\mathbb{N}}}$.

Example XIII.6.4. By Theorem IX.5.11 and Theorem XIII.5.2, $(\mathbb{U}\mathbb{N}, \beta)$ is the Čech-Stone compactification $\beta\mathbb{N}$ of $\mathbb{N}$. We could also see that via Theorem XIII.6.3, because on a discrete space, ultrafilters and ultrafilters on functionally closed sets coincide. As $\mathbb{N}$ is a Hausdorff locally compact topological space, it is open in $\beta\mathbb{N}$ by Proposition IX.3.18. Thus $\beta\mathbb{N} \setminus \mathbb{N}$ is compact, as already noted for $\mathbb{U}^*(\mathbb{N})$. We already know that $\beta\mathbb{N}$ is zero-dimensional. This can also be seen from Theorem XIII.5.1.

Example XIII.6.5. The cardinality of $\beta\mathbb{R} \setminus \mathbb{R}$ is $2^{\mathfrak{c}}$. In fact, we know from Theorem II.7.1 and Example XIII.6.4 that $\mathrm{card}(\beta\mathbb{N}) = 2^{\mathfrak{c}}$. Moreover,

$$\mathrm{card}(\beta\mathbb{R}) = \mathrm{card}(\beta\mathbb{N}).$$

To see this, let $f : \mathbb{N} \to \mathbb{Q}$ be a bijection. As $\mathbb{N}$ is discrete, $f \in C(\mathbb{N}, \beta\mathbb{R})$ and thus there exists an extension $\tilde{f} \in C(\beta\mathbb{N}, \beta\mathbb{R})$ of f.

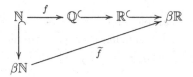

Hence

$$\tilde{f}(\beta\mathbb{N}) = \mathrm{cl}_{\beta\mathbb{R}} \, \mathbb{Q} = \beta\mathbb{R},$$

and thus $\mathrm{card}(\beta\mathbb{N}) \geq \mathrm{card}(\beta\mathbb{R})$. On the other hand, $\mathrm{card}(\beta\mathbb{N}) \leq \mathrm{card}(\beta\mathbb{R})$, as every bounded function on $\mathbb{N} \subset \mathbb{R}$ has a continuous extension on $\beta\mathbb{R}$, thus $\beta\mathbb{N} \subset \beta\mathbb{R}$.

As each ultrafilter containing a bounded set converges in $\mathbb{R}$, the set $\beta\mathbb{R} \setminus \mathbb{R}$ consists of unbounded $\mathcal{Z}$-ultrafilters, that is, those containing either $\mathbb{R}_{\geq r} := \{x \in \mathbb{R} : x \geq r\}$ for all r or $\mathbb{R}_{\leq r} := \{x \in \mathbb{R} : x \leq r\}$ for all r. In

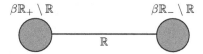

Figure XIII.1   The Čech-Stone compactification $\beta\mathbb{R}$ of $\mathbb{R}$.

other words, if $y \in \beta\mathbb{R} \setminus \mathbb{R}$, then either $V \cap \mathbb{R}_{\geq r} \neq \varnothing$ for each $V \in \mathcal{V}(y)$ and every $\mathbb{R}_{\geq r}$, or $V \cap \mathbb{R}_{\leq r} \neq \varnothing$ for each $V \in \mathcal{V}(y)$ and every $\mathbb{R}_{\leq r}$. Hence $\beta\mathbb{R} \setminus \mathbb{R}$ is composed of two clusters around $+\infty$ and $-\infty$.

As each continuous bounded function on $\mathbb{R}_{\geq r}$ has a continuous extension to $\beta\mathbb{R}$ (by the definition of the Čech-Stone compactification), we can consider $\beta\mathbb{R}_{\geq r}$ as a part of $\beta\mathbb{R}$, so that $\beta\mathbb{R}_{\geq r} \setminus \mathbb{R}_{\geq r} \subset \beta\mathbb{R} \setminus \mathbb{R}$. As the neighborhoods of the elements of $\beta\mathbb{R} \setminus \mathbb{R}$ are not bounded, $\beta\mathbb{R}_{\geq r} \setminus \mathbb{R}_{\geq r}$ is independent of $r \in \mathbb{R}$ and is compact, because $\mathbb{R}_{\geq r}$ is locally compact, thus open in $\beta\mathbb{R}_{\geq r}$ by Proposition IX.3.18. In particular,

$$\beta\mathbb{R}_{\geq 0} \setminus \mathbb{R} = \bigcap\nolimits_{r \in \mathbb{R}} \beta\mathbb{R}_{\geq r},$$

so that $\beta\mathbb{R}_{\geq 0} \setminus \mathbb{R}$ is connected by Exercise XII.1.24. An analogous argument holds for the homeomorphic space $\beta\mathbb{R}_{\leq 0} \setminus \mathbb{R}$. The compactification $\beta\mathbb{R}$ of $\mathbb{R}$ is connected as the closure of the connected space $\mathbb{R}$. However,

$$\beta\mathbb{R} \setminus \mathbb{R} = (\beta\mathbb{R}_{\geq 0} \setminus \mathbb{R}) \cup (\beta\mathbb{R}_{\leq 0} \setminus \mathbb{R})$$

is disconnected. For example $\arctan : \mathbb{R} \to \left[-\frac{\pi}{2}, \frac{\pi}{2} \right]$ has an extension $f \in C(\beta\mathbb{R}, \left[-\frac{\pi}{2}, \frac{\pi}{2} \right])$. As the neighborhoods of the elements of $\beta\mathbb{R} \setminus \mathbb{R}$ are not bounded, $f(\beta\mathbb{R} \setminus \mathbb{R})$ is equal to $\{ -\frac{\pi}{2}, \frac{\pi}{2} \}$.

Example XIII.6.6. Since $\mathbb{Q}$ is zero-dimensional, the Čech-Stone compactification $\beta\mathbb{Q}$ of $\mathbb{Q}$ is zero-dimensional by Theorem XIII.5.1. For each $y \in \beta\mathbb{Q}$ and every $V \in \mathcal{V}(y)$, there exists a compact set $W \subset V$ such that $W \in \mathcal{V}(y)$. As $\mathbb{Q}$ is not locally compact, $W \cap (\beta\mathbb{Q} \setminus \mathbb{Q}) \neq \varnothing$, that is, $\beta\mathbb{Q} \setminus \mathbb{Q}$ is dense in $\beta\mathbb{Q}$.

As for $\beta\mathbb{R} \setminus \mathbb{R}$, the difference $\beta\mathbb{Q} \setminus \mathbb{Q}$ consists of two (disconnected) clusters about $-\infty$ and $+\infty$. On the other hand, each continuous bounded function $\mathbb{Q}$ has a unique continuous extension to $\mathbb{R}$. It follows that $\beta\mathbb{Q} \setminus \mathbb{Q}$ includes $\mathbb{R} \setminus \mathbb{Q}$.

As card $\mathbb{Q} = $ card $\mathbb{N}$ and $\mathbb{N}$ is discrete, there exists a bijection $f : \mathbb{N} \to \mathbb{Q}$ and $f \in C(\mathbb{N}, \beta\mathbb{Q})$. Hence, there exists $\tilde{f} \in C(\beta\mathbb{N}, \beta\mathbb{Q})$ and $\tilde{f}(\beta\mathbb{N}) = \mathrm{cl}_{\beta\mathbb{Q}} f(\mathbb{N}) = \beta\mathbb{Q}$, because $\tilde{f}(\beta\mathbb{N})$ is compact, which implies card$(\beta\mathbb{N}) \geq$ card$(\beta\mathbb{Q})$. Indeed,

$$\mathrm{card}(\beta\mathbb{Q}) = \mathrm{card}(\beta\mathbb{N}),$$

because $\mathbb{N} \subset \mathbb{Q}$, thus $\beta\mathbb{N} \subset \beta\mathbb{Q}$ and so card$(\beta\mathbb{N}) \leq$ card$(\beta\mathbb{Q})$.

Chapter XIV

Classification of spaces

XIV.1 Modifiers, projectors, and coprojectors

Let $\mathbf{I}$ denote the class of *all* convergences and let $\mathbf{I}(X)$ be the *set* of convergences on a set X. As observed before in Section III.6, $\mathbf{I}(X)$ is a complete lattice whose greatest element is the discrete convergence on X and least element is the chaotic convergence on X. If $\mathbf{C}$ is a subclass of $\mathbf{I}$ we also write $\mathbf{C}(X)$ for the convergences of the class $\mathbf{C}$ with underlying set X.

So far, the term *map* has been used in the framework of sets. Its extension to the context of classes is intuitively obvious: it associates with every object of one class (its domain) a unique object of another class (its range).

Definition XIV.1.1. A map $F : \mathbf{I} \to \mathbf{I}$ is called a *modifier* if

$$|F\xi| = |\xi|$$
$$\xi \le \tau \implies F\xi \le F\tau.$$

In other words, if ξ is a convergence on X, then $F\xi$ is also a convergence on X. Since we are using the notion of map for modifiers, we may define it via its restrictions to sets $F : \mathbf{I}(X) \to \mathbf{I}(X)$.

We have already encountered a number of modifiers (The table below lists some of the most important modifiers among those already introduced.) Additionally, any finite composite of modifiers is clearly also a modifier.

Modifiers are ordered pointwise, that is, if F and M are two modifiers, then

$$F \ge M \iff \bigvee_{\xi} F\xi \ge M\xi.$$

In order to investigate special classes of modifiers, we first need to discuss certain properties of classes of convergences.

F	modifier
L	finitely deep modifier
S	pseudotopological modifier
S$_0$	pretopological modifier
T	topological modifier
D	diagonal modifier
Seq	sequentially based modifier
K	locally compact modifier

Definition XIV.1.2. A class of convergences is *projective* if it is closed under suprema and *coprojective* if it is closed under infima.

We use the convention that the supremum of the empty family of convergences on X is the chaotic convergence on X and the infimum of the empty family of convergences on X is the discrete convergence on X.

In other words, a class **C** of convergences is projective if

$$\Xi \subset \mathbf{C}(X) \Longrightarrow \bigvee \Xi \in \mathbf{C}(X), \qquad \text{(projective)}$$

and coprojective if

$$\Xi \subset \mathbf{C}(X) \Longrightarrow \bigwedge \Xi \in \mathbf{C}(X) \qquad \text{(coprojective)}$$

for every set X.

Of course, the class **I** of all convergences is both projective and coprojective.

Let F be a modifier. Consider the classes $\mathbf{F}_+$ and $\mathbf{F}_-$ defined by

$$\mathbf{F}_+(X) := \{\xi \in \mathbf{I}(X) : F\xi \geq \xi\}$$
$$\mathbf{F}_-(X) := \{\xi \in \mathbf{I}(X) : \xi \geq F\xi\}$$

for each set X.

Proposition XIV.1.3. *If F is a modifier, then $\mathbf{F}_+$ is projective and $\mathbf{F}_-$ is coprojective.*

Proof. Since

$$\bigvee \Xi \geq \xi \Longrightarrow F(\bigvee \Xi) \geq F\xi$$

for each $\xi \in \Xi$, because F is a modifier, we conclude that, whenever $F\xi \geq \xi$ for all $\xi \in \Xi$,

$$F(\bigvee \Xi) \geq \bigvee_{\xi \in \Xi} F\xi \geq \bigvee \Xi.$$

This shows that $\mathbf{F}_+$ is projective. $\qquad\square$

Exercise XIV.1.4. Show that if F is a modifier then $\mathbf{F}_-$ is coprojective.

Example XIV.1.5. If I is the *identity modifier*, that is, $\mathrm{I}\xi = \xi$ for all ξ, then $\mathbf{I}_+ = \mathbf{I}_- = \mathbf{I}$.

Definition XIV.1.6. A modifier F is called *expansive* (respectively, *contractive* and *idempotent*) if

$$F \geq \mathrm{I}, \qquad\qquad \text{(expansive)}$$
$$F \leq \mathrm{I}, \qquad\qquad \text{(contractive)}$$
$$F = F \circ F. \qquad\qquad \text{(idempotent)}$$

We denote by fix F the class of convergences that are left unchanged by the modifier F:

$$\operatorname{fix} F := \{\xi \in \mathbf{I} : F\xi = \xi\}.$$

Of course,

Proposition XIV.1.7. *If F is a modifier, then*

$$F \text{ is contractive} \iff \mathbf{F}_+ = \operatorname{fix} F \iff \mathbf{F}_- = \mathbf{I},$$
$$F \text{ is expansive} \iff \mathbf{F}_- = \operatorname{fix} F \iff \mathbf{F}_+ = \mathbf{I},$$
$$F \text{ is idempotent} \iff F(\mathbf{I}) = \operatorname{fix} F.$$

Example XIV.1.8. Let L, S, S_0, D and T denote the finitely deep, pseudotopological, pretopological, diagonal, and topological modifiers respectively. These modifiers are all contractive, so that

$$\mathbf{L}_- = \mathbf{S}_- = \mathbf{S}_{0-} = \mathbf{D}_- = \mathbf{T}_- = \mathbf{I}$$

by Proposition XIV.1.7, and

$$\mathbf{L}_+ = \text{class of finitely deep convergences}$$
$$\mathbf{S}_+ = \text{class of pseudotopologies}$$
$$\mathbf{S}_{0+} = \text{class of pretopologies}$$
$$\mathbf{D}_+ = \text{class of diagonal convergences}$$
$$\mathbf{T}_+ = \text{class of topologies}.$$

Example XIV.1.9. The sequentially based modification Seq and the locally compact modification K are expansive. By Proposition XIV.1.7, $\mathbf{Seq}_+ = \mathbf{K}_+ = \mathbf{I}$ and

$$\mathbf{Seq}_- = \text{class of sequentially based convergences}$$
$$\mathbf{K}_- = \text{class of locally compact convergences}.$$

As a consequence of these observations and Proposition XIV.1.3, fix F is projective whenever F is contractive, and coprojective whenever F is expansive.

Definition XIV.1.10. An idempotent modifier is called a *projector* if it is contractive, and a *coprojector* if it is expansive.

Therefore, if P is a projector then the class

$$\text{fix}\, P = \mathbf{P}_+ = P(\mathbf{I})$$

is projective. Moreover, $P\xi$ is the finest convergence of fix P that is coarser than ξ. Indeed, $P\xi \leq \xi$ because P is contractive, $P\xi \in \text{fix}\, P$ because P is idempotent, and if $\sigma = P\sigma \leq \xi$ then $P\sigma \leq P\xi$ so that $\sigma \leq P\xi$. In summary:

Proposition XIV.1.11. *If P is a projector then* fix P *is projective and $P\xi$ is the finest convergence of* fix P *that is coarser than ξ.*

Proposition XIV.1.12. *If C is a coprojector then* fix C *is coprojective, and $C\xi$ is the coarsest convergence of* fix C *that is finer than ξ.*

Moreover, each (co)projective class is the image of a unique (co)projector:

Theorem XIV.1.13. *Let $\mathbf{P}$ be a projective class of convergences. For each convergence ξ there is the finest convergence $P\xi$ in $\mathbf{P}(|\xi|)$ that is coarser than ξ. Moreover, P is a projector.*

Proof. Let $\Xi := \{\tau \in \mathbf{P}(|\xi|) : \tau \leq \xi\}$. The set Ξ is non-empty, because, in view of Definition XIV.1.2, the chaotic convergence on $|\xi|$ belongs to Ξ. Moreover,

$$P\xi := \bigvee_{\tau \in \Xi} \tau \in \Xi$$

because $\mathbf{P}$ is closed under suprema, and is by definition the finest convergence in $\mathbf{P}(|\xi|)$ coarser than ξ. Moreover, P is a projector: it is by definition contractive, and it is idempotent because $P\xi \in \mathbf{P}$ is itself the finest convergence of $\mathbf{P}$ coarser than ξ. $\square$

Theorem XIV.1.14. *Let $\mathbf{C}$ be a coprojective class of convergences. For each convergence ξ there is the coarsest convergence $C\xi$ in $\mathbf{C}(|\xi|)$ that is finer than ξ. Moreover C is a coprojector.*

Definition XIV.1.15. If $\mathbf{P}$ is a projective class, then the convergence $P\xi$ is called *projection of ξ on* $\mathbf{P}$, and P is *the projector on* $\mathbf{P}$.

If $\mathbf{C}$ is a coprojective class, then the convergence $C\xi$ is called *coprojection of ξ on* $\mathbf{C}$, and C is *the coprojector on* $\mathbf{C}$.

Proposition XIV.1.16. *Let $\mathbf{P}$ be a projective class. The projector P on $\mathbf{P}$ is the smallest element of the class of modifiers F satisfying $\mathbf{F}_+ = \mathbf{P}$.*

Proof. Let F be a modifier such that $\mathbf{F}_+ = \mathbf{P}$. The projector P is contractive so that $\xi \geq P\xi$ for each convergence ξ. Hence $F\xi \geq F(P\xi)$ and $F(P\xi) \geq P\xi$ because $P\xi \in \mathbf{P}(|\xi|) = \mathbf{F}_+(|\xi|)$. Hence $F\xi \geq P\xi$ for each ξ. $\qquad\qquad\square$

Proposition XIV.1.17. *Let $\mathbf{C}$ be a coprojective class. The coprojector C on $\mathbf{C}$ is the largest element of the class of modifiers F with $\mathbf{F}_- = \mathbf{C}$.*

Proposition XIV.1.18. *Let $\mathbf{P}$ be a projective class. If F is a contractive modifier with $\mathbf{F}_+ = \mathbf{P}$ then $P = \bigwedge_{\alpha \in \mathrm{Ord}} F^\alpha$ where F^α is defined by transfinite induction by $F^1 = F$ and $F^\alpha = F(\bigwedge_{\beta < \alpha} F^\beta)$.*

Proof. By Proposition XIV.1.16 $F \geq P$. We show by transfinite induction that $F^\alpha \geq P$ for all α. Assume that $F^\beta \geq P$ for each $\beta < \alpha$. Then $\bigwedge_{\beta < \alpha} F^\beta \geq P$ and

$$F\Big(\bigwedge_{\beta < \alpha} F^\beta \xi \Big) \geq F(P\xi)$$

for each ξ and $F(P\xi) \geq P\xi$ because $P\xi \in \mathbf{P}(|\xi|) = \mathbf{F}_+(|\xi|)$. Hence $F^\alpha \xi \geq P\xi$.

Moreover, for each ξ, the transfinite sequence $\{F^\alpha \xi\}_{\alpha \in \mathrm{Ord}}$ is decreasing. As $\mathbf{I}(\xi)$ is a complete lattice, it cannot be strictly decreasing, so that there is α such that $F^\alpha \xi = F^{\alpha+1}\xi$ so that $F^\alpha \xi \in \mathbf{F}_+(|\xi|) = \mathbf{P}(|\xi|)$ and $\xi \geq F^\alpha \xi$. Since $P\xi$ is by definition the finest convergence of $\mathbf{P}(|\xi|)$ coarser than ξ, we conclude that $P\xi \geq F^\alpha \xi$ and we have equality, which completes the proof. $\qquad\qquad\square$

Proposition XIV.1.19. *Let $\mathbf{C}$ be a coprojective class. If F is an expansive modifier with $\mathbf{F}_- = \mathbf{C}$ then $C = \bigvee_{\alpha \in \mathrm{Ord}} F^\alpha$ where F^α is defined by transfinite induction by $F^1 = F$ and $F^\alpha = F(\bigvee_{\beta < \alpha} F^\beta)$.*

We have seen that L, S, S$_0$ and T are projectors, while K and Seq are coprojectors. Another property that we have checked in each instance is that these modifiers preserve continuity.

XIV.2   Functors, reflectors and coreflectors

Definition XIV.2.1. A modifier F is a (concrete) *functor* if

$$C(\xi,\sigma) \subset C(F\xi,F\sigma).$$

Example XIV.2.2. The modifiers L, S, S$_0$, T, K and Seq are all functors. On the other hand, Example VI.2.23 shows that the diagonal modifier D is not a functor.

Proposition XIV.2.3. *Let F be a modifier. The following are equivalent:*

(1) F is a functor;
(2) for each map f and each convergence τ on its codomain,

$$F(f^-\tau) \geq f^-(F\tau); \qquad\qquad \text{(XIV.2.1)}$$

(3) for each map f and each convergence ξ on its domain,

$$f(F\xi) \geq F(f\xi). \qquad\qquad \text{(XIV.2.2)}$$

Proof. (1) $\implies$ (2): The map $f : |f^-\tau| \to |\tau|$ is continuous and F is a functor, hence $f : |F(f^-\tau)| \to |F\tau|$ is continuous, that is, $F(f^-\tau) \geq f^-(F\tau)$ by Proposition IV.2.9.

(2) $\implies$ (3): Let $f : X \to Y$ and let ξ be a convergence on X. Let $\tau := f\xi$. By (2), we have

$$F(f^- f\xi) \geq f^-(F(f\xi)),$$

and $\xi \geq f^- f\xi$ by (IV.2.5), so that $F\xi \geq f^-(F(f\xi))$. Hence

$$f(F\xi) \geq f f^-(F(f\xi)) \geq F(f\xi),$$

by (IV.2.4).

(3) $\implies$ (1): By Proposition IV.2.9, continuity of a map $f : |\xi| \to |\tau|$ means $f\xi \geq \tau$, so that $F(f\xi) \geq F\tau$. By (3), we conclude that $f(F\xi) \geq F\tau$, which amounts to the continuity of $f : |F\xi| \to |F\tau|$ by Proposition IV.2.9. $\qquad\qquad\square$

Note that even if a collection of modifiers $\mathfrak{F}$ is formally a proper class, for each convergence ξ, $\{F\xi : F \in \mathfrak{F}\}$ is a subset of the set $\mathbf{I}(|\xi|)$ of convergences on $|\xi|$. Hence, we will be able to consider collections of functors without taking care whether it is indexed by a set or a class.

Proposition XIV.2.4. *Suppose that $\mathfrak{F}$ is a collection of functors. Then $\bigvee_{F\in\mathfrak{F}} F$ and $\bigwedge_{F\in\mathfrak{F}} F$ are also functors.*

Proof. Let $f \in C(\xi, \tau)$. For each $F \in \mathfrak{F}$, $f \in C(F\xi, F\tau)$. Thus

$$f \in \bigcap_{F \in \mathfrak{F}} C(F\xi, F\tau) \subset \bigcap_{F \in \mathfrak{F}} C(\bigvee_{G \in \mathfrak{F}} G\xi, F\tau),$$

and, in view of Exercise IV.1.2,

$$\bigcap_{F \in \mathfrak{F}} C(\bigvee_{G \in \mathfrak{F}} G\xi, F\tau) = C(\bigvee_{G \in \mathfrak{F}} G\xi, \bigvee_{F \in \mathfrak{F}} F\tau).$$

Thus $\bigvee_{F \in \mathfrak{F}} F$ is a functor.

On the other hand, if $x \in \lim_{\bigwedge_{F \in \mathfrak{F}} F\xi} \mathcal{F}$, there is $F_0 \in \mathfrak{F}$ with $x \in \lim_{F_0 \xi} \mathcal{F}$ and $f \in C(F_0\xi, F_0\tau)$ so that

$$f(x) \in \lim_{F_0 \tau} f[\mathcal{F}] \subset \lim_{\bigwedge_{F \in \mathfrak{F}} F\tau} f[\mathcal{F}].$$

Thus, $\bigwedge_{F \in \mathfrak{F}} F$ is a functor. $\qquad\square$

Definition XIV.2.5. We call a class of convergences (concretely) *reflective* if it is projective and closed under initial convergences, and *coreflective* if it is coprojective and closed under final convergences.

Proposition XIV.2.6. *A projective class is reflective if and only if its projector is a functor. A coprojective class is coreflective if and only if its coprojector is a functor.*

Proof. If **P** is a projective class, whose projector P is a functor, then for each $\tau \in \mathbf{P}$ and map f with codomain $|\tau|$, we have

$$P(f^-\tau) \geq f^-(P\tau)$$

by Proposition XIV.2.3. Since $P\tau = \tau$ and P is contractive, we conclude that $f^-\tau = P(f^-\tau)$ so that $f^-\tau \in \mathbf{P}$. Therefore the class **P** is reflective.

Conversely, if a projective class **P** is reflective, then $f^-\tau \in \mathbf{P}$ whenever $\tau \in \mathbf{P}$. In particular, $f^-(P\tau) \in \mathbf{P}$ so that

$$f^-(P\tau) \leq P(f^-(P\tau)) \leq P(f^-\tau)$$

because P is contractive. By Proposition XIV.2.3, P is a functor. $\qquad\square$

Definition XIV.2.7. A projector (respectively, coprojector) that is also a functor is called a *reflector* (respectively, *coreflector*).

Accordingly, L, S, S_0, T are reflectors, K and Seq are coreflectors and D is an example of a projector that is not a reflector.

In the case of a functor F, Proposition XIV.1.3 rephrases as:

Proposition XIV.2.8. *If F is a concrete functor, then the class $\mathbf{F}_+$ is reflective and the class $\mathbf{F}_-$ is coreflective.*

Proof. We already know that $\mathbf{F}_+$ is projective. Let P denote the associated projector. For any map f and convergence τ on the codomain, $P\tau \in \mathbf{F}_+$ so that $FP\tau \geq P\tau$. Thus

$$F(f^-(P\tau)) \geq f^-(FP\tau) \geq f^-(P\tau),$$

because F is a functor and thus satisfies (XIV.2.1). As a result, $f^-(P\tau) \in \mathbf{F}_+$ so that

$$P(f^-\tau) \geq P(f^-(P\tau)) \geq f^-(P\tau),$$

and P is a functor, hence a reflector. $\qquad\square$

Proposition XIV.2.9. *If $\mathfrak{F}$ is a collection of reflectors, then $\bigvee_{F\in\mathfrak{F}} F$ is a reflector. If $\mathfrak{F}$ is a collection of coreflectors, then $\bigwedge_{F\in\mathfrak{F}} F$ is a coreflector.*

Proof. We only prove the first statement. In view of Proposition XIV.2.4, and the fact that $\bigvee_{F\in\mathfrak{F}} F$ is contractive whenever each F is, we only need to show that $\bigvee_{F\in\mathfrak{F}} F$ is idempotent. For each $G \in \mathfrak{F}$,

$$\bigvee_{F\in\mathfrak{F}} F\left(\bigvee_{F\in\mathfrak{F}} F\xi\right) \geq \bigvee_{F\in\mathfrak{F}} F(G\xi) \geq G(G\xi) = G\xi.$$

Thus, $\bigvee_{F\in\mathfrak{F}} F\left(\bigvee_{F\in\mathfrak{F}} F\xi\right) \geq \bigvee_{F\in\mathfrak{F}} F\xi$, and, as $\bigvee_{F\in\mathfrak{F}} F$ is contractive, it is also idempotent. $\qquad\square$

Theorem XIV.1.13, Proposition XIV.1.16 and Proposition XIV.1.18 remain valid if one substitutes reflective for projective and reflector for projector. Similarly, Theorem XIV.1.14, Proposition XIV.1.17 and Proposition XIV.1.19 remain valid if one substitutes coreflective for coprojective and coreflector for coprojector.

Proposition XIV.2.10. *Let F be a functor. For every ξ and τ,*

$$F(\xi \times \tau) \geq F\xi \times F\tau.$$

Proof. Since $\xi \times \tau = p_{|\xi|}^- \xi \vee p_{|\tau|}^- \tau,$

$$F(\xi \times \tau) \geq F(p_{|\xi|}^- \xi) \text{ and } F(\xi \times \tau) \geq F(p_{|\tau|}^- \tau)$$

so that, in view of (XIV.2.1),

$$F(\xi \times \tau) \geq F(p_{|\xi|}^- \xi) \vee F(p_{|\tau|}^- \tau) \geq p_{|\xi|}^-(F\xi) \vee p_{|\tau|}^-(F\tau) = F\xi \times F\tau.$$

$\qquad\square$

On the other hand, the reverse inequality may fail. For instance, we have seen that the reflector S_0 on pretopologies does not commute with products (Example V.2.27), nor does the reflector T on topologies (Example V.4.40). On the other hand, the reflectors L on convergences of finite depth, and S on pseudotopologies do commute with finite product, as show Corollary IV.4.3 and Theorem VIII.3.9.

XIV.3 Adherence-determined convergences

If $\mathbb{D}$ denotes a class of filters, then $\mathbb{D}\xi$ stands for the set of filters on a convergence space (X, ξ) that are in $\mathbb{D}$. If $\mathbb{D}\xi$ is the same for every convergence ξ on X, we use the notation $\mathbb{D}X$. We list below several important instances of $\mathbb{D}$, each of which is independent of the convergence.

Let $\mathbb{F}_\kappa$ denote the class of filters of character less than $\aleph_\kappa$, that is, that admit a filter-base of cardinality less than $\aleph_\kappa$. In particular, $\mathbb{F}_0$ is the class of filters with a finite filter-base, that is, principal filters, and $\mathbb{F}_1$ is the class of countably based filters.

class of filters	$\mathbb{D}$
principal	$\mathbb{F}_0$
countably based	$\mathbb{F}_1$
of character less than $\aleph_\kappa$	$\mathbb{F}_\kappa$
all	$\mathbb{F}$

The class $\mathbb{K}$ of filters that contain a compact set is an example of a class of filters that depends on the convergence. Explicitly,

$$\mathbb{K}\xi = \{\mathcal{F} \in \mathbb{F}|\xi| : \mathcal{F} \cap \mathcal{K}(\xi) \neq \varnothing\}.$$

Given two classes $\mathbb{D}$ and $\mathbb{J}$ of filters, we write $\mathbb{D} \subset \mathbb{J}$ if $\mathbb{D}\xi \subset \mathbb{J}\xi$ for each convergence ξ. For instance, for each $\kappa \geq 1$,

$$\mathbb{F}_0 \subset \mathbb{F}_1 \subset \mathbb{F}_\kappa \subset \mathbb{F}.$$

By convention, we assume that any class of filters considered contains the degenerate filter on each set.

Recall that a convergence ξ is a pseudotopology if and only if

$$\bigcap_{\mathcal{D}\#\mathcal{F}} \operatorname{adh}_\xi \mathcal{D} \subset \lim_\xi \mathcal{F}.$$

and a pretopology if and only if

$$\bigcap_{D \# \mathcal{F}} \mathrm{adh}_\xi \, D \subset \lim_\xi \mathcal{F}.$$

More generally:

Definition XIV.3.1. Given a class $\mathbb{D}$ of filters, we say that a convergence is *determined by adherences of* $\mathbb{D}$*-filters* (or $\mathbb{D}$*-adherence determined*) if

$$\bigcap_{\mathbb{D} \xi \ni \mathcal{D} \# \mathcal{F}} \mathrm{adh}_\xi \, \mathcal{D} \subset \lim_\xi \mathcal{F}.$$

Note that the reverse inclusion in Definition XIV.3.1 is always true.

XIV.3.1 *Reflective classes*

Recall that the reflector S on pseudotopologies and the reflector S_0 on pretopologies are given by

$$\lim_{\mathrm{S} \xi} \mathcal{F} = \bigcap_{\mathcal{H} \# \mathcal{F}} \mathrm{adh}_\xi \, \mathcal{H} \text{ and } \lim_{\mathrm{S}_0 \xi} \mathcal{F} = \bigcap_{A \# \mathcal{F}} \mathrm{adh}_\xi \, A.$$

More generally, we define the map $\mathrm{A}_\mathbb{D}$ by

$$\lim_{\mathrm{A}_\mathbb{D} \xi} \mathcal{F} = \bigcap_{\mathbb{D} \xi \ni \mathcal{D} \# \mathcal{F}} \mathrm{adh}_\xi \, \mathcal{D}. \qquad (\text{XIV.3.1})$$

Accordingly, the reflectors S on pseudotopologies and S_0 on pretopologies satisfy:

$$\mathrm{S} = \mathrm{A}_\mathbb{F}$$
$$\mathrm{S}_0 = \mathrm{A}_{\mathbb{F}_0}.$$

We will now investigate under what conditions on $\mathbb{D}$ the map $\mathrm{A}_\mathbb{D}$ is a projector or a reflector on convergences determined by adherences of $\mathbb{D}$-filters, like in the cases of $\mathbb{F}$ and $\mathbb{F}_0$.

We shall consider the following properties of a class $\mathbb{D}$ of filters (for each $\xi \in \mathbf{I}$ and every $f : X \to |\tau|$)

$$\sigma \le \xi \Longrightarrow \mathbb{D}\sigma \subset \mathbb{D}\xi \qquad (\text{isotone})$$
$$\mathbb{D}\,(\mathrm{A}_\mathbb{D}\,\xi) = \mathbb{D}\xi \qquad (\text{adherence-stable})$$
$$\mathcal{D} \in \mathbb{D}\tau \Longrightarrow f^-[\mathcal{D}] \in \mathbb{D}(f^-\tau). \qquad (\text{initial})$$

Note that if a class $\mathbb{D}$ does not depend on convergence, then (isotone) and (adherence-stable) are satisfied.

Proposition XIV.3.2. *Let* $\mathbb{D}$ *be a class of filters.*

(1) If $\mathbb{D}$ *is isotone then* $A_{\mathbb{D}}$ *is a contractive modifier;*

(2) If $\mathbb{D}$ *is isotone and adherence-stable then* $A_{\mathbb{D}}$ *is a projector and*

$$\mathrm{adh}_{\xi}\,\mathcal{D} = \mathrm{adh}_{A_{\mathbb{D}\xi}}\,\mathcal{D}, \qquad (\mathrm{XIV.3.2})$$

for all $\mathcal{D} \in \mathbb{D}\xi$;

(3) If $\mathbb{D}$ *is isotone and initial then* $A_{\mathbb{D}}$ *is a functor.*

Proof. (1): Let $\sigma \leq \xi$. Then $\mathbb{D}\sigma \subset \mathbb{D}\xi$ and $\mathrm{adh}_{\xi}\,\mathcal{D} \subset \mathrm{adh}_{\sigma}\,\mathcal{D}$ for every filter $\mathcal{D}$, so that

$$\begin{aligned} \lim_{A_{\mathbb{D}}\xi}\mathcal{F} &= \bigcap_{\mathbb{D}\xi \ni \mathcal{D}\#\mathcal{F}} \mathrm{adh}_{\xi}\,\mathcal{D} \\ &\subset \bigcap_{\mathbb{D}\sigma \ni \mathcal{D}\#\mathcal{F}} \mathrm{adh}_{\xi}\,\mathcal{D} \\ &\subset \bigcap_{\mathbb{D}\sigma \ni \mathcal{D}\#\mathcal{F}} \mathrm{adh}_{\sigma}\,\mathcal{D} = \lim_{A_{\mathbb{D}}\sigma}\mathcal{F}. \end{aligned}$$

In other words, $A_{\mathbb{D}}\,\sigma \leq A_{\mathbb{D}}\,\xi$ and $A_{\mathbb{D}}$ is a modifier. Moreover, $A_{\mathbb{D}}\,\xi \leq \xi$ by definition.

(2): We only need to show that $A_{\mathbb{D}}$ is idempotent, which is clear if $\mathbb{D}(A_{\mathbb{D}}\,\xi) = \mathbb{D}\xi$ and $\mathrm{adh}_{\xi}\,\mathcal{D} = \mathrm{adh}_{A_{\mathbb{D}}\,\xi}\,\mathcal{D}$ for every $\mathcal{D} \in \mathbb{D}\xi$. To see the latter, first note that $\mathrm{adh}_{\xi}\,\mathcal{D} \subset \mathrm{adh}_{A_{\mathbb{D}}\,\xi}\,\mathcal{D}$ because $A_{\mathbb{D}}\,\xi \leq \xi$. On the other hand, if $x \in \mathrm{adh}_{A_{\mathbb{D}}\,\xi}\,\mathcal{D}$ then there is $\mathcal{F}\#\mathcal{D}$ with $x \in \lim_{A_{\mathbb{D}}\,\xi}\mathcal{F}$ so that $x \in \mathrm{adh}_{\xi}\,\mathcal{H}$ for each $\mathcal{H} \in \mathbb{D}$ with $\mathcal{H}\#\mathcal{F}$. In particular, $x \in \mathrm{adh}_{\xi}\,\mathcal{D}$.

(3): If $\mathbb{D}$ satisfies (isotone), we have seen that $A_{\mathbb{D}}$ is a modifier. By Proposition XIV.2.3, we only need to show that $A_{\mathbb{D}}(f^{-}\tau) \geq f^{-}(A_{\mathbb{D}}\,\tau)$ for every map f and convergence τ on its codomain. Let $x \in \lim_{A_{\mathbb{D}}(f^{-}\tau)}\mathcal{F}$. We need to show that $f(x) \in \lim_{A_{\mathbb{D}}\,\tau}f[\mathcal{F}]$. To this end, let $\mathcal{D} \in \mathbb{D}\tau$ with $\mathcal{D}\#f[\mathcal{F}]$, equivalently, $f^{-}[\mathcal{D}]\#\mathcal{F}$. By (initial), $f^{-}[\mathcal{D}] \in \mathbb{D}(f^{-}\tau)$ so that $x \in \mathrm{adh}_{f^{-}\tau}\,f^{-}[\mathcal{D}]$. In view of (VIII.1.10), $f(x) \in \mathrm{adh}_{\tau}\,f[f^{-}[\mathcal{D}]]$ and $f[f^{-}[\mathcal{D}]] \geq \mathcal{D}$ so that $x \in \mathrm{adh}_{\tau}\,\mathcal{D}$. $\qquad\square$

Corollary XIV.3.3. *If* $\mathbb{D}$ *is isotone, adherence-stable, and initial, then* $A_{\mathbb{D}}$ *is a reflector.*

Note that (XIV.3.2) characterizes the projection on convergences determined by adherences of $\mathbb{D}$-filters in the following sense:

Proposition XIV.3.4. *If* $\mathbb{D}$ *is isotone and adherence-stable, then* $\xi \geq A_{\mathbb{D}}\,\theta$ *if and only if*

$$\mathrm{adh}_{\xi}\,\mathcal{D} \subset \mathrm{adh}_{\theta}\,\mathcal{D}$$

for every $\mathcal{D} \in \mathbb{D}\theta$.

Proof. If $\xi \geq A_{\mathbb{D}} \theta$ then $\mathrm{adh}_\xi \mathcal{D} \subset \mathrm{adh}_{A_{\mathbb{D}} \theta} \mathcal{D} = \mathrm{adh}_\theta \mathcal{D}$ for each $\mathcal{D} \in \mathbb{D}\theta$, by (adherence-stable). Conversely, if $\mathrm{adh}_\xi \mathcal{D} \subset \mathrm{adh}_\theta \mathcal{D}$ for each $\mathcal{D} \in \mathbb{D}\theta$ and $x \in \lim_\xi \mathcal{F}$, then $x \in \mathrm{adh}_\xi \mathcal{D}$ for each $\mathcal{D} \in \mathbb{D}\theta$ with $\mathcal{D} \# \mathcal{F}$. Thus $x \in \mathrm{adh}_\theta \mathcal{D}$ and $x \in \lim_{A_{\mathbb{D}} \theta} \mathcal{F}$. $\qquad\qquad\qquad\qquad\qquad\qquad\qquad\qquad\qquad\qquad\square$

XIV.3.2   *Composable classes of filters*

We assume that every class of filters contains the degenerate filter of every set.

Definition XIV.3.5. A class $\mathbb{D}$ of filters is *non-empty* if for every set X, $\mathbb{D}X$ contains a non-degenerate filter.

Definition XIV.3.6. Let $\mathbb{D}$ and $\mathbb{J}$ be two classes of filters. We say that $\mathbb{D}$ is $\mathbb{J}$-*composable* if

$$\left.\begin{array}{r} \mathcal{D} \in \mathbb{D}X \\ \mathcal{J} \in \mathbb{J}(X \times Y) \end{array}\right\} \Longrightarrow \mathcal{J}[\mathcal{D}] \in \mathbb{D}Y,$$

for every pair of sets X and Y. A class $\mathbb{D}$ of filters is called *composable* if it is $\mathbb{D}$-composable.

In particular, a class $\mathbb{D}$ of filters is $\mathbb{F}_0$-composable if the image $R[\mathcal{D}]$ of a $\mathbb{D}$-filter $\mathcal{D}$ on X under a relation $R \subset X \times Y$ is a $\mathbb{D}$-filter on Y. In this case, $\mathbb{D}$ is in particular initial.

Lemma XIV.3.7. *Let $\mathbb{J}$ be a class of filters that is $\mathbb{F}_0$-composable. If $\mathbb{D}$ is a non-empty $\mathbb{J}$-composable class of filters, then*

(1) $\mathbb{F}_0 \subset \mathbb{J} \subset \mathbb{D}$;
(2) $\mathbb{D}$ does not depend on convergence and therefore is isotone and adherence-stable;
(3) If $\mathcal{D} \in \mathbb{D}(X)$ and $y \notin X$ then $\mathcal{D} \wedge \{y\} \in \mathbb{D}(X \cup \{y\})$;
(4) If $\mathcal{D} \in \mathbb{D}(X)$ and $\mathcal{J} \in \mathbb{J}$ then $\mathcal{D} \vee \mathcal{J} \in \mathbb{D}(X)$;
(5) If $\mathcal{D} \in \mathbb{D}(X)$ and $\mathcal{J} \in \mathbb{J}(Y)$ then $\mathcal{D} \times \mathcal{J} \in \mathbb{D}(X \times Y)$.

Proof. (1). Let $A \in \mathbb{F}_0(X)$ and let $\mathcal{J}$ be a non-degenerate $\mathbb{J}$-filters on X. Let $R := X \times A$. Then

$$R[\mathcal{J}] = A \in \mathbb{J}$$

because $\mathbb{J}$ is $\mathbb{F}_0$-composable, so that $\mathbb{F}_0 \subset \mathbb{J}$. Moreover, if $\mathcal{J} \in \mathbb{J}(X)$, $\mathcal{D} \in \mathbb{D}(X)$, and $p_X : X \times X \to X$ is a projection, then $p_X^-[\mathcal{J}] \in \mathbb{J}(X \times X)$ by $\mathbb{F}_0$-composability of $\mathbb{J}$, so that

$$(p_X^-[\mathcal{J}])[\mathcal{D}] = \mathcal{J} \in \mathbb{D}$$

because $\mathbb{D}$ is $\mathbb{J}$-composable. Thus $\mathbb{J} \subset \mathbb{D}$.

(2). By (1), $\mathbb{D}$ is in particular $\mathbb{F}_0$-composable. If σ and ξ are two convergences on the same set X then

$$\mathcal{D} \in \mathbb{D}\sigma \implies i_X[\mathcal{D}] = \mathcal{D} \in \mathbb{D}\xi$$

and similarly $\mathbb{D}\xi \subset \mathbb{D}\sigma$, so that we obtain equality.

(3). Let

$$R := \{(x,x) : x \in X\} \cup (X \times \{y\}) \subset X \times (X \cup \{y\}).$$

Then $R[\mathcal{D}] = \mathcal{D} \wedge \{y\}^\uparrow \in \mathbb{D}(X \cup \{y\})$ because $\mathbb{D}$ is $\mathbb{F}_0$-composable.

(4). Given $\mathcal{D} \in \mathbb{D}(X)$ and $\mathcal{J} \in \mathbb{J}(X)$, consider $d : X \to X \times X$ defined by $d(x) = (x,x)$. The filter $d[\mathcal{J}] \in \mathbb{J}(X \times X)$, so that

$$(d[\mathcal{J}])[\mathcal{D}] = \mathcal{D} \vee \mathcal{J} \in \mathbb{D}(X).$$

(5). Since $\mathcal{D} \times \mathcal{J} = p_X^-[\mathcal{D}] \vee p_Y^-[\mathcal{J}]$, this follows from (4) and the fact that both $\mathbb{D}$ and $\mathbb{J}$ are $\mathbb{F}_0$-composable. $\qquad\square$

Corollary XIV.3.8. *If $\mathbb{D}$ is an $\mathbb{F}_0$-composable class, then $A_\mathbb{D}$ is a reflector satisfying*

$$A_\mathbb{D}(f^-\tau) = f^-(A_\mathbb{D}\,\tau),$$

for every map f and convergence τ on its codomain.

Proof. That $A_\mathbb{D}$ is a reflector follows immediately from Corollary XIV.3.3 and the subsequent considerations. In particular,

$$A_\mathbb{D}(f^-\tau) \geq f^-(A_\mathbb{D}\,\tau).$$

To see that the reverse inequality holds, let $x \in \lim_{f^-(A_\mathbb{D}\,\tau)} \mathcal{F}$ and let $\mathcal{D} \in \mathbb{D}$ with $\mathcal{D}\#\mathcal{F}$. We only need to show that $x \in \mathrm{adh}_{f^-\tau} \mathcal{D}$. The fact that $x \in \lim_{f^-(A_\mathbb{D}\,\tau)} \mathcal{F}$ means that $f(x) \in \lim_{A_\mathbb{D}\,\tau} f[\mathcal{F}]$. Moreover, $f[\mathcal{D}] \in \mathbb{D}$ by $\mathbb{F}_0$-composability, and $f[\mathcal{D}]\#f[\mathcal{F}]$ because $\mathcal{D}\#\mathcal{F}$. Therefore

$$f(x) \in \mathrm{adh}_{A_\mathbb{D}\,\tau} f[\mathcal{D}] = \mathrm{adh}_\tau f[\mathcal{D}],$$

by (XIV.3.2). Moreover $f^-(\mathrm{adh}_\tau f[\mathcal{D}]) = \mathrm{adh}_{f^-\tau} \mathcal{D}$ by (VIII.1.10) so that $x \in \mathrm{adh}_{f^-\tau} \mathcal{D}$. $\qquad\square$

Corollary XIV.3.9. *If* $\mathbb{D}$ *is an* $\mathbb{F}_0$-*composable class of filters, then*

$$A_{\mathbb{D}}(\xi_{|A}) = (A_{\mathbb{D}}\,\xi)_{|A}$$

for every convergence ξ *and* $A \subset |\xi|$.

In particular, the classes of convergences determined by adherence of all filters (pseudotopologies), of countably based filters, of principal filters (pretopologies), are all reflective, and their reflectors commute with initial constructions, in particular subspaces.

Definition XIV.3.10. A convergence determined by adherences of countably based filters is called a *paratopology*. Paratopologies form a reflective class. We denote the corresponding reflector by

$$S_1 := A_{\mathbb{F}_1}\,.$$

XIV.3.3 *Conditional compactness*

We have already considered a variant of compactness relative to a class of filters, namely countable compactness or $\mathbb{F}_1$-compactness (see Definition IX.8.4), as well as compactness for families (Definition IX.7.1), in particular filters. Moreover, Proposition IX.7.10 characterizes the convergence of $\mathcal{F}$ in $S\,\xi$ in terms of the compactness of $\mathcal{F}$ for ξ. More generally:

Definition XIV.3.11. Let $\mathcal{A}$ and $\mathcal{B}$ be two families of subsets of a convergence space (X, ξ), and let $\mathbb{D}$ be a class of filters. We say that $\mathcal{A}$ is $\mathbb{D}$-*compact at* $\mathcal{B}$ if

$$\big(\mathcal{D} \in \mathbb{D}X \text{ and } \mathcal{D}\#\mathcal{A}\big) \Longrightarrow \mathrm{adh}_\xi\,\mathcal{D}\#\mathcal{B}. \qquad (\text{XIV.3.3})$$

In case $\mathcal{B} = \{B\}$ or $\mathcal{B} = \{B\}^\uparrow$ then $\mathcal{B}$ can equivalently be replaced by B in (XIV.3.3), and we say that $\mathcal{A}$ is $\mathbb{D}$-compact at B. Similarly, if $\mathcal{A} = \{A\}$ or $\mathcal{A} = \{A\}^\uparrow$ then $\mathcal{A}$ can equivalently be replaced by A in (XIV.3.3), and we say that A is $\mathbb{D}$-compact at $\mathcal{B}$.

We say that a family $\mathcal{A}$ is *relatively* $\mathbb{D}$-*compact*, or $\mathbb{D}$-*compactoid*, if it is $\mathbb{D}$-compact at X and $\mathbb{D}$-*compact* if it is $\mathbb{D}$-compact at $\mathcal{A}$.

Accordingly, a family $\mathcal{A}$ is compact at $\mathcal{B}$ if it is $\mathbb{F}$-compact at $\mathcal{B}$, and a subset A of (X, ξ) is countably compact at B (in the sense of Definition IX.7.1) if and only if it is $\mathbb{F}_1$-compact at B (in the sense of Definition XIV.3.11).

Proposition XIV.3.12. *Let $\mathbb{D}$ be a class of filters and let ξ be a convergence. Then*

$$x \in \lim_{A_{\mathbb{D}\xi}} \mathcal{F} \iff \mathcal{F} \text{ is } \mathbb{D}\text{-compact at } \{x\} \text{ (for } \xi\text{)}.$$

Corollary XIV.3.13. *A filter $\mathcal{F}$ is $\mathbb{F}_0$-compact at $\{x\}$ for ξ if and only if $\mathcal{F} \geq \mathcal{V}_\xi(x)$.*

Exercise XIV.3.14. Show that in a T_1-convergence space, A is $\mathbb{F}_0$-compact at B if and only if $A \subset B$.

Remark XIV.3.15. Exercise XIV.3.14 shows that $\mathbb{F}_0$-compactness is essentially trivial for sets, while Corollary XIV.3.13 shows that it is far from trivial for general filters.

Exercise XIV.3.16. Let $\mathcal{A}$ and $\mathcal{B}$ be two isotone families of subsets of a convergence space (X, ξ). Show that $\mathcal{A}$ is $\mathbb{F}_0$-compact at $\mathcal{B}$ if and only if

$$\mathcal{A} \geq \mathcal{V}_\xi(\mathcal{B}) = \bigcup_{B \in \mathcal{B}} \bigcap_{x \in B} \mathcal{V}_\xi(x).$$

$\mathbb{D}$-compactness is preserved by continuous maps, provided that $\mathbb{D}$ is $\mathbb{F}_0$-composable:

Proposition XIV.3.17. *Let $\mathcal{A}$ and $\mathcal{B}$ be two families of subsets of a convergence space (X, ξ) and let $\mathbb{D}$ be an $\mathbb{F}_0$-composable class of filters. If $f : |\xi| \to |\tau|$ is continuous and $\mathcal{A}$ is $\mathbb{D}$-compact at $\mathcal{B}$ then $f[\mathcal{A}]$ is $\mathbb{D}$-compact at $f[\mathcal{B}]$.*

Proof. If $\mathcal{D} \in \mathbb{D}Y$ with $\mathcal{D} \# f[\mathcal{A}]$ then $f^-[\mathcal{D}] \in \mathbb{D}X$ by $\mathbb{F}_0$-composability and $f^-[\mathcal{D}] \# \mathcal{A}$ by (II.2.9). Therefore $\operatorname{adh}_\xi f^-[\mathcal{D}] \# \mathcal{B}$ by $\mathbb{D}$-compactness of $\mathcal{A}$. In other words, for each $B \in \mathcal{B}$ there is $\mathcal{U}_B \in \beta(f^-[\mathcal{D}])$ and $x_B \in B$ with $x_B \in \lim_\xi \mathcal{U}_B$. By continuity of f, $f(x_B) \in f(B) \cap \lim_\tau f[\mathcal{U}_B]$ and $f[\mathcal{U}_B] \# \mathcal{D}$. Therefore, $\operatorname{adh}_\tau \mathcal{D} \# f[\mathcal{B}]$. $\qquad\square$

Corollary XIV.3.18. *Let $\mathbb{D}$ be an $\mathbb{F}_0$-composable class of filters. If $f : |\xi| \to |\tau|$ is continuous and $K \subset |\xi|$ is $\mathbb{D}$-compact, then $f(K)$ is $\mathbb{D}$-compact in τ.*

Remark XIV.3.19. Note that in view of Proposition XIV.3.12, the fact that $A_{\mathbb{D}}$ is a functor if $\mathbb{D}$ is $\mathbb{F}_0$-composable reflects preservation of $\mathbb{D}$-compactness under continuous maps.

In view of Corollary XIV.3.8, it is relevant to identify other examples of $\mathbb{F}_0$-composable classes of filters, as they determine reflective classes of convergences, and, as we will see shortly in the next section, also coreflective classes. We do so in Section XIV.5. In view of Proposition XIV.3.17, such classes also determine variants of compactness that are preserved by continuous maps.

XIV.4   Convergences based in a class of filters

Definition XIV.4.1. Given a class $\mathbb{D}$ of filters, we say that a convergence ξ is $\mathbb{D}$-*based* if ξ admits at each point a pavement composed of elements of $\mathbb{D}\xi$.

In other words, ξ is $\mathbb{D}$-based if and only if whenever $x \in \lim_\xi \mathcal{F}$ there is $\mathcal{D} \in \mathbb{D}\xi$ with $\mathcal{D} \leq \mathcal{F}$ and $x \in \lim_\xi \mathcal{D}$. Hence, ξ is $\mathbb{D}$-based if and only if

$$\lim\nolimits_\xi \mathcal{F} \subset \bigcup_{\mathbb{D}\xi \ni \mathcal{D} \leq \mathcal{F}} \lim\nolimits_\xi \mathcal{D}.$$

The inclusion above is in fact an equality, as the reverse inclusion is true for every convergence.

Accordingly, a convergence has countable character if and only if it is $\mathbb{F}_1$-based, and is locally compact if and only if it is $\mathbb{K}$-based.

Coreflective classes

Given a class $\mathbb{D}$ of filters, we define a map $B_\mathbb{D}$ by

$$\lim\nolimits_{B_\mathbb{D}\, \xi} \mathcal{F} = \bigcup_{\mathbb{D}\xi \ni \mathcal{D} \leq \mathcal{F}} \lim\nolimits_\xi \mathcal{D}.$$

If $\mathbb{E}$ denotes the class of sequential filters, and $\mathbb{K}$ that of filters containing a compact set, then

$$\mathrm{Seq} = B_\mathbb{E} \;\text{ and }\; \mathrm{K} = B_\mathbb{K}$$

are coreflectors.

More generally, to see under what condition on $\mathbb{D}$ the map $B_\mathbb{D}$ is a coprojector or a coreflector, consider the following properties of a class $\mathbb{D}$ of filters (for each $\xi \in \mathbf{I}$ and $f : |\xi| \to Y$):

$$\sigma \leq \xi \Longrightarrow \mathbb{D}\sigma \supseteq \mathbb{D}\xi \qquad\qquad \text{(antitone)}$$

$$\mathbb{D}(B_\mathbb{D}\, \xi) = \mathbb{D}\xi \qquad\qquad \text{(base-stable)}$$

$$\mathcal{D} \in \mathbb{D}\xi \Longrightarrow f[\mathcal{D}] \in \mathbb{D}(f\xi). \qquad\qquad \text{(final)}$$

Note that if the class $\mathbb{D}$ does not depend on convergence, then (antitone) and (base-stable) are satisfied.

Note also that if $\mathbb{D}X$ does not contain filters of the form $\{x\}^{\uparrow}$ then $\mathrm{B}_{\mathbb{D}}\,\iota$ ([1]) is not a convergence, because it is not centered. We say that a class of filters *contains points* if it contains all principal ultrafilters.

Proposition XIV.4.2. *Let $\mathbb{D}$ be a class of filters that contains points.*

(1) If $\mathbb{D}$ is antitone then $\mathrm{B}_{\mathbb{D}}$ is an expansive modifier;

(2) If $\mathbb{D}$ is antitone and base-stable then $\mathrm{B}_{\mathbb{D}}$ is a coprojector and

$$\lim{}_{\xi}\,\mathcal{D} = \lim{}_{\mathrm{B}_{\mathbb{D}}\,\xi}\,\mathcal{D}, \qquad\qquad (\mathrm{XIV.4.1})$$

for all $\mathcal{D} \in \mathbb{D}\xi$;

(3) If $\mathbb{D}$ is antitone and final then $\mathrm{B}_{\mathbb{D}}$ is a functor.

Proof. (1): Let $\sigma \le \xi$. Then $\mathbb{D}\xi \subset \mathbb{D}\sigma$ and $\lim_{\xi} \mathcal{D} \subset \lim_{\sigma} \mathcal{D}$ for every filter $\mathcal{D}$, so that

$$\lim{}_{\mathrm{B}_{\mathbb{D}}\,\xi}\,\mathcal{F} = \bigcup_{\mathcal{D}\xi \ni \mathcal{D} \le \mathcal{F}} \lim{}_{\xi}\,\mathcal{D} \subset \bigcup_{\mathcal{D}\sigma \ni \mathcal{D} \le \mathcal{F}} \lim{}_{\sigma}\,\mathcal{D} = \lim{}_{\mathrm{B}_{\mathbb{D}}\,\sigma}\,\mathcal{F}.$$

In other words, $\mathrm{B}_{\mathbb{D}}\,\sigma \le \mathrm{B}_{\mathbb{D}}\,\xi$ and $\mathrm{B}_{\mathbb{D}}$ is a modifier. Moreover, $\mathrm{B}_{\mathbb{D}}\,\xi \ge \xi$ by definition.

(2): We only need to show that $\mathrm{B}_{\mathbb{D}}$ is idempotent, which is clear if $\mathbb{D}(\mathrm{B}_{\mathbb{D}}\,\xi) = \mathbb{D}\xi$ and $\lim_{\xi} \mathcal{D} = \lim_{\mathrm{B}_{\mathbb{D}}\,\xi} \mathcal{D}$ for every $\mathcal{D} \in \mathbb{D}\xi$. To see the latter, first note that $\lim_{\xi} \mathcal{D} \supseteq \lim_{\mathrm{B}_{\mathbb{D}}\,\xi} \mathcal{D}$ because $\mathrm{B}_{\mathbb{D}}\,\xi \ge \xi$. On the other hand, $\lim_{\xi} \mathcal{D} \subset \lim_{\mathrm{B}_{\mathbb{D}}\,\xi} \mathcal{D}$ is clear if $\mathbb{D}(\mathrm{B}_{\mathbb{D}}\,\xi) = \mathbb{D}\xi$ because $\mathcal{D} \le \mathcal{D}$.

(3): If $\mathbb{D}$ satisfies (antitone), we have seen that $\mathrm{B}_{\mathbb{D}}$ is a modifier. By Proposition XIV.2.3, we only need to show that $f(\mathrm{B}_{\mathbb{D}}\,\xi) \ge \mathrm{B}_{\mathbb{D}}(f\xi)$ for every map f and convergence ξ on its domain. Let $f : X \to Y$ and ξ be a convergence on X. Since $\mathbb{D}$ contains points, the restrictions of $f(\mathrm{B}_{\mathbb{D}}\,\xi)$ and $\mathrm{B}_{\mathbb{D}}(f\xi)$ to $Y \setminus f(X)$ are both discrete, and therefore coincide.

If $y \in f(X)$ and $y \in \lim_{f(\mathrm{B}_{\mathbb{D}}\xi)} \mathcal{F}$ then there is $\mathcal{D} \in \mathbb{D}\xi$ with $x \in \lim_{\xi} \mathcal{D} \cap f^{-}(y)$ and $f[\mathcal{D}] \le \mathcal{F}$. Since $f[\mathcal{D}] \in \mathbb{D}(f\xi)$ by (final) and $y \in \lim_{f\xi} f[\mathcal{D}]$, we conclude that $y \in \lim_{\mathrm{B}_{\mathbb{D}}(f\xi)} \mathcal{F}$. $\qquad\square$

Corollary XIV.4.3. *If $\mathbb{D}$ is a class of filters that contains points and is antitone, base-stable and final, then $\mathrm{B}_{\mathbb{D}}$ is a coreflector.*

Corollary XIV.4.4. *If $\mathbb{D}$ is an $\mathbb{F}_0$-composable class of filters, then $\mathrm{B}_{\mathbb{D}}$ is a coreflector. If moreover $\mathbb{D}$ is composable, then $\mathrm{B}_{\mathbb{D}}$ commutes with finite products.*

---

[1] where ι is, as usual, the discrete convergence on a set X.

Proof. In view of Lemma XIV.3.7 with $\mathbb{J} = \mathbb{F}_0$, if $\mathbb{D}$ is $\mathbb{F}_0$-composable, it is independent of the convergence and therefore is antitone and base-stable. Moreover, it contains $\mathbb{F}_0$, hence contains points, and it is final. Thus $B_{\mathbb{D}}$ is a coreflector by Corollary XIV.4.3. To show that it commutes with finite products we only need to show that

$$B_{\mathbb{D}}\,\xi \times B_{\mathbb{D}}\,\tau \geq B_{\mathbb{D}}(\xi \times \tau)$$

by Proposition XIV.2.10. But this follows from the definition of $B_{\mathbb{D}}$ and the fact that a product of two $\mathbb{D}$-filters is a $\mathbb{D}$-filter, by Lemma XIV.3.7 applied with $\mathbb{J} = \mathbb{D}$. □

Example XIV.4.5. $\mathbb{F}_\kappa$ is $\mathbb{F}_0$-composable for each κ so that the class of convergences of character less than κ is coreflective. On the other hand, the class $\mathbb{E}$ of sequential filters does not depend on convergence and therefore is antitone and adherence-stable. Moreover, it is final because the image of a sequence under a function is a sequence. Thus Seq is a coreflector. But $\mathbb{E}$ is not $\mathbb{F}_0$-composable.

We use I_κ as a shorthand for

$$I_\kappa := B_{\mathbb{F}_\kappa}\,.$$

In particular, I_0 denotes the coreflector on $\mathbb{F}_0$-based convergences, that is, the coreflector on *finitely generated* convergences, while I_1 denotes the coreflector on $\mathbb{F}_1$-based convergences, that is, the coreflector on *first-countable* convergences.

Example XIV.4.6 (directed graphs). We say that a convergence ξ is a *directed graph* (with loops) if all its convergent filters are principal ultrafilters, in other words, if it is $(\mathbb{U} \cap \mathbb{F}_0)$-based. The terminology comes from the fact that a directed graph, that is, a set of vertices and directed edges, can be identified with such a convergence (on the set of vertices) by interpreting each directed edge $x \to y$ of the graph as convergence: $y \in \lim\{x\}^\uparrow$. Of course, the graph needs to have a loop at each vertex to ensure that $x \in \lim\{x\}^\uparrow$ for every x, but we may add these loops as a convention, or consider preconvergences that are $(\mathbb{U} \cap \mathbb{F}_0)$-based.

Conversely, given a convergence in which the only convergent filters are principal ultrafilters, we can represent the convergence as a directed graph whose set of vertices is the underlying set of the convergence, and a directed edge $x \to y$ is traced if and only if $y \in \lim\{x\}^\uparrow$. Since in a convergence $x \in \lim\{x\}^\uparrow$ for every x, the resulting graph has a loop at each vertex. We could drop this requirement on the graph by considering preconvergences instead.

Definition XIV.4.7. A convergence space X is *reciprocal* if for every x, y in X,

$$\lim\{x\}^\uparrow \cap \lim\{y\}^\uparrow = \varnothing \text{ or } \lim\{x\}^\uparrow = \lim\{y\}^\uparrow.$$

Example XIV.4.8 ((undirected) graphs). We say that a convergence ξ is *a graph* or *an undirected graph* if it is a reciprocal directed graph in the sense of Example XIV.4.6, that is, a reciprocal $(\mathbb{U} \cap \mathbb{F}_0)$-based convergence. The interpretation of such a graph as a $(\mathbb{U} \cap \mathbb{F}_0)$-based convergence on the set of its vertices, and of a $(\mathbb{U} \cap \mathbb{F}_0)$-based convergence as a graph with set of vertices its underlying set, is done just like in Example XIV.4.6. The only difference is that starting from a reciprocal $(\mathbb{U} \cap \mathbb{F}_0)$-based convergence, the resulting graph has the property that $y \to x$ whenever $x \to y$, so that we consider this situation as a non-directed edge joining x and y. Conversely, we interpret a non-directed edge joining x and y in a graph as $y \in \lim\{x\}^\uparrow$ and $x \in \lim\{y\}^\uparrow$ in the corresponding $(\mathbb{U} \cap \mathbb{F}_0)$-based convergence on the set of vertices, which is then reciprocal.

XIV.5 Other $\mathbb{F}_0$-composable classes of filters

Recall (from page 30) that a filter $\mathcal{F}$ is *countably deep* if $\bigcap_{A \in \mathcal{A}} A \in \mathcal{F}$ whenever $\mathcal{A}$ is a countable family with $\mathcal{A} \subset \mathcal{F}$. The class of filters of countable depth is denoted by $\mathbb{F}_{\wedge 1}$.

Principal filters are countably deep, that is, $\mathbb{F}_0 \subset \mathbb{F}_{\wedge 1}$. On the other hand, by Exercise II.2.22, a free filter $\mathcal{F}$ on an uncountable set X is countably deep if and only if

$$\mathcal{F} \geq (X)_1,$$

where $(X)_1$ is the cocountable filter on X.

Exercise XIV.5.1. Show that a free countably deep filter has no countable element.

Lemma XIV.5.2. *The class $\mathbb{F}_{\wedge 1}$ is composable; in particular it is $\mathbb{F}_0$-composable.*

Proof. Let $\mathcal{H} \in \mathbb{F}_{\wedge 1}(X \times Y)$ and $\mathcal{F} \in \mathbb{F}_{\wedge 1}X$, and let $\{A_i : i \in \mathbb{N}\} \subset \mathcal{H}[\mathcal{F}]$. For each $i \in I$, there is $H_i \in \mathcal{H}$ and $F_i \in \mathcal{F}$ such that $H_i F_i \subset A_i$. Since $\bigcap_{i \in \mathbb{N}} H_i \in \mathcal{H}$, $\bigcap_{i \in \mathbb{N}} F_i \in \mathcal{F}$ and

$$\left(\bigcap_{i \in \mathbb{N}} H_i \right) \left(\bigcap_{i \in \mathbb{N}} F_i \right) \subset \bigcap_{i \in \mathbb{N}} H_i F_i \subset \bigcap_{i \in \mathbb{N}} A_i,$$

we conclude that $\bigcap_{i\in\mathbb{N}} A_i \in \mathcal{H}[\mathcal{F}]$.   □

Recall (from Definition II.8.5) that a filter $\mathcal{F}$ on a set X is called *Fréchet* if

$$\mathcal{F} = \bigwedge_{\mathcal{E}\in\mathbb{E},\,\mathcal{E}\geq\mathcal{F}} \mathcal{E},$$

where $\mathbb{E}$ denotes the class of sequential filters.

We know from Exercise II.8.10 that a filter $\mathcal{F}$ is Fréchet if and only if

$$A \in \mathcal{F}^{\#} \implies \underset{\mathcal{H}\in\mathbb{F}_1}{\exists}\ \mathcal{H} \geq \mathcal{F} \vee A. \tag{XIV.5.1}$$

Let $\mathbb{F}_0^{\triangle}$ denote the class of Fréchet filters.

Lemma XIV.5.3. *The class $\mathbb{F}_0^{\triangle}$ of Fréchet filters is $\mathbb{F}_0$-composable.*

Proof. If $\mathcal{F} \in \mathbb{F}_0^{\triangle}X$, $R \subset X \times Y$ and $A\#R[\mathcal{F}]$ then $R^-(A)\#\mathcal{F}$ and there is $\mathcal{H} \in \mathbb{F}_1 X$ with $\mathcal{H} \geq \mathcal{F} \vee R^-(A)$. Since $\mathbb{F}_1$ is $\mathbb{F}_0$-composable, $R[\mathcal{H}] \in \mathbb{F}_1 Y$. Moreover, $R[\mathcal{H}] \geq R[\mathcal{F}]$ and $R[\mathcal{H}]\#A$, so that $R[\mathcal{H}] \vee A$ is countably based and $R[\mathcal{H}] \vee A \geq R[\mathcal{F}] \vee A$. Hence $R[\mathcal{F}]$ is Fréchet.   □

Definition XIV.5.4. A filter $\mathcal{F}$ on a set X is *countably tight* if

$$\underset{A\in\mathcal{F}^{\#}}{\forall}\ \underset{B\subset A}{\exists}\ \ \operatorname{card} B \leq \aleph_0,\ B \in \mathcal{F}^{\#}.$$

More generally a filter $\mathcal{F}$ has *tightness at most κ* if

$$\underset{A\in\mathcal{F}^{\#}}{\forall}\ \underset{B\subset A}{\exists}\ \ \operatorname{card} B \leq \kappa,\ B \in \mathcal{F}^{\#}.$$

We denote by $\mathbb{F}_{\#1}$ the class of countably tight filters.

Lemma XIV.5.5. *The class $\mathbb{F}_{\#1}$ of filters is $\mathbb{F}_1$-composable, hence $\mathbb{F}_0$-composable.*

Proof. Let $\mathcal{F} \in \mathbb{F}_{\#1}X$, $\mathcal{R} \in \mathbb{F}_1(X\times Y)$ and $A\#\mathcal{R}[\mathcal{F}]$. Let $\{R_n : n \in \mathbb{N}\}$ be a decreasing filter-base of $\mathcal{R}$. For each n, $A\#R_n[\mathcal{F}]$, equivalently, $R_n^-(A)\#\mathcal{F}$. As $\mathcal{F}$ has countable tightness, there is a countable subset $B_n = \{b_{k,n} : k \in \mathbb{N}\}$ of $R_n^-(A)$ with $B_n\#\mathcal{F}$. For each $k \in \mathbb{N}$, there is $a_{k,n} \in A$ with $(b_{k,n}, a_{k,n}) \in R_n$. Then $A_n := \{a_{k,n} : k \in \mathbb{N}\}$ is a countable subset of A with $A_n\#R_n[\mathcal{F}]$ because for each $F \in \mathcal{F}$, there is $b_{n,k} \in B_n \cap F$ so that $a_{n,k} \in A_n \cap R(F)$. Hence $\bigcup_{n\in\mathbb{N}} A_n$ is a countable subset of A that meshes with $\mathcal{R}[\mathcal{F}]$.   □

Exercise XIV.5.6. Verify that
$$\mathbb{F}_0 \subset \mathbb{F}_1 \subset \mathbb{F}_0^{\triangle} \subset \mathbb{F}_{\#1},$$
and that
$$\mathbb{F}_{\#1} \cap \mathbb{F}_{\wedge 1} = \mathbb{F}_0.$$

Thus, we have the following inclusions of classes, represented by arrows:

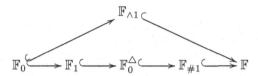

These inclusions are all strict:

(1) the cocountable filter on an uncountable set is in $\mathbb{F}_{\wedge 1} \setminus \mathbb{F}_0$;
(2) the filter generated by a free sequence on an infinite set is in $\mathbb{F} \setminus \mathbb{F}_{\wedge 1}$ and in $\mathbb{F}_1 \setminus \mathbb{F}_0$;
(3) $\mathcal{V}_{f\xi}(\infty)$ in Example IV.9.25 is in $\mathbb{F}_0^{\triangle} \setminus \mathbb{F}_1$ by Exercise V.2.22;
(4) the filter $\mathcal{F} \times \bigwedge_{i \in \mathbb{N}} \mathcal{G}_i$ in Example II.8.17 is in $\mathbb{F}_{\#1} \setminus \mathbb{F}_0^{\triangle}$ by Exercise XIV.5.7 below;
(5) a uniform ultrafilter on an uncountable set is in $\mathbb{F} \setminus \mathbb{F}_{\#1}$.

Exercise XIV.5.7. Show that the filter $\mathcal{F} \times \bigwedge_{i \in \mathbb{N}} \mathcal{G}_i$ in Example II.8.17 is in $\mathbb{F}_{\#1} \setminus \mathbb{F}_0^{\triangle}$.

Figure XIV.1 below is a diagram of reflective classes associated with the classes of filters considered so far.

XIV.6 Functorial inequalities and classification of spaces

Here are some important examples of $\mathbb{D}$-based convergences:

class $\mathbb{D}$ of filters	$\mathbb{D}$-filters	class of $\mathbb{D}$-based convergences
$\mathbb{K}$	locally compact	*locally compact*
$\mathbb{F}_0$	principal	*finitely generated*
$\mathbb{F}_1$	countably based	*first-countable*
$\mathbb{F}_{\wedge 1}$	countably deep	*P-space*
$\mathbb{F}_0^{\triangle}$	Fréchet	*Fréchet*
$\mathbb{F}_{\#1}$	countably tight	*countably tight*

$$T \leq S_0 \leq S_1 \leq A_{\mathbb{F}_0^{\triangle}} \leq A_{\mathbb{F}_{\#1}} \leq S$$

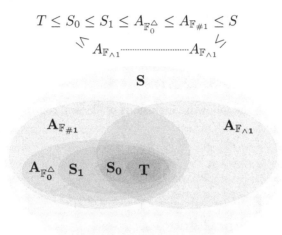

Figure XIV.1   Reflective classes of convergences: the order between the reflectors corresponds to the inclusions between these classes.

Remark XIV.6.1. Fréchet convergences are also called *Fréchet-Urysohn*.

In view of Corollary XIV.4.4, each of the classes of convergences considered above is coreflective. Several other important coreflective classes of convergences can be characterized via inequalities of the form

$$\xi \geq JE\xi, \tag{XIV.6.1}$$

where J is a reflector and E is a coreflector.

Example XIV.6.2 (sequential spaces). Recall from Definition V.6.14 on page 154 that a convergence is *sequential* if every sequentially closed set is closed, and that we have characterized sequential convergences ξ via a functorial inequality (V.6.6) of the type (XIV.6.1), namely,

$$\xi \geq \mathrm{T\,Seq}\,\xi. \tag{XIV.6.2}$$

Example XIV.6.3 (Fréchet pretopologies). We have defined Fréchet convergences as $\mathbb{F}_0^{\triangle}$-based convergences. If ξ is a Fréchet convergence, $x \in \lim_\xi \mathcal{F}$ and $A \# \mathcal{F}$, then $A \# \mathcal{L}$ where $\mathcal{L} \leq \mathcal{F}$, $x \in \lim_\xi \mathcal{L}$ and $\mathcal{L} \in \mathbb{F}_0^{\triangle}$. By definition of Fréchet filters, there is a sequential filter $\mathcal{E}$ such that $\mathcal{E} \geq \mathcal{L} \vee A$. Hence $x \in \lim_\xi \mathcal{E}$ and $A \in \mathcal{E}$ so that $x \in \mathrm{adh}_{\mathrm{Seq}\,\xi} A$. In other words,

$$\xi \geq S_0 \,\mathrm{Seq}\,\xi. \tag{XIV.6.3}$$

Conversely, if ξ is a pretopology satisfying (XIV.6.3) then $\mathcal{V}_\xi(x) \in \mathbb{F}_0^\triangle$ because if $A \# \mathcal{V}_\xi(x)$ then $x \in \mathrm{adh}_\xi A \subset \mathrm{adh}_{S_0 \, \mathrm{Seq}\,\xi} A = \mathrm{adh}_{\mathrm{Seq}\,\xi} A$, so that there is a sequence on A finer than $\mathcal{V}_\xi(x)$. Hence ξ is $\mathbb{F}_0^\triangle$-based, that is, Fréchet.

Example XIV.6.4 (strongly Fréchet pretopologies). A pretopology ξ is called *strongly Fréchet* if whenever $x \in \bigcap_{n \in \mathbb{N}} \mathrm{adh}_\xi A_n$ for a decreasing sequence $\{A_n\}_{n=1}^\infty$ of subsets of $|\xi|$, there is a sequential filter $(x_n)_n^\uparrow$ with $x_n \in A_n$ for each n, and $x \in \lim_\xi (x_n)_n$. Note that $\{A_n : n \in \mathbb{N}\}$ is a filter-base for a filter in $\mathbb{F}_1$ and that every filter in $\mathbb{F}_1$ has such a decreasing filter-base (see Exercise II.8.2). Moreover, in a pretopology

$$\mathrm{adh}\,\mathcal{H} = \bigcap_{H \in \mathcal{H}} \mathrm{adh}\,H.$$

Therefore, a pretopology is strongly Fréchet if and only if

$$\mathrm{adh}_\xi \,\mathcal{H} \subset \mathrm{adh}_{\mathrm{Seq}\,\xi}\,\mathcal{H},$$

for every $\mathcal{H} \in \mathbb{F}_1$. This, in turn, is equivalent to

$$\xi \geq S_1 \,\mathrm{Seq}\,\xi.$$

Definition XIV.6.5. A filter $\mathcal{F}$ is *strongly Fréchet* if and only if for every $\mathcal{H} \in \mathbb{F}_1$ such that $\mathcal{H} \# \mathcal{F}$, there is a sequential filter $\mathcal{E} \geq \mathcal{H} \vee \mathcal{F}$. We denote by $\mathbb{F}_1^\triangle$ the class of strongly Fréchet filters.

Exercise XIV.6.6. Show that:

(1) $\mathrm{adh}_\xi \,\mathcal{H} \subset \mathrm{adh}_{\mathrm{Seq}\,\xi}\,\mathcal{H}$ for every $\mathcal{H} \in \mathbb{F}_1$ if and only if $\xi \geq S_1 \,\mathrm{Seq}\,\xi$.
(2) If moreover ξ is a pretopology, $\xi \geq S_1 \,\mathrm{Seq}\,\xi$ if and only if ξ is $\mathbb{F}_1^\triangle$-based.

Proposition XIV.6.7. *If $\mathbb{D}$ is a class of filters included in $\mathbb{F}_1^\triangle$ then*

$$A_\mathbb{D}\,\mathrm{Seq} = A_\mathbb{D}\,I_1\,.$$

Proof. Since $I_1 \leq \mathrm{Seq}$, we only need to show that $A_\mathbb{D}\,I_1 \geq A_\mathbb{D}\,\mathrm{Seq}$. If $x \in \lim_{A_\mathbb{D}\,I_1\,\xi} \mathcal{F}$ and $\mathcal{D} \in \mathbb{D}$ with $\mathcal{D} \# \mathcal{F}$, then $x \in \mathrm{adh}_{I_1\,\xi}\,\mathcal{D}$, that is, there is $\mathcal{L} \in \mathbb{F}_1$ with $\mathcal{L} \# \mathcal{D}$ and $x \in \lim_\xi \mathcal{L}$. But since $\mathcal{D} \in \mathbb{F}_1^\triangle$, there is a sequential filter $\mathcal{E} \geq \mathcal{L} \vee \mathcal{D}$ so that $x \in \lim_\xi \mathcal{E}$ and $x \in \mathrm{adh}_{\mathrm{Seq}\,\xi}\,\mathcal{D}$. Thus $x \in \lim_{A_\mathbb{D}\,\mathrm{Seq}\,\xi}\mathcal{F}$. $\square$

In particular a pretopology ξ is

$$\text{sequential} \iff \xi \geq T\,I_1\,\xi$$
$$\text{Fréchet} \iff \xi \geq S_0\,I_1\,\xi$$
$$\text{strongly Fréchet} \iff \xi \geq S_1\,I_1\,\xi.$$

Proposition XIV.6.8. *The class* $\mathbb{F}_1^\triangle$ *of strongly Fréchet filters is* $\mathbb{F}_1$-*composable, hence* $\mathbb{F}_0$-*composable.*

Exercise XIV.6.9. Prove Proposition XIV.6.8.

On the other hand, the class of bisequential spaces is defined in terms of countably based filters rather than sequences, resulting in a formulation via the following functorial inequality:

Example XIV.6.10 (bisequential spaces). A convergence space ξ is *bisequential* if whenever $x \in \lim_\xi \mathcal{U}$ for an ultrafilter $\mathcal{U}$, there is a countably based filter $\mathcal{H} \leq \mathcal{U}$ with $x \in \lim_\xi \mathcal{H}$. In other words, ξ is bisequential if $\lim_\xi \mathcal{U} \subset \lim_{\mathrm{I}_1 \xi} \mathcal{U}$ for every ultrafilter $\mathcal{U}$, that is, if

$$\xi \geq \mathrm{S}\,\mathrm{I}_1\,\xi. \tag{bisequential}$$

Since $\mathrm{S} \geq \mathrm{S}_1 \geq \mathrm{S}_0 \geq \mathrm{T}$, every bisequential pretopology is strongly Fréchet, every strongly Fréchet pretopology is in particular Fréchet, and every Fréchet pretopology is sequential.

Definition XIV.6.11. A filter $\mathcal{F}$ is *bisequential* if and only if for every filter $\mathcal{H}$ such that $\mathcal{H}\#\mathcal{F}$, there is $\mathcal{L} \in \mathbb{F}_1$ such that $\mathcal{L} \geq \mathcal{F}$ and $\mathcal{L}\#\mathcal{H}$. We denote by $\mathbb{F}^\triangle$ the class of bisequential filters.

Exercise XIV.6.12. Show that:

(1) $\mathrm{adh}_\xi\,\mathcal{H} \subset \mathrm{adh}_{\mathrm{I}_1\,\xi}\,\mathcal{H}$ for every filter $\mathcal{H}$ if and only if $\xi \geq \mathrm{S}\,\mathrm{I}_1\,\xi$.
(2) If moreover ξ is a pretopology, $\xi \geq \mathrm{S}\,\mathrm{I}_1\,\xi$ if and only if ξ is $\mathbb{F}^\triangle$-based.

Proposition XIV.6.13. *The class* $\mathbb{F}^\triangle$ *of bisequential filters is* $\mathbb{F}_1$-*composable, hence* $\mathbb{F}_0$-*composable.*

Exercise XIV.6.14. Prove Proposition XIV.6.13.

Example XIV.6.15 (*k*-spaces). A topology ξ is called a *k-topology* if a subset whose intersection which each compact subset is closed (in the relative topology of that compact set), is closed in ξ. It follows from the Claim below that a topology ξ is a *k*-topology if and only if

$$\xi \geq \mathrm{T}\,\mathrm{K}\,\xi.$$

Claim. $C \subset |\xi|$ is closed in $\mathrm{K}\,\xi$ if and only if $C \cap K$ is closed in K for every compact subset K of ξ.

Proof. If C is $\mathrm{K}\,\xi$-closed and $C \cap K \in \mathcal{F}$ with $x \in \lim_\xi \mathcal{F} \cap K$, then $K \in \mathcal{F}$ so that $x \in \lim_{\mathrm{K}\,\xi} \mathcal{F}$. Since $C \in \mathcal{F}$ and C is $\mathrm{K}\,\xi$-closed, $x \in C$.

Conversely, assume that $C \cap K$ is closed in K for every compact subset K of ξ, and let $\mathcal{F}$ be a filter with $x \in \lim_{\mathrm{K}\,\xi} \mathcal{F}$ and $C \in \mathcal{F}$. Since $\mathcal{F}$ converges for $\mathrm{K}\,\xi$, there is a compact subset K_0 of ξ in $\mathcal{F}$. Moreover, $K_1 = K \cup \{x\}$ is also compact, and belongs to $\mathcal{F}$ as well. Hence $K_1 \cap C \in \mathcal{F}$ and $x \in K_1 \cap \lim_\xi \mathcal{F}$. Since $K_1 \cap C$ is closed, $x \in C$. Therefore, C is closed for $\mathrm{K}\,\xi$. $\qquad\square$

Example XIV.6.16 (k'-spaces). A topology ξ is called a *k'-topology* if whenever $x \in \mathrm{cl}_\xi A$, there is a compact set K with $x \in \mathrm{cl}_\xi(A \cap K)$. It follows from the Claim below that a topology ξ is a k'-topology if and only if

$$\xi \geq \mathrm{S}_0 \,\mathrm{K}\, \xi.$$

Claim. The following are equivalent in a convergence space ξ:

(1) If $x \in \mathrm{adh}_\xi A$, there exists a compact set K with $x \in \mathrm{adh}_\xi(A \cap K)$;
(2) $\xi \geq \mathrm{S}_0 \,\mathrm{K}\, \xi$.

Proof. In view of Proposition XIV.3.4, the functorial inequality $\xi \geq \mathrm{S}_0 \,\mathrm{K}\, \xi$ is equivalent to $\mathrm{adh}_\xi A \subset \mathrm{adh}_{\mathrm{K}\,\xi} A$ for every subset A of $|\xi|$. Moreover,

$$x \in \mathrm{adh}_{\mathrm{K}\,\xi} A \iff x \in \mathrm{adh}_\xi(A \cap K) \text{ for some compact set } K.$$

Indeed, if $x \in \mathrm{adh}_{\mathrm{K}\,\xi} A$ then there is an ultrafilter $\mathcal{U}$ with $A \in \mathcal{U}$ and $x \in \lim_{\mathrm{K}\,\xi} \mathcal{U}$, that is, $x \in \lim_\xi \mathcal{U}$ and there is a ξ-compact set $K \in \mathcal{U}$. Thus $K \cap A \in \mathcal{U}$ and $x \in \mathrm{adh}_\xi(A \cap K)$. Conversely, if $x \in \mathrm{adh}_\xi(A \cap K)$ for some compact set K, there is an ultrafilter $\mathcal{U}$ with $A \cap K \in \mathcal{U}$ and $x \in \lim_\xi \mathcal{U}$. Moreover, $x \in \lim_{\mathrm{K}\,\xi} \mathcal{U}$ because $K \in \mathcal{U}$ and thus $x \in \mathrm{adh}_{\mathrm{K}\,\xi} A$ because $A \in \mathcal{U}$. $\qquad\square$

Example XIV.6.17 (strongly k'-spaces). A topology ξ is *strongly k'* if whenever a decreasing sequence $\{A_n\}_n$ of subsets satisfies $x \in \bigcap_{n \in \mathbb{N}} \mathrm{cl}_\xi A_n$ there is a compact set K such that $x \in \bigcap_{n \in \mathbb{N}} \mathrm{cl}_\xi(A_n \cap K)$. Note that $\{A_n : n \in \mathbb{N}\}$ is a filter-base for a filter in $\mathbb{F}_1$ and that every filter in $\mathbb{F}_1$ has such a decreasing filter-base (see Exercise II.8.2). In view of the characterization (VIII.1.5) of the adherence of a filter in a (pre)topological space, a topology is strongly k' if and only if

$$x \in \mathrm{adh}_\xi \mathcal{H} \implies \underset{K \in \mathcal{K}(\xi)}{\exists} x \in \mathrm{adh}_\xi(\mathcal{H} \vee K), \qquad (\text{XIV.6.4})$$

for every $\mathcal{H} \in \mathbb{F}_1$. Moreover, (XIV.6.4) is equivalent to $\mathrm{adh}_\xi \, \mathcal{H} \subset \mathrm{adh}_{\mathrm{K}\,\xi} \, \mathcal{H}$, so that, in view of Proposition XIV.3.4, ξ is strongly k' if and only if

$$\xi \geq \mathrm{S}_1 \, \mathrm{K} \, \xi.$$

Exercise XIV.6.18. Show that if ξ is a pseudotopology then $\xi \geq \mathrm{K} \, \xi$ if and only if $\xi \geq \mathrm{S} \, \mathrm{K} \, \xi$.

Similarly, it can be shown (the interested reader may for instance consult [32, Proposition 3]) that if ξ is a pseudotopology then

$$\xi \geq \mathrm{Seq}\, \xi \iff \xi \geq \mathrm{S}\, \mathrm{Seq}\, \xi.$$

Table XIV.1 below gathers the examples of this section of topological properties (of a topology) characterized by a functorial inequality of the type

$$\xi \geq JE\xi$$

for a reflector J and a coreflector E.

	I_1	Seq	K
I	first-countable $\xi \geq \mathrm{I}_1 \, \xi$	sequentially based $\xi \geq \mathrm{Seq}\, \xi$	locally compact $\xi \geq \mathrm{K} \, \xi$
S	bisequential $\xi \geq \mathrm{S}\, \mathrm{I}_1 \, \xi$	sequentially based $\xi \geq \mathrm{S}\, \mathrm{Seq}\, \xi$	locally compact $\xi \geq \mathrm{S}\, \mathrm{K} \, \xi$
S_1	strongly Fréchet $\xi \geq \mathrm{S}_1 \, \mathrm{I}_1 \, \xi$	strongly Fréchet $\xi \geq \mathrm{S}_1 \, \mathrm{Seq}\, \xi$	strongly k' $\xi \geq \mathrm{S}_1 \, \mathrm{K} \, \xi$
S_0	Fréchet $\xi \geq \mathrm{S}_0 \, \mathrm{I}_1 \, \xi$	Fréchet $\xi \geq \mathrm{S}_0 \, \mathrm{Seq}\, \xi$	k' $\xi \geq \mathrm{S}_0 \, \mathrm{K} \, \xi$
T	sequential $\xi \geq \mathrm{T}\, \mathrm{I}_1 \, \xi$	sequential $\xi \geq \mathrm{T}\, \mathrm{Seq}\, \xi$	k $\xi \geq \mathrm{T}\, \mathrm{K} \, \xi$

Table XIV.1   Classes of spaces and functorial inequalities.

XIV.7   Reflective and coreflective hulls

In Section VII.3 we have already studied convergences that are initial with respect to all the continuous maps to a single convergence. If σ is a convergence, then

$$R_\sigma \xi := \bigvee_{f \in C(\xi,\sigma)} f^- \sigma$$

defines a concrete functor, as in (VII.3.1). Since a map $f : |\xi| \to |\sigma|$ is continuous if and only if $\xi \geq f^- \sigma$,

$$\xi \geq R_\sigma \xi$$

for every σ and ξ, that is, R_σ is contractive. Similarly,

$$C_\sigma \xi := \bigwedge_{f \in C(\sigma,\xi)} f \sigma$$

also defines a concrete functor, and since $f : |\sigma| \to |\xi|$ is continuous if and only if $f \sigma \geq \xi$,

$$\xi \leq C_\sigma \xi$$

for every σ and ξ, that is, C_σ is expansive. In this section, we examine when the reverse inequalities hold.

Definition XIV.7.1. Given a convergence σ, a convergence ξ is said to be *σ-initial* if $\xi = \bigvee_{f \in C(\xi,\sigma)} f^- \sigma$, that is, if

$$\xi \leq R_\sigma \xi,$$

and *σ-final* if $\xi = \bigwedge_{f \in C(\sigma,\xi)} f \sigma$, that is, if

$$\xi \geq C_\sigma \xi.$$

By definition, the class of σ-initial convergences is the largest class of convergences in which $\{\sigma\}$ is initially dense, while the class of σ-final convergences is the largest class of convergences in which $\{\sigma\}$ is finally dense. The choice of notations stems from the fact that the class $\mathbf{R}_\sigma$ of convergences fixed by R_σ is, as we will see below, the smallest reflective class of convergences that contains σ, and the coreflective class $\mathbf{C}_\sigma$ of convergences fixed by C_σ is the smallest coreflective class of convergences that contains σ. By definition, $\mathbf{R}_\sigma$ is simple (in the sense of Definition VIII.5.1), for $\{\sigma\}$ is initially dense in $\mathbf{R}_\sigma$.

Theorem XIV.7.2. *Given two convergences ξ and σ,*

(1) $C(R_\sigma \xi, \sigma) = C(\xi, \sigma)$;
(2) $R_\sigma \xi$ is the finest σ-initial convergence coarser than ξ;
(3) R_σ is a reflector.

Proof. (1). $C(\mathrm{R}_\sigma\,\xi,\sigma) \subset C(\xi,\sigma)$ because $\mathrm{R}_\sigma\,\xi \leq \xi$, and if $f \in C(\xi,\sigma)$, then $f \in C(\mathrm{R}_\sigma\,\xi,\sigma)$ because by definition $\mathrm{R}_\sigma\,\xi \geq f^-\sigma$.

(2). As a result of (1), $\mathrm{R}_\sigma\,\xi$ is a σ-initial convergence $\bigvee_{f\in C(\mathrm{R}_\sigma\,\xi,\sigma)} f^-\sigma = \bigvee_{f\in C(\xi,\sigma)} f^-\sigma = \mathrm{R}_\sigma\,\xi$, and it is coarser than ξ. If now $\theta \leq \xi$ and θ is σ-initial, then

$$\theta = \bigvee_{f\in C(\theta,\sigma)} f^-\sigma \leq \bigvee_{f\in C(\xi,\sigma)} f^-\sigma = \mathrm{R}_\sigma\,\xi,$$

for $C(\theta,\sigma) \subset C(\xi,\sigma)$.

(3). In view of (2) and Proposition XIV.1.11, R_σ is a projector. It is also a functor, hence a reflector, for if $h \in C(\xi,\tau)$ then for every $f \in C(\tau,\sigma)$, $f \circ h \in C(\xi,\sigma)$ and thus

$$\mathrm{R}_\sigma\,\xi = \bigvee_{g\in C(\xi,\sigma)} g^-\sigma \geq \bigvee_{f\in C(\tau,\sigma)} (f \circ h)^-\,\sigma$$

and

$$\bigvee_{f\in C(\tau,\sigma)} (f \circ h)^-\,\sigma = h^-\left(\bigvee_{f\in C(\tau,\sigma)} f^-\sigma \right)$$

by (IV.3.5). Thus $\mathrm{R}_\sigma\,\xi \geq h^-(\mathrm{R}_\sigma\,\tau)$ so that $h \in C(\mathrm{R}_\sigma\,\xi, \mathrm{R}_\sigma\,\tau)$. $\square$

We leave the proof of the dual statement below as an exercise.

We call $\mathrm{R}_\sigma\,\xi$ the σ-*initial modification* of ξ, and $\mathrm{C}_\sigma\,\xi$ the σ-*final modification* of ξ.

Theorem XIV.7.3. *Given two convergences ξ and σ,*

(1) $C(\sigma, \mathrm{C}_\sigma\,\xi) = C(\sigma,\xi)$;
(2) $\mathrm{C}_\sigma\,\xi$ *is the coarsest σ-final convergence finer than ξ;*
(3) C_σ *is a coreflector.*

Exercise XIV.7.4. Prove Theorem XIV.7.3.

Example XIV.7.5 (topologies). Let $\sigma = \$$ be the Sierpiński space. Proposition V.4.18 states that the $\$$-initial convergences are exactly the topologies, so that $\mathrm{R}_\$ = \mathrm{T}$ is the topologizer.

Example XIV.7.6. By Theorem V.1.14, the class $\mathbf{R}_{\maltese}$, where $\maltese$ is the Bourdaud convergence, is the class of pretopologies and $\mathrm{R}_{\maltese} = \mathrm{S}_0$ is the pretopologizer.

Example XIV.7.7. Let $\sigma = \nu$ be the standard topology of the reals. By definition VII.6.1, ξ is functionally regular if and only if it is ν-initial. In particular, the class of functionally regular convergences is simple and reflective.

Now, given a class $\mathbf{D}$ of convergences, for every convergence ξ, we have:

$$\xi \geq \bigvee_{\sigma \in \mathbf{D}} R_\sigma \xi := R_\mathbf{D} \xi$$

and

$$C_\mathbf{D} \xi := \bigwedge_{\sigma \in \mathbf{D}} C_\sigma \xi \geq \xi.$$

We call a convergence $\mathbf{D}$-*initial* if $R_\mathbf{D} \xi \geq \xi$ and $\mathbf{D}$-*final* if $\xi \geq C_\mathbf{D} \xi$.

Theorem XIV.7.8. *Given a class $\mathbf{D}$ of convergences, for every convergences ξ,*

(1) $C(R_\mathbf{D} \xi, \sigma) = C(\xi, \sigma)$ *for every $\sigma \in \mathbf{D}$;*
(2) $C(R_\mathbf{D} \xi, \sigma) = C(\xi, \sigma)$ *for every $\sigma = R_\mathbf{D} \sigma$;*
(3) $R_\mathbf{D} \xi$ *is the finest $\mathbf{D}$-initial convergence coarser than ξ;*
(4) $R_\mathbf{D}$ *is a reflector;*
(5) *If $\mathbf{W} \subset \mathbf{D}$ is initially dense in $\mathbf{D}$, then $R_\mathbf{W} = R_\mathbf{D}$.*

Proof. (1) follows from the fact that for every ξ and every $\sigma \in \mathbf{D}$,

$$\xi \geq R_\mathbf{D} \xi \geq R_\sigma \xi,$$

and $C(R_\sigma \xi, \sigma) = C(\xi, \sigma)$.

The proofs of (3) and (4) follows that of Theorem XIV.7.2 with virtually no change, but the insertion of the appropriate suprema over $\sigma \in \mathbf{D}$.

(5) follows from the observation that if

$$(f : |\xi| \to |\sigma|)_{f \in C(\xi, \sigma), \sigma \in \mathbf{D}}$$

is initial and, for each $\sigma \in \mathbf{D}$,

$$(g : |\sigma| \to |\tau|)_{g \in C(\sigma, \tau), \tau \in \mathbf{W}}$$

is also initial, then the family of composites $g \circ f : |\xi| \to |\tau|$ is initial, so that, *a fortiori*,

$$(h : \xi| \to |\tau|)_{h \in C(\xi, \tau), \tau \in \mathbf{W}}$$

is initial.

Finally, (2) follows from (4) for $R_\mathbf{D} \xi \leq \xi$ ensures $C(R_\mathbf{D} \xi, \sigma) \subset C(\xi, \sigma)$ for every σ, and the fact that $R_\mathbf{D}$ is a functor ensures $C(\xi, \sigma) \subset C(R_\mathbf{D} \xi, R_\mathbf{D} \sigma)$. $\square$

In particular, $R_\mathbf{D}$ is the smallest reflective class of convergences containing $\mathbf{D}$.

Corollary XIV.7.9. *Let F be an idempotent modifier and let $\mathbf{D}$ be an initially dense subclass of* fix F. *The following are equivalent:*

(1) F *is a reflector;*
(2) $\mathrm{R}_{\mathrm{fix}\,F} = F$;
(3) $C(\xi,\sigma) = C(F\xi,\sigma)$ *for every convergence ξ and every $\sigma \in$ fix F;*
(4) $C(\xi,\sigma) = C(F\xi,\sigma)$ *for every convergence ξ and every $\sigma \in \mathbf{D}$.*

Proof. $(1) \Longrightarrow (2)$ for if F is a reflector then fix $F = \mathbf{F}$ is a reflective class and fix $\mathrm{R}_{\mathrm{fix}\,F}$ is the smallest reflective class containing fix F, hence is fix F.

$(2) \Longrightarrow (3)$ by Theorem XIV.7.8 (1).

$(3) \Longrightarrow (1)$: Because F is idempotent, $F\xi \in$ fix F and the identity map $i : |F\xi| \to |F\xi|$ belongs to $C(F\xi, F\xi) \subset C(\xi, F\xi)$, so that $\xi \geq F\xi$ and F is contractive, hence a projector. Moreover, F is a functor, hence a reflector. Indeed, for every ξ and τ, $C(\xi,\tau) \subset C(\xi, F\tau)$ because F is contractive. By (3),

$$C(\xi,\tau) \subset C(\xi, F\tau) \subset C(F\xi, F\tau)$$

and the conclusion follows.

Finally, $(3) \Longrightarrow (4)$ because $\mathbf{D} \subset$ fix F, and $(4) \Longrightarrow (3)$ because if $\mathbf{D}$ is initially dense in fix F, then $\mathrm{R}_{\mathbf{D}} = \mathrm{R}_{\mathrm{fix}\,F}$ and the result follows from Theorem XIV.7.8. $\qquad\qquad\square$

Thus, for a reflector F, the reflection $F\xi$ is determined by the sets $C(\xi,\sigma)$ where σ ranges over $\mathbf{F}$, equivalently over an initially dense subclass $\mathbf{D}$ of $\mathbf{F}$. In other words:

Corollary XIV.7.10. *Let F be a reflector, let $\mathbf{D}$ be an initially dense subclass of $\mathbf{F}$, and let ξ and τ be two convergences on the same underlying set. The following are equivalent:*

(1) $\tau \geq F\xi$;
(2) *for all $\sigma = F\sigma$, $C(\xi,\sigma) \subset C(\tau,\sigma)$;*
(3) *for all $\sigma \in \mathbf{D}$, $C(\xi,\sigma) \subset C(\tau,\sigma)$.*

Proof. $(1) \Longrightarrow (2)$ because $C(\xi,\sigma) = C(F\xi,\sigma) \subset C(\tau,\sigma)$ and $(2) \Longrightarrow (1)$ for the identity map $i : |\xi| \to |F\xi|$ is continuous, so that $i : |\tau| \to |F\xi|$ is continuous, that is, $\tau \geq F\xi$. Finally, $(2) \Longrightarrow (3)$ is obvious, and $(3) \Longrightarrow (2)$ because $\mathrm{R}_{\mathbf{D}} = F$. $\qquad\qquad\square$

Dually,

Theorem XIV.7.11. *Given a class* **D** *of convergences, for every convergences* ξ,

(1) $C(\sigma, \mathrm{C_D}\,\xi) = C(\sigma, \xi)$ *for every* $\sigma \in \mathbf{D}$;
(2) $C(\sigma, \mathrm{C_D}\,\xi) = C(\sigma, \xi)$ *for every* $\sigma = \mathrm{C_D}\,\sigma$;
(3) $\mathrm{C_D}\,\xi$ *is the coarsest* **D**-*final convergence finer than* ξ;
(4) $\mathrm{C_D}$ *is a coreflector;*
(5) *If* $\mathbf{W} \subset \mathbf{D}$ *is a finally dense in* **D**, *then* $\mathrm{C_W} = \mathrm{C_D}$.

Corollary XIV.7.12. *Let* F *be an idempotent modifier and let* **D** *be a finally dense subclass of* fix F. *The following are equivalent:*

(1) F *is a coreflector;*
(2) $\mathrm{C_{\mathrm{fix}\,F}} = F$;
(3) $C(\sigma, \xi) = C(\sigma, F\xi)$ *for every convergence* ξ *and every* $\sigma \in$ fix F;
(4) $C(\sigma, \xi) = C(\sigma, F\xi)$ *for every convergence* ξ *and every* $\sigma \in \mathbf{D}$.

Corollary XIV.7.13. *Let* F *be a coreflector, let* **D** *be a finally dense subclass of* **F**, *and let* ξ *and* τ *be two convergences on the same underlying set. The following are equivalent:*

(1) $\tau \leq F\xi$;
(2) *for all* $\sigma = F\sigma$, $C(\sigma, \xi) \subset C(\sigma, \tau)$;
(3) *for all* $\sigma \in \mathbf{D}$, $C(\sigma, \xi) \subset C(\sigma, \tau)$.

XIV.8   Conditional compactness and cover-compactness

Let us examine $\mathbb{D}$-compactness, that is, the conditional compactness with respect to some of the classes $\mathbb{D}$ introduced in the previous sections.

In view of the definitions of the classes $\mathbb{F}_0^\triangle$ of Fréchet filters, $\mathbb{F}_1^\triangle$ of strongly Fréchet filters, and $\mathbb{F}^\triangle$ of bisequential filters, it is natural to associate to a class $\mathbb{D}$ of filters, the class $\mathbb{D}^\triangle$ of filters $\mathcal{F}$ that satisfy

$$\mathcal{D} \in \mathbb{D},\ \mathcal{D}\#\mathcal{F} \implies \underset{\mathcal{L} \in \mathbb{F}_1}{\exists}\ \mathcal{L} \geq \mathcal{F},\ \mathcal{L}\#\mathcal{D}. \tag{XIV.8.1}$$

Note that by definition of the relation $\triangle$, regardless of what the class $\mathbb{D}$ is,

$$\mathbb{F}_1 \subset \mathbb{D}^\triangle$$

for we can take $\mathcal{F} = \mathcal{L}$ in (XIV.8.1) if $\mathcal{F} \in \mathbb{F}_1$.

Proposition XIV.8.1. *A filter* $\mathcal{D} \in \mathbb{D}(X)$ *is* $\mathbb{D}^\triangle$-*compact (at* $\mathcal{A} \subset 2^X$) *if and only if it is* $\mathbb{F}_1$-*compact (at* $\mathcal{A} \subset 2^X$).

Proof. If $\mathcal{D}$ is $\mathbb{D}^{\triangle}$-compact, then it is in particular $\mathbb{F}_1$-compact because $\mathbb{F}_1 \subset \mathbb{D}^{\triangle}$. Conversely, assume that $\mathcal{D}$ is $\mathbb{F}_1$-compact at $\mathcal{A}$ and let $\mathcal{H} \in \mathbb{D}^{\triangle}$ with $\mathcal{H} \# \mathcal{D}$. Then by definition of $\mathbb{D}^{\triangle}$, there is $\mathcal{L} \in \mathbb{F}_1$ with $\mathcal{L} \geq \mathcal{H}$ and $\mathcal{L} \# \mathcal{D}$. Thus $\operatorname{adh} \mathcal{L} \# \mathcal{A}$ because $\mathcal{D}$ is $\mathbb{F}_1$-compact at $\mathcal{A}$. But $\operatorname{adh} \mathcal{L} \subset \operatorname{adh} \mathcal{H}$, so that $\operatorname{adh} \mathcal{H} \# \mathcal{A}$ and $\mathcal{D}$ is $\mathbb{D}^{\triangle}$-compact. $\qquad\square$

Corollary XIV.8.2. *A subset of a convergence space is $\mathbb{F}_0^{\triangle}$-compact if and only if it is $\mathbb{F}_1$-compact. A countably based filter is $\mathbb{F}_1^{\triangle}$-compact if and only if it is $\mathbb{F}_1$-compact.*

On the other hand, a countably based filter may be $\mathbb{F}_1$-compact at a set, but not $\mathbb{F}_0^{\triangle}$-compact at that set:

Example XIV.8.3 (A countably based filter that is $\mathbb{F}_1$-compact, but not $\mathbb{F}_0^{\triangle}$-compact). As we have seen in Theorem XVII.8.5, there are Fréchet filters that are not strongly Fréchet. Let $\mathcal{F}$ be such a filter in $\mathbb{F}_0^{\triangle}(X) \setminus \mathbb{F}_1^{\triangle}(X)$, that is also free. Then there is $\mathcal{H} \in \mathbb{F}_1(X)$ with $\mathcal{H} \# \mathcal{F}$ such that

$$\mathcal{L} \in \mathbb{F}_1(X),\ \mathcal{L} \geq \mathcal{H} \Longrightarrow \mathcal{L} \not\geq \mathcal{F}.$$

Define on $Y := X \cup \mathbb{U}X$ the (pre)topology in which every point of X and every point of $\mathbb{U}X \setminus (\beta(\mathcal{H}) \setminus \beta(\mathcal{F}))$ is isolated and, for every $\mathcal{U} \in \beta(\mathcal{H}) \setminus \beta(\mathcal{F})$, $\mathcal{V}(\{\mathcal{U}\}) := \mathcal{U}^{\uparrow} \wedge \{\mathcal{U}\}$ where $\mathcal{U}^{\uparrow}$ denotes the filter generated on Y by $\mathcal{U} \in \mathbb{F}X$. Then $\mathcal{H}^{\uparrow}$ is not (relatively) $\mathbb{F}_0^{\triangle}$-compact, for $\mathcal{F}^{\uparrow}$ is a Fréchet filter that meshes with $\mathcal{H}$ and $\operatorname{adh}_Y \mathcal{F}^{\uparrow} = \varnothing$. On the other hand, if $\mathcal{G} \in \mathbb{F}_1$ and $\mathcal{G} \# \mathcal{H}$, then $\mathcal{G} \vee \mathcal{H} \in \mathbb{F}_1$ so that $\beta(\mathcal{G} \vee \mathcal{H}) \cap \beta(\mathcal{F}) = \varnothing$ and $\operatorname{adh}_Y (\mathcal{G} \vee \mathcal{H}) = \beta(\mathcal{G} \vee \mathcal{H})$. Thus, $\mathcal{H}$ is $\mathbb{F}_1$-compact at $\beta(\mathcal{H})$.

Definition XIV.8.4. A set (or a family) is called *Lindelöf* if it is $\mathbb{F}_{\wedge 1}$-compact.

Covers were introduced in Section IX.10, where the duality between ideal covers and non-adherent filters was established in Corollary IX.10.6. Via this duality, cover-compactness (Definition IX.11.2) could be characterized in terms of filters by (IX.11.1), and by Definition IX.11.12 for families. We can extend the approach to compactness relative to classes:

Definition XIV.8.5. Given two classes $\mathbb{D}$ and $\mathbb{J}$ of filters, a family $\mathcal{A}$ of subsets of $|\xi|$ is *cover-$(\mathbb{D}/\mathbb{J})$-compact* at a family $\mathcal{B}$ if for every $\mathcal{D} \in \mathbb{D}$

$$\left(\bigvee_{\mathcal{J} \in \mathbb{J}} \mathcal{J} \leq \mathcal{D} \Longrightarrow \operatorname{adh} \mathcal{J} \# \mathcal{A} \right) \Longrightarrow \operatorname{adh} \mathcal{D} \# \mathcal{B}.$$

We have seen (Example IX.11.9) that even cover-$(\mathbb{F}/\mathbb{F}_0)$-compactness of sets is not preserved by continuous maps, in sharp contrast to Proposition XIV.3.17.

We will denote by

$$\operatorname{adh}_{\xi}^{\natural} \mathbb{D}(\xi) \subset \mathbb{D}(\xi)$$

the condition that $\operatorname{adh}_{\xi}^{\natural} \mathcal{D} \in \mathbb{D}$ whenever $\mathcal{D}$ is a filter of the class $\mathbb{D}$ on $|\xi|$.

Theorem XIV.8.6. *Let $\mathbb{D}$ be an $\mathbb{F}_0$-composable class such that $\operatorname{adh}_{\xi}^{\natural} \mathbb{D}(\xi) \subset \mathbb{D}(\xi)$ and assume that $\xi = \mathrm{A}_{\mathbb{D}}\, \xi$. Then ξ is a topology if and only if every filter that is $\mathbb{D}$-compact (at a set) is cover-$(\mathbb{D}/\mathbb{F}_0)$-compact (at that set).*

Exercise XIV.8.7. Following the example of Theorem IX.11.15, prove Theorem XIV.8.6.

Chapter XV

Classification of maps

XV.1 Various types of quotient maps

XV.1.1 *Remarks on the quotient convergence*

In Section IV.2, we constructed the quotient of a set under an equivalence relation, or equivalently a partition, and have seen how each surjection can be interpreted as the canonical surjection for such a quotient.

We have seen that given a convergence space (X, ξ) and a surjection $f : X \to Y$, the set Y, interpreted as a quotient of X, can be canonically endowed with the *quotient convergence*, namely the final convergence $f\xi$ for the canonical surjection f. It is by definition the finest convergence on the quotient set Y that makes the canonical surjection f continuous from (X, ξ). Accordingly, a continuous surjection $f : |\xi| \to |\tau|$ is called a *convergence quotient map* if $\tau = f\xi$ (see Definition IV.8.2). We have seen in Theorem IV.7.3 that this construction commutes with products. Thus:

Corollary XV.1.1. *If F is a family of convergence quotient maps $f : |\xi_f| \to |\tau_f|$, then the product map*

$$\bigotimes F : \left| \prod_{f \in F} \xi_f \right| \to \left| \prod_{f \in F} \tau_f \right|$$

is a convergence quotient map.

Additionally, the construction of the quotient convergence is *hereditary* in the following sense: Let $f : (X, \xi) \to (Y, \tau)$ be a convergence quotient map, that is, f is onto and $\tau = f\xi$. On a subset B of Y, there are two convergences to consider in this situation: the convergence $(f\xi)_{|B}$ induced on B by $f\xi$ via the inclusion map $j : B \to Y$, and the quotient convergence $f(\xi_{|f^- B})$ for the convergence $\xi_{|f^- B}$ induced on $f^-(B)$ by ξ via the inclusion map $i : f^- B \to X$:

$$
\begin{array}{ccc}
f^-(B) & \xrightarrow{\ i\ } & (X, \xi) \\
{\scriptstyle f}\downarrow & & \downarrow{\scriptstyle f} \\
B & \xrightarrow{\ j\ } & (Y, f\xi)
\end{array}
$$

We say that convergence quotients are hereditary, or compatible with subsets, because these two convergences coincide, as the following proposition shows.

Proposition XV.1.2. *Let $f : |\xi| \to |\tau|$ be a convergence quotient map, let B be a subset of Y and let $j : B \to Y$ and $i : f^- B \to X$ denote inclusion maps. Then*

$$
f\xi_{|B} = f(\xi_{|f^- B}),
$$

so that the map $f_{|f^- B} : \left|\xi_{|f^- B}\right| \to \left|f\xi_{|B}\right|$ is a convergence quotient map.

Proof. That $f(\xi_{|f^- B}) \geq f\xi_{|B}$ follows from Proposition IV.2.15, as the restriction of a continuous map is continuous. On the other hand, if $y \in \lim_{f\xi_{|B}} \mathcal{F}$ then $y \in B \in \mathcal{F}$ and, in view of (IV.2.3), there is a filter $\mathcal{G} \in \mathbb{F}X$ with $x \in \lim_\xi \mathcal{G}$, $f(x) = y$ and $f[\mathcal{G}] \leq \mathcal{F}$. Since $B \in \mathcal{F}$, $f^- B \in \mathcal{G}^\#$ and

$$
f[\mathcal{G} \vee f^- B] \leq \mathcal{F}.
$$

As $x \in f^- B \cap \lim_\xi (\mathcal{G} \vee f^- B)$, we conclude that $y \in \lim_{f(\xi_{|f^- B})} \mathcal{F}$, which proves the reverse inequality. $\qquad\square$

More generally:

Definition XV.1.3. A class $\mathcal{M}$ of onto maps is called *hereditary* if whenever $f : |\xi| \to |\tau|$ is a map in $\mathcal{M}$ and B is a subset of Y, the restriction

$$
f_{|f^- B} : \left|\xi_{|f^- B}\right| \to \left|\tau_{|B}\right|
$$

is also in $\mathcal{M}$.

XV.1.2 *Topologically quotient maps*

Exercise XV.1.4. Show that if $f : X \to Y$ and ξ is a convergence on X, then $\mathrm{T}(f\xi)$ is the finest topology on Y making f continuous (from ξ).

Note that if $f : |\xi| \to |\tau|$ is a continuous map and τ is a topology then

$$\tau \leq f\xi \Longrightarrow \mathrm{T}\tau = \tau \leq \mathrm{T}(f\xi).$$

Thus τ *is* the finest topology making f continuous if and only if

$$\tau \geq \mathrm{T}(f\xi).$$

Thus the following definition:

Definition XV.1.5. A continuous surjection $f : |\xi| \to |\tau|$ is called a T-*quotient map, or simply quotient map* if $\tau \geq \mathrm{T}(f\xi)$.

Remark XV.1.6. We could call such maps *topologically quotient* maps to contrast them with quotient maps in the sense of convergences. Because of its traditionally topological use, we prefer to say *convergence quotient map* for the latter, and simply *quotient map* for the former.

Of course, every convergence quotient map is quotient because $f\xi \geq \mathrm{T}(f\xi)$. On the other hand, the map $f : |\xi| \to |\,\mathrm{S}_0(f\xi)|$ in Example IV.9.25, as represented on the right-hand side of Figure V.2 (page 126), is quotient but not convergence quotient.

Proposition XV.1.7. *A continuous surjection* $f : |\xi| \to |\tau|$ *is a quotient map if and only if for every* $U \subset |\tau|$,

$$U \in \mathcal{O}_\tau \iff f^-(U) \in \mathcal{O}_\xi, \tag{XV.1.1}$$

if and only if for every $U \subset |\tau|$,

$$U \in \mathcal{C}_\tau \iff f^-(U) \in \mathcal{C}_\xi. \tag{XV.1.2}$$

Proof. Of course, we only need to prove one equivalence. Since f is continuous, $\tau \leq f\xi$, so that f is quotient if and only if

$$\mathrm{T}\tau = \mathrm{T}f\xi,$$

which in turn is equivalent to (XV.1.2) by Proposition V.3.26. $\square$

Given an onto map $f : X \to Y$, we call a subset A of X *saturated (for f)* if $f^-(f(A)) = A$. In view of Proposition XV.1.7:

Corollary XV.1.8. *A quotient map sends saturated closed (resp. open) sets to closed (resp. open) sets.*

Corollary XV.1.9. *If a continuous surjection is either closed or open then it is quotient.*

Exercise XV.1.10. Prove Corollary XV.1.9.

Unlike convergence quotient maps, quotient maps are neither productive nor hereditary. More specifically, even the product of an identity map with a quotient map need not be a quotient map:

Example XV.1.11 (Two quotient maps whose product is not quotient (e.g., [90])). Let $q : \mathbb{R} \to \mathbb{R}/\mathbb{Z}$ be the canonical surjection, where $\mathbb{R}/\mathbb{Z}$ is endowed with the quotient topology ([1]).

We claim that $q \times i_{\mathbb{Q}} : \mathbb{R} \times \mathbb{Q} \to \mathbb{R}/\mathbb{Z} \times \mathbb{Q}$ is onto and continuous (as a product of continuous maps), but not quotient. Otherwise, it would map closed saturated subsets of $\mathbb{R} \times \mathbb{Q}$ to closed subsets of $\mathbb{R}/\mathbb{Z} \times \mathbb{Q}$, but this is not the case.

Indeed, let $\{a_n\}_{n \in \mathbb{N}}$ be a sequence of irrationals converging to 0 in $\mathbb{R}$ and for each n, let $\{r_m^n\}_{m \in \mathbb{N}}$ be a sequence of rationals converging to a_n. The set

$$A = \left\{ \left(n + \frac{1}{m}, r_m^n\right) : n, m \in \mathbb{N}, m > 1 \right\}$$

is a closed saturated subset of $\mathbb{R} \times \mathbb{Q}$. But $(q \times i_{\mathbb{Q}})(A) = A$ is not closed in $\mathbb{R}/\mathbb{Z} \times \mathbb{Q}$ because $(q(0), 0) \notin A$ but each neighborhood of $q(0)$ in $\mathbb{R}/\mathbb{Z}$ contains $q(n + \frac{1}{m}) = q(\frac{1}{m})$ for m sufficiently large, and each neighborhood of 0 in $\mathbb{Q}$ contains a set of the form $\{r_m^n : m > k(n), n > p_0\}$ where $k : \mathbb{N} \to \mathbb{N}$. Therefore each neighborhood of $(q(0), 0)$ in $\mathbb{R}/\mathbb{Z} \times \mathbb{Q}$ meshes with A and $(q(0), 0) \in \mathrm{cl}_{\mathbb{R}/\mathbb{Z} \times \mathbb{Q}} A$.

Example XV.1.12 (A non-hereditary quotient). (e.g., [90]) Let $X = \{0, 1, 2, 3\}$ endowed with the topology $\{\varnothing, \{0, 2\}, \{1, 3\}, \{0, 1, 2, 3\}\}$ and let $Y = \{0, 1, 2\} = X/\{2, 3\}$ endowed with the quotient topology. Let $f : X \to Y$ be the associated canonical surjection. Notice that none of the proper subsets of Y has an open pre-image, so that the topology of Y is chaotic. Consider the subset $B = \{0, 1\}$ of Y and the restriction of f

$$f_{|f^- B} : f^- B = \{0, 1\} \to B = \{0, 1\}.$$

Here $f^- B$ carries the topology induced by that of X and is therefore discrete, while B carries the topology induced by Y and is therefore chaotic. The map $f_{|f^- B}$ is not quotient, because for instance $\{0\} \subset B$ is not open in B even though $f_{|f^- B}^-\{0\} = \{0\}$ is open in $f^- B$.

---

[1]Here $\mathbb{R}/\mathbb{Z}$ denotes the quotient $\mathbb{R}/\sim$, where $\sim$ denotes the equivalence relation on $\mathbb{R}$ defined by $x \sim y$ if $x = y$ or $\{x, y\} \subset \mathbb{Z}$.

XV.1.3 *Hereditarily quotient maps*

Exercise XV.1.13. Show that if $f : X \to Y$ and ξ is a convergence on X, then $S_0(f\xi)$ is the finest pretopology on Y making f continuous (from ξ).

Note that if $f : |\xi| \to |\tau|$ is a continuous map and τ is a pretopology then

$$\tau \leq f\xi \implies S_0\, \tau = \tau \leq S_0(f\xi).$$

Thus τ *is* the finest pretopology making f continuous if and only if

$$\tau \geq S_0(f\xi).$$

Thus the following definition:

Definition XV.1.14. A continuous surjection $f : |\xi| \to |\tau|$ is called S_0-*quotient* or *hereditarily quotient* if $\tau \geq S_0(f\xi)$.

Of course, every convergence quotient map is S_0-quotient because $f\xi \geq S_0(f\xi)$ and every S_0-quotient map is quotient because $S_0(f\xi) \geq \mathrm{T}(f\xi)$. On the other hand, the map $f : |\xi| \to |S_0(f\xi)|$ in Example IV.9.25, as represented on the right-hand side of Figure V.2 (page 126), is S_0-quotient but not convergence quotient.

Proposition XV.1.15. *A closed continuous surjection between topological spaces is hereditarily quotient, and thus quotient.*

Exercise XV.1.16. Show Proposition XV.1.15.

Example XV.1.17 (A quotient map that is not S_0-quotient). If π is a non-topological pretopology, such as in Example V.4.6, then the identity map $i : |\pi| \to |\mathrm{T}\,\pi|$ is quotient but not S_0-quotient, because $i\pi = \pi$ and $S_0\,\pi = \pi > \mathrm{T}\,\pi$.

The use of the terminology *hereditarily quotient* map for S_0-quotient map is justified by the following facts:

Proposition XV.1.18. *The class of S_0-quotient maps is hereditary.*

Proof. If $f : |\xi| \to |\tau|$ is S_0-quotient, then $\tau \geq S_0(f\xi)$. Let $B \subset |\tau|$. In view of Proposition XV.1.2,

$$f\left(\xi_{|f^-B}\right) = f\xi_{|B}.$$

By Corollary V.2.24,

$$S_0\left(f(\xi_{|f^-B})\right) = S_0((f\xi)_{|B}) = (S_0(f\xi))_{|B}$$

so that

$$\tau_{|B} \geq (\mathrm{S}_0(f\xi))_{|B} = \mathrm{S}_0\left(f(\xi_{|f^- B})\right)$$

and $f : \left(f^- B, \xi_{f-B}\right) \to (B, \tau_{|B})$ is S_0-quotient.   □

Theorem XV.1.19. *Let* $f : |\xi| \to |\tau|$ *be a* T*-quotient map. Then the quotient is hereditary, that is,* $f_{|f-B} : |\xi_{f-B}| \to |\tau_{|B}|$ *is quotient for every* $B \subset |\tau|$, *if and only if* f *is* S_0*-quotient.*

Proof. If f is S_0-quotient, it is hereditarily S_0-quotient by Proposition XV.1.18, hence hereditarily T-quotient.

Conversely, if f is not S_0-quotient, there is $\mathcal{G} \in \mathbb{F}Y$ with

$$y \in \lim_\tau \mathcal{G} \subset \lim_{\mathrm{T}\, f\xi} \mathcal{G},$$

but $y \notin \lim_{\mathrm{S}_0\, f\xi} \mathcal{G}$. Thus there is $A \in \mathcal{G}^\#$ with $y \notin \mathrm{adh}_{f\xi} A$, but $y \in \mathrm{cl}_{f\xi} A$ because $y \in \lim_{\mathrm{T}\, f\xi} \mathcal{G}$. Let $B := A \cup \{y\}$. Then

$$f_{|f-B} : |\xi_{|f-B}| \to |\tau_{|B}|$$

is not T-quotient. Indeed, A is not closed in $(B, \tau_{|B})$ because $y \in \mathrm{cl}_\tau A$. On the other hand, in view of (V.2.7), we conclude that

$$\mathrm{adh}_\xi f^- A \cap f^- y = \varnothing,$$

so that A is closed in $\xi_{|f-B}$ because

$$f^- B = f^- A \cup f^- y.$$

□

However a product of S_0-quotient map does not even need to be quotient:

Example XV.1.20 (A product of S_0-quotient maps that is not T-quotient). In Example V.2.26, $\sigma \vee \tau$ is the discrete topology and τ is a prime pretopology, hence a topology, so that we have in fact $\tau = \mathrm{T}\,\tau$ and

$$\mathrm{S}_0\, \sigma \vee \tau \not\geq \mathrm{T}(\sigma \vee \tau).$$

Let f and g be the following two identity-carried maps $f : |\tau| \to |\tau|$ and $g : |\sigma| \to |\mathrm{S}_0\, \sigma|$. They are both S_0-quotient, but their product

$$g \times f : |\sigma \times \tau| \to |\mathrm{S}_0\, \sigma \times \tau|$$

is not even quotient, because Example V.2.27 shows that

$$\mathrm{S}_0\, \sigma \times \tau \not\geq \mathrm{T}(\sigma \times \tau).$$

So far, we have seen that the class of T-quotient map is neither productive nor hereditary, that the smaller class of S_0-quotient maps is the largest hereditary class of T-quotient maps, but is not productive. We will now look for the largest class of quotient maps that is both productive and hereditary.

XV.1.4 *Quotient maps relative to a reflector*

Both quotient and hereditarily quotient maps are defined by the following common scheme:

Definition XV.1.21. Let J be a reflector. A continuous surjection $f : |\xi| \to |\tau|$ is *quotient relative to J*, or *J-quotient* if

$$\tau \geq J(f\xi).$$

Quotient maps corresponds to $J = \mathrm{T}$ and hereditarily quotient maps corresponds to $J = \mathrm{S}_0$.

If J is a reflector, $f : X \to Y$, and ξ is a convergence on X, then $J(f\xi)$ is the finest convergence of the class $\mathbf{J}$ on Y making f continuous (from ξ). Moreover, that if $f : |\xi| \to |\tau|$ is a continuous map and $\tau = J\tau$ then

$$\tau \leq f\xi \Longrightarrow J\tau = \tau \leq J(f\xi).$$

Thus τ *is* the finest convergence of the class $\mathbf{J}$ making f continuous if and only if $\tau \geq J(f\xi)$.

As we have seen in Section XIV.3, an $\mathbb{F}_0$-composable class $\mathbb{D}$ of filters determines a reflector $\mathrm{A}_\mathbb{D}$ that commutes with initial constructions (Corollary XIV.3.8). As a result:

Proposition XV.1.22. *If $\mathbb{D}$ is a non-empty $\mathbb{F}_0$-composable class of filters then the class of $\mathrm{A}_\mathbb{D}$-quotient maps is hereditary.*

Exercise XV.1.23. Following the model of Proposition XV.1.18 and using Corollary XIV.3.8, prove Proposition XV.1.22.

Exercise XV.1.24. Show that

(1) if J is a reflector that commutes with finite products, that is,

$$J(\xi \times \tau) = J\xi \times J\tau$$

for all ξ and τ, then the class of J-quotient maps is productive, that is, if f and g are two J-quotient maps, so is $f \times g$;

(2) If J is a reflector that commutes with arbitrary products, that is,

$$J\Big(\prod_{\xi \in \Xi} \xi\Big) = \prod_{\xi \in \Xi} J\xi$$

for any set Ξ of convergences, then an arbitrary product of J-quotient maps is J-quotient.

XV.1.5 *Biquotient maps*

Definition XV.1.25. A continuous surjection $f : |\xi| \to |\tau|$ is called S-*quotient* or *biquotient* if $\tau \geq S(f\xi)$.

Since $S \geq S_0 \geq T$, every biquotient map is hereditarily quotient, in particular quotient.

In view of Proposition XV.1.22, the class of biquotient map is hereditary, because $S = A_\mathbb{F}$ and $\mathbb{F}$ is composable. Moreover, the reflector S commutes with products (Theorem VIII.3.9). By Exercise XV.1.24, we conclude:

Proposition XV.1.26. *A product of biquotient maps is a biquotient map.*

Thus, the map g in Example XV.1.20 is hereditarily quotient but not biquotient, for otherwise, its product with the identity map f would be biquotient.

As a consequence of the observations above, the class of biquotient maps is a class of quotient maps that is both hereditary and productive. Moreover, it is the largest class with these properties:

Theorem XV.1.27. *Let $f : |\xi| \to |\tau|$ be a continuous surjection. The following are equivalent:*

(1) f is biquotient;
(2) $f \times g$ is biquotient for every biquotient map g;
(3) For every convergence σ,

$$f \times i_\sigma : |\xi \times \sigma| \to |\tau \times \sigma|$$

is biquotient;
(4) For every prime topology σ,

$$f \times i_\sigma : |\xi \times \sigma| \to |\tau \times \sigma|$$

is hereditarily quotient.

Proof. $(1) \implies (2) \implies (3) \implies (4)$ follow from the productivity of biquotient map, and the fact that biquotient maps are in particular hereditarily quotient.

$(4) \implies (1)$: Assume that f is not biquotient, that is, there is $\mathcal{G} \in \mathbb{F}|\tau|$ and

$$y \in \lim{}_\tau \mathcal{G} \setminus \lim{}_{S(f\xi)} \mathcal{G}.$$

In other words, there is $\mathcal{U} \in \mathbb{U}(\mathcal{G})$ with $y \notin \lim_{f\xi} \mathcal{U}$. Consider on $|\tau|$ the prime topology defined by $\mathcal{N}_\sigma(y) = \mathcal{U} \wedge \{y\}^\uparrow$. Then $f \times i_\sigma$ is not S_0-quotient. Indeed, $(y,y) \in \lim_{\tau \times \sigma} \mathcal{U} \times \mathcal{U}$ while the set

$$\Delta := \{(t,t) \in |\tau| : t \neq y\}$$

meshes with $\mathcal{U} \times \mathcal{U}$ but

$$(y,y) \notin \mathrm{adh}_{(f \times i_\sigma)(\xi \times \sigma)} \Delta = \mathrm{adh}_{f\xi \times \sigma} \Delta.$$

Indeed, if $(t,y) \in \lim_{f\xi \times \sigma} \mathcal{H}$ for $\Delta \in \mathcal{H}$, then $p_{|\sigma|}\mathcal{H} = \mathcal{U}$. Thus $\Delta[p_{|\sigma|}\mathcal{H}] \# p_{|f\xi|}\mathcal{H}$ and $\Delta[p_{|\sigma|}\mathcal{H}] = \Delta\mathcal{U} = \mathcal{U}$ so that $\mathcal{U} \geq p_{|f\xi|}\mathcal{H}$. Therefore, $t \neq y$ because $y \notin \lim_{f\xi} \mathcal{U}$. Thus $(\tau \times \sigma) \not\geq S_0 (f\xi \times \sigma)$. $\qquad\square$

Theorem XV.1.27 suggests a pair of dual classes of questions:

(1) Given a class of convergences **D**, what are the (continuous onto) maps satisfying

$$f \times i_\sigma$$

is (hereditarily) quotient for every σ in **D**;

(2) Given a class of (onto continuous) maps $\mathcal{M}$, can we characterize maps of the class $\mathcal{M}$ with a result of the form: $f \in \mathcal{M}$ if and only if $f \times i_\sigma$ is (hereditarily) quotient for every σ in **D**; for a certain class **D** of convergences.

We will see more examples of results of this kind in Section XV.4.

Before turning to the next section, let us examine two important instances of J-quotient maps.

XV.1.6 *Almost open maps*

Definition XV.1.28. I-quotient maps, or convergence quotient maps, are also called *almost open*.

Since $I \geq S$, every almost open map is biquotient. We already have seen that almost open maps, aka convergence quotient maps, are productive and hereditary.

To shed light on this terminology, recall that a map $f : |\xi| \to |\tau|$ is *open* if $f(U)$ is τ-open whenever U is ξ-open. By definition, f is open if and only if $f : |T\xi| \to |T\tau|$ is open.

Exercise XV.1.29. Let ξ be a topology, and let $f : |\xi| \to |\tau|$ be an onto map. Show that f is open if and only if for every $y \in |\tau|$ and $\mathcal{G} \in \mathbb{F}|\tau|$ with $y \in \lim_\tau \mathcal{G}$:

$$\mathcal{G} \geq f[\mathcal{N}(x)]$$

for all $x \in f^- y$.

Corollary XV.1.30. *An onto map $f : |\xi| \to |\tau|$, where $\xi = \mathrm{T}\,\xi$, is*

(1) open if and only if for each $y \in |\tau|$ and $\mathcal{G} \in \mathbb{F}|\tau|$ with $y \in \lim_\tau \mathcal{G}$ and for all $x \in f^- y$, there is $\mathcal{F}_x \in \mathbb{F}|\xi|$ with $x \in \lim_\xi \mathcal{F}_x$ and $\mathcal{G} \geq f[\mathcal{F}_x]$;

(2) almost open if and only if for each $y \in |\tau|$ and $\mathcal{G} \in \mathbb{F}|\tau|$ with $y \in \lim_\tau \mathcal{G}$ there is $x \in f^- y$ and there is $\mathcal{F} \in \mathbb{F}|\xi|$ with $x \in \lim_\xi \mathcal{F}$ and $\mathcal{G} \geq f[\mathcal{F}]$.

Proof. (1) follows from Exercise XV.1.29 and (2) follows directly from (IV.2.3). $\qquad\qquad\square$

In particular, open onto maps are almost open, hence biquotient.

XV.1.7 *Countably biquotient map*

Definition XV.1.31. S_1-quotient map are also called *countably biquotient*.

We will see in the next section that such maps play an important role for preservation and reconstruction of various classes of spaces characterized by inequalities of the form

$$\xi \geq S_1\, E\xi$$

where E is a coreflector, such as the classes of strongly Fréchet and of strongly k' spaces.

Moreover, we will see in Section XV.4 that they can be characterized in terms similar to Theorem XV.1.27.

XV.2 Interactions between maps and spaces

Table XIV.1 gathers characterizations of various topological properties in terms of functorial inequalities of the form

$$\xi \geq JE\xi,$$

where J is a reflector, and E a coreflector. On the other hand, various kinds of quotient maps are characterized as J-quotient maps $f : |\xi| \to |\tau|$, that is, maps for which

$$\tau \geq J(f\xi).$$

There are natural preservation and reconstruction results linking these classes of maps and of spaces.

Theorem XV.2.1. *Let J be a reflector and E a coreflector. If $\xi \geq JE\xi$ and $f : |\xi| \to |\tau|$ is J-quotient, then*

$$\tau \geq JE\tau.$$

Proof. Because f is J-quotient, we have $\tau \geq J(f\xi)$ and $\xi \geq JE\xi$ so that

$$\tau \geq J(f(JE\xi)) \geq JJE(f\xi)$$

by Proposition XIV.2.3, because JE is a functor. Since J is idempotent and $f\xi \geq \tau$ by continuity of f, we conclude that $\tau \geq JE\tau$. $\quad\square$

Proposition XV.2.2. *Let J be a reflector and E a coreflector. A convergence τ satisfies $\tau \geq JE\tau$ if and only if there is a J-quotient map $f : |\sigma| \to |\tau|$ with $\sigma = E\sigma$.*

Proof. $\tau \geq JE\tau$ if and only if $i_{|\tau|} : |E\tau| \to |\tau|$ is J-quotient. $\quad\square$

Remark XV.2.3. Note that if the topology ξ built from σ in Proposition IV.8.5 satisfies $\xi = E\xi$ whenever $\sigma = E\sigma$, then we can assume σ to be a topology in Proposition XV.2.2. For instance, if $E = B_{\mathbb{D}}$ for a $\mathbb{F}_0$-compsoable class $\mathbb{D}$ of filters, then this condition is satisfied and we can assume σ to be a topology.

Reproducing Table XIV.1 to include classes of quotient, we obtain:

reflector \ coreflector	I_1	Seq	K
almost open I	first-countable $\xi \geq I_1\,\xi$	sequentially based $\xi \geq \mathrm{Seq}\,\xi$	locally compact $\xi \geq K\,\xi$
biquotient S	bisequential $\xi \geq S\,I_1\,\xi$	sequentially based $\xi \geq S\,\mathrm{Seq}\,\xi$	locally compact $\xi \geq S\,K\,\xi$
countably biquotient S_1	strongly Fréchet $\xi \geq S_1\,I_1\,\xi$	strongly Fréchet $\xi \geq S_1\,\mathrm{Seq}\,\xi$	strongly k' $\xi \geq S_1\,K\,\xi$
hereditarily quotient S_0	Fréchet $\xi \geq S_0\,I_1\,\xi$	Fréchet $\xi \geq S_0\,\mathrm{Seq}\,\xi$	k' $\xi \geq S_0\,K\,\xi$
quotient T	sequential $\xi \geq T\,I_1\,\xi$	sequential $\xi \geq T\,\mathrm{Seq}\,\xi$	k $\xi \geq T\,K\,\xi$

Table XV.1 By Theorem XV.2.1, every property in a given row is preserved by the type of maps in that row, or the rows above. By Proposition XV.2.2, a class of spaces in a given row is characterized as the class of images under the class of maps in that row of spaces in the top row of the same column.

XV.3 Compact relations

Recall that we can consider relations $R \subset X \times Y$ as multivalued maps $R : X \rightrightarrows Y$ and $R^- : Y \rightrightarrows X$.

We will now see that many important kinds of maps can be seen as relations that preserve one type of compactness or another.

Definition XV.3.1. A relation $R : |\xi| \rightrightarrows |\tau|$ is $\mathbb{D}$-*compact* if for every subset A of X and every filter $\mathcal{F}$ that is ξ-$\mathbb{D}$-compact at A, the filter $R[\mathcal{F}]$ is $\mathbb{D}$-compact at $R(A)$ in τ.

Proposition XV.3.2. *If $\mathbb{D}$ is $\mathbb{F}_0$-composable, then $R : |\xi| \rightrightarrows |\tau|$ is $\mathbb{D}$-compact if and only if $R[\mathcal{F}]$ is τ-$\mathbb{D}$-compact at $R(x)$ whenever $x \in \lim_\xi \mathcal{F}$.*

Proof. If $x \in \lim_\xi \mathcal{F}$ then $\mathcal{F}$ is compact at $\{x\}$, so $R[\mathcal{F}]$ is $\mathbb{D}$-compact at $R(x)$ whenever R is a $\mathbb{D}$-compact relation.

Conversely, assume that $R[\mathcal{F}]$ is $\mathbb{D}$-compact at $R(x)$ whenever $x \in \lim_\xi \mathcal{F}$, and consider a filter $\mathcal{G}$ on X which is $\mathbb{D}$-compact at A. Let $\mathcal{D}\#R[\mathcal{G}]$ be a $\mathbb{D}$-filter on $|\tau|$. Then $R^-[\mathcal{D}]\#\mathcal{G}$ and $R^-[\mathcal{D}] \in \mathbb{D}|\xi|$ because $\mathbb{D}$ is $\mathbb{F}_0$-composable. Thus there exists $x \in A \cap \mathrm{adh}_\xi R^-[\mathcal{D}]$. Therefore, there exists $\mathcal{U}\#R^-[\mathcal{D}]$ such that $x \in \lim_\xi \mathcal{U}$. By assumption, $R[\mathcal{U}]$ is $\mathbb{D}$-compact at $R(x) \subset R(A)$. Since $\mathcal{D}\#R[\mathcal{U}]$, the filter $\mathcal{D}$ has adherent points in $R(x)$

hence in $R(A)$. □

Corollary XV.3.3. *Let* $\mathbb{D}$ *be an* $\mathbb{F}_0$*-composable class of filters and let* $f : |\xi| \to |\tau|$ *with* $\tau = \mathrm{A}_{\mathbb{D}}\,\tau$. *The following are equivalent:*

(1) f is continuous;
(2) f is a compact relation;
(3) f is a $\mathbb{D}$-compact relation.

Proof. $(1 \implies 2)$. If $x \in \lim_{\xi} \mathcal{F}$, then $f(x) \in \lim_{\tau} f[\mathcal{F}]$ so that $f[\mathcal{F}]$ is compact at $f(x)$ and f is a compact relation by Proposition XV.3.2. $(2 \implies 3)$ is obvious, and $(3 \implies 1)$ follows from Proposition XIV.3.12 and $\tau = \mathrm{A}_{\mathbb{D}}\,\tau$. □

In particular, $\mathbb{F}_0$-compact (equivalently compact) maps between pretopological spaces are exactly the continuous ones.

Theorem XV.3.4. *Let* $\mathbb{D}$ *be an* $\mathbb{F}_0$*-composable class of filters. Let* $f : |\xi| \to |\tau|$ *be a surjection. Then* $f : |\xi| \to |\tau|$ *is* $\mathrm{A}_{\mathbb{D}}$*-quotient if and only if* $f : |f^-\tau| \to |f\xi|$ *is a $\mathbb{D}$-compact relation.*

Proof. Assume f is $\mathrm{A}_{\mathbb{D}}$-quotient and let $x \in \lim_{f^-\tau} \mathcal{F}$. Then

$$f(x) \in \lim_{\tau} f[\mathcal{F}] \subset \lim_{\mathrm{A}_{\mathbb{D}}\,f\xi} f[\mathcal{F}]$$

so that $f(x) \in \mathrm{adh}_{f\xi} \mathcal{D}$ whenever $\mathcal{D} \in \mathbb{D}|\tau|$ and $\mathcal{D}\#f[\mathcal{F}]$. By Proposition XV.3.2, we conclude that $f : |f^-\tau| \to |f\xi|$ is a $\mathbb{D}$-compact relation.

Conversely, assume that $f : |f^-\tau| \to |f\xi|$ is $\mathbb{D}$-compact and let $y \in \lim_{\tau} \mathcal{G}$. For every $x \in f^-(y)$, $x \in \lim_{f^-\tau} f^-[\mathcal{G}]$. By $\mathbb{D}$-compactness of $f : |f^-\tau| \to |f\xi|$, $f[f^-[\mathcal{G}]]$ is $\mathbb{D}$-compact at $f(x) = y$ for $f\xi$. Since f is onto, $f[f^-[\mathcal{G}]] = \mathcal{G}$, and in view Proposition XIV.3.12, we conclude that $y \in \lim_{\mathrm{A}_{\mathbb{D}}\,f\xi} \mathcal{G}$, that is, $\tau \geq \mathrm{A}_{\mathbb{D}}\,f\xi$. □

Note that the following, while not obvious from the definition, is an immediate consequence of Proposition XV.3.2:

Corollary XV.3.5. *If* $\mathbb{D}$ *is* $\mathbb{F}_0$*-composable and* $\mathbb{J} \subset \mathbb{D}$ *then a $\mathbb{D}$-compact relation is also $\mathbb{J}$-compact.*

In particular, when $\mathbb{D}$ contains the class of principal filters, then a $\mathbb{D}$-compact relation R is $\mathbb{F}_0$-compact. Moreover $R(x)$ is $\mathbb{D}$-compact for each x in the domain of R, because $\{x\}^{\uparrow}$ is $\mathbb{D}$-compact at $\{x\}$.

We will now see that when the cover and filter versions of $\mathbb{D}$-compactness coincide in the range of the relation (note that by Theorem XIV.8.6, this means, under mild assumptions, that the range of the relation is topological), the converse is true.

Proposition XV.3.6. *Let $\mathbb{D}$ be an $\mathbb{F}_0$-composable class of filters. If $R : |\xi| \rightrightarrows |\tau|$ is an $\mathbb{F}_0$-compact relation and if $R(x)$ is cover-$(\mathbb{D}/\mathbb{F}_0)$-compact in τ for every $x \in X$, then R is $\mathbb{D}$-compact.*

Proof. Using Proposition XV.3.2, we need to show that $R[\mathcal{F}]$ is $\mathbb{D}$-compact at $R(x)$ whenever $x \in \lim_\xi \mathcal{F}$. Consider a $\mathbb{D}$-filter $\mathcal{D}\#R[\mathcal{F}]$. Then, $\text{adh}_\tau \mathcal{D}\#R(x)$ for every $D \in \mathcal{D}$ because R is $\mathbb{F}_0$-compact. By cover-$(\mathbb{D}/\mathbb{F}_0)$-compactness of $R(x)$, we conclude that $\text{adh}_\tau \mathcal{D}\#R(x)$. $\qquad\square$

In view of Theorem XIV.8.6, we conclude:

Corollary XV.3.7. *Let $\mathbb{D}$ be a non-empty $\mathbb{F}_0$-composable class of filters such that $\text{adh}_\tau^\natural \mathbb{D}(\tau) \subset \mathbb{D}(\tau)$ and let τ be a topology. Then $R : |\xi| \rightrightarrows |\tau|$ is $\mathbb{D}$-compact if and only if it is $\mathbb{F}_0$-compact and $R(x)$ is $\mathbb{D}$-compact in τ for every $x \in X$.*

We now examine what it means for the inverse relation $f^- : |\tau| \rightrightarrows |\xi|$ of a map $f : |\xi| \rightarrow |\tau|$ to be $\mathbb{D}$-compact.

Proposition XV.3.8. *Let $f : |\xi| \rightarrow |\tau|$. If $f^- : |\tau| \rightrightarrows |\xi|$ is $\mathbb{F}_0$-compact, then f is a closed map. If moreover $S_0\,\xi = T\,\xi$, the converse is true.*

Proof. Assume that $f^- : |\tau| \rightrightarrows |\xi|$ is $\mathbb{F}_0$-compact and that $f(H)$ is not τ-closed, for some $H \subset |\xi|$. Then there exists a filter $\mathcal{G}$ on $|\tau|$ with $f(H)\#\mathcal{G}$, $y \in \lim_\tau \mathcal{G}$ and $y \notin f(H)$. By $\mathbb{F}_0$-compactness of f^-, the filter $f^-[\mathcal{G}]$ is $\mathbb{F}_0$-compact at $f^-(y)$ for ξ. Since $H\#f^-[\mathcal{G}]$, we conclude that $\text{adh}_\xi H \cap f^-(y) \neq \varnothing$. Thus H is not ξ-closed, for $y \notin f(H)$ means $f^-(y) \cap H = \varnothing$. Therefore f is a closed map.

Conversely, assume that f is a closed map and let $y \in \lim_\tau \mathcal{G}$. We only need to show that $f^-[\mathcal{G}]$ is $\mathbb{F}_0$-compact at $f^-(y)$. Let $H\#f^-[\mathcal{G}]$. Then $\text{adh}_\xi H$ is closed if $S_0\,\xi = T\,\xi$, and $\text{adh}_\xi H$ meshes with $f^-[\mathcal{G}]$. Thus $f(\text{adh}_\xi H)$ is closed and meshes with $\mathcal{G}$, so that $y \in f(\text{adh}_\xi H)$, that is, $f^-(y) \cap \text{adh}_\xi H \neq \varnothing$. $\qquad\square$

Remark XV.3.9. Maps with a $\mathbb{F}_0$-compact inverse relation have been called *adherent map* [26]. Proposition XV.3.8 states that adherent maps are closed and that both classes of maps coincide among maps whose domain has closed adherences of sets, such as a topological space.

Traditionally, closed maps (between topological spaces) with compact fibers have been called *perfect maps*. We have also introduced a notion of perfect map in Definition XI.6.3.

Exercise XV.3.10. Show that a map $f : |\xi| \to |\tau|$ is perfect in the sense of Definition XI.6.3 if and only if $f^- : \left|\tau_{|f(|\xi|)}\right| \rightrightarrows |\xi|$ is a compact relation.

More generally:

Definition XV.3.11. Let $\mathbb{D}$ be a class of filters. A map $f : |\xi| \to |\tau|$ is called a $\mathbb{D}$-*perfect* if it is a closed map and each fiber $f^-(y)$ is $\mathbb{D}$-compact (in ξ).

class $\mathbb{D}$	$\mathbb{D}$-perfect maps
$\mathbb{F}$	perfect
$\mathbb{F}_{\wedge 1}$	inversely Lindelöf
$\mathbb{F}_1$	countably perfect
$\mathbb{F}_0$	closed

We have seen that if $f^- : |\tau| \rightrightarrows |\xi|$ is $\mathbb{D}$-compact then $f : |\xi| \to |\tau|$ is $\mathbb{D}$-perfect. Combining Proposition XV.3.8 and Corollary XV.3.7, we obtain the converse:

Corollary XV.3.12. *Let ξ be a topology, let $\mathbb{D}$ be a non-empty $\mathbb{F}_0$-composable class of filters such that $\mathrm{adh}_\xi^\natural \mathbb{D}(\xi) \subset \mathbb{D}(\xi)$, and let $f : |\xi| \to |\tau|$. Then $f^- : |\tau| \rightrightarrows |\xi|$ is a $\mathbb{D}$-compact relation if and only if f is $\mathbb{D}$-perfect.*

Exercise XV.3.13. Prove Corollary XV.3.12.

We have seen that continuous maps, $\mathrm{A}_\mathbb{D}$-quotient maps and $\mathbb{D}$-perfect maps are all instances of $\mathbb{D}$-compact relations. Note that if f is a map between two topological spaces, these characterizations remain valid for continuous maps and $\mathbb{D}$-perfect maps without leaving the realm of topological spaces. However, it is necessary to consider general convergence spaces to characterize $\mathrm{A}_\mathbb{D}$-quotient maps in those terms as in Theorem XV.3.4, as $f\xi$ would generally fail to be topological, even if ξ is.

Recall from Corollary XV.1.9 that continuous onto closed maps are quotient maps. Moreover:

Proposition XV.3.14. *A $\mathbb{D}$-perfect onto map is $\mathrm{A}_\mathbb{D}$-quotient.*

Proof. We show that $f : |f^- \tau| \to |f\xi|$ is $\mathbb{D}$-compact if $f^- : |\tau| \rightrightarrows |\xi|$ is. To this end, let $x \in \lim_{f^- \tau} \mathcal{F}$, that is, $f(x) \in \lim_\tau f[\mathcal{F}]$. Since f^- is $\mathbb{D}$-compact, $f^-[f[\mathcal{F}]]$ is $\mathbb{D}$-compact at $f^-(f(x))$ for ξ. By continuity of $f : |\xi| \to |f\xi|$ and Proposition XIV.3.17, we conclude that $f[f^-[f[\mathcal{F}]]] = f[\mathcal{F}]$ is $\mathbb{D}$-compact at $f(f^-(f(x))) = f(x)$ for $f\xi$, which completes the proof. $\square$

class $\mathbb{D}$	reflector J	$\mathbb{D}$-perfect maps $\Longrightarrow$	J-quotient
$\mathbb{F}$	S	perfect	biquotient
$\mathbb{F}_{\wedge 1}$	$A_{\mathbb{F}_{\wedge 1}}$	inversely Lindelöf	weakly biquotient
$\mathbb{F}_1$	S_1	countably perfect	countably biquotient
$\mathbb{F}_0$	S_0	closed	hereditarily quotient
$\mathbb{F}_0$	T	closed	quotient

XV.4 Product of spaces and of maps

In view of the previous results, the concept of $\mathbb{D}$-compactness for families is versatile enough to apply to compactness of sets (principal filters), convergence (Proposition XIV.3.12), and various classes of maps (Theorem XV.3.4 and Corollary XV.3.12). Thus, product theorems for this notion will yield corollaries for product of $\mathbb{D}$-compact spaces, for product of local properties, for product of maps, and will naturally link spaces and maps.

Here is a simple example of such a result, which will illustrate the unifying nature of the notion of $\mathbb{D}$-compact filters. See [81] for more general results.

Theorem XV.4.1. *Let $\mathbb{D}$ be a composable class of filters containing $\mathbb{F}_0$. Let $\mathcal{F} \in \mathbb{F}X$ and $A \subset X$ where (X, ξ) is a convergence space. The following are equivalent:*

(1) $\mathcal{F}$ is $\mathbb{D}$-compact at A;
(2) for every convergence space (Y, τ), every $\mathcal{G} \in \mathbb{D}(Y)$ and every $B \subset Y$

$$\mathcal{G} \text{ compact at } B \Longrightarrow \mathcal{F} \times \mathcal{G} \text{ is } \mathbb{D}\text{-compact at } A \times B;$$

(3) for every prime $\mathbb{D}$-based topological space (Y, τ) with non-isolated point ∞_Y,

$$\mathcal{F} \times \mathcal{N}_\tau(\infty_Y) \text{ is } \mathbb{F}_0\text{-compact at } A \times \{\infty_Y\}.$$

Proof. (1) $\implies$ (2): Let $\mathcal{G} \in \mathbb{D}(Y)$ be compact at $B \subset Y$ and let $\mathcal{D} \in \mathbb{D}(X \times Y)$ with $\mathcal{D}\#(\mathcal{F} \times \mathcal{G})$. We then have $\mathcal{D}^-[\mathcal{G}]\#\mathcal{F}$ by (II.2.10), and $\mathcal{D}^-[\mathcal{G}] \in \mathbb{D}(X)$ because $\mathbb{D}$ is composable. By $\mathbb{D}$-compactness of $\mathcal{F}$ at A, there is $\mathcal{W}\#\mathcal{D}^-[\mathcal{G}]$ with $\lim_\xi \mathcal{W} \cap A \neq \varnothing$. Thus $\mathcal{D}[\mathcal{W}]\#\mathcal{G}$ and, as $\mathcal{G}$ is compact at B, there is $\mathcal{U}\#\mathcal{D}[\mathcal{W}]$ with $\lim_\tau \mathcal{U} \cap B \neq \varnothing$. Then $(\mathcal{W} \times \mathcal{U})\#\mathcal{D}$ and $\lim_{\xi \times \tau}(\mathcal{W} \times \mathcal{U}) \cap (A \times B) \neq \varnothing$, and the conclusion follows.

(2) $\implies$ (3) is obvious. (3) $\implies$ (1): Assume that $\mathcal{F}$ is not $\mathbb{D}$-compact at A. Then there is $\mathcal{D} \in \mathbb{D}(X)$ with $\mathcal{D}\#\mathcal{F}$ but $\operatorname{adh}_\xi \mathcal{D} \cap A = \varnothing$. Consider $Y := X \cup \{\infty_Y\}$, where $\infty_Y \notin X$ and define on Y the prime topology τ defined by $\mathcal{N}_\tau(\infty_Y) := \mathcal{D} \wedge \{\infty_Y\}^\uparrow$. In view of Lemma XIV.3.7, $\mathcal{N}_\tau(\infty_Y) \in \mathbb{D}(Y)$ so that τ is $\mathbb{D}$-based.

Since $\mathcal{D}\#\mathcal{F}$ the subset

$$\Delta := \{(x,x) : x \in X\}$$

of $X \times Y$ meshes with $\mathcal{F} \times \mathcal{N}_\tau(\infty_Y)$, but $\operatorname{adh}_{\xi \times \tau} \Delta \cap (A \times \{\infty_Y\}) = \varnothing$, and thus $\mathcal{F} \times \mathcal{N}_\tau(\infty_Y)$ is not $\mathbb{F}_0$-compact at $A \times \{\infty_Y\}$.

Indeed, if $\Delta \in \mathcal{H} \in \mathbb{F}(X \times Y)$ then $p_X[\mathcal{H}] = p_Y[\mathcal{H}]$. If $\lim_\xi p_X[\mathcal{H}] \cap A \neq \varnothing$ then $p_X[\mathcal{H}]$ does not mesh with $\mathcal{D}$ because $\operatorname{adh}_\xi \mathcal{D} \cap A = \varnothing$. But then $p_Y[\mathcal{H}]$ does not mesh with $\mathcal{N}_\tau(\infty_Y)$ and therefore does not converge to ∞_Y in τ. $\qquad\square$

The case $\mathcal{F} = \{A\}^\uparrow = \{|\xi|\}^\uparrow$ and $\mathcal{G} = \{B\}^\uparrow = \{|\tau|\}^\uparrow$ yields:

Corollary XV.4.2. *Let $\mathbb{D}$ be a composable class of filters containing $\mathbb{F}_0$. If ξ is $\mathbb{D}$-compact and τ is compact, then $\xi \times \tau$ is $\mathbb{D}$-compact.*

In particular, a product of a compact space with a countably compact (respectively Lindelöf) space is countably compact (respectively Lindelöf).

Corollary XV.4.3. *Let $\mathbb{D}$ be a composable class of filters containing $\mathbb{F}_0$ and let ξ be a topology. The following are equivalent:*

(1) ξ is $\mathbb{D}$-compact;
(2) for every $\mathbb{D}$-based convergence τ, $p_{|\tau|} : |\xi \times \tau| \to |\tau|$ is $\mathbb{D}$-perfect;
(3) for every $\mathbb{D}$-based prime topological space, $p_{|\tau|} : |\xi \times \tau| \to |\tau|$ is closed.

Exercise XV.4.4. Deduce Corollary XV.4.3 from Theorem XV.4.1 and Corollary XV.3.12.

In particular,

$\mathbb{D}$	a topology ξ is	iff $p_{|\tau|} : |\xi \times \tau| \to |\tau|$ is closed for every
$\mathbb{F}$	compact	topology τ
$\mathbb{F}_{\wedge 1}$	Lindelöf	P-topology τ
$\mathbb{F}_1$	countably compact	first-countable topology τ

Additionally, if ξ is a compact topology and τ a convergence, the projection $p_{|\tau|}$ is not only closed, but perfect. If ξ is Lindelöf and τ is a P-convergence, then $p_{|\tau|}$ is not only closed, but inversely Lindelöf. If ξ is countably compact and τ is a first-countable convergence, then $p_{|\tau|}$ is not only closed, but countably perfect.

Applying Theorem XV.4.1 when A is a singleton, we obtain via Proposition XIV.3.12:

Corollary XV.4.5. *Let $\mathbb{D}$ be a composable class of filters containing $\mathbb{F}_0$. Let ξ and θ be two convergences on the same set. The following are equivalent:*

(1) $\theta \geq \mathrm{A}_{\mathbb{D}}\,\xi$;
(2) for every convergence τ,

$$\theta \times \mathrm{B}_{\mathbb{D}}\,\tau \geq \mathrm{A}_{\mathbb{D}}(\xi \times \tau);$$

(3) for every $\mathbb{D}$-based prime topology τ,

$$\theta \times \tau \geq \mathrm{S}_0(\xi \times \tau).$$

Exercise XV.4.6. Prove Corollary XV.4.5.

Corollary XV.4.7. *Let $\mathbb{D}$ be a composable class of filters containing $\mathbb{F}_0$ and let $\mathbb{J}$ be a class of filters of closed under finite product such that $\mathbb{D} \subset \mathbb{J}$. The following are equivalent:*

(1) $\xi \geq \mathrm{A}_{\mathbb{D}}\,\mathrm{B}_{\mathbb{J}}\,\xi$;
(2) $\xi \times \tau \geq \mathrm{A}_{\mathbb{D}}\,\mathrm{B}_{\mathbb{J}}(\xi \times \tau)$ *for every* $\tau \geq \mathrm{S}\,\mathrm{B}_{\mathbb{D}}\,\tau$;
(3) $\xi \times \tau \geq \mathrm{S}_0\,\mathrm{B}_{\mathbb{J}}(\xi \times \tau)$ *for every $\mathbb{D}$-based prime topology τ.*

Exercise XV.4.8. Deduce Corollary XV.4.7 from Corollary XV.4.5.

For instance, if $\mathbb{J} = \mathbb{D} = \mathbb{F}_1$, we obtain, in view of Table XIV.1:

Corollary XV.4.9. *A convergence space is strongly Fréchet if and only if its product with every bisequential convergence (equivalently every prime first-countable topology) is strongly Fréchet (equivalently Fréchet).*

If $\mathbb{D} = \mathbb{F}_0$ and $\mathbb{J} = \mathbb{F}_1$, Corollary XV.4.7 becomes:

Corollary XV.4.10. *A convergence space is Fréchet if and only if its product with every finitely generated convergence (equivalently finitely generated prime topology) is Fréchet.*

Theorem XV.4.1 also applies to products of maps via Theorem XV.3.4 and Corollary XV.3.12. For instance:

Exercise XV.4.11. Deduce Theorem XV.1.27 from Theorems XV.4.1 and XV.3.4.

More generally, we can consider products of relations, and consider $\mathbb{D}$-compactness of a product relation. If $R : X \rightrightarrows Y$ and $S : W \rightrightarrows Z$ then

$$R \times S : (X \times W) \rightrightarrows (Y \times Z)$$

is defined by $(R \times S)(x, w) = R(x) \times S(w)$.

Theorem XV.4.12. *Let $\mathbb{D}$ be a composable class of filters that contains $\mathbb{F}_0$ and let $R : |\xi| \rightrightarrows |\tau|$. The following are equivalent:*

(1) R is a $\mathbb{D}$-compact relation;

(2) $R \times S : |\xi \times \sigma| \rightrightarrows |\tau \times \theta|$ is a $\mathbb{D}$-compact relation, for every $\mathbb{D}$-based convergence σ and compact relation $S : |\sigma| \rightrightarrows |\theta|$;

(3) $R \times i_{|\sigma|} : |\xi \times \sigma| \rightrightarrows |\tau \times \sigma|$ is a $\mathbb{F}_0$-compact relation, for every $\mathbb{D}$-based prime topology σ.

Proof. (1) $\implies$ (2) Let $x \in \lim_\xi \mathcal{F}$ and $y \in \lim_\sigma \mathcal{G}$. We can assume $\mathcal{G} \in \mathbb{D}$ because σ is $\mathbb{D}$-based. Since R is $\mathbb{D}$-compact and S is compact, $R[\mathcal{F}]$ is $\mathbb{D}$-compact at $R(x)$ and $S[\mathcal{G}]$ is compact at $S(y)$. Moreover, $S[\mathcal{G}] \in \mathbb{D}$ because $\mathcal{G} \in \mathbb{D}$ and $\mathbb{D}$ is composable, in particular $\mathbb{F}_0$-composable. By Theorem XV.4.1,

$$R[\mathcal{F}] \times S[\mathcal{G}] = (R \times S)[\mathcal{F} \times \mathcal{G}]$$

is $\mathbb{D}$-compact at $R(x) \times S(y) = (R \times S)(x, y)$. Thus $R \times S$ is $\mathbb{D}$-compact.

(2) $\implies$ (3) is obvious and (3) $\implies$ (1) follows from the fact that if $x \in \lim_\xi \mathcal{F}$ and for every prime $\mathbb{D}$-based topology σ with non-isolated point ∞_σ,

$$(R \times i_{|\sigma|})[\mathcal{F} \times \mathcal{N}_\sigma(\infty_\sigma)] = R[\mathcal{F}] \times \mathcal{N}_\sigma(\infty_\sigma)$$

is $\mathbb{F}_0$-compact at $R(x) \times \{\infty_\sigma\}$, we conclude from Theorem XV.4.1 that $R[\mathcal{F}]$ is $\mathbb{D}$-compact at $R(x)$. $\square$

Corollary XV.4.13. *Let $\mathbb{D}$ be a composable class of filters that contains $\mathbb{F}_0$, and let $f : |\xi| \to |\tau|$ be a continuous surjection. The following are equivalent:*

(1) f *is* $A_\mathbb{D}$-*quotient;*

(2) $f \times g : |\xi \times \sigma| \to |\tau \times \theta|$ *is* $A_\mathbb{D}$-*quotient for every* $\mathbb{D}$-*based convergence* θ *and biquotient map* $g : |\sigma| \to |\theta|$;

(3) $f \times i_{|\sigma|} : |\xi \times \sigma| \to |\tau \times \sigma|$ *is hereditarily quotient for every* $\mathbb{D}$-*based prime topology* σ.

Corollary XV.4.14. *Let* $\mathbb{D}$ *be a composable class of filters that contains* $\mathbb{F}_0$, *and let* $f : |\xi| \to |\tau|$ *be a surjection, where* $\xi = \mathrm{T}\,\xi$. *The following are equivalent:*

(1) f *is* $\mathbb{D}$-*perfect;*

(2) $f \times g : |\xi \times \sigma| \to |\tau \times \theta|$ *is* $\mathbb{D}$-*perfect for every* $\mathbb{D}$-*based convergence* σ *and perfect map* $g : |\sigma| \to |\theta|$;

(3) $f \times i_{|\sigma|} : |\xi \times \sigma| \to |\tau \times \sigma|$ *is closed for every* $\mathbb{D}$-*based prime topology* σ.

Exercise XV.4.15. Use Theorem XV.3.4 and Corollary XV.3.12 to deduce Corollaries XV.4.13 and XV.4.14 from Theorem XV.4.12.

Thus, we can summarize the instances of this pair of results for a continuous surjection $f : |\xi| \to |\tau|$ between topological spaces as follows:

$\mathbb{D}$	f is	iff $f \times g$ is	for every map g	with	iff $f \times i_{|\sigma|}$ is	for every prime top. σ that is
$\mathbb{F}$	biquotient	biquotient	biquotient		hereditarily quotient	
	perfect	perfect	perfect		closed	
$\mathbb{F}_1$	countably biquotient	countably biquotient	biquotient	first-countable range	hereditarily quotient	first-countable
	countably perfect	countably perfect	perfect	first-countable domain	closed	
$\mathbb{F}_{\wedge 1}$	weakly biquotient	weakly biquotient	biquotient	P-space range	hereditarily quotient	a P-space
	inversely Lindelöf	inversely Lindelöf	perfect	P-space domain	closed	
$\mathbb{F}_0$	hereditarily quotient	hereditarily quotient	biquotient	finitely generated range	hereditarily quotient	finitely generated
	closed	closed	perfect	finitely generated domain	closed	

Chapter XVI

Spaces of maps

Among all convergences on a set $C(\xi,\sigma)$ of maps that are continuous from ξ to σ we individuate those particularly related to the duality between ξ and $C(\xi,\sigma)$.

The *pointwise convergence* $p(\xi,\sigma)$ on $C(\xi,\sigma)$ amounts to the convergence of the values of maps, separately at each element of $|\xi|$ ([1]). Therefore the pointwise convergence $p(\xi,\sigma)$ depends on a convergence ξ merely because of its domain $C(\xi,\sigma)$.

In contrast, various properties of σ are preserved by $p(\xi,\sigma)$. In particular, if σ is a topological space, then $p(\xi,\sigma)$ is topological, too. One of the consequences of this little dependence of the pointwise convergence on the domain convergence is that the pointwise limit in $|\sigma|^{|\xi|}$ ([2]) of a filter on $C(\xi,\sigma)$ is not, in general, included in $C(\xi,\sigma)$.

The *natural convergence* $[\xi,\sigma]$, also called the *continuous convergence* or the *convergence of joint continuity* (as in Example IV.5.3) does not have this defect: if a filter on $C(\xi,\sigma)$ converges to f with respect to $[\xi,\sigma]$ (defined on $|\sigma|^{|\xi|}$) then $f \in C(\xi,\sigma)$ provided that σ be regular. Our observation can be restated: $C(\xi,\sigma)$ is closed in $|\sigma|^{|\xi|}$ with respect to $[\xi,\sigma]$ but not with respect to $p(\xi,\sigma)$.

---

[1]The *uniform convergence* represents an opposite extreme (provided that a uniform structure is defined on Z). We do not study uniform structures systematically in this work.

[2]A priori, the pointwise convergence is defined in a natural way on the set of all maps, in our case $|\sigma|^{|\xi|}$.

XVI.1 Evaluation and adjoint maps

If X and Z are sets, then Z^X stands for the set of maps from X to Z. We assume from now on that X and Z are non-empty.

Accordingly, $Z^{X \times Y}$ denotes the set of all maps from the product $X \times Y$ to Z. On the other hand, $(Z^X)^Y$ stands for the set of all the maps from Y to the set Z^X of all the maps from X to Z. It turns out that there is a canonical bijection between them, called the *exponential law* and denoted by

$$Z^{X \times Y} \stackrel{\mathrm{exp}}{\cong} (Z^X)^Y. \tag{XVI.1.1}$$

Indeed, if $g : X \times Y \to Z$, then $x \mapsto g(x, y)$ is a map from X to Z for each $y \in Y$. This defines the *exponential map* $\exp : Z^{X \times Y} \to (Z^X)^Y$, defined by

$$(\exp g)(y)(x) := g(x, y). \tag{XVI.1.2}$$

The *natural coupling* (called also the *evaluation*) is a map

$$e = \langle \cdot, \cdot \rangle : X \times Z^X \to Z \tag{XVI.1.3}$$

defined by $e(x, f) = \langle x, f \rangle := f(x)$.

As we shall see, in general, the map $\exp g : Y \to Z^X$ is neither injective nor surjective.

Conversely, if $h : Y \to Z^X$, then the *transposition*

$$\widehat{h}(x, y) := h(y)(x) \tag{XVI.1.4}$$

maps $(Z^X)^Y$ into $Z^{X \times Y}$.

Clearly,

$$\widehat{\exp g} = g \text{ and } \exp \widehat{h} = h,$$

so that exp is a bijection between $Z^{X \times Y}$ and $(Z^X)^Y$.

In particular, if $Y := Z^X$ and $g\,(x, y) := \langle x, y \rangle$ is the evaluation, then $(\exp g)\,(y) = y$ for each $y \in Z^X$, that is, $\exp g$ is in this case the identity map of Z^X.

If $f : X \to Y$, then the *upper adjoint* map $f^* : Z^Y \to Z^X$ of f (with respect to Z) is defined by

$$f^*(h) := h \circ f,$$

that is,

$$f^*(h)(x) = h\,(f(x))$$

for every $h : Y \to Z$ and $x \in X$. On using the evaluation, we get the following equivalent definition (^{3})

$$\langle x, f^*(h) \rangle := \langle f(x), h \rangle. \tag{XVI.1.5}$$

Therefore $f^{**} : Z^{(Z^X)} \to Z^{(Z^Y)}$ is given by $f^{**}(w) = (f^*)^*(w) = w \circ f^*$ for every $w : Z^X \to Z$. Of course, $f^{**}(w) : Z^Y \to Z$, hence for each $h : Y \to Z$,

$$f^{**}(w)(h) = (w \circ f^*)(h) = w(h \circ f) \in Z. \tag{XVI.1.6}$$

Of course,

$$f : X \to Y,$$
$$f^* : Z^Y \to Z^X,$$
$$f^{**} : Z^{(Z^X)} \to Z^{(Z^Y)}.$$

The *point-evaluation map* $j_X : X \to Z^{(Z^X)}$ is defined by

$$j_X\,(x)\,(h) := h\,(x) \tag{XVI.1.7}$$

for each $h \in Z^X$. If X is a singleton, then obviously j_X is injective.

Example XVI.1.1. Let us consider the particular case of the identity $i : X \to X$. Then for every map $h : X \to Z$,

$$i^*\,(h) = h \circ i = h,$$

that is, $i^* : Z^X \to Z^X$ is an identity, too.

Proposition XVI.1.2. *If* card $X > 1$, *then the point-evaluation map with respect to Z is injective if and only if* card $Z > 1$.

---

^{3}On the left-hand, $\langle \cdot, \cdot \rangle : X \times Z^X \to Z$ and on the right-hand $\langle \cdot, \cdot \rangle : Y \times Z^Y \to Z$.

Proof. Let z_0, z_1 be two distinct elements of Z and let $j_X(x_0) = j_X(x_1)$. If $x_0 \neq x_1$, then there exists a map $h : X \to Z$ such that $h(x_0) = z_0$ and $h(x_1) = z_1$, thus $j_X(x_0)(h) = z_0$ and $j_X(x_1)(h) = z_1$, a contradiction. $\square$

Proposition XVI.1.3. *Let* $j_X : X \to Z^{(Z^X)}$ *and* $j_Y : Y \to Z^{(Z^Y)}$ *be the point-evaluation maps and let* $f : X \to Y$. *Then*

$$
\begin{array}{ccc}
X & \xrightarrow{\ j_X\ } & Z^{(Z^X)} \\
{\scriptstyle f}\big\downarrow & & \big\downarrow{\scriptstyle f^{**}} \\
Y & \xrightarrow[\ j_Y\ ]{} & Z^{(Z^Y)}
\end{array}
$$

is commutative, that is,

$$f^{**} \circ j_X = j_Y \circ f. \tag{XVI.1.8}$$

Proof. Indeed, $f^{**}(j_X(x)) = j_X(x) \circ f^*$ is a map from Z^Y to Z, so if $h \in Z^Y$ then $(j_X(x) \circ f^*)(h) = j_X(x)(f^*(h)) = j_X(x)(h \circ f)$. Now, $h \circ f \in Z^X$ and $j_X(x) : Z^X \to Z$ defined by (XVI.1.7). Hence $j_X(x)(h \circ f) = h(f(x)) = j_Y(f(x))(h) = (j_Y \circ f)(x)(h)$. $\square$

If $g : Z \to W$, then the *lower adjoint* map $g_* : Z^X \to W^X$ of g (with respect to X) is defined by

$$g_X(h) = g_*(h) := g \circ h,$$

that is,

$$g_*(h)(x) := g(h(x))$$

for every $h : X \to Z$ and each $x \in X$, or equivalently,

$$\langle x, g_*(h) \rangle = \langle h(x), g \rangle. \tag{XVI.1.9}$$

If $A \subset X$ and $F \subset Z^X$, then

$$\langle A, F \rangle := \{ \langle x, f \rangle : x \in A, f \in F \}.$$

If now $\mathcal{A}$ is a family of subsets of X and $\mathcal{F}$ is a family of subsets of Z^X, then

$$\langle \mathcal{A}, \mathcal{F} \rangle := \{ \langle A, F \rangle : A \in \mathcal{A}, F \in \mathcal{F} \}.$$

This notation is used along with another, in which

$$F(A) := \langle A, F \rangle \text{ and } \mathcal{F}(\mathcal{A}) := \langle \mathcal{A}, \mathcal{F} \rangle.$$

XVI.2   Adjoint maps on spaces of continuous maps

Let us enrich our considerations by endowing sets with convergences. Let ξ be a convergence on X and σ a convergence on Z. Then $C(\xi, \sigma)$ denotes the subset of Z^X consisting of the maps that are continuous from ξ to σ.

Recall from Proposition IV.2.9 that $f \in C(\xi, \sigma)$ if and only if $f\xi \geq \sigma$ if and only if $\xi \geq f^- \sigma$. Therefore

$$\zeta \geq \xi \text{ and } \sigma \geq \tau \Longrightarrow C(\xi, \sigma) \subset C(\zeta, \tau).$$

In fact, if $f \in C(\xi, \sigma)$ then $f\zeta \geq f\xi \geq \sigma \geq \tau$ implies that $f \in C(\zeta, \tau)$.

In some instances, each map is continuous. For example, for every convergences ξ on X and σ on Z,

$$C(\xi, o) = Z^X,$$
$$C(\iota, \sigma) = Z^X,$$

where ι is the discrete topology (on X) and o is the chaotic topology (on Z).

The evaluation (XVI.1.3) will be often restricted to $|\xi| \times C(\xi, \sigma)$

$$\langle \cdot, \cdot \rangle : |\xi| \times C(\xi, \sigma) \to |\sigma|. \tag{XVI.2.1}$$

Similarly, the point-evaluation map $j_{|\xi|} : |\xi| \to |\sigma|^{\left(|\sigma|^{|\xi|}\right)}$ can be restricted to

$$j_{\xi} : |\xi| \to |\sigma|^{C(\xi, \sigma)} \tag{XVI.2.2}$$

by $j_{\xi}(x)(h) = h(x)$ for all $h \in C(\xi, \sigma)$.

Recall that $F \subset Y^X$ *separates points of* X if for every x and t in X

$$x \neq t \Longrightarrow \underset{f \in F}{\exists} \ f(x) \neq f(t).$$

Proposition XVI.2.1. *The restriction* j_{ξ} *of the point-evaluation map to* $C(\xi, \sigma)$ *is injective if and only if* $C(\xi, \sigma)$ *separates points of* $|\xi|$.

Exercise XVI.2.2. Prove Proposition XVI.2.1.

In the same vein, if $f \in C(\xi, \tau)$ and $h \in C(\tau, \sigma)$, then $h \circ f \in C(\xi, \sigma)$. Consequently, the upper adjoint

$$f^* : |\sigma|^{|\tau|} \to |\sigma|^{|\xi|}$$

maps $C(\tau, \sigma)$ into $C(\xi, \sigma)$, and thus we can consider the restriction

$$f^{\sigma} : C(\tau, \sigma) \to C(\xi, \sigma).$$

Example XVI.2.3. If $\xi \geq \tau$ then the identity map $i \in C(\xi, \tau)$ and thus $i^\sigma : C(\tau, \sigma) \to C(\xi, \sigma)$. By Example XVI.1.1, $i^* : |\sigma|^{|\tau|} \to |\sigma|^{|\xi|}$ is the identity, hence its restriction i^σ is an *injection* and $C(\tau, \sigma) \subset C(\xi, \sigma)$.

Similarly, if $g \in C(\sigma, \theta)$ and $h \in C(\xi, \sigma)$ then $g \circ h \in C(\xi, \theta)$, hence

$$g_* : |\sigma|^{|\xi|} \to |\theta|^{|\xi|}$$

maps $C(\xi, \sigma)$ into $C(\xi, \theta)$, and we can consider the restriction

$$g_\xi : C(\xi, \sigma) \to C(\xi, \theta).$$

At this stage, the spaces of continuous maps are not endowed with any convergence. We shall later consider convergences on the spaces of continuous maps that make the adjoint maps continuous.

XVI.3   Fundamental convergences on spaces of continuous maps

Definition XVI.3.1. The *natural convergence* $[\xi, \sigma]$ is the coarsest convergence on $C(\xi, \sigma)$, for which the evaluation (XVI.2.1) is jointly continuous.

As we will see, the natural convergence on sets of continuous maps satisfies the *exponential law*

$$[\xi \times \tau, \sigma] \stackrel{\mathrm{exp}}{\equiv} [\tau, [\xi, \sigma]] \tag{XVI.3.1}$$

analogous to (XVI.1.1) on sets of continuous functions. This is to be compared with:

Definition XVI.3.2. The *pointwise convergence* $p(\xi, \sigma)$ is the coarsest convergence on $C(\xi, \sigma)$, for which for every $x \in |\xi|$, the point evaluation $j_\xi(x) : |p(\xi, \sigma)| \to |\sigma|$ is continuous, that is, for every $x \in |\xi|$, the restriction of (XVI.2.1) to $\{x\} \times C(\xi, \sigma)$ is continuous.

Among several intermediate convergences between the pointwise and the natural ones is the so-called compact-open convergence.

Definition XVI.3.3. The *compact-open convergence* $k(\xi, \sigma)$ is the coarsest convergence on $C(\xi, \sigma)$, for which for every compact subset K of $|\xi|$, the restriction of the evaluation (XVI.2.1) to $K \times C(\xi, \sigma)$ is continuous.

By definition

$$p(\xi, \sigma) \leq k(\xi, \sigma) \leq [\xi, \sigma]. \tag{XVI.3.2}$$

Moreover:

Proposition XVI.3.4. *Let ξ and σ be two convergences. If ξ is discrete, then $[\xi, \sigma] = p(\xi, \sigma)$. If ξ is hereditarily locally compact, then $[\xi, \sigma] = k(\xi, \sigma)$.*

Exercise XVI.3.5. Prove Proposition XVI.3.4.

In the case where the convergence σ is induced by a metric d, that is, in the case where $\sigma = \tilde{d}$, *uniform convergence* $u(\xi, d)$ (as in Definition VII.4.2) is also available.

Exercise XVI.3.6. Show that if (Y, d) is a metric space and (X, ξ) is a convergence space then

$$[\xi, \tilde{d}] \leq u(\xi, d)$$

and that the equality holds whenever ξ is compact.

The mentioned advantages of the natural convergence over the pointwise convergence more than compensate what may be considered a defect, that is, the topologicity of ξ and σ does not imply that of $[\xi, \sigma]$. The compact-open convergence can be seen as a compromise: it is topological whenever ξ and σ are, and, while it does not satisfies the exponential law in general, it does in some important cases.

XVI.4 Pointwise convergence

We have examined an instance of pointwise convergence in Example III.4.6 and observed in Exercise IV.5.1 that the pointwise convergence on $\mathbb{R}^{\mathbb{R}}$ coincides with the product convergence on $\prod_{x \in \mathbb{R}} \mathbb{R}$.

More generally, if ξ and σ are convergence spaces, then a filter $\mathcal{F}$ on $|\sigma|^{|\xi|}$ converges to $f_0 \in |\sigma|^{|\xi|}$ in the *pointwise convergence* $p(|\xi|, \sigma)$, in symbols,

$$f_0 \in \lim_{p(|\xi|, \sigma)} \mathcal{F}$$

whenever $f_0(x) \in \lim_\sigma \mathcal{F}(x)$ for every $x \in X$, where

$$\mathcal{F}(x) := \{\{f(x) : f \in F\} : F \in \mathcal{F}\}$$
$$= j_X(x)[\mathcal{F}].$$

Exercise XVI.4.1. Verify that $p(|\xi|, \sigma)$ and $\prod_{x \in |\xi|} \sigma_x$ are homeomorphic.

Note that $p(|\xi|, \sigma)$ does not depend on the convergence ξ but merely on its underlying set $|\xi|$. We denote by

$$p(\xi, \sigma) := p(|\xi|, \sigma)_{|C(\xi,\sigma)}$$

its restriction to the set $C(\xi, \sigma)$ of continuous functions. We notice that if σ is a topological space, then $p(|\xi|, \sigma)$, hence its subspace $p(\xi, \sigma)$, are topological as well. More generally,

Proposition XVI.4.2. *If* **F** *is a (concretely) reflective class of convergences and* $\sigma \in$ **F** *then* $p(|\xi|, \sigma)$ *and* $p(\xi, \sigma)$ *are in* **F**.

Proof. By definition, $p(|\xi|, \sigma)$ is the coarsest convergence on $|\sigma|^{|\xi|}$, for which, for every $x \in |\xi|$, the point-evaluation $j_{|\xi|}(x) : |\sigma|^{|\xi|} \to |\sigma|$ is continuous. In other words, $p(|\xi|, \sigma)$ is the initial convergence with respect to a set of maps valued in $|\sigma|$:

$$p(|\xi|, \sigma) = \bigvee_{x \in |\xi|} j_{|\xi|}(x)^{-}\sigma.$$

The conclusion follows from the fact that a reflective class is closed under initial constructions, in particular subspace, and suprema (see Section XIV.2). □

Corollary XVI.4.3. *If* σ *is topological (respectively regular, functionally regular), so are* $p(|\xi|, \sigma)$ *and* $p(\xi, \sigma)$.

Note that $p(\xi, \sigma)$ is *not* closed in $p(|\xi|, \sigma)$, as shows the following example.

Example XVI.4.4 (A pointwise limit of continuous functions may fail to be continuous). Let ν denote the standard convergence on $\mathbb{R}$, and consider, for each $n \in \mathbb{N}$, the function $f_n : \mathbb{R} \to \mathbb{R}$ defined by

$$f_n(x) = \begin{cases} 0 & \text{if } x < -\frac{1}{n} \text{ or } x > \frac{1}{n} \\ nx + 1 & \text{if } -\frac{1}{n} \leq x \leq 0 \\ -nx + 1 & \text{if } 0 < x \leq \frac{1}{n} \end{cases}.$$

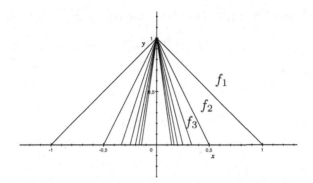

It is easily verified that the functions f_n are all continuous on $\mathbb{R}$, and that

$$\lim_{p(|\nu|,\nu)}(f_n)_{n\in\mathbb{N}} = \begin{cases} 0 & \text{if } x \neq 0 \\ 1 & \text{if } x = 0 \end{cases} \qquad (\text{XVI.4.1})$$

is a discontinuous function.

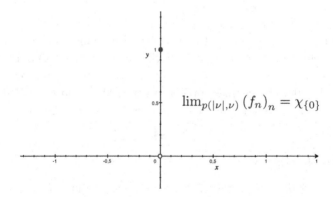

Exercise XVI.4.5. Show that the functions f_n of Example XVI.4.4 are all continuous and that (XVI.4.1) holds.

Consider the point evaluation map (XVI.2.2), where $C(\xi,\sigma)$ is endowed with the pointwise convergence $p(\xi,\sigma)$. For each $x \in |\xi|$, the map $j_\xi(x) : |p(\xi,\sigma)| \to |\sigma|$ is continuous, for $f \in \lim_{p(\xi,\sigma)} \mathcal{F}$ means that $f(x) = j_\xi(x)(f) \in \lim_\sigma j_\xi(x)[\mathcal{F}]$ for each x. Thus j_ξ maps $|\xi|$ to $C(p(\xi,\sigma),\sigma)$. Moreover:

Proposition XVI.4.6. *For every convergences ξ and σ,*

$$j_\xi : |\xi| \to |p\,(p(\xi,\sigma),\sigma)\,|$$

is continuous, and

$$C(\xi,\sigma) = C(j_\xi^-\,(p(p(\xi,\sigma),\sigma)),\sigma).$$

Proof. Let $x \in \lim_\xi \mathcal{F}$. To show $j_\xi(x) \in \lim_{p(p(\xi,\sigma),\sigma)} j_\xi[\mathcal{F}]$, it is enough to show that

$$j_\xi(x)(f) \in \lim_\sigma j_\xi[\mathcal{F}](f) \iff f(x) \in \lim_\sigma f[\mathcal{F}] \qquad (\text{XVI.4.2})$$

for every $f \in C(\xi,\sigma)$, which is true by continuity of f. Thus $j_\xi : |\xi| \to |p\,(p(\xi,\sigma),\sigma)\,|$ is continuous, so that $\xi \geq j_\xi^-\,(p(p(\xi,\sigma),\sigma))$ and thus

$$C(j_\xi^-\,(p(p(\xi,\sigma),\sigma)),\sigma) \subset C(\xi,\sigma).$$

To see the reverse inclusion, note that by (XVI.4.2) $x \in \lim_{j_\xi^-\,(p(p(\xi,\sigma),\sigma))} \mathcal{F}$ is equivalent to $f(x) \in \lim_\sigma f[\mathcal{F}]$ for every $f \in C(\xi,\sigma)$. $\square$

XVI.5   Natural convergence

Consider the preconvergence of joint continuity defined on $\mathbb{R}^{\mathbb{R}}$ in Example IV.5.3, which defines a convergence $[\mathbb{R},\mathbb{R}]$ on the set $C(\mathbb{R},\mathbb{R})$ of real-valued continuous functions on $\mathbb{R}$. More generally, given two convergence spaces ξ and σ, consider the coarsest preconvergence σ^ξ on $|\sigma|^{|\xi|}$, called *natural preconvergence* or *preconvergence of joint continuity*, making the evaluation

$$\langle \cdot,\cdot \rangle : |\xi \times \sigma^\xi| \to |\sigma|$$

continuous. In other words,

$$f \in \lim_{\sigma^\xi} \mathcal{F} \iff \underset{x \in |\xi|}{\forall}\ \underset{\mathcal{G} \in \lim_\xi^-(x)}{\forall}\ f(x) \in \lim_\sigma \langle \mathcal{G},\mathcal{F} \rangle\,.$$

By definition:

Proposition XVI.5.1. *The largest subset of $|\sigma|^{|\xi|}$ on which σ^ξ defines a convergence is the set $C(\xi,\sigma)$ of continuous functions:*

$$f \in \lim_{\sigma^\xi} \{f\}^\uparrow \iff f \in C(\xi,\sigma).$$

The convergence induced on $C(\xi,\sigma)$ by σ^ξ is the *natural convergence* or *convergence of joint continuity* $[\xi,\sigma]$ introduced in Section VIII.6. It is the coarsest among the convergences θ on $C(\xi,\sigma)$, for which the evaluation

$e : |\xi \times \theta| \to |\sigma|$ is (jointly) continuous, that is, the coarsest among the convergences θ on $C(\xi, \sigma)$, for which

$$\xi \times \theta \geq e^-\sigma, \qquad (\text{XVI.5.1})$$

where $e^-\sigma$ is the initial convergence of σ by e in (XVI.1.3). Given $f \in C(\xi, \sigma)$ and a filter $\mathcal{F}$ on $C(\xi, \sigma)$, the convergence is characterized by (VIII.6.2), that is,

$$f \in \lim_{[\xi,\sigma]} \mathcal{F} \iff \underset{x \in |\xi|}{\forall} \ \underset{\mathcal{G} \in \lim_\xi^-(x)}{\forall} \ f(x) \in \lim_\sigma \langle \mathcal{G}, \mathcal{F} \rangle .$$

XVI.5.1 *Continuity of limits*

In contrast to pointwise convergence, $[\xi, \sigma]$ is a closed subspace of σ^ξ, as long as σ is regular. In fact:

Theorem XVI.5.2. *A convergence σ is regular if and only if for every convergence ξ (equivalently, every topology ξ), every filter $\mathcal{F}$ on $|\sigma|^{|\xi|}$ and every $f \in |\sigma|^{|\xi|}$,*

$$f \in \lim_{\sigma^\xi} \mathcal{F} \Longrightarrow f \in C(\xi, \sigma).$$

Proof. Assume that σ is regular and that $f \in \lim_{\sigma^\xi} \mathcal{F}$. We want to show that $f(x_0) \in \lim_\sigma f[\mathcal{G}]$ whenever $x_0 \in \lim_\xi \mathcal{G}$. Since $f \in \lim_{\sigma^\xi} \mathcal{F}$,

$$f(x_0) \in \lim_\sigma \langle \mathcal{G}, \mathcal{F} \rangle \subset \lim_\sigma \text{adh}_\sigma^\natural \langle \mathcal{G}, \mathcal{F} \rangle \qquad (\text{XVI.5.2})$$

by regularity.

Moreover,

$$f[\mathcal{G}] \geq \text{adh}_\sigma^\natural \langle \mathcal{G}, \mathcal{F} \rangle . \qquad (\text{XVI.5.3})$$

Indeed, for every $G \in \mathcal{G}$ and $x \in G$, $f(x) \in \lim_\sigma \langle x, \mathcal{F} \rangle$ because $x \in \lim_\xi \{x\}^\uparrow$, and $\langle G, F \rangle \in \langle x, \mathcal{F} \rangle$ for every $F \in \mathcal{F}$. Thus $f(G) \subset \text{adh}_\sigma \langle G, F \rangle$. In view of (XVI.5.2) and (XVI.5.3), $f(x_0) \in \lim_\sigma f[\mathcal{G}]$.

Assume conversely that σ is not regular and let $Y := |\sigma|$. In view of Theorem VI.3.18, there are $A \neq \varnothing$, maps $l : A \to Y$ and $\mathcal{S} : A \to \mathbb{F}Y$ with $l(a) \in \lim_\sigma \mathcal{S}(a)$, for every $a \in A$, and $y_0 \in \lim_\sigma \mathcal{S}(\mathcal{F}) \setminus \lim_\sigma l[\mathcal{F}]$.

We define the pretopology ξ on the set

$$X := (Y \times A) \cup A \cup \{x_\infty\}$$

where $x_\infty \notin Y \cup A$ by $\mathcal{V}_\xi((y, a)) := \{(y, a)\}^\uparrow$ for every $(y, a) \in Y \times A$, $\mathcal{V}_\xi(a) := (\mathcal{S}(a) \times \{a\}^\uparrow) \wedge \{a\}^\uparrow$ for every $a \in A$, and

$$\mathcal{V}_\xi(x_\infty) := \{x_\infty\}^\uparrow \wedge \bigcup_{F \in \mathcal{F}} \bigcap_{a \in F} (\mathcal{S}(a) \times \{a\}^\uparrow) \wedge \{a\}^\uparrow.$$

By construction, this pretopology is diagonal, hence a topology.

Consider the map $f : |\xi| \to |\sigma|$ defined by $f(a) = l(a)$ for all $a \in A$, $f((y,a)) = y$ for all $(y,a) \in Y \times A$, and $f(x_\infty) = y_0$. The map f is not continuous, for $x_\infty \in \lim_\xi \mathcal{F}$ but $y_0 \notin \lim_\sigma f[\mathcal{F}]$ because $f[\mathcal{F}] = l[\mathcal{F}]$.

However, we can construct a filter $\mathcal{H}$ on Y^X with $f \in \lim_{\sigma\xi} \mathcal{H}$. To this end, consider

$$P := \{h \in Y^X : h_{|Y \times A} = p_Y \text{ and } h(x_\infty) = y_0\},$$

$$\mathcal{A} := \bigcup_{a \in A} \left\{P \cap j_X(a)^{-1}(S) : S \in \mathcal{S}(a)\right\},$$

$$\mathcal{B} := \bigcup_{F \in \mathcal{F}} \left\{\bigcap_{a \in F} \left(P \cap j_X(a)^{-1}\left(S_a^H\right)\right) : \left(S_a^H\right)_{a \in H} \in \prod_{a \in H} \mathcal{S}(a)\right\},$$

where j_X is as in (XVI.1.7).

Then $\mathcal{A} \cup \mathcal{B}$ has the finite intersection property (See Exercise XVI.5.3 below) and thus, by Proposition II.2.14, generates a filter $\mathcal{H}$ on Y^X.

Remains to show that $f \in \lim_{\sigma\xi} \mathcal{H}$. This is the case because

$$\langle \mathcal{V}_\xi(x_\infty), \mathcal{H}\rangle \geq \mathcal{S}(\mathcal{F}) \wedge \{y_0\}^\uparrow, \tag{XVI.5.4}$$

$$\langle \mathcal{V}_\xi(a), \mathcal{H}\rangle \geq \mathcal{S}(a) \text{ for all } a \in A, \tag{XVI.5.5}$$

$$\langle (y,a), \mathcal{H}\rangle = \{y\}^\uparrow \text{ for all } (y,a) \in Y \times A, \tag{XVI.5.6}$$

so that

$$f(x_\infty) = y_0 \in \lim_\sigma \langle \mathcal{V}_\xi(x_\infty), \mathcal{H}\rangle,$$

$$f(a) = l(a) \in \lim_\sigma \langle \mathcal{V}_\xi(a), \mathcal{H}\rangle \text{ for all } a \in A,$$

$$f((y,a)) = y \in \lim_\sigma \langle (y,a), \mathcal{H}\rangle \text{ for all } (y,a) \in Y \times A.$$

Indeed, (XVI.5.4) follows from the fact that for every $B \in \mathcal{S}(\mathcal{F})$, there is $F \in \mathcal{F}$, and for each $a \in F$, there is $S_a^F \in \mathcal{S}(a)$ such that $\bigcup_{a \in F} S_a^F \subset B$, and

$$\left\langle \bigcup_{a \in F}\left(\left(S_a^F \times \{a\}\right) \cup \{a\}\right), \bigcap_{a \in F} j_X(a)^{-1}\left(S_a^F\right) \cap P\right\rangle \subset \bigcup_{a \in F} S_a^F.$$

(XVI.5.5) follows from the observation that, by definition of P,

$$\left\langle (S \times \{a\}) \cup \{a\}, j_X(a)^{-1}(S) \cap P\right\rangle \subset S$$

for every $S \in \mathcal{S}(a)$. Finally, (XVI.5.6) follows from $P \in \mathcal{F}$. $\square$

Exercise XVI.5.3. Show that the family $\mathcal{A} \cup \mathcal{B}$ of subsets of Y^X defined in the proof of Theorem has the finite intersection property.

XVI.5.2 *Exponential law*

We have seen that $[\xi, \sigma]$ has finite depth whenever σ does (Remark IV.5.5), and that $[\xi, \sigma]$ is pseudotopological whenever σ is (Theorem VIII.6.2), while $[\xi, \sigma]$ may fail to be a topology even when both ξ and σ are topological (Theorem IX.4.1). Thus there is no direct analog of Proposition XVI.4.2 for $[\xi, \sigma]$. We will see in Sections XVII.1 and XVII.2 how the results we have just mentioned are part of a larger theory linking the reflective properties of the natural convergence with the behavior of functors with respect to products.

If $f \in C(\xi \times \tau, \sigma)$ then for every $y \in |\tau|$, the map $x \mapsto f(x, y)$ is continuous from ξ to σ, that is, $(\exp f)(y) \in C(\xi, \sigma)$. Moreover,

$$e \circ (i_\xi \times \exp f) = f, \qquad (\text{XVI.5.7})$$

because

$$e \circ (i_\xi \times \exp f)(x, y) = \langle x, (\exp f)(y) \rangle = (\exp f)(y)(x) = f(x, y):$$

Additionally:

Proposition XVI.5.4. *If $f \in C(\xi \times \tau, \sigma)$ then*

$$\exp f : |\tau| \to |[\xi, \sigma]|$$

is continuous.

Proof. (XVI.5.7) ensures that

$$\xi \times (\exp f)\tau \geq e^- \sigma$$

by continuity of f, and $[\xi, \sigma]$ is the coarsest convergence satisfying (XVI.5.1). Therefore, $(\exp f)\tau \geq [\xi, \sigma]$, that is, $\exp f : |\tau| \to |[\xi, \sigma]|$ is continuous. $\qquad\square$

Now, if $g \in C(\tau, [\xi, \sigma])$ then $i_\xi \times g \in C(\xi \times \tau, \xi \times [\xi, \sigma])$, and

$$e \circ (i_\xi \times g) = \widehat{g}$$

because

$$e \circ (i_\xi \times g)(x, y) = \langle x, g(y) \rangle = \widehat{g}(x, y)$$

by definition of the transposition $\widehat{g}$ of g. Thus

$$\widehat{g} \in C(\xi \times \tau, \sigma) \tag{XVI.5.8}$$

for it is a composite of continuous functions, as shown on the diagram:

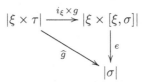

Moreover:

Theorem XVI.5.5. *For every convergences ξ, τ, and σ, the exponential map is an homeomorphism:*

$$[\xi \times \tau, \sigma] \overset{\exp}{\cong} [\tau, [\xi, \sigma]].$$

Proof. Proposition XVI.5.4 and (XVI.5.8) show that the bijection $\exp :$ $|\xi \times \tau|^{|\sigma|} \to \left(|\sigma|^{|\xi|}\right)^{|\tau|}$ of (XVI.1.1) restricts to a bijection from $C(\xi \times \tau, \sigma)$ to $C(\tau, [\xi, \sigma])$.

Moreover, $\exp$ is continuous. To see this, consider $f \in \lim_{[\xi \times \tau, \sigma]} \mathcal{H}$. To show that $\exp f \in \lim_{[\tau, [\xi, \sigma]]} \exp[\mathcal{H}]$, we need to show that for every $y \in |\tau|$ and $\mathcal{G}$ with $y \in \lim_\tau \mathcal{G}$,

$$(\exp f)(y) \in \lim_{[\xi, \sigma]} \langle \mathcal{G}, \exp[\mathcal{H}] \rangle.$$

To this end, let $x \in \lim_\xi \mathcal{F}$ and note that

$$\langle \mathcal{F}, \langle \mathcal{G}, \exp[\mathcal{H}] \rangle \rangle = \langle \mathcal{F} \times \mathcal{G}, \mathcal{H} \rangle,$$

for

$$\langle t, \langle s, \exp h \rangle \rangle = (\exp h)(s)(t) = h(t, s) = \langle (t, s), h \rangle$$

for every $h \in C(\xi \times \tau, \sigma)$, $t \in |\xi|$ and $s \in |\tau|$. Thus, since $f \in \lim_{[\xi \times \tau, \sigma]} \mathcal{H}$,

$$f(x, y) \in \lim_\sigma \langle \mathcal{F} \times \mathcal{G}, \mathcal{H} \rangle,$$

we conclude that $(\exp f)(y)(x) \in \lim_\sigma \langle \mathcal{F}, \langle \mathcal{G}, \exp[\mathcal{H}] \rangle \rangle$, and $\exp$ is continuous.

Moreover, the transposition $\widehat{\cdot} : |[\tau, [\xi, \sigma]]| \to |[\xi \times \tau, \sigma]|$ is also continuous: let $h \in \lim_{[\tau, [\xi, \sigma]]} \mathcal{H}$, and let $x \in \lim_\xi \mathcal{F}$ and $y \in \lim_\tau \mathcal{G}$. Note that

$$\left\langle \mathcal{F} \times \mathcal{G}, \widehat{\mathcal{H}} \right\rangle = \langle \mathcal{F}, \langle \mathcal{G}, \mathcal{H} \rangle \rangle$$

and $h(y) \in \lim_{[\xi, \sigma]} \langle \mathcal{G}, \mathcal{H} \rangle$ because $h \in \lim_{[\tau, [\xi, \sigma]]} \mathcal{H}$, so that

$$h(y)(x) \in \lim_\sigma \langle \mathcal{F}, \langle \mathcal{G}, \mathcal{H} \rangle \rangle, \text{ that is, } \widehat{h}(x, y) \in \lim_\sigma \left\langle \mathcal{F} \times \mathcal{G}, \widehat{\mathcal{H}} \right\rangle,$$

which shows that $\widehat{h} \in \lim_{[\xi \times \tau, \sigma]} \widehat{\mathcal{H}}$, that is, the transposition is continuous. $\square$

XVI.5.3 *Finer subspaces and natural convergence*

By definition of the natural preconvergence,

$$\xi \le \theta \text{ and } \sigma \ge \tau \Longrightarrow \sigma^\xi \ge \tau^\theta.$$

However, while $|\sigma|^{|\xi|} = |\tau|^{|\theta|}$, the inclusion $C(\xi, \sigma) \subset C(\theta, \tau)$ may be strict, so that $[\xi, \sigma]$ and $[\theta, \tau]$ may not have the same underlying set. Yet, the inclusion map $i_{C(\theta,\tau)}^{C(\xi,\sigma)} : |[\xi, \sigma]| \to |[\theta, \tau]|$ is continuous. This leads to the following definition:

Definition XVI.5.6. If τ and υ are convergences, then we write

$$\tau \trianglerighteq \upsilon$$

and say that τ *is a finer subspace of* υ whenever $|\tau| \subset |\upsilon|$ and $i_{|\upsilon|}^{|\tau|} \in C(\tau, \upsilon)$, where $i_{|\upsilon|}^{|\tau|} : |\tau| \to |\upsilon|$ stands for the inclusion map.

In particular, if $\tau \ge \upsilon$, then $\tau \trianglerighteq \upsilon$, in which case the inclusion map is the identity. The relation $\trianglerighteq$ on the class of convergences is that of order (see Section A.6). We denote by $\bigvee^\trianglerighteq$ and $\bigwedge^\trianglerighteq$ the supremum and the infimum with respect to $\trianglerighteq$.

Proposition XVI.5.7. *Let Ξ be a set of convergences. Then*

(1) $\bigvee^\trianglerighteq \Xi$ *exists if and only if* $\bigcap_{\xi \in \Xi} |\xi| \ne \varnothing$. *In this case,* $|\bigvee^\trianglerighteq \Xi| = \bigcap_{\xi \in \Xi} |\xi|$ *and* $\bigvee^\trianglerighteq \Xi$ *is the initial convergence for the inclusion maps of* $\bigcap_{\xi \in \Xi} |\xi|$ *in each* $|\xi|$:

$$\bigvee^\trianglerighteq \Xi = \bigvee_{\xi \in \Xi} \xi_{|\bigcap_{\xi \in \Xi} |\xi|}.$$

(2) $|\bigwedge^\trianglerighteq \Xi| = \bigcup_{\xi \in \Xi} |\xi|$ *and* $\bigwedge^\trianglerighteq \Xi$ *is the final convergence for the inclusion maps of each* $|\xi|$ *in* $\bigcup_{\xi \in \Xi} |\xi|$:

$$\bigwedge^\trianglerighteq \Xi = \bigwedge_{\xi \in \Xi} i_{\bigcup_{\xi \in \Xi} |\xi|}^{|\xi|} \xi.$$

Exercise XVI.5.8. Prove Proposition XVI.5.7.

The discussion at the beginning of this subsection leads to:

Proposition XVI.5.9. *If $\theta \ge \xi$ and $\sigma \ge \tau$ then $[\xi, \sigma] \trianglerighteq [\theta, \tau]$.*

Proof. Under these conditions, if $f \in C(\xi, \sigma)$ then all the maps in the diagram below are continuous:

$$
\begin{array}{ccc}
|\xi| & \xrightarrow{f} & |\sigma| \\
\uparrow{\scriptstyle i_{|\xi|}} & & \downarrow{\scriptstyle i_{|\sigma|}} \\
|\theta| & \xrightarrow{f} & |\tau|
\end{array}
$$

so that $f \in C(\theta, \tau)$. Moreover, the inclusion map $i_C : |[\xi, \sigma]| \to |[\theta, \tau]|$ is continuous, for if $f \in \lim_{[\xi, \sigma]} \mathcal{F}$ and $x \in \lim_{\theta} \mathcal{G} \subset \lim_{\xi} \mathcal{G}$ then $f(x) \in \lim_{\sigma} \langle \mathcal{G}, \mathcal{F} \rangle \subset \lim_{\tau} \langle \mathcal{G}, \mathcal{F} \rangle$. $\qquad \square$

Proposition XVI.5.10. *For each ξ and each set Σ of convergences on a common underlying set,*

$$
[\xi, \bigvee \Sigma] = \bigvee_{\sigma \in \Sigma}^{\unrhd} [\xi, \sigma]
$$

and

$$
[\bigwedge \Sigma, \xi] = \bigvee_{\sigma \in \Sigma}^{\unrhd} [\sigma, \xi].
$$

Proof. By Exercise IV.1.2,

$$
C(\xi, \bigvee \Sigma) = \bigcap_{\sigma \in \Sigma} C(\xi, \sigma) \text{ and } C(\bigwedge \Sigma, \xi) = \bigcap_{\sigma \in \Sigma} C(\sigma, \xi),
$$

and, by Proposition XVI.5.9, for every $\sigma \in \Sigma$,

$$
[\xi, \bigvee \Sigma] \unrhd [\xi, \sigma] \text{ and } [\bigwedge \Sigma, \xi] \unrhd [\sigma, \xi].
$$

Thus, in view of Proposition XVI.5.7, $\bigvee_{\sigma \in \Sigma}^{\unrhd} [\xi, \sigma]$ is a convergence on $C(\xi, \bigvee \Sigma)$ coarser or equal to $[\xi, \bigvee \Sigma]$, and $\bigvee_{\sigma \in \Sigma}^{\unrhd} [\sigma, \xi]$ is a convergence on $C(\bigwedge \Sigma, \xi)$ coarser or equal to $[\bigwedge \Sigma, \xi]$.

Let θ be a convergence on $C(\xi, \bigvee \Sigma)$ with $\theta \unrhd [\xi, \sigma]$ for every $\sigma \in \Sigma$, let $h \in \lim_{\theta} \mathcal{H}$, and $x \in \lim_{\xi} \mathcal{F}$. We need to show that $h(x) \in \lim_{\bigvee \Sigma} \langle \mathcal{F}, \mathcal{H} \rangle$ in order to conclude that $\theta \geq [\xi, \bigvee \Sigma]$. For every $\sigma \in \Sigma$, let $i_{|\sigma|} := i_{|[\xi, \sigma]|}^{|\theta|}$ denote the inclusion map. By continuity, $i_{|\sigma|}(h) \in \lim_{[\xi, \sigma]} i_{|\sigma|}[\mathcal{H}]$, so that $(i_{|\sigma|}(h))(x) \in \lim_{\sigma} \langle \mathcal{F}, i_{|\sigma|}[\mathcal{H}] \rangle$ for every $\sigma \in \Sigma$. Thus $h(x) \in \lim_{\bigvee \Sigma} \langle \mathcal{F}, \mathcal{H} \rangle$.

Similarly, if θ is a convergence on $C(\bigwedge \Sigma, \xi)$ with $\theta \unrhd [\sigma, \xi]$ for every $\sigma \in \Sigma$, we need to show that $\theta \geq [\bigwedge \Sigma, \xi]$. To this end, let $h \in \lim_{\theta} \mathcal{H}$ and $s \in \lim_{\bigwedge \Sigma} \mathcal{F}$, that is, there is $\sigma \in \Sigma$ with $s \in \lim_{\sigma} \mathcal{F}$. Let $i := i_{|[\sigma, \xi]|}^{|\theta|}$ denote the inclusion map. By continuity, $i(h) \in \lim_{[\sigma, \xi]} i[\mathcal{H}]$ so that $(i(h))(s) \in \lim_{\xi} \langle \mathcal{F}, i[\mathcal{H}] \rangle$, that is, $h(s) \in \lim_{\xi} \langle \mathcal{F}, \mathcal{H} \rangle$, and the conclusion follows. $\qquad \square$

On the other hand, if $\theta \unrhd \xi$ and $\sigma \unrhd \tau$ and $f \in C(\xi, \sigma)$ then all the maps below are continuous

$$
\begin{array}{ccc}
|\xi| & \xrightarrow{\;f\;} & |\sigma| \\
i^{|\theta|}_{|\xi|} \big\uparrow & \times & \big\downarrow i^{|\sigma|}_{|\tau|} \\
|\theta| & & |\tau|
\end{array}
\quad .
$$

Let us then call *restriction map* the map $r : \|[\xi, \sigma]\| \to \|[\theta, \sigma]\|$ defined by

$$
r(f) = f \circ i^{|\theta|}_{|\xi|}
$$

and *corestriction map* the map $c : \|[\xi, \sigma]\| \to \|[\xi, \tau]\|$ defined by

$$
c(f) = i^{|\sigma|}_{|\tau|} \circ f.
$$

Remark XVI.5.11. In other words, the restriction map is the upper adjoint

$$
r = \left(i^{|\theta|}_{|\xi|} \right)^{\sigma}
$$

while the corestriction map is the lower adjoint

$$
c = \left(i^{|\sigma|}_{|\tau|} \right)_{\xi}.
$$

Proposition XVI.5.12. *If $\theta \unrhd \xi$ and $\sigma \unrhd \tau$ then the restriction map $r : \|[\xi, \sigma]\| \to \|[\theta, \sigma]\|$ and corestriction map $c : \|[\xi, \sigma]\| \to \|[\xi, \tau]\|$ are continuous.*

In view of Remark XVI.5.11, Proposition XVI.5.12 is in fact an instance of Theorem XVI.5.13 below.

XVI.5.4 *Continuity of adjoint maps*

As we have seen before if $f : |\xi| \to |\tau|$ is continuous then

$$
f^{\sigma} : \|[\tau, \sigma]\| \to \|[\xi, \sigma]\| \text{ and } f_{\sigma} : \|[\sigma, \xi]\| \to \|[\sigma, \tau]\|.
$$

Moreover:

Theorem XVI.5.13. *If $f \in C(\xi, \tau)$, then, for every σ, $f^{\sigma} \in C([\tau, \sigma], [\xi, \sigma])$ and $f_{\sigma} \in C([\sigma, \xi], [\sigma, \tau])$.*

Proof. Let $h \in \lim_{[\tau,\sigma]} \mathcal{H}$. In order to prove that $f^\sigma(h) \in \lim_{[\xi,\sigma]} f^\sigma[\mathcal{H}]$, one must establish that $\langle x, f^\sigma(h) \rangle \in \lim_\sigma \langle \mathcal{F}, f^\sigma[\mathcal{H}] \rangle$, that is, in view of (XVI.1.5),

$$\langle f(x), h \rangle \in \lim_\sigma \langle f[\mathcal{F}], \mathcal{H} \rangle \qquad\qquad (\text{XVI.5.9})$$

for every $x \in |\xi|$, and each filter $\mathcal{F}$ such that $x \in \lim_\xi \mathcal{F}$. Because f is continuous $f(x) \in \lim_\tau f[\mathcal{F}]$, and since by assumption, $h \in \lim_{[\tau,\sigma]} \mathcal{H}$, (XVI.5.9) holds.

Let now $h \in \lim_{[\sigma,\xi]} \mathcal{H}$. To show that $f_\sigma(h) \in \lim_{[\sigma,\tau]} f_\sigma[\mathcal{H}]$, one must establish that $\langle s, f_\sigma(h) \rangle \in \lim_\tau \langle \mathcal{F}, f_\sigma[\mathcal{H}] \rangle$, for every $s \in |\sigma|$ and each filter $\mathcal{F}$ with $s \in \lim_\sigma \mathcal{F}$. In view of (XVI.1.9),

$$\langle s, f_\sigma(h) \rangle \in \lim_\tau \langle \mathcal{F}, f_\sigma[\mathcal{H}] \rangle \iff \langle h(s), f \rangle \in \lim_\tau f\left[\langle \mathcal{F}, \mathcal{H} \rangle \right].$$

The latter follows from $h(s) \in \lim_\xi \langle \mathcal{F}, \mathcal{H} \rangle$ and the continuity of $f : |\xi| \to |\tau|$. $\square$

Example XVI.5.14. We have seen in Example XVI.2.3 that if $\xi \geq \tau$, then, for every σ, $C(\tau,\sigma) \subset C(\xi,\sigma)$ and

$$i^\sigma : C(\tau,\sigma) \to C(\xi,\sigma)$$

is the inclusion map. By Theorem XVI.5.13, $i^\sigma \in C([\tau,\sigma],[\xi,\sigma])$, that is, $[\tau,\sigma] \geq [\xi,\sigma]$, as already established by Proposition XVI.5.9. Again by Theorem XVI.5.13, $i^{\sigma\sigma} = (i^\sigma)^\sigma$ is a continuous map

$$i^{\sigma\sigma} : |[[\xi,\sigma],\sigma]| \to |[[\tau,\sigma],\sigma]|$$

which, in view of Remark XVI.5.11, corresponds to the restriction of maps from $C(\xi,\sigma)$ to $C(\tau,\sigma)$.

Note that the point-evaluation map (XVI.2.2) takes each $x \in |\xi|$ to a map

$$j_\xi(x) : |[\xi,\sigma]| \to |\sigma|$$
$$h \mapsto h(x)$$

which is continuous (because $[\xi,\sigma] \geq p(\xi,\sigma)$). Thus j_ξ maps $|\xi|$ to $C([\xi,\sigma],\sigma)$. Moreover:

Proposition XVI.5.15. *For every convergences ξ and σ, the point evaluation map*

$$j_\xi : |\xi| \to |[[\xi,\sigma],\sigma]|$$

is continuous.

Proof. Indeed, if $h \in \lim_{[\xi,\sigma]} \mathcal{H}$ and $x \in \lim_{\xi} \mathcal{F}$, then by definition,

$$\langle x, h \rangle \in \lim_{\sigma} \langle \mathcal{F}, \mathcal{H} \rangle = \lim_{\sigma} j_{\xi}[\mathcal{F}][\mathcal{H}].$$

$\square$

In view of Theorem XVI.5.13 and Proposition XVI.5.15, Proposition XVI.1.3 finds an analogue on spaces of continuous functions endowed with the natural convergence:

Theorem XVI.5.16. *For every continuous map $f \in C(\xi, \tau)$ and every convergence σ, the following diagram is commutative and only contains continuous maps:*

$$
\begin{array}{ccc}
|\xi| & \xrightarrow{\quad j_{\xi} \quad} & |[[\xi,\sigma],\sigma]| \\
f \downarrow & & \downarrow f^{\sigma\sigma} \\
|\tau| & \xrightarrow{\quad j_{\tau} \quad} & |[[\tau,\sigma],\sigma]|
\end{array}
$$

that is,

$$j_{\tau} \circ f = f^{\sigma\sigma} \circ j_{\xi}.$$

XVI.5.5 *Initial structures for adjoint maps*

Note that given a function f with domain $|\xi|$, $f \in C(\xi, f\xi)$, so that, by Theorem XVI.5.13, $f^{\sigma} \in C([f\xi,\sigma],[\xi,\sigma])$, for every σ. In other words,

$$[f\xi,\sigma] \geq (f^{\sigma})^{-}[\xi,\sigma]. \tag{XVI.5.10}$$

In fact:

Proposition XVI.5.17. *For every convergences ξ and σ, and every onto map f with domain $|\xi|$,*

$$[f\xi,\sigma] = (f^{\sigma})^{-}[\xi,\sigma].$$

Proof. To show the reverse inequality of (XVI.5.10), let $h \in \lim_{(f^{\sigma})^{-}[\xi,\sigma]} \mathcal{H}$, that is, in view of (IV.2.1), $f^{\sigma}(h) \in \lim_{[\xi,\sigma]} f^{\sigma}[\mathcal{H}]$. Let $y \in \lim_{f\xi} \mathcal{G}$. Because f is onto, by (IV.2.3) there is $x \in f^{-}(y)$ and $\mathcal{F}$ with $x \in \lim_{\xi} \mathcal{F}$ and $\mathcal{G} \geq f[\mathcal{F}]$. Thus

$$\langle x, f^{\sigma}(h) \rangle \in \lim_{\sigma} \langle \mathcal{F}, f^{\sigma}[\mathcal{H}] \rangle$$

because $f^{\sigma}(h) \in \lim_{[\xi,\sigma]} f^{\sigma}[\mathcal{H}]$, and, in view of (XVI.1.5), this rephrases as

$$\langle f(x), h \rangle \in \lim_{\sigma} \langle f[\mathcal{F}], \mathcal{H} \rangle$$

so that $h(y) \in \lim_{\sigma} \langle \mathcal{G}, \mathcal{H} \rangle$ and $h \in \lim_{[f\xi,\sigma]} \mathcal{H}$. $\square$

Similarly, given a map g with codomain $|\sigma|$, the map $g : |g^- \sigma| \to |\sigma|$ is continuous, so that, for every ξ,

$$g_\xi : |[\xi, g^- \sigma]| \to |[\xi, \sigma]|$$

is continuous, by Theorem XVI.5.13. In other words,

$$[\xi, g^- \sigma] \geq g_\xi^- [\xi, \sigma]. \qquad\qquad (\text{XVI.5.11})$$

In fact:

Proposition XVI.5.18. *For every convergences ξ and σ and map g with codomain $|\sigma|$,*

$$[\xi, g^- \sigma] = g_\xi^- [\xi, \sigma].$$

Proof. To show the reverse inequality of (XVI.5.11), let $h \in \lim_{g_\xi^- [\xi, \sigma]} \mathcal{H}$, that is, $g_\xi(h) \in \lim_{[\xi, \sigma]} g_\xi[\mathcal{H}]$ by (IV.2.1). Then, for every $x \in |\xi|$ and filter $\mathcal{F}$ with $x \in \lim_\xi \mathcal{F}$,

$$\langle x, g_\xi(h) \rangle \in \lim_\sigma \langle \mathcal{F}, g_\xi[\mathcal{H}] \rangle ,$$

which, in view of (XVI.1.9), rephrases as

$$\langle h(x), g \rangle \in \lim_\sigma g\, [\langle \mathcal{F}, \mathcal{H} \rangle] .$$

By (IV.2.1), this means $h(x) \in \lim_{g^- \sigma} \langle \mathcal{F}, \mathcal{H} \rangle .$ Thus, $h \in \lim_{[\xi, g^- \sigma]} \mathcal{H}.$ $\square$

Combining Propositions XVI.5.10 and XVI.5.18 yields:

Corollary XVI.5.19. *If ξ is a convergence, and for each $i \in I$, g_i is a map with codomain $|\sigma_i|$, then*

$$[\xi, \bigvee_{i \in I} g_i^- \sigma_i] = \bigvee_{i \in I}^{\rhd} ((g_i)_\xi)^- [\xi, \sigma_i].$$

Similarly, combining Propositions XVI.5.10 and XVI.5.17 yields:

Corollary XVI.5.20. *If σ is a convergence, and for each $i \in I$, f_i is an onto map with domain $|\xi_i|$, then*

$$[\bigwedge_{i \in I} f_i \xi_i, \sigma] = \bigvee_{i \in I}^{\rhd} (f_i^\sigma)^- [\xi_i, \sigma].$$

The next two sections will develop a theory of duality between a convergence ξ and its σ-*duals* $[\xi, \sigma]$. In particular, we will study reflective properties of duals $[\xi, \sigma]$ in terms of the *primal convergence* ξ. It turns out that, because of (XVI.3.1), this quest is intimately intertwined with problems of stability under product of various properties. Section XVII.7 will explore the consequences of these two more theoretical sections in terms of product theorems.

XVI.6   Compact subsets of function spaces (Ascoli-Arzelà)

In its original form, the Ascoli Theorem provides conditions for a sequence of continuous real-valued functions on a closed interval to have a *uniformly* convergent subsequence. To this end, Ascoli and Arzelà independently introduced the notion of an equicontinuous sequence: the sequence $\{f_n\}_n$ is *equicontinuous at* $x \in [a, b]$ if for every $\varepsilon > 0$ there is $\delta > 0$ such that

$$|x - y| < \delta \implies |f_n(x) - f_n(y)| < \varepsilon$$

for every $n \in \mathbb{N}$, and $\{f_n\}_{n \in \mathbb{N}}$ is called *equicontinuous* if it is equicontinuous at every x in $[a, b]$. The sequence is *uniformly bounded* if there is an M such that $|f_n(x)| < M$ for all $x \in [a, b]$ and all $n \in \mathbb{N}$. The combination of both conditions yields the desired property [4]:

Theorem XVI.6.1 (Ascoli). *If a sequence of functions is equicontinuous and uniformly bounded, then it has a uniformly convergent subsequence.*

In other words, under these conditions, the set $\{f_n : n \in \mathbb{N}\}$ has sequentially compact closure in the space $C([a, b], \mathbb{R})$ of real-valued continuous functions on $[a, b]$ endowed with the topology of uniform convergence. Since $C([a, b], \mathbb{R})$ is metrizable in this topology (by the uniform norm), sequential compactness and compactness are equivalent, by Corollary IX.8.31. Arzelà extended Ascoli's theorem to general sets of functions [3], obtaining what in modern terms would be called a criterion of compactness of a set of real-valued continuous functions. Therefore, the *abstract Arzelà-Ascoli quest* is to find sufficient (and hopefully, also necessary) conditions on subsets H of a space $C(X, Y)$ of continuous functions between two spaces X and Y to have compact closure, that is, to be *relatively compact*, for some appropriate analogue of the topology of uniform convergence. The literature is rich in results of that type. Theorems of the Ascoli-Arzelà type have become an ubiquitous and efficient tool in a variety of contexts.

Since we want to characterize (relatively) compact subsets of spaces of continuous maps, pseudotopologies seem to be the natural context. However, we will first consider a particular case where equicontinuity still makes sense:

Definition XVI.6.2. Let ξ be a Kent space ([4]) and (Y, d) be a metric space. A set $F \subset C(\xi, \tilde{d})$ is *equicontinuous at* $x \in |\xi|$ if for every $\varepsilon > 0$ and

---

[4]that is, if $x \in \lim_{\xi} \mathcal{G}$ then $x \in \lim_{\xi} \mathcal{G} \wedge \{x\}^{\uparrow}$. In this context, we may restrict ourselves to $\mathcal{G} \leq \{x\}^{\uparrow}$ when considering $\mathcal{G} \in \lim_{\xi}^{-}(x)$.

every $\mathcal{G} \in \lim_{\xi}^{-}(x)$, there is $G \in \mathcal{G}$ such that

$$(t \in G, f \in F) \implies d(f(t), f(x)) < \varepsilon$$

and *equicontinuous* if it is equicontinuous at every $x \in |\xi|$.

We need to generalize equicontinuity to a notion that we can use for general convergence spaces.

Definition XVI.6.3. Let (X, ξ) and (Y, σ) be two convergences spaces. A set $H \subset Y^X$ is *evenly continuous at* $x \in |\xi|$ if for every filter $\mathcal{F}$ on Y^X such that $H \in \mathcal{F}$,

$$y \in \lim_{\sigma} \langle x, \mathcal{F} \rangle, \ x \in \lim_{\xi} \mathcal{G} \implies y \in \lim_{\sigma} \langle \mathcal{G}, \mathcal{F} \rangle. \qquad \text{(XVI.6.1)}$$

H is *evenly continuous* if it is evenly continuous at every $x \in |\xi|$.

Proposition XVI.6.4. *Let ξ be a Kent space and (Y, d) be a metric space. If $H \subset C(\xi, \tilde{d})$ is equicontinuous (at x) then it is evenly continuous (at x). Moreover, if $\langle x, H \rangle$ is relatively compact (in $\tilde{d}$), the converse is true.*

Proof. Let $\mathcal{F}$ be a filter on $C(\xi, \tilde{d})$ with $H \in \mathcal{F}$, and assume that $y \in \lim_{\tilde{d}} \langle x, \mathcal{F} \rangle$. It suffices to show that $y \in \lim_{\tilde{d}} \langle \mathcal{G}, \mathcal{F} \rangle$ for every $\mathcal{G} \in \lim_{\xi}^{-}(x)$ with $\mathcal{G} \le \{x\}^{\uparrow}$.

For every $\varepsilon > 0$, there is $F \in \mathcal{F}$ with $d(f(x), y) < \frac{\varepsilon}{2}$ for every $f \in F$. On the other hand, since H is equicontinuous, there is $G \in \mathcal{G}$ such that $d(f(t), f(x)) < \frac{\varepsilon}{2}$ for every $t \in G$ and $f \in F \cap H$. Since d is a metric

$$d(f(t), y) \le d(f(t), f(x)) + d(f(x), y) < \varepsilon,$$

for every $t \in G$ and $f \in F \cap H$, so that $\langle G, F \cap H \rangle \subset B_d(y, \varepsilon)$, and the conclusion follows.

Assume now that $\langle x, H \rangle$ is relatively compact. We proceed by contrapositive, assuming that H is not equicontinuous at x, that is, there is $\varepsilon_0 > 0$ and $\mathcal{G}_0 \in \lim_{\xi}^{-}(x)$ such that for every $G \in \mathcal{G}_0$ there is $t \in G$ and $f_G \in H$ with $d(f_G(t), f(x)) > \varepsilon_0$. Consider an ultrafilter $\mathcal{U}$ containing the following filter-base:

$$\{\{f_G : G \subset G_1\} : G_1 \in \mathcal{G}_0\}.$$

Since $H \in \mathcal{U}$, $\langle x, \mathcal{U} \rangle$ is an ultrafilter on the relatively compact set $\langle x, H \rangle$ and is thus convergent to some point y for $\tilde{d}$. But $\langle \mathcal{G}_0, \mathcal{U} \rangle$ does not converge, because for every $G_1 \in \mathcal{G}$ and $U \in \mathcal{U}$,

$$\text{diam} \langle G_1, U \cap \{f_G : G \subset G_1\} \rangle \ge \varepsilon_0.$$

Thus, H is not evenly continuous at x. $\qquad\square$

It is immediate from Definition XVI.6.3 that:

Corollary XVI.6.5. *If ξ and σ are convergences, $H \subset |\sigma|^{|\xi|}$ is evenly continuous and $H \in \mathcal{F}$ then*

$$f \in \lim_{p(|\xi|,\sigma)} \mathcal{F} \Longrightarrow f \in \lim_{\sigma^\xi} \mathcal{F}.$$

In other words, if H is evenly continuous, then the topology of pointwise convergence and the natural (pre)convergence coincide on H.

As a result, an evenly continuous set of functions must be a set of continuous functions, which justifies the terminology:

Corollary XVI.6.6. *If $H \subset |\sigma|^{|\xi|}$ is evenly continuous then $H \subset C(\xi, \sigma)$.*

Proof. For each $f \in H$, $f \in \lim_{p(|\xi|,\sigma)}\{f\}^\uparrow$ so that, by Corollary XVI.6.5, $f \in \lim_{\sigma^\xi}\{f\}^\uparrow$, and, by Proposition XVI.5.1, $f \in C(\xi, \sigma)$. □

Theorem XVI.6.7. *Let ξ be a convergence, and let σ be a regular convergence. If $H \subset C(\xi, \sigma)$ is evenly continuous and $\langle x, H \rangle$ is relatively compact in σ for each x in $|\xi|$, then H is relatively compact in $[\xi, \sigma]$ (and thus, also in the compact-open topology $k(\xi, \sigma)$).*

Proof. Let $\mathcal{U}$ be an ultrafilter on $C(\xi, \sigma)$ containing H. We need to show that $\lim_{[\xi,\sigma]} \mathcal{U} \neq \varnothing$. Since $\langle x, H \rangle$ belongs to the ultrafilter $\langle x, \mathcal{U} \rangle$ and $\langle x, H \rangle$ is relatively compact, there is $y(x) \in \lim_\sigma \langle x, \mathcal{U} \rangle$, for each $x \in |\xi|$. Therefore, $y(\cdot) \in \lim_{p(|\xi|,\sigma)} \mathcal{U}$ and $H \in \mathcal{U}$ so that $y(\cdot) \in \lim_{\sigma^\xi} \mathcal{U}$ because of Corollary XVI.6.5. In view of Theorem XVI.5.2, $y(\cdot)$ is continuous because σ is regular, which concludes the proof. The parenthetical statement follows from (XVI.3.2). □

Conversely:

Theorem XVI.6.8. *Let ξ be a convergence, and let σ be a Hausdorff pseudotopology. If H is relatively compact in $[\xi, \sigma]$ then $\langle x, H \rangle$ is relatively compact in σ for each x in $|\xi|$, and H is evenly continuous.*

Proof. If H is (relatively) compact in $[\xi, \sigma]$ then $\langle x, H \rangle$ is (relatively) compact in σ for each x in $|\xi|$ because each map $\langle x, \cdot \rangle$ is continuous and therefore preserves (relative) compactness.

To show that H is evenly continuous, assume that $H \in \mathcal{F}$, that $y \in \lim_\sigma \langle x, \mathcal{F} \rangle$ and that $x \in \lim_\xi \mathcal{G}$. Since σ is a pseudotopology, to show that

$y \in \lim_\sigma \langle \mathcal{G}, \mathcal{F} \rangle$ we only need to show that $y \in \lim_\sigma \mathcal{W}$ for each ultrafilter $\mathcal{W}$ that contains $\langle \mathcal{G}, \mathcal{F} \rangle$. Note that $\mathcal{W} \# e [\mathcal{G} \times \mathcal{F}]$. By (II.2.9)

$$e^- [\mathcal{W}] \# (\mathcal{G} \times \mathcal{F}),$$

which, in view of (II.2.10) amounts to

$$(e^- [\mathcal{W}]) [\mathcal{G}] \# \mathcal{F}.$$

In particular, the filters $(e^- [\mathcal{W}]) [\mathcal{G}]$ and $\mathcal{F}$ have a least upper bound $\mathcal{H}$ that contains H. By relative compactness of H, there is an ultrafilter $\mathcal{U}$ finer than $\mathcal{H}$ and a function $f \in \lim_{[\xi,\sigma]} \mathcal{U}$. In particular, $f(x) \in \lim_\sigma \langle x, \mathcal{U} \rangle$ and

$$\langle x, \mathcal{U} \rangle \supseteq \langle x, \mathcal{H} \rangle \supseteq \langle x, \mathcal{F} \rangle,$$

and $y \in \lim_\sigma \langle x, \mathcal{F} \rangle$ so that $y \in \lim_\sigma \langle x, \mathcal{U} \rangle$. As σ is Hausdorff, $y = f(x)$. Now, $y = f(x) \in \lim_\sigma \langle \mathcal{G}, \mathcal{U} \rangle$ because $f \in \lim_{[\xi,\sigma]} \mathcal{U}$. Moreover,

$$\mathcal{U} \# (e^- [\mathcal{W}]) [\mathcal{G}],$$

so that, in view of (II.2.10),

$$e [\mathcal{U} \times \mathcal{G}] \# \mathcal{W}.$$

As $\mathcal{W}$ is an ultrafilter, $\mathcal{W} \supseteq \langle \mathcal{G}, \mathcal{U} \rangle$ and $y \in \lim_\sigma \mathcal{W}$, which completes the proof. $\qquad \square$

Corollary XVI.6.9. *If ξ is a convergence, and σ is a Hausdorff regular pseudotopology, then a subset H of $[\xi, \sigma]$ is relatively compact if and only if $\langle x, H \rangle$ is relatively compact in σ for each x in $|\xi|$, and H is evenly continuous.*

In particular, since a metric induces a (pseudo)topology that is Hausdorff and regular, Proposition XVI.6.4 and Corollary XVI.6.9 yield:

Corollary XVI.6.10. *If ξ is a Kent space and (Y, d) is a metric space, then a subset H of $[\xi, \tilde{d}]$ is relatively compact if and only if $\langle x, H \rangle$ is relatively compact in $\tilde{d}$ for each x in $|\xi|$, and H is equicontinuous.*

In view of Proposition XVI.3.4:

Corollary XVI.6.11. *If ξ is a hereditarily locally compact convergence, and σ is a Hausdorff regular pseudotopology, then a subset H of $k(\xi, \sigma)$ is relatively compact if and only if $\langle x, H \rangle$ is relatively compact in σ for each x in $|\xi|$, and H is evenly continuous.*

Corollary XVI.6.12. *If ξ is a hereditarily locally compact Kent space and (Y, d) is a metric space, then a subset H of $k(\xi, \tilde{d})$ is relatively compact if and only if $\langle x, H \rangle$ is relatively compact in $\tilde{d}$ for each x in $|\xi|$, and H is equicontinuous.*

Extension to filters

Everything in this section can be extended from subsets (identified to principal filters) of function spaces to arbitrary filters on $C(X, Y)$. Namely:

Definition XVI.6.13. Let (X, ξ) and (Y, σ) are two convergence spaces and $\mathcal{H} \in \mathbb{F}(Y^X)$. The filter $\mathcal{H}$ is called *evenly continuous at x* if for every $\mathcal{F} \in \beta(\mathcal{H})$, (XVI.6.1) holds. $\mathcal{H}$ is *evenly continuous* if it is evenly continuous at every $x \in X$.

Definition XVI.6.14. If ξ is a Kent space and (Y, d) is a metric space, then $\mathcal{F} \in \mathbb{F}(C(\xi, \tilde{d}))$ is *equicontinuous at $x \in |\xi|$* if for every $\varepsilon > 0$ and every $\mathcal{G} \in \lim_{\xi}^{-}(x)$ there is $G \in \mathcal{G}$ and $F \in \mathcal{F}$ with

$$(t \in G, f \in F) \Longrightarrow d(f(t), f(x)) < \varepsilon,$$

and *equicontinuous* if it is equicontinuous at every $x \in |\xi|$.

On the other hand, the notion of relative compactness for filters has already been introduced. The results of the section are then easily extended by adapting the proofs above accordingly.

Proposition XVI.6.15. *Let ξ be a Kent space and (Y, d) be a metric space. If $\mathcal{H} \in \mathbb{F}(C(\xi, \tilde{d}))$ is equicontinuous (at x) then it is evenly continuous (at x). Moreover, if $\langle x, \mathcal{H} \rangle$ is compactoid (in $\tilde{d}$), the converse is true.*

Exercise XVI.6.16. Prove Proposition XVI.6.15.

Corollary XVI.6.17. *If ξ and σ are convergences, $\mathcal{H} \in \mathbb{F}\left(|\sigma|^{|\xi|}\right)$ is evenly continuous and $\mathcal{F} \geq \mathcal{H}$ then*

$$f \in \lim_{p(|\xi|, \sigma)} \mathcal{F} \Longrightarrow f \in \lim_{\sigma \xi} \mathcal{F}.$$

Theorem XVI.6.18. *Let ξ be a convergence, and let σ be a regular convergence. If $\mathcal{H}$ is an evenly continuous filter on $C(\xi, \sigma)$ and $\langle x, \mathcal{H} \rangle$ is compactoid in σ for each x in $|\xi|$, then $\mathcal{H}$ is compactoid in $[\xi, \sigma]$ (and thus, also in the compact-open topology $k(\xi, \sigma)$).*

Exercise XVI.6.19. Prove Theorem XVI.6.18.

Theorem XVI.6.20. *Let ξ be a convergence, and let σ be a Hausdorff pseudotopology. If $\mathcal{H}$ is a relatively compact filter in $[\xi, \sigma]$ then $\langle x, \mathcal{H} \rangle$ is relatively compact in σ for each x in $|\xi|$, and $\mathcal{H}$ is evenly continuous.*

Exercise XVI.6.21. Prove Theorem XVI.6.20.

Chapter XVII

Duality

XVII.1   Natural duality

The previous section presented some of the convenient features of the natural convergence $[\xi, \sigma]$ on the set $C(\xi, \sigma)$ of continuous maps from ξ to σ. In particular, Proposition XVI.5.15 states that the point-evaluation map

$$j_\xi : |\xi| \to |[[\xi, \sigma], \sigma]|$$

is continuous, that is, $\xi \geq j_\xi^- [[\xi, \sigma], \sigma]$. In other words, defining the σ-*bidual modification of* ξ as

$$\text{Epi}^\sigma \, \xi := j_\xi^- [[\xi, \sigma], \sigma],$$

we have

$$\xi \geq \text{Epi}^\sigma \, \xi. \tag{XVII.1.1}$$

By definition, $x \in \lim_{\text{Epi}^\sigma \, \xi} \mathcal{F}$ means that $j_\xi(x) \in \lim_{[[\xi, \sigma], \sigma]} j_\xi[\mathcal{F}]$, that is, for every $f \in C(\xi, \sigma)$ and $\mathcal{G} \in \lim_{[\xi, \sigma]}^-(f)$,

$$\langle f, j_\xi(x) \rangle \in \lim_\sigma \langle \mathcal{G}, j_\xi[\mathcal{F}] \rangle \iff f(x) \in \lim_\sigma \langle \mathcal{F}, \mathcal{G} \rangle.$$

Thus,

$$x \in \lim_{\text{Epi}^\sigma \, \xi} \mathcal{F} \iff \underset{f \in C(\xi, \sigma)}{\forall} \underset{\mathcal{G} \in \lim_{[\xi, \sigma]}^-(f)}{\forall} f(x) \in \lim_\sigma \langle \mathcal{F}, \mathcal{G} \rangle. \tag{XVII.1.2}$$

The σ-bidual map turns out to have remarkable properties of interest in the study of the duality between a convergence ξ and its σ-*dual* $[\xi, \sigma]$:

Theorem XVII.1.1. *For every convergence* σ, Epi^σ *is a reflector. Moreover, for every convergence* ξ, $\text{Epi}^\sigma \, \xi$ *is the coarsest among convergences* θ *on* $|\xi|$ *such that*

$$[\theta, \sigma] = [\xi, \sigma].$$

Proof. We first show that Epi^σ is a functor. Let $f : |\xi| \to |\tau|$ be continuous, and apply Theorem XVI.5.16 to the effect that the diagram of continuous maps

$$
\begin{array}{ccc}
|\xi| & \xrightarrow{\;j_\xi\;} & |[[\xi,\sigma],\sigma]| \\
{\scriptstyle f}\big\downarrow & & \big\downarrow{\scriptstyle f^{\sigma\sigma}} \\
|\tau| & \xrightarrow{\;j_\tau\;} & |[[\tau,\sigma],\sigma]|
\end{array}
$$

is commutative. Thus

$$j_\tau \circ f : |\,\text{Epi}^\sigma\,\xi| \to |[[\tau,\sigma],\sigma]|$$

is continuous because $f^{\sigma\sigma} \circ j_\xi : |\,\text{Epi}^\sigma\,\xi| \to |[[\tau,\sigma],\sigma]|$ is the composite of two continuous maps. Therefore,

$$(j_\tau \circ f)^-\,[[\tau,\sigma],\sigma] \le \text{Epi}^\sigma\,\xi$$

and, in view of (IV.2.8),

$$(j_\tau \circ f)^-\,[[\tau,\sigma],\sigma] = f^-\,(\text{Epi}^\sigma\,\tau)$$

so that $f^-\,(\text{Epi}^\sigma\,\tau) \le \text{Epi}^\sigma\,\xi$, that is, $f \in C(\text{Epi}^\sigma\,\xi, \text{Epi}^\sigma\,\tau)$. The case where f is the identity map $i : |\xi| \to |\tau|$ where $\xi \ge \tau$ shows that Epi^σ is isotone.

Thus Epi^σ is a functor, that is contractive, by (XVII.1.1). To show that it is also idempotent, hence a reflector, it is enough to show that

$$[\text{Epi}^\sigma\,\xi, \sigma] = [\xi,\sigma]$$

for every ξ.

Since

$$j_\xi^-\,(p(p(\xi,\sigma),\sigma)) \le j_\xi^-\,[[\xi,\sigma],\sigma] = \text{Epi}^\sigma\,\xi \le \xi,$$

Proposition XVI.4.6 ensures that $C(\text{Epi}^\sigma\,\xi, \sigma) = C(\xi,\sigma)$. Moreover, Proposition XVI.5.9 and (XVII.1.1) yield $[\text{Epi}^\sigma\,\xi,\sigma] \trianglerighteq [\xi,\sigma]$, which therefore means $[\text{Epi}^\sigma\,\xi,\sigma] \ge [\xi,\sigma]$.

To see the reverse inequality, let $f \in \lim_{[\xi,\sigma]} \mathcal{F}$, that is,

$$f(x) \in \lim_\sigma \langle \mathcal{G}, \mathcal{F} \rangle,$$

for every $x \in |\xi|$ and $\mathcal{G} \in \lim_\xi^-(x)$. In view of (XVII.1.2), $f \in \lim_{[\text{Epi}^\sigma\,\xi,\sigma]} \mathcal{F}$, and we have

$$[\text{Epi}^\sigma\,\xi,\sigma] = [\xi,\sigma].$$

To see that $\text{Epi}^\sigma\,\xi$ is the coarsest convergence on $|\xi|$ with this property, assume that θ satisfies $|\theta| = |\xi|$ and $[\theta,\sigma] = [\xi,\sigma]$. Then $[[\theta,\sigma],\sigma] = [[\xi,\sigma],\sigma]$, so that

$$\theta \ge \text{Epi}^\sigma\,\theta = \text{Epi}^\sigma\,\xi.$$

$\square$

Now if R is a reflector,

$$\mathrm{Epi}^R := \bigvee_{\sigma = R\sigma} \mathrm{Epi}^\sigma,$$

is the *R-bidual modification*.

Proposition XVII.1.2. *If R is a reflector, so is Epi^R. Moreover, $\mathrm{Epi}^R \xi$ is the coarsest among convergences θ on $|\xi|$ satisfying*

$$[\theta, \sigma] = [\xi, \sigma] \text{ for every } \sigma = R\sigma.$$

Proof. Epi^R is a contractive modifier, because each Epi^σ is. Moreover, for every $\sigma = R\sigma$,

$$\xi \geq \mathrm{Epi}^R \xi \geq \mathrm{Epi}^\sigma \xi,$$

and $[\mathrm{Epi}^\sigma \xi, \sigma] = [\xi, \sigma]$ by Theorem XVII.1.1, so that

$$[\mathrm{Epi}^R \xi, \sigma] = [\xi, \sigma] \text{ for every } \sigma = R\sigma.$$

Moreover, by Proposition XIV.2.9, Epi^R is a reflector because each functor Epi^σ is.

If now θ is a convergence on $|\xi|$ with $[\theta, \sigma] = [\xi, \sigma]$ for every $\sigma = R\sigma$, then $\theta \geq \mathrm{Epi}^\sigma \xi$ for every $\sigma = R\sigma$ by Theorem XVII.1.1, so that $\theta \geq \mathrm{Epi}^R \xi$. $\square$

Recall (from Section XIV.7) that R_σ defined by

$$R_\sigma \xi = \bigvee_{f \in C(\xi, \sigma)} f^- \sigma$$

is the reflector on the smallest reflective class of convergences that contains σ, and more generally, if $\mathbf{D}$ is a class of convergences, $R_{\mathbf{D}} := \bigvee_{\sigma \in \mathbf{D}} R_\sigma$ is the reflector on the smallest reflective class containing $\mathbf{D}$.

By definition, $\mathrm{Epi}^{R_\sigma} \geq \mathrm{Epi}^\sigma$. Moreover, if $\tau = R_\sigma \tau$, then, by Corollary XVI.5.19,

$$[\mathrm{Epi}^\sigma \xi, \tau] = [\mathrm{Epi}^\sigma \xi, \bigvee_{f \in C(\tau, \sigma)} f^- \sigma] = \bigvee^{\trianglerighteq}_{f \in C(\tau, \sigma)} ((f)_\xi)^- [\mathrm{Epi}^\sigma \xi, \sigma]$$

$$= \bigvee^{\trianglerighteq}_{f \in C(\tau, \sigma)} ((f)_\xi)^- [\xi, \sigma] \text{ by Theorem XVII.1.1}$$

$$= [\xi, \tau].$$

In view of Proposition XVII.1.2, $\mathrm{Epi}^\sigma \geq \mathrm{Epi}^{R_\sigma}$, so that

$$\mathrm{Epi}^\sigma = \mathrm{Epi}^{R_\sigma}. \tag{XVII.1.3}$$

More generally,

$$\mathrm{Epi}^{R_{\mathbf{D}}} = \bigvee_{\sigma \in \mathbf{D}} \mathrm{Epi}^\sigma. \tag{XVII.1.4}$$

Exercise XVII.1.3. Prove (XVII.1.4) following the example (XVII.1.3).

In view of (XVII.1.2), given a reflector R, $x \in \lim_{\text{Epi}^R \xi} \mathcal{F}$ if and only if for every $\sigma = R\sigma$, every $f \in C(\xi, \sigma)$ and every $\mathcal{G}$ with $f \in \lim_{[\xi,\sigma]} \mathcal{G}$, $f(x) \in \lim_{\sigma} \langle \mathcal{F}, \mathcal{G} \rangle$. In particular, if $x \in \lim_{\text{Epi}^R \xi} \mathcal{F}$, then $f(x) \in \lim_{\sigma} \langle \mathcal{F}, \{f\}^{\uparrow} \rangle = \lim_{\sigma} f[\mathcal{F}]$ for every $\sigma = R\sigma$ and every $f \in C(\xi, \sigma)$, that is, $x \in \lim_{R_R \xi} \mathcal{F}$. Since $R_R = R$, we conclude that

$$\text{Epi}^R \geq R. \tag{XVII.1.5}$$

Lemma XVII.1.4. *Let R be a reflector, let $\mathbf{W}$ be finally dense in a class $\mathbf{C}$ of convergences, and let $\mathbf{D}$ be initially dense in $\mathbf{R}$. The following are equivalent:*

(1) for every $\tau \in \mathbf{C}$,

$$\theta \times \tau \geq R(\xi \times \tau);$$

(2) for every $\tau \in \mathbf{C}$,

$$\sigma = R\sigma \implies C(\xi \times \tau, \sigma) \subset C(\theta \times \tau, \sigma);$$

(3) for every $\tau \in \mathbf{W}$,

$$\sigma \in \mathbf{D} \implies C(\xi \times \tau, \sigma) \subset C(\theta \times \tau, \sigma);$$

(4) for every $\tau \in \mathbf{W}$,

$$\theta \times \tau \geq R(\xi \times \tau).$$

Proof. (1) $\iff$ (2) and (3) $\iff$ (4) are instances of Corollary XIV.7.10, and (2) $\implies$ (3) is obvious. Finally, to see that (3) $\implies$ (2) first note that if $\mathbf{W}$ is finally dense in $\mathbf{C}$ then

$$\xi \times \mathbf{W} := \{\xi \times \tau : \tau \in \mathbf{W}\}$$

is finally dense in $\xi \times \mathbf{C}$ and $\theta \times \mathbf{W}$ is finally dense in $\theta \times \mathbf{C}$ because for any ξ and $\Lambda \subset \mathbf{I}$,

$$\xi \times \bigwedge_{\lambda \in \Lambda} \lambda = \bigwedge_{\lambda \in \Lambda} (\xi \times \lambda).$$

Thus, Corollary XIV.7.12 with $F = I$ the identity coreflector applies to the effect $C(\xi \times \tau, \sigma) \subset C(\theta \times \tau, \sigma)$ for every $\tau \in \mathbf{W}$ implies the $C(\xi \times \tau, \sigma) \subset C(\theta \times \tau, \sigma)$ for every $\tau \in \mathbf{C}$. $\square$

Theorem XVII.1.5. *Let R be a reflector, $\mathbf{D}$ be an initially dense subclass of $\mathbf{R}$, $\mathbf{F}$ a finally dense subclass of the class $\mathbf{I}$ of convergences, and let $\xi \geq \theta$. The following are equivalent:*

(1) $\theta \geq \mathrm{Epi}^R \xi$;

(2) for every convergence τ,

$$\theta \times \tau \geq \mathrm{Epi}^R(\xi \times \tau);$$

(3) for every convergence $\tau \in \mathbf{F}$,

$$\theta \times \tau \geq R(\xi \times \tau);$$

(4) $[\xi, \sigma] \geq [\theta, \sigma]$ *for every $\sigma = \mathrm{Epi}^R \sigma$ (in particular, every $\sigma = R\sigma$);*

(5) $[\xi, \sigma] \geq [\theta, \sigma]$ *for every $\sigma \in \mathbf{D}$.*

Proof. (1) $\implies$ (2): By the exponential law (XVI.3.1), for every $\sigma = R\sigma$, $[\mathrm{Epi}^R \xi \times \tau, \sigma] = [\tau, [\mathrm{Epi}^R \xi, \sigma]]$, and $[\mathrm{Epi}^R \xi, \sigma] = [\xi, \sigma]$ by Proposition XVII.1.2. Thus,

$$[\mathrm{Epi}^R \xi \times \tau, \sigma] = [\tau, [\xi, \sigma]] \overset{(XVI.3.1)}{=} [\xi \times \tau, \sigma],$$

and Proposition XVII.1.2 applies to the effect that

$$\mathrm{Epi}^R \xi \times \tau \geq \mathrm{Epi}^R(\xi \times \tau).$$

(2) $\implies$ (3) follows from (XVII.1.5). (2) $\implies$ (4): Applying (2) with $|\tau|$ a singleton, we obtain that $\theta \geq \mathrm{Epi}^R \xi$ so that $\xi \geq \theta \geq \mathrm{Epi}^R \xi$. By Corollary XIV.7.9 and Proposition XVII.1.2, $[\xi, \sigma] = [\theta, \sigma]$ for every $\sigma = \mathrm{Epi}^R \sigma$. Recall that for every σ, $\xi \times [\xi, \sigma] \geq ev^- \sigma$, so that, if $\sigma = \mathrm{Epi}^R \sigma$,

$$\mathrm{Epi}^R(\xi \times [\xi, \sigma]) \geq \mathrm{Epi}^R(ev^- \sigma) \geq ev^-(\mathrm{Epi}^R \sigma) = ev^- \sigma.$$

Apply (2) with $\tau = [\xi, \sigma]$ to the effect that

$$\theta \times [\xi, \sigma] \geq \mathrm{Epi}^R(\xi \times [\xi, \sigma]) \geq ev^- \sigma.$$

Thus, $[\xi, \sigma]$ is a convergence on $C(\theta, \sigma)$ whose product with θ is finer than $ev^- \sigma$. In view of (VIII.6.1), $[\theta, \sigma]$ is the coarsest such convergence, so that $[\xi, \sigma] \geq [\theta, \sigma]$.

(3) $\implies$ (5) is shown like (2) $\implies$ (4) after noting that (3) is equivalent to $\theta \times \tau \geq R(\xi \times \tau)$ for all τ by Lemma XVII.1.4, because $\sigma = R\sigma$ for every $\sigma \in \mathbf{D}$.

(4) $\implies$ (5) is obvious. (5) $\implies$ (1) : For each $\sigma \in \mathbf{D}$, $[\xi, \sigma] = [\theta, \sigma]$, for Proposition XVI.5.9 and $\xi \geq \theta$ gives $[\theta, \sigma] \geq [\xi, \sigma]$. Thus, in view of (XVII.1.4), $\mathrm{Epi}^R \xi = \mathrm{Epi}^R \theta \leq \theta$. $\square$

Note in particular that the equivalence (1) $\Longleftrightarrow$ (4) above applied for $\theta = \mathrm{Epi}^R \xi$ yields

$$\sigma = \mathrm{Epi}^R \sigma \Longrightarrow [\mathrm{Epi}^R \xi, \sigma] = [\xi, \sigma], \qquad \text{(XVII.1.6)}$$

because $[\mathrm{Epi}^R \xi, \sigma] \geq [\xi, \sigma]$ follows from Proposition XVI.5.9.

Recall from Proposition XIV.2.10 that every functor F satisfies

$$F(\xi \times \tau) \geq F\xi \times F\tau$$

for every convergences ξ and τ. If the reverse inequality is true, we will say that F *commutes with finite products*.

Corollary XVII.1.6. *If R is a reflector, then Epi^R is a reflector that commutes with finite products.*

Proof. By Proposition XVII.1.2, Epi^R is a reflector. Moreover, by Theorem XVII.1.5,

$$\mathrm{Epi}^R \xi \times \tau \geq \mathrm{Epi}^R(\xi \times \tau)$$

for every ξ and τ. Applying Epi^R again,

$$\mathrm{Epi}^R \xi \times \mathrm{Epi}^R \tau \geq \mathrm{Epi}^R \left(\mathrm{Epi}^R \xi \times \tau \right) \geq \mathrm{Epi}^R(\xi \times \tau).$$

$\square$

On the other hand,

Proposition XVII.1.7. *If M is a functor that commutes with finite products, then for every ξ and σ,*

$$M[\xi, \sigma] \unrhd [M\xi, M\sigma].$$

Proof. Applying M to $\xi \times [\xi, \sigma] \geq e^- \sigma$, we obtain, taking (XIV.2.1) into account,

$$M\xi \times M[\xi, \sigma] \geq M(\xi \times [\xi, \sigma]) \geq M \left(e^- \sigma \right) \geq e^- \left(M\sigma \right).$$

As $C\left(\xi, \sigma \right) \subset C\left(M\xi, M\sigma \right)$, (VIII.6.1) yields $M[\xi, \sigma] \unrhd [M\xi, M\sigma]$. $\square$

Corollary XVII.1.8. *Let R be a reflector and ξ be a convergence. For every $\sigma = \mathrm{Epi}^R \sigma$, $[\xi, \sigma] = \mathrm{Epi}^R[\xi, \sigma]$.*

Proof. In view of Corollary XVII.1.6 and Proposition XVII.1.7,

$$\mathrm{Epi}^R[\xi, \sigma] \unrhd [\mathrm{Epi}^R \xi, \mathrm{Epi}^R \sigma] = [\xi, \sigma]$$

because $\sigma = \mathrm{Epi}^R \sigma$, and $[\mathrm{Epi}^R \xi, \sigma] = [\xi, \sigma]$ for every such convergence σ, by (XVII.1.6). $\square$

Remark XVII.1.9. In categorical terms, the (full) reflective subcategory **Epi**^{R} of the category of convergence spaces and continuous maps is *Cartesian-closed* because the exponential law (XVI.3.1) is satisfied in this category, as a result of Corollary XVII.1.8. We prefer to use the term *exponential* for classes of convergences that satisfy the exponential law. Moreover, if **R** is surjectively finally dense in **Epi**^{R}, **Epi**^{R} is the smallest reflective class of convergence spaces that contains **R** and has this property (of being closed for the formation of function spaces $[\xi, \sigma]$, and thus satisfying (XVI.3.1)). We say that **Epi**^{R} is the *exponential hull* ([1]) of **R**.

Indeed, if **C** is another such reflective class, it must contain every convergence of the form $[\tau, \sigma]$ where τ and σ are in **R**. If now ξ is any convergence, $[\xi, \sigma] = \text{Epi}^R[\xi, \sigma]$ for $\sigma \in \mathbf{R} \subset \mathbf{Epi}^R$. By final density, $[\xi, \sigma] = \bigwedge_{i \in I} f_i \tau_i$ for onto maps $f_i : |\tau_i| \to |[\xi, \sigma]|$, where every $\tau_i \in \mathbf{R}$. By Corollary XVI.5.20,

$$[[\xi, \sigma], \sigma] = [\bigwedge_{i \in I} f_i \tau_i, \sigma] = \bigvee_{i \in I}^{\triangleright}(f_i^\sigma)^-[\tau_i, \sigma],$$

and each $[\tau_i, \sigma]$ must be in **C**, so that $[[\xi, \sigma], \sigma]$ is in **C** because **C** is reflective. Thus **C** contains $\text{Epi}^\sigma \xi$ for every ξ and every $\sigma \in \mathbf{R}$, and thus it contains **Epi**^{R}.

XVII.2 Modified duality

This section may seem very abstract, but is motivated by its applications in Section XVII.7.

If now F is a (concrete) functor, let

$$\text{Epi}_F^\sigma \xi := i^-[F[\xi, \sigma], \sigma]$$

and if R is a reflector, let

$$\text{Epi}_F^R \xi := \bigvee_{\sigma = R\sigma} \text{Epi}_F^\sigma \xi.$$

Proposition XVII.2.1. *If F is a functor, so is Epi_F^σ, and thus, so is Epi_F^R, for each reflector R.*

Proof. By Theorem XVI.5.13, if $f \in C(\xi, \tau)$ then $f^\sigma \in C([\tau, \sigma], [\xi, \sigma])$, and, since F is a functor, $f^\sigma \in C(F[\tau, \sigma], F[\xi, \sigma])$. Applying Theorem XVI.5.13 again, $f^{\sigma\sigma} \in C([F[\xi, \sigma], \sigma], [F[\tau, \sigma], \sigma])$. Thus, the proof that Epi^σ is a functor in Theorem XVII.1.1 applies virtually unchanged to Epi_F^σ. That Epi_F^R is a functor follows from Proposition XIV.2.4. $\square$

---

[1] *Cartesian-closed hull.* in classical terms.

Let Dis denote the *discretization functor*, which associates to each convergence ξ the discrete convergence on $|\xi|$.

Lemma XVII.2.2. *Let F be a functor, and let ξ and σ be two convergences.*

(1) $x \in \lim_{\mathrm{Epi}_F^\sigma \xi} \mathcal{F}$ *if and only if for every $f \in C(\xi, \sigma)$ and every $\mathcal{G} \in \mathbb{F}(C(\xi, \sigma))$,*

$$f \in \lim_{F[\xi, \sigma]} \mathcal{G} \Longrightarrow f(x) \in \lim_\sigma \langle \mathcal{F}, \mathcal{G} \rangle.$$

(2)

$$\mathrm{Epi}_{\mathrm{Dis}}^\sigma = \mathrm{R}_\sigma.$$

(3) If R is a reflector,

$$\mathrm{Epi}_{\mathrm{Dis}}^R = R.$$

Proof. (1) is proved just like (XVII.1.2). Thus, in the case where $F = \mathrm{Dis}$, $x \in \lim_{\mathrm{Epi}_{\mathrm{Dis}}^\sigma \xi} \mathcal{F}$ if and only if for every $f \in C(\xi, \sigma)$, $f(x) \in \lim_\sigma f[\mathcal{F}]$, that is, if and only if $x \in \lim_{\mathrm{R}_\sigma \xi} \mathcal{F}$. □

Lemma XVII.2.3. *If F and G are two functors, then*

$$G \geq F \Longrightarrow \mathrm{Epi}_F^\sigma \geq \mathrm{Epi}_G^\sigma \geq \mathrm{R}_\sigma$$

and if R is a reflector

$$G \geq F \Longrightarrow \mathrm{Epi}_F^R \geq \mathrm{Epi}_G^R \geq R.$$

Similarly, if R_1, R_2 are reflectors and F is a functor, then

$$R_2 \geq R_1 \Longrightarrow \mathrm{Epi}_F^{R_2} \geq \mathrm{Epi}_F^{R_1}.$$

Proof. By Proposition XVI.5.9, $[F[\xi, \sigma], \sigma] \unrhd [G[\xi, \sigma], \sigma]$ because $G[\xi, \sigma] \geq F[\xi, \sigma]$, so that $\mathrm{Epi}_F^\sigma \geq \mathrm{Epi}_G^\sigma$. Thus, since $G \leq \mathrm{Dis}$ for every functor G, $\mathrm{Epi}_G^\sigma \geq \mathrm{Epi}_{\mathrm{Dis}}^\sigma = \mathrm{R}_\sigma$, by Lemma. The version with a reflector R is similar. If $R_2 \geq R_1$ then fix $R_1 \subset$ fix R_2 and the inequality follows. □

Theorem XVII.2.4. *Let R be a reflector and let F be a functor. Assume $\theta \geq R\xi$. The following are equivalent:*

(1) $\theta \geq \mathrm{Epi}_F^R \xi$;
(2) $\theta \times F\tau \geq R(\xi \times \tau)$ *for every τ;*
(3) $F[\xi, \sigma] \unrhd [\theta, \sigma]$ *for every $\sigma = R\sigma$.*

Note that the assumption that $\theta \geq R\xi$ ensures that $C(\theta, \sigma) \supseteq C(\xi, \sigma)$ for every $\sigma = R\sigma$. If additionally $\xi \geq \theta$, then $C(\theta, \sigma) = C(\xi, \sigma)$ for every $\sigma = R\sigma$ and $\trianglerighteq$ can be replaced by $\geq$ in (3).

Proof. (1) $\implies$ (2): Because R is a reflector, it is enough, by Corollary XIV.7.10, to show that if $\sigma = R\sigma$ and $f : |\xi \times \tau| \to |\sigma|$ is continuous, then $f : |\theta \times F\tau| \to |\sigma|$ is continuous. By (VIII.6.1), $\exp f : |\tau| \to |[\xi, \sigma]|$ is continuous, so is $\exp f : |F\tau| \to |F[\xi, \sigma]|$ because F is a functor. Hence $(\exp f)^\sigma : |[F[\xi, \sigma], \sigma]| \to |[F\tau, \sigma]|$ is continuous as well, by Theorem XVI.5.13. Since $\theta \geq \mathrm{Epi}_F^\sigma \xi$, the map $(\exp f)^\sigma \circ i : |\theta| \to |[F\tau, \sigma]|$ is continuous, by definition of Epi^F. But that is equivalent to $f : |\theta \times F\tau| \to |\sigma|$ continuous.

(2) $\implies$ (3): The evaluation map $ev : |\xi \times [\xi, \sigma]| \to |\sigma|$ is continuous so that

$$\xi \times [\xi, \sigma] \geq e^- \sigma.$$

As R is a functor,

$$R(\xi \times [\xi, \sigma]) \geq R(e^- \sigma) \geq e^-(R\sigma) = e^- \sigma.$$

By (2), with $\tau := [\xi, \sigma]$,

$$\theta \times F[\xi, \sigma] \geq e^- \sigma.$$

But $[\theta, \sigma]$ is the coarsest convergence on $C(\theta, \sigma) \supseteq C(\xi, \sigma)$ making the evaluation continuous. Thus,

$$F[\xi, \sigma] \trianglerighteq [\theta, \sigma].$$

(3) $\implies$ (1): It is enough to show that $j : |\theta| \to |[F[\xi, \sigma], \sigma]|$ defined by $j(x)(f) = f(x)$ is continuous whenever $\sigma = R\sigma$. The map $j \in C(\theta, [[\theta, \sigma], \sigma])$ is continuous by Proposition XVI.5.15. Moreover, by Proposition XVI.5.9 $F[\xi, \sigma] \trianglerighteq [\theta, \sigma]$ ensures that and that

$$[[\theta, \sigma], \sigma] \geq [F[\xi, \sigma], \sigma]$$

and the conclusion follows. $\qquad\square$

As in the case of natural duality, where $F = I$ is the identity functor, we only need consider convergences σ in an initially dense subclass:

Lemma XVII.2.5. *If F is a functor and σ a convergence*

$$\mathrm{Epi}_F^\sigma = \mathrm{Epi}_F^{R\sigma} .$$

If $\mathbf{D}$ is an initially dense subclass of $\mathbf{R}$, then

$$\mathrm{Epi}_F^R = \bigvee_{\sigma \in \mathbf{D}} \mathrm{Epi}_F^\sigma .$$

Proof. We show the first part. By definition, $\text{Epi}_F^{R_\sigma} \geq \text{Epi}_F^\sigma$. On the other hand, for every $\tau = R_\sigma \tau$, $\tau = \bigvee_{f \in C(\tau,\sigma)} f^- \sigma$ so that, in view of Corollary XVI.5.19,

$$[\text{Epi}_F^\sigma \xi, \tau] = [\text{Epi}_F^\sigma \xi, \bigvee_{f \in C(\tau,\sigma)} f^- \sigma] = \bigvee_{f \in C(\tau,\sigma)}^{\trianglerighteq} ((f)_\xi)^- [\text{Epi}_F^\sigma \xi, \sigma].$$

Moreover, in view of Theorem XVII.2.4, $[\text{Epi}_F^\sigma \xi, \sigma] \leq F[\xi, \sigma]$ so that, using (XIV.2.1),

$$[\text{Epi}_F^\sigma \xi, \tau] \leq \bigvee_{f \in C(\tau,\sigma)}^{\trianglerighteq} ((f)_\xi)^- F[\xi, \sigma]$$

$$\leq \bigvee_{f \in C(\tau,\sigma)}^{\trianglerighteq} F((f)_\xi)^- [\xi, \sigma] \leq F\left(\bigvee_{f \in C(\tau,\sigma)}^{\trianglerighteq} ((f)_\xi)^- [\xi, \sigma]\right)$$

$$\leq F[\xi, \tau].$$

Thus, by Theorem XVII.2.4 with $\text{Epi}_F^\sigma \xi = \theta$ and $R = R_\sigma$, $\text{Epi}_F^\sigma \xi \geq \text{Epi}_F^{R_\sigma} \xi$.

The second part is similar. $\qquad\square$

Corollary XVII.2.6. *Let R be a reflector and let F be a functor. Let* **D** *be an initially dense subclass of* **R**. *Assume $\theta \geq R\xi$. Then the equivalent conditions of Theorem XVII.2.4 are equivalent to*

$$F[\xi, \sigma] \trianglerighteq [\theta, \sigma] \text{ for every } \sigma \in \mathbf{D}.$$

If F is a coreflector, then Theorem XVII.2.4 can be refined as follows.

Corollary XVII.2.7. *Let R be a reflector and let F be a coreflector. Let* **D** *be an initially dense subclass of* **R**, *and let* **W** *be a finally dense subclass of* **F**. *Assume $\theta \geq R\xi$. The following are equivalent:*

(1) $\theta \geq \text{Epi}_F^R \xi$;
(2) $\theta \times F\tau \geq R(\xi \times \tau)$ for every τ;
(3) $\theta \times \tau \geq R(\xi \times \tau)$ for $\tau \geq F\tau$;
(4) $\theta \times \tau \geq R(\xi \times \tau)$ for $\tau \in \mathbf{W}$;
(5) $F[\xi, \sigma] \trianglerighteq [\theta, \sigma]$ for every $\sigma = R\sigma$;
(6) $F[\xi, \sigma] \trianglerighteq [\theta, \sigma]$ for every $\sigma \in \mathbf{D}$.

Proof. (1) $\Longleftrightarrow$ (2) $\Longleftrightarrow$ (5) $\Longleftrightarrow$ (6) is Theorem XVII.2.4 and Corollary XVII.2.6. Moreover, (2) $\Longrightarrow$ (3) $\Longrightarrow$ (4) are immediate. (4) $\Longrightarrow$ (3) follows from Lemma XVII.1.4, and (3) $\Longrightarrow$ (2) because if $\tau \in \mathbf{I}$ and F is a coreflector,

$$R(\xi \times \tau) \leq R(\xi \times F\tau) \underset{(3)}{\leq} \theta \times F\tau.$$

$\qquad\square$

A crucial observation for applications of our scheme to product theorems is the following generalization of Corollary XVII.1.6:

Theorem XVII.2.8. *If F and B is are two functors verifying*
$$F\xi \times B\tau \geq F(\xi \times \tau)$$
and R is a reflector, then
$$\mathrm{Epi}_F^R \xi \times F\tau \geq \mathrm{Epi}_B^R(\xi \times \tau)$$
for every ξ and τ.

Proof. In view of Theorem XVII.2.4, it is enough to show that $(\mathrm{Epi}_F^R \xi \times F\tau) \times B\sigma \geq R((\xi \times \tau) \times \sigma)$ for every σ, but this follows immediately from
$$\mathrm{Epi}_F^R \xi \times (F\tau \times B\sigma) \geq \mathrm{Epi}_F^R \xi \times F(\tau \times \sigma) \geq R(\xi \times (\tau \times \sigma)).$$
$\square$

In particular, if F is commutes with finite products, we can take $F = B$, and obtain
$$\mathrm{Epi}_F^R \xi \times F\tau \geq \mathrm{Epi}_F^R(\xi \times \tau),$$
for every ξ and τ. Corollary XVII.1.6 is recovered when $F = B = \mathrm{I}$ is the identity functor.

We are now in a position to fully generalize Theorem XVII.1.5:

Corollary XVII.2.9. *Let R be a reflector and let F and B be functors satisfying*
$$F\xi \times B\tau \geq F(\xi \times \tau).$$
Assume $\theta \geq R\xi$, and let $\mathbf{D}$ be an initially dense subclass of $\mathbf{R}$. The following are equivalent:

(1) $\theta \geq \mathrm{Epi}_F^R \xi$;
(2) $\theta \times F\tau \geq \mathrm{Epi}_B^R(\xi \times \tau)$ for every τ;
(3) $\theta \times F\tau \geq R(\xi \times \tau)$ for every τ;
(4) $F[\xi, \sigma] \trianglerighteq [\theta, \sigma]$ for every $\sigma \leq \mathrm{Epi}_B^R \sigma$;
(5) $F[\xi, \sigma] \trianglerighteq [\theta, \sigma]$ for every $\sigma = R\sigma$;
(6) $F[\xi, \sigma] \trianglerighteq [\theta, \sigma]$ for every $\sigma \in \mathbf{D}$.

Proof. (1) $\Longrightarrow$ (2) is Theorem XVII.2.8. We have seen that $\mathrm{Epi}_B^R \geq R$, so that (2) $\Longrightarrow$ (3). Finally (3) $\Longrightarrow$ (5) $\Longrightarrow$ (6) $\Longrightarrow$ (1) follows from Theorem XVII.2.4 and Corollary XVII.2.6. Now (2) $\Longrightarrow$ (4) corresponds to (2) $\Longrightarrow$ (3) in Theorem XVII.2.4, where R is replaced by Epi_B^R, whose proof only depends on the fact that R is a functor. Finally, (4) $\Longrightarrow$ (5) because $R \leq \mathrm{Epi}_B^R$. $\square$

In particular, if F is a coreflector that commutes with finite products, we can then take $F = B$ and combine Corollaries XVII.2.7 and XVII.2.9 to the effect that:

Corollary XVII.2.10. *Let R be a reflector and let F be a coreflector that commutes with finite products. Assume $\theta \geq R\xi$, and let* $\mathbf{D}$ *be an initially dense subclass of* $\mathbf{R}$ *and* $\mathbf{W}$ *is a finally dense subclass of* $\mathbf{F}$. *The following are equivalent:*

(1) $\theta \geq \mathrm{Epi}_F^R \xi$;
(2) $\theta \times F\tau \geq \mathrm{Epi}_F^R(\xi \times \tau)$ for every τ;
(3) $\theta \times \tau \geq R(\xi \times \tau)$ for every $\tau \in \mathbf{W}$;
(4) $F[\xi, \sigma] \trianglerighteq [\theta, \sigma]$ for every $\sigma \leq \mathrm{Epi}_F^R \sigma$;
(5) $F[\xi, \sigma] \trianglerighteq [\theta, \sigma]$ for every $\sigma = R\sigma$;
(6) $F[\xi, \sigma] \trianglerighteq [\theta, \sigma]$ for every $\sigma \in \mathbf{D}$.

Proposition XVII.2.11. *If F is a coreflector, then Epi_F^σ is a reflector, and Epi_F^R is a reflector whenever R is a reflector.*

Proof. Since $F \geq \mathrm{I}$, Lemma XVII.2.3 yields

$$\mathrm{Epi}_F^\sigma \leq \mathrm{Epi}_\mathrm{I}^\sigma = \mathrm{Epi}^\sigma \leq \mathrm{I}$$

so that Epi_F^σ is contractive. Since Epi_F^σ is a functor, we only to show that it is idempotent. Consider the equivalence between (1) and (3) in Theorem XVII.2.4, for $\theta = \mathrm{Epi}_F^\sigma(\mathrm{Epi}_F^\sigma \xi)$. We have

$$F[\xi, \sigma] \trianglerighteq F[\mathrm{Epi}_F^\sigma \xi, \sigma] \trianglerighteq [\mathrm{Epi}_F^\sigma(\mathrm{Epi}_F^\sigma \xi), \sigma],$$

so that $\mathrm{Epi}_F^\sigma(\mathrm{Epi}_F^\sigma \xi) \geq \mathrm{Epi}_F^\sigma \xi$. Since Epi_F^σ is contractive, it is idempotent. $\square$

Therefore, if F is a coreflector that commutes with finite products, we can use Epi_F^R instead of R in Corollary XVII.2.10 to the effect that:

Corollary XVII.2.12. *If F is a coreflector that commutes with finite products and R is a reflector, then*

$$\mathrm{Epi}_F^{\mathrm{Epi}_F^R} = \mathrm{Epi}_F^R.$$

In particular, when F is the identity functor, we get

$$\mathrm{Epi}^{\mathrm{Epi}^R} = \mathrm{Epi}^R.$$

XVII.3 Concrete characterizations of bidual reflectors

The applicability of modified duality depends in part on the internal characterization of instances of functors Epi_F^R. The first observation is that Corollary XV.4.5 can be seen as an instance of the equivalence $(1) \iff (2) \iff (3)$ in Corollary XVII.2.10, so that:

Theorem XVII.3.1. *Let $\mathbb{D}$ be a composable class of filters containing $\mathbb{F}_0$. Then*

$$\mathrm{Epi}_{\mathrm{B}_\mathbb{D}}^{\mathrm{S}_0} = \mathrm{A}_\mathbb{D}.$$

Proof. By Corollary XIV.4.4, $\mathrm{B}_\mathbb{D}$ is a coreflector that commutes with finite products. Thus, Corollary XVII.2.10 applies with $F = \mathrm{B}_\mathbb{D}$ and $R = \mathrm{S}_0$ to the effect that

$$\underset{\tau}{\forall} \, (\theta \times \mathrm{B}_\mathbb{D}\,\tau \geq \mathrm{S}_0(\xi \times \tau)) \iff \theta \geq \mathrm{Epi}_{\mathrm{B}_\mathbb{D}}^{\mathrm{S}_0}\,\xi$$

while Corollary XV.4.5 yields

$$\underset{\tau}{\forall} \, (\theta \times \mathrm{B}_\mathbb{D}\,\tau \geq \mathrm{S}_0(\xi \times \tau)) \iff \theta \geq \mathrm{A}_\mathbb{D}\,\xi,$$

which completes the proof. ☐

On the other hand, to describe $\mathrm{Epi}_{\mathrm{B}_\mathbb{D}}^{\mathrm{T}}$ explicitly, we need to introduce notations. If $A \subset |\xi|$, then

$$A_\bullet := \bigcup_{a \in A} \mathrm{cl}_\xi\{a\}, \qquad (\text{XVII.3.1})$$

and given a filter $\mathcal{F} \in \mathbb{F}|\xi|$,

$$\mathcal{F}_\bullet := \{F_\bullet : F \in \mathcal{F}\}^\uparrow,$$

and finally, given a class $\mathbb{D}$ of filters

$$\mathbb{D}_\bullet := \{\mathcal{D} \in \mathbb{D} : \mathcal{D} = \mathcal{D}_\bullet\}.$$

Theorem XVII.3.2. *Let $\mathbb{D}$ be an $\mathbb{F}_0$-composable class of filters. Then*

$$\lim_{\mathrm{Epi}_{\mathrm{B}_\mathbb{D}}^{\mathrm{T}}\xi} \mathcal{F} = \bigcap_{\mathbb{D}_\bullet \ni \mathcal{D} \# \mathcal{F}} \mathrm{cl}_\xi(\mathrm{adh}_\xi\,\mathcal{D}).$$

Proof. Since $\$_0$ is initially dense in topologies, $\mathrm{Epi}_{B_\mathbb{D}}^{\$_0} = \mathrm{Epi}_{B_\mathbb{D}}^T$, so that $x \in \lim_{\mathrm{Epi}_{B_\mathbb{D}}^T \xi} \mathcal{F}$ if and only if for every $\mathfrak{D} \in \mathbb{D}([\xi, \$_0])$ and $A \in \lim_{[\xi, \$_0]} \mathfrak{D}$,

$$\chi_A(x) \in \lim_{\$_0} \langle \mathcal{F}, \mathfrak{D} \rangle,$$

which is an empty condition if $\chi_A(x) = 1$. Thus, $x \in \lim_{\mathrm{Epi}_{B_\mathbb{D}}^T \xi} \mathcal{F}$ if and only if $\{0\} \in \langle \mathcal{F}, \mathfrak{D} \rangle$ for every closed subset A of ξ with $x \notin A$, and every $\mathfrak{D} \in \mathbb{D}([\xi, \$_0])$ with $A \in \lim_{[\xi, \$_0]} \mathfrak{D}$. But $\{0\} \in \langle \mathcal{F}, \mathfrak{D} \rangle$ means that $\mathcal{F}$ and the reduced filter $\mathrm{rdc}(\mathfrak{D})$ (as defined in (VIII.7.2)) do not mesh. In other words, $x \in A$ whenever $\mathrm{rdc}(\mathfrak{D}) \# \mathcal{F}$ and $\mathrm{adh}_\xi \mathrm{rdc}(\mathfrak{D}) \subset A$, that is, $x \in \mathrm{cl}_\xi(\mathrm{adh}_\xi \mathrm{rdc}(\mathfrak{D}))$ whenever $\mathrm{rdc}(\mathfrak{D}) \# \mathcal{F}$. But $\mathcal{D} \in \mathbb{D}_\bullet(|\xi|)$ if and only if there is $\mathfrak{D} \in \mathbb{D}([\xi, \$_0])$ with $\mathrm{rdc}(\mathfrak{D}) = \mathcal{D}$, which will complete the proof.

To see this, note that $\mathrm{cl}_\xi\{x\} \in |[\xi, \$_0]|$ for each $x \in |\xi|$ and thus, if $\mathcal{D} \in \mathbb{D}_\bullet(|\xi|)$, the filter

$$\mathfrak{D} := \{\{\mathrm{cl}_\xi\{x\} : x \in D\} : D \in \mathcal{D}\}^\uparrow$$

is a $\mathbb{D}$-filter on $[\xi, \$_0]$ for it is the image of $\mathcal{D}$ under the relation $\{(x, \mathrm{cl}_\xi\{x\}) : x \in |\xi|\}$, and $\mathrm{rdc}(\mathfrak{D}) = \mathcal{D}$. On the other hand, if $\mathfrak{D} \in \mathbb{D}([\xi, \$_0])$, then $\mathrm{rdc}(\mathfrak{D}) \in \mathbb{D}$ because it is the inverse image of $\mathfrak{D}$ under the relation $\{(x, A) \in |\xi| \times [\xi, \$_0]| : x \in A\}$. Moreover, $\mathrm{rdc}(\mathfrak{D}) \in \mathbb{D}_\bullet$ for $A = A_\bullet$ whenever A is a union of closed sets. $\square$

XVII.4   Epitopologies

In particular, for $\mathbb{D} = \mathbb{F}$, $B_\mathbb{D}$ is the identity functor and Theorems XVII.3.1 and XVII.3.2 particularize to

$$\mathrm{Epi}^{S_0} = A_\mathbb{F} = S$$

and

$$\lim_{\mathrm{Epi}^T \xi} \mathcal{F} = \bigcap_{\mathbb{F}_\bullet \ni \mathcal{H} \# \mathcal{F}} \mathrm{cl}_\xi(\mathrm{adh}_\xi \mathcal{F}). \tag{XVII.4.1}$$

Definition XVII.4.1. A convergence ξ is an *epitopology* if $\xi = \mathrm{Epi}^T \xi$.

The formula (XVII.4.1) above gives an explicit characterization of epitopologies. Here is a reformulation in terms of the dual closure to the $\bullet$ closure (XVII.3.1): If $A \subset |\xi|$ then

$$A^* := \{x \in |\xi| : \mathrm{cl}\{x\} \cap A \neq \varnothing\},$$

and accordingly, if $\mathcal{H}$ is a filter,

$$\mathcal{H}^* := \{H^* : H \in \mathcal{H}\}^\uparrow.$$

Of course,

$$\mathcal{H}_\bullet \# \mathcal{F} \iff \mathcal{H} \# \mathcal{F}^*, \qquad \text{(XVII.4.2)}$$

for $H_\bullet \cap F \neq \varnothing$ means that there is $x \in H$ with $\mathrm{cl}\{x\} \cap F \neq \varnothing$, that is, $x \in F^*$.

Proposition XVII.4.2. *A convergence is an epitopology if and only if it is a *-regular pseudotopology with closed limits.*

Proof. If ξ is an epitopology, then

$$\xi \leq \mathrm{Epi}^{\mathrm{T}}\,\xi \leq \mathrm{Epi}^{\mathrm{S}_0}\,\xi = \mathrm{S}\,\xi$$

by Lemma XVII.2.3, so that ξ is a pseudotopology. By (XVII.4.1), limit sets for ξ are closed. Moreover, ξ is *-regular: if $x \in \lim_\xi \mathcal{F}$ and $\mathcal{H} \in \mathbb{F}_\bullet$ with $\mathcal{H} \# \mathcal{F}^*$, then $\mathcal{H} \# \mathcal{F}$ by (XVII.4.2), so that $x \in \mathrm{cl}_\xi(\mathrm{adh}_\xi\,\mathcal{H})$ and $x \in \lim_\xi \mathcal{F}^*$ by (XVII.4.1).

If now ξ is a *-regular pseudotopology with closed limits, and $x \in \lim_{\mathrm{Epi}^{\mathrm{T}}\,\xi} \mathcal{F}$, we only need to show that $x \in \lim_\xi \mathcal{U}$ whenever $\mathcal{U} \in \beta(\mathcal{F}^*)$ to show that $x \in \lim_\xi \mathcal{F}$, because ξ is a *-regular pseudotopology. By (XVII.4.2), $\mathcal{U}_\bullet \# \mathcal{F}$ so that $x \in \mathrm{cl}_\xi(\mathrm{adh}_\xi\,\mathcal{U}_\bullet)$. Moreover,

$$\mathrm{adh}_\xi\,\mathcal{U}_\bullet = \mathrm{adh}_\xi\,\mathcal{U} = \lim_\xi \mathcal{U},$$

where the second equality follows from the fact that $\mathcal{U}$ is an ultrafilter. The first equality follows from

$$\mathcal{H} \# \mathcal{U}_\bullet \iff \mathcal{H}^* \# \mathcal{U}$$

and $\lim_\xi \mathcal{H} = \lim_\xi \mathcal{H}^*$ by *-regularity. Since limit sets are closed,

$$\mathrm{cl}_\xi(\mathrm{adh}_\xi\,\mathcal{U}_\bullet) = \mathrm{cl}_\xi(\lim_\xi \mathcal{U}) = \lim_\xi \mathcal{U},$$

and the $x \in \lim_\xi \mathcal{F}$ follows. We have shown that $\mathrm{Epi}^{\mathrm{T}}\,\xi \geq \xi$, that is, ξ is an epitopology. $\qquad\square$

XVII.5 Functionally embedded convergences

Proposition XVII.5.1. *If σ is a regular convergence and E is a coreflector, then $\mathrm{Epi}_E^\sigma\,\xi$ is $\mathrm{R}_\sigma\,\xi$-regular for every convergence ξ.*

Proof. $x \in \lim_{\mathrm{Epi}_E^\sigma\,\xi} \mathcal{F}$ if and only if $f(x) \in \lim_\sigma \langle \mathcal{F}, \mathcal{G} \rangle$ for every $f \in C(\xi, \sigma)$ and every filter $\mathcal{G}$ on $C(\xi, \sigma)$ with $f \in \lim_{E[\xi,\sigma]} \mathcal{G}$. Since σ is regular, $f(x) \in \lim_\sigma \mathrm{adh}_\sigma^\natural \langle \mathcal{F}, \mathcal{G} \rangle$. Moreover,

$$\langle \mathrm{adh}_{\mathrm{R}_\sigma\,\xi}^\natural \mathcal{F}, \mathcal{G} \rangle \geq \mathrm{adh}_\sigma^\natural \langle \mathcal{F}, \mathcal{G} \rangle, \qquad \text{(XVII.5.1)}$$

and the conclusion follows. To see (XVII.5.1), note that $x \in \mathrm{adh}_{\mathrm{R}_\sigma\,\xi}\, F$ if there is a filter $\mathcal{H}$ with $F \in \mathcal{H}$ and $f(x) \in \lim_\sigma f[\mathcal{H}]$ for every $f \in C(\xi, \sigma)$, so that in particular, $f(x) \in \mathrm{adh}_\sigma f(F)$ for every $f \in C(\xi, \sigma)$. Thus for every $F \in \mathcal{F}$ and $G \in \mathcal{G}$, $\langle \mathrm{adh}_{\mathrm{R}_\sigma\,\xi}\, F, G \rangle \subset \mathrm{adh}_\sigma \langle F, G \rangle$. $\qquad\square$

Note that $\mathrm{Epi}^{S_0} = \mathrm{Epi}^{¥}$ and $\mathrm{Epi}^{T} = \mathrm{Epi}^{\$_0}$, but neither $\$_0$ nor $¥$ is regular, so that Proposition XVII.5.1 does not apply. On the other hand, the standard topology ν of the real line is regular. We call a convergence ξ *functionally embedded* if

$$\mathrm{Epi}^\nu\, \xi = \xi.$$

Corollary XVII.5.2. *The convergence* $\mathrm{Epi}^\nu\, \xi$ *is* $R_\nu\xi$ *regular, for every convergence* ξ. *In particular, a functionally embedded convergence is regular with respect to its functionally regular modification.*

We have already observed that fixed points for $\mathrm{Epi}^{¥}$ and $\mathrm{Epi}^{\$_0}$ are pseudotopologies. In fact:

Proposition XVII.5.3. *If* $\sigma = S\sigma$ *then* $\mathrm{Epi}^\sigma \leq S$ *so that every* $\xi = \mathrm{Epi}^\sigma\, \xi$ *is pseudotopological.*

Proof. By Theorem VIII.6.2, $[[\xi, \sigma], \sigma]$ is pseudotopological, hence $j_{|\xi|}^-[[\xi, \sigma], \sigma] = \mathrm{Epi}^\sigma\, \xi$ is pseudotopological too. Thus, $\mathrm{Epi}^\sigma \leq S$. $\qquad\square$

Lemma XVII.5.4. *If* S *is a cover of a convergence* ξ *then the filter-base*

$$\mathcal{S}^{=0} := \big\{ \{ f \in C(\xi, \nu) : S \subset f^-(0) \} : S \in \S^\cup \big\}$$

converges to the zero function $\overline{0}$ *in* $[\xi, \nu]$.

Proof. $\mathcal{S}^{=0}$ is easily seen to be a filter-base because $\mathcal{S}^\cup$ is an ideal base. If $\mathcal{G} \in \mathbb{F}|\xi|$ with $x \in \lim_\xi \mathcal{G}$ then there is $S \in \S \cap \mathcal{G}$, so that there is $F \in \S^{=0}$ with $\langle S, F \rangle = \{0\}$ and thus $\overline{0} \in \lim_{[\xi, \nu]} \mathcal{S}^{=0}$. $\qquad\square$

Theorem XVII.5.5. *A Hausdorff convergence is functionally embedded if and only if it is a pseudotopology that is regular with respect to its functionally regular modification.*

Proof. In view of Corollary XVII.5.2 and Proposition XVII.5.3, a functionally embedded convergence is a pseudotopology that is regular with respect to its functionally regular modification. Assume conversely, that $\xi = S\xi$ is Hausdorff and $\mathrm{R}_\nu\, \xi$-regular. Since $\xi = S\xi \geq \mathrm{Epi}^\nu\, \xi$, it is enough to show that for an ultrafilter $\mathcal{U}$ with $x \in \lim_{\mathrm{Epi}^\nu \, \xi} \mathcal{U}$ we have $x \in \lim_\xi \mathcal{U}$. Assume to the contrary that $x \notin \lim_\xi \mathcal{U}$. Because ξ is Hausdorff and $\xi \geq \mathrm{Epi}^\nu \, \xi$, $\lim_\xi \mathcal{U} = \varnothing$. By $\mathrm{R}_\nu \, \xi$-regularity, $\mathcal{U} \not\geq \mathrm{adh}^{\natural}_{\mathrm{R}_\nu \, \xi} \mathcal{F}$ whenever $\mathcal{F}$ is ξ-convergent. Thus, for every ξ-convergent filter $\mathcal{F}$, there is a $\mathrm{R}_\nu \, \xi$-closed set $F_\mathcal{F} \in \mathcal{F}$ with $F_\mathcal{F}^c \in \mathcal{U}$. Let $\mathcal{S} := \{F_\mathcal{F} : \mathcal{F} \in \xi^-(X)\}$. In view of Lemma XVII.5.4, the associated filter-base $\mathcal{S}^{=0}$ satisfies $\overline{0} \in \lim_{[\xi, \nu]} \mathcal{S}^{=0}$. Since $x \in \lim_{\mathrm{Epi}^\nu \, \xi} \mathcal{U}$, $0 \in \lim_\nu \langle \mathcal{U}, \mathcal{S}^{=0} \rangle$, that is, for every $\varepsilon \in (0, 1)$, there is $U_\varepsilon \in \mathcal{U}$ and $G_\varepsilon \in \S^{=0}$ with $\langle U_\varepsilon, G_\varepsilon \rangle \subset (-\varepsilon, \varepsilon)$. Moreover There is $S \in \S^\cup$ such that

$$\{f \in C(\xi, \nu) : S \subset f^-(0)\} \subset G_\varepsilon.$$

By definition $S^c \in \mathcal{U}$. Pick $t \in S^c \cap U_\varepsilon$. Since S is $\mathrm{R}_\nu \, \xi$-closed, there is, in view of Corollary VII.6.9, a function $g \in C(\xi, \nu)$ with $g(t) = 1$ and $g(S) = 0$. By definition, $g \in G_\varepsilon$ and $\langle U_\varepsilon, G_\varepsilon \rangle \subset (-\varepsilon, \varepsilon)$ so that $g(t) \in (-\varepsilon, \varepsilon)$, in contradiction to $g(t) = 1$. $\qquad \Box$

XVII.6   Exponential hulls and exponential objects

In view of Remark XVII.1.9:

Corollary XVII.6.1. *The exponential hull of the category of*

(1) pretopological spaces (and continuous maps) is that of pseudotopological spaces (and continuous maps);

(2) topological spaces (and continuous maps) is that of epitopological spaces (and continuous maps);

(3) functionally regular topological spaces (and continuous maps) is that of functionally embedded spaces (and continuous maps).

In looking for the exponential hull, we focus on the smallest (reflective) class of convergence containing a given reflective class, in which (XVI.3.1) is satisfied for all triplet ξ, τ, σ. On the other hand, in a class where (XVI.3.1) is not satisfied for all triplets, it is often useful to know which objects ξ ensures that (XVI.3.1) for all τ and σ in the class. More specifically, given a reflective class $\mathbf{R}$ of convergences, the part

$$C(\xi \times \tau, \sigma) \equiv C(\tau, [\xi, \sigma])$$

of (XVI.3.1) is a statement in $\mathbf{R}$ for every $\tau = R\tau$ and $\sigma = R\sigma$ if and only if

$$\sigma = R\sigma \implies [\xi, \sigma] = R[\xi, \sigma]. \qquad (\text{XVII.6.1})$$

Definition XVII.6.2. A convergence ξ satisfying (XVII.6.1) *exponential in* **R**, or *R-dual*.

Note that in this definition, we do not insist (in contrast to the traditional categorical view) on $\xi = R\xi$. However, since by Proposition XVII.1.2 $[\xi, \sigma] = [\mathrm{Epi}^R \xi, \sigma]$ for every $\sigma = R\sigma$, *we may assume that* $\xi = \mathrm{Epi}^R \xi$ when considering R-dual convergences.

In view of Theorem XVII.2.4 with $R = F$ and $\theta = \xi$ and Corollary XVII.2.6:

Proposition XVII.6.3. *Let* **R** *be a reflective class of convergences, let* **D** *be an initially dense subclass of* **R**, *and let* $\xi = \mathrm{Epi}^R \xi$. *The following are equivalent:*

(1) ξ is R-dual;
*(2) $[\xi, \sigma] = R[\xi, \sigma]$ for all $\sigma \in$ **D***;*
(3)

$$\xi \geq \mathrm{Epi}_R^R \xi = \bigvee_{\sigma \in \mathbf{D}} \mathrm{Epi}_R^\sigma \xi;$$

(4) for every convergence τ,

$$\xi \times R\tau \geq R(\xi \times \tau);$$

(5) for every R-quotient map f, $i_\xi \times f$ is R-quotient.

Proof. The only needed clarification is the equivalence (4) $\Longleftrightarrow$ (5). Assume (4) and $f : |\tau| \to |\sigma|$ an onto map with $\sigma \geq Rf\tau$. Then

$$\xi \times \sigma \geq \xi \times Rf\tau \underset{(4)}{\geq} R(\xi \times f\tau) \underset{(IV.7.2)}{=} R(i_\xi \times f)(\xi \times \tau),$$

so that $i_\xi \times f$ is R-quotient. Conversely, assuming (5), we obtain (4) by using the identity carried R-quotient map $f = i_{|\tau|} : |\tau| \to |R\tau|$. $\qquad \square$

Proposition XVII.6.4. *Let* **R** *be a reflective class of convergences. Then the restriction* $\mathrm{Epi}_R^R : \mathbf{Epi^R} \to \mathbf{Epi^R}$ *is an expansive functor.*

Proof. By Proposition XVII.2.1, $\mathrm{Epi}_R^R : \mathbf{I} \to \mathbf{I}$ is a functor. In view of Lemma XVII.2.3, $\mathrm{Epi}_R^R \geq \mathrm{Epi}^R$ for $\mathbf{I} \geq R$, so that the restriction of Epi_R^R to $\mathbf{Epi^R}$ is expansive. It is moreover valued in $\mathbf{Epi^R}$. Indeed, $\mathrm{Epi}_R^R \xi$ is the coarsest θ such that

$$\theta \times R\tau \geq R(\xi \times \tau) \qquad\qquad (XVII.6.2)$$

for all τ, and Epi^R commutes with product by Corollary XVII.1.6, and is finer than R, so that applying Epi^R to (XVII.6.2) yields:

$$\text{Epi}^R \theta \times \text{Epi}^R R\tau \geq \text{Epi}^R R(\xi \times \tau) \iff \text{Epi}^R \theta \times R\tau \geq R(\xi \times \tau),$$

which implies in particular (for $\theta = \text{Epi}_R^R \xi$) that $\text{Epi}^R \text{Epi}_R^R \xi \geq \text{Epi}_R^R \xi$, that is, $\text{Epi}_R^R \xi \in \mathbf{Epi^R}$. $\qquad\square$

Corollary XVII.6.5. *The class*

$$\mathbf{Epi_{R-}^R} = \{\xi \in \mathbf{Epi^R} : \xi \geq \text{Epi}_R^R \xi\}$$

of R-dual convergences is a coreflective subclass of $\mathbf{Epi^R}$ *and the coreflector* D_R *is obtained by transfinite iteration of* Epi_R^R *as in Proposition XIV.1.19. Moreover,* D_R *commutes with finite products.*

Proof. We only need to show that D_R commutes with finite products. To this end, it is enough to show that a product of two R-dual convergences is R-dual, for then $\mathrm{D}_R \xi \times \mathrm{D}_R \tau$ is an R-dual convergence finer than $\xi \times \tau$, so that $\mathrm{D}_R \xi \times \mathrm{D}_R \tau \geq \mathrm{D}_R(\xi \times \tau)$. The reverse inequality is Proposition XIV.2.10.

Assume that ξ and τ are R-dual, and let $\sigma = R\sigma$. Then by (XVI.3.1)

$$[\xi \times \tau, \sigma] \equiv [\tau, [\xi, \sigma]],$$

and $[\xi, \sigma] = R[\xi, \sigma]$ because ξ is R-dual, so that $[\tau, [\xi, \sigma]] = R[\tau, [\xi, \sigma]]$ because τ is R-dual. Thus $[\xi \times \tau, \sigma] = R[\xi \times \tau, \sigma]$ and $\xi \times \tau$ is R-dual. $\qquad\square$

We have seen in Proposition XVII.2.11 that if R is a reflector and C a coreflector, then Epi_C^R is a reflector. Additionally, in view of Lemma XVII.2.2(2), every reflector R can be represented this way:

$$R = \text{Epi}_{\text{Dis}}^R.$$

We may ask if such a representation can be obtained for coreflectors other than Dis.

Proposition XVII.6.6. *Let* $\mathbf{R}$ *be a reflective class and let C be a coreflector in* $\mathbf{Epi^R}$ *that commutes with finite products. Then*

$$R = \text{Epi}_C^R \iff C \geq \mathrm{D}_R.$$

Proof. If $C \geq \mathrm{D}_R$, then by Lemma XVII.2.3,

$$\text{Epi}_{\mathrm{D}_R}^R \geq \text{Epi}_C^R \geq R.$$

Moreover, $\text{Epi}^R_{D_R} \leq R$ for

$$[D_R[\xi, \sigma], \sigma] = R[D_R[\xi, \sigma], \sigma]$$

for every $\sigma = R\sigma$, since $D_R[\xi, \sigma]$ is R-dual. Thus $R = \text{Epi}^R_C$.

Conversely, if $R = \text{Epi}^R_C$ then every $\tau = C\tau$ is R-dual. Indeed, if $\tau = C\tau$, Theorem XVII.2.8 with $F = B = C$ yields

$$\text{Epi}^R_C \xi \times \tau \geq \text{Epi}^R_C(\xi \times \tau) \iff R\xi \times \tau \geq R(\xi \times \tau)$$

for every ξ. Thus, for every $\tau = \text{Epi}^R \tau$, $C\tau$ is an R-dual convergence of **Epi^R** finer than τ. By definition of D_R, $C\tau \geq D_R \tau$. $\qquad\square$

In view of Theorem IX.4.1:

Theorem XVII.6.7. *A Hausdorff regular topology is* T*-dual if and only if it is locally compact* (^{2}).

Proof. Indeed, since $\$_0$ is initially dense in the class **T** of topologies, a convergence ξ is T-dual if and only if $[\xi, \$_0] = \text{T}[\xi, \$_0]$, by Corollary XVII.2.6, and the result follows from Theorem IX.4.1. $\qquad\square$

A full characterization of (non-necessarily topological) T-dual convergences is not available to our knowledge. However, while we will not provide proofs here to simplify the discussion, it was shown in [33] and further discussed in [58] that

$$\text{core compact} \implies \text{T-dual} \implies \text{T-core compact}, \qquad (\text{XVII.6.3})$$

where a *convergence space* is called *core compact* if whenever $x \in \lim \mathcal{F}$, there is $\mathcal{G} \leq \mathcal{F}$ with $x \in \lim \mathcal{G}$ and for every $G \in \mathcal{G}$ there is $G' \in \mathcal{G}$ such that G' is compact at G; and a convergence space is called T-*core compact* if whenever $x \in \lim \mathcal{F}$ and $U \in \mathcal{O}_X(x)$, there is $F \in \mathcal{F}$ that is compact at U.

In particular, finitely generated convergences are clearly core compact, so that $I_0 \geq D_T$ and Proposition XVII.6.6 applies to the effect that

$$\text{Epi}^T_{I_0} = \text{T}. \qquad (\text{XVII.6.4})$$

Proposition XVII.6.8. *If ξ is $R_\nu\,\xi$-regular and $[\xi, \nu]$ is pretopological, then ξ is locally compact.*

---

^{2}The assumptions of separation (Hausdorff regular) are inessential and only made to simplify formulations. A variant of local compactness called *core compactness* characterizes T-dual topologies. Characterizing T-dual general convergences is more challenging.

Proof. If $[\xi, \nu]$ is a pretopology, then for every convergent filter $\mathcal{F}$ on $|\xi|$, there is $F_0 \in \mathcal{F}$ and $V_0 \in \mathcal{V}_{[\xi,\nu]}(\overline{0})$ with

$$\langle F_0, V_0 \rangle \subset (-\tfrac{1}{2}, \tfrac{1}{2}). \qquad (XVII.6.5)$$

Moreover, F_0 can be assumed to be $R_\nu \xi$-closed, hence closed, by $R_\nu \xi$-regularity. We show that F_0 is compact. If not, as F_0 is closed, there is a filter $\mathcal{U}$ with $F_0 \in \mathcal{U}$ and $\mathrm{adh}_\xi \mathcal{U} = \varnothing$. Because ξ is $R_\nu \xi$-regular, we can assume that $\mathcal{U} = \mathcal{O}_{R_\nu \xi}(\mathcal{U})$ by Proposition VI.5.2. By Theorem III.5.12, $\mathcal{U}_c$ is a cover of ξ, and moreover $\mathcal{U}_c = \mathrm{cl}^\natural_{R_\nu \xi} \mathcal{U}_c$.

By Lemma XVII.5.4, $\mathcal{H} := (\mathcal{U}_c)^{=0}$ is a filter on $C(\xi, \nu)$ with $\overline{0} \in \lim_{[\xi,\nu]} \mathcal{H}$. Since $[\xi, \nu]$ is pretopological, $V_0 \in \mathcal{H}$, that is, there is an $R_\nu \xi$-open set $U \in \mathcal{U}$ with

$$\{f \in C(\xi, \nu) : U^c \subset f^-(0)\} \subset V_0. \qquad (XVII.6.6)$$

Let $x \in U = \mathrm{int}_{R_\nu \xi} U \subset F_0$. Since ξ is $R_\nu \xi$-regular and U^c is $R_\nu \xi$-closed, there is $h \in C(\xi, \nu)$ with $h(U^c) = \{0\}$ and $h(x) = 1$. By (XVII.6.6), $h \in V_0$, in contradiction to (XVII.6.5), for $x \in F_0$. $\qquad \square$

Theorem XVII.6.9. *Let ξ be a Hausdorff functionally regular topology. The following are equivalent:*

(1) ξ is locally compact;
(2) $[\xi, \nu] = S_0[\xi, \nu]$;
(3) $[\xi, \nu] = T[\xi, \nu]$;
(4) $[\xi, \nu] = R_\nu[\xi, \nu]$.

Proof. If ξ is a Hausdorff functionally regular locally compact topology, it is in particular T-dual by virtue of Theorem XVII.6.7, so that $[\xi, \nu]$ is topological, Hausdorff, and functionally embedded, hence functionally regular[3]. This shows (1) $\Longrightarrow$ (4). That (4) $\Longrightarrow$ (3) $\Longrightarrow$ (2) is obvious. Finally, (2) $\Longrightarrow$ (1) by Proposition XVII.6.8. $\qquad \square$

In other words,

Corollary XVII.6.10. *A Hausdorff functionally regular topology is R_ν-dual if and only if it is locally compact.*

On the other hand:

---

[3]While there are topological functionally embedded spaces that are not functionally regular [13], it is easily seen that such phenomenon does not occur for $[\xi, \nu]$.

Theorem XVII.6.11. *Let $\mathbb{D}$ be a composable class of filters containing $\mathbb{F}_0$. Then a pseudotopology τ is $A_\mathbb{D}$-dual if and only if*

$$\tau \geq S B_\mathbb{D} \tau.$$

Proof. By Theorem XVII.3.1, $A_\mathbb{D} = \mathrm{Epi}_{B_\mathbb{D}}^{S_0}$ and Theorem XVII.2.8 applies with $F = B = B_\mathbb{D}$ to the effect that

$$A_\mathbb{D}\, \xi \times B_\mathbb{D}\, \tau \geq A_\mathbb{D}(\xi \times \tau).$$

Since S commutes with product, we apply S to this inequality to the effect that

$$A_\mathbb{D}\, \xi \times S B_\mathbb{D}\, \tau \geq A_\mathbb{D}(\xi \times \tau),$$

so that if $\tau \geq S B_\mathbb{D}\, \tau$ then

$$A_\mathbb{D}\, \xi \times \tau \geq A_\mathbb{D}(\xi \times \tau)$$

for every ξ, and τ is $A_\mathbb{D}$-dual.

Conversely, if $\tau \not\geq S B_\mathbb{D}\, \tau$, then there is an ultrafilter $\mathcal{U}$ and

$$x_0 \in \lim_\tau \mathcal{U} \setminus \lim_{B_\mathbb{D}\, \tau} \mathcal{U}$$

so that $x_0 \notin \lim_\tau \mathcal{D}$ whenever $\mathcal{D} \in \mathbb{D}$ and $\mathcal{D} \# \mathcal{U}$. Since τ is a pseudotopology, for each such $\mathcal{D}$, there is an ultrafilter $\mathcal{W}_\mathcal{D}$ finer than $\mathcal{D}$ with $x_0 \notin \lim_\tau \mathcal{W}_\mathcal{D}$.

Let ξ be the prime convergence on $|\tau|$ with non-isolated point x_0 defined by

$$x \in \lim_\xi \mathcal{F} \iff \underset{\mathcal{D} \in \mathbb{D}}{\exists}\ \mathcal{D} \# \mathcal{U} : \mathcal{F} \geq \mathcal{W}_\mathcal{D} \wedge \{x_0\}.$$

Of course, $(x_0, x_0) \in \lim_{A_\mathbb{D}\, \xi \times \tau}(\mathcal{U} \times \mathcal{U})$ but $(x_0, x_0) \notin \lim_{A_\mathbb{D}(\xi \times \tau)}(\mathcal{U} \times \mathcal{U})$ (in fact, $(x_0, x_0) \notin \lim_{S_0(\xi \times \tau)}(\mathcal{U} \times \mathcal{U})$). Indeed,

$$\Delta := \{(x, x) \in |\xi \times \tau| : x \neq x_0\} \# (\mathcal{U} \times \mathcal{U})$$

but $(x_0, x_0) \notin \mathrm{adh}_{\xi \times \tau} \Delta$, for if $\Delta \in \mathcal{H}$ and $x_0 \in \lim_\xi p_{|\xi|} \mathcal{H}$ then $\mathcal{H} = \mathcal{W}_\mathcal{D}$ for some $\mathcal{D} \in \mathbb{D}$ with $\mathcal{D} \# \mathcal{U}$. But then $p_{|\tau|} \mathcal{H} = p_{|\xi|} \mathcal{H}$ does not converge to $x_0 \notin \lim_\tau \mathcal{W}_\mathcal{D}$. $\qquad\square$

Corollary XVII.6.12. *The S_0-dual pseudotopologies are exactly the finitely generated ones. The S_1-dual pseudotopologies are exactly the bisequential ones.*

Proof. By Theorem XVII.6.11 for $\mathbb{D} = \mathbb{F}_0$, S_0-dual pseudotopologies are those satisfying $\xi = S I_0 \xi$. To see that ξ is finitely generated, that is, $\xi = I_0 \xi$, note that if $x \in \lim_{S I_0 \xi} \mathcal{F}$ then every ultrafilter $\mathcal{U} \in \beta(\mathcal{F})$ contains an elements $U_\mathcal{U}$ with $x \in \lim_\xi U_\mathcal{U}$. By Proposition II.6.5, there is a finite subset $\mathbb{S}$ of $\beta(\mathcal{F})$ with $\bigcup_{\mathcal{U} \in \mathbb{S}} U_\mathcal{U} \in \mathcal{F}$. As ξ is a pseudotopologies, it has finite depth, so that

$$x \in \lim_\xi \bigwedge_{\mathcal{U} \in \mathbb{S}} U_\mathcal{U}^\uparrow = \lim_\xi \left(\bigcup_{\mathcal{U} \in \mathbb{S}} U_\mathcal{U} \right)^\uparrow,$$

and $x \in \lim_{I_0 \xi} \mathcal{F}$.

Similarly, By Theorem XVII.6.11 for $\mathbb{D} = \mathbb{F}_1$, S_1-dual pseudotopologies are those satisfying $\xi = S I_1 \xi$, that is, bisequential ones as defined in Example XIV.6.10. $\qquad\square$

XVII.7   Duality and product theorems

In this section, we will mostly focus on product theorems in subclasses of sequential spaces, as a way to illustrate the method of modified duality. Other examples can be found in [33, 77–80, 83].

Recall from Section XIV.6 that a convergence ξ is sequential if $\xi \geq T I_1 \xi$, and that its subclasses of Fréchet, strongly Fréchet, and bisequential convergences come in the form of two variants each. For instance, a convergence is Fréchet-based if it is $\mathbb{F}_0^\triangle$-based and Fréchet if $\xi \geq S_0 I_1 \xi$. A Fréchet-based convergence is Fréchet, and the two notions coincide among pretopologies. Similarly, the variants of strong Fréchetness and bisequentiality in terms of base are stronger than the variants in terms of functorial inequalities, and the two coincide among pretopologies. To summarize ([4]):

Corollary XVII.2.10 yields the following analogue of Corollary XV.4.7:

Proposition XVII.7.1. *Let R be a reflector and let $\mathbf{W}$ be a finally dense subclass of $B_\mathbb{D}$. Let $\mathbb{D}$ be a composable class of filters containing $\mathbb{F}_0$ and let $\mathbb{J}$ be a class of filters of closed under finite product such that $\mathbb{D} \subset \mathbb{J}$. The following are equivalent:*

(1) $\xi \geq \mathrm{Epi}_{B_\mathbb{D}}^R B_\mathbb{J} \xi$;

(2) $\xi \times \tau \geq \mathrm{Epi}_{B_\mathbb{D}}^R B_\mathbb{J} (\xi \times \tau)$ *for every $\mathbb{D}$-based convergence τ;*

(3) $\xi \times \tau \geq R B_\mathbb{J} (\xi \times \tau)$ *for every $\tau \in \mathbf{W}$.*

---

[4]For completeness, we add a row on productively Fréchet convergences to be defined and studied below.

$\mathbb{D}$	$\mathbb{D}$-based	inequality version
		sequential: $\xi \geq T\,I_1\,\xi$
$\mathbb{F}_0^{\triangle}$	Fréchet-based	Fréchet: $\xi \geq S_0\,I_1\,\xi$
$\mathbb{F}_1^{\triangle}$	strongly Fréchet-based	strongly Fréchet: $\xi \geq S_1\,I_1\,\xi$
$(\mathbb{F}_1^{\triangle})^{\triangle}$	productively Fréchet-based	productively Fréchet: $\xi \geq A_{\mathbb{F}_1^{\triangle}}\,I_1\,\xi$
$\mathbb{F}^{\triangle}$	bisequential-based	bisequential: $\xi \geq S\,I_1\,\xi$
$\mathbb{F}_1$	first-countable	$\xi \geq I_1\,\xi$
$\mathbb{F}_0$	finitely generated	$\xi \geq I_0\,\xi$

Table XVII.1 Subclasses of sequential convergences introduced so far. The lower row, the narrower the class of convergence. Variants on the left are stronger than those on the right, and they coincide among pretopologies.

Proof. Corollary XVII.2.10 (1) $\Longleftrightarrow$ (2) $\Longleftrightarrow$ (3) applied with $F = B_{\mathbb{D}}$, θ replaced by ξ, and ξ replaced by $B_{\mathbb{J}}\,\xi$ yields

$$\xi \geq \mathrm{Epi}_{B_{\mathbb{D}}}^{R}\,B_{\mathbb{J}}\,\xi \iff \underset{\tau}{\forall}\,\left(\xi \times B_{\mathbb{D}}\,\tau \geq \mathrm{Epi}_{B_{\mathbb{D}}}^{R}(B_{\mathbb{J}}\,\xi \times \tau)\right)$$

$$\iff \underset{\tau \in \mathbf{W}}{\forall}\,(\xi \times \tau \geq R(B_{\mathbb{J}}\,\xi \times \tau)).$$

Since $\mathbb{D} \subset \mathbb{J}$, we have $B_{\mathbb{J}} \leq B_{\mathbb{D}}$. Thus,

$$\underset{\tau}{\forall}\,\left(\xi \times B_{\mathbb{D}}\,\tau \geq \mathrm{Epi}_{B_{\mathbb{D}}}^{R}(B_{\mathbb{J}}\,\xi \times \tau)\right) \Longrightarrow \underset{\tau = B_{\mathbb{D}}\,\tau}{\forall}\,(\xi \times \tau \geq \mathrm{Epi}_{B_{\mathbb{D}}}^{R}(B_{\mathbb{J}}\,\xi \times B_{\mathbb{D}}\,\tau))$$

$$\geq \mathrm{Epi}_{B_{\mathbb{D}}}^{R}(B_{\mathbb{J}}\,\xi \times B_{\mathbb{J}}\,\tau)$$

$$\geq \mathrm{Epi}_{B_{\mathbb{D}}}^{R}\,B_{\mathbb{J}}(\xi \times \tau).$$

This proves (1) $\Longrightarrow$ (2), and (2) $\Longrightarrow$ (3) is obvious. Moreover, (3) $\Longrightarrow$ (1) follows from (3) $\Longrightarrow$ (1) in Corollary XVII.2.10 and the observation that if $\xi \times \tau \geq RB_{\mathbb{J}}(\xi \times \tau)$ for every $\tau \in \mathbf{W}$, then

$$\xi \times \tau \geq R(B_{\mathbb{J}}\,\xi \times B_{\mathbb{J}}\,\tau) \geq R(B_{\mathbb{J}}\,\xi \times \tau).$$

$\square$

In particular, if $R = S_0$ and $\mathbf{W}$ is the class of prime $\mathbb{D}$-based topologies, Corollary XV.4.7 is recovered, and thus so are its instances Corollary XV.4.9 and Corollary XV.4.10, which characterize strongly Fréchet and Fréchet convergences in terms of product.

Remark XVII.7.2. It is interesting to note that the fact that the product only need be checked for prime $\mathbb{D}$-based topologies is not the result of an ad-hoc construction like in Corollary XV.4.5 but of the structural role of prime topologies as a finally dense subclass of finitely deep convergences, which is good enough, as we can restrict ourselves to pseudotopologies in this context, for $\mathrm{Epi}^{\mathrm{S}_0} = \mathrm{S}$.

While a product of a strongly Fréchet convergence with a bisequential convergence is (strongly) Fréchet, a product of two strongly Fréchet topologies may fail to be Fréchet. To illustrate this fact, the next section provides a construction of a compact Fréchet topology whose square is not Fréchet (Theorem XVII.8.5). Moreover, a compact Fréchet convergence is strongly Fréchet, as can easily be seen directly, or as we will see as a byproduct of our study in Corollary XVII.7.15.

But modified duality is flexible enough to characterize convergences whose product with every strongly Fréchet-based convergence is Fréchet. We call such convergences *productively Fréchet*. By Corollary XV.4.9, every bisequential convergence is productively Fréchet.

Note first that the fact that the product of a strongly Fréchet convergence with a first-countable one is strongly Fréchet extends to the $\mathbb{D}$-based variants:

Lemma XVII.7.3. *A product of a strongly Fréchet filter with a countably based filter is a strongly Fréchet filter, so that,*

$$\underset{\xi,\tau}{\forall} \ \mathrm{B}_{\mathbb{F}_1^{\triangle}} \, \xi \times \mathrm{I}_1 \, \tau \geq \mathrm{B}_{\mathbb{F}_1^{\triangle}}(\xi \times \tau). \qquad (\mathrm{XVII.7.1})$$

Proof. If $\mathcal{B} \in \mathbb{F}_1(X)$, $\mathcal{F} \in \mathbb{F}_1^{\triangle}(Y)$ and $\mathcal{H} \in \mathbb{F}_1(X \times Y)$ with $\mathcal{H}\#(\mathcal{B} \times \mathcal{F})$, then $\mathcal{H}[\mathcal{B}]\#\mathcal{F}$ by (II.2.10) and $\mathcal{H}[\mathcal{B}] \in \mathbb{F}_1(Y)$ because $\mathbb{F}_1$ is composable. Thus, there is $\mathcal{L} \in \mathbb{F}_1(Y)$ such that $\mathcal{L} \geq \mathcal{H}[\mathcal{B}] \vee \mathcal{F}$ because $\mathcal{F} \in \mathbb{F}_1^{\triangle}$. As a consequence,

$$(\mathcal{B} \times \mathcal{L}) \vee \mathcal{H} \geq (\mathcal{B} \times \mathcal{F}).$$

Since $\mathcal{B} \times \mathcal{L} \in \mathbb{F}_1$, we conclude that $(\mathcal{B} \times \mathcal{F}) \in \mathbb{F}_1^{\triangle}(X \times Y)$. The formula (XVII.7.1) follows at once. $\qquad\square$

Corollary XVII.7.4. *The following are equivalent:*

(1) ξ is productively Fréchet;

(2) $\xi \geq \mathrm{A}_{\mathbb{F}_1^{\triangle}} \, \mathrm{I}_1 \, \xi$;

(3) $\mathrm{adh}_\xi \, \mathcal{H} \subset \mathrm{adh}_{\mathrm{Seq}\,\xi} \, \mathcal{H}$ for every strongly Fréchet filter $\mathcal{H}$;

(4) $\xi \times \tau$ is strongly Fréchet for every strongly Fréchet-based convergence τ;

(5) $\xi \times \tau$ is Fréchet for every prime strongly Fréchet topology τ.

Proof. In view of (XVII.7.1), we can use $R = S_0$, $F = B_{\mathbb{F}_1^\triangle}$ and $B = I_1$ in Corollary XVII.2.9 with θ replaced by ξ and ξ replaced by $I_1 \xi$ to the effect that

$$\xi \geq A_{\mathbb{F}_1^\triangle} I_1 \xi \iff \underset{\tau}{\forall} \; \xi \times B_{\mathbb{F}_1^\triangle} \tau \geq S_1(I_1 \xi \times \tau)$$
$$\iff \underset{\tau}{\forall} \; \xi \times B_{\mathbb{F}_1^\triangle} \tau \geq S_0(I_1 \xi \times \tau),$$

which, in view of Corollary XVII.2.7 with $\mathbf{W}$ the class of prime $\mathbb{F}_1^\triangle$-based topologies, rephrases as

$$\xi \geq A_{\mathbb{F}_1^\triangle} I_1 \xi \iff \underset{\tau}{\forall} \; \xi \times B_{\mathbb{F}_1^\triangle} \tau \geq S_1(I_1 \xi \times \tau) \qquad \text{(XVII.7.2)}$$
$$\iff \underset{\tau \in \mathbf{W}}{\forall} \; \xi \times \tau \geq S_0(I_1 \xi \times \tau).$$

Therefore, if $\xi \geq A_{\mathbb{F}_1^\triangle} I_1 \xi$ and $\tau \geq B_{\mathbb{F}_1^\triangle} \tau$ then

$$\xi \times \tau \geq S_1(I_1 \xi \times \tau) \geq S_1(I_1 \xi \times B_{\mathbb{F}_1^\triangle} \tau) \underset{(XVII.7.1)}{\geq} S_1 B_{\mathbb{F}_1^\triangle}(\xi \times \tau).$$

Moreover, as a strongly Fréchet based convergence is strongly Fréchet

$$B_{\mathbb{F}_1^\triangle}(\xi \times \tau) \geq S_1 I_1 (B_{\mathbb{F}_1^\triangle}(\xi \times \tau)).$$

Since $\mathbb{F}_1 \subset \mathbb{F}_1^\triangle$, $B_{\mathbb{F}_1^\triangle} \leq I_1$ and $I_1 \circ B_{\mathbb{F}_1^\triangle} = I_1$. Thus

$$\xi \times \tau \geq S_1 B_{\mathbb{F}_1^\triangle}(\xi \times \tau) \geq S_1 S_1 I_1 B_{\mathbb{F}_1^\triangle}(\xi \times \tau) = S_1 I_1(\xi \times \tau)$$

and $\xi \times \tau$ is strongly Fréchet. This shows (2) $\implies$ (4). Moreover, (4) $\iff$ (1) by definition, and (2) $\iff$ (3) by Proposition XIV.6.7. That (4) $\implies$ (5) is obvious. Remains to show (5) $\implies$ (2). This follows easily from (XVII.7.2), for if $\xi \times \tau \geq S_0 I_1(\xi \times \tau)$ for every $\tau \in \mathbf{W}$, then in particular

$$\underset{\tau \in \mathbf{W}}{\forall} \; \xi \times \tau \geq S_0(I_1 \xi \times I_1 \tau) \geq S_0(I_1 \xi \times \tau).$$

$\square$

The name *productively Fréchet* is justified by the following:

Corollary XVII.7.5. *A finite product of productively Fréchet pretopologies is productively Fréchet.*

Proof. Let ξ and τ be two productively Fréchet pretopologies. To show that $\xi \times \tau$ is too, in view of Corollary XVII.7.4, it is enough to show that its product with an arbitrary strongly Fréchet topology is Fréchet. But

$$(\xi \times \tau) \times \sigma = \xi \times (\tau \times \sigma)$$

and $\tau \times \sigma$ is a strongly Fréchet pretopology because τ is a productively Fréchet pretopology, so that $\xi \times (\tau \times \sigma)$ is also a (strongly) Fréchet pretopology because ξ is productively Fréchet. $\square$

Exercise XVII.7.6. Show that $\mathcal{F} \in \left(\mathbb{F}_1^\triangle\right)^\triangle$ (as defined via (XIV.8.1)) if and only if $\mathcal{F} \times \mathcal{G} \in \mathbb{F}_1^\triangle$ (equivalently, $\mathcal{F} \times \mathcal{G} \in \mathbb{F}_0^\triangle$) for every $\mathcal{G} \in \mathbb{F}_1^\triangle$. We call filters of $\mathbb{F}_1^{\triangle\triangle} := \left(\mathbb{F}_1^\triangle\right)^\triangle$ *productively Fréchet*. Deduce that a convergence is productively Fréchet-based if and only if its product with every strongly Fréchet-based convergence (equivalently prime topology) is strongly Fréchet-based (equivalently Fréchet-based).

On the other hand, if $R = \mathrm{T}$, $\mathbb{D} = \mathbb{F}_0$, and $\mathbb{J} = \mathbb{F}_1$, then $\mathrm{Epi}_{\mathbb{B}_\mathbb{D}}^\mathrm{T} = \mathrm{T}$ by (XVII.6.4), so that Proposition XVII.7.1 yields:

Corollary XVII.7.7. *A convergence is sequential if and only if its product with every finitely generated convergence (equivalently, finitely generated prime topology) is sequential.*

Definition XVII.7.8. A convergence is called *strongly sequential* if

$$\xi \geq \mathrm{Epi}_{\mathrm{I}_1}^\mathrm{T} \, \mathrm{I}_1 \, \xi.$$

Since $\mathrm{S}_0 \geq \mathrm{T}$, $\mathrm{Epi}_F^{\mathrm{S}_0} \geq \mathrm{Epi}_F^\mathrm{T}$ for any functor F. In particular $\mathrm{Epi}_{\mathrm{I}_1}^{\mathrm{S}_0} = \mathrm{S}_1 \geq \mathrm{Epi}_{\mathrm{I}_1}^\mathrm{T}$. Hence,

Proposition XVII.7.9. *Every strongly Fréchet convergence is strongly sequential.*

In view of Theorem XVII.3.2 and Proposition XIV.6.7, we have an internal characterization of strongly sequential convergences:

Proposition XVII.7.10. $\xi = \mathrm{Epi}_{\mathrm{I}_1}^\mathrm{T} \, \xi$ *is strongly sequential if and only if*

$$\mathrm{adh}_\xi \, \mathcal{H} \subset \mathrm{cl}_{\mathrm{Seq}\,\xi}(\mathrm{adh}_{\mathrm{Seq}\,\xi} \, \mathcal{H})$$

for every countably based filter $\mathcal{H} \in (\mathbb{F}_1)_\bullet$.

We now consider a simpler reformulation in the context of regular Hausdorff sequential topologies. We call a convergence a *Tanaka* convergence if

$$\mathcal{H} \in \mathbb{F}_1, \ \mathrm{adh}_\xi \, \mathcal{H} \neq \varnothing \implies \mathrm{adh}_{\mathrm{Seq}\,\xi} \, \mathcal{H} \neq \varnothing. \qquad (\mathrm{XVII.7.3})$$

Of course, in view of Proposition XVII.7.10, a strongly sequential convergence is a Tanaka convergence. Moreover,

Proposition XVII.7.11. *A regular Hausdorff sequential topology is strongly sequential if and only if it is a Tanaka convergence.*

Proof. The convergence is in particular T_1 so $(\mathbb{F}_1)_\bullet = \mathbb{F}_1$. Let $\mathcal{H} \in \mathbb{F}_1$ with a decreasing countable base $(H_n)_n$, and let $x \in \mathrm{adh}_\xi \mathcal{H}$. Because ξ is topological, $\mathrm{adh}_\xi \mathcal{H} = \bigcap_{n \in \mathbb{N}} \mathrm{cl}_\xi H_n$, and thus

$$x \in \bigcap_{n \in \mathbb{N}} \mathrm{cl}_\xi(H_n \cap W)$$

for each closed neighborhood W of x. Since ξ is a Tanaka convergence, there is a sequence $(x_n^W)_n \# (\mathcal{H} \vee W)$ with $\lim(x_n^W)_n \cap W \neq \varnothing$, for W is closed. Since ξ is regular, $\mathcal{N}_\xi(x) = \mathrm{cl}_\xi^\natural \mathcal{N}_\xi(x)$ so that $\bigcup_{W = \mathrm{cl}\, W \in \mathcal{N}_\xi(x)} \lim(x_n^W)_n \# \mathcal{N}_\xi(x)$ and

$$x \in \mathrm{cl}_\xi(\mathrm{adh}_{\mathrm{Seq}\,\xi} \mathcal{H}) = \mathrm{cl}_{\mathrm{Seq}\,\xi}(\mathrm{adh}_{\mathrm{Seq}\,\xi} \mathcal{H})$$

by sequentiality of ξ. $\square$

Corollary XVII.7.12. *A regular Hausdorff Fréchet topology is strongly sequential if and only if it is strongly Fréchet.*

Exercise XVII.7.13. Prove Corollary XVII.7.12.

Corollary XVII.7.14. *A regular Hausdorff locally countably compact sequential topology is strongly sequential.*

Proof. By Proposition XVII.7.11, it is enough to show that the topology is a Tanaka topology. If $\mathcal{H} \in \mathbb{F}_1$ and $\mathrm{adh}_\xi \mathcal{H} \neq \varnothing$, there is x with $\mathcal{N}_\xi(x) \# \mathcal{H}$ so that, by local compactness, there is a countably compact set K with $K \# \mathcal{H}$. Let $(H_n)_{n \in \mathbb{N}}$ be a decreasing filter-base for $\mathcal{H}$. Pick $x_n \in H_n \cap K$ for each n. In view of Proposition IX.8.27, there is a convergent subsequence of $(x_n)_n$ so that $\mathrm{adh}_{\mathrm{Seq}\,\xi} \mathcal{H} \neq \varnothing$. $\square$

In view of Corollary XVII.7.12,

Corollary XVII.7.15. *A regular Hausdorff locally countably compact Fréchet topology is strongly Fréchet.*

If $R = \mathrm{T}$ and $\mathbb{D} = \mathbb{J} = \mathbb{F}_1$, Proposition XVII.7.1 gives:

Corollary XVII.7.16. *A convergence is strongly sequential if and only if its product with every first-countable convergence (equivalently, every prime first-countable topology) is strongly sequential (equivalently, sequential).*

To summarize instances of Corollary XVII.2.9 in this context:

ξ is	iff for all τ	$\xi \times \tau$ is
Fréchet	finitely generated	Fréchet
$\xi \geq S_0 I_1 \xi$	finitely generated prime topology	Fréchet
strongly Fréchet	bisequential	strongly Fréchet
$\xi \geq S_1 I_1 \xi$	metrizable prime topology	Fréchet
productively Fréchet	strongly Fréchet-based	strongly Fréchet
$\xi \geq A_{F_1^\triangle} I_1 \xi$	strongly Fréchet prime topology	Fréchet
sequential	finitely generated	sequential
$\xi \geq T I_1 \xi$	finitely generated prime topology	sequential
strongly sequential	first-countable	strongly sequential
$\xi \geq \mathrm{Epi}_{I_1}^T I_1 \xi$	metrizable prime topology	sequential

In view of Proposition XVII.6.3 and Theorem XVII.6.7:

Theorem XVII.7.17. *A product of a Hausdorff regular locally compact sequential topology and a sequential convergence is sequential* ([5]).

Proof. Let $\xi = T I_1 \xi$ be a Hausdorff regular and locally compact, and let $\tau \geq T I_1 \tau$. Then

$$T I_1 (\xi \times \tau) = T(I_1 \xi \times I_1 \tau)$$

because I_1 commutes with finite products by Corollary XIV.4.4. Moreover,

$$T(I_1 \xi \times I_1 \tau) \leq \mathrm{Epi}_{I_1}^T I_1 \xi \times I_1 \tau$$

and ξ is strongly sequential by Corollary XVII.7.14. Thus

$$T I_1 (\xi \times \tau) \leq \xi \times I_1 \tau$$

and applying the topologizer T on both sides

$$T I_1 (\xi \times \tau) \leq T(\xi \times I_1 \tau) \leq \xi \times T I_1 \tau \leq \xi \times \tau$$

because ξ is T-dual. Thus $\xi \times \tau$ is sequential. $\qquad\square$

---

[5] We could refine this result with a little more effort: a Hausdorff regular topology is locally countably compact if and only if its product with every sequential convergence (equivalently, topology) is sequential, and this can also be generalized to a version without separation. See [33].

XVII.8   Non-Fréchet product of two Fréchet compact topologies

We know already from Theorem II.8.4 that if $\mathcal{E}_n$ is a free sequential filter such that $\mathcal{E}_n \leq \mathcal{E}_{n+1}$ for each $n \in \mathbb{N}$, and if $\bigvee_{n \in \mathbb{N}} \mathcal{E}_n$ is a sequential filter, then there is n_0 such that $\mathcal{E}_{n_0} = \mathcal{E}_n$ for every $n \geq n_0$.

If $\{\mathcal{E}_n\}_n$ is a sequence of sequential filters such that $\mathcal{E}_n \leq \mathcal{E}_{n+1}$ for each n, then $\bigvee_{n < \omega} \mathcal{E}_n$ is countably based, and thus there exists a sequential filter $\mathcal{E} \geq \bigvee_{n < \omega} \mathcal{E}_n$. The following theorem improves this fact.

Theorem XVII.8.1 (Dočkálková). *Let $\mathcal{A}$ be a MAD family on an infinite set and let $\{E_n\}_n$ be a sequence of countable infinite sets such that $E_{n+1} \setminus E_n$ is finite for every n, and*

$$\{A \in \mathcal{A} : \operatorname{card}(A \cap E_n) = \aleph_0\} \qquad (\text{XVII.8.1})$$

is infinite for each n. Then there is an infinite set H such that $H \setminus E_n$ is finite for each n, and

$$\{A \in \mathcal{A} : \operatorname{card}(A \cap H) = \aleph_0\} \qquad (\text{XVII.8.2})$$

is infinite.

Proof. Let $\{x_n^0\}_n$ be a sequence of distinct terms such that $x_n^0 \in E_n$ for each n. As $\mathcal{A}$ is *MAD*, there exists $A_1 \in \mathcal{A}$ such that $H_0 := \{x_n^0 : n \in \mathbb{N}\}$ has infinitely many elements in common with A_1. We shall construct by induction a sequence $\{A_k\}_k$ of distinct elements of $\mathcal{A}$, and $H_k := \{x_n^k : n \geq k\}$ such that for each n and k,

$$x_n^k \in E_n \setminus \bigcup_{1 \leq j \leq k} A_j,$$

$$\operatorname{card}(H_k \cap A_{k+1}) = \aleph_0.$$

The formula above is the k-th step of the induction. We have already proved it for $k = 0$. Suppose that $l > 0$ and that the claim holds for $k < l$. By assumption, the set (XVII.8.1) is infinite, hence there is a sequence $\{x_n^l\}_n$ of distinct terms such that $x_n^l \in E_n \setminus \bigcup_{1 \leq j \leq l} A_j$ for each n and l. As H_l is infinite, there exists $A_{l+1} \in \mathcal{A}$ such that $H_l \cap A_{l+1}$ is infinite, and since $H_l \cap \bigcup_{1 \leq j \leq l} A_j = \varnothing$, the set A_{l+1} is distinct from $A_1, \ldots, A_l$. Therefore, the set $H := \bigcup_{k \in \mathbb{N}} H_k$ fulfills (XVII.8.2) and $H \setminus E_n$ is finite for every n. $\qquad\square$

Recall that if $\mathcal{A}$ is a family of infinite subsets of X and $B \subset X$ then

$$\mathcal{A} \vee_\infty B := \{A \cap B : A \in \mathcal{A}, \operatorname{card}(A \cap B) \geq \aleph_0\}.$$

Theorem XVII.8.2 (Simon). *There exists a MAD family $\mathcal{A}$ on a countably infinite set and two disjoint subfamilies $\mathcal{A}_0$ and $\mathcal{A}_1$ of $\mathcal{A}$ such that if $\mathcal{A}_k \vee_\infty B$ is maximal then $\mathcal{A}_k \vee_\infty B$ is finite, for each infinite set B and for $k = 0, 1$.*

Proof. Let X be a countably infinite set. Suppose that the claim is not true and let $\mathcal{A}$ be a *MAD* family on X such that $\operatorname{card}(\mathcal{A}) = c$. Let us order $\mathcal{A}$ by $2^{\mathbb{N}}$ as follows

$$\mathcal{A} = \{A_D : D \subset \mathbb{N}\}.$$

Let $\mathcal{A}_{n,0} = \{A_D : n \notin D\}$ and $\mathcal{A}_{n,1} = \{A_D : n \in D\}$. Then $\mathcal{A} = \mathcal{A}_{n,0} \cup \mathcal{A}_{n,1}$ and $\mathcal{A}_{n,0} \cap \mathcal{A}_{n,1} = \varnothing$. By assumption, there exist $k_0 \in \{0,1\}$ and an infinite set E_0 such that $\mathcal{A}_{0,k_0} \vee_\infty E_0$ is infinite and maximal on E_0. Suppose that we have already found $k_0, k_1, \ldots, k_p \in \{0,1\}$ and infinite sets $E_0 \supset E_1 \supset \ldots \supset E_p$ such that $\mathcal{A}_{n,k_n} \vee_\infty E_n$ is an infinite *MAD* family on E_n for every $0 \le n \le p$. Then by assumption, there exist $k_{p+1} \in \{0,1\}$ and an infinite subset E_{p+1} of E_p such that $\mathcal{A}_{p+1,k_{p+1}} \vee_\infty E_{p+1}$ is infinite and maximal on E_{p+1}. It follows in particular that $\{A \in \mathcal{A} : \operatorname{card}(A \cap E_p) = \infty\}$ is infinite for every $p < \omega$, hence by Theorem XVII.8.1, there exists an infinite set H such that $H \backslash E_p$ is finite for each p and $\{A \in \mathcal{A} : \operatorname{card}(A \cap H) = \infty\}$ is infinite. As $\mathcal{A} = \{A_D : D \subset N\}$, there exists $D \ne \{n : k_n = 1\}$ such that $A_D \cap H$ is infinite, and since $H \setminus E_p$ is finite, $A_D \cap E_p$ is infinite for every p. If $p \notin D \cap \{n : k_n = 1\}$, then $A_D \notin \mathcal{A}_{p,k_p}$, and because $\mathcal{A}_{p,k_p}$ is *MAD* on E_p, there exists $A \in \mathcal{A}_{p,k_p}$ such that $A_D \cap A \cap E_p$ is infinite. But A and A_D are distinct elements of the *MAD* family $\mathcal{A}$, hence $A \cap A_D$ is finite: a contradiction. $\qquad\square$

Recall that a filter $\mathcal{F}$ is Fréchet if and only if

$$\mathcal{F} \supset \bigcap_{E \ni \mathcal{E} \ge \mathcal{F}} \mathcal{E}. \qquad\qquad \text{(XVII.8.3)}$$

Theorem XVII.8.3. *A filter $\mathcal{F}$ on a countable set N is Fréchet if and only if*

$$\beta^*(\mathcal{F}) \subset \operatorname{cl}_\beta \operatorname{int}_\beta \beta^*(\mathcal{F}).$$

Proof. By applying the Stone transform to (XVII.8.3), we obtain

$$\beta^*(\mathcal{F}) \subset \operatorname{cl}_\beta (\bigcup_{E \ni \mathcal{E} \ge \mathcal{F}} \beta^*(\mathcal{E})).$$

On the other hand,

$$\operatorname{int}_\beta \beta^*(\mathcal{F}) = \bigcup_{E \ni \mathcal{E} \ge \mathcal{F}} \beta^*(\mathcal{E}),$$

because $\mathcal{U} \in \text{int}_\beta \, \beta^*(\mathcal{F})$ if and only if there exists $E \subset N$ such that $\mathcal{U} \in \beta^*(E) \subset \beta^*(\mathcal{F})$. Now $\beta^*(E) = \beta^*((E)_0)$ for the cofinite filter $(E)_0$ of E, which is sequential by Proposition II.2.28. $\qquad\square$

In view of Proposition IX.6.11, it follows that if $\mathcal{A}$ is a *MAD* family of subsets of N, then its residual filter $\mathcal{F}_\mathcal{A}$ is not Fréchet. On the other hand,

Proposition XVII.8.4. *If $\mathcal{A}$ is an* AD *family on N such that for every infinite subset B of N, if $\mathcal{A} \vee_\infty B$ is maximal then it is finite, then the residual filter $\mathcal{F}_\mathcal{A}$ of $\mathcal{A}$ is Fréchet.*

Proof. Let $B \in \mathcal{F}_\mathcal{A}^\#$, that is, $\beta(B) \cap \beta(\mathcal{F}_\mathcal{A}) \neq \varnothing$. If there is no sequential filter finer than $\mathcal{F}_\mathcal{A} \vee B$, then $\beta(B) \cap \text{int}\,\beta(\mathcal{F}_\mathcal{A}) = \text{int}(\beta(B) \cap \beta(\mathcal{F}_\mathcal{A})) = \varnothing$, hence $\mathcal{A} \vee_\infty B$ is maximal, and by the assumption, finite. Consequently, $\beta(B) \cap \beta(\mathcal{F}_\mathcal{A}) = \varnothing$, contrary to the assumption. $\qquad\square$

Theorem XVII.8.5 (Simon). [101] *There exists a Hausdorff compact Fréchet topology such that its square is not Fréchet.*

Proof. By Lemma XVII.8.2, there exists a *MAD* family $\mathcal{A}$ on N and its two disjoint subfamilies $\mathcal{A}_0$ and $\mathcal{A}_1$, such that if $\mathcal{A}_k \vee_\infty B$ is maximal then $\mathcal{A}_k \vee_\infty B$ is finite, for each infinite set B and for $k = 0, 1$. By Proposition XVII.8.4, the residual filters $\mathcal{F}_k := \mathcal{F}_{\mathcal{A}_k}$ are Fréchet. Thus the corresponding AD compact topologies $\overline{\omega\zeta_k}$ (as in Example XIII.2.10) are Hausdorff, compact, and Fréchet because $\mathcal{N}_{\overline{\omega\zeta_k}}(\infty_{Y_k}) \vee N = \mathcal{F}_k$ and $\mathcal{N}_{\overline{\omega\zeta_k}}(\infty_{Y_k}) \vee \mathcal{A}_k = (\mathcal{A}_k)_0$.

We shall see that the product $\overline{\omega\zeta_0} \times \overline{\omega\zeta_1}$ is not Fréchet. Its restriction to $\{(\infty_0, \infty_1)\} \cup \{(n, n) : n \in \mathbb{N}\}$ is a prime topology in which the trace $\mathcal{F}$ of the neighborhood filter of (∞_0, ∞_1) on $\{(n, n) : n \in \mathbb{N}\}$ is $(\mathcal{F}_0 \times \mathcal{F}_1) \vee \{(n, n) : n \in \mathbb{N}\}$. In other words, $F \in \mathcal{F}$ whenever there exist $F_0 \in \mathcal{F}_0$ and $F_1 \in \mathcal{F}_1$ such that

$$F \supset (F_0 \times F_1) \cap \{(n, n) : n \in \mathbb{N}\} = \{(n, n) : n \in F_0 \cap F_1\}.$$

Therefore $\mathcal{F} \approx \mathcal{F}_0 \vee \mathcal{F}_1$. On the other hand, we have the equalities of disjoint unions $\mathbb{U}^*(N) = \bigcup_{A \in \mathcal{A}_0} \beta^*(A) \cup \beta^*(\mathcal{F}_0) = \bigcup_{A \in \mathcal{A}_1} \beta^*(A) \cup \beta^*(\mathcal{F}_1)$. As a result, $\beta^*(\mathcal{F}_0) \cap \beta^*(\mathcal{F}_1) = \mathbb{U}(N) \setminus \bigcup_{A \in \mathcal{A}} \beta^*(A) = \beta^*(\mathcal{F}_\mathcal{A})$, that is, $\mathcal{F}_\mathcal{A} = \mathcal{F}_0 \vee \mathcal{F}_1$. The filter $\mathcal{F}_\mathcal{A}$ is not Fréchet, as the residual filter of a *MAD* family, and so $\overline{\omega\zeta_0} \times \overline{\omega\zeta_1}$ is not Fréchet, because it admits a non-Fréchet subspace. To conclude the proof consider the simple sum $\zeta = \overline{\omega\zeta_0} \oplus \overline{\omega\zeta_1}$. As its components are Fréchet, it is Fréchet. The square ζ^2 has $\overline{\omega\zeta_0} \times \overline{\omega\zeta_1}$ as a subspace, and thus is not Fréchet. $\qquad\square$

XVII.9 Spaces of real-valued continuous functions

XVII.9.1 *Cauchy completeness*

Recall from Section VII.4 that binary operations are lifted pointwise from $\mathbb{R}$ to $\mathbb{R}^X$. That induces a natural notion of Cauchy filters (or fundamental filters) on $\mathbb{R}^X$. Given subsets F and G of $\mathbb{R}^X$, let

$$F - G := \{f - g : f \in F, g \in G\},$$

and given two filters $\mathcal{F}$ and $\mathcal{G}$ on $\mathbb{R}^X$, let

$$\mathcal{F} - \mathcal{G} := \{F - G : F \in \mathcal{F}, G \in \mathcal{G}\}^{\uparrow}.$$

Definition XVII.9.1. Let τ be a (pre)convergence on Y^X. A filter $\mathcal{F}$ on $\mathbb{R}^X$ is τ-*Cauchy* if

$$\overline{0} \in \lim_{\tau}(\mathcal{F} - \mathcal{F}).$$

We will say that τ is *Cauchy complete* if every Cauchy filter converges.

By definition, if a filter is τ-Cauchy and $\tau \geq \theta$, then it is also θ-Cauchy.

Lemma XVII.9.2. *If ξ is a convergence, $\tau \geq p(\xi, \sigma)$ and $\mathcal{F}$ is a τ-Cauchy filter on $C(\xi, \nu)$, then $\mathcal{F}$ converges pointwise in $\mathbb{R}^{|\xi|}$.*

Proof. Under these assumptions, $0 \in \lim_{\nu}\langle x, \mathcal{F} - \mathcal{F}\rangle$ for every $x \in |\xi|$, so that $\langle x, \mathcal{F}\rangle$ is a Cauchy filter in ν (in the metric sense), and is thus convergent. □

Corollary XVII.9.3. $p(X, \nu)$ *is Cauchy-complete.*

In contrast, $p(\nu, \nu)$ is not Cauchy-complete. Indeed, the sequential filter associated with the sequence $\{f_n\}_n$ of Example XVI.4.4 is Cauchy but not convergent in $p(\nu, \nu)$.

On the other hand, the natural convergence is always Cauchy complete:

Theorem XVII.9.4. *If ξ is a convergence then $[\xi, \nu]$ is Cauchy complete.*

Proof. If $\mathcal{F}$ is a $[\xi, \nu]$-Cauchy filter, then $\mathcal{F}$ converges pointwise to a function $f \in \mathbb{R}^{|\xi|}$ by Lemma XVII.9.2. Moreover, $\mathcal{F}$ is equicontinuous (Exercise XVII.9.5), hence evenly continuous by Proposition XVI.6.15. In view of Corollary XVI.6.17, $f \in \lim_{\nu^{\xi}} \mathcal{F}$, and, since ν is regular, $f \in \lim_{[\xi, \nu]} \mathcal{F}$ by Theorem XVI.5.2. □

Exercise XVII.9.5. Show that $\mathcal{F}$ in the proof of Theorem XVII.9.4 is equicontinuous.

Remark XVII.9.6. The notion of Cauchy completeness that we have introduced for a (pre)convergence on $|\xi|^{|\nu|}$ or on $C(\xi, \nu)$ can be defined more generally on a *convergence group*, that is, a group endowed with a convergence making the group operation and inversion continuous. If the convergence of a convergence group is a topology, then this becomes a *topological group*. In this context, Cauchy complete topological groups are often called *Raikov complete*. We do not study convergence and topological groups in this book. Let us only mention that a countably complete Hausdorff topological group is Raikov complete (e.g., [2, Theorem 4.3.7]), but not conversely, and that the two notions of completeness coincide among metrizable topological groups. In fact, a Cauchy complete pretopological group of countable character is also countably complete. Indeed, a pretopological convergence group is a topological group [40, Theorem 5], and a first-countable topological group is metrizable (e.g., [2, Theorem 3.3.12]).

XVII.9.2 *Completeness number*

Since the pointwise convergence $p(X, \mathbb{R})$ is homeomorphic with the product convergence $\prod_{x \in X} \mathbb{R}$, it is evident that the results of Section XI.8 established for product convergences still hold for $p(X, \mathbb{R})$. In particular,

Corollary XVII.9.7. $p(X, \mathbb{R})$ *is* $\max(\mathrm{compl}(\mathbb{R}), \mathrm{card}\,X)$*-complete.*

To relate completeness of $p(\xi, \mathbb{R})$ and of $\mathbb{R}$ is more difficult, because $p(\xi, \mathbb{R})$ is the restriction of $p(|\xi|, \mathbb{R})$ to $C(\xi, \mathbb{R})$, which, as we have seen, is generally not closed in $\mathbb{R}^X$.

Lemma XVII.9.8. *If ξ is Hausdorff and functionally regular topology, then $C(\xi, \mathbb{R})$ is dense in $\mathbb{R}^X$.*

Proof. To prove that $C(\xi, \mathbb{R})$ is dense for the pointwise convergence of $\mathbb{R}^\xi$, it is enough to show that for each finite subset F of $|\xi|$ and every $f \in |\mathbb{R}|^{|\xi|}$, there exists $f_0 \in C(\xi, \mathbb{R})$ such that f_0 and f coincide on F. If $F = \{x_0, x_1\}$ then, by the functional regularity of ξ, there is $g \in C(\xi, \mathbb{R})$ such that $g(x_0) = 0$ and $g(x_1) = 1$. Hence

$$f_0(x) := (f(x_1) - f(x_0))\, g(x) + f(x_0)$$

coincides with f on F. By induction on the cardinality of F, if there is $f_0 \in C(\xi, \mathbb{R})$ that coincides with f on F and if $x_F \notin F$, then there is $h \in C(\xi, \mathbb{R})$

such that $h(F) = \{0\}$ and $h(x_F) = 1$. Then $f_1(x) := f_0(x) + f(x_F)h(x)$ is continuous and coincides with f on $F \cup \{x_F\}$. $\square$

Proposition XVII.9.9. *If ξ is Hausdorff and functionally regular topology, and $p(\xi, \mathbb{R})$ is λ-complete, then ξ is a discrete space of cardinality not greater than λ.*

Proof. Under the assumptions, $C(\xi, \mathbb{R})$ is dense in $|\mathbb{R}|^{|\xi|}$ and $\prod_{x \in |\xi|} \mathbb{R}$ is weakly diagonal, so that $C(\xi, \mathbb{R})$ is a G_λ-subset of $|\mathbb{R}|^{|\xi|}$.

We shall show that $C(\xi, \mathbb{R}) = \prod_{x \in |\xi|} \mathbb{R}$. Let $g \in \prod_{x \in |\xi|} \mathbb{R}$. Define

$$T_g(f) := f + g$$

for each $f \in \prod_{x \in |\xi|} \mathbb{R}$. Then T_g is a homeomorphism, thus $T_g(C(\xi, \mathbb{R}))$ is a dense $G_{\aleph_0}$ in a Baire space $\prod_{x \in |\xi|} \mathbb{R}$ and thus $T_g(C(\xi, \mathbb{R})) \cap C(\xi, \mathbb{R})$ is dense. In particular, there exists $f \in C(\xi, \mathbb{R})$ such that $T_g(f) \in T_g(C(\xi, \mathbb{R})) \cap C(\xi, \mathbb{R})$, hence $g = T_g(f) - f \in C(\xi, \mathbb{R})$. $\square$

Corollary XVII.9.10. *If ξ is a Hausdorff functionally regular topology, then $p(\xi, \nu)$ is countably complete if and only if ξ is a countable discrete space.*

In particular, if ξ is discrete and uncountable then $p(\xi, \nu) = [\xi, \nu]$ is not countably complete.

XVII.9.3 *Character and weight*

Recall (from Definition XIV.8.5) that, given two classes $\mathbb{D}$ and $\mathbb{J}$ of filters, a convergence ξ is cover-$(\mathbb{D}/\mathbb{J})$-compact if for every $\mathcal{D} \in \mathbb{D}$,

$$\left(\bigvee_{\mathcal{J} \in \mathbb{J}} \mathcal{J} \leq \mathcal{D} \implies \text{adh}_\xi \mathcal{J} \neq \varnothing \right) \implies \text{adh}_\xi \mathcal{D} \neq \varnothing.$$

By Theorem III.5.12, ξ is cover-$(\mathbb{D}/\mathbb{J})$-compact if and only if for every cover $\mathcal{S} \in \mathbb{D}_\neg$ of ξ (which is then an ideal cover) there is a subfamily $\mathcal{A} \subset \mathcal{S}$ with $\mathcal{A} \in \mathbb{J}_\neg$ and $\mathcal{A}$ is a cover of ξ.

Exercise XVII.9.11. Show that the following are equivalent:

(1) ξ is cover-$(\mathbb{F}/\mathbb{F}_\kappa)$-compact;
(2) for every cover $\mathcal{S}$ of ξ there is a subfamily $\mathcal{A} \subset \mathcal{S}$ of cardinality less than κ such that $\mathcal{A}^\cup$ is a cover of ξ;
(3) for every ideal cover $\mathcal{S}$ of ξ there is a subfamily $\mathcal{A} \subset \mathcal{S}$ of cardinality less than κ such that $\mathcal{A}$ is a cover of ξ.

Definition XVII.9.12. The least cardinal κ such that a convergence ξ is cover-$(\mathbb{F}/\mathbb{F}_\kappa)$-compact is called *cover-Lindelöf number of* ξ and is denoted $\mathfrak{l}(\xi)$.

Theorem XVII.9.13. [39] *If ξ is a functionally embedded convergence, then*

$$\chi([\xi,\nu]) = \mathfrak{l}(\xi).$$

Proof. Since translations T_g are homeomorphisms of $[\xi,\nu]$, this convergence is homogeneous and it is enough to show that $\chi([\xi,\nu],\overline{0}) = \mathfrak{l}(\xi)$.

Assume that $\mathfrak{l}(\xi) = \kappa$ and let $\mathcal{F}$ be a filter on $C(\xi,\nu)$ with $\overline{0} \in \lim_{[\xi,\nu]} \mathcal{F}$. For every convergent filter $\mathcal{G}$ on $|\xi|$, and every $n \in \mathbb{N}$, there is $F_{n,\mathcal{G}} \in \mathcal{F}$ and $G_{n,\mathcal{G}} \in \mathcal{G}$ with

$$\langle G_{n,\mathcal{G}}, F_{n,\mathcal{G}} \rangle \subset \left(-\tfrac{1}{2^n}, \tfrac{1}{2^n}\right). \qquad \text{(XVII.9.1)}$$

Then, for each $n \in \mathbb{N}$, the family $\mathcal{S}_n := \{G_{n,\mathcal{G}} : \lim_\xi \mathcal{G} \neq \varnothing\}$ is a cover of ξ, and thus, in view of Exercise XVII.9.11, has a subfamily $\mathcal{C}_n$ of cardinality less than κ such that $\mathcal{C}_n^\cup$ is a cover. For each $C \in \mathcal{C}_n$, there is $F_{C,n} \in \mathcal{F}$ corresponding to C via (XVII.9.1). Moreover,

$$\left\{F_{C,n} : C \in \bigcup_{n \in \mathbb{N}} \mathcal{C}_n\right\}$$

has cardinality less than κ and has the finite intersection property, hence generates a filter $\mathcal{H} \leq \mathcal{F}$ with $\mathcal{H} \in \mathbb{F}_\kappa$. It remains to see that $\overline{0} \in \lim_{[\xi,\nu]} \mathcal{H}$. For each n and $\mathcal{G}$ with $\lim_\xi \mathcal{G} \neq \varnothing$, there is a finite subset $\mathcal{D}$ of $\mathcal{C}_n$ with $\bigcup_{C \in \mathcal{D}} C \in \mathcal{G}$. The corresponding intersection $\bigcap_{C \in \mathcal{D}} F_{C,n}$ belongs to $\mathcal{H}$, and $\langle \bigcup_{C \in \mathcal{D}} C, \bigcap_{C \in \mathcal{D}} F_{C,n} \rangle \subset (-\tfrac{1}{2^n}, \tfrac{1}{2^n})$ so that $\overline{0} \in \lim_\nu \langle \mathcal{G}, \mathcal{H} \rangle$. We conclude that $\chi([\xi,\nu],\overline{0}) \leq \kappa$.

Assume conversely that $\chi([\xi,\nu],\overline{0}) = \kappa$, and let $\mathcal{S}$ be a cover of ξ. Since ξ is functionally embedded, ξ is $R_\nu\,\xi$-regular, and thus the family $\mathrm{cl}^{\natural}_{R_\nu\,\xi}\mathcal{S}$ is also a cover. Therefore, we may assume that $\mathcal{S} = \mathrm{cl}^{\natural}_{R_\nu\,\xi}\mathcal{S}$.

By Lemma XVII.5.4, $\mathcal{S}^{=0}$ converges to $\overline{0}$ in $[\xi,\nu]$ so that there is $\mathcal{H} \in \mathbb{F}_\kappa \cap \lim_{[\xi,\nu]}^{-}(\overline{0})$ with $\mathcal{H} \leq \mathcal{S}^{=0}$. Let $\mathcal{B} \subset \mathcal{H}$ be a filter-base of cardinality at most κ. For every $\mathcal{G}$ with $\lim_\xi \mathcal{G} \neq \varnothing$, there is $B_\mathcal{G} \in \mathcal{B}$ and $G_\mathcal{G} \in \mathcal{G}$ with $\langle G_\mathcal{G}, B_\mathcal{G} \rangle \subset (-1,1)$. For each $B \in \mathcal{B}$, let $R_B := \{x \in |\xi| : \langle x, B \rangle \subset (-1,1)\}$. By construction $\mathcal{R} := \{R_B : B \in \mathcal{B}\}$ is a cover of ξ of cardinality at most κ. Moreover, for each $B \in \mathcal{B}$, there is a finite subset $\mathcal{D}_B$ of $\mathcal{S}$ with

$$\bigcap_{S \in \mathcal{D}_B} \{g : S \subset g^-(0)\} \subset B.$$

We claim that

$$R_B \subset \bigcup_{S \in \mathcal{D}_B} S, \qquad\qquad (XVII.9.2)$$

so that $\{\bigcup_{S \in \mathcal{D}_B} S : B \in \mathcal{B}\}$ is a cover of cardinality at most κ included in $\mathcal{S}^{\cup}$, which will conclude the proof. To see (XVII.9.2), assume to the contrary that there is $x \in R_B \setminus \bigcup_{S \in \mathcal{D}_B} S$. Since $\bigcup_{S \in \mathcal{D}_B} S$ is closed in $R_\nu \xi$, there is $f \in C(\xi, \nu)$ with $f(x) = 1$ and $f(\bigcup_{S \in \mathcal{D}_B} S) = \{0\}$. Because $f(\bigcup_{S \in \mathcal{D}_B} S) = \{0\}$,

$$f \in \bigcap_{S \in \mathcal{D}_B} \{g : S \subset g^-(0)\} \subset B,$$

but this contradicts that if $f \in B$ and $x \in R_B$ then $f(x) \in (-1, 1)$. $\qquad \square$

Corollary XVII.9.14. *If ξ is a functionally embedded convergence then*

$$\chi(\xi) \leq \mathfrak{l}([\xi, \nu]).$$

Proof. By Theorem XVII.9.13, $\mathfrak{l}([\xi, \nu]) = \chi([[\xi, \nu], \nu])$ and ξ is homeomorphic to a subspace of $[[\xi, \nu], \nu]$. Thus $\chi(\xi) \leq \chi([[\xi, \nu], \nu])$ and the conclusion follows. $\qquad \square$

Theorem XVII.9.15. *If ξ is a functionally embedded convergence then*

$$\mathrm{w}(\xi) = \mathrm{w}([\xi, \nu]).$$

Proof. It is enough to prove that $\mathrm{w}([\xi, \nu]) \leq \mathrm{w}(\xi)$ for any functionally embedded convergence ξ, for then $\mathrm{w}([[\xi, \nu], \nu]) \leq \mathrm{w}([\xi, \nu])$ and ξ is homeomorphic to a subspace of $[[\xi, \nu], \nu]$, so that

$$\mathrm{w}(\xi) \leq \mathrm{w}([[\xi, \nu], \nu]) \leq \mathrm{w}([\xi, \nu]).$$

To this end, let $\mathcal{B}$ be a base for the convergence ξ of cardinality $\kappa = \mathrm{w}(\xi)$. For $B \in \mathcal{B}$, $r \in \mathbb{Q}$ and $n \in \mathbb{N}_1$, let

$$M_{B,r,n} := \left\{ f \in C(\xi, \nu) : f(B) \subset \left(r - \tfrac{1}{n}, r + \tfrac{1}{n}\right) \right\}.$$

Let $\mathcal{M} := \{M_{B,r,n} : B \in \mathcal{B}, r \in \mathbb{Q}, n \in \mathbb{N}_1\}$. Clearly, $\mathcal{M}$ and $\mathcal{A} := \mathcal{M}^{\cap}$ have cardinality κ. We claim that $\mathcal{A}$ is a base for $[\xi, \nu]$, which will complete the proof.

Let $f \in \lim_{[\xi, \nu]} \mathcal{F}$. For every $x \in |\xi|$ and every filter $\mathcal{G}$ with $x \in \lim_\xi \mathcal{G}$, there is $\mathcal{H} \leq \mathcal{G}$ with $x \in \lim_\xi \mathcal{H}$ and $\mathcal{H}$ has a filter-base composed of elements of $\mathcal{B}$. Thus, for every $n \in \mathbb{N}_1$, there is $B \in \mathcal{B} \cap \mathcal{G}$ and $F \in \mathcal{F}$ with

$$\langle B, F \rangle \subset \left(f(x) - \tfrac{1}{2n}, f(x) + \tfrac{1}{2n}\right).$$

Now pick $r \in \mathbb{Q}$ with $|f(x) - r| \leq \frac{1}{2n}$. For $g \in M_{B,r,2n}$ and $t \in B$, we then have

$$|g(t) - f(x)| \leq |g(t) - r| + |r - f(x)| \leq 2\frac{1}{2n} = \frac{1}{n},$$

that is

$$\langle B, M_{B,r,2n} \rangle \subset \left(f(x) - \frac{1}{n}, f(x) + \frac{1}{n} \right).$$

If now $g \in F$ and $t \in B$,

$$|g(t) - r| \leq |g(t) - f(x)| + |f(x) - r| \leq \frac{1}{2n} + \frac{1}{2n} = \frac{1}{n},$$

so that $F \subset M_{B,r,n}$. Thus, by construction, $\mathcal{F} \cap \mathcal{A}$ generates a filter coarser than $\mathcal{F}$ with $f \in \lim_{[} \xi, \nu](\mathcal{F} \cap \mathcal{A})^{\uparrow}$, that is, $\mathcal{A}$ is a base of $[\xi, \nu]$. $\square$

Recall from Theorem IV.9.40 that a separable and metrizable topology is second countable. Moreover, the converse is true by Theorem VII.7.14: a regular T_1 topology of countable weight is metrizable (and of course separable). Hausdorff functionally regular topologies are in particular T_1, regular, and functionally embedded. Hence, in view of Theorem XVII.9.15:

Corollary XVII.9.16. *A Hausdorff functionally regular topology ξ is metrizable and separable if and only if $[\xi, \nu]$ has countable weight.*

Combining Corollary XVII.9.16 and Theorem XVII.6.9, we have:

Corollary XVII.9.17. *Let ξ be a Hausdorff functionally regular topology. Then $[\xi, \nu]$ is a separable metrizable topology if and only if ξ is separable, metrizable, and locally compact.*

Chapter XVIII

Functional partitions and metrization

XVIII.1 Introduction

In this chapter we often use the abbreviations spelled out in (I.1.1).

So far we used the term *family* for a set of subsets of a given set. We shall extend this usage to sets of real-valued functions on a given set, because we shall often pass from *families of functions* to some related *families of sets*.

Recall that a convergence space (X, ξ) is said to be *metrizable* if there exists a metric d on X such that $\xi = \tilde{d}$; in other words, if $x \in \lim_\xi \mathcal{F}$ is equivalent to

$$\underset{r>0}{\forall}\ B_d(x, r) \in \mathcal{F}. \qquad (\text{XVIII.1.1})$$

Such a metric is said to be *compatible* with the convergence.

Our approach to metrization is based on the theory of partitions developed by Jerzy Dydak [36], [5].

How can one say that a convergence space is metrizable? We already know several conditions that a metrizable convergence space should satisfy. In particular, in view of Proposition VII.7.10:

Each metrizable space is a (Hausdorff) hereditarily normal topology of countable character. In particular, we restrict ourselves to topologies in this chapter.

We shall see that metrizable topologies have a property that is (strictly) stronger than hereditary normality; it is called perfect normality.

XVIII.2 Perfect normality

Recall that, by the Urysohn Lemma (Theorem VII.7.14), a topological space X is normal if for every couple F_0, F_1 of disjoint closed subsets of X, there exists $f \in C(X, [0, 1])$ such that

$$F_0 \subset f^{-1}(0) \text{ and } F_1 \subset f^{-1}(1).$$

A topological space X is called *perfectly normal* if for every couple F_0, F_1 of disjoint closed subsets of X, there exists $f \in C(X, [0, 1])$ such that

$$f^{-1}(0) = F_0 \text{ and } f^{-1}(1) = F_1.$$

Proposition XVIII.2.1. *Metrizable spaces are perfectly normal.*

Proof. If F is a closed subset of a metrizable space X, then in view of Exercise V.3.8, dist_F is a continuous function and $F = \{\text{dist}_F = 0\}$. If $F = \varnothing$, then $\varnothing = \{\bar{1} = 0\}$. It follows that if F_0, F_1 are disjoint non-empty closed subsets of a metrizable space X, and

$$f(x) := \frac{\text{dist}_{F_0}(x)}{\text{dist}_{F_0}(x) + \text{dist}_{F_1}(x)},$$

then $f^{-1}(0) = F_0$ and $f^{-1}(1) = F_1$. □

It turns out that perfect normality is equivalent to a formally simpler condition.

Recall (from Section VII.5) that a subset U of a topological space X is *functionally open* if there exists a positive ([1]) continuous function f such that $U = \{f > 0\}$, equivalently if there exists $f \in C(X, [0, 1])$ such that $U = \{f > 0\}$. Recall also that a subset of a topological space is *functionally closed* if it is of the form $\{f = 0\}$ for some continuous real-valued function f, equivalently if its complement is functionally open.

Theorem XVIII.2.2 (Vedenisov). *A T_1 topology is perfectly normal if and only if every open set is functionally open (equivalently, each closed set is functionally closed).*

Proof. Let X be perfectly normal and let U be an open subset of X. If $U = X$, then $U = \{\bar{1} > 0\}$; if $U = \varnothing$, then $U = \{\bar{0} > 0\}$. So let U be a proper open subset of X. Let $u \in U$ be arbitrary. Since X is T_1, the set $\{u\}$ is closed. Hence there exists $f \in C(X, [0, 1])$ such that $f^{-1}(1) = \{u\}$ and $f^{-1}(0) = \{f = 0\} = X \setminus U$. Obviously, $\{f > 0\} = U$.

Suppose that each open set is functionally open and let F_0, F_1 be disjoint closed subsets of X. By hypothesis, there exist $f_0, f_1 \in C(X, [0, 1])$ such that $X \setminus F_0 = \{f_0 > 0\}$ and $X \setminus F_1 = \{f_1 > 0\}$. In other words, $F_0 = \{f_0 = 0\}$ and $F_1 = \{f_1 = 0\}$.

---

[1]Recall that we say that $f : X \to \mathbb{R}$ is *positive* if $f(x) \geq 0$ for every $x \in X$.

Then, for each $x \in X$, let

$$f(x) := \frac{f_0(x)}{f_0(x) + f_1(x)}.$$

Because F_0, F_1 are disjoint, $f_0(x) + f_1(x) > 0$ for every $x \in X$, so that f is well defined for every $x \in X$ and continuous. Clearly, the range of f is included in $[0,1]$. Moreover, if $x \in F_0$ then $f(x) = 0$, if $x \in F_1$, then $f(x) = 1$ and if $x \notin F_0 \cup F_1$, then $0 < f_0(x)$ and $0 < f_1(x)$, thus $0 < f(x) < 1$. $\qquad\square$

Exercise XVIII.2.3. Show that a space is perfectly normal if and only if it is normal and if each closed set is $G_{\aleph_0}$.

Definition XVIII.2.4. A family $\mathcal{A}$ of subsets of a topological space is called *locally finite* if each point admits a neighborhood that meets at most a finite number elements of $\mathcal{A}$. It is called *σ-locally finite* if it is a countable union of locally finite families.

Exercise XVIII.2.5. Show that if $\mathcal{A}$ is a locally finite family, then

$$\mathrm{cl}\left(\bigcup_{A \in \mathcal{A}} A \right) \subset \bigcup_{A \in \mathcal{A}} \mathrm{cl}\, A.$$

Lemma XVIII.2.6. *Each Hausdorff regular topology with a σ-locally finite base is perfectly normal.*

Proof. Let X be a Hausdorff regular topological space and let $(\mathcal{B}_n)_n$ be a sequence of locally finite families of open sets such that $\bigcup_{n \in \mathbb{N}} \mathcal{B}_n$ is a base of open sets of X. Let F_0 and F_1 be disjoint open subsets of X. For each $x \in X \setminus F_1$ there exists $n(x) \in \mathbb{N}$ and $W(x) \in \mathcal{B}_n$ such that $x \in W(x) \subset \mathrm{cl}\, W(x) \subset X \setminus F_1$, because X is regular. For every $n \in \mathbb{N}$ let $W_n := \bigcup \{W(x) : n(x) = n\}$. By Exercise XVIII.2.5, $\mathcal{B}_n$ is locally finite for each $n \in \mathbb{N}$, so that $\mathrm{cl}\, W_n = \bigcup \{\mathrm{cl}\, W(x) : n(x) = n\}$ and thus $F_0 \subset \bigcup_{n \in \mathbb{N}} W_n \subset \bigcup_{n \in \mathbb{N}} \mathrm{cl}\, W_n = X \setminus F_1$, which implies that X is normal. Moreover, this shows that each open set is a countable union of closed sets, hence each closed set is $G_{\aleph_0}$, hence X is perfectly normal by Exercise XVIII.2.3. $\qquad\square$

There exist normal topologies that are not perfectly normal.

Example XVIII.2.7. Let X be an uncountable set endowed with the *prime cofinite topology* and let x_∞ be its non-isolated point. Then X is normal by Exercise VII.7.8.

We see that X is not perfectly normal, for $\{x_\infty\}$ is closed, but not a $G_{\aleph_0}$ subset. Indeed, if $\{x_\infty\} = \bigcap_{n \in \mathbb{N}} G_n$, where G_n is open for every $n \in \mathbb{N}$, then $X \setminus G_n$ is finite, hence $\bigcup_{n \in \mathbb{N}} (X \setminus G_n) = X \setminus \bigcap_{n \in \mathbb{N}} G_n = X \setminus \{x_\infty\}$ is countable, which contradicts that X is uncountable.

In view of Proposition XVIII.2.1, each metrizable convergence space is a T_1 perfectly normal topology of countable character, but the converse is false. Indeed, recall from Theorem IV.9.40 that a separable metrizable topology has countable weight, and from Corollary IV.9.41 that the Sorgenfrey line is an example of a separable topology of countable character and uncountable weight, hence is not metrizable. It turns out that the Sorgenfrey line is also perfectly normal, thus providing the desired example.

Proposition XVIII.2.8. *The Sorgenfrey line from Example IV.9.24 is perfectly normal.*

Proof. We already have shown in Example VII.7.11 that the Sorgenfrey line is (hereditarily) normal. In view of Exercise XVIII.2.3, it is enough to show that every closed subset is a G_δ-set, equivalently that every open subset is an F_σ-set. To this end, let O be an open subset of the Sorgenfrey line. Let P be the interior of O in the usual topology of the real line. Then P is a countable union of open intervals (a, b), and each (a, b) is a countable union of intervals of the form $[a_n, b)$, where $(a_n)_n$ is a decreasing sequence of elements of (a, b) such that $\lim_{n \to \infty} a_n = a$. Of course, each $[a_n, b)$ is closed and open. If x is an element of $O \setminus P$, then there is t_x such that $[x, t_x) \subset O$. If y is another element of $O \setminus P$ and r_y is such that $[y, r_y) \subset O$, then $[x, t_x) \cap [y, r_y) = \varnothing$, for otherwise x or y would belong to P (²). Since each interval $[x, t_x)$ contains a rational number, $O \setminus P$ is countable. $\quad\square$

XVIII.3   Pseudometrics

A map $d : X \times X \to \mathbb{R}_+$ is said to be a *pseudometric* if for each $x, y, z \in X$,

$$d(x, y) = d(y, x), \tag{XVIII.3.1}$$

$$d(x, x) = 0, \tag{XVIII.3.2}$$

$$d(x, z) \leq d(x, y) + d(y, z). \tag{XVIII.3.3}$$

---

²If $[x, t_x) \cap [y, r_y) \neq \varnothing$ and, for instance, $x < y$, then $(x, r_y) \subset O$, hence $y \in (x, r_y) \subset P$, which is a contradiction.

Of course, a pseudometric is a metric if $d(x, y) = 0$ implies $x = y$. If d is a pseudometric on X and $r > 0$, then

$$B_d(x, r) := \{y \in X : d(x, y) < r\}$$

is called a *ball* centered at x with radius r.

Example XVIII.3.1. The function $d : \mathbb{R} \times \mathbb{R} \to \mathbb{R}_+$ given by $d(x, y) := |y|$ is a pseudometric, but not a metric.

Each family $\mathcal{D}$ of pseudometrics on X defines a convergence $\widetilde{\mathcal{D}}$ on X to the effect that

$$x \in \lim_{\widetilde{\mathcal{D}}} \mathcal{F} \iff \underset{d \in \mathcal{D}}{\forall} \underset{r > 0}{\forall} B_d(x, r) \in \mathcal{F}.$$

Exercise XVIII.3.2. Let $\mathcal{D}$ be a family of pseudometrics. Show that

(1) $\widetilde{\mathcal{D}}$ is the coarsest topology, for which each $d \in \mathcal{D}$ is continuous on $X \times X$.
(2) $x \in \lim_{\widetilde{\mathcal{D}}} \{x_k\}_k$ if and only if $\lim_{k \to \infty} d(x_k, x) = 0$ for each $d \in \mathcal{D}$.
(3) $\widetilde{\mathcal{D}}$ is a topology. A base of neighborhoods of x for $\widetilde{\mathcal{D}}$ is given by

$$\{B_d(x, r) : \varepsilon > 0, d \in \mathcal{D}\}^{\cap}.$$

(4) $\mathcal{D}_1 := \{\min(d, \overline{1}) : d \in \mathcal{D}\}$ is a family of pseudometrics such that $\widetilde{\mathcal{D}} = \widetilde{\mathcal{D}_1}$.
(5) If $d_1, d_2, \ldots, d_n$ are pseudometrics on X, then

$$d(x, y) := \max_{1 \leq k \leq n} d_k(x, y)$$

is a pseudometric on X. Moreover,

$$B_d(x, r) = \bigcap_{k=1}^{n} B_{d_k}(x, r).$$

A family $\mathcal{D}$ of pseudometrics is called *saturated* if $\max_{d \in \mathcal{D}_0} d \in \mathcal{D}$ for each finite subfamily $\mathcal{D}_0$ of $\mathcal{D}$. The least saturated family of pseudometrics including $\mathcal{D}$ is denoted by $\mathcal{D}^{\vee}$. It is obvious that $\widetilde{\mathcal{D}^{\vee}} = \widetilde{\mathcal{D}}$.

We say that a family $\mathcal{D}$ of pseudometrics *distinguishes points* of X, if for every $x_0, x_1 \in X$

$$x_0 \neq x_1 \implies \underset{d \in \mathcal{D}}{\exists} d(x_0, x_1) \neq 0;$$

it *distinguishes points from closed sets* if for each closed subset F of X and $x \notin F$, there exists $d \in \mathcal{D}$ such that $d(x, F) > 0$.

Exercise XVIII.3.3. The topology $\widetilde{\mathcal{D}}$ is Hausdorff if and only if $\mathcal{D}$ distinguishes points.

Proposition XVIII.3.4. *If $\mathcal{D}$ is a countable family of pseudometrics distinguishing points, then $\widetilde{\mathcal{D}}$ is metrizable.*

Proof. If $\mathcal{D} := \{d_n : n \in \mathbb{N}\}$ is a countable family of pseudometrics, all bounded by 1, then $\widetilde{\mathcal{D}}$ is of countable character. It is straightforward that

$$d(x,y) := \sum_{n \in \mathbb{N}} \frac{1}{2^n} d_n(x,y) \qquad \text{(XVIII.3.4)}$$

is a pseudometric, that is also bounded by 1. If $\mathcal{D}$ distinguishes points, then (XVIII.3.4) is a metric, for if $d(x,y) = 0$, then $d_n(x,y) = 0$ for each $n \in \mathbb{N}$, hence $x = y$. Because both $\widetilde{\mathcal{D}}$ and d are of countable character, to see that d is a metric compatible with $\widetilde{\mathcal{D}}$, it is enough to see that the convergent sequences coincide for $\widetilde{\mathcal{D}}$ and d.

If $\{x_k\}_k$ is such that $0 = \lim_{k \to \infty} d(x_k, x)$, then $\lim_{k \to \infty} d_n(x_k, x) = 0$ for each $n \in \mathbb{N}$.

Conversely, let $\lim_{k \to \infty} d_n(x_k, x) = 0$ for each $n \in \mathbb{N}$. Let $\varepsilon > 0$. Then there exists $n_\varepsilon \in \mathbb{N}$ such that

$$\sum_{n=n_\varepsilon}^{\infty} \frac{1}{2^n} d_n(x,y) \leq \sum_{n=n_\varepsilon}^{\infty} \frac{1}{2^n} < \frac{\varepsilon}{2}$$

for every $x, y \in X$. On the other hand, for each $n < n_\varepsilon$ there exists $k_n(\varepsilon) \in \mathbb{N}$ such that $d_n(x_k, x) < \frac{\varepsilon}{4}$ for $k > k_n(\varepsilon)$. Hence, for $k > \max_{n < n_\varepsilon} k_n(\varepsilon)$,

$$d(x_k, x) \leq \frac{\varepsilon}{4} \sum_{n < n_\varepsilon} \frac{1}{2^n} + \frac{\varepsilon}{2} < \varepsilon.$$

$\square$

Of course, (XVIII.3.4) is a compatible metric for $\widetilde{\mathcal{D}}$.

Corollary XVIII.3.5. *Each countable product of metrizable convergences is metrizable.*

Proof. If $x \in \prod_{m \in \mathbb{N}} X_m$, then we denote by $x(m)$ the m-th factor of x. Let d_n be a metric (bounded by 1) on X_n for each $n \in \mathbb{N}$. Because each metric induces a topological convergence, their product is also topological.

We observe that

$$d_n^*(x,y) := d_n(x(n), y(n))$$

is a pseudometric on $\prod_{m \in \mathbb{N}} X_m$. Moreover the neighborhood filter of each $x \in \prod_{m \in \mathbb{N}} X_m$ in the product topology has a subbase of the form

$$\{\{y : d_n^*(x,y) < \varepsilon\} : n \in \mathbb{N}, \varepsilon > 0\}.$$

Therefore, the product topology is that determined by the countable family of pseudometrics $\{d_n^* : n \in \mathbb{N}\}$. This family distinguishes points, for if $x_0 \neq$

x_1, then there is $n \in \mathbb{N}$ such that $x_0(n) \neq x_1(n)$, thus $d_n^*(x_0, x_1) > 0$. Therefore, the product topology is metrizable, by Proposition XVIII.3.4.

$\square$

Theorem XVIII.3.6. *If X_j is a Hausdorff convergence space such that card $X_j \geq 2$ for each $j \in J$, but J is not countable, then $\prod_{j \in J} X_j$ is not metrizable.*

Proof. Let x_j, y_j be two distinct elements of X_j for each $j \in J$. Because X_j is Hausdorff, $\{x_j, y_j\}$ with the convergence induced from X_j is discrete. To conclude, it is enough to show that $\prod_{j \in J} \{x_j, y_j\}$ has uncountable character. The family $\{p_j^{-1}(f(j)) : j \in J\}^{\cap}$ is a neighborhood base of $f \in \prod_{j \in J} X_j$.

If $\{V_n : n \in \mathbb{N}\}$ were a base of neighborhoods of f, then, for each $n \in \mathbb{N}$, there would exist a finite subset F_n of J such that

$$\bigcap_{j \in F_n} p_j^{-1}(f(j)) \subset V_n.$$

Therefore if $F := \bigcup_{n \in \mathbb{N}} F_n$, then $\bigcap_{j \in F} p_j^{-1}(f(j)) \subset V_n$ for each $n \in \mathbb{N}$. On the other hand, $J \setminus F \neq \varnothing$, because F is countable, but J is not. If $j_0 \notin F$, then

$$\bigcap_{j \in F} p_j^{-1}(f(j)) \setminus p_j^{-1}(f(j_0)) \neq \varnothing,$$

and if $g \in \prod_{j \in J} X_j$ is such that $g(j) = f(j)$ for $j \in F$ and $g(j_0) \neq f(j_0)$, then

$$g \in \bigcap_{j \in F} p_j^{-1}(f(j)) \setminus p_j^{-1}(f(j_0)).$$

Accordingly, V_n is not a subset of $p_j^{-1}(f(j_0))$ for $n \in \mathbb{N}$, so that $\{V_n : n \in \mathbb{N}\}$ is not a neighborhood base of f.

$\square$

XVIII.4 Functional covers and partitions

If $\mathcal{F}$ is a family of real positive functions on a set X, then the family

$$\text{supp}\, \mathcal{F} := \{\{f > 0\} : f \in \mathcal{F}\}$$

is called the *support* of $\mathcal{F}$.

Definition XVIII.4.1. A family $\mathcal{F}$ of real positive continuous functions on a topological space X is called a *functional cover* if for each $x \in X$ there exists $f \in \mathcal{F}$ such that $f(x) > 0$ (³).

---

³In other words $\sup_{f \in \mathcal{F}} f(x) > 0$ for every $x \in X$.

If moreover, $h : X \to (0, \infty]$ and for each $x \in X$,

$$h(x) = \sum_{f \in \mathcal{F}} f(x),$$

then $\mathcal{F}$ is called a *partition* of h. In particular a partition $\mathcal{F}$ is called a *partition of unity* if for every $x \in X$,

$$\sum_{f \in \mathcal{F}} f(x) = 1.$$

Recall that (^{4})

$$\sum_{f \in \mathcal{F}} f(x) := \sup \left\{ \sum_{f \in \mathcal{T}} f(x) : \mathcal{T} \in [\mathcal{F}]^{<\omega} \right\}. \qquad \text{(XVIII.4.1)}$$

These definitions entail

Proposition XVIII.4.2. *A family $\mathcal{F} \subset C(X, \mathbb{R}_+)$ is a functional cover on X if and only if* supp $\mathcal{F}$ *is a cover* (^{5}) *of X.*

This proposition enables us to define properties of functional covers by analogous properties of covers. For instance, in view of Theorem XVIII.2.2,

Proposition XVIII.4.3. *Let X be T_1. Then every open cover is the support of a functional cover if and only if X is perfectly normal.*

Example XVIII.4.4. A family $\mathcal{A}$ of subsets of a topological space X is called *discrete* if each point of X has a neighborhood that intersects at most one element of $\mathcal{A}$.

A functional cover $\mathcal{F}$ on a topological space X is called *discrete* if for each $x \in X$ there exists a neighborhood V of x such that

$$\{ f \in \mathcal{F} : \sup_V f > 0 \}$$

is a singleton. Of course, a functional cover is discrete if and only if its support is a discrete family subsets.

Example XVIII.4.5. A functional cover $\mathcal{F}$ of a topological space X is called *locally finite* if for each $x \in X$ there exists a neighborhood V of x such that card $\{ f \in \mathcal{F} : \sup_V f > 0 \} < \infty$. Sure enough, a functional cover is locally finite if and only if its support is a locally finite family of sets.

---

[4] If $A \subset \mathbb{R}_+$, then $\sum_{r \in A} r := \sup \{ \sum_{r \in T} r : T \in [A]^{<\omega} \}$.

[5] Of course, such a cover consists of open sets.

A functional cover $\mathcal{F}$ is called a *refinement* of a cover $\mathcal{U}$ (traditionally, $\mathcal{F}$ is called *subordinated* to $\mathcal{U}$) if the support of $\mathcal{F}$ is a refinement of $\mathcal{U}$. A functional cover $\mathcal{F}$ is called a *refinement* of a functional cover $\mathcal{H}$ if (the support of) $\mathcal{F}$ is a refinement of the support of $\mathcal{H}$. A functional cover $\mathcal{F}$ on X is called *bounded* if there exists $c > 0$ such that

$$\sup\{f(x) : x \in X, f \in \mathcal{F}\} \le c.$$

It turns out that each open cover of a topological normal space admits a finer bounded functional cover. An important property, stronger than normality, is the existence of a finer partition of unity for each open cover. Let us characterize the functional covers, the sum of which is continuous.

A function $h : X \to (0, \infty)$ is called *finite*.

Proposition XVIII.4.6. *If $\mathcal{F}$ is a partition of a finite function h on a topological space X, then h is continuous if and only if for each $\varepsilon > 0$ and each $x \in X$, there exists $U \in \mathcal{N}(x)$ and a finite subset $\mathcal{T}$ of $\mathcal{F}$ such that*

$$\sup_{u \in U} \sum_{f \in \mathcal{F} \setminus \mathcal{T}} f(u) < \varepsilon.$$

Proof. Suppose that h is finite and the condition is fulfilled. If $\varepsilon > 0$ and $x \in X$, then by (XVIII.4.1), there exists a finite subset $\mathcal{T}$ of $\mathcal{F}$ such that $\sum_{f \in \mathcal{T}} f(x) > h(x) - \frac{\varepsilon}{2}$. As $\sum_{f \in \mathcal{T}} f$ is continuous, there exists a neighborhood U of x such that

$$h(x) - \varepsilon < \sum_{f \in \mathcal{T}} f(x) - \frac{\varepsilon}{2} < \sum_{f \in \mathcal{T}} f(u) \le h(u)$$

for each $u \in U$.

If $\varepsilon > 0$ and $x \in X$, then by the condition, there exists a finite subset $\mathcal{T}$ of $\mathcal{F}$ and a neighborhood U of x such that

$$h(u) - \sum_{f \in \mathcal{T}} f(u) = \sum_{f \in \mathcal{F} \setminus \mathcal{T}} f(u) < \frac{\varepsilon}{2}$$

for each $u \in U$. As $\sum_{f \in \mathcal{T}} f$ is continuous, there exists a neighborhood W of x, included in U, and such that, for each $u \in W$,

$$h(x) + \varepsilon \ge \sum_{f \in \mathcal{T}} f(x) + \varepsilon,$$

$$\sum_{f \in \mathcal{T}} f(u) + \frac{\varepsilon}{2} = h(u) - \sum_{f \in \mathcal{F} \setminus \mathcal{T}} f(u) + \frac{\varepsilon}{2} > h(u).$$

Conversely, let h be continuous. If $x \in X$ and $\varepsilon > 0$, then by (XVIII.4.1) there exists a finite subset $\mathcal{T}$ of $\mathcal{F}$ such that $h(x) - \sum_{f \in \mathcal{T}} f(x) < \frac{\varepsilon}{2}$. As h

is continuous, $\sum_{f \in \mathcal{F} \setminus \mathcal{T}} f = h - \sum_{f \in \mathcal{T}} f$ is continuous, hence there exists $U \in \mathcal{N}(x)$ such that

$$\sup_{u \in U} \sum_{f \in \mathcal{F} \setminus \mathcal{T}} f(u) < \varepsilon.$$

$\square$

If $\mathcal{F}$ is a functional cover, then

$$\mathcal{F}_{fin} := \left\{ \sum_{f \in \mathcal{T}} f : \mathcal{T} \in [\mathcal{F}]^{<\omega} \right\}$$

is a functional cover. Let us observe that

Proposition XVIII.4.7. *If $\sum_{f \in \mathcal{F}} f$ is finite, then it is continuous if and only if $\mathcal{F}_{fin}$ is equicontinuous.*

Proof. If $\sum_{f \in \mathcal{F}} f$ is continuous, then, by Proposition XVIII.4.6, for every $x \in X$ and each $\varepsilon > 0$ there exists $U \in \mathcal{V}(x)$ and a finite subset $\mathcal{T}$ of $\mathcal{F}$ such that

$$\sup_{u \in U} \sum_{f \in \mathcal{F} \setminus \mathcal{T}} f(u) < \frac{\varepsilon}{4}.$$

In particular, if $\mathcal{S}$ is a finite subset of $\mathcal{F}$, then $\sup_{u \in U} \sum_{f \in \mathcal{S} \setminus \mathcal{T}} f(u) < \frac{\varepsilon}{4}$, hence

$$\sup_{u \in U} \left| \sum_{f \in \mathcal{S} \setminus \mathcal{T}} f(u) - \sum_{f \in \mathcal{S} \setminus \mathcal{T}} f(x) \right| < \frac{\varepsilon}{2}.$$

On the other hand, the family of continuous functions $\{ \sum_{f \in \mathcal{P}} f : \mathcal{P} \subset \mathcal{T} \}$ is finite, hence equicontinuous, and then there exists $V \in \mathcal{V}(x)$ such that

$$\left| \sum_{f \in \mathcal{P}} f(x) - \sum_{f \in \mathcal{P}} f(v) \right| < \frac{\varepsilon}{2} \qquad \text{(XVIII.4.2)}$$

for each $v \in V$. Therefore

$$\left| \sum_{f \in \mathcal{R}} f(x) - \sum_{f \in \mathcal{S}} f(v) \right| < \varepsilon$$

for each finite subset $\mathcal{S}$ of $\mathcal{F}$ and each $v \in U \cap V$, proving that $\mathcal{F}_{fin}$ is equicontinuous.

Conversely, if $\mathcal{F}_{fin}$ is equicontinuous on X, then for each $x \in X$ and each $\varepsilon > 0$ there exists a neighborhood V of x such that for each finite subset $\mathcal{P}$ of $\mathcal{F}$,

$$\sup_{v \in V} \left| \sum_{f \in \mathcal{P}} (f(v) - f(x)) \right| < \frac{\varepsilon}{2}.$$

On the other hand, by (XVIII.4.1), there exists a finite subset $\mathcal{P}$ of $\mathcal{F}$ such that

$$\sum_{f\in\mathcal{F}\setminus\mathcal{P}} f(x) = \sum_{f\in\mathcal{F}} f(x) - \sum_{f\in\mathcal{P}} f(x) < \frac{\varepsilon}{2},$$

hence, by (XVIII.4.2), for each $v \in V$,

$$\sum_{f\in\mathcal{F}\setminus\mathcal{P}} f(v) \le \sum_{f\in\mathcal{F}\setminus\mathcal{P}} f(x) + \frac{\varepsilon}{2} < \varepsilon.$$

$\square$

Definition XVIII.4.8. A family $\mathcal{G}$ of real continuous positive functions is called *strongly equicontinuous* if for each $x \in X$ and each $\varepsilon > 0$ there exist a neighborhood V of x and a finite subfamily $\mathcal{T}$ of $\mathcal{G}$ such that

$$\sup_{u\in V} g(u) < \varepsilon$$

for each $g \in \mathcal{G} \setminus \mathcal{T}$.

Lemma XVIII.4.9. *Each strongly equicontinuous family is equicontinuous. If $\mathcal{G}$ is equicontinuous and $\sum_{g\in\mathcal{G}} g(x) < \infty$ for each $x \in X$, then $\mathcal{G}$ is strongly equicontinuous.*

Proof. Let $\varepsilon > 0$ and $x \in X$. Let $\mathcal{G}$ be strongly equicontinuous and let $\mathcal{T}$ and V be as in the definition. Then

$$|g(u) - g(x)| < 2\varepsilon \qquad\qquad \text{(XVIII.4.3)}$$

for each $u \in V$ and $g \in \mathcal{G}\setminus\mathcal{T}$. As $\mathcal{T}$ is a finite family of continuous functions, $\mathcal{T}$ is equicontinuous, that is, there exists a neighborhood U of x such that (XVIII.4.3) holds for each $u \in U$ and $g \in \mathcal{T}$. We conclude that (XVIII.4.3) holds for every $g \in \mathcal{G}$ and $u \in V \cap U$, that is, $\mathcal{G}$ is equicontinuous.

Conversely, let $x \in X$ and $\varepsilon > 0$. By (XVIII.4.1) there is a finite subfamily $\mathcal{T}$ of $\mathcal{G}$ such that $\sum_{g\in\mathcal{G}} g(x) < \sum_{g\in\mathcal{T}} g(x) + \varepsilon$ and by continuity, there exists a neighborhood V of x such that $\sum_{g\in\mathcal{G}} g(x) < \sum_{g\in\mathcal{T}} g(u) + \varepsilon$ for each $u \in V$. It follows that for every $g \in \mathcal{G} \setminus \mathcal{T}$,

$$g(x) \le \sum_{g\in\mathcal{G}\setminus\mathcal{T}} f(x) < \varepsilon,$$

hence by equicontinuity, there exists a neighborhood U of x such that $g(u) < \varepsilon$ for each $g \in \mathcal{G} \setminus \mathcal{T}$ and every $u \in V \cap U$, that is, $\mathcal{G}$ is strongly equicontinuous. $\square$

Proposition XVIII.4.10. *Let $\mathcal{G}$ be a functional cover of a topological space X such that $\sum_{g\in\mathcal{G}} g$ is finite. Then the following conditions are equivalent:*

(1) $\mathcal{G}$ is equicontinuous,
(2) $\left\{\max(\overline{0}, g - h) : g \in \mathcal{G}\right\}$ is locally finite for each $h \in C(X, (0,\infty))$,
(3) $\left\{\max(\overline{0}, g - \overline{\varepsilon}) : g \in \mathcal{G}\right\}$ is locally finite for each $\varepsilon > 0$.

Proof. (1) implies (2) : As $\sum_{g\in\mathcal{G}} g$ is finite, equicontinuity is equivalent to strong equicontinuity in view of Lemma XVIII.4.9. Let $\varepsilon > 0$ and $x \in X$. As $h \in C(X, (0,\infty))$, there exists a neighborhood U of x such that $h(u) > \frac{1}{2}h(x)$ for each $u \in U$. By the strong equicontinuity of $\mathcal{G}$, there exists a finite subset $\mathcal{T}$ of $\mathcal{G}$ and a neighborhood W of x such that

$$\sum_{g\in\mathcal{G}\setminus\mathcal{T}} g(u) < \frac{1}{2}h(x)$$

for each $u \in W$. Therefore if $g \in \mathcal{G} \setminus \mathcal{T}$ then $g(u) - h(u) < 0$ for each $u \in V \cap W$, proving (2).

(2) implies (3) for $h = \overline{\varepsilon}$ for $\varepsilon > 0$.

(3) implies (1): for each $\varepsilon > 0$ and $x \in X$, there exists a finite subset $\mathcal{T}$ of $\mathcal{G}$ and a neighborhood V of x such that $\left\{\max(\overline{0}, g - \overline{\varepsilon}) > 0\right\} \cap V = \varnothing$ for each $g \in \mathcal{G} \setminus \mathcal{T}$. This implies that $\sup_{u\in V} g(u) < \varepsilon$ for each $g \in \mathcal{G} \setminus \mathcal{T}$, that is, $\mathcal{G}$ is strongly equicontinuous, hence equicontinuous. □

Lemma XVIII.4.11. *If $\mathcal{F} \subset C(X, \mathbb{R})$ is equicontinuous and $\sup \{f(x) : f \in \mathcal{F}\} < \infty$ for each $x \in X$, then*

$$\left\{\sup_{f\in\mathcal{T}} f : \mathcal{T} \subset \mathcal{F}\right\}$$

is equicontinuous.

Proof. Let $f_{\mathcal{T}} := \sup_{f\in\mathcal{T}} f$. Let $x \in X$ and $\varepsilon > 0$. Then $f_{\mathcal{T}}(x) + \varepsilon \geq f(x) + \varepsilon$ for each $f \in \mathcal{T}$ and for each $\mathcal{T} \subset \mathcal{F}$. By the equicontinuity of $\mathcal{F}$ there exists $V \in \mathcal{V}(x)$ such that for each $v \in V$ and each $f \in \mathcal{T}$,

$$f_{\mathcal{T}}(x) + \varepsilon \geq f(x) + \varepsilon \geq f(v),$$

hence $f_{\mathcal{T}}(x) + \varepsilon \geq \sup_{f\in\mathcal{T}} f(v) = f_{\mathcal{T}}(v)$ for each $v \in V$. On the other hand, for each $\mathcal{T} \subset \mathcal{F}$, there exists $g_{\mathcal{T}} \in \mathcal{T}$ such that $g_{\mathcal{T}}(x) > f_{\mathcal{T}}(x) - \varepsilon$ and as $\mathcal{F}$ is equicontinuous, there exists $V \in \mathcal{V}(x)$ such that

$$f_{\mathcal{T}}(v) \geq g_{\mathcal{T}}(v) \geq f_{\mathcal{T}}(x) - \varepsilon$$

for each $v \in V$ and each $\mathcal{T} \subset \mathcal{F}$. □

Recall (see page 299) that a family $\mathcal{W}$ is called a *regular refinement* of a family $\mathcal{U}$ if for each $W \in \mathcal{W}$ there exists $U \in \mathcal{U}$ such that $\operatorname{cl} W \subset U$.

Proposition XVIII.4.12. *Each equicontinuous partition of a finite function on a topological space admits a regular refinement that is a locally finite functional cover.*

Proof. If $\mathcal{F}$ is an equicontinuous partition of a finite function on a topological space X, then so is $\{\min(f, \overline{1}) : f \in \mathcal{F}\}$. Hence it can be assumed that $\mathcal{F} \subset C(X, [0, 1])$. By Lemma XVIII.4.11 and Proposition XVIII.4.7,

$$F(x) := \sup_{f \in \mathcal{F}} f(x)$$

is strictly positive and continuous. Therefore

$$\mathcal{G} := \{\max(\overline{0}, f - \tfrac{1}{2}F) : f \in \mathcal{F}\}$$

is a functional cover, because $\sup_{g \in \mathcal{G}} g$ is strictly positive for each $g \in \mathcal{G}$. As

$$\{f > \tfrac{1}{2}F\} \subset \operatorname{cl}\{f > \tfrac{1}{2}F\} \subset \{f \geq \tfrac{1}{2}F\} \subset \{f > 0\}$$

for each $f \in \mathcal{F}$, the functional cover $\mathcal{G}$ is a regular refinement of $\mathcal{F}$. Finally, thanks to Proposition XVIII.4.10, $\mathcal{G}$ is locally finite. $\square$

Corollary XVIII.4.13. *Every partition of unity on a topological space admits a locally finite refining partition of unity.*

Proof. If $\mathcal{F}$ is a partition of unity (hence of a continuous function), then in particular $\mathcal{F}$ is equicontinuous. By Proposition XVIII.4.12, there exists a locally finite functional cover $\mathcal{G}$ that refines $\mathcal{F}$, hence $h := \sum_{g \in \mathcal{G}} g$ is continuous and strictly positive. Therefore $\{g/h : g \in \mathcal{G}\}$ is a locally finite partition of unity. $\square$

Corollary XVIII.4.14. *If an open cover of a topological space admits a refining partition of unity, then it admits a refining locally finite partition of unity.*

Proposition XVIII.4.15. *A Hausdorff topological space X is normal if and only if each finite open cover of X admits a refining finite functional cover by a partition of unity.*

Proof. If X is normal, then we proceed by induction on the cardinality of cover. If $n = 1$, then $\{X\}$ is an open cover of X of cardinality 1. If $n = 2$, then let U_0, U_1 be open sets such that $U_0 \cup U_1 = X$, that is, $X \setminus U_0$ and $X \setminus U_1$ are disjoint and closed. By the Urysohn Lemma VII.7.14, there exists $f \in C(X, [0, 1])$ such that $X \setminus U_0 \subset f^{-1}(0)$ and $X \setminus U_1 \subset f^{-1}(1)$, equivalently, $\{f > 0\} \subset U_0$ and $\{\overline{1} - f > 0\} \subset U_1$, that is, $\{f, \overline{1} - f\}$ is a partition of unity refining $\{U_0, U_1\}$.

Suppose that the condition for covers of cardinality n and let

$$\{U_0, U_1, \ldots, U_{n-1}, U_n\}$$

be an open cover of X. Let $\{g_0, g_1, \ldots, g_{n-1}\}$ be a partition of unity refining $\{U_0, U_1, \ldots, U_{n-1} \cup U_n\}$ and let $\{h_0, h_1\}$ be a partition of unity refining $\{U_0 \cup U_1 \cup \ldots \cup U_{n-1}, U_n\}$. Then

$$\overline{1} = (h_0 + h_1) \sum_{k=0}^{n-1} g_k$$

$$= \sum_{k=0}^{n-2} (h_0 g_k) + (h_0 + h_1) g_{n-1} + h_1 g_n,$$

and consequently,

$$f_k := h_0 g_k, \text{ if } 0 \le k < n - 1,$$
$$f_{n-1} := (h_0 + h_1) g_{n-1},$$
$$f_n := h_1 g_n$$

is partition of unity refining $\{U_0, U_1, \ldots, U_{n-1}, U_n\}$. $\qquad\square$

XVIII.5 Paracompactness

A topology is called *paracompact* if each open cover admits an open locally finite refining cover.

It follows that every compact topology is paracompact.

Lemma XVIII.5.1. *If F_0, F_1 are closed subsets of a paracompact space X such that for each $x \in F_0$, there exist $U_x \in \mathcal{N}(x)$ and $V_x \in \mathcal{V}(F_1)$ with $U_x \cap V_x = \varnothing$, then $\mathcal{V}(F_0)$ and $\mathcal{V}(F_1)$ do not mesh.*

Proof. The family $\mathcal{U} := \{U_x : x \in F_0\} \cup \{X \setminus F_0\}$ is an open cover of X. Let $\mathcal{W}$ be a locally finite refinement of $\mathcal{U}$.

If we set $\mathcal{W}_0 := \{W \in \mathcal{W} : W \cap F_0 \neq \varnothing\}$, then $\operatorname{cl} W \cap F_1 = \varnothing$ for each $W \in \mathcal{W}_0$ and

$$F_0 \subset \bigcup_{W \in \mathcal{W}_0} W \subset \operatorname{cl}(\bigcup_{W \in \mathcal{W}_0} W) = \bigcup_{W \in \mathcal{W}_0} \operatorname{cl} W,$$

by Exercise XVIII.2.5. Hence $\bigcup_{W \in \mathcal{W}_0} W \in \mathcal{V}(F_0)$ and $X \setminus \bigcup_{W \in \mathcal{W}_0} \operatorname{cl} W \in \mathcal{N}(F_1)$. $\square$

If F_0 is closed and $x \notin F_0$, then by Lemma XVIII.5.1 with $F_1 := \{x\}$,

Corollary XVIII.5.2. *Each Hausdorff paracompact space is regular.*

Theorem XVIII.5.3. *Each Hausdorff paracompact topology is normal.*

Proof. If F_0, F_1 are closed subsets of a paracompact space X. By Corollary XVIII.5.2, for each $x \in F_0$ there exist $U_x \in \mathcal{V}(x)$ and $V_x \in \mathcal{V}(F_1)$ such that $U_x \cap V_x = \varnothing$. By Lemma XVIII.5.1, $\mathcal{V}(F_0)$ and $\mathcal{V}(F_1)$ do not mesh. $\square$

Proposition XVIII.5.4. *If X is a Hausdorff topological space, then the following conditions are equivalent:*

(1) X is paracompact;
(2) for each open cover of X there exists a refining locally finite partition of unity;
(3) for each open cover of X there exists a refining partition of unity.

Proof. (1) implies (2) : If X is Hausdorff and paracompact, then each open cover has a locally finite refinement $\mathcal{W}$ that covers X. By Theorem XVIII.5.3, X is normal, hence regular, and thus there exists a cover $\{F_W : W \in \mathcal{W}\}$ such that $\operatorname{cl} F_W \subset W$ for each $W \in \mathcal{W}$. As X is normal, by the Urysohn Lemma VII.7.14, for each $W \in \mathcal{W}$ there exists a continuous function $f_W : X \to [0,1]$ such that $f_W(F_W) = \{1\}$ and $f_W(X \setminus W) = \{0\}$. Since $\mathcal{W}$ and $\{F_W : W \in \mathcal{W}\}$ are locally finite covers, $\{f_W : W \in \mathcal{W}\}$ is a locally finite functional cover, hence $\sum_{V \in \mathcal{W}} f_V$ is a continuous strictly positive function. Hence $\{f_W / \sum_{V \in \mathcal{W}} f_V : W \in \mathcal{W}\}$ is a locally finite partition of unity that refines $\mathcal{W}$.

Sure enough (2) implies (3).

(3) implies (2) : If $\mathcal{U}$ is an open cover and $\mathcal{F}$ is a partition of unity refining $\mathcal{U}$, then by Corollary XVIII.4.14, there exists a locally finite partition of unity $\mathcal{G}$ refining $\mathcal{F}$, hence $\mathcal{U}$. $\square$

Lemma XVIII.5.5. *For each bounded equicontinuous functional cover of a topological space, there exists a refining equicontinuous partition of a bounded function.*

Proof. Let $\mathcal{F}$ be a functional cover bounded by $c > 0$. Let $\mathcal{F} = \{f_s : s \in S\}$, where S is well ordered. Let

$$g_s(x) := \max(\overline{0}, f_s(x) - \sup_{t < s} f_t(x)),$$

for each $x \in X$ and $s \in S$. By Lemma XVIII.4.11, $\{f_s - \sup_{t < s} f_t : s \in S\}$ is equicontinuous, hence $\{g_s : s \in S\}$ is equicontinuous.

Since each f_s is positive,

$$t < s \Longrightarrow g_s(x) \leq f_s(x) - f_t(x) \tag{XVIII.5.1}$$

for each $x \in X$. Let $s(x)$ be the least s such that $f_s(x) > 0$. Then $f_{s(x)}(x) = g_{s(x)}(x)$, hence $\sum_{s \in S} g_s(x) > 0$.

In view of (XVIII.5.1), for each increasing finite sequence $\{t_1, t_2, \ldots, t_n\}$ of $\{s \in S : g_s(x) > 0\}$,

$$\sum_{t \in T} g_t(x) \leq f_{t_1}(x) + (f_{t_2}(x) - f_{t_1}(x)) + \ldots + (f_{t_n}(x) - f_{t_{n-1}}(x)),$$

so that

$$\sum_{t \in T} g_t(x) \leq f_{t_n}(x) \leq c.$$

Hence $g(x) := \sum_{s \in S} g_s(x) \leq c$ by (XVIII.4.1). Therefore $\{g_s : s \in S\}$ is an equicontinuous partition of a bounded function refining $\mathcal{F}$. □

Theorem XVIII.5.6. *Let X be a topological normal space. An open cover $\mathcal{U}$ of X admits a refining partition of unity provided that there exists an equicontinuous functional cover $\mathcal{F}$ refining $\mathcal{U}$.*

Proof. Let $\mathcal{U}$ be an open cover of a normal space X and let $\mathcal{F}$ be an equicontinuous functional cover refining $\mathcal{U}^{\cup}$. Hence $\mathcal{F}_1 := \{\min(1, f) : f \in \mathcal{F}\}$ is bounded and has all the properties of $\mathcal{F}$ mentioned above. By Lemma XVIII.5.5, there exists a partition $\mathcal{G}$ of a bounded function refining $\mathcal{F}$. By Proposition XVIII.4.12, there exists an equicontinuous locally finite functional cover $\mathcal{H}$ refining $\mathcal{U}$. Therefore $\sum_{h \in \mathcal{H}} h$ is continuous and strictly positive, hence by dividing each element of $\mathcal{H}$ by $\sum_{h \in \mathcal{H}} h$, we get a partition of unity refining $\mathcal{U}$. □

Theorem XVIII.5.7 (A. H. Stone). *Each metrizable space is para-compact.*

Proof. If d is a metric on X bounded by 1 and if $\mathcal{U}$ is an open cover of X, then $\mathcal{F} = \left\{ \text{dist}_{X \backslash U} : U \in \mathcal{U} \right\}$ is an equicontinuous bounded functional cover refining $\mathcal{U}$. By virtue of Theorem XVIII.5.6, there exists a partition of unity refining $\mathcal{U}$, hence thanks to Proposition XVIII.5.4, X is paracompact. $\square$

Corollary XVIII.5.8. *Each metrizable space admits a σ-locally finite base.*

Proof. If X is a metrizable space and d is a compatible metric, then by Theorem XVIII.5.7, for each $n \in \mathbb{N}_1$, the cover $\mathcal{B}_n := \{B_d(x, \frac{1}{n}) : x \in X\}$ of X admits a σ-locally finite refining cover $\mathcal{U}_n$. Hence $\mathcal{U} := \bigcup_{n \in \mathbb{N}_1} \mathcal{U}_n$ is a σ-locally finite open base. $\square$

XVIII.6 Fragmentations of partitions of unity

Let $\mathcal{F}$ be a family of positive real functions on a set X. As before, we denote by $[\mathcal{F}]^{<\omega}$ the set of finite subfamilies of $\mathcal{F}$ and by

$$[\mathcal{F}]^{<\omega}(f)$$

the set of finite subfamilies of $\mathcal{F}$ containing f.

A family of positive real functions $\left\{ \varphi_\mathcal{T} : \mathcal{T} \in [\mathcal{F}]^{<\omega} \right\}$, is called a *fragmentation* of $\mathcal{F}$ provided that for each $f \in \mathcal{F}$,

$$f = \sum_{\mathcal{T} \in [\mathcal{F}]^{<\omega}(f)} \frac{1}{\text{card}\, \mathcal{T}}\, \varphi_\mathcal{T} \qquad (\text{XVIII.6.1})$$

and for each $x \in X$, the set $\{\mathcal{T} : \varphi_\mathcal{T}(x) > 0\}$ is totally ordered.

Therefore if $\left\{ \varphi_\mathcal{T} : \mathcal{T} \in [\mathcal{F}]^{<\omega} \right\}$ is a fragmentation of $\mathcal{F}$, then

$$\sum_{f \in \mathcal{F}} f = \sum_{f \in \mathcal{F}} \sum_{\mathcal{T} \in [\mathcal{F}]^{<\omega}(f)} \frac{1}{\text{card}\, \mathcal{T}}\, \varphi_\mathcal{T} = \sum_{\mathcal{T} \in [\mathcal{F}]^{<\omega}} \varphi_\mathcal{T}, \qquad (\text{XVIII.6.2})$$

because in the double sum above each finite subfamily $\mathcal{T}$ of $\mathcal{F}$ appears $\text{card}\, \mathcal{T}$ times. Hence, $\mathcal{F}$ is a partition of unity if and only if $\left\{ \varphi_\mathcal{T} : \mathcal{T} \in [\mathcal{F}]^{<\omega} \right\}$ is a partition of unity.

Theorem XVIII.6.1. *For each partition of unity $\mathcal{F}$ on X, the family*

$$\mathcal{F}' = \left\{ \varphi_\mathcal{T} : \mathcal{T} \in [\mathcal{F}]^{<\omega} \right\}, \qquad (\text{XVIII.6.3})$$

where

$$\varphi_{\mathcal{T}}(x) := \text{card}\, \mathcal{T} \cdot \max\{0, \min_{f \in \mathcal{T}} f(x) - \sup_{f \in \mathcal{F} \setminus \mathcal{T}} f(x)\} \qquad \text{(XVIII.6.4)}$$

is a unique fragmentation of $\mathcal{F}$.

Proof. Let us show that $\{\mathcal{T} : \varphi_{\mathcal{T}}(x) > 0\}$ is totally ordered for each $x \in X$. If $\varphi_{\mathcal{T}}(x) > 0$, then, by (XVIII.6.4),

$$\min_{f \in \mathcal{T}} f(x) > \sup_{f \in \mathcal{F} \setminus \mathcal{T}} f(x).$$

Thus if $f \in \mathcal{T}$, then $f(x) > g(x)$ for each $g \in \mathcal{F} \setminus \mathcal{T}$. Consequently, if $\mathcal{T}_0, \mathcal{T}_1$ are two finite subfamilies of $\mathcal{F}$ such that $\varphi_{\mathcal{T}_0}(x)\varphi_{\mathcal{T}_1}(x) > 0$, $f_1 \in \mathcal{T}_1 \setminus \mathcal{T}_0$ and $f_0 \in \mathcal{T}_0 \setminus \mathcal{T}_1$, then $f_1(x) > f_0(x)$ and $f_0(x) > f_1(x)$, which is impossible. We infer that $\mathcal{T}_0 \subset \mathcal{T}_1$ or $\mathcal{T}_1 \subset \mathcal{T}_0$.

It remains to show (XVIII.6.1). Since $\mathcal{F}$ is a partition of unity, for each $x \in X$ the set

$$\mathcal{F}_x := \{f \in \mathcal{F} : f(x) > 0\}$$

is countable (possibly finite). Let $\mathcal{F}_x = \{g_n : n \in N\}$, where N is an initial segment or the whole of $\mathbb{N}$ and $g_n(x) \geq g_{n+1}(x)$ for each $n \in N$ such that $n \in N$.

If $\mathcal{T}$ is a finite subset of $\mathcal{F}$ such that $\varphi_{\mathcal{T}}(x) > 0$, then by (XVIII.6.4), there exists $k \in N$ such that $\mathcal{T} = \mathcal{T}_k := \{g_n : n \in N, n \geq k\}$ and thus

$$\frac{1}{\text{card}\, \mathcal{T}_k} \varphi_{\mathcal{T}_k}(x) = \begin{cases} g_k(x) - g_{k+1}(x), & \text{if } k + 1 \in N, \\ g_k(x), & \text{if } k = \max N. \end{cases}$$

If $f \in \mathcal{T} \subset \mathcal{F}_x$, then there exists $n_0 \in N$ such that $f = g_{n_0}$. Therefore,

$$f(x) = \sum_{n_0 \leq k \in N} (g_k(x) - g_{k+1}(x))$$

$$= \sum_{n_0 \leq k \in N} \frac{1}{\text{card}\, \mathcal{T}_k} \varphi_{\mathcal{T}_k}(x),$$

that is, (XVIII.6.1).

To prove uniqueness, consider a fragmentation

$$\mathcal{A} = \{\alpha_{\mathcal{T}} : \mathcal{T} \in [\mathcal{F}]^{<\omega}\}$$

of $\mathcal{F}$. For each $x \in X$, the set

$$\mathcal{A}_x := \{\alpha_{\mathcal{T}} \in \mathcal{A} : \alpha_{\mathcal{T}}(x) > 0\}$$

is (possibly finite) countable, because $\mathcal{A}$ is a partition of unity. As $\mathcal{A}$ is a fragmentation of $\mathcal{F}$, the set $\{\mathcal{T} : \alpha_{\mathcal{T}}(x) > 0\}$ is totally ordered, thus there exists an initial segment N of $\mathbb{N}_1$ such that $\mathcal{T}_n \subset \mathcal{T}_{n+1}$ for each $n \in N$ with $n + 1 \in N$ and

$$\mathcal{A}_x := \{\alpha_{\mathcal{T}_n} : n \in N\}.$$

If

$$f_n(x) := \sum_{n \le k \in N} \frac{1}{\operatorname{card} \mathcal{T}_k} \alpha_{\mathcal{T}_k}(x), \qquad (\text{XVIII.6.5})$$

then $f(x) = f_{n+1}(x)$ if $f \in \mathcal{T}_{n+1} \setminus \mathcal{T}_n$ and $f(x) = 0$ if $f \notin \bigcup_{n \in N} \mathcal{T}_n$.

We shall show that $\mathcal{A} = \mathcal{F}'$. Fix $x \in X$ and consider a finite subset $\mathcal{T}$ of $\mathcal{F}$. If $\mathcal{T} \setminus \bigcup_{n \in N} \mathcal{T}_n \ne \varnothing$, then $\alpha_{\mathcal{T}}(x) = 0$ and $f(x) = 0$ for each $f \in \mathcal{T} \setminus \bigcup_{n \in N} \mathcal{T}_n$. Therefore $\varphi_{\mathcal{T}}(x) = 0$, so that $\alpha_{\mathcal{T}}(x) = \varphi_{\mathcal{T}}(x)$.

Otherwise, let n be the least natural number such that $\mathcal{T} \subset \mathcal{T}_n$. If $f \in \mathcal{T}_n \setminus \mathcal{T}$ then $\alpha_{\mathcal{T}_k}(x) = 0$ and, by (XVIII.6.5), $f(x) \le g(x)$ for all $g \in \mathcal{T}$, because $\mathcal{T} = \mathcal{T} \cap \bigcup_{n \ge k \in N} \mathcal{T}_k$. Hence $\gamma_{\mathcal{T}}(x) = 0$ and $\gamma_{\mathcal{T}}(x) = \varphi_{\mathcal{T}}(x)$. $\mathcal{T} = \mathcal{T}_n$. There remains the case $\mathcal{T} = \mathcal{T}_n$.

Set $\mathcal{T}_0 := \varnothing$. If $f \in \mathcal{T}_n \setminus \mathcal{T}_{n-1}$ for $n \in \mathbb{N}_1$, then by (XVIII.6.5), $f(x) = f_n(x) = \min_{h \in \mathcal{T}_n} h(x)$. On the other hand, $f_{n+1}(x) = \sup_{g \notin \mathcal{T}_n} g(x)$, so that

$$0 < f_n(x) - f_{n+1}(x) = \min_{h \in \mathcal{T}_n} h(x) - \sup_{g \notin \mathcal{T}_n} g(x) = \frac{1}{\operatorname{card} \mathcal{T}_n} \gamma_{\mathcal{T}_n}(x),$$

what proves $\varphi_{\mathcal{T}_n}(x) = \gamma_{\mathcal{T}_n}(x)$. $\qquad \square$

Theorem XVIII.6.2. *Every open cover of a metrizable space has an open locally finite and σ-discrete refinement.*

Proof. If X is metrizable and d is a compatible metric bounded by 1, then for each open cover $\mathcal{U}$ of X, the family

$$\{\operatorname{dist}_{X \setminus U} : U \in \mathcal{U}\}$$

is an equicontinuous and bounded functional cover. By Lemma XVIII.5.5, there exists a refining equicontinuous partition $\mathcal{F}$ of a bounded function. By Proposition XVIII.4.10, the fragmentation $\mathcal{F}'$ of $\mathcal{F}$ is an equicontinuous partition of unity refining $\mathcal{F}$ thanks to (XVIII.6.1).

For each $n \in \mathbb{N}$, the elements of $\mathcal{U}_n := \{\{\varphi_{\mathcal{T}} > 0\} : \mathcal{T} \subset \mathcal{F}, \operatorname{card} \mathcal{T} = n\}$ are disjoint. Indeed, if $\{\varphi_{\mathcal{T}} > 0\} \cap \{\varphi_{\mathcal{S}} > 0\} \ne \varnothing$ then $\mathcal{T} \subset \mathcal{S}$ or $\mathcal{S} \subset \mathcal{T}$, hence $\mathcal{T} = \mathcal{S}$, because $\operatorname{card} \mathcal{T} = \operatorname{card} \mathcal{S}$.

By virtue of XVIII.6.1, for each $n \in \mathbb{N}$ and $k \in \mathbb{N}_1$, the family

$$\mathcal{U}_{n,k} := \{\{\varphi_{\mathcal{T}} > \tfrac{1}{k}\} : \mathcal{T} \subset \mathcal{F}, \operatorname{card} \mathcal{T} = n\}$$

is a regular locally finite refinement of $\mathcal{U}_n$. Hence $\mathcal{U}_{n,k}$ is discrete, because if $x \in X$ and V is a neighborhood intersecting finitely many elements of $\mathcal{U}_{n,k}$, then there exists at most one $\mathcal{T}$ with $\operatorname{card} \mathcal{T} = n$ such that $x \in \operatorname{cl}\{\varphi_{\mathcal{T}} > \tfrac{1}{k}\}$. Thus there exists a neighborhood W of x such that $W \cap \operatorname{cl}\{\varphi_{\mathcal{S}} > \tfrac{1}{k}\} = \varnothing$ for each $\mathcal{S} \subset \mathcal{F}$ with $\operatorname{card} \mathcal{S} = n$. $\qquad \square$

XVIII.7   Metrization theorems

Theorem XVIII.7.1 (Dydak). *A Hausdorff topological space X is metrizable if and only if there exists a partition of unity $\mathcal{F}$ on X such that*

$$\{\{f > 0\} : f \in \mathcal{F}\}$$

is a base of open subsets of X.

Proof. Suppose that X is metrizable and fix a compatible metric on X. Then $\mathcal{B}_n := \{B(x, \frac{1}{n}) : x \in X\}$ is an open cover of X for each $n \in \mathbb{N}_1$ and $\mathcal{B} := \bigcup_{n \in \mathbb{N}_1} \mathcal{B}_n$ is a base of open subsets of X. If $\{f_{x,n} : x \in X\}$ is a partition of unity refining $\mathcal{B}_n$, then $\{\frac{1}{2^n} f_{x,n} : n \in \mathbb{N}_1, x \in X\}$ is a partition of unity refining $\mathcal{B}$.

Conversely, let τ be a topology on X and let $\mathcal{F}$ be a partition of unity on X such that $\{\{f > 0\} : f \in \mathcal{F}\}$ is a base of open subsets for τ. Then the function

$$d(x,y) = \sum_{f \in \mathcal{F}} |f(x) - f(y)|$$

is a compatible metric on X. For each $x \in X$, the map $y \longmapsto d(x,y)$ is τ-continuous, because $\mathcal{F}_{\mathrm{fin}}$ is equicontinuous as a partition of a continuous function (Proposition XVIII.4.7), hence $\{|f(x) - f| : f \in \mathcal{F}\}_{\mathrm{fin}}$ is equicontinuous for each $x \in X$. It follows that the topology $\widetilde{d}$ (defined by d) is coarser than τ. To see that $\widetilde{d} \geq \tau$, let O be a τ-open set and let $x \in O$. By assumption, there is $f_0 \in \mathcal{F}$ such that $x \in \{f_0 > 0\} \subset O$. Therefore $B_d(x, f_0(x))$ is $\widetilde{d}$-open and $x \in B_d(x, f_0(x)) \subset O$. Indeed, if $y \notin O$ then $f_0(y) = 0$ and $d(x,y) \geq |f_0(x) - f_0(y)| = f_0(x) > 0$, which means that $y \notin B_d(x, f_0(x))$. $\square$

Theorem XVIII.7.2 (Nagata-Smirnov). *A topological space is metrizable if and only if it is regular and admits a σ-locally finite base of open sets.*

Proof. Necessity follows from Corollary XVIII.5.8. Conversely, let X be a regular topological space and $\mathcal{B} = \bigcup_{n \in \mathbb{N}} \mathcal{B}_n$ be an open base of X such that $\mathcal{B}_n$ is locally finite for each $n \in \mathbb{N}_1$.

By adding an open set including $\bigcup_{B \in \mathcal{B}_n} B$ if necessary, we can suppose that $\mathcal{B}_n$ is an open cover of X for each $n \in \mathbb{N}_1$. By Lemma XVIII.2.6, X is perfectly normal, hence for each $B \in \mathcal{B}$ there is $f_B \in C(X, [0,1])$ such that $B = \{f_B > 0\}$. As $\mathcal{B}_n$ is a locally finite cover of X, $f_n := \sum_{B \in \mathcal{B}_n} f_B$ is continuous and strictly positive for every $n \in \mathbb{N}_1$. Therefore $\{f_B/f_n : B \in \mathcal{B}_n\}$ is a partition of unity for each $n \in \mathbb{N}_1$, hence

$$\left\{ \frac{f_B}{2^n f_n} : B \in \mathcal{B}_n, n \in \mathbb{N}_1 \right\}$$

is a partition of unity, and its support $\mathcal{B}$ is a base of open subsets of X. By virtue of Theorem XVIII.7.1 of Dydak, X is metrizable. $\qquad\square$

Since each σ-discrete family is σ-locally finite, we infer from Theorems XVIII.6.2 and XVIII.7.2 that

Theorem XVIII.7.3 (Bing). *A topological space is metrizable if and only if it is regular and admits a σ-discrete base of open sets.*

We omit several other famous metrization theorems, because their presentation would require an introduction of a series of notions. Let us mention the metrization criteria of Bing and Alexandrov, and metrization theorems of Moore, Alexandrov-Urysohn, and Arhangel'skii. Details can be found for instance in [37].

Appendix A

Set theory

A.1 Axiomatic set theory

An intuitive idea of a set is that of a collection of elements. Moreover, it is intuitive that we can define a set associated with a given property ψ, that is, the set of those x that fulfill a formula $\psi(x)$.

If no other rules were imposed, then intuitively we would be tempted to consider a set U of all sets. However if such a set existed, then

$$S := \{x \in U : x \notin x\}$$

would be a set corresponding to the formula $x \notin x$. As Bertrand Russell pointed out in his famous paradox, as each set is determined by its elements, we should be able to decide if S belongs to S. However, it follows from the definition of S that

$$S \in S \iff S \notin S,$$

which is paradoxical.

Axiomatic approach can disambiguate a theory. It consists in a finite stock of *primitive* (that is, undefined) *notions*, and a finite stock of *axioms* (that is, of propositions that are declared true). A meaning of a primitive notion is specified by the axioms. All other notions of a theory are defined from primitive notions with the aid of logical concepts and procedures. All propositions other than axioms are deduced from axioms with the aid of logical inferential rules.

Indeed, one defines a notion with the aid of other notions, so that there should be some notions that are undefined (primitive notions). Alike, to prove that a proposition is true, one needs to use other propositions that are known to be true. Therefore, some propositions (axioms) should be declared true from the very beginning.

An axiomatic theory is *consistent* if the axioms are not contradictory (^{6}).

The most commonly adopted axiomatic set theory is that of *Zermelo and Fraenkel* with the *axiom of choice* (*ZFC*), while (*ZF*) denotes the Zermelo-Fraenkel set theory without the axiom of choice. The notion of *set* is primitive. It is determined by its *elements* via the elementary relation $\in$. We say that x *belongs* to (or, is an *element* of) X if

$$x \in X.$$

We define *inclusion* $\subset$ by

$$A \subset B \iff \underset{x}{\forall}(\, x \in A \implies x \in B),$$

and we say that A is a *subset* of B.

Without presenting systematically ZFC (^{7}), we list its axioms. They give the meaning to the concept of set. They tell us which new sets can be obtained from the already existing sets (^{8}).

(1) *Axiom of extensionality* states that $X = Y$ if and only if X and Y have the same elements, that is,

$$z \in X \iff z \in Y.$$

(2) *Axiom of replacement.* Let $\varphi(x, y)$ be a formula such that

$$\varphi(x, y_0) = \varphi(x, y_1) \implies y_0 = y_1,$$

then for every set X there exists a set Y such that $Y = \{y : x \in X \wedge \varphi(x, y)\}$.

Each formula with two variables fulfilling the condition above, defines a map Φ (on the class of all sets) such that $y = \Phi(x)$ if and only if $\varphi(x, y)$. Therefore, the axiom of separation can rephrased: For each map Φ and every set X,

$$\Phi(X) := \{\Phi(x) : x \in X\}$$

is a set.

If a formula $\psi(x) = \varphi(x, y)$ is independent of y, then the axiom of replacement and the existence of a set (that we will see later) imply

---

^{6}It was shown by Gödel that it is not possible to prove consistency of a theory at least as complex as the axiomatic arithmetic within that theory. This is also the case of the axiomatic set theory that we present here.

^{7}A detailed outline of the theory can be found, for instance, in [56].

^{8}The axiom of *existence of a set* is not mentioned, because it is a consequence of the axiom of infinity.

(3) *Axiom of separation* stipulates that for every formula $\psi(x)$ (with truth value true or false) and every set X there is a set
$$\{x \in X : \psi(x)\}.$$

(4) *Axiom of union* states that for every set X (of sets) there is a set $\bigcup X$ defined by
$$y \in \bigcup X \iff \underset{A \in X}{\exists} y \in A.$$

(5) *Axiom of power set* states that for every set X there is the set 2^X of all sets included in X:
$$Y \in 2^X \iff Y \subset X.$$

(6) *Axiom of regularity* states that each non-empty set has an $\in$-minimal element.

   The axiom of regularity implies that no set is an element of itself. In other words, $x \notin x$ for every set x. In particular, $x \neq \{x\}$ every x, for if $x = \{x\}$ then $x \notin x = \{x\}$. It follows from the axioms of regularity and separation that the set of all sets cannot exist (see Russell's paradox).

(7) *Axiom of infinity* postulates existence of an infinite set.

   The axiom of infinity implies the existence of a set, which implies the existence of empty set, denoted by $\varnothing$. In fact, if X is a set, then $\{x \in X : x \neq x\}$ is a set by the axiom of separation and is empty by the axiom of extensionality.

   In order to define *infinity*, we postulate the existence of an *inductive set*, that is a set X such that $\varnothing \in X$ and
$$x \in X \implies \{x\} \in X.$$

   If an inductive set X were finite, then the operation $x \longmapsto \{x\}$ would have a fixed point $x = \{x\}$, which contradicts the axiom of regularity.

(8) *Axiom of choice.* For every set X of non-empty sets, there is a function $f : X \to \bigcup X$ with $f(A) \in A$ for all $A \in X$.

A.2 Basic set theory

Let us construct some sets with the aid of these axioms. From the axioms of replacement, of power set and of extensionality, for any a and b there is a set $\{a, b\}$ that contains exactly a and b. A *singleton* is defined by $\{a\} := \{a, a\}$, and an *ordered pair* (a, b) is formally defined by $(a, b) := \{\{a\}, \{a, b\}\}$ to the effect that
$$(a_0, b_0) = (a_1, b_1) \iff ((a_0 = a_1) \wedge (b_0 = b_1)),$$

where $\wedge$ stands for the logical conjunction (*and*). The *product* $X \times Y$ is the set of all ordered (x, y) with $x \in X$ and $y \in Y$.

A subset R of $X \times Y$ can be seen as a binary relation R from X to Y. Define as usual

$$R(x) := \{y \in Y : (x, y) \in R\}$$

and

$$R(F) := \bigcup_{x \in F} R(x),$$

and, similarly, the inverse relation R^- from Y to X by

$$R^-(y) = \{x \in X : (x, y) \in R\}.$$

Thus we will sometimes consider relations as *multivalued maps* and use the notation

$$R : X \rightrightarrows Y \text{ and } R^- : Y \rightrightarrows X,$$

rather than

$$R : X \to 2^Y \text{ and } R^- : Y \to 2^X.$$

A *map* $f : X \to Y$ is determined by its *graph*, that is, a binary relation $\widehat{f} \subset X \times Y$ such $(\{x\} \times Y) \cap \widehat{f}$ is a singleton for every $x \in X$, as follows

$$y = f(x) \iff (x, y) \in \widehat{f}. \tag{A.2.1}$$

A *partial map* f from X to Y

$$f : X \rightarrowtail Y$$

is defined by (A.2.1), where $(\{x\} \times Y) \cap \widehat{f}$ is either a singleton or empty for every $x \in X$. The *domain* $\mathrm{dom}\, f$ of a partial map f consists of those $x \in X$, for which $(\{x\} \times Y) \cap \widehat{f} \neq \varnothing$. Accordingly, a partial map f restricted to $\mathrm{dom}\, f$ is a map.

As we have said in Section I.1, there exists a one-to-one correspondence between each map $f : X \to Y$ and a relation $\widehat{f}$, called the *graph of* f, namely,

$$\widehat{f} := \{(x, y) \in X \times Y : y = f(x)\}.$$

Of course, $\widehat{f}(x) = \{f(x)\}$. Actually, in axiomatic set theory, a *map* (*function, mapping*) is defined as a particular relation. A relation R from X to Y is called *injective* if $R(x) \cap R(t) = \varnothing$ whenever $x \neq t$, and *surjective* (or *onto*) if $R(X) = Y$. Clearly, a function $f : X \to Y$ is injective (respectively, onto) if and only if its graph is an injective (respectively, onto) relation.

A set X is said to be finite *finite* if each injection $f : X \to X$ is surjective; a set is infinite if it is not finite.

Exercise A.2.1. Show that a relation $R \subset X \times Y$ is the graph of a map from X to Y if and only if the inverse relation R^- is both injective and surjective.

If $\{Y_x : x \in X\}$ is a *family*, that is a set of sets, then the *Cartesian product*

$$\prod_{x \in X} Y_x$$

is defined as the set of all maps $\varphi : X \to \bigcup_{x \in X} Y_x$ such that $\varphi(x) \in Y_x$ for each $x \in X$. It follows from this definition that if $Y_x = Y$ for each $x \in X$, then

$$\prod_{x \in X} Y = Y^X.$$

The projection $p_{x_0} : \prod_{x \in X} Y_x \to Y_{x_0}$ is defined by

$$p_{x_0}(\varphi) := \varphi(x_0).$$

Sometimes one denotes elements of $\prod_{x \in X} Y_x$ by $(y_x)_{x \in X}$, that is, $y_x := \varphi(x)$ and implicitly $y_x \in Y_x$.

We are now in a position to reformulate the axiom of choice.

Axiom of choice. If X is a set and $Y_x \neq \varnothing$ for each $x \in X$, then $\prod_{x \in X} Y_x \neq \varnothing$.

A.3 Natural numbers

The *natural numbers* were defined, axiomatically by Giuseppe Peano in [87, 1889], as a set $\mathbb{N}$ with a (*shift*) function $s : \mathbb{N} \to \mathbb{N}$ such that $\mathbb{N} \backslash s(\mathbb{N}) = \{0\}$. This description includes already the *induction principle*, which says that if a subset A of $\mathbb{N}$ is such that $0 \in A$ and $s(A) \subset A$, then $A = \mathbb{N}$. For each $n \in \mathbb{N}$, the element $s(n)$ of $\mathbb{N}$ is called a *successor* of n, and is also denoted by $s(n) = n + 1$.

With the aid of $s : \mathbb{N} \to \mathbb{N}$, of a zero element 0 and of the induction principle, one defines the usual order, addition and multiplication on $\mathbb{N}$. For each $m \in \mathbb{N}$, let

$$\mathbb{N}_m := \{n \in \mathbb{N} : n \geq m\}.$$

Of course, $\mathbb{N}_0 = \mathbb{N}$.

A.4   Cardinality

Two sets X, Y are said to be *equipotent* $(X \sim Y)$ if there exists a bijection $f : X \to Y$. If X, Y, Z are sets then

(1) $X \sim X$ (reflexivity),
(2) if $X \sim Y$ then $Y \sim X$ (symmetry), and
(3) $X \sim Y$ and $Y \sim Z$ entails $X \sim Z$ (transitivity).

Therefore *equipotency* is an equivalence relation on the class of all sets. An equivalence class of equipotent sets is called a *cardinal number*. We denote card X (or $|X|$) the equivalence class of sets that are equipotent to X, that is, the *cardinality* of X.

Definition A.4.1. A set X is *finite* if each injection from X to X is a surjection.

Example A.4.2. The empty set $\varnothing$ is finite. If $R \subset \varnothing \times \varnothing$, then $R = \varnothing$. As $\varnothing^- (\varnothing) = \varnothing$, the inverse relation of $\varnothing$ is surjective. On the other hand, $\varnothing^- = \varnothing$ is injective, because the image of each element is empty. It follows that $\varnothing$ is the graph of a map. This is the only map from $\varnothing$ to $\varnothing$ and, by the same argument that we used for $\varnothing^-$, it is a bijection.

Proposition A.4.3. *If X is finite, then $X \cup \{y\}$ is finite for each y.*

Proof. It is enough to show this when $y \notin X$. Let $f : X \cup \{y\} \to X \cup \{y\}$ be injective. If $f(X) \subset X$, then $f(X) = X$, because X is finite, hence $f(y) = y$, and thus f is surjective.

If $y \in f(X)$, then there exists a single $x_0 \in X$ such that $f(x_0) = y$. Let $x_1 := f(y)$ and let $h : X \cup \{y\} \to X \cup \{y\}$ be a bijection for which all the points are fixed except for x_0 and x_1. Then $g := h \circ f$ is an injection injection such that $g(x_0) = y$ and $g(y) = x_0$. Accordingly, $g(X \setminus \{x_0\}) \subset X \setminus \{x_0\}$ and, since $X \setminus \{x_0\}$ is finite, $g(X \setminus \{x_0\}) = X \setminus \{x_0\}$. On the other hand, $g(\{x_0, y\}) = \{x_0, y\}$, what proves that g is surjective. It follows that $f = h^{-1} \circ g$ is surjective. $\square$

We conclude, that for each $n \in \mathbb{N}$ the set $\{x_1, \dots, x_n\}$ is finite. Conversely,

Proposition A.4.4. *If X is finite, then there is $n \in \mathbb{N}$ such that $X \sim \{x_1, \dots, x_n\}$.*

Proof. If $X = \varnothing$, then $\operatorname{card} X = 0$, otherwise there exists $x_1 \in X$. By induction, we construct a sequence of elements of X such that

$$x_k \in X \setminus \{x_1, \ldots, x_{k-1}\}.$$

Then there exists $n \in \mathbb{N}$ such that $X \setminus \{x_1, \ldots, x_n\} = \varnothing$, that is, $X \sim \{1, \ldots, n\}$. Otherwise, X would contain an injective sequence $(x_n)_{n \in \mathbb{N}}$ and thus $\operatorname{card} \mathbb{N} \leq \operatorname{card} X$, which is a contradiction. $\square$

Proposition A.4.5. $\operatorname{card} X \leq \operatorname{card} Y$ *if and only if there exists an injective map $f : X \to Y$.*

Proof. If $\operatorname{card} X \leq \operatorname{card} Y$ then by definition, there exist sets X_0 and Y_0 such that $X_0 \subset Y_0, X_0 \sim X$ and $Y_0 \sim Y$: in other words, there exist bijections $h : X \to X_0$ and $k : Y_0 \to Y$. Consequently, $k \circ h$ is an injection from X to Y. Conversely, if $f : X \to Y$ is an injection, then $X \sim f(X)$ and $f(X) \subset Y$. $\square$

Proposition A.4.6. *If X, Y are non-empty sets, then there exists an injective map $f : X \to Y$ if and only if there exists a surjective map $g : Y \to X$.*

Proof. If $f : X \to Y$ is injective, then let $g(y) := f(x)$ if $y \in f(X)$. If $y \in Y \setminus f(X)$ then $g(y)$ can be any element of X (as, by assumption there exists $x_0 \in X$, we can set $g(y) := x_0$ for each $y \in Y \setminus f(X)$.

Conversely, if $g : Y \to X$ is surjective, then for each $x \in X$ there exists an element $f(x) \in g^{-}(x)$. If $x_0 \neq x_1$ then $g^{-}(x_0) \cap g^{-}(x_1) = \varnothing$, hence $f(x_0) \neq f(x_1)$. This proves that f is injective. $\square$

Let κ and λ be cardinal numbers. We write

$$\kappa \leq \lambda$$

if there exist sets $Y \subset X$ such that $\operatorname{card} Y = \kappa$ and $\operatorname{card} X = \lambda$.

Obviously, $\leq$ is a relation on cardinals which is reflexive and transitive. We shall show that $\leq$ is also antisymmetric, that is, an order.

Theorem A.4.7 (Cantor-Bernstein). *If $\kappa \leq \lambda$ and $\lambda \leq \kappa$, then $\lambda = \kappa$.*

Proof. Let κ be the cardinality of X and λ be the cardinality of Y. Since $\kappa \geq \lambda$ we can assume without loss of generality that $Y \subset X$. On the other hand, $\kappa \leq \lambda$, there exists an injective map $f : X \to Y$.

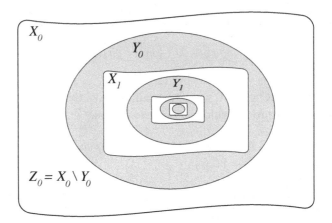

Let $X_0 := X, Y_0 := Y, Z_0 := X_0 \setminus Y_0$ and

$$X_{n+1} := f(X_n), Y_{n+1} := f(Y_n) \text{ and } Z_{n+1} := f(Z_n).$$

As f is injective,

$$Z_1 = f(Z_0) = f(X_0 \setminus Y_0) = f(X_0) \setminus f(Y_0) = X_1 \setminus Y_1,$$

and, more generally, $Z_n = X_n \setminus Y_n$. Let $Z := \bigcup_{n=0}^{\infty} Z_n$. Hence

$$f(Z) = \bigcup_{n=0}^{\infty} f(Z_n) = \bigcup_{n=1}^{\infty} Z_n = Z \setminus Z_0.$$

Moreover, $Z_n \cap Z_k = \varnothing$ if $n \neq k$.

Let us define $h : X \to Y$ by

$$h(x) := \begin{cases} f(x), & \text{if } x \in Z, \\ x, & \text{if } x \notin Z. \end{cases}$$

The map h is injective, because $f(Z) \subset Z$ and the restriction of f to Z is injective, and the identity is injective on $X \setminus Z$. The map h is surjective on Y. Indeed, as h is injective,

$$h(X) = h(Z \cup (X \setminus Z)) = f(Z) \cup X \setminus Z$$
$$= (Z \setminus Z_0) \cup (X \setminus Z) = X \setminus Z_0 = Y.$$

We conclude that $\kappa = \lambda$. $\qquad\qquad\qquad\qquad\qquad\qquad\qquad\qquad\square$

Consequently the relation $\leq$ on cardinal numbers is antisymmetric. We write

$$\kappa < \lambda \text{ if } \kappa \leq \lambda \text{ and } \kappa \neq \lambda.$$

The cardinal number *zero* is the cardinality of the empty set

$$\operatorname{card} \varnothing = 0,$$

the cardinality of one-element sets is 1, and so on. We call *aleph zero* the cardinality of $\mathbb{N}$,

$$\operatorname{card} \mathbb{N} = \aleph_0.$$

A cardinal is called *countable* if it is the image of a sequence. A countable cardinal is either finite or infinite.

Exercise A.4.8. A countable cardinal is infinite if and only if it is $\aleph_0$.

Exercise A.4.9. $\aleph_0$ is the least infinite cardinal.

The *sum* $\kappa + \lambda$ of two cardinals κ, λ is defined as the cardinality of the disjoint union $X \cup Y$, where $\operatorname{card} X = \kappa$ and $\operatorname{card} Y = \lambda$. The *product* $\kappa \cdot \lambda$ of two cardinals κ, λ is defined as the cardinality of $X \times Y$, where $\operatorname{card} X = \kappa$ and $\operatorname{card} Y = \lambda$. The *power* λ^κ is defined as the cardinality of Y^X, where $\operatorname{card} X = \kappa$ and $\operatorname{card} Y = \lambda$.

Exercise A.4.10. Show that

$$\kappa \leq \lambda \implies \kappa \cdot \mu \leq \lambda \cdot \mu, \ \kappa^{\lambda+\mu} = \kappa^\lambda \cdot \kappa^\mu, \ \left(\kappa^\lambda\right)^\mu = \kappa^{\lambda \cdot \mu}. \tag{A.4.1}$$

On the other hand,

Proposition A.4.11. $\aleph_0 = \aleph_0 + \aleph_0 = \aleph_0 \cdot \aleph_0$.

Proof. Of course, $\aleph_0 \leq \aleph_0 + \aleph_0 \leq \aleph_0 \cdot \aleph_0$. As $\operatorname{card}(\mathbb{N}) = \aleph_0$, it is enough to show that $\operatorname{card}(\mathbb{N} \times \mathbb{N}) \leq \operatorname{card}(\mathbb{N})$. In fact, we see that $\operatorname{card}(\mathbb{N} \times \mathbb{N}) \leq \operatorname{card}(\mathbb{N})$, because the map

$$f(p,q) := \frac{(p+q)(p+q+1)}{2} + p \tag{A.4.2}$$

is injective from $\mathbb{N} \times \mathbb{N}$ onto $\mathbb{N}$. Indeed, if

$$S(n) := \sum_{k=0}^n k = \frac{n(n+1)}{2},$$

then $f(p,q) = S(p,q) + p$. In particular, $n < S(n+1) - S(n)$ for each $n \in \mathbb{N}$. But $(p_0, q_0) < (p_1, q_1)$ if and only if

$$p_0 + q_0 < p_1 + q_1, \text{ or } p_0 + q_0 = p_1 + q_1 \text{ and } p_0 < p_1.$$

In the first case $p_0 + q_0 < S(p_1 + q_1) - S(p_0 + q_0)$, hence

$$f(p_0, q_0) = S(p_0 + q_0) + p_0 < S(p_1 + q_1) \leq f(p_1, q_1).$$

As for the second, $f(p_1, q_1) - f(p_0, q_0) = p_1 - p_0 > 0$. We have proved injectivity. $\square$

In fact, we can show that (A.4.2) is bijective.

Exercise A.4.12. Prove that (A.4.2) is surjective.

Corollary A.4.13. $\operatorname{card} \mathbb{Q} = \aleph_0$.

Proof. By Proposition A.4.11, the cardinality of $\mathbb{Q}$ is equal to $\mathbb{Q}_+ := \{q \in \mathbb{Q} : q \geq 0\}$. On the other hand, each $q \in \mathbb{Q}$ can be written as $q = \frac{k}{n}$, where $k \in \mathbb{N}$ and $n \in \mathbb{N}_1$, thus the map $(k, n) \longmapsto \frac{k}{n}$ is surjective from $\mathbb{N} \times \mathbb{N}_1$ onto $\mathbb{Q}_+$, thus $\aleph_0 \geq \operatorname{card} \mathbb{Q}_+ \geq \operatorname{card} \mathbb{N} = \aleph_0$, because $\mathbb{N} \subset \mathbb{Q}_+$. $\qquad\square$

We denote by $\mathfrak{c}$ the cardinality of the *real line* $\mathbb{R}$, and we call it the *continuum*.

Proposition A.4.14. *If κ is a cardinal, then*

$$\kappa^0 = 1,$$
$$1^\kappa = 1.$$
$$\kappa > 0 \Longrightarrow 0^\kappa = 0.$$

Proof. By definition, $\kappa^0 = \operatorname{card} Y^\varnothing$ where $\operatorname{card} Y = \kappa$. The set $\varnothing \times Y = \varnothing$ has only one subset $\varnothing$ ($\operatorname{card} 2^{\varnothing \times Y} = 1$), that is, there is the only relation $\varnothing$ in $\varnothing \times Y$. This relation is an application, because $\varnothing^{-1} y = \varnothing$ for each $y \in Y$, hence in particular $\varnothing^{-1} y_0 \cap \varnothing^{-1} y_1 = \varnothing$ for every $y_0 \neq y_1$. On the other hand, $\varnothing^{-1}(Y) = \varnothing$.

By definition, $1^\kappa = \operatorname{card}(\{0\}^X)$, where $\operatorname{card} X = \kappa$. There is the only relation R in $X \times \{0\}$ for which $R^{-1}\{0\} = X$, that is, $R := X \times \{0\}$. This is an application, because $R^{-1} y_0 \cap R^{-1} y_1 = \varnothing$ if $y_0 \neq y_1$, hence $1^\kappa = 1$.

Finally $0^\kappa = \operatorname{card}(\varnothing^X)$, where $\operatorname{card} X > 0$. There is only one relation in $\varnothing = X \times \varnothing$, that is $\varnothing$. This however is not an application, because $\varnothing = \varnothing^{-1}\varnothing \neq X$. It follows that $0^\kappa = 0$. $\qquad\square$

Theorem A.4.15 (Cantor). *If κ is a cardinal, then $\kappa < 2^\kappa$.*

Proof. Let X be a set of cardinality κ. The map $h : X \to 2^X$ given by $h(x) := \{x\} \in 2^X$ is injective. Therefore $\kappa \leq 2^\kappa$. If $f : X \to 2^X$ is any application, then

$$\{x \in X : x \notin f(x)\}$$

does not belong to $f(X)$, that is, f is not surjective. Indeed, if there were $y \in X$ such that $f(y) = \{x \in X : x \notin f(x)\}$, then $y \in f(y)$ if and only if $y \notin f(y)$, which is a contradiction. $\qquad\square$

It follows that there exist infinitely many infinite cardinals. In fact, if κ is an infinite cardinal, for example, $\kappa = \aleph_0$, then the cardinals

$$\kappa < 2^\kappa < 2^{2^\kappa} < 2^{2^{2^\kappa}} < \ldots$$

are all different.

A.5 Continuum

By definition, $\{0,1\}^{\mathbb{N}} := \prod_{n \in \mathbb{N}} \{0,1\}$ is the set of all applications f from the set of natural numbers to $\{0,1\}$:

$$f \in \{0,1\}^{\mathbb{N}} \Longleftrightarrow f : \mathbb{N} \to \{0,1\}.$$

Therefore, it can be identified with the set of all subsets of $\mathbb{N}$ with the aid of the bijection

$$f \mapsto \{n \in \mathbb{N} : f(n) = 1\}$$

from $\{0,1\}^{\mathbb{N}}$ onto $2^{\mathbb{N}}$. This set is called the *Cantor cube*. It follows from the definition that $\mathrm{card}(\{0,1\}^{\mathbb{N}}) = 2^{\aleph_0}$.

The *Cantor set* C consists of the reals r, the tertiary representation of which fulfills

$$r = \sum_{n=1}^{\infty} \frac{r^{[n]}}{3^n} \quad \text{and} \quad \underset{n \in \mathbb{N}_1}{\forall} \ r^{[n]} \in \{0,2\}. \tag{A.5.1}$$

Consequently $C \subset [0,1]$.

Proposition A.5.1. *Each element of the Cantor set C admits a unique representation (A.5.1).*

Proof. Let r, s be two elements of C. Of course, $r^{[n]}$ and $s^{[n]}$ are the n-th terms of the tertiary representations (A.5.1) of r and s, respectively. Let $\#(r,s)$ be the first $n \in \mathbb{N}_1$ such that $r^{[n]} \neq s^{[n]}$ and $\#(r,s) = \infty$ and let $r = s$. As a consequence, if $r, s \in C$ and $r < s$, then there exists n such that $1 \leq n < \infty$ and $n = \#(r,s)$. Therefore $r^{[n]} = 0$ and $s^{[n]} = 2$. We shall prove that

$$\min \{s - r : \#(r,s) = n\} = \frac{1}{3^n}.$$

As $r < s$ and $n = \#(r,s)$,

$$s - r = \frac{2}{3^n} + \sum_{k=n+1}^{\infty} \frac{s^{[k]} - r^{[k]}}{3^k} \geq \frac{2}{3^n} - \frac{2}{3^{n+1}} \sum_{k=0}^{\infty} \frac{1}{3^k} = \frac{1}{3^n}. \tag{A.5.2}$$

This lower bound is attained if $r^{[n]} = 0$ and $r^{[k]} = 2$ for each $k > n$ and $s^{[n]} = 2$ and $s^{[k]} = 0$ for each $k > n$. $\square$

The Cantor set C is the intersection of the following sets:

$$[0,1] \qquad \text{(level 0)}$$

$$\left[0,\tfrac{1}{3}\right] \cup \left[\tfrac{2}{3},1\right] \qquad \text{(level 1)}$$

$$\left[0,\tfrac{1}{9}\right] \cup \left[\tfrac{2}{9},\tfrac{1}{3}\right] \cup \left[\tfrac{2}{3},\tfrac{7}{9}\right] \cup \left[\tfrac{8}{9},1\right] \qquad \text{(level 2)}$$

$$\cdots$$

$$\left[0,\tfrac{1}{3^n}\right] \cup \left[\tfrac{2}{3^n},\tfrac{3}{3^n}\right] \cup \ldots \cup \left[\tfrac{3^n-3}{3^n},\tfrac{3^n-2}{3^n}\right] \cup \left[\tfrac{3^n-1}{3^n},1\right] \qquad \text{(level } n\text{)}$$

For example, if $\#(r,s) = 1$, that is, $r^{[1]} \neq s^{[1]}$, then r and s belong to two different factor of the first level; if $\#(r,s) = 2$, then $r^{[1]} = s^{[1]}$ and $r^{[2]} \neq s^{[2]}$, hence r and s belong to the same level of the first level and to two different levels of the second levels.

Proposition A.5.2. *The Cantor set is equipotent to the Cantor cube.*

Proof. The application of $\{0,1\}^{\mathbb{N}}$ to C given by

$$F(f) := \sum_{n=1}^{\infty} \frac{2f(n)}{3^n}$$

is surjective, by the definition of the Cantor set (A.5.1), is injective, thanks to Proposition A.5.1. □

Theorem A.5.3. $\mathfrak{c} = 2^{\aleph_0}$.

Proof. Let $F : \mathbb{R} \to 2^{\mathbb{Q}}$ by

$$F(r) := \{f_r(n) : n \in \mathbb{N}\},$$

where $f_r : \mathbb{N} \to \mathbb{Q}$ is injective and such that $r = \lim_{n \to \infty} f_r(n)$ ([9]). If $r_0 \neq r_1$, then

$$\{f_{r_0}(n) : n \in \mathbb{N}\} \cap \{f_{r_1}(n) : n \in \mathbb{N}\}$$

is finite, and therefore $f_{r_0} \neq f_{r_1}$. Hence $F : \mathbb{R} \to 2^{\mathbb{Q}}$ is injective, thus $\mathfrak{c} = \operatorname{card} \mathbb{R} \leq \operatorname{card} 2^{\mathbb{Q}} = 2^{\aleph_0}$. As the Cantor set C is a subset of $\mathbb{R}$, $2^{\aleph_0} = \operatorname{card} C \leq \operatorname{card} \mathbb{R} = \mathfrak{c}$. Hence, by Theorem A.4.7 of Cantor-Bernstein, $\mathfrak{c} = 2^{\aleph_0}$. □

Corollary A.5.4. $\mathfrak{c} = \mathfrak{c} + \mathfrak{c} = \aleph_0 \cdot \mathfrak{c} = \mathfrak{c} \cdot \mathfrak{c} = \mathfrak{c}^{\aleph_0}$ ([10]).

Proof. Of course, by (A.4.1),

$$\mathfrak{c} \leq \mathfrak{c} + \mathfrak{c} \leq \aleph_0 \cdot \mathfrak{c} \leq \mathfrak{c} \cdot \mathfrak{c},$$

because $1 \leq 2 \leq \aleph_0 \leq \mathfrak{c}$, and $\mathfrak{c} \cdot \mathfrak{c} = \mathfrak{c}^2 \leq \mathfrak{c}^{\aleph_0}$, because $2 \leq \aleph_0$.

On the other hand, by Proposition A.4.11 and by Theorem A.5.3, $\mathfrak{c}^{\aleph_0} = (2^{\aleph_0})^{\aleph_0} = 2^{\aleph_0 \cdot \aleph_0} = 2^{\aleph_0} = \mathfrak{c}$. □

---

[9]Of course, there are many sequences that verify this condition

[10]This corollary is a consequence of Theorem A.11.6 to be discussed later.

A.6 Order

Let X be a set ordered by $\leq$, that is, a binary relation on X fulfilling

$$x \leq x, \tag{A.6.1}$$

$$x_0 \leq x_1 \text{ and } x_1 \leq x_2 \Longrightarrow x_0 \leq x_2, \tag{A.6.2}$$

$$x_0 \leq x_1 \text{ and } x_1 \leq x_0 \Longrightarrow x_0 = x_1. \tag{A.6.3}$$

The strict order $<$ associated with $\leq$ is defined by

$$x_0 < x_1 \Longleftrightarrow x_0 \leq x_1 \text{ and } x_0 \neq x_1.$$

If X and Y are ordered sets and $f : X \to Y$ is a map, then f is *increasing* if

$$x_0 \leq x_1 \Longrightarrow f(x_0) \leq f(x_1),$$

and *strictly increasing* if

$$x_0 < x_1 \Longrightarrow f(x_0) < f(x_1).$$

Of course, every strictly increasing map is increasing.

Let A be a subset of a partially ordered set $(X, \leq)$. The *supremum of A*, denoted $\sup A$ or $\bigvee_{a \in A} a$, is the smallest element of $\left\{ x \in X : \underset{a \in A}{\forall} \, a \leq x \right\}$ provided that it exists. Similarly, the *infimum of A*, denoted $\inf A$ or $\bigwedge_{a \in A} a$ is the greatest element of $\left\{ x \in X : \underset{a \in A}{\forall} \, x \leq a \right\}$ if it exists.

If f is increasing and bijective and if the inverse map f^{-1} is increasing, then f is called an *order isomorphism*. Two ordered sets X, Y are said to be *(order) isomorphic* $(X \simeq Y)$, if there exists an order isomorphism between them. Let

$$\text{Min } A := \left\{ a \in A : \underset{x \in A}{\forall} \, a \leq x \right\},$$

$$\text{Max } A := \left\{ a \in A : \underset{x \in A}{\forall} \, x \leq a \right\}.$$

An ordered set X is called *totally ordered* if each couple $x_0, x_1 \in X$, either $x_0 \leq x_1$ or $x_0 \geq x_1$. If X is totally ordered and $A \subset X$, then Min A is either empty or a singleton, and then we write $\text{Min } A = \{\min A\}$. Similarly, Max A is either empty or a singleton, in which case we write $\text{Max } A = \{\max A\}$. Then the supremum of A is $\min \left\{ x \in X : \underset{a \in A}{\forall} \, a \leq x \right\}$ and the infimum of A is $\max \left\{ x \in X : \underset{a \in A}{\forall} \, x \leq a \right\}$ provided they exist.

A.7 Lattice

If $(X, \leq)$ is a partially ordered set such that

$$\sup\{x_1, x_2\} = x_1 \vee x_2$$

and

$$\inf\{x_1, x_2\} = x_1 \wedge x_2$$

exist for each x_1 and x_2 in X, then X is called a *lattice*.

A *complete lattice* is a lattice in which

$$\sup A = \bigvee A$$

and

$$\inf A = \bigwedge A$$

exist for each non-empty subset. In particular, a complete lattice has a least and a greatest element.

The real line $(\mathbb{R}, \leq)$ is a lattice that is not complete, but it can easily be completed: Let

$$\overline{\mathbb{R}} := \{-\infty\} \cup \mathbb{R} \cup \{\infty\}$$

denote the *extended real line*, in which $-\infty \leq x \leq \infty$ for all $x \in \mathbb{R}$. Then $(\overline{\mathbb{R}}, \leq)$ is a complete lattice.

A lattice X is called *distributive* if

$$x \wedge (y \vee z) = (x \wedge y) \vee (x \wedge z) \tag{A.7.1}$$

for all $x, y, z \in X$. In a complete lattice, stronger forms of distributivity can be considered:

A complete lattice $(X, \leq)$ is called a *frame* if it satisfies the following infinite distributive law:

$$x \wedge \bigvee A = \bigvee_{a \in A} (x \wedge a), \tag{A.7.2}$$

for all $x \in X$ and $A \subset X$.

An even stronger form of distributivity in a complete lattice is that arbitrary inf distribute over arbitrary sup. We will formulate this property in different terms:

If X is a complete lattice, then for every non-degenerate family $\mathcal{A}$ of subsets of X,

$$\bigvee_{A \in \mathcal{A}} \bigwedge A = \sup_{A \in \mathcal{A}} \inf_{x \in A} x \leq \inf_{H \in \mathcal{A}^{\#}} \sup_{x \in H} x = \bigwedge_{H \in \mathcal{A}^{\#}} \bigvee H. \tag{A.7.3}$$

Indeed, let $H \in \mathcal{A}^{\#}$, that is, for each $A \in \mathcal{A}$ there exists $x_A \in H \cap A$. Hence $\inf_{x \in A} x \leq x_A$ and $H \supset \{x_A : A \in \mathcal{A}\} \in \mathcal{A}^{\#}$. Therefore

$$\sup_{A \in \mathcal{A}} \inf_{x \in A} x \leq \sup_{A \in \mathcal{A}} x_A \leq \sup_{x \in H} x,$$

and, since this holds for every $H \in \mathcal{A}^{\#}$, (A.7.3) holds.

A complete lattice X is said to be *completely distributive* if for every for every non-degenerate family $\mathcal{A}$ of subsets of X,

$$\bigvee_{A \in \mathcal{A}} \bigwedge A = \sup_{A \in \mathcal{A}} \inf_{x \in A} x \geq \inf_{H \in \mathcal{A}^{\#}} \sup_{x \in H} x = \bigwedge_{H \in \mathcal{A}^{\#}} \bigvee H, \qquad (A.7.4)$$

that is, in view of (A.7.3), equality holds.

Example A.7.1. The power set 2^X of any set X is a completely distributive lattice for the inclusion order.

Indeed, if $\mathfrak{A}$ is a family of subsets of 2^X, and

$$x \notin \bigcup_{A \in \mathfrak{A}} \bigcap_{A \in \mathcal{A}} A$$

then for each $\mathcal{A} \in \mathfrak{A}$ there is $A_\mathcal{A} \in \mathcal{A}$ with $x \notin A_\mathcal{A}$. Then $\mathcal{H} := \{A_\mathcal{A} : \mathcal{A} \in \mathfrak{A}\} \in \mathfrak{A}^{\#}$ and $x \notin \bigcup_{H \in \mathcal{H}} H$ so that

$$x \notin \bigcap_{\mathcal{H} \in \mathfrak{A}^{\#}} \bigcup_{H \in \mathcal{H}} H,$$

which proves (A.7.4).

Proposition A.7.2. $\overline{\mathbb{R}}$ *is a completely distributive lattice.*

Proof. Let $\mathcal{A}$ be a family of subsets of $\overline{\mathbb{R}}$ and let $\varepsilon > 0$. For each $A \in \mathcal{A}$, there is $a_A \in A$ such that $a_A \leq \bigwedge A + \varepsilon$. Moreover, $\{a_A : A \in \mathcal{A}\} \in \mathcal{A}^{\#}$ so that

$$\bigwedge_{H \in \mathcal{A}^{\#}} \bigvee H \leq \bigvee_{A \in \mathcal{A}} a_A \leq \bigvee_{A \in \mathcal{A}} (\bigwedge A + \varepsilon) = \varepsilon + \bigvee_{A \in \mathcal{A}} \bigwedge A$$

for every $\varepsilon > 0$, which yields (A.7.4). □

Exercise A.7.3. Show that

(1) If V is completely distributive lattice, then for every non-degenerate family $\mathcal{A}$ of subsets of a set X and each map $f : X \to V$,

$$\sup_{H \in \mathcal{A}^{\#}} \inf_{v \in H} f(v) = \inf_{A \in \mathcal{A}} \sup_{v \in A} f(v).$$

(2) A two-point set $\{0, 1\}$ with $0 \leq 1$ is a completely distributive lattice.

(3) If α is a propositional function on V, then for every non-degenerate family $\mathcal{A}$ of subsets of V,

$$\underset{A \in \mathcal{A}}{\forall} \underset{v \in A}{\exists} \alpha(v) \equiv \underset{H \in \mathcal{A}^{\#}}{\exists} \underset{v \in H}{\forall} \alpha(v).$$

A.8   Well ordered sets

An ordered set X is called *well ordered* if $\operatorname{Min} A \neq \varnothing$ for every $\varnothing \neq A \subset X$. A subset of a well ordered set is well ordered.

Exercise A.8.1. Each well ordered set is totally ordered.

Exercise A.8.2. A sequence on a well-ordered set has a non-decreasing subsequence.

Example A.8.3. The real line $\mathbb{R}$ with the natural order, as well as its subsets $\mathbb{Z}$ and $\mathbb{Q}$ are not well ordered, while $\mathbb{N}$ is well ordered.

Proposition A.8.4. *If X is well ordered and $f : X \to X$ strictly increasing, then for each $x \in X$,*

$$x \leq f(x).$$

Proof. Otherwise,

$$A := \{x \in X : x > f(x)\} \neq \varnothing.$$

As X is well ordered, there exists $x_0 = \min A$. Therefore $x_1 := f(x_0) < x_0$ and, since f is strictly increasing, $f(x_1) < f(x_0) = x_1$, that is, $x_1 \in A$ and $x_1 < \min A$, which is a contradiction. ☐

Proposition A.8.5. *If X is a well ordered set and $f : X \to X$ is an order isomorphism, then f is the identity: $f(x) = x$ for each $x \in X$.*

Proof. If f is an isomorphism, then by Proposition A.8.4, $x \leq f(x)$ and, on the other hand, $y \leq f^{-1}(y)$ for each $y \in X$, hence $x \leq f(x) \leq f^{-1}(f(x)) = x$. ☐

Proposition A.8.6. *If X and Y are well ordered and isomorphic, then there exists a unique isomorphism.*

Proof. If $f : X \to Y$ and $g : X \to Y$ are isomorphisms, then $g^{-1} \circ f : X \to X$ is an isomorphism, hence by Proposition A.8.5, $g^{-1} \circ f = i_X$ and thus $g = f$. ☐

Proposition A.8.7. *If X, Y are totally ordered and $f : X \to Y$ is a strictly increasing bijection, then f is an isomorphism.*

Proof. Let us show that $f^{-1} : Y \to X$ is strictly increasing. If $y_0 < y_1$ but $f^{-1}(y_0) < f^{-1}(y_1)$, then $f^{-1}(y_0) \geq f^{-1}(y_1)$, because the order of X is total. Hence

$$y_0 = f\left(f^{-1}(y_0)\right) \geq f\left(f^{-1}(y_1)\right) = y_1,$$

which is a contradiction. □

This proposition is no longer valid if the order of the domain is not total.

Example A.8.8. Let $X := 2^{\{0,1\}}$ with the order given by the inclusion, that is, $\varnothing \subset \{0\} \subset \{0,1\}$ and $\varnothing \subset \{1\} \subset \{0,1\}$. On the hand, let $\varnothing \prec \{0\} \prec \{1\} \prec \{0,1\}$. Then the identity $i : (X, \subset) \to (X, \prec)$ is strictly increasing, but not an isomorphism.

If W is well ordered, then for each $x \in W$ the set

$$W(x) := \{w \in W : w < x\}$$

is called the *initial segment* of W corresponding to x.

Proposition A.8.9. *A well ordered set is not isomorphic to any of its initial segments.*

Proof. If W is well ordered, $x \in W$ and $f : W \to W(x)$ is strictly increasing, then $f(x) \in W(x)$, hence $f(x) < x$, in contradiction with Proposition A.8.4. □

Theorem A.8.10 (Trichotomy). *If X, Y are well ordered, then exactly one of the following situations holds:*

(1) $X \simeq Y$,
(2) there exists $y_0 \in Y$ such that $X \simeq Y(y_0)$,
(3) there exists $x_0 \in X$ such that $Y \simeq X(x_0)$.

Proof. Let

$$F := \{(x,y) \in X \times Y : X(x) \simeq Y(y)\}.$$

If $y_0, y_1 \in F(x)$ then $Y(y_0) \simeq X(x) \simeq Y(y_1)$, hence $y_0 = y_1$, that is, F is single-valued. For the same reason, F^{-1} is single-valued. Therefore there exists a map $f : F^{-1}(Y) \to F(X)$ such that $F(x) = \{f(x)\}$ for each $x \in F^{-1}(Y)$.

Let $(x_0, y_0) \in F$ and $h : X(x_0) \to Y(y_0)$ is an isomorphism, then $X(x) \simeq Y(h(x))$ for each $x < x_0$, hence $(x, h(x)) \in F$ and $h(x) < y_0$. This shows that f is strictly increasing and that if $x < x_0$ and $x_0 \in F^{-1}(Y)$, then $x \in F^{-1}(Y)$. Symmetrically, if $y < y_0 \in F(X)$ then $y \in F(X)$.

If $F(X) = Y$ then $X \simeq Y$ by Proposition A.8.7, because f is strictly increasing and bijective.

If $F(X) \neq Y$ then there exists $y_0 = \min(Y \setminus F(X))$ and thus $F(X) = Y(y_0)$. It follows $F^{-1}(Y) = X$, for otherwise there would exist $x_0 = \min(X \setminus F^{-1}(Y))$ and an application $f : X(x_0) \to Y(y_0)$ is bijective and monotone, thus an isomorphism, so that $(x_0, y_0) \in F$ in contradiction with $x_0 = \min(X \setminus F^{-1}(Y))$. Therefore $X \simeq Y(y_0)$. Symmetrically, if $F(X) = Y$, but $F^{-1}(Y) \neq X$, then there exists $x_0 = \min\left(X \setminus F^{-1}(Y)\right)$ and thus $X(x_0) \simeq Y$.   □

Two ordered sets have the same *order type* if they are isomorphic. The order type is an equivalence relation on the class of ordered sets. Let us denote by $t(X)$ the equivalence class of X for this relation.

A.9   Ordinal numbers

An *ordinal (number)* is an order type of well ordered sets. The class of all ordinal numbers is denoted by Ord. This class is not a set.

We denote by 0 the order type of the empty set, by n the type of a well ordered set of cardinality n. The natural order of $\mathbb{N}$ is denoted by ω_0 or ω.

Exercise A.9.1. Let $n \in \mathbb{N}$. Show that

(1) all the well ordered sets of cardinality n are isomorphic,
(2) there are $n!$ ways of well ordering each set of cardinality n.

The class $\{\alpha \in \text{Ord} : \alpha < \beta\}$ is a set for each ordinal β.

By Theorem A.8.10, the class of ordinals is well ordered by the following relation: $\alpha < \beta$ if there exist X, Y such that $t(X) = \alpha$, $t(Y) = \beta$ and there exists $y \in Y$ such that $X \simeq Y(y)$. In particular,

$$0 < 1 < \ldots < n \in \mathbb{N}_0$$

for each natural number n.

Example A.9.2. Let X be a countably infinite set. Let $x_\infty \in X$ and $X_0 := X \setminus \{x_\infty\} = \{x_n : n \in \mathbb{N}\}$, where all the terms are distinct. Let

$$\underset{0 \leq n \in \mathbb{N}}{\forall} \quad x_n < x_{n+1} < x_\infty.$$

Note that X is well ordered.

Proposition A.9.3. *For each ordinal* α,

$$\alpha = t\left(\{\beta \in \mathrm{Ord} : \beta < \alpha\}\right).$$

Proof. Let X be such that $t(X) = \alpha$. By Theorem A.8.10, $\beta < \alpha$ if and only if there exists $x \in X$ such that $\beta = t(X(x))$. Conversely, $t(X(x)) < \alpha$ for each $x \in X$. Hence $X \simeq \{\beta \in \mathrm{Ord} : \beta < \alpha\}$. $\qquad\square$

Theorem A.9.4 (Zermelo). *The axiom of choice implies that each set can be well ordered.*

Proof. The empty set $\varnothing$ is well ordered by 0. If $X \neq \varnothing$ then by the axiom of choice (Axiom 8), there exists a map $f : 2^X \setminus \{\varnothing\} \to X$, such that $f(A) \in A$ for every $\varnothing \neq A \subset X$. Let

$$a_0 = f(X) \text{ and } a_\beta := f(X \setminus \{a_\xi : \xi < \beta\})$$

provided that $X \setminus \{a_\xi : \xi < \beta\}$ is not empty. We set

$$\tau := \min\{\beta : X = \{a_\xi : \xi < \beta\}\}.$$

Then $X = \{a_\xi : \xi < \tau\}$ is well ordered by the corresponding $\leq$. $\qquad\square$

Proposition A.9.5 (Successor). *For each ordinal* α, *there exists the least ordinal* β *such that* $\alpha < \beta$.

Proof. Let X be a well ordered set such that $t(X) = \alpha$. By Theorem A.9.4 of Zermelo, 2^X can be well ordered. If β is the type of this order, then $\alpha < \beta$, hence there exists a least ordinal in $\{\beta \in \mathrm{Ord} : \alpha < \beta\}$. $\qquad\square$

Denote by $S(\alpha)$ such ordinal and call it the *successor of* α. An ordinal β is said to be an *ordinal successor* if there exists α such that $S(\alpha) = \beta$; if an ordinal is not successor, then it is called a *limit ordinal*. For instance, each natural number $n > 0$ is successor, while 0, ω_0 are limit ordinals. We adopt the convention that $\sup \varnothing = 0$.

Proposition A.9.6. *An ordinal* β *is a limit ordinal if and only if* $\beta = \sup\{\alpha : \alpha < \beta\}$.

Proof. If β is an ordinal, then by definition,

$$\sup\{\alpha : \alpha < \beta\} = \min\{\gamma \in \mathrm{Ord} : \underset{\alpha < \beta}{\forall} \ \alpha \leq \gamma\}.$$

Of course, $\min\{\gamma \in \mathrm{Ord} : \forall_{\alpha < \beta} \ \alpha \leq \gamma\} \leq \beta$. If β is a successor ordinal, then $S(\min\{\gamma \in \mathrm{Ord} : \forall_{\alpha < \beta} \ \alpha \leq \gamma\}) = \beta$, for otherwise

$$\min\{\gamma \in \mathrm{Ord} : \underset{\alpha < \beta}{\forall} \ \alpha \leq \gamma\} = \beta.$$

$\qquad\square$

Of course, if X is uncountable, then every well ordered type of X is strictly greater than ω_0. We denote

$$\omega_1 := \min \left\{ t\left(X\right) : \operatorname{card} X > \aleph_0 \right\}.$$

Let α be an ordinal and let X be a set. A *sequence of length α* on X is defined as a map $\xi \mapsto x_\xi$ from $\{\xi \in \mathrm{Ord} : \xi < \alpha\}$ to X. It is denoted by

$$\left(x_\xi\right)_{\xi < \alpha}.$$

Of course, a sequence $(\gamma_n)_n$ in the usual sense is a special case: it is a sequence of length ω_0.

Proposition A.9.7. *Each decreasing sequence of ordinals is stationary.*

Proof. Let $\{\gamma_\xi : \xi < \alpha\}$ be such that $\zeta < \xi$ implies $\gamma_\zeta \geq \gamma_\xi$. Suppose that it is not stationary, that is, there exists a sequence $\{\xi_n : n \in \mathbb{N}\}$ such that $\{\gamma_{\xi_n} : n \in \mathbb{N}\}$ has all distinct terms. As $\{\gamma_{\xi_n} : n \in \mathbb{N}\}$ is well ordered, there exists $n_0 < \omega$ such that $\gamma_{\xi_{n_0}} \leq \gamma_{\xi_n}$ for each $n \in \mathbb{N}$. Hence $\gamma_{\xi_{n_0}} = \gamma_{\xi_n}$ for each $n_0 \leq n \in \mathbb{N}$. A contradiction. $\qquad\square$

The *limit* of an increasing sequence $(\gamma_\xi)_{\xi < \alpha}$ is defined by

$$\lim_{\xi \to \alpha} \gamma_\xi := \sup \{\gamma_\xi : \xi < \alpha\}.$$

A sequence $(\gamma_\xi)_{\xi < \beta}$ is said to be *continuous* if

$$\lim_{\xi \to \delta} \gamma_\xi = \gamma_\delta$$

for each limit ordinal $\delta < \beta$.

Proposition A.9.8. *For each ordinal ξ, let γ_ξ be an ordinal. Suppose that $\xi < \zeta$ implies $\gamma_\xi < \gamma_\zeta$. If for each ordinal β, the sequence $(\gamma_\xi)_{\xi < \beta}$ is continuous, then*

$$\underset{\alpha}{\forall} \, \underset{\beta}{\exists} \, (\alpha < \beta \text{ and } \gamma_\beta = \beta).$$

Proof. Let $\alpha_0 := \alpha$ and $\alpha_{n+1} := \gamma_{\alpha_n}$ for $n > 0$. Since for each η the sequence $(\gamma_\xi)_{\xi < \eta}$ is strictly increasing, $\alpha_{n+1} = \gamma_{\alpha_n} \geq \alpha_n > \alpha$ for each $n > 0$. Hence $\beta := \lim_{n \to \omega} \alpha_{n+1} = \lim_{n \to \omega} \gamma_{\alpha_n} = \gamma_\beta$. $\qquad\square$

Theorem A.9.9 (Transfinite induction). *Let γ be an ordinal and P be a subset of $\{\alpha \in \mathrm{Ord} : \alpha < \gamma\}$ such that*

$$0 \in P, \tag{A.9.1}$$

$$\underset{\alpha < \beta < \gamma}{\forall} \, \alpha \in P \Longrightarrow \beta \in P, \tag{A.9.2}$$

then $P = \{\alpha \in \mathrm{Ord} : \alpha < \gamma\}$.

Proof. Otherwise $\{\alpha \in \text{Ord} : \alpha < \gamma\} \setminus P \neq \varnothing$ and thus there exists $\beta = \min(\{\alpha \in \text{Ord} : \alpha < \gamma\} \setminus P)$. By (A.9.1) and (A.9.2), $\beta \in P$, which yields a contradiction. $\qquad\square$

Notice that (A.9.2) implies that $\alpha \in P \implies S(\alpha) \in P$ if $S(\alpha) < \gamma$.

Theorem A.9.10 (Zorn-Kuratowski Lemma). *Let X be a set ordered by $\leq$. If for each totally ordered subset L of X there exists $w \in X$ such that $l \leq w$ for each $l \in L$, then for each $x \in X$ there exists a maximal element $u \in X$ such that $x \leq u$.*

Proof. Let $x \in X$ and $x_0 := x$. Suppose we have constructed a strictly increasing $(x_\xi)_{\xi < \alpha}$. Let

$$A_\alpha := \Big\{ x \in X : \underset{\xi < \alpha}{\forall}\ x_\xi < x \Big\}.$$

If $A_\alpha \neq \varnothing$, then by Axiom 8 of choice, there exists $x_\alpha \in A_\alpha$. There exists an ordinal β such that $A_\beta = \varnothing$, because for a sequence $(x_\xi)_{\xi < \beta}$ of distinct terms on X, the cardinality of β is not greater than card X. As $\{x_\xi : \xi < \beta\}$ is totally ordered, there exists $u \in X$ such that $x_\xi \leq u$ for each $\xi < \beta$. But $A_\beta = \varnothing$ entails $u \in \{x_\xi : \xi < \beta\}$, hence $u = \max\{x_\xi : \xi < \beta\}$ so that there is no $y \in X$ with $y > u$. $\qquad\square$

A.10 Ordinal arithmetic

If α is the type of X ordered by $\leq_X$ and β is the type of Y ordered by $\leq_Y$, then the *sum* $\alpha + \beta$ is the type of the disjoint union $X \cup Y$ ordered by $\leq$, the strict part of which is

$$v < w \iff \begin{cases} v \in X, w \in Y \text{ or} \\ v, w \in X, v <_X w \text{ or} \\ v, w \in Y, v <_Y w. \end{cases}$$

Observe that

$$1 + \omega = \omega < \omega + 1 = S(\omega).$$

Thus the addition is not commutative.

Proposition A.10.1. *Let α, β, γ ordinals.*

(1) If $\alpha < \beta$, then $\gamma + \alpha < \gamma + \beta$.
(2) If $\alpha \leq \beta$, then $\alpha + \gamma \leq \beta + \gamma$.

The *product* $\alpha\beta$ is the type of $X \times Y$ ordered lexicographically on the right:

$$(x_0, y_0) < (x_1, y_1) \Longleftrightarrow \begin{cases} y_0 < y_1, \\ y_0 = y_1 \text{ and } x_0 < x_1. \end{cases}$$

By Proposition A.9.6, the ordinal 2 is the type of the natural order of $\{0, 1\}$. It follows from the definition above that $\omega_0 2$ is the order type of

$$(0,0) < (1,0) < \ldots < (n,0), \ldots < (0,1) < (1,1) < \ldots < (n,1), \ldots$$

for each $1 < n \in \mathbb{N}_0$, while 2ω is the order type of

$$(0,0) < (0,1) < (1,0) < (1,1) \ldots < (n,0) < (n,1), \ldots$$

In particular,

$$2\omega_0 = \omega_0 < \omega_0 2 = \omega_0 + \omega_0.$$

It is easy to see that:

Proposition A.10.2. *Let* α, β, γ *be ordinals.*

(1) If $\alpha < \beta$, *then* $\gamma\alpha < \gamma\beta$.
(2) If $\alpha \le \beta$, *then* $\alpha\gamma \le \beta\gamma$.

The *power* is defined by induction, as follows

$$\alpha^0 = 1,$$
$$\alpha^{\beta+1} := \alpha^\beta \cdot \alpha,$$
$$\beta \text{ limit} \Longrightarrow \alpha^\beta := \sup_{\xi < \beta} \alpha^\xi.$$

In particular,

$$0 < \omega_0 < \omega_0 2 := \omega_0 + \omega_0 < \ldots < \omega_0^2 := \omega_0 \omega_0$$
$$< \ldots < \omega_0^{\omega_0} < \ldots < \omega_1 < \omega_1 + 1 < \omega_1 + \omega_0 \ldots$$

For each ordinal α, one defines a *canonical order* of the set $\alpha \times \alpha := \{(x, y) : x, y < \alpha\}$ by

$$(x_0, y_0) \prec (x_1, y_1) \Longleftrightarrow \begin{cases} \max(x_0, y_0) < \max(x_1, y_1) \text{ or} \\ \max(x_0, y_0) = \max(x_1, y_1) \text{ and } x_0 < x_1 \text{ or} \\ \max(x_0, y_0) = \max(x_1, y_1) \text{ and } x_0 = x_1, y_0 < y_1. \end{cases}$$

For instance,

$$(0,0) \prec (0,1) \prec (1,0) \prec (1,1) \prec (0,2) \prec (1,2) \prec (2,0) \prec (2,1) \prec (2,2) \prec \ldots$$

Proposition A.10.3. *For each ordinal θ, the set $\theta \times \theta$ ordered canonically is well ordered.*

Proof. If $\varnothing \neq A \subset \theta \times \theta$, then let

$$\alpha_0 := \min \left\{ \max(\beta, \delta) : (\beta, \delta) \in A \right\}.$$

It exists, because $\varnothing \neq \{\max(\beta, \delta) : (\beta, \delta) \in A\} \subset \mathrm{Ord}$. Let

$$B := \{(\beta, \delta) \in A : \alpha_0 = \max(\beta, \delta)\}.$$

If $\beta = \alpha_0$ for each $(\beta, \delta) \in B$, then $\delta_0 := \min \{\delta : (\alpha_0, \delta) \in B\}$ fulfills $(\alpha_0, \delta_0) = \min A$. If there exists $(\beta, \delta) \in B$ such that $\beta < \alpha_0$, then $(\beta, \delta) = (\beta, \alpha_0)$, and thus $\beta_0 := \min \{\beta : (\beta, \alpha_0) \in B\}$ fulfills $(\beta_0, \alpha_0) = \min A$. $\qquad\square$

Observe that $\alpha \times \alpha := \{(\xi, \eta) : \xi < \alpha, \eta < \alpha\}$ is the initial segment of $(0, \alpha)$.

Let $\Gamma(\alpha, \beta)$ be the type ordinal of $\{(\xi, \eta) : (\xi, \eta) \prec (\alpha, \beta)\}$. If $(\alpha_0, \beta_0) \prec (\alpha_1, \beta_1)$ then $\Gamma(\alpha_0, \beta_0) \leq \Gamma(\alpha_1, \beta_1)$, because $\Gamma(\alpha_0, \beta_0)$ is an initial segment of $\Gamma(\alpha_1, \beta_1)$.

Proposition A.10.4. *For every ordinal θ there exists ordinals α and β such that $\theta = \Gamma(\alpha, \beta)$.*

Proof. Sure enough, $0 = \Gamma(0, 0)$. Suppose that for each $\xi < \delta$ there exist ordinals x_ξ, y_ξ such that $\Gamma(x_\xi, y_\xi) = \xi$. Therefore

$$\sup \{\Gamma(x_\xi, y_\xi) + 1 : \xi < \delta\} = \sup \{\xi + 1 : \xi < \delta\} = \delta.$$

We conclude thanks to Theorem A.9.9. $\qquad\square$

Let

$$\gamma(\alpha) := \Gamma(\alpha, \alpha). \qquad (\mathrm{A}.10.1)$$

As the map γ is strictly increasing, $\alpha \leq \gamma(\alpha)$. Observe that $\gamma(\omega_0) = \omega_0$. We will see later that for each ordinal α the ordinal ω_α is a fixed point of γ.

A.11   Ordinal-cardinal numbers

For every ordinal number α we denote by $\mathrm{card}\,\alpha$ the cardinality of a well ordered set of the type α. An ordinal α is said to be *cardinal* if $\mathrm{card}\,\beta < \mathrm{card}\,\alpha$ for each ordinal $\beta < \alpha$.

We denote by ω_1 the first uncountable ordinal. Of course,

$$\omega_1 = \min \{\alpha \in \mathrm{Ord} : \aleph_0 < \mathrm{card}\,\alpha\}.$$

Example A.11.1. All the finite ordinals, ω_0 and ω_1 are ordinals-cardinals, while $\omega_0 + 1$, $\omega_0 2$, ω_0^2 are not, because they have the same cardinality as ω_0 that is strictly smaller than them.

There is a bijection between ordinals-cardinals and the cardinals studied in Section A.4. There is however an important difference between ordinals-cardinals and cardinals. The first ones are particular ordinals numbers, hence the addition is not commutative for them. For example, ω_0 and ω_1 are ordinals-cardinals. Notice that $\omega_1 < \omega_1 + \omega_0$, but $\operatorname{card} \omega_1 = \operatorname{card} \omega_1 + \operatorname{card} \omega_0$, as we will see in a moment.

Therefore we need distinguish arithmetic operations on ordinals-cardinals and on cardinals.

Proposition A.11.2. *For every ordinal β there exists an ordinal-cardinal that is greater than β.*

Proof. For every set X we define the *Hartogs number*

$$h(X) = \min \{ \alpha : \operatorname{card} \alpha > \operatorname{card} X \}.$$

As all well orders on 2^X constitute a set and for each $\alpha \in \mathrm{Ord}$ such that $\operatorname{card} \alpha = 2^{\operatorname{card} X} > \operatorname{card} X$, the number $h(X)$ exists. If $\beta \in \mathrm{Ord}$, then $\operatorname{card} \beta < h(\operatorname{card} \beta)$, hence $\beta < h(\operatorname{card} \beta)$. $\qquad \square$

Corollary A.11.3. *For every ordinal β, there exists the least ordinal-cardinal λ such that $\beta < \lambda$. Its cardinality is denoted by $\operatorname{card} (\beta)^+$.*

As we have already noticed, ω_0 is the least countably infinite ordinal-cardinal, that is, $\operatorname{card} \omega_0 = \aleph_0$.

Let

$$\aleph_{\alpha+1} := \aleph_\alpha^+,$$

$$\alpha \text{ limit } \implies \aleph_\alpha := \sup \{ \aleph_\gamma : \gamma < \alpha \}.$$

The ordinal-cardinal number of $\aleph_\alpha$ is denoted by ω_α. The least uncountable cardinal is $\aleph_1$. The least cardinal strictly greater than $\aleph_1$ is denoted by $\aleph_2$, and so on. Therefore, (for each natural n),

$$0 < n < \aleph_0 < \aleph_1 < \ldots < \aleph_n < \aleph_{\omega_0} < \aleph_{\omega_0+1} < \cdots$$

Corollary A.11.4. *Each infinite cardinal is an aleph.*

As $\mathfrak{c} = 2^{\aleph_0}$ is uncountable,

$$\aleph_0 < \aleph_1 \leq \mathfrak{c}.$$

The proposition $\aleph_1 = \mathfrak{c}$ is called the *Continuum Hypothesis* ([11]). It is, as well as its contrary $\aleph_1 < \mathfrak{c}$, independent of the axioms of Set Theory. One can add it as a supplementary axiom without affecting its consistency, as one can add its negation $\aleph_1 < \mathfrak{c}$. There arise two consistent theories with a lot of common theorems.

Theorem A.11.5. $\aleph_\alpha \cdot \aleph_\alpha = \aleph_\alpha$ *for each ordinal* α.

Proof. Let $\Gamma(\alpha, \beta)$ be the type of the canonical order of

$$\{(\xi, \eta) : (\xi, \eta) \prec (\alpha, \beta)\}$$

and let $\gamma(\alpha) := \Gamma(\alpha, \alpha)$ as in (A.10.1). We will show that $\gamma(\omega_\alpha) := \Gamma(\omega_\alpha, \omega_\alpha) = \omega_\alpha$ for each ordinal α.

This is true for $\alpha = 0$. If $\{\alpha : \omega_\alpha \prec \gamma(\omega_\alpha)\} \neq \varnothing$, then in this set there exists a least element α_0. Hence ω_{α_0} is an initial segment of $\Gamma(\omega_{\alpha_0}, \omega_{\alpha_0})$. Let $x, y < \omega_{\alpha_0}$ such $\Gamma(x, y) = \omega_\alpha$. Since ω_{α_0} is a limit ordinal, there exists a δ such that $x, y < \delta < \omega_{\alpha_0}$. Now, $\delta \times \delta$ is an initial segment (given by $(0, \delta)$) containing x and y, so that $\omega_{\alpha_0} \subset \Gamma(\delta, \delta)$, thus card $(\delta \times \delta) \geq \aleph_{\alpha_0}$. But card $(\delta \times \delta) = $ card $\delta \cdot$ card δ and, by hypothesis, card $\delta \cdot$ card $\delta = $ card $\delta < \aleph_{\alpha_0}$, which is a contradiction. $\qquad \square$

This theorem enables one to establish several important facts on the cardinal arithmetic.

Theorem A.11.6. *(AC) Let* κ, λ *be infinite cardinal numbers. Then*

$$\kappa = \kappa + \kappa = \kappa \cdot \kappa, \tag{A.11.1}$$

$$\kappa + \lambda = \kappa \cdot \lambda = \max(\kappa, \lambda), \tag{A.11.2}$$

$$\kappa \leq \lambda \implies \kappa^\lambda = 2^\lambda. \tag{A.11.3}$$

Proof. (A.11.1) On one hand, $\kappa \leq \kappa + \kappa \leq \kappa \cdot \kappa$, and, on the other, there exists an ordinal α such that $\kappa = \aleph_\alpha$. Thus $\kappa \cdot \kappa = \aleph_\alpha \cdot \aleph_\alpha = \aleph_\alpha = \kappa$, by Theorem A.11.5.

(A.11.2) If, for instance, $\kappa = \max(\kappa, \lambda)$, then $\kappa \leq \kappa + \lambda \leq \kappa \cdot \lambda$, and by (A.11.1), $\kappa \cdot \lambda \leq \kappa \cdot \kappa = \kappa$.

(A.11.3) Let $\kappa = \aleph_\alpha$ and $\lambda = \aleph_\beta$, where $\alpha \leq \beta$. Hence $2^{\aleph_\beta} \leq \aleph_\alpha^{\aleph_\beta} \leq \left(2^{\aleph_\alpha}\right)^{\aleph_\beta} = 2^{\aleph_\alpha \aleph_\beta} = 2^{\max(\aleph_\alpha \aleph_\beta)} = 2^{\aleph_\beta}$. $\qquad \square$

---

[11] The continuum hypothesis was formulated in the 1880 by Georg Cantor who attempted to prove it unsuccessfully of course. Kurt Gödel proved in 1940 that it is not contradictory with **ZFC**, that is, that one cannot prove that it is false. Finally, in 1963 Paul J. Cohen, aged 29, that it cannot be deduced from **ZFC**, and thus is independent.

Proposition A.11.7. *If κ is an infinite cardinal and $\lambda_\alpha > 0$ for each $\alpha < \kappa$, then*

$$\sum_{\alpha < \kappa} \lambda_\alpha = \kappa \cdot \sup_{\alpha < \kappa} \lambda_\alpha.$$

Proof. Let $\lambda := \sup_{\alpha < \kappa} \lambda_\alpha$. Then $\sum_{\alpha < \kappa} \lambda_\alpha \leq \sum_{\alpha < \kappa} \lambda = \kappa \cdot \lambda$. On the other hand, as $\lambda_\alpha \geq 1$ for each $\alpha < \kappa$,

$$\kappa = \sum_{\alpha < \kappa} 1 \leq \sum_{\alpha < \kappa} \lambda_\alpha.$$

As $\lambda = \sup_{\alpha < \kappa} \lambda_\alpha \leq \sum_{\alpha < \kappa} \lambda_\alpha$,

$$\lambda \cdot \kappa \leq \lambda \cdot \sum_{\alpha < \kappa} \lambda_\alpha \leq \left(\sum_{\alpha < \kappa} \lambda_\alpha \right)^2 = \sum_{\alpha < \kappa} \lambda_\alpha,$$

because the square of each infinite cardinal is equal to that cardinal. $\qquad\square$

Bibliography

[1] Antoine, P. (1966). Etude élémentaire des catégories d'ensembles structurés, *Bull. Soc. Math. Belge* **18**.

[2] Arhangel'skii, A. and Tkachenko, M. (2008). *Topological Groups and Related Structures, An Introduction to Topological Algebra.*, Vol. 1 (Springer Science & Business Media).

[3] Arzelà, C. (1889). Funzioni di linee, *Atti della Reale Accademia dei Lincei. Serie quarta* **5**, pp. 342–348.

[4] Ascoli, G. (1884). Le curve limiti di una varietà data di curve, *Atti della R. Accad. Dei Lincei. Memorie della Cl. Sci. Fis. Mat. Nat.* **18**, 3, pp. 521–586.

[5] Austin, K. and Dydak, J. (2014). Partitions of unity and coverings, *Topology and its Applications* **173**, pp. 74–82.

[6] Bartsh, R., Dencker, P. and Poppe, H. (1996). Ascoli-Arzelà-theory based on continuous convergence in an (almost) non-Hausdorff setting. in E. Giuli (ed.), *Categorical Topology* (Kluwer Academic), pp. 221–240.

[7] Bastiani, A. (1964). Applications différentiables et variétés différentiables de dimension infinie, *J. Analyse Math.* **13**, pp. 1–114.

[8] Beattie, R. and Butzmann, H. P. (2002). *Convergence Structures and Applications to Functional Analysis* (Kluwer Academic).

[9] Binz, E. (1975). *Continuous Convergence in $C(X)$* (Springer-Verlag), lecture Notes Math. 469.

[10] Bourdaud, G. (1975a). Espaces d'Antoine et semi-espaces d'Antoine, *Cahiers de Topologies et Géométrie Différentielle* **16**, pp. 107–133.

[11] Bourdaud, G. (1975b). Some Cartesian closed topological categories of convergence spaces, in *Categorical Topology* (Springer-Verlag), pp. 93–108, lecture Notes in Math 540.

[12] Brito, W. (2011). *El Teorema de Categoría de Baire y Aplicaciones* (Universidad de los Andes).

[13] Butzmann, H. and Müller, B. (1976). Topological c-embedded spaces, *General Topology and its Applications* **6**, pp. 17–20.

[14] Cantor, G. (1884). Über unendliche, lineare Punktmannigfaltigkeiten, *Mathematische Annalen* **23**, pp. 453–488.

[15] Cartan, H. (1937a). Filtres et ultrafiltres, *C. R. Acad. Sc. Paris* **205**, pp. 777–779.

[16] Cartan, H. (1937b). Théorie des filtres, *C. R. Acad. Sc. Paris* **205**, pp. 595–598.

[17] Choquet, G. (1947). Sur les notions de filtre et de grille, *C. R. Acad. Sci. Paris* **224**, pp. 171–173.

[18] Choquet, G. (1947-48). Convergences, *Ann. Univ. Grenoble* **23**, pp. 55–112.

[19] Christenson, C. and Voxman, W. (1998). *Aspects of Topology* (Bcs Assoc.).

[20] Comfort, W. W. and Negrepontis, S. (1974). *The Theory of Ultrafilters* (Springer-Verlag, New York-Heidelberg).

[21] Cook, C. (1968). On continuous extensions, *Mathematische Annalen* **176**, 4, pp. 302–304.

[22] Cook, C. H. (1973). Compact pseudo-convergences, *Math. Ann.* **202**, pp. 193–202.

[23] Cook, C. H. and Fischer, H. R. (1965). On equicontinuity and continuous convergence, *Math. Ann.* **159**, pp. 94–104.

[24] Cook, C. H. and Fischer, H. R. (1967). Regular convergence spaces, *Math. Ann.* **174**, pp. 1–7.

[25] Dolecki, S. (1996). Convergence-theoretic methods in quotient quest, *Topology Appl.* **73**, pp. 1–21.

[26] Dolecki, S. (2002). Convergence-theoretic characterizations of compactness, *Topology and its Applications* **125**, pp. 393–417.

[27] Dolecki, S. (2004). Elimination of covers in completeness, *Topology Proceedings* **28**, pp. 445–465.

[28] Dolecki, S. (2009). An initiation into convergence theory, in F. Mynard and E. Pearl (eds.), *Beyond Topology, Contemporary Mathematics 486*, Vol. Beyond Topology (A.M.S.), pp. 115–161.

[29] Dolecki, S. (2016). Completeness number of families of subsets of convergence spaces, *Topology and its Applications*. To appear.

[30] Dolecki, S., Greco, G. H. and Lechicki, A. (1985). Compactoid and compact filters, *Pacific J. Math.* **117**, pp. 69–98.

[31] Dolecki, S., Greco, G. H. and Lechicki, A. (1995). When do the upper Kuratowski topology (homeomorphically, Scott topology) and the cocompact topology coincide? *Trans. Amer. Math. Soc.* **347**, pp. 2869–2884.

[32] Dolecki, S., Jordan, F. and Mynard, F. (2007). Reflective classes of sequentially based convergence spaces, sequential continuity and sequence-rich filters, *Topology Proceedings* **31**, 2, pp. 457–479.

[33] Dolecki, S. and Mynard, F. (2000). Convergence-theoretic mechanisms behind product theorems, *Topology and its Applications* **104**, pp. 67–99.

[34] Dolecki, S. and Mynard, F. (2010). When is the Isbell topology a group topology? *Topology and its Applications* **157**, 10, pp. 1370–1378.

[35] Van Dowen, E. K. (1984). The integers and topology, in K. Kunnen and J. E. Vaughan (eds.), *Handbook of set-theoretic topology* (North-Holland), pp. 111–167.

[36] Dydak, J. (2003). Partitions of unity, *Topology Proceedings* **27**, 1, pp. 125–171.

[37] Engelking, R. (1989). *General Topology* (Heldermann Verlag).

[38] Escardó, M., Lawson, J. and Simpson, A. (2004). Comparing Cartesian closed categories of (core) compactly generated spaces, *Topology and its Applications* **143**, 1-3, pp. 105–145.

[39] Feldman, W. A. (1973). Axioms of countability and the algebra $C(X)$, *Pacific J. Math.* **47**, pp. 81–89.

[40] Fischer, H. (1959). Limesraüme, *Math. Ann.* **137**, pp. 269–303.

[41] Fréchet, M. (1906). Sur quelques points du calcul fonctionnel, *Rend. Circ. Mat. Palermo* **22**, pp. 1–74.

[42] Frič, R. and Kent, D. C. (1992). Regularity and extension of maps, *Mathematica Slovaca* **42**, 3, pp. 349–357.

[43] Frolík, Z. (1967). Sums of ultrafilters, *Bull. Amer. Math. Soc.* **73**, pp. 87–91.

[44] Gähler, W. (1977). *Grundstrukturen der Analysis* (Akademie-Verlag).

[45] Gauld, D. and Mynard, F. (2005/2006). Differentiability as continuity, *Real Analysis Exchange* **31**, 2, pp. 425–430.

[46] Gazik, R. J., Park, B. H. and Richardson, G. D. (1984). A Wallman-type compactification for convergence spaces, *Proc. Amer. Math. Soc.* **92**, 2, pp. 301–304.

[47] Gierz, G., Hofmann, K., Keimel, K., Lawson, J., Mislove, M. and Scott, D. (2003). *Continuous Lattices and Domains, Encyclopedia of Mathematics*, Vol. 93 (Cambridge University Press).

[48] Gillman, L. and Jerison, M. (1960). *Rings of Continuous Functions* (Van Nostrand).

[49] Greco, G. (1984). Decomposizioni di semifiltri e Γ-limiti sequenziali in reticoli completamente distributivi, *Ann. Mat. pura e appl.* **137**, pp. 61–82.

[50] Greco, G. H. (1983). Limitoidi e reticoli completi, *Ann. Univ. Ferrara* **29**, pp. 153–164.

[51] Greco, G. H. (1985). The sequential defect of the cross topology is ω_1, *Topology and its Applications* **19**, pp. 91–94.

[52] Greco, G. H. (2012). *Analisi Matematica Uno, Funzioni di una variabile reale (calcolo differenziale e integrale)* (Trento).

[53] Hausdorff, F. (1914). *Grundzüge der Mengenlehre* (Veit & Comp.).

[54] Herrlich, H. (1965). Wann sind alle stetigen Abbildungen in Y konstant, *Math. Zeitschrift.* **90**, pp. 152–154.

[55] Herrlich, H. (1968). *Topologische Reflexionen und Coreflexionen*, Lecture Notes in Mathematics, No. 78 (Springer-Verlag, Berlin-New York).

[56] Jech, T. (1978). *Set Theory* (Academic Press).

[57] Jordan, F. and Mynard, F. (2006). Compatible relations on filters and stability of local topological properties under supremum and product, *Top. Appl.* **153**, pp. 2386–2412.

[58] Jordan, F. and Mynard, F. (2011). Core compactness and diagonality in spaces of open sets, *Applied General Topology* **12**, 2, pp. 143–162.

[59] Keller, H. (1968). Die Limes-uniformisierbarkeit der Limesraüme, *Math. Ann.* **176**, pp. 334–341.

[60] Kent, D. C. (1969). Convergence quotient maps, *Fund. Math.* **65**, pp. 197–205.

[61] Kent, D. C., McKennon, K., Richardson, G. and Schroder, M. (1974). Continuous convergence in $C(X)$, *Pacific J. Math.* **52**, pp. 457–465.

[62] Kent, D. C. and Richardson, G. (1972). Regular compactification of convergence spaces, *Proc. Amer. Math. Soc.* **31**, pp. 571–573.

[63] Kent, D. C. and Richardson, G. (1975). Locally compact convergence spaces, *Michigan Math. J.* **22**, pp. 353–360.

[64] Kent, D. C. and Richardson, G. (1979a). Compactifications of convergence spaces, *Internat. J. Math. and Math. Sci.* **2**, 3, pp. 345–368.

[65] Kent, D. C. and Richardson, G. (1979b). Some product theorems for convergence spaces, *Math. Nachr.* **87**, pp. 43–51.

[66] Kent, D. C. and Richardson, G. (1996). Convergence spaces and diagonal conditions, *Topology Appl.* **70**, pp. 167–174.

[67] Kent, D. C. and Richardson, G. D. (1973a). The decomposition series of a convergence space, *Czechoslovak Math. J.* **23(98)**, pp. 437–446.

[68] Kent, D. C. and Richardson, G. D. (1973b). Open and proper maps between convergence spaces, *Czechoslovak Math. J.* **23(98)**, pp. 15–23.

[69] Kent, D. C. and Richardson, G. D. (1981). Completely regular and ω-regular spaces, *Proc. Amer. Math. Soc.* **82**, 4, pp. 649–652.

[70] Kent, D. C. and Richardson, G. D. (1990). p-regular convergence spaces, *Math. Nachr.* **149**, pp. 215–222.

[71] Kowalsky, H. J. (1954). Limesräume und Kompletierung, *Math. Nach.* **12**, pp. 302–340.

[72] Lowen-Colebunders, E. (1977). Completeness properties for convergence spaces, *Pacific J. Math.* **70**, 2, pp. 401–411.

[73] Lowen-Colebunders, E. (1978). An internal and an external characterization of convergence spaces in which adherences of filters are closed, *Proc. Amer. Math. Soc.* **72**, pp. 205–210.

[74] Lowen-Colebunders, E. (1980). The Choquet hyperspace structure for convergence spaces, *Math. Nachr.* **95**, pp. 17–26.

[75] Lowen-Colebunders, E. (1981). Open and proper maps characterized by continuous set-valued maps, *Canad. J. Math.* **33**, 4, pp. 929–936.

[76] Michael, E. (1972). A quintuple quotient quest, *Gen. Topology Appl.* **2**, pp. 91–138.

[77] Mynard, F. (2002a). Coreflectively modified continuous duality applied to classical product theorems, *Applied General Topology* **2 (2)**, pp. 119–154.

[78] Mynard, F. (2002b). When are compact, countably compact and Lindelöf convergences topological? *Topology Proceedings* **26**, 1, pp. 283–296.

[79] Mynard, F. (2003). First-countability, sequentiality and tightness of the upper Kuratowski convergence, *Rocky Mountain J. of Math.* **33**, 3, pp. 1011–1038.

[80] Mynard, F. (2004). Coreflectively modified duality, *Rocky Mountain J. of Math.* **34**, 2, pp. 733–758.

[81] Mynard, F. (2007a). Products of compact filters and applications to classical product theorems, *Topology and its Applications* **154**, 4, pp. 953–968.

[82] Mynard, F. (2007b). Relations that preserve compact filters, *Applied Gen. Top.* **8**, 2, pp. 171–185.

[83] Mynard, F. (2007c). Unified characterization of exponential objects in TOP, PRTOP and PARATOP, *Cahiers de Géométrie et Topologie Différentielle Catégorique* **48**, 70-80.

[84] Mynard, F. (2013). A convergence-theoretic viewpoint on the Arzelà-Ascoli theorem, *Real Analysis Exchange* **38**, 2, pp. 431–444.

[85] Oxtoby, J. C. (1960/1961). Cartesian products of Baire spaces, *Fund. Math.* **49**, pp. 157–166.

[86] Peano, G. (1887). *Applicazioni geometriche del calcolo infinitesimale* (Fratelli Bocca Editori).

[87] Peano, G. (1889). *Arithmetices principia, nova methodo exposita* (Bocca, Augustae Taurinorum).

[88] Poppe, H. (1974). *Compactness in general function spaces* (Deutscher Verlag der Wissenschaften).

[89] Preuss, G. (1988). *Theory of Topological Structures, Mathematics and its Applications*, Vol. 39 (Reidel Publishing, Dordrecht).

[90] Preuss, G. (2002). *Convenient Topology* (Kluwer).

[91] Preuss, G. (2009). Semiuniform convergence spaces and filter spaces, in E. Pearl and F. Mynard (eds.), *Beyond Topology*, Contemporary Mathematics 486 (A.M.S.), pp. 333–373.

[92] Richardson, G. (1970). A Stone-Čech compactification for limit spaces, *Proc. Amer. Math. Soc.* **25**, 403-404.

[93] Richardson, G. D. (1980). Applications of convergence spaces, *Bull. Austral. Math. Soc.* **21**, 1, pp. 107–123.

[94] Richardson, G. D. and Kent, D. C. (1975). A note on $C_c(X)$, *Proc. Amer. Math. Soc.* **49**, pp. 441–445.

[95] Riesz, F. (1907). Die Genesis des Raumbegriffs, *Math. und Naturwiss. Berichte aus Ungarn* **24**, pp. 309–353.

[96] Schroder, M. (1974). Compactness theorems, in *Categorical Topology* (Springer-Verlag), pp. 566–577.

[97] Schwarz, F. (1983a). Powers and exponential objects in initially structured categories and application to categories of limits spaces, *Quaest. Math.* **6**, pp. 227–254.

[98] Schwarz, F. (1983b). Product compatible reflectors and exponentiality, in L. Bentley et al. (ed.), *Proceedings of the International Conference held at the University of Toledo* (Heldermann), pp. 505–522.

[99] Schwarz, F. (1984). Topological continuous convergence, *Manuscripta Math.* **49**, pp. 79–89.

[100] Schwarz, F. (1992). Hulls of classes of "Gestüfte Raüme", in W. Gähler et al (ed.), *Recent Developments of General Topology and its Applications* (Akademie Verlag), pp. 299–302.

[101] Simon, P. (1980). A compact Fréchet space whose square is not Fréchet, *Comment. Math. Univ. Carolin.* **21**, pp. 749–753.

[102] Urysohn, P. (1925). Über die Mächtigkeit der zusammenhängenden Mengen, *Mathematische Annalen* **94**, pp. 262–295.

List of symbols

This list of symbols is sectioned into the following groups: **S**ets and spaces, **F**amilies of sets, **C**lasses and collections of families, **M**aps and map operations, **R**elations, **C**ardinality and cardinal invariants, **C**onvergence, and **O**ther. Symbols are ordered automatically within a section, and thus do not follow any particular logical order. A symbol may appear under more than one header.

Sets and spaces

2^X	set of all subsets of X, page 4
$[\mathbb{R}, \mathbb{R}]$	set of real-valued continuous functions on $\mathbb{R}$ endowed with the convergence of joint continuity, or natural convergence, page 93
$[\xi, \$_0]$	upper Kuratowski convergence on $\mathcal{C}_\xi$, page 228
$[\xi, \$_0]^\#$	lower Kuratowski topology, page 228
$[\xi, \$_1]$	Scott convergence on $\mathcal{O}_\xi$, page 228
$[\xi, \sigma]$	set of continuous functions from ξ to σ endowed with the natural convergence (see also pages 416, 420), page 225
$\$_0$	Sierpiński convergence on $\{0,1\}$ ($\{0\}$ open), page 61
$\$_1$	Sierpiński convergence on $\{0,1\}$ ($\{1\}$ open), page 61
$\beta\mathbb{N}$	Čech-Stone compactification of $\mathbb{N}$, page 351
$\beta\mathbb{Q}$	Čech-Stone compactification of $\mathbb{Q}$, page 351
$\beta\mathbb{R}$	Čech-Stone compactification of $\mathbb{R}$, page 351
$\operatorname{conv} F$	convex hull of a set F, page 182
$\ker\{x_n\}_n$	kernel of a sequence $\{x_n\}_n$, page 35
$\ker \mathcal{F}$	kernel of a filter $\mathcal{F}$, page 33
$\mathbb{C}$	set of complex numbers, page 99
$\mathbb{N}_k$	set of natural numbers greater or equal to k, page 6

$\mathbb{R}_+$	set of positive real numbers, page 5
$\mathbb{N}$	set of natural numbers, page 6
$\mathbb{R}$	set of real numbers, page 5
$\mathbb{U}X$	set of ultrafilters on X, page 44
$\mathbf{I}(X)$	set of convergences on X, page 70
$\mathbf{J}(X)$	set of preconvergences on X, page 70
$\mathbf{L}(X)$	set of finitely deep convergences on X, page 71
ν	standard convergence on the real line, page 56
ω_1	first uncountable ordinal (see also Appendix A.9), page 138
ω_α	ordinal-cardinal of $\aleph_\alpha$, page 520
$\overline{\beta\xi}$	Čech-Stone compactification of ξ, page 341
$\overline{\omega\xi}$	one-point compactification of ξ, page 336
$\overline{j\xi}$	compactification of ξ with embedding j, page 339
$\pi[x,(X)_0]$	prime cofinite convergence with distinguished point x, page 59
$\pi[x,\mathcal{G}]$	prime pretopology with distinguished point x and $\mathcal{V}(x) = \mathcal{G} \wedge \{x\}^{\uparrow}$, page 74
$\pi[x,\mathbb{G}]$	prime convergence with distinguished point x and pavement $\mathbb{G}$ at x, page 74
$\prod_{X\in\mathcal{X}} X$	Cartesian product of the set of sets $\mathcal{X}$, page 39
$\S X$	set of families of subsets of X, page 53
σ^ξ	natural preconvergence on $|\sigma|^{|\xi|}$, page 420
$\varnothing$	empty set, page 6
$|\xi|$	underlying set of a convergence ξ, page 56
$\xi_{\$_0}$	Kuratowski convergence on $\mathcal{C}_\xi$, page 229
$¥$	Bourdaud pretopology, page 120
$\zeta(\mathcal{A})$	AD topology associated to the AD family $\mathcal{A}$, page 280
$\{f < r\}$	set of points x where $f(x) < r$, page 5
$\{f = g\}$	set of points x where $f(x) = g(x)$, page 5
$\{f = r\}$	set of points x where $f(x) = r$, page 5
$\{f > r\}$	set of points x where $f(x) > r$, page 5
$\{f \geq g\}$	set of points x where $f(x) \geq g(x)$, page 5
$\{f \geq r\}$	set of points x where $f(x) \geq r$, page 5
$\{f \leq g\}$	set of points x where $f(x) \leq g(x)$, page 5
$\{f \leq r\}$	set of points x where $f(x) \leq r$, page 5
A^c	complement of A, page 4
$A_\bullet$	$\bigcup_{a\in A} \mathrm{cl}_\xi\{a\}$, page 449
$C(\xi, \$_0)$	set of closed subsets of $|\xi|$ seen as a function space, page 227
$C(\xi, \tau)$	set of continuous functions $f : |\xi| \to |\tau|$, page 79

$C(X,Y)$	set of continuous functions $f : X \to Y$, page 79
$C_b(X,Y)$	set of bounded continuous maps from X to Y, page 290
$F - G$	$\{f - g : f \in F, g \in G\}$, page 469
$h(S,f)$	hull of extensionability of S for f, page 206
$H(\xi,\sigma)$	set of homeomorphisms from ξ to σ, page 81
$k(\xi,\sigma)$	compact-open convergence on $C(\xi,\sigma)$, page 416
O^+	$\{C \in \mathcal{C}_\xi : C \cap O \neq \varnothing\}$, page 229
$p(\xi,\sigma)$	pointwise convergence on $C(\xi,\sigma)$, page 418
$R(F)$	image of the set F under the relation R, page 37
S_1	unit circle in the complex plane, page 99
$u(\xi,d)$	uniform convergence on $C(\xi,\tilde{d})$, page 417
$W(x)$	initial segment of the well-ordered set W corresponding to x, page 513
$X/\sim$	quotient set of X by the equivalence relation $\sim$, page 98
$X \times Y$	product of the sets X and Y, page 5
Y^X	set of functions from X to Y, page 4

Families of sets

$\mathcal{A} \leq \mathcal{D}$	$\mathcal{D}$ is finer than $\mathcal{A}$ (families), page 12
$(X)_1$	cocountable filter on X, page 30
$(x_n)_n$	family of tails of the sequence $\{x_n\}_n$, page 10
$[X]^{<\omega}$	set of finite subsets of the set X, page 6
$\mathcal{A} \times \mathcal{B}$	family of products of elements of $\mathcal{A}$ and $\mathcal{B}$, page 5
$\mathcal{A} \vee_\infty \mathcal{B}$	$\{A \cap B : A \in \mathcal{A}, \operatorname{card}(A \cap B) \geq \aleph_0\}$, page 466
$\mathcal{A}^\cap$	family of finite intersections of elements of $\mathcal{A}$, page 32
$\mathcal{A}^\cup$	family of finite unions of elements of $\mathcal{A}$, page 32
$\mathcal{A}^{\#\#}$	$(\mathcal{A}^\#)^\#$, page 39
$\beta(\mathcal{F})$	set of ultrafilters finer than the filter(-base) $\mathcal{F}$, page 44
$\beta(A)$	set of ultrafilters on X containing $A \subset X$, page 247
$\beta^*(A)$	set of free ultrafilters on X containing $A \subset X$, page 247
$\beta^*(\mathcal{F})$	set of free ultrafilters finer than $\mathcal{F}$, page 247
$\bigvee_{\mathcal{F} \in \mathbb{D}} \mathcal{F}$	supremum of $\mathbb{D}$ in $(\mathbb{F}X, \leq)$, page 33
$\bigwedge_{\mathcal{F} \in \mathbb{D}} \mathcal{F}$	infimum of $\mathbb{D}$ in $(\mathbb{F}X, \leq)$, page 30
$\mathcal{C}_\xi$	family of ξ-closed sets, page 129
$\mathcal{F} - \mathcal{G}$	$\{F - G : F \in \mathcal{F}, G \in \mathcal{G}\}^\uparrow$, page 469
$\mathcal{F}_\mathcal{A}$	residual filter of the family $\mathcal{A}$, page 254
$\mathcal{K}(\xi)$	family of ξ-compact sets, page 231
$\mathcal{K}^o(\xi)$	family of ξ-compactoid sets , page 231
$(X)_0$	family of cofinite subsets of X, page 13

$(x_n)_n^{\uparrow}$ sequential filter associated to $\{x_n\}_n$. See also, $\mathcal{A}^{\uparrow}$, page 10

$[A]_0$ family of sets that are almost equal to A, page 17

$\mathrm{adh}_\xi^{\natural} \mathcal{F}$ $\{\mathrm{adh}_\xi F : F \in \mathcal{F}\}$, page 172

$\mathrm{int}_\xi^{\natural} \mathcal{B}$ $\{\mathrm{int}_\xi B : B \in \mathcal{B}\}$, page 139

$\mathrm{supp}\, \mathcal{F}$ support of the family $\mathcal{F}$ of functions, page 481

$\mathbb{I}X$ set of isotone families on X, page 28

$\mathcal{A}^{\#}$ grill of a family $\mathcal{A}$, page 39

$\mathcal{B}(x)$ family $\{B(x,r) : r > 0\}$ of balls centered at x, page 7

$\mathcal{F}^*$ free part of the filter $\mathcal{F}$, page 34

$\mathcal{R}^-[\mathcal{G}]$ (possibly degenerate) filter generated by the sets $R^-(G)$ for $R \in \mathcal{R}$ and $G \in \mathcal{G}$, page 38

$\mathcal{A}\#\mathcal{B}$ $\mathcal{A}$ and $\mathcal{B}$ mesh, page 31

$\mathcal{A} \approx \mathcal{D}$ $\mathcal{A}$ and $\mathcal{D}$ are equivalent families, page 12

$\mathcal{A}(\mathcal{G})$ contour of $\mathcal{A}(\cdot)$ along $\mathcal{G}$ (see also page 161), page 53

$\mathcal{A}^{\downarrow}$ family of subsets of elements of $\mathcal{A}$, page 28

$\mathcal{A}^{\downarrow x}$ family of subsets of elements of $\mathcal{A}$ in X, page 28

$\mathcal{A}^{\uparrow}$ isotonization of the family $\mathcal{A}$, page 28

$\mathcal{A}^{\uparrow x}$ isotonization of the family $\mathcal{A}$ in X, page 28

$\mathcal{A}_c$ family of complements of elements of the family $\mathcal{A}$, page 28

$\mathcal{C}_\xi^\nu$ family of functionally ξ-closed sets, page 189

$\mathcal{F} \vee \mathcal{G}$ supremum of $\mathcal{F}$ and $\mathcal{G}$ in $(\mathbb{F}X, \leq)$, page 31

$\mathcal{F} \vee A$ abbreviation for $\mathcal{F} \vee \{A\}^{\uparrow}$, page 32

$\mathcal{F}^{\bullet}$ principal part of the filter $\mathcal{F}$, page 34

$\mathcal{G}(F)$ contour of $\mathcal{G}(\cdot)$ along the principal filter F, that is, $\bigcap_{a \in F} \mathcal{G}(a)$, page 161

$\mathcal{G}^{\Uparrow}$ extension of $\mathcal{G}$ to $\S X$, page 53

$\mathcal{H}^{\star}$ contour of $\mathcal{H} \in \S\S X$, page 53

$\mathcal{K}_c^o(\xi)$ cocompactoid filter of ξ, page 232

$\mathcal{N}(\mathcal{N}(x))$ contour of $\mathcal{N}(\cdot)$ along $\mathcal{N}(x)$, page 131

$\mathcal{O}_\xi^\nu$ family of functionally ξ-open sets, page 189

$\mathcal{P} \vee \mathcal{R}$ family of pairwise intersections of elements of the covers $\mathcal{P}$ and $\mathcal{R}$, page 69

$\mathcal{R}[\mathcal{F}]$ (possibly degenerate) filter generated by the sets $R(F)$ for $R \in \mathcal{R}$ and $F \in \mathcal{F}$, page 38

$\mathcal{S}^{=0}$ $\{\{f \in C(\xi,\nu) : S \subset f^-(0)\} : S \in \S^{\cup}\}$, page 452

$\mathcal{V}_\xi(x)$ vicinity filter of x for ξ, page 59

$\mathcal{Z}(\cdot)$ assignment to a convergence of a meet-semilattice of subsets, page 183

$\mathcal{N}_\xi(x)$	neighborhood filter of x for ξ, page 131
$\mathcal{O}_\xi$	family of ξ-open sets, page 128
$\prod_{X \in \mathcal{X}} \mathcal{F}_X$	polyhedral filter, page 39
$\mathrm{rdc}(\mathcal{F})$	reduction of a family $\mathcal{F}$ of subsets of $\mathcal{C}_\xi$, page 227
$\{\varnothing\}^\uparrow$	degenerate filter, page 34
$\mathcal{A}\#\mathcal{B}$	abbreviation for $\{A\}\#\mathcal{B}$, page 31
$a^\natural\mathcal{A}$	$\{a(A) : A \in \mathcal{A}\}$ where $a : 2^X \to 2^X$ and $\mathcal{A} \subset 2^X$, page 171
$A^\uparrow$	abbreviation for $\{A\}^\uparrow$, page 33
$f[\mathcal{A}]$	$\{f(A) : A \in \mathcal{A}\}$, page 14
$f^-[\mathcal{B}]$	$\{f^-(B) : B \in \mathcal{B}\}$, page 14
$f^\to[\mathcal{F}]$	$f[\mathcal{F}]^\uparrow \cap \mathcal{C}^\nu_\tau$, page 346
$G^{\Uparrow}$	$\{\mathcal{A} \in \S X : G \in \mathcal{A}\}$, page 54
$H^\star$	$\bigcap_{\mathcal{A} \in H} \mathcal{A}$ where $H \subset \S X$, page 53
$R[\mathcal{F}]$	image of the family $\mathcal{F}$ under the relation R, page 38

Classes and collections of families

$\beta(\mathcal{F})$	set of ultrafilters finer than the filter(-base) $\mathcal{F}$, page 44
$\beta(A)$	set of ultrafilters on X containing $A \subset X$, page 247
$\beta^*(A)$	set of free ultrafilters on X containing $A \subset X$, page 247
$\beta^*(\mathcal{F})$	set of free ultrafilters finer than $\mathcal{F}$, page 247
$\mathbb{A}$	class of antidistributive families, page 43
$\mathbb{D}_\bullet$	subclass of the class of filters $\mathbb{D}$ formed by filters with a filter-base closed for $\cdot_\bullet$, page 449
$\mathbb{F}$	class of all filters, page 33
$\mathbb{F}_\# X$	set of filter-grills on X, page 41
$\mathbb{F}_{\mathcal{Z}} X$	set of $\mathcal{Z}$-filters on X, page 344
$\mathbb{I} X$	set of isotone families on X, page 28
$\mathbb{E}$	class of sequential filters, page 35
$\mathbb{F}^\triangle$	class of bisequential filters, page 378
$\mathbb{F}_0$	class of principal filters, page 33
$\mathbb{F}_0^\triangle$	class of Fréchet filters, page 374
$\mathbb{F}_1$	class of countably based filters, page 48
$\mathbb{F}_1^{\triangle\triangle}$	class of productively Fréchet filters, page 463
$\mathbb{F}_1^\triangle$	class of strongly Fréchet filters, page 377
$\mathbb{F}_{\#1}$	class of countably tight filters, page 374
$\mathbb{F}_*$	class of free filters, page 33
$\mathbb{F}_\kappa$	class of filters of character less than $\aleph_\kappa$, page 363
$\mathbb{F}_{\wedge 1}$	class of countably deep filters, page 373
$\mathbb{F} X$	set of filters on X, page 29

$\mathbb{G}_\neg$	$\{\mathcal{G}_c : \mathcal{G} \in \mathbb{G}\}$, page 301
$\mathbb{U}$	class of ultrafilters, page 44
$\mathbb{U}^\cap$	class of finite infima of ultrafilters, page 104
$\mathbb{U}_{\mathcal{Z}}(X)$	set of all $\mathcal{Z}$-ultrafilters on X, page 344
$\mathbb{U}_\xi^\nu$	the set of all $\mathcal{C}_\xi^\nu$-ultrafilters, page 349
$\mathbb{U}X$	set of ultrafilters on X, page 44
$\overline{\mathbb{F}}_\#$	class of possibly degenerate filter-grills (class of distributive families), page 43
$\overline{\mathbb{F}}X$	set of possibly degenerate filters on X, page 32
$f\,[\![\mathbb{G}]\!]$	$\{f\,[\mathcal{G}] : \mathcal{G} \in \mathbb{G}\}$, page 307
$f^-\,[\![\mathbb{H}]\!]$	$\{f^-\,[\mathcal{H}] : \mathcal{H} \in \mathbb{H}\}$, page 307

Maps and map operations

$\bigotimes F$	product map of elements of the family F of maps, page 95
χ_A	characteristic function of a set A, page 4
ΔF	diagonal map for the family F of maps $f : X \to Y_f$, page 94
$\exp$	exponential map, page 412
$\frac{f}{g}$	(pointwise) quotient of real valued functions, page 187
$\hat{h}$	transposition of $h : Y \to Z^X$, page 412
$\langle \cdot, \cdot \rangle$	evaluation map (see also page 420), page 92
$[\mathcal{F}]^{<\omega}(f)$	set of finite subfamilies of $\mathcal{F}$ containing f, page 491
$\{x_n\}_{n\in N}$	sequence $f : N \to X$ where $x_n := f(n)$, page 10
dist_A	$\mathrm{dist}_A(x) = \inf_{a\in A} d(x,a)$, page 131
$\mathrm{dom}\,f$	domain of the (partial) map f, page 4
$\mathcal{D}^\vee$	least saturated family of pseudometric including $\mathcal{D}$, page 479
$\mathcal{F}'$	fragmentation of the partition of unity $\mathcal{F}$, page 491
$\mathcal{I}$	identity map of $\mathbb{U}X$, page 248
$\mathcal{Z}(\cdot)$	assignment to a convergence of a meet-semilattice of subsets, page 183
$\overline{r}$	real-valued function identically equal to r, page 5
ψ_A	indicator function of a set A, page 4
e	evaluation map, page 92
e_x	point-evaluation defined by $e_x(f) = f(x)$, page 92
$f(A)$	$\{f(x) : x \in A\}$, page 5
$f + g$	(pointwise) sum of real valued functions, page 187
$f : X \rightarrowtail Y$	partial map from X to Y, page 4
$f : |\xi| \to |\tau|$	See notational convention, page 79
$f : X \to Y$	map from X to Y, page 4
$f \cdot g$	(pointwise) product of real valued functions, page 187

$f[\mathcal{A}]$	$\{f(A) : A \in \mathcal{A}\}$, page 14
$f[\![\mathbb{G}]\!]$	$\{f[\mathcal{G}] : \mathcal{G} \in \mathbb{G}\}$, page 307
$f \times g$	product map $f \times g : X \times W \to Y \times Z$ of $f : X \to Y$ and $g : W \to Z$, page 95
$f \vee g$	(pointwise) sup of real valued functions, page 187
$f \wedge g$	(pointwise) inf of real valued functions, page 187
f^*	upper adjoint map, page 413
$f^-(B)$	$\{x : f(x) \in B\}$, page 5
$f^-[\![\mathbb{H}]\!]$	$\{f^-[\mathcal{H}] : \mathcal{H} \in \mathbb{H}\}$, page 307
f^σ	restriction of the adjoint map f^* to the set of continuous maps, page 415
$f^-[\mathcal{B}]$	$\{f^-(B) : B \in \mathcal{B}\}$, page 14
$f^{|B}$	restriction of $f : X \to Y$ to $B \subset Y$, page 85
$f_{|A}$	restriction of $f : X \to Y$ to $A \subset X$, page 85
$g \succ f$	g is a subquence of f, page 15
g_*	lower adjoint map, page 414
g_ξ	lower adjoint map of g restricted to continuous functions, page 416
i_X^A	inclusion map of A into X, page 64
j_X	point-evaluation map $(j_X(x)(f) := f(x))$, page 413
j_ξ	point-evaluation of continuous functions, page 415
p_X	projection from a product onto the component X, page 5
$R : X \rightrightarrows Y$	R is a relation between X and Y considered as a multivalued map from X to Y, page 500
s_m	right shift of $\mathbb{N}$ by m, page 10
T_g	translation by g in a set of real-valued functions, page 471

Relations

$\mathcal{A} \le \mathcal{D}$	$\mathcal{D}$ is finer than $\mathcal{A}$ (families), page 12
$\approx_0$	almost equality of sets, page 17
$\bigvee A$	supremum of the subset A of a lattice, page 70
$\bigvee^{\unrhd}$	supremum for $\unrhd$, page 425
$\bigwedge A$	infimum of the subset A of a lattice, page 70
$\bigwedge^{\unrhd}$	infimum for $\unrhd$, page 425
$\mathcal{F} - \mathcal{G}$	$\{F - G : F \in \mathcal{F}, G \in \mathcal{G}\}^\uparrow$, page 469
$\kappa \le \lambda$	κ is not greater than λ (order on cardinal numbers), page 6
$[A]_0$	family of sets that are almost equal to A, page 17
$[x]$	equivalence class of x for a given equivalence relation, page 97
$\mathcal{A} \# \mathcal{B}$	$\mathcal{A}$ and $\mathcal{B}$ mesh, page 31

$\mathcal{A} \approx \mathcal{D}$	$\mathcal{A}$ and $\mathcal{D}$ are equivalent families, page 12
$\overline{j_0\xi} \approx \overline{j_1\xi}$	the compactifications $\overline{j_0\xi}$ and $\overline{j_1\xi}$ are equivalent, page 339
$\overline{j_0\xi} \gtrsim \overline{j_1\xi}$	the compactification $\overline{j_0\xi}$ is bigger than $\overline{j_1\xi}$, page 339
$\prec_\xi$	$\mathcal{P} \prec_\xi A$ ($\mathcal{P}$ is a ξ-cover of A), page 68
$\sim_f$	equivalence relation determined by the fibers of a map f, page 98
$\subset_0$	almost inclusion, page 17
$\tau \trianglerighteq \upsilon$	τ is a finer subspace of υ, page 425
$\triangle$	relation between filters, page 385
$\triangleleft$	refinement relation on families of sets, page 28
$\zeta \geq \xi$	ζ is finer than ξ (order on convergences), page 57
$A \# \mathcal{B}$	abbreviation for $\{A\} \# \mathcal{B}$, page 31
$F - G$	$\{f - g : f \in F, g \in G\}$, page 469
$g \succ f$	g is a subquence of f, page 15
$R(F)$	image of the set F under the relation R, page 37
$R : X \rightrightarrows Y$	R is a relation between X and Y considered as a multivalued map from X to Y, page 500
$R[\mathcal{F}]$	image of the family $\mathcal{F}$ under the relation R, page 38
R^-	inverse relation of the relation R, page 37
$x \vee y$	supremum in a lattice, page 70
$x \wedge y$	infimum in a lattice, page 70

Cardinality and cardinal invariants (for a convergence ξ)

$\aleph_0$	smallest infinite cardinal number, page 6
$\aleph_\alpha$	cardinal number aleph α, page 520
$\operatorname{card} X$	cardinality of a set X, page 6
$\chi(\mathcal{F})$	character of the filter $\mathcal{F}$, page 105
$\chi(\mathbb{G})$	character of the collection $\mathbb{G}$ of filters, page 106
$\chi(\xi)$	character of ξ, page 106
$\chi(\xi, x)$	character of ξ at x, page 106
$\chi_*(\xi)$	strong character of ξ, page 106
$\chi_*(\xi, x)$	strong character of ξ at x, page 106
$\kappa \leq \lambda$	κ is not greater than λ (order on cardinal numbers), page 6
λ^κ	cardinality of the set of functions from a set of cardinality κ to a set of cardinality λ, page 6
$\|\mathcal{F}\|$	robustness of a filter $\mathcal{F}$, page 48
$\operatorname{compl}(\xi)$	completeness number of ξ, page 304
$\operatorname{d}(\xi)$	density of ξ, page 109

$d^*(\xi)$	strong density of ξ, page 110
$so(\xi)$	sequential order of ξ, page 154
$t(\xi)$	topological defect of ξ, page 151
$w(\xi)$	weight of ξ, page 105
$\mathfrak{l}(\xi)$	cover Lindelöf number of ξ, page 472
$\mathfrak{c}$	cardinality of the real line, page 6
$\mathfrak{p}(\xi)$	paving number of ξ, page 72
$\mathfrak{p}(\xi, x)$	paving number of ξ at x, page 72

Convergence

$\lim_\xi^-(x)$	set of filters converging to x for ξ, page 72
(X, ξ)	convergence space with underlying set X and convergence structure ξ, page 56
$[\mathbb{R}, \mathbb{R}]$	set of real-valued continuous functions on $\mathbb{R}$ endowed with the convergence of joint continuity, or natural convergence, page 93
$[\xi, \$_0]$	upper Kuratowski convergence on $\mathcal{C}_\xi$, page 228
$[\xi, \$_0]^\#$	lower Kuratowski topology, page 228
$[\xi, \$_1]$	Scott convergence on $\mathcal{O}_\xi$, page 228
$[\xi, \sigma]$	set of continuous functions from ξ to σ endowed with the natural convergence (see also pages 416, 420), page 225
$\mathrm{adh}\, A$	adherence of the set A, page 121
$\mathrm{adh}_\xi\, A$	adherence of the set A, page 121
β	Stone convergence on $\mathbb{U}X$, page 248
$\bigoplus_{\xi \in \Xi} \xi$	sum or coproduct of the set Ξ of convergences, page 89
$\boldsymbol{FX}$	final convergence for $\{(f, \xi_f) : f \in \boldsymbol{F}\}$, where $\boldsymbol{F}$ is a collection of maps $f : |\xi_f| \to Y$, page 87
$\boldsymbol{F}^-\boldsymbol{Y}$	initial convergence for $\{(f, \upsilon_f) : f \in \boldsymbol{F}\}$, where $\boldsymbol{F}$ is a collection of maps $f : X \to |\upsilon_f|$, page 86
$\mathcal{C}_\xi$	family of ξ-closed sets, page 129
$\mathrm{cl}_\xi\, A$	the closure of the set A for ξ, page 130
$\mathrm{int}_\xi\, A$	interior of the set A for ξ, page 130
ι	discrete convergence, page 57
ι_X	discrete convergence on X, page 57
$\lim f$	limit set of a quence f on a premetric space, page 16
$\lim(x_n)_n$	limit set of $\{x_n\}_n$, equivalently of its sequential filter, page 13
$\lim_{n \in N} x_n$	set of limit points of a sequence $\{x_n\}_{n \in N}$ in a premetric space, page 11

$\lim_p (f_n)_n$	set of pointwise limits of the sequence of functions $\{f_n\}_{n \in N}$, page 19
$\lim_{s^*}$	limit for the upper convergence or dual Scott convergence, page 75
$\lim_s$	limit for the Scott convergence or lower convergence on a lattice, page 75
$\lim_\beta$	limit in the Stone convergence β on $\mathbb{U}X$, page 248
$\lim_\xi^{-1}(x)$	set of filters converging to x, page 55
$\lim_{ae}$	convergence almost everywhere, page 113
$\lim_{n \to \infty}$	limit of a sequence in the usual sense, page 2
$\mathrm{adh}_\xi \mathcal{A}$	adherence of the family $\mathcal{A}$, page 67
$\mathrm{adh}_\xi^\gamma A$	γ-iterated adherence of A (γ is an ordinal), page 150
$\mathrm{adh}_\xi^\natural \mathcal{F}$	$\{\mathrm{adh}_\xi F : F \in \mathcal{F}\}$, page 172
$\mathrm{inh}_\xi \mathcal{A}$	inherence of the family $\mathcal{A}$ for ξ, page 214
$\mathrm{inh}_\xi A$	inherence of the set A, page 123
$\mathrm{int}_\xi^\natural \mathcal{B}$	$\{\mathrm{int}_\xi B : B \in \mathcal{B}\}$, page 139
$\mathrm{Seq}\,\nu$	sequentially based modification of ν, page 60
$\liminf \mathcal{F}$	lower limit of a filter $\mathcal{F}$ in a lattice, page 76
$\limsup \mathcal{F}$	upper limit of a filter $\mathcal{F}$ in a lattice, page 76
$\lim_p \mathcal{F}$	set of pointwise limits of a filter $\mathcal{F}$, page 22
$\lim_\xi \mathcal{F}$	limit set of the filter $\mathcal{F}$ for the convergence ξ, page 55
$\mathbf{I}$	class of convergences, page 355
$\mathbf{I}(X)$	set of convergences on X, page 70
$\mathbf{J}(X)$	set of preconvergences on X, page 70
$\mathbf{L}(X)$	set of finitely deep convergences on X, page 71
$\mathcal{V}_\xi(x)$	vicinity filter of x for ξ, page 59
$\mathcal{V}_p(f)$	family of vicinities of f for the pointwise convergence, page 20
$\mathcal{N}(x)$	neighborhood filter of x (convergence is implicit), page 131
$\mathcal{N}_\xi(x)$	neighborhood filter of x for ξ, page 131
ν	standard convergence on the real line, page 56
$\mathcal{O}_\xi$	family of ξ-open sets, page 128
$\mathcal{O}_\xi(x)$	family of ξ-open sets containing x, page 128
$\overline{\beta\xi}$	Čech-Stone compactification of ξ, page 341
$\overline{\omega\xi}$	one-point compactification of ξ, page 336
$\overline{j\xi}$	compactification of ξ with embedding j, page 339
$\pi[x, (X)_0]$	prime cofinite convergence with distinguished point x, page 59

$\pi[x, \mathbb{G}]$	prime convergence with distinguished point x and pavement $\mathbb{G}$ at x, page 74
$\prod_{\xi \in \Xi} \xi$	product convergence for the set Ξ of convergences, page 89
$\mathcal{V}(x)$	vicinity filter of x, page 9
$\mathcal{V}_\xi(x)$	vicinity filter of x for ξ, page 115
$\varnothing_X$	empty convergence on X, page 56
$|\xi|$	underlying set of a convergence ξ, page 56
$\widehat{f\xi}$	final preconvergence for (f, ξ), page 83
$\widetilde{\mathcal{D}}$	topology generated by the family $\mathcal{D}$ of pseudometrics, page 479
$\tilde{d}$	convergence induced by the (pre)metric d, page 56
ξ^γ	pretopology whose principal adherence is adh_ξ^γ, page 163
$\xi_{|A}$	convergence induced on A by ξ, page 64
$\xi \times \tau$	product convergence (finite product), page 89
$\xi_{\$_0}$	Kuratowski convergence on $\mathcal{C}_\xi$, page 229
$\zeta \geq \xi$	ζ is finer than ξ (order on convergences), page 57
$A_\bullet$	$\bigcup_{a \in A} \mathrm{cl}_\xi\{a\}$, page 449
$f\xi$	final convergence for (f, ξ), page 83
$f^-\tau$	initial (pre)convergence for (f, τ), page 82
$h(S, f)$	hull of extensionability of S for f, page 206
$H(\xi, \sigma)$	set of homeomorphisms from ξ to σ, page 81
$k(\xi, \sigma)$	compact-open convergence on $C(\xi, \sigma)$, page 416
o	chaotic convergence, page 57
o_X	chaotic convergence on X, page 57
$p(\xi, \sigma)$	pointwise convergence on $C(\xi, \sigma)$, page 418
T_0	property of convergence: points are distinguished by the convergence, page 58
T_2	property of convergence: limits are unique, page 57
T_1	property of convergence: singletons are closed, page 58
$T_{3\frac{1}{2}}$	property of convergence: Hausdorff functionally regular, page 191
$u(\xi, d)$	uniform convergence on $C(\xi, \tilde{d})$, page 417

Modifiers

Epi^σ	initial convergence for the bidual with respect to σ, page 437
Epi^R	R-bidual modifier, page 439
Epi_F^σ	F-modified σ-bidual modifier, page 443
Epi_F^R	F-modified R-bidual modifier, page 443

$A_\mathbb{D}$ projection on convergences determined by adherences of filters of $\mathbb{D}$, page 364

$B_\mathbb{D}$ coprojection on $\mathbb{D}$-based convergences, page 370

C_σ σ-final modifier, page 381

Dis discretization functor, page 444

D diagonal modifier, page 170

D_R coreflector on R-dual convergences, page 455

I identity functor, page 219

I_κ coreflector on $\mathbb{F}_\kappa$-based convergences, page 372

Kh hereditarily locally compact modifier, page 244

K locally compact modifier, page 242

L finitely deep modifier, page 71

R_ν functionally regular modifier, page 191

R_υ υ-initial modifier(see also page 381) , page 184

Seq ν sequentially based modification of ν, page 60

S pseudotopologizer, page 219

S_0 pretopologizer, page 117

S_1 paratopologizer, page 262

T topologizer, page 140

Reg regularizer, page 172

Seq sequentially based modifier, page 154

Other

$\{x_n\}_{n \in N}$ sequence $f : N \to X$ where $x_n := f(n)$, page 10

diam A diameter of a subset A of a (pre)metric space, page 7

dom f domain of the (partial) map f, page 4

$\mathcal{D}^\vee$ least saturated family of pseudometric including $\mathcal{D}$, page 479

$B(x, r)$ ball centered at x of radius r for a (pre)metric, page 7

$f(A)$ $\{f(x) : x \in A\}$, page 5

$f^-(B)$ $\{x : f(x) \in B\}$, page 5

G_δ intersection of countably many open sets, page 310

G_κ intersection of κ many open sets, page 310

O^+ $\{C \in \mathcal{C}_\xi : C \cap O \neq \varnothing\}$, page 229

$S_\Pi(f)$ Riemann Sum of f for the marked partition Π, page 24

Index

Printed in the United States
By Bookmasters